Guide to

Oracle 10g

Rocky Conrad
San Antonio College

Joline Morrison
Mike Morrison
University of Wisconsin-Eau Claire

COURSE TECHNOLOGY
CENGAGE Learning

Australia • Brazil • Japan • Korea • Mexico • Singapore • Spain • United Kingdom • United States

COURSE TECHNOLOGY
CENGAGE Learning

Guide to Oracle 10g
Rocky Conrad, Mike Morrison and Joline Morrison

Senior Vice President, Publisher: Kristen Duerr

Vice President, Technology Product Strategy: Mac Mendelsohn

Executive Editor: Bob Woodbury

Senior Product Manager: Eunice Yeates

Senior Acquisitions Editor: Maureen Martin

Development Editor: Lynne Raughley

Production Editor: Kristen Guevara

Associate Product Manager: Jennifer Smith

Senior Marketing Manager: Karen Seitz

Manufacturing Coordinator: Justin Palmeiro

Compositor: GEX Publishing Services

For product information and technology assistance, contact us at
Cengage Learning Customer & Sales Support, 1-800-354-9706

For permission to use material from this text or product, submit all requests online at **cengage.com/permissions**
Further permissions questions can be emailed to
permissionrequest@cengage.com

ISBN-13: 978-0-619-21629-0

ISBN: 0-619-21629-8

Course Technology
25 Thomson Place
Boston, Massachusetts 02210
USA

Cengage Learning is a leading provider of customized learning solutions with office locations around the globe, including Singapore, the United Kingdom, Australia, Mexico, Brazil, and Japan. Locate your local office at: **international.cengage.com/region**

Cengage Learning products are represented in Canada by Nelson Education, Ltd.

For your lifelong learning solutions, visit **course.cengage.com**

Purchase any of our products at your local college store or at our preferred online store **www.ichapters.com**

Oracle is a registered trademark, and Oracle 10g, Oracle9i, Oracle8i, SQL*Plus,iSQL*Plus and PL/SQL, are trademarks or registered trademarks of Oracle Corporation and/or its affiliates.

Microsoft Windows NT, Microsoft Windows 2000, Microsoft Windows 2003, and Microsoft Windows XP are either trademarks or registered trademarks of Microsoft Corporation in the United States and/or other countries. Microsoft is a registered trademark of Microsoft Corporation in the United States and/or other countries. Course Technology is an independent entity from the Microsoft Corporation and not affiliated with Microsoft in any manner.

Printed in Canada
5 6 7 8 9 10 09

BRIEF
Contents

TABLE OF

Contents

CHAPTER FIVE
Introduction to Forms Builder

Preface

The Oracle Database 10*g* software provides a rich environment for illustrating multi-user and client/server database concepts, such as managing concurrent users and sharing database resources, and allows users to develop database applications in a production environment using the Oracle Developer 10*g* utilities. Building on the success of the Morrisons' approach in previous editions, *Guide to Oracle 10*g provides a comprehensive guide for developing database applications using the Oracle 10*g* object relational database and the Developer 10*g* application development utilities. As in previous editions, this book addresses database development activities including using SQL commands to create tables and insert, update, delete, and view data values. It provides an overview of PL/SQL, explores the Developer 10*g* application development tools, and describes how to create an integrated database application. New to this edition is coverage of how to display application components in a Web browser and an overview of database administration activities using the browser-based version of Enterprise Manager.

The Intended Audience

This book is intended to support individuals in a database course in which the instructor has chosen to illustrate database development concepts using an Oracle 10*g* object relational database. The book describes database system concepts and application design techniques, but that is not its primary focus. The book assumes students have already used a programming language such as Visual Basic, Java, C++, or COBOL, or a scripting language such as Perl or JavaScript. No prior database knowledge is necessary, but a basic understanding of relational database concepts is helpful.

The Approach

This book integrates content and theory with tutorial exercises that help you put into practice what you are reading. Each chapter contains one to three lessons. Each lesson concludes with a Chapter Summary and Review Questions that highlight and reinforce the major concepts. Problem-solving Cases appear at the end of each lesson and give you the opportunity to practice and reinforce the techniques presented in the chapter. *Guide to Oracle 10*g emphasizes sound database design and development techniques and GUI design skills in the Microsoft Windows environment. It addresses real-world considerations for creating and managing realistic database development projects.

*Guide to Oracle 10*g distinguishes itself from other Oracle books because it is designed specifically for users and instructors in educational environments. It is written in a clear and easy-to-read style. Each chapter is divided into separate lessons that break down the concepts into manageable sections. The Review Questions challenge students' understanding of the lesson concepts, and provide an excellent source of quiz and examination questions. The Problem-solving Cases at the end of each lesson emphasize the issues that must be considered when developing applications using the Oracle 10*g* software, and are easy to follow, are comprehensive, and have been extensively tested.

Overview and Organization of This Book

The text, examples, tutorials, Review Questions, and Problem-solving Cases in *Guide to Oracle 10*g help you achieve the following objectives:

- Become familiar with relational database concepts and terms, and understand the architecture of a client-server database.

- Understand the Oracle 10*g* SQL commands used to create and manage database tables and to insert, update, and delete database records.

- Understand how to use SQL commands to retrieve data from a relational database.

- Learn how to create PL/SQL programs.

- Learn how to use the Developer 10*g* Forms Builder and Reports Builder tools to create an integrated database application.

- Create PL/SQL named program units that are stored in the database and in the workstation file system.

- Become familiar with the tasks that database administrators perform, and understand how to perform many of these tasks using the Web browser-based Oracle Enterprise Manager utility.

In **Chapter 1**, you learn about client/server and relational database concepts. This chapter also introduces two case study databases that are used throughout the book. These sample databases contain a variety of realistic data values and relationships to enhance your learning. In **Chapter 2**, you learn how to create database objects such as tables and views using SQL*Plus, Oracle's command-line SQL utility. **Chapter 3** provides a comprehensive overview of inserting, updating, deleting, and retrieving data using SQL*Plus. You learn how to join multiple tables, create nested queries, and perform arithmetic operations on retrieved data. In **Chapter 4**, you learn how to write programs in PL/SQL, Oracle's procedural programming language. You will use PL/SQL programs throughout the rest of the book to create database applications.

Chapters 5 through 8 provide instructions for using the Oracle Developer 10*g* application set, which includes Forms Builder and Reports Builder. In **Chapter 5**, you are introduced to Forms Builder, and you create simple database forms to manipulate data values. In **Chapter 6**, you learn how to use Forms Builder to create more complex and flexible forms. In **Chapter 7**, you learn how to create database reports using the Developer 10*g* Reports Builder tool. **Chapter 8** integrates the form and report components by teaching you how to create an integrated database application. New to this edition is coverage on how to display forms and reports in a Web browser window, and how to use the Report Server utility to generate dynamic reports and display them in Web browsers.

In **Chapter 9**, you learn how to create named PL/SQL programs that allow you to store and reuse PL/SQL code. You create PL/SQL programs using both SQL*Plus commands and the Forms Builder utility. In **Chapter 10**, you explore advanced form concepts, including storing and retrieving image data in form applications and creating form applications for databases that contain many records. **Chapter 11** provides you with the opportunity to learn about database administration tasks, and provides hands-on instructions for performing a variety of administrative support tasks, such as creating tablespaces, user accounts, and roles, using Oracle Enterprise Manager. Chapter 11 also provides an overview of the database installation and configuration process, and describes the architecture and components of an Oracle 10*g* database.   Brand new to this edition is an **Appendix** covering *i*SQL*Plus. It explains how to navigate the *i*SQL*Plus interface and clearly demonstrates the differences between *i*SQL*Plus and SQLPlus.

Features

- **Lesson Objectives**. Each lesson begins with a list of the important concepts that you will master in the lesson. This list provides a quick reference to the contents of the lesson as well as a useful study aid.

- **Step-By-Step Methodology**. Each chapter introduces new concepts, explains why the concepts are important, and provides tutorial exercises to illustrate the concepts. When you are learning a new concept, detailed instructions lead you through each step, and the numerous illustrations include labels that direct your attention to how your screen should look. As you proceed through the tutorials, less detailed instructions are provided for familiar tasks, and more detailed instructions are provided for new concepts.

- **Case Approach**. Two running cases present database-related problems that you could reasonably expect to encounter in business, followed by a demonstration of an application that could be used to solve the problem. The sample databases referenced in the two ongoing cases represent realistic client/server applications with several database tables, and require supporting multiple users simultaneously at different physical locations. The Clearwater Traders database represents a standard sales order and inventory system. The Northwoods University database illustrates a student registration system.

- **Data Disks**. Data Disks provide files for creating the sample databases used in the chapters, as well as files needed to complete the tutorials and Problem-solving Cases.

- **HELP?**. These paragraphs anticipate problems you may encounter and help you to resolve them on your own. This feature facilitates independent learning and frees the instructor to focus on substantive issues rather than on common procedural errors.

- **Tips**. These notes provide additional information about a procedure—for example, an alternative method of performing the procedure.

- **CAUTION**. These paragraphs anticipate common errors and provide strategies to help you avoid these errors before they occur.

- **Summaries**. Following each lesson is a Summary that recaps the programming concepts and commands covered in the lesson.

- **Review Questions and Problem-solving Cases**. Each lesson concludes with Review Questions that test your understanding of the concepts you learned. Problem-solving Cases provide you with additional practice of the skills and concepts you learned in the lesson.

In addition to these book-based features, *Guide to Oracle 10*g provides two software components:

- The Oracle 10*g* Developer Suite, (Version 9.0.4) CDs, which are in the envelope bound into this book, enable users to install this software on their own computers at home. Users can then connect to either an Oracle 10*g* Enterprise Edition or Personal Edition database. (The database is not part of the Developer 10*g* software, and must be obtained, installed, and configured separately.) You can use the Developer 10*g* software with Microsoft Windows NT, Windows 2000 Professional or Server, Windows 2003 Server, and Windows XP Professional operating systems. The installation and configuration instructions for Developer 10*g* are available at *www.course.com/cdkit*. Look for this book's title and front cover, and click the link to access the information specific to this book.

Before proceeding to use the software, **you must** register the software and agree to the Oracle Technology Network Developer License Terms in order to receive the key code to unlock the software. Please go to *http://otn.oracle.com/books/*. Upon registering the software, you agree that Oracle may contact you for marketing purposes. You also agree that any information you provide Oracle may be used for marketing purposes.

- The Oracle Database 10*g* Software is available when purchased as a separate bundle with this book. It provides the Oracle 10*g* Enterprise Edition, Standard Edition, and Personal Edition database software, (Version 10.1.0), on a CD. You can use the software included in the kit with Microsoft Windows NT, Windows 2000 Professional or Server, Windows 2003 Server, and Windows XP Professional operating systems. The installation instructions for Oracle 10*g* and the log in procedures are available at *www.course.com/cdkit*. Look for this book's title and front cover, and click the link to access the information specific to this book.

The Oracle Server and Client Software

This book was written and tested using the following software:

- Database Server: Oracle 10g Enterprise Edition Server, Version 10.1.0.2.0, installed on a Windows 2003 Server.

- Oracle 10g Developer Suite, Version 9.0.4.0.1, installed on a Windows XP Professional workstation.

Supplemental Materials

This book is accompanied by the following materials:

- **Electronic Instructor's Manual**. The Instructor's Manual assists in class preparation by providing suggestions and strategies for teaching the text, chapter outlines, technical notes, quick quizzes, discussion topics, and key terms.

- **Solutions Manual**. The Solutions Manual contains answers to end-of-chapter questions and exercises.

- **Sample Syllabi and Course Outline**. The sample syllabi and course outlines are provided as a foundation to begin planning and organizing your course.

- **ExamView Test Bank**. ExamView® allows instructors to create and administer printed, computer (LAN-based), and Internet exams. The Test Bank includes hundreds of questions that correspond to the topics covered in this text, enabling students to generate detailed study guides that include page references for further review. The computer-based and Internet testing components allow students to take exams at their computers, and also save the instructor time by grading each exam automatically. The Test Bank is also available in Blackboard and WebCT versions posted online at *www.course.com*.

- **PowerPoint Presentations**. Microsoft PowerPoint slides for each chapter are included as a teaching aid for classroom presentation, to make available to students on the network for chapter review, or to be printed for classroom distribution. Instructors can add their own slides for additional topics they introduce to the class.

- **Figure Files**. Figure and table files from each chapter are provided for your use in the classroom.

- **Data Files**. Data Files, containing all of the data necessary for steps within the chapters and Problem-solving Cases, are provided through the Course Technology Web site at *www.course.com*, and are also available on the Instructor's Resources CD-ROM.

ACKNOWLEDGMENTS

This book rests on a sound foundation, the proven approach that Joline and Mike Morrison established in previous editions. Many thanks also to the individuals involved in the development and production of this book, including Eunice Yeates, Senior Product Manager; Lynne Raughley, Development Editor; Kristen Guevara, Production Editor; and Quality Assurance Testers: Chris Scriver, Serge Palladino, Shawn Day, and Susan Whalen. I am grateful to the many reviewers who provided helpful and insightful comments during the development of this book. They include; Angela Mattia, J. Sargeant Reynolds Community College; Reni Abraham, Houston Community College; Varsha Leiseth, Dunwoody College of Technology; Barbara Nicolai, Purdue University Calumet; and David Spaisman, Katharine Gibbs School. Special thanks to Angela Mattia for her help with the Appendix. I would like to thank my wife Donna and my three children, Katrina, Cathlena, and Clayton for their loving support.

Rocky Conrad

Read This Before You Begin

TO THE USER

Data Files

To complete the steps and projects in this book, you need data files that have been created for this book. Your instructor will provide the data files to you. You also can obtain the files electronically from the Course Technology Web site by connecting to *www.course.com*, and then searching for the title of this book. Each chapter in the book has its own set of data files. These include files used in the tutorials, and also include any files that may be needed for end-of-lesson Problem-solving Cases.

In the book, you are usually asked to open files from the C:\OraData drive and folder path. Create a folder named OraData in the root of your C: drive, and copy the data files to the OraData folder. You can store your data files in an alternative location, however, and you are usually reminded that your drive letter (and path) may be different. Throughout, keep in mind that it is always possible to use a different drive. You are free to save files wherever it is convenient for you.

Solution Files

Throughout this book you will be instructed to save files to your Solution Disk (for example, "save the file as Ch10Inventory.fmb in the Chapter10\Tutorials folder on your Solution Disk"). Therefore, it is important to make sure you are using the correct Solution Disk when you begin working on each chapter. To improve application performance, we recommend that you save the files on a hard drive rather than on floppy disks. You should create a folder named OraSolutions on the C: drive of your computer (or an alternate drive on which you wish to store your solution files.) Then create the following folders within the OraSolutions folder: Chapter1, Chapter2, Chapter3, Chapter4, Chapter5, Chapter6, Chapter7, Chapter8, Chapter9, Chapter10, and Chapter11. Within each of these folders create two subfolders. Name the first subfolder Tutorials and the second subfolder Cases.

Using Your Own Computer

You can use your own computer to complete the chapter exercises and end-of-chapter Problem-solving Cases in Chapters 2 through 10. To use your own computer, you must have the following:

- **350MHz Pentium II or faster computer.** This computer must have at least 256 MB of RAM and at least 2 GB of free hard disk space to install the Developer 10*g* tools. If you choose to install the Oracle 10*g* Personal Database, you need 3 GB of free hard disk space. If you choose to install both Developer 10*g* and Oracle 10*g* Personal Edition, you need 5 GB of free hard disk space.

- **Microsoft Windows 2000 Server or Professional with Service Pack 1 or higher, or Windows XP Professional.** You cannot install Developer 10*g* or the Oracle 10*g* database on Windows XP Home Edition, Windows ME, or Windows 98 operating systems.

- **Oracle 10*g* Developer Suite tools (Version 9.0.4).** The Oracle Developer 10*g* (Version 9.0.4) tools are included on two CDs in an envelope bound into this book. The installation and configuration instructions for Developer 10*g* are available at *course.com/cdkit*. Look for this book's title and front cover, and click the link to access the information specific to this book.

Before proceeding to use the software, **you must** register the software and agree to the Oracle Technology Network Developer License Terms in order to receive the key code to unlock the software. Please go to *http://otn.oracle.com/books/*. Upon registering the software, you agree that Oracle may contact you for marketing purposes. You also agree that any information you provide Oracle may be used for marketing purposes.

- **Oracle 10*g* Database (Version 10.1.0).** You must install either an Oracle 10*g* Personal Edition database (Version 10.1.0.2.0 or later) on your computer, or you must be able to access an Oracle 10*g* Enterprise database server that has been set up for you by your instructor. The Course Technology Kit for Oracle 10*g* Software contains the database software. Your instructor will provide instructions on how to access the Oracle 10*g* database server.

- The Oracle 10*g* Database Software, which is available as a separate bundle with this book, contains the database software necessary to perform all the Oracle 10*g* database tasks shown in this textbook. Detailed installation, configuration, and log on information are provided at *course.com/cdkit*. Look for this book's title and front cover, and click the link to access the information specific to this book.

- **Data Disk files.** You will not be able to complete in-chapter tutorial exercise and cases in this book, using your own computer, unless you have the Data Disk files. You can get the Data Disk files from your instructor, or you can obtain them electronically from the Course Technology Web site at *course.com* and then searching for this book title.

A Note on Syntax

This book contains many examples of code statements. Sometimes these statements are too long to fit on a single line in this book. A single line of code that must appear on two lines in the book due to its length appears with an arrow in the right margin as follows:

```
extracted_string := SUBSTR(original_string, ↵
starting_point, number_of_characters);
```

Visit Our World Wide Web Site

Additional materials designed especially for you might be available for your course on the World Wide Web. Go to *course.com* periodically, and search this site for more details.

TO THE INSTRUCTOR

To complete the chapters in this book, your users must use a set of data files. These files are included in the Instructor's Resources. The Instructor's Resources also contains solutions to all of the Tutorials and Problem-solving Cases used in the book. You may also obtain the data and solution files electronically through the Course Technology Web site at *course.com*. Follow the instructions in the Help file to copy the data files to your server or standalone computer. You can view the Help file using a text editor such as WordPad or Notepad. Once the files are copied, you should instruct your users how to copy the data files to their own computers or workstations.

The Tutorials and Problem-solving Cases in this book were tested on Windows XP Professional workstations running Oracle Developer 10*g* (Version 9.0.4). On the database side, the Tutorials and Problem-solving Cases were tested using Oracle 10*g* Enterprise Edition, (Version 10.1.0.2.0), running on Windows 2003 Server.

To complete the chapter exercises and Problem-solving Cases in Chapters 2 through 10, your users must have access to the Developer 10*g* tools and be able to connect to either an Oracle 10*g* (version 10.1.0) Enterprise Edition database or Personal Oracle 10*g* database. If your users are connecting to an Enterprise Edition database, you must create a separate account for each user, and you must configure these accounts so they have specific privileges that enable the students to perform the tasks in each chapter. Chapter 11 contains detailed instructions for creating user accounts. User accounts need to be granted the following privileges:

- CREATE SESSION
- CREATE TABLE
- CREATE SEQUENCE
- CREATE VIEW
- CREATE PROCEDURE
- CREATE TRIGGER

In addition, to complete the exercises in Chapter 11, students must be granted the DBA role.

To complete the chapter exercises and Problem-solving Cases in Chapter 11, your users must have access to a Web browser. Some of the exercises in Chapter 11 require users to perform configuration tasks. If your users are connecting to an Enterprise Edition database, you must supply the users with the global database name and SID. Some of the exercises in Chapter 11 require the users to shut down and start the database, so you may need to make special provisions to allow your students to complete these exercises.

Course Technology Data Files

You are granted a license to copy the data files to any computer or computer network used by individuals who have purchased this book.

1

CLIENT/SERVER DATABASES AND THE ORACLE 10g RELATIONAL DATABASE

Objectives

After completing this chapter, you should be able to:

♦ Develop an understanding of the purpose of database systems

♦ Describe the purpose and contents of an entity-relationship model

♦ Explain the steps in the normalization process

♦ Describe the differences between personal and client/server databases

♦ Understand the Northwoods University student registration data-base and the Clearwater Traders sales order database

---

Organizations use databases to store and maintain data in an organized way. If an organization needs to store and maintain a large amount of data that must be viewed and updated by many users at the same time, it often uses a client/server database such as Oracle 10g. The Oracle 10g database environment includes a client/server database management system (DBMS) and utilities for developing and managing database applications. In this book, you will learn how to use these utilities and practice using them to build database applications. You will also learn how to perform database administration tasks, and how to configure a Web server that allows users to view and update data using Web-based applications.

TIP

This book focuses on using Oracle 10*g* to create database tables, retrieve database data, and create database applications. Determining database requirements and designing database tables is significantly different from the more mechanical task of creating them. This book provides general guidelines for determining database requirements and designing database tables; however, it does not provide a rigorous treatment of the subject.

In this chapter, you will learn about database systems and become familiar with relational database concepts and terms. You will learn about personal and client/server databases, and understand the differences between these two types of databases. And, you will become familiar with the two case study databases you will use throughout the book: the Northwoods University student registration database and the Clearwater Traders sales database.

DATABASE SYSTEMS

When organizations first began converting from manual to computerized data-processing systems, each individual application had its own set of data files that were used only for that application. Figure 1-1 shows an example of a data file that contains information about students. This file contains individual data items, or **fields**, that describe characteristics about students, such as each student's ID, last name, first name, middle initial, address, city, state, and zip code. Fields are also called **columns**. In the data file in Figure 1-1, a comma separates each field. A **record** is a collection of related fields that contain related information. In Figure 1-1, each student record appears on a separate line in the file.

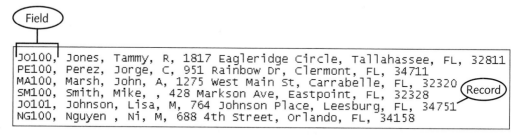

Figure 1-1   Example student data file

As applications became more complex, organizations encountered problems with managing data stored in data files. One problem was the proliferation of data management programs. For example, in a student registration system, the system to enable students to enroll in courses would have its own data file, the system to create class lists would have its own file, and the system to create student transcripts would have its own file. Because each data file contained different data fields, programmers had to write and maintain a separate program to insert, update, delete, and retrieve data that was stored in each different data file, as shown in Figure 1-2. For a large data-processing system with many data files, this involved a good many programs that all performed essentially the same tasks.

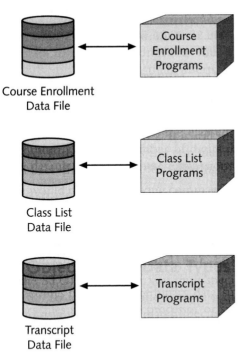

Figure 1-2   Proliferation of data management programs in file-based data processing

Another problem with data files was the presence of redundant data. Figure 1-3 shows an example of a data file that contains information about the courses in which each student has enrolled.

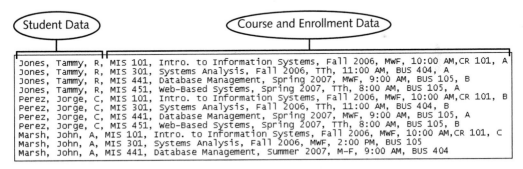

Figure 1-3   Enrollment data file containing redundant data

This file contains fields representing the student's last name, first name, and middle initial. The file also contains enrollment information for the student, including the course name and associated course number (such as MIS 101), the term, the days, time, and location of the class, and the student's course grade, if assigned. Note that the student names repeat for each course in which each student is enrolled. Because each student will probably take many courses during his or her academic career, the file will store the student's last name, first name, and middle initial many times. For a system that contains tens of thousands of students who enroll in several courses over their academic careers, the disk space required to store this redundant data becomes significant.

Also, note that the information for each course section repeats for each student who enrolls in the course. If data about a course needs to be updated, such as the day, time, or location, it must be updated for each student/enrollment record, which makes the application for updating enrollment records more complex.

Yet another problem with data files is that they might contain inconsistent data. If a student changes his or her name, the name data must be updated in both the student file shown in Figure 1-1 and the enrollment file shown in Figure 1-3. If different files contain different data values for the same student, the system is inconsistent and errors occur.

To address these problems with data files, programmers developed databases to store and manage application data. A **database** stores all organizational data in a central location. A good database design strives to eliminate redundant data to reduce the possibility of inconsistent data. In a database system, a single application called the **database management system** (**DBMS**) performs all routine data-handling operations. The DBMS provides a central set of common functions for managing a database, which include inserting, updating, retrieving, and deleting data values. For example, all applications in the student registration system, such as the ones to support the enrollment, class list, and transcript systems, interface with the DBMS to access the database data. The person who is responsible for installing, administering, and maintaining the database is called the **database administrator**, or **DBA**.

Figure 1-4 illustrates the database approach to data processing.

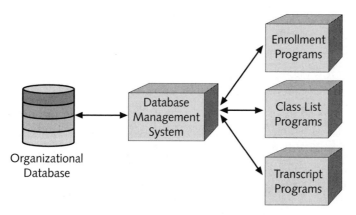

Figure 1-4   Database approach to data processing

OVERVIEW OF RELATIONAL DATABASES

Most modern databases are **relational databases**, which store data in a tabular format. A relational database organizes data in **tables**, or matrixes with columns and rows. Columns represent different data fields—the characteristics or **attributes** about the entity. The rows contain individual records—all the attributes about a specific **instance** of an entity. Figure 1-5 shows an example of two relational database tables.

(Column/Field)

STUDENT

S_ID	S_LAST	S_FIRST	S_MI	S_ADDRESS	S_STATE	S_ZIP
JO100	Jones	Tammy	R	1817 Eagleridge Circle	FL	32811
PE100	Perez	Jorge	C	951 Rainbow Dr	FL	34711
MA100	Marsh	John	A	1275 West Main St	FL	32320
SM100	Smith	Mike		428 Markson Ave	FL	32328
JO101	Johnson	Lisa	M	764 Johnson Place	FL	34751
NG100	Nguyen	Ni	M	688 4th Street	FL	34158

FACULTY

F_ID	F_LAST	F_FIRST	F_MI	F_PHONE	F_RANK
1	Marx	Teresa	J	4075921695	Associate
2	Zhulin	Mark	M	4073875682	Full
3	Langley	Colin	A	4075928719	Assistant
4	Brown	Jonnel	D	4078101155	Full

(Row/Record)

Figure 1-5 Examples of relational database tables

NOTE

In this book, database table and field names appear in all capital letters, such as STUDENT and FACULTY. Individual words are separated by underscores, such as S_ID for student ID and S_ADDRESS for student address.

In database terminology, an **entity** is an object about which you want to store data (that is, students, faculty members, courses, and so on). In a relational database, a different table stores data about each different entity. For example, in Figure 1-5, one table stores data about students, and a second table stores data about faculty members. A third table might store data about courses. Storing data about different entities in different tables minimizes storing redundant data. To connect information about different entities, you must create **relationships**, which are links that show how different records are related. For example, a student can enroll in many courses, and each course can be composed of many different students. Therefore, you must create a relationship between the course record and the student record to determine which students are in which courses. Relationships among records in different tables are established through **key fields**, values in database tables that help to either identify an individual row or to link data from different tables. There are five main types of key fields in a relational database: primary keys, candidate keys, surrogate keys, foreign keys, and composite keys. The following sections describe these types of key fields in more detail.

Primary Keys

A **primary key** is a column in a relational database table whose value must be unique for each row. The primary key serves to identify the individual occurrence of an entity. Every row must have a primary key, and the primary key cannot be NULL. **NULL** is not the equivalent of a blank space or a zero. NULL means that a value is absent or unknown so no entry is made for that data element. In the STUDENT table in Figure 1-5, the S_ID column is a good choice for the primary key because a unique value can be assigned for each student. The S_ADDRESS (student address) column might be another choice for the primary key, but there are three problems with using this column. First, two different students might have the same address (as in roommates or members of the same family). Second, the column contains text data that is prone to data entry errors, such as spelling errors and variations in capitalization. Being prone to data entry errors is a problem because database developers use primary key values to create relationships with other database tables, and the values must *exactly* match in the different tables. Third, the student's address might change, which could result in multiple records being included in the table for the same student. The best choice for a primary key is a column that contains numerical data or some type of abbreviated code such as the last two letters of the student's last name plus a numeric value, as shown in the S_ID column of the STUDENT table in Figure 1-5.

TIP

Ideally, you should use only numeric values to identify each record in a table because there are fewer chances of data entry errors, and data processing is faster with numbers than with characters. However, for a table that contains thousands, possibly millions, of records, searches for a specific individual tend to be quicker if the primary key field contains one to three letters of the person's name (for example, a driver's license number starting with the first letter of the driver's last name) and the data is sorted in order of the primary key. In addition, sometimes the primary key field is designed to also be part of a coding scheme (for example, VIN for an automobile), so that characteristics about the item can be quickly identified.

Candidate Keys

When you are designing a relational database table, many columns might be possible choices for the table's primary key. A **candidate key** is any column that could be used as the primary key. In contrast, a primary key is the candidate key that you actually select to be the table's unique identifier. A candidate key should be a column that is unique for each record and does not change.

TIP

Good candidate key choices include identification numbers such as product stock-keeping units (SKUs), book ISBN numbers, and student identification numbers. Because of increases in identity theft and privacy issues, Social Security numbers are not considered good candidate keys. For these reasons, even if the database contains an individual's Social Security number, that number should never be considered the main method of identifying a person within the database. If there is no alternate data available to uniquely identify a person, create a surrogate key as discussed in the next section.

Surrogate Keys

What do you do when a table does not contain any suitable candidate keys? Figure 1-6 shows a FACULTY table with this problem.

FACULTY

F_LAST	F_FIRST	F_MI	F_PHONE	F_RANK
Marx	Teresa	J	4075921695	Associate
Zhulin	Mark	M	4073875682	Full
Langley	Colin	A	4075928719	Assistant
Brown	Jonnel	D	4078101155	Full

Figure 1-6   Table lacking suitable candidate keys

The F_LAST or F_FIRST columns, or even the combination of the two columns, are not suitable candidate keys because many people have the same name. Telephone numbers are not good candidate keys because people often change their telephone numbers and several people can share the same telephone (for example, members of the same household). To address the problem of a table with no suitable candidate keys, a good database design practice is to create a surrogate key. A **surrogate key** is a column that you create to be the record's primary key identifier. A surrogate key has no real relationship to the row to which it is assigned, other than to identify the row uniquely. Usually, developers configure the database to generate surrogate key values automatically.

In an Oracle 10*g* database, you can automatically generate surrogate key values using a sequence. **Sequences** are sequential lists of numbers the database automatically generates and that guarantee each primary key value is unique. For the FACULTY table in Figure 1-6, you could create a new surrogate key field named F_ID, and you might start the F_ID values at 1 or 100. The F_ID numbers will not change, and every faculty member gets a unique number. Surrogate keys are always numerical columns, because the database generates surrogate key values automatically by incrementing the previous value by one.

NOTE

The Oracle DBMS does include the option of increasing the sequence values by a specific integer. In the case of a financial institution, you would not want each account number to be separated by only one number. Instead, to decrease the chances of posting an entry to the wrong account due to transposed numbers, you could have Oracle 10g increase the value by 1307 for each account number generated. This prevents the sequence from generating two account numbers that are only different by one digit.

Foreign Keys

How do you use key fields to create relationships among records? In Figure 1-5, no relationships exist between the rows in the STUDENT table and the rows in the FACULTY table. However, every student has a faculty advisor, so there must be a way to connect students with faculty members. One way to represent this relationship is to combine the FACULTY and STUDENT fields into a single table, and store the data for each student's faculty advisor directly in each student row, as shown in Figure 1-7.

Redundant faculty data

STUDENT

S_ID	S_LAST	S_FIRST	S_MI	S_ADDRESS	S_STATE	S_ZIP	F_LAST	F_FIRST	F_MI	F_PHONE	F_RANK
JO100	Jones	Tammy	R	1817 Eagleridge Circle	FL	32811	Zhulin	Mark	M	4073875682	Full
PE100	Perez	Jorge	C	951 Rainbow Dr	FL	34711	Langley	Colin	A	4075928719	Assistant
MA100	Marsh	John	A	1275 West Main St	FL	32320	Zhulin	Mark	M	4073875682	Full
SM100	Smith	Mike		428 Markson Ave	FL	32328	Marx	Teresa	J	4075921695	Associate
JO101	Thomley	Lisa	M	764 Johnson Place	FL	34751	Zhulin	Mark	M	4073875682	Full
NG100	Nguyen	Ni	M	688 4th Street	FL	34158	Brown	Jonnel	D	4078101155	Full

Figure 1-7   Creating relationships by repeating data

The problem with this approach is that it stores redundant data. Note that the data for faculty member Mark Zhulin is repeated three times, once for each student he advises. If Mark advises 50 or 100 students, his data is repeated 50 or 100 times. Recall that storing redundant data is undesirable because it occupies valuable storage space. And, when the same data is stored in multiple places, it can become inconsistent.

Relational databases provide a better way to create relationships by using foreign keys. A **foreign key** is a column in a table that is a primary key in another table. The foreign key creates a relationship between the two tables. Rather than storing all of the faculty member's data in each associated student's record, you can simply store the F_ID, which is the primary key of the FACULTY table, in each student record. Figure 1-8 shows the STUDENT table, now containing the F_ID field for each student's advisor as a foreign key in each student record. This creates a relationship between the STUDENT table and the FACULTY table. Notice that the 4 stored in the F_ID column for Ni Nguyen identifies Jonnel Brown in the FACULTY table as his faculty advisor.

STUDENT

S_ID	S_LAST	S_FIRST	S_MI	S_ADDRESS	S_STATE	S_ZIP	F_ID
100	Jones	Tammy	R	1817 Eagleridge Circle	FL	32811	2
101	Perez	Jorge	C	951 Rainbow Dr	FL	34711	3
102	Marsh	John	A	1275 West Main St	FL	32320	2
103	Smith	Mike		428 Markson Ave	FL	32328	1
104	Johnson	Lisa	M	764 Johnson Place	FL	34751	2
105	Nguyen	Ni	M	688 4th Street	FL	34158	4

FACULTY

F_ID	F_LAST	F_FIRST	F_MI	F_PHONE	F_RANK
1	Marx	Teresa	J	4075921695	Associate
2	Zhulin	Mark	M	4073875682	Full
3	Langley	Colin	A	4075928719	Assistant
4	Brown	Jonnel	D	4078101155	Full

Figure 1-8   Creating relationships using foreign keys

By using foreign keys to create relationships, you repeat only the foreign key values. If the data values for a faculty member change, you update the values only in the FACULTY table. The primary key of the FACULTY table, which is the F_ID value, does not change, so the STUDENT table does not need to be updated if the faculty data changes.

A foreign key value must exist in the table where it is a primary key. For example, suppose you have a new record for S_ID JO100 that specifies that the student's F_ID (advisor ID) value is 5. Currently, there is no record for F_ID 5 in the FACULTY table, so the student record is invalid. Foreign key values must match the value in the primary key table *exactly*. So unless the primary key is being used for coding purposes, it is best to use number values for primary keys rather than text values that are prone to typographical, punctuation, spelling, and case-variation errors. If the concept of where to place a foreign key seems a little confusing, the normalization process discussed in a later section should help clarify how to organize all the data elements.

Composite Keys

Sometimes database designers have to combine multiple columns to create a unique primary key. The ENROLLMENT table in Figure 1-9 shows an example of this situation. In this table, the S_ID column is not a candidate key, because one student may enroll in many courses. Similarly, the COURSE_NO column is not a candidate key, because one course may have many students enrolled. The GRADE column is not a candidate key, because many students receive the same grade in courses.

As an alternative to adding a surrogate column for the primary key, a combination of the current columns may create a value that can uniquely identify each row in the table. A **composite key** is a unique key that you create by combining two or more columns. For example, because each student enrolls in each course only once, the combination of the S_ID column and the COURSE_NO column represents a unique identifier for each record. The combination of the S_ID and GRADE columns is not unique unless a student could never receive two As, Bs, and so on while completing coursework at the university. In addition, the combination of the COURSE_NO and GRADE columns is not unique because more than one student can earn the same grade in the same class. Therefore, the combination of values in the S_ID and COURSE_NO columns is the most practical method of uniquely identifying each row in the ENROLLMENT table.

A composite key usually comprises fields that are primary keys in other tables. As shown in Figure 1-9, the S_ID field in the ENROLLMENT table is part of the table's primary key, and is also a foreign key that references an S_ID value in the STUDENT table. Similarly, the COURSE_NO field in the ENROLLMENT table is part of the primary key, and is also a foreign key that references a COURSE_NO value in the COURSE table.

STUDENT

S_ID	S_LAST	S_FIRST	S_MI	S_ADDRESS	S_STATE	S_ZIP	F_ID
JO100	Jones	Tammy	R	1817 Eagleridge Circle	FL	32811	2
PE100	Perez	Jorge	C	951 Rainbow Dr	FL	34711	3
MA100	Marsh	John	A	1275 West Main St	FL	32320	2
SM100	Smith	Mike		428 Markson Ave	FL	32328	1
JO101	Johnson	Lisa	M	764 Johnson Place	FL	34751	2
NG100	Nguyen	Ni	M	688 4th Street	FL	34158	4

ENROLLMENT

S_ID	COURSE_NO	GRADE
JO100	CGS 270	C
NG100	CGS 270	B
MA100	MIS 100	A
SM100	COP 174	B
PE100	CGS 270	A
JO100	COP 174	B
JO101	CGS 210	B
MA100	CGS 210	C

COURSE

COURSE_NO	COURSE_NAME	CREDITS
MIS 100	Information Systems	3
CGS 210	Computer Applications	3
CGS 270	Database Management	3
COP 174	Introduction to Java	3

Figure 1-9   Example of a composite key

DATABASE DESIGN

In the preceding section, it is fairly simple to identify which data elements should be stored in each table and how each table should be linked. However, in the real world, things are usually not so simple. In some cases, you are told only that the organization needs to store or track certain data, and you have to decide the best way to sort the data into tables. This can be a particularly daunting task if the database is required to store information on thousands of different data elements for numerous entities. Although the focus of this textbook is not to teach you everything about database design, you should have at least a basic understanding of how to structure the database tables. There are two main tasks involved with the design of a database: developing an entity-relationship (ER) model and normalizing the database tables. Each of these is discussed in the following subsections.

Entity-relationship Model

An ER model is designed to help you identify which entities need to be included in the database. A basic ER model is composed of squares (or rectangles) and lines. The squares represent the entities, whereas the lines depict the relationship between the entities. Three types of relationships can exist between two entities: one-to-one, one-to-many, and many-to-many.

In a **one-to-one (1:1) relationship,** each occurrence of a specific entity, called an **instance**, is found only once in each set of data. For example, if you worked for the Bureau of Vital Statistics and were defining the relationship between birth certificates and individuals, you would use a one-to-one relationship. Why? Because legally each birth certificate is issued for only one person, and each person has only one birth certificate. Figure 1-10 shows an ER model for this 1:1 relationship. Notice that the line drawn between the two entities is a simple straight line. A straight line is used to represent a one-to-one relationship.

Figure 1-10   Example of a one-to-one relationship

One-to-one relationships are rare in a relational database; usually you work with one-to-many relationships. In a **one-to-many (1:M) relationship**, an instance can only appear once in one entity, but one or more times in the other entity. For example, a publisher may publish several different books, but each book sold at a bookstore has only one publisher. This means that after the database is created, each publisher would be listed in a PUBLISHER table only once, but each publisher could have several books listed in the BOOK table. Each of these books would have a foreign key field that links back to a specific record in the PUBLISHER table. Figure 1-11 shows the 1:M relationship between the two entities. The ER model uses a straight line with a crow's foot on the BOOK portion of the relationship to identify the "many" side of the relationship.

Figure 1-11   Example of a one-to-many relationship

The final type of relationship is a **many-to-many (N:M) relationship**. In a many-to-many relationship, an instance can occur multiple times in each entity. For example, a student can take many different classes in the same term, and each class can be composed of many different students. A crow's foot is included at both ends of the relationship line to indicate an N:M relationship, as shown in Figure 1-12.

Figure 1-12   Example of a many-to-many relationship

An N:M relationship is of particular importance to a database designer. This type of relationship cannot be represented in the physical database (i.e., in the actual database tables). Recall the data displayed in Figure 1-3. The data represents an N:M relationship because many students were enrolled in the same course and each course had more than one student. If you tried to convert that data set directly into two database tables, one table being the STUDENT table and the other being the COURSE table, there would be no way to correctly link the data in the tables together. This basically means it would be nearly impossible to correctly retrieve data from the database.

So, how do you eliminate the problem caused by an N:M relationship? During the design of the actual database tables, the N:M relationship is broken down into a series of two or more 1:M relationships through the use of a **linking table** in the process of normalization. In many cases, the linking table is simply a table that contains the primary key of each of the tables being connected. The following section on normalization demonstrates use of a linking table.

Normalization

Normalization is a step-by-step process used by database designers to determine which data elements should be stored in which tables. The purpose of normalization is to eliminate data redundancy. Beginning with unnormalized data, the designer can complete a series of steps to convert the data to a normalized form. Although there are several normalized forms, most designers are concerned only with first, second, and third normal forms. These forms are presented in the following sections.

Unnormalized Data

Unnormalized data does not have a primary key identified and/or contains repeating groups. Look at the data shown in Figure 1-13.

STUDENT

S_ID	S_LAST	S_ADDRESS	S_STATE	S_ZIP	COURSE_NO	COURSE_NAME	CREDITS	GRADE	F_ID	F_LAST
JO100	Jones	1817 Eagleridge Circle	FL	32811	CGS 270	Database Management	3	C	1	Marx
					COP174	Introduction to Java	3	B		
PE100	Perez	951 Rainbow Dr	FL	34711	CGS 270	Database Management	3	A	3	Langley
MA100	Marsh	1275 West Main St	FL	32320	MIS 100	InformationSystems	3	A	4	Brown
					CGS 210	Computer Applications	3	C		
SM100	Smith	428 Markson Ave	FL	32328	COP 174	Introduction to Java	3	B	2	Zhulin
JO101	Johnson	764 Johnson Place	FL	34751	CGS 210	Computer Applications	3	B	4	Brown
NG100	Nguyen	688 4th Street	FL	34158	CGS 270	Database Management	3	B	3	Langley

Figure 1-13   Example of unnormalized data

Some of the data, such as the student's first name, faculty member's first name, and so on, have been removed to conserve space.

TIP

Notice in Figure 1-13 that students JO100 and MA100 are both taking more than one course. The multiple entries in the COURSE_NO, COURSE_NAME, CREDITS, and GRADE fields for these two students are known as repeating groups. The easiest way to remove repeating groups is to create a separate record for each value in the repeating group. As shown in Figure 1-14, creation of a separate record simply involves placing each value in a separate row and then duplicating the information from the remaining fields.

STUDENT

S_ID	S_LAST	S_ADDRESS	S_STATE	S_ZIP	COURSE_NO	COURSE_NAME	CREDITS	GRADE	F_ID	F_LAST
JO100	Jones	1817 Eagleridge Circle	FL	32811	CGS 270	Database Management	3	C	1	Marx
JO100	Jones	1817 Eagleridge Circle	FL	32811	COP174	Introduction to Java	3	B	1	Marx
PE100	Perez	951 Rainbow Dr	FL	34711	CGS 270	Database Management	3	A	3	Langley
MA100	Marsh	1275 West Main St	FL	32320	MIS 100	InformationSystems	3	A	4	Brown
MA100	Marsh	1275 West Main St	FL	32320	CGS 210	Computer Applications	3	C	4	Brown
SM100	Smith	428 Markson Ave	FL	32328	COP 174	Introduction to Java	3	B	2	Zhulin
JO101	Johnson	764 Johnson Place	FL	34751	CGS 210	Computer Applications	3	B	4	Brown
NG100	Nguyen	688 4th Street	FL	34158	CG 2270	Database Management	3	B	3	Langley

Figure 1-14   Example of unnormalized data with repeating groups removed

Now that the repeating groups have been removed, a primary key field must be identified to convert the data into **first normal form (1NF)**. First normal form means that the data has been organized in such a manner that it has a primary key and no repeating groups. Recall that a primary key is a field that uniquely identifies a record within a table. Previously, the S_ID field was used to identify each student. However, in Figure 1-14, there is more than one entry for each student, so the S_ID field does not contain unique values. As you examine the table, notice that no single field can be used to identify a record. Therefore, a composite key is required. Remember from the previous section that a composite key is created from multiple fields. In this example, the S_ID and COURSE_NO fields can be used as a composite primary key because no two records have identical entries in both fields. Now the table is in 1NF.

A shorthand method of identifying a table and its contents is to give the name of the table followed by a list of the field names, separated by commas, within a set of parentheses. The first normal form of the data given previously in Figure 1-13 is: STUDENT(S_ID, S_LAST, S_ADDRESS, S_STATE, S_ZIP, COURSE NO, COURSE_NAME, CREDITS, GRADE, F_ID, F_LAST). Notice that the S_ID and COURSE_NO fields are underlined in the field list. The underlining indicates that these two columns represent the primary key for the table.

Now that the table is in 1NF, it is time to convert to **second normal form (2NF)**. A table is in 2NF if it fulfills these two conditions: it is in 1NF, and it has no partial dependencies. Because of the previous steps performed, the table meets the first condition, so it is time to examine the second condition—partial dependencies. A **partial dependency** means that the fields within the table are dependent only on part of the primary key. This problem can only exist if there is a composite primary key, which the STUDENT table currently has in the form of the S_ID and COURSE_NO fields.

The basic procedure for identifying a partial dependency is to look at each field that is not part of the composite primary key and make certain you are required to have *both* parts of the composite field to determine the value of the data element and not just one part of the composite field. For example, is the value in the COURSE_NAME field determined by the combination of values in both the S_ID and COURSE_ID fields? No—the name of course depends only on the value in the COURSE_ID field. Therefore, there is at least one partial dependency currently in the STUDENT table.

To remove partial dependencies from a table, list each part of the composite key, as well as the entire composite key, as separate entries, as shown in Figure 1-15. Then examine the remaining fields, and determine which attribute, or characteristic, is determined by each portion of the composite primary key.

```
S_ID
COURSE_NO
S_ID + COURSE_NO
```

Figure 1-15 Listing of the components of the composite primary key

The first field to examine is S_LAST, which contains the student's last name. In this case, the last name is determined by which student is being referenced, not by the course. Therefore, the S_LAST field should be assigned to the list of fields determined by the student's ID, as shown in Figure 1-16.

```
S_ID, S_LAST
COURSE_NO
S_ID + COURSE_NO
```

Figure 1-16   Field assignment to eliminate partial dependency

As you begin examining each of the remaining fields from Figure 1-14, you should notice that the value S_ADDRESS, S_STATE, S_ZIP and information regarding the student's advisor (F_ID and F_LAST) all depend on which student is being examined rather than the course. Therefore, each of these fields is dependent only on the S_ID portion of the composite primary key. In addition, the name of the course and the credits earned for course completion are all based on which course number is stored in the COURSE_NO field, and not which value is assigned to the S_ID field. But what about the value stored in the GRADE field? In this particular case, the grade *is* based on the combination of values in the S_ID and COURSE_NO fields (i.e., a grade is entered for a particular student in a particular course). The completed assignment of fields based on the primary key dependency is shown in Figure 1-17.

```
S_ID, S_LAST, S_ADDRESS, S_STATE, S_ZIP, F_ID, F_LAST
COURSE_NO, COURSE_NAME, CREDITS
S_ID + COURSE_NO, GRADE
```

Figure 1-17   Completed field assignment example

After you have figured out which field is determined by which portion of the composite primary key, in effect you have divided the data into different tables. Data regarding the student is placed in one table, data regarding the courses is placed in another table, and the student's grade is placed in a third table. Of course, each table is given a name to reflect the data it contains, as shown in Figure 1-18. Are the tables' contents starting to resemble the tables previously shown in Figure 1-9?

```
STUDENT(S_ID, S_LAST, S_ADDRESS, S_STATE, S_ZIP, F_ID, F_LAST)
COURSE(COURSE_NO, COURSE_NAME, CREDITS)
ENROLLMENT(S_ID, COURSE_NO, GRADE)
```

Figure 1-18   Results of converting the STUDENT table to second
                   normal form

Before assuming the tables are now in 2NF, go back and reexamine each table. First, is each table in 1NF? In this case, yes—each table has a primary key and no repeating groups. Now, are the tables in 2NF? Again, yes—there are no partial dependencies in the

ENROLLMENT table. The STUDENT and COURSE tables are automatically in 2NF because they do not have composite primary keys *and* they are in 1NF. In other words, a table is automatically in 2NF if its primary key consists of only one field.

The final step in the normalization process is to convert the tables to **third normal form (3NF)**. A table is considered to be in 3NF if it is in 2NF and does not have any **transitive dependencies**. A transitive dependency means that a field is dependent on another field within the table that is *not* the primary key field. The ENROLLMENT table is in 3NF because the value in the GRADE field is dependent on the primary key, the S_ID, and COURSE_NO fields. In addition, the COURSE table is in 3NF because the course name and credit hours are dependent on which course is identified by the COURSE_NO field, which is the table's primary key. But what about the STUDENT table? Although the student's last name, address information, and advisor values depend on which student ID is being referenced, the advisor's last name is actually dependent on which advisor ID is included in the student record, not on the student. This would be especially obvious if the table also included data such as the advisor's phone number, office location, and so on. Therefore, the advisor's last name should be removed from the STUDENT table. Because the advisor's last name is not relevant to the primary keys of the COURSE and ENROLLMENT tables, it is placed in a separate table named FACULTY, along with the F_ID field to uniquely identify each entry in the new FACULTY table. The new table is shown in Figure 1-19, along with the revised STUDENT table. Now that all fields in the STUDENT table are entirely dependent on the primary key (S_ID), the STUDENT table is now in 3NF.

```
FACULTY(F_ID, F_LAST)
STUDENT(S_ID, S_LAST, S_ADDRESS, S_STATE, S_ZIP, F_ID)
```

Figure 1-19   Results of removing transitive dependency from the STUDENT table

Notice in Figure 1-18 that the F_ID field is still included in the STUDENT table. The F_ID field is the foreign key that will be used to link the student to his or her advisor's name, which is stored in the FACULTY table. The normal procedure to follow after converting all tables in the database to 3NF is to double-check each table and make certain that all tables representing entities that have a relationship are linked through the use of foreign keys. If you are not certain whether there should be a link between two tables, simply refer to your ER model.

The ER model previously shown in Figure 1-12 displays the original relationship between the STUDENT and COURSE entities. Recall during the discussion of the many-to-many relationship that such a relationship needs to be broken down into a series of one-to-many relationships before the actual database tables are created. Notice what happened during the normalization process—the relationship between the STUDENT and COURSE entities now consists of three tables: STUDENT, COURSE, and

ENROLLMENT. The ENROLLMENT table is the linking table that now has a one-to-many relationship with the STUDENT table, and a one-to-many relationship with the COURSE table. Because the FACULTY entity was identified during the normalization process, this entity has been added to the revised ER model shown in Figure 1-20.

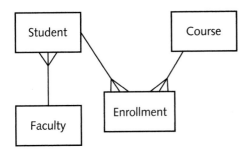

Figure 1-20   Revised entity-relationship model

During practical application of the normalization process in which there may be hundreds of tables, you would normally convert each table to 3NF and then make certain the necessary foreign keys have been included to link the tables after all the tables have been normalized.

DATABASE SYSTEMS

A database system consists of the DBMS, which manages the physical storage and data retrieval, and **database applications**, which provide the interface that allows users to interact with the database. When organizations first started using databases, the database DBMS and the database applications both ran on large centralized mainframe computers that users accessed using terminals. **Terminals** are devices that do not perform any processing, but only send keyboard input and receive video output from a central computer. As distributed computing and microcomputers became popular during the 1980s, two new kinds of database management systems emerged: personal databases and client/server databases.

Before you can understand the differences between personal and client/server databases, you must understand server and client processes. When computers are connected through networks, some computers act as servers, and other computers act as clients. A **server** is a computer that shares its resources with other computers. Examples of server resources include data, such as files; hardware, such as printers; or shared server-based programs, such as databases and e-mail. A **server process** is a program that listens for requests for resources from clients and responds to those requests. A **client** is a program that requests and uses server resources. For example, when you start an e-mail program, the e-mail program starts a client process that connects to an e-mail server. The e-mail server process notifies you if you have new messages, sends you your new messages when you request them, and forwards your outgoing messages to the recipient's e-mail server. Today, almost any computer can be a server and can run more than one server process. Therefore, it is more accurate to refer to a server as a particular process that enables other computers to share the resources of the computer on which the server process is running.

TIP

The same computer can act as both a client and a server. For example, your workstation could share files with other users, and it could also request files from other servers.

Personal Database Management Systems

With a **personal database**, the DBMS and the database applications run on the same workstation and appear to the user as a single integrated application. People use personal databases primarily for creating single-user database applications. For example, you might use a personal database to create an address book or track information about your personal finances.

NOTE

Microsoft Access is the dominant personal database in the marketplace today, so this discussion of personal databases focuses on Microsoft Access. Although Microsoft Access and Oracle 10*g* Personal Edition are both considered personal databases, they differ extensively in their approach to managing data. The personal edition of Oracle 10*g* derives its features from Oracle's client/server database, which the next section describes. Conversely, Access has no client/server features. Oracle Corporation has imposed limits on Oracle 10*g* Personal Edition to prevent people from using it for large multiuser database applications.

Organizations sometimes use personal databases to create simple database applications that multiple people use at the same time. For example, a computer laboratory manager might use a personal database to create a multiuser database application to track laboratory equipment or employee schedules. Personal databases such as Microsoft Access support small multiuser database applications by storing the database application files on a file server instead of on a single user's workstation and then transmitting the files or the parts of files containing the desired data to various users across a network, as shown in Figure 1-21.

Client workstation
running Access

File server
(stores .mdb file)

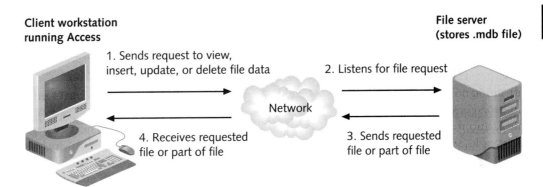

1. Sends request to view,
insert, update, or delete file data

2. Listens for file request

Network

4. Receives requested
file or part of file

3. Sends requested
file or part of file

Figure 1-21 Using a personal database for a multiuser application

The Microsoft Access personal database stores all data for a database in a single file with an .mdb extension. To run a multiuser personal database application, the database administrator stores the .mdb file on a central file server. Each client workstation loads the entire DBMS into main memory, along with the database applications to view, insert, update, or print data. As an individual user interacts with the database, his or her workstation sends requests to the file server for all or part of the database .mdb file. A request for a small amount of data sometimes requires the file server to transmit the entire file, which may contain hundreds of megabytes of data, to the client's workstation. The client workstation then filters the requested data. As a result, Microsoft Access and other similar personal database management systems sometimes impose a heavy load on client workstations and on the network. The network must be fast enough to handle the traffic generated when transferring database files to the client workstation and sending them back to the server for database additions and updates.

As a general rule, database developers should use a personal database only for applications that are not mission critical. In a personal database system, when a client workstation requests a data file, data within the file is locked and unavailable to other users. If a client workstation fails because of a software malfunction or power failure during a database operation, the data file that is locked remains unavailable to other users, and sometimes becomes damaged. The central database file might be repairable, but all users must quit using the database during the repair process, which could take several hours. Updates, deletions, and insertions taking place at the time of the failure often cannot be reconstructed. If repair is not possible, the DBA can restore the database to its state at the time of the last regular backup, but transactions that occurred since the backup are lost.

Another problem with using personal databases involves how they support transaction processing. **Transaction processing** refers to grouping related database changes into units of work that must either all succeed or all fail. For example, assume a customer writes a check to deposit money from a checking account into a money market account. The bank must ensure that the checking account is debited for the amount and the money market account is credited for the same amount. If any part of the transaction fails, then neither account balance should change. When you perform a transaction using a Microsoft Access personal database, the personal database records the related changes in the client workstation's main memory in a record called a **transaction log**. If you decide to abort the transaction, or if some part of the transaction does not succeed, the DBMS can use the transaction log to reverse, or **roll back**, the changes. However, recall that the transaction log is in the client workstation's main memory. If the client workstation fails in the middle of the transaction, the transaction log is lost, so the changes cannot be rolled back. Depending on the order of the transactions, a failed client workstation could result in a depleted checking account and unchanged money market account, or an enlarged money market account and unchanged checking account.

Personal databases such as Microsoft Access are excellent for creating small database systems, but are not really suitable for creating large, mission-critical database systems. To create large, mission-critical systems, database developers often use client/server database management systems.

Client/Server Database Management Systems

Today's organizations often need databases that support tens of thousands of simultaneous users and support mission-critical tasks. To accomplish this, they use client/server databases, which take advantage of distributed processing and networked computers by distributing processing across multiple computers. The following sections describe client/server database concepts and the Oracle 10*g* client/server DBMS.

Client/Server Database Concepts

In a **client/server database**, the DBMS server process runs on one workstation, and the database applications run on separate client workstations across the network. Figure 1-22 illustrates the client/server database architecture.

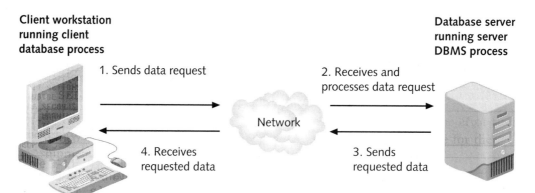

Figure 1-22   Client/server database architecture

The **client database process** sends data requests across the network. The **server DBMS process** runs on the database server and listens for requests from clients. When the server DBMS process receives a data request, it retrieves the data from the database, performs the requested functions on the data (retrieving, inserting, updating, sorting, and so on), and sends *only* the requested data, rather than the entire database, back to the client. As a result, client/server databases generate less network traffic than personal databases and are less likely to get bogged down by an overloaded network.

Another important difference between client/server and personal databases is how each type of database handles hardware and software malfunctions. Most client/server DBMS server processes have extra features to minimize the chance of failure, and when they do fail, they have powerful recovery mechanisms that often operate automatically. Client/server systems also differ from personal database systems in how they handle transaction processing. Client/server systems maintain a file-based transaction log on the database server. If a client workstation fails before finishing a transaction, the database server automatically rolls back all the transaction changes. If the server crashes, the file-based transaction log allows transactions to be rolled back when the server is restarted.

In summary, client/server databases are preferred for database applications that retrieve and manipulate small amounts of data from databases containing large numbers of records because they minimize network traffic and improve response times. Client/server systems are essential for mission-critical applications because of their failure handling, recovery mechanisms, and transaction-management control. Client/server databases also have a rich set of database management and administration tools for handling large numbers of users. Client/server databases are also ideal for Web-based database applications that require increased security and fault tolerance. Organizations use a client/server database if the database generally has more than 10 simultaneous users and if the database is mission critical.

The Oracle 10g Client/Server Database

Oracle 10*g* is the latest release of Oracle Corporation's relational database. Oracle 10*g* is a client/server database, so you must consider its environment in terms of server-side programs and client-side programs. On the server side, the DBA installs and configures the Oracle 10*g* DBMS server process. After the DBA installs and configures the DBMS on the database server, a process on the database server listens for incoming user requests and commands from client database processes. These client database processes may be Oracle utilities for executing database commands, or specific applications that support accounting, finance, production, and sales business functions. All Oracle server- and client-side programs use **Oracle Net**, which is a utility that enables network communication between the client and the server. Whenever you start an Oracle client-side process, an Oracle Net process automatically starts on the client workstation. The client

Oracle Net process communicates with an Oracle Net process running on the database server. The process running on the server forwards user requests to the Oracle DBMS. Figure 1-23 illustrates how Oracle 10g uses Oracle Net to enable communication between server-side and client-side database processes.

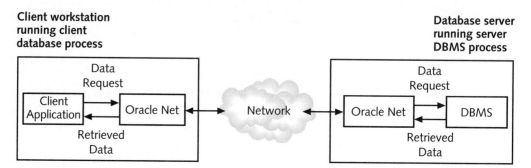

Figure 1-23   Client/server architecture for Oracle 10g DBMS

 If you are using the Oracle 10g Personal Edition database on your computer, you will install and configure the database yourself, and run both the server DBMS process and the client database programs on the same workstation.

Along with the server DBMS, another Oracle 10g server-side product is the **Oracle Application Server**, which creates a World Wide Web site that allows users to access Oracle databases and create dynamic Web pages that interact with an Oracle database. Usually, a DBA installs and configures the Oracle server programs.

On the client side, Oracle 10g provides not only utilities for creating commands to retrieve and manipulate database data, but also development tools for designing and creating custom applications to support user tasks. In this book, you will use client-side applications to learn how to interact with an Oracle database, create and manage database tables, and create database applications. Specifically, you will use the following Oracle client products:

- **SQL*Plus**—Used for creating and testing command-line SQL queries and executing PL/SQL procedural programs. (PL/SQL is Oracle's sequential programming language that is used to create programs that process database data.)

- **Oracle 10g Developer Suite**—Used for developing database applications. This book describes how to use the following Developer tools:

  - **Forms Builder**—Used for creating custom user applications

  - **Reports Builder**—Used for creating reports for displaying, printing, and distributing summary data

- **Enterprise Manager**—Used for performing database administration tasks such as creating new user accounts and configuring how the DBMS stores and manages data

THE DATABASE CASES

In this book, you will encounter tutorial exercises and end-of-chapter cases that illustrate how to use Oracle 10*g* using databases developed for Clearwater Traders and Northwoods University, two fictional organizations. As previously mentioned, the focus of this book is on database development rather than on database design, so the rationale behind the design of all the database tablesis not described. Nevertheless, it is important that you remember a few general database design principles. You should follow these principles when creating database tables:

- Unless instructed otherwise, convert all tables to third normal form.

- To link tables, make certain you include the primary key as the foreign key in the table on the "many" side of the relationship.

- When creating a database and inserting data values, you must specify the data type for each column. Different database management systems have specific names for their data types, but in general, data types include numbers, text strings, date/time values, time interval values, or binary data such as images or sounds. In general, you should use only a number data type for columns that store numerical values that are involved in calculations. For example, you would store inventory quantity-on-hand values in number columns, but you would not store telephone numbers in number columns. You store data values that contain numeric characters that are not used in calculations, such as telephone numbers and postal codes, as text strings.

Client/server DBMSs are appropriate for the case study databases because these systems have many simultaneous users accessing the system from different locations. Each database has data examples to illustrate database development tasks such as creating and maintaining database tables, retrieving data, and developing data entry and maintenance forms and output reports.

The Clearwater Traders Sales Order Database

Clearwater Traders markets a line of clothing and sporting goods through mail-order catalogs. Clearwater Traders currently accepts customer orders through telephone, mail, and fax, and wants to begin accepting orders using its Web site. The company recently experienced substantial growth and, as a result, has decided to offer 24-hour customer order service. The existing microcomputer-based database system cannot handle the current transaction volume generated by sales representatives processing incoming orders. Managers are also concerned that the current database does not have the failure-handling and recovery capabilities needed for an ordering system that cannot tolerate failures or downtime.

When a customer orders an item, the system must confirm that the ordered item is in stock. If the item is in stock, the system must update the available quantity on hand to reflect that the item has been sold. If the item is not in stock, the system needs to advise the customer when the item will be available. The customer must log onto the system using a username and password, and then complete the ordering process. Customers should also be able to log onto the system and view information about the status of their orders.

When Clearwater Traders receives new inventory shipments, a receiving clerk at Clearwater Traders must update the inventory to show the new quantities on hand. The system must produce invoices that can be included with customer shipments and must print reports showing inventory levels. Marketing managers want to be able to track each order's source in terms of the catalog number or whether it originates on the Web site.

These processes require the following data items:

- Customer name, address, daytime and evening telephone numbers, usernames, and passwords

- Order date, payment method (check or credit card), order source (catalog description or Web site), and associated item numbers, sizes, colors, and quantities ordered

- Item descriptions and photo images, as well as item categories (women's clothing, outdoor gear, and so on), prices, and quantities on hand. Many clothing items are available in multiple sizes and colors. Sometimes the same item has different prices depending on the item size.

- Information about incoming product shipments

Figure 1-24 shows sample data for Clearwater Traders. Each database table displays the table name, column names, and the type of data (Number, String, or Date/Time) the column stores. In Figure 1-24 and later figures, the primary key or keys will be positioned as the first columns in the figure. The CUSTOMER table displays six customer records. Customer 1 is Neal Graham, who lives at 9815 Circle Dr., Tallahassee, FL, and his zip code is 32308. His daytime telephone number is 904-555-1897, and his evening telephone number is 904-555-8599. C_ID has been designated as the table's primary key.

CUSTOMER

C_ID	C_LAST	C_FIRST	C_MI	C_DOB	C_ADDRESS	C_CITY	C_STATE	C_ZIP
Number	String	String	String	Date/Time	String	String	String	String
1	Graham	Neal	R	12/10/1967	9815 Circle Dr.	Tallahassee	FL	32308
2	Sanchez	Myra	T	8/14/1958	172 Alto Park	Seattle	WA	42180
3	Smith	Lisa	M	4/12/1960	850 East Main	Santa Ana	CA	51875
4	Phelp	Paul		1/18/1981	994 Kirkman Rd.	Northpoint	NY	11795
5	Lewis	Sheila	A	8/30/1978	195 College Blvd.	Newton	GA	37812
6	James	Thomas	E	6/01/1973	348 Rice Lane	Radcliff	WY	87195

CUSTOMER (continued)

C_DPHONE	C_EPHONE	C_USERID	C_PASSWORD
String	String	String	String
9045551897	9045558599	grahamn	barbiecar
4185551791	4185556643	sanchezmt	qwert5
3075557841	3075559852	smithlm	joshua5
4825554788	4825558219	phelpp	hold98er
3525554972	3525551811	lewissa	125pass
7615553485	7615553319	jamest	nok$tell

ORDER_SOURCE

OS_ID	OS_DESC
Number	String
1	Winter 2005
2	Spring 2006
3	Summer 2006
4	Outdoor 2006
5	Children's 2006
6	Web Site

ORDERS

O_ID	O_DATE	O_METHPMT	C_ID	OS_ID
Number	Date/Time	String	Number	Number
1	5/29/2006	CC	1	2
2	5/29/2006	CC	5	6
3	5/31/2006	CHECK	2	2
4	5/31/2006	CC	3	3
5	6/01/2006	CC	4	6
6	6/01/2006	CC	4	3

CATEGORY

CAT_ID	CAT_DESC
Number	String
1	Women's Clothing
2	Children's Clothing
3	Men's Clothing
4	Outdoor Gear

ITEM

ITEM_ID	ITEM_DESC	CAT_ID	ITEM_IMAGE
Number	String	Number	String
1	Men's Expedition Parka	3	parka.jpg
2	3-Season Tent	4	tents.jpg
3	Women's Hiking Shorts	1	shorts.jpg
4	Women's Fleece Pullover	1	fleece.jpg
5	Children's Beachcomber Sandals	2	sandals.jpg
6	Boy's Surf Shorts	2	surfshorts.jpg
7	Girl's Soccer Tee	2	girlstee.jpg

Figure 1-24   Clearwater Traders database

ORDER_LINE

O_ID	INV_ID	OL_QUANTITY
Number	Number	Number
1	1	1
1	14	2
2	19	1
3	24	1
3	26	1
4	12	2
5	8	1
5	13	1
6	2	1
6	7	3

SHIPMENT

SHIP_ID	SHIP_DATE_EXPECTED
Number	Date/Time
1	09/15/2006
2	11/15/2006
3	06/25/2006
4	06/25/2006
5	08/15/2006

INVENTORY

INV_ID	ITEM_ID	COLOR	INV_SIZE	INV_PRICE	INV_QOH
Number	Number	String	String	Number	Number
1	2	Sky Blue		259.99	16
2	2	Light Grey		259.99	12
3	3	Khaki	S	29.95	150
4	3	Khaki	M	29.95	147
5	3	Khaki	L	29.95	0
6	3	Navy	S	29.95	139
7	3	Navy	M	29.95	137
8	3	Navy	L	29.95	115
9	4	Eggplant	S	59.95	135
10	4	Eggplant	M	59.95	168
11	4	Eggplant	L	59.95	187
12	4	Royal	S	59.95	0
13	4	Royal	M	59.95	124
14	4	Royal	L	59.95	112
15	5	Turquoise	10	15.99	121
16	5	Turquoise	11	15.99	111
17	5	Turquoise	12	15.99	113
18	5	Turquoise	1	15.99	121
19	5	Bright Pink	10	15.99	148
20	5	Bright Pink	11	15.99	137
21	5	Bright Pink	12	15.99	134
22	5	Bright Pink	1	15.99	123
23	1	Spruce	S	199.95	114
24	1	Spruce	M	199.95	17
25	1	Spruce	L	209.95	0
26	1	Spruce	XL	209.95	12
27	6	Blue	S	15.95	50
28	6	Blue	M	15.95	100
29	6	Blue	L	15.95	100
30	7	White	S	19.99	100
31	7	White	M	19.99	100
32	7	White	L	19.99	100

Figure 1-24   Clearwater Traders database (continued)

SHIPMENT_LINE

SHIP_ID	INV_ID	SL_QUANTITY	SL_DATE_RECEIVED
Number	Number	Number	Date/Time
1	1	25	09/10/2006
1	2	25	09/10/2006
2	2	25	
3	5	200	
3	6	200	
3	7	200	
4	12	100	08/15/2006
4	13	100	08/25/2006
5	23	50	08/15/2006
5	24	100	08/15/2006
5	25	100	08/15/2006

COLOR

COLOR
String
Sky Blue
Light Grey
Khaki
Navy
Royal
Eggplant
Blue
Red
Spruce
Turquoise
Bright Pink
White

Figure 1-24 Clearwater Traders database (continued)

The ORDER_SOURCE table has OS_ID as the primary key and contains a column named OS_DESC, which describes the order source as a specific catalog or the company Web site. The ORDERS table shows six customer orders. The table includes the O_ID column, which is a surrogate key that is the table's primary key; the O_DATE column, showing the date the customer places the order; the O_METHPMT column, indicating the payment method: CC (credit card) or CHECK; the C_ID column, a foreign key that creates a relationship to the CUSTOMER table; and the OS_ID column, a foreign key that creates a relationship to the ORDER_SOURCE table. The first record shows information for O_ID 1, dated 5/29/2006, method of payment CC (credit card), and ordered by customer 1, Neal Graham. The OS_ID foreign key indicates that the Spring 2006 catalog was the source for the order.

The CATEGORY table displays different product categories: Women's Clothing, Children's Clothing, Men's Clothing, and Outdoor Gear. The CAT_ID column is this table's primary key. The ITEM table contains seven different items. ITEM_ID is the table's primary key. The CAT_ID column is a foreign key that creates a relationship with the CATEGORY table. Item 1, Men's Expedition Parka, is in the Men's Clothing category. The ITEM_IMAGE column contains a text string that represents the name of the JPEG image file that stores an image of each item.

The INVENTORY table contains specific inventory numbers for specific merchandise item sizes and colors. It also shows the price and quantity on hand (QOH) for each item. Items that are not available in different sizes contain NULL, or undefined, values in their INV_SIZE columns. Notice that some items have different prices for different sizes. For example, for ITEM_ID 1 (Men's Expedition Parka), the small (S) and medium (M) inventory items are priced at $199.95, whereas the large (L) and extra large (XL) items are priced at $209.95. INV_ID is the primary key of this table, and ITEM_ID is a foreign key that creates a relationship with the ITEM table.

The ORDER_LINE table represents the individual inventory items in a customer order. The first line of O_ID 1 specifies one Sky Blue 3-Season Tent, and the second line of this order specifies two large Royal-colored Women's Fleece Pullovers. This information is used to create the printed customer order invoice and to calculate sales revenues. Note that the primary key of this table is not O_ID, because more than one record might have the same O_ID. The primary key is a composite key made up of the combination of O_ID and INV_ID. An order might have several different inventory items, but it will never have the same inventory item listed more than once. Along with being part of the primary key, O_ID and INV_ID are also foreign keys because they create relationships to the ORDERS and INVENTORY tables.

The COLOR column of the INVENTORY table is also a foreign key, which references the COLOR table. The COLOR table is a lookup table. A **lookup table** is also sometimes called a pick list. It contains a list of legal values for a column in another table. Notice the variety of colors shown in the INVENTORY table (Sky Blue, Light Grey, Khaki, Navy, Royal, and so on). If users are allowed to type these colors each time an inventory item is added to the table, data entry errors might occur. For example, a query looking for sales of items with the Light Grey color will not find instances if Light Grey is spelled *Light Gray*, or is specified with a different combination of upper- and lowercase letters, such as *Light grey* or *LIGHT GREY*. Typically, when a Clearwater Traders employee user enters a new inventory item, he or she selects a color from a pick list that displays values from the COLOR table. Thus the user need not type the color directly, which reduces errors. Small lists that are unlikely to change over time might be coded directly into an application, but large lists to which items might be added over time are usually stored in a separate lookup table.

The SHIPMENT table contains a schedule of expected shipments and the date each shipment is expected. The primary key of the SHIPMENT table is SHIP_ID. A shipment can include multiple inventory items, so the SHIPMENT_LINE table records the corresponding shipment ID, the inventory ID, the quantity of each item, and the date each item was received. If a shipment line has not been received, the date value is NULL. The first SHIPMENT_LINE record shows that the first line of SHIP_ID 1 is for 25 Sky Blue 3-Season Tents. The second line of SHIP_ID 1 is for 25 Light Grey 3-Season Tents. Both items were received on 09/10/2006. Notice that the primary key of this table cannot be SHIP_ID, because some shipments have multiple lines. The primary key cannot be INV_ID, because the same inventory item might be in several shipments. The primary key must be a composite key comprising SHIP_ID and INV_ID, because each shipment might consist of multiple inventory items, but each inventory item is listed only once per shipment. Therefore, each SHIP_ID/INV_ID combination is unique for each record. SHIP_ID and INV_ID are also foreign keys in this table, because they reference records in the SHIPMENT and INVENTORY tables.

Figure 1-25 shows a visual representation of the Clearwater Traders database tables. In this representation, the primary key for each table appears in boldface, and foreign key columns appear in italics. Relationships between tables are represented by join lines. For example, a join line connects the C_ID column in the CUSTOMER table to the C_ID column in the ORDERS table. Note that for tables that contain a composite primary key, the columns that compose the primary key appear in boldface italic type. The italics are added because the composite columns are also foreign keys. For example, in the ORDER_LINE table in Figure 1-25, the columns that make up the table's composite primary key (O_ID and INV_ID) appear in boldface italics. This indicates that the columns are part of a primary key and that the columns are also foreign keys.

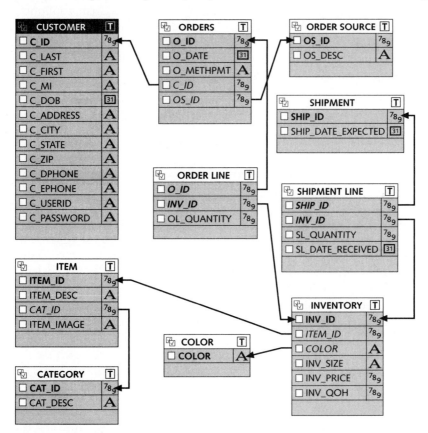

Figure 1-25   Visual representation of the Clearwater Traders database

In the visual database representation, the icon beside each column indicates the type of data that the column stores. The Number icon [789] indicates that the column stores numerical values, the Text icon A indicates that the column stores text values, and the Date icon [31] indicates that the column stores date values.

The Northwoods University Student Registration Database

Northwoods University has decided to replace its aging mainframe-based student registration system with a more modern client/server database system. School officials want students to be able to retrieve course availability information, register for courses, and print transcripts using personal computers located in the student computer labs. In addition, faculty members must be able to retrieve student course lists, drop and add students, and record course grades. Faculty members must also be able to view records for the students they advise. Security is a prime concern, so student and course records must be protected by password access.

The data items for the Northwoods database are:

- Student name, address, telephone number, class (freshman, sophomore, junior, or senior), date of birth, PIN (personal identification number), and advisor ID

TIP

In a production system, databases usually store PINs or passwords as encrypted values.

- Course call number (such as MIS 101), course name, credits, location, duration, maximum enrollment, instructor, and term offered

- Instructor name, office location, telephone number, rank, and PIN

- Student enrollment and grade information

Figure 1-26 shows sample data for the Northwoods database. The LOCATION table identifies building codes, room numbers, and room capacities. LOC_ID is the primary key. The FACULTY table describes five faculty members. The first record shows faculty member Teresa Marx, whose office is located at BUS 424, and whose telephone number is 407-592-1695. She has the rank of ASSOCIATE (associate professor), and her PIN is 6338. This F_PIN (faculty PIN) column will be used as a password to determine whether a faculty member can update specific student or course records. F_ID is the primary key, and LOC_ID is a foreign key that references the LOCATION table. The F_IMAGE column contains the binary data of a photo image for each faculty member. Figure 1-26 shows the name of the JPEG image file that contains the image data.

LOCATION

LOC_ID	BLDG_CODE	ROOM	CAPACITY
Number	String	String	Number
1	CR	101	150
2	CR	202	40
3	CR	103	35
4	CR	105	35
5	BUS	105	42
6	BUS	404	35
7	BUS	421	35
8	BUS	211	55
9	BUS	424	1
10	BUS	402	1
11	BUS	433	1
12	LIB	217	2
13	LIB	222	1

FACULTY

F_ID	F_LAST	F_FIRST	F_MI	LOC_ID	F_PHONE	F_RANK	F_SUPER	F_PIN	F_IMAGE
Number	String	String	String	Number	String	String	Number	String	Binary
1	Marx	Teresa	J	9	4075921695	Associate	4	6338	marx.jpg
2	Zhulin	Mark	M	10	4073875682	Full		1121	zhulin.jpg
3	Langley	Colin	A	12	4075928719	Assistant	4	9871	langley.jpg
4	Brown	Jonnel	D	11	4078101155	Full		8297	brown.jpg
5	Sealy	James	L	13	4079817153	Associate	2	6089	sealy.jpg

STUDENT

S_ID	S_LAST	S_FIRST	S_MI	S_ADDRESS	S_CITY	S_STATE	S_ZIP
String	String	String	String	String	String	String	String
JO100	Jones	Tammy	R	1817 Eagleridge Circle	Tallahasse	FL	32811
PE100	Perez	Jorge	C	951 Rainbow Dr	Clermont	FL	34711
MA100	Marsh	John	A	1275 West Main St	Carrabelle	FL	32320
SM100	Smith	Mike		428 Markson Ave	Eastpoint	FL	32328
JO101	Johnson	Lisa	M	764 Johnson Place	Leesburg	FL	34751
NG100	Nguyen	Ni	M	688 4th Street	Orlando	FL	34158

STUDENT (continued)

S_PHONE	S_CLASS	S_DOB	S_PIN	F_ID	TIME_ENROLLED
String	String	Date/Time	String	Number	Interval
7155559876	SR	07/14/85	8891	1	3 YEARS 2 MONTHS
7155552345	SR	08/19/85	1230	1	4 YEARS 6 MONTHS
7155553907	JR	10/10/82	1613	1	3 YEARS 0 MONTHS
7155556902	SO	9/24/86	1841	2	2 YEARS 2 MONTHS
7155558899	SO	11/20/86	4420	4	1 YEAR 11 MONTHS
7155554944	FR	12/4/87	9188	3	0 YEARS 4 MONTHS

Figure 1-26   Northwoods University database

TERM

TERM_ID	TERM_DESC	STATUS	START_DATE
Number	*String*	*String*	*String*
1	Fall 2005	CLOSED	29-AUG-05
2	Spring 2006	CLOSED	09-JAN-06
3	Summer 2006	CLOSED	15-MAY-05
4	Fall 2006	CLOSED	28-AUG-06
5	Spring 2007	CLOSED	08-JAN-07
6	Summer 2007	OPEN	07-MAY-07

COURSE

COURSE_ID	COURSE_NO	COURSE_NAME	CREDITS
Number	*String*	*String*	*Number*
1	MIS 101	Intro. to Info. Systems	3
2	MIS 301	Systems Analysis	3
3	MIS 441	Database Management	3
4	CS 155	Programming in C++	3
5	MIS 451	Web-Based Systems	3

COURSE_SECTION

C_SEC_ID	COURSE_ID	TERM_ID	SEC_NUM	F_ID	C_SEC_DAY	C_SEC_TIME	C_SEC_DURATION	LOC_ID	MAX_ENRL
Number	*Number*	*Number*	*Number*	*Number*	*String*	*Date/Time*	*Interval*	*Number*	*Number*
1	1	4	1	2	MWF	10:00 AM	50 MINUTES	1	140
2	1	4	2	3	TR	9:30 AM	1 HOUR 15 MINUTES	7	35
3	1	4	3	3	MWF	8:00 AM	50 MINUTES	2	35
4	2	4	1	4	TR	11:00 AM	1 HOUR 15 MINUTES	6	35
5	2	5	2	4	TR	2:00 PM	1 HOUR 15 MINUTES	6	35
6	3	5	1	1	MWF	9:00 AM	50 MINUTES	5	30
7	3	5	2	1	MWF	10:00 AM	50 MINUTES	5	30
8	4	5	1	5	TR	8:00 AM	1 HOUR 15 MINUTES	3	35
9	5	5	1	2	MWF	2:00 PM	50 MINUTES	5	35
10	5	5	2	2	MWF	3:00 PM	50 MINUTES	5	35
11	1	6	1	1	MTWRF	8:00 AM	1 HOUR 30 MINUTES	1	50
12	2	6	1	2	MTWRF	8:00 AM	1 HOUR 30 MINUTES	6	35
13	3	6	1	3	MTWRF	9:00 AM	1 HOUR 30 MINUTES	5	35

ENROLLMENT

S_ID	C_SEC_ID	GRADE
Number	*Number*	*String*
1	1	A
1	4	A
1	6	B
1	9	B
2	1	C
2	5	B
2	6	A
2	9	B
3	1	C
3	12	
3	13	
4	11	
4	12	
5	1	B
5	5	C
5	9	C
5	11	
5	13	
6	11	
6	12	

Figure 1-26   Northwoods University database (continued)

The STUDENT table displays six student records. Its primary key column is S_ID. The first record displays data for S_ID JO100, student Tammy Jones, who lives at 1817 Eagleridge Circle, in Tallahassee, Florida. Her telephone number is 715-555-9876, she is a senior, her date of birth is 7/14/85, and her faculty advisor is Teresa Marx. Note that the S_PIN (student PIN) column stores her student personal identification number to control data access. F_ID (faculty ID) is a foreign key that refers to the F_ID column in the FACULTY table. The TIME_ENROLLED column stores the time interval that Tammy has been enrolled at Northwoods University, which is 3 years and 2 months.

The TERM table has an ID number that links terms to different course offerings. TERM_ID is the primary key. The table contains a text description of each term and a STATUS column that shows whether enrollment is open or closed. The first record shows a TERM_ID of 1 for the Fall 2005 term, with enrollment status as CLOSED. The START_DATE column contains the date that the term begins.

The COURSE table shows records for five courses. The first course record, COURSE_ID 1, has the course number MIS 101 and is named "Intro. to Info. Systems." It provides three credits. COURSE_ID is the primary key, and the table contains no foreign keys.

The COURSE_SECTION table shows the course offerings for specific terms and includes columns that display the course ID, section number, ID of the instructor teaching the section, and day and time. The C_SEC_DURATION column stores a time interval that represents how long the course lasts each day. The LOC_ID column stores a foreign key reference to the course location, and the MAX_ENRL column specifies each course section's maximum allowable enrollment. C_SEC_ID is the primary key, and COURSE_ID, TERM_ID, F_ID, and LOC_ID are all foreign key columns. The first record shows that C_SEC_ID 1 is section 1 of MIS 101. It is offered in the Fall 2006 term and is taught by Mark Zhulin. The section meets on Mondays, Wednesdays, and Fridays at 10:00 AM in room CR 101. The class lasts for 50 minutes, and the course has a maximum enrollment of 140 students.

The ENROLLMENT table shows students who are currently enrolled in each course section and their associated grade if one has been assigned. The primary key for this table is a composite key composed of S_ID and C_SEC_ID.

Figure 1-27 shows a visual representation of the tables in the Northwoods University database. As with the visual representation of the Clearwater Traders database in Figure 1-25, primary key columns appear in boldface, foreign keys appear in italics, and foreign key relationships appear as join lines. The interval icon indicates that the column stores time interval data.

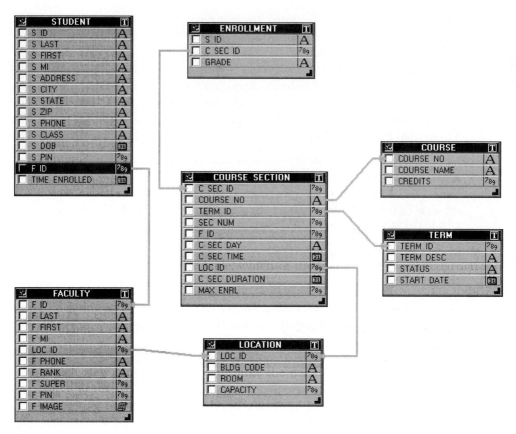

Figure 1-27   Visual representation of the Northwoods University database

---

CHAPTER SUMMARY

- When organizations first began converting from manual to computerized data-processing systems, each individual application had its own set of data files that were used only for that application. As applications became more complex, data files became difficult to manage, because each data file required a separate program to maintain its contents and because data files for related applications contained redundant data. Redundant data takes up valuable file space and can become inconsistent. To address these problems, organizations began using databases to store and manage application data.

❐ A database stores all organizational data in a central location. In a database system, the database management system (DBMS) provides a central set of common functions for managing a database, which include inserting, updating, retrieving, and deleting data values. The person who is responsible for installing, administering, and maintaining the database is called the database administrator (DBA).

❐ Most modern databases are relational databases, which store data in a tabular format. Columns represent different data columns, and rows contain individual data records. In a relational database, data about different entities is stored in separate tables. You create relationships that link related data using key columns.

❐ A primary key is a column that uniquely identifies a specific record in a database table. Primary key values must be unique within a table and cannot be NULL. Primary key columns should contain numeric data rather than text data, because text data is more prone to data entry errors. A candidate key is any column that could be used as the primary key.

❐ When a database table does not contain any suitable candidate keys, you must create a surrogate key, which has no intrinsic relation to the record and is used for the sole purpose of identifying the record. An Oracle DBMS automatically generates surrogate key values using sequences, which are sequential lists of numbers that guarantee that each surrogate key value is unique.

❐ A foreign key is a column in a table that is also a primary key in another table and that creates a relationship between the two tables.

❐ A composite key is a primary key composed of the combination of two or more columns. The combination of the two column values must be unique for every record.

❐ An entity-relationship (ER) model is used to describe the types of relationships between entities. There are three possible types of relationships: one-to-one, one-to-many, and many-to-many.

❐ In the rare one-to-one (1:1) relationship, an instance occurs only once in each entity.

❐ In a one-to-many (1:M) relationship, an instance can only occur once in one entity, but multiple times in the other entity.

❐ In a many-to-many (N:M) relationship, an instance can occur multiple times in each entity. A many-to-many relationship cannot be included in the physical database and must be broken down into a series of one-to-many relationships using a linking table. The linking table is identified during the normalization process.

❐ The normalization process is used to determine which fields belong in which tables.

- Unnormalized data contains repeating groups and/or no primary key.

- Data in first normal form (1NF) has a primary key and no repeating groups.

- Data in second normal form (2NF) is in 1NF and contains no partial dependencies.

- Data in third normal form (3NF) is in 2NF and contains no transitive dependencies.

- A database system consists of the DBMS, which manages the physical data storage, and database applications, which provide the user interface to the database. Today, many database systems use a client/server architecture, in which the DBMS runs as a server process, and the database applications run as client processes.

- With a personal database, the DBMS and the database applications run on the same workstation and appear to the user as a single integrated application. Personal database systems are best suited for single-user database applications. When organizations use personal databases for a multiuser application, the personal database downloads to the user's client workstation some or all of the data that a user needs. This process can be slow and cause network congestion.

- Client/server databases divide the database into a server process that runs on a network server and user application processes that run on individual client workstations. Client/server databases send data requests to the server and return the results of data requests to the client workstation. This process minimizes network traffic and congestion. Client/server databases have better failure recovery mechanisms than personal databases and automatically handle competing user transactions.

- In an Oracle 10*g* client/server database, Oracle 10*g* provides client-side utilities for executing SQL commands and designing and creating custom applications, and a server-side DBMS. All Oracle server- and client-side programs use Oracle Net, which is a utility that enables network communication between the client and the server.

- The Oracle database development environment has programs that run on the database server and on the client workstation. Server-side programs include the database and the Oracle Application Server.

- Client-side programs include utilities for creating database tables and queries; for developing data forms, reports, and graphics based on database table data; and for performing database administration tasks.

- To avoid creating database tables that contain redundant data, you should group related items that describe a single entity together in a common table. If a table seems to describe more than one entity, then it probably needs to be split into two tables, and one table needs to contain a foreign key that references records in the other table.

1

❐ You must specify a data type for each column in a database (e.g., numbers, strings, date/time values). In general, you should use only a number data type for columns that store values that are involved in calculations and for the primary key. You store data values that contain numeric characters that are not used in calculations, such as telephone numbers and postal codes, as text strings.

REVIEW QUESTIONS

1. What is a database?

2. A(n) _____ within a table is used to represent an attribute being collected about an entity, whereas a(n) _____ represents the specific occurrence of an entity.

3. List the five main types of key fields available in a relational database.

4. A person's Social Security number should always be used as a primary key because each Social Security number can only be assigned to one person. True or False?

5. When linking two tables together, the foreign key is always stored in the _____ side of a one-to-many relationship.

6. A partial dependency can only occur if you have what type of primary key?

7. If a table has a primary key, has no repeating groups, and no partial dependencies, the table is in _____ normal form.

8. Explain what type of relationship exists between these two entities: automobiles and automobile owners.

9. Oracle Net is a utility that enables network communication between a client and the server. True or False?

10. A client is a program designed to listen for, and respond to, requests made by another computer. True or False?

MULTIPLE CHOICE

1. Which of the following can occur when there is redundant data stored in the database?

a. It is impossible to uniquely identify a record.

b. Updates or changes can result in inconsistent data.

c. Linking tables cannot be created.

d. all of the above

2. Which of the following identifies the type of application used to manage a database?

    a. client

    b. server

    c. database management system

    d. transaction-processing system

3. Which of the following is *not* correct in regard to a primary key?

    a. A primary key cannot be NULL.

    b. A primary key must appear as a foreign key in another table.

    c. The value assigned to a primary key must be unique for each record.

    d. None. All of the above are correct statements.

4. A transitive dependency cannot exist if the database table is _____.

    a. unnormalized

    b. in first normal form

    c. in second normal form

    d. in third normal form

5. A database table can only be normalized if it contains a _____.

    a. foreign key

    b. surrogate key

    c. composite key

    d. none of the above

6. Which of the following is normally used to generate a surrogate key in Oracle 10*g*?

    a. sequences

    b. key fields

    c. candidate keys

    d. the normalization process

7. Which of the following is a client-side tool used for displaying, printing, and distributing summary data?

    a. SQL*Plus

    b. Enterprise Manager

    c. Forms Builder

    d. Reports Builder

8. Which of the following statements describes a transitive dependency?

   a. A field is dependent on only a portion of a composite primary key.

   b. A field is dependent on a field within the table that is not the table's primary key.

   c. The primary key is composed of more than one field.

   d. There are multiple values assigned to a field for a particular record.

9. A field is also called a _____ in a relational database.

   a. column

   b. row

   c. record

   d. key

10. A(n) _____ contains the changes that have been made to the database.

   a. rollback log

   b. transaction log

   c. server process

   d. application server

PROBLEM-SOLVING CASES

1. The manager for the Clearwater Traders wants to collect the following data for each order placed by a customer: customer's name and address, item(s) ordered, quantity of each item, item's size or color—if applicable—and the retail price of each item.

   a. Create an entity-relationship model representing the data that the manager wants to store in the database, based on the following assumptions:

   ❑ Each customer can place multiple orders.

   ❑ Each order can only belong to a single customer.

   ❑ Different items can be ordered on the same order.

   ❑ Each item on an order can have a different size and/or color, and some items may not have a size or color.

   b. Based on the data elements the Clearwater Traders' manager wants you to include in the database, assign an appropriate field name to each of the data elements. Make certain the name of each field is descriptive so it will be easily recognizable in the final version of the database.

   c. Take the named data elements from the previous step and convert the data to first normal form.

d. Convert the first normal form of the data elements to second normal form. Use the shorthand method to specify which elements belong to each entity and underline the primary key for each element.

e. Convert each of the previously identified entities to third normal form. Make certain that the necessary foreign keys have been added to the final tables to support the relationships shown in your initial ER model.

2. Identify the tables and fields that must be referenced in the Clearwater Traders database to answer the following questions:

a. How many 3-Season Tents are left in stock?

b. Which customer(s) lives in Georgia?

c. How many customers placed an order in May 2006?

d. Who was the last customer to order a Men's Expedition Parka?

e. Which items of women's clothing were sold between May 29 and June 1?

3. Using the Clearwater Traders data set presented in Figure 1-24, determine the answer for each of the five questions asked in Problem 2.

4. Based on the Clearwater Traders data set presented in Figure 1-24, draft an ER model representing the relationships in the database.

5. Using the ER model created in Problem 4, create a list of each table using the shorthand method shown in Figure 1-19. Underline the primary key of each table once and underline any foreign keys the table may contain twice.

2

CREATING AND MODIFYING DATABASE TABLES

Objectives

After completing this chapter, you should be able to:

♦ Use Structured Query Language (SQL) commands to create, modify, and drop database tables

♦ Explain Oracle 10g user schemas

♦ Define Oracle 10g database tables

♦ Create database tables using SQL*Plus

♦ Debug Oracle 10g SQL commands and use online help resources available through the Oracle Technology Network (OTN)

♦ View information about your database tables using Oracle 10g data dictionary views

♦ Modify and delete database tables using SQL*Plus

With Oracle 10*g* you use Structured Query Language (SQL) for data entry, retrieval, and manipulation. SQL is the standard query language for relational databases. In this chapter, you learn how to use SQL*Plus, Oracle's command-line SQL utility to issue SQL commands. Because the first step in creating the database is to make tables, this chapter focuses on the SQL commands required to create and modify tables.

INTRODUCTION TO SQL

Users interact with relational databases using high-level query languages. Query languages have commands that contain standard English words such as CREATE, ALTER, INSERT, UPDATE, and DELETE. The standard query language for relational databases is **Structured Query Language (SQL)**. Basic SQL consists of about 30 commands that enable users to create database objects and manipulate and view data. The American National Standards Institute (ANSI) has published standards for SQL. The most recent version was published in 1999, and is called SQL-99. Most relational database vendors do not yet fully comply with SQL-99, but almost all relational database applications support the SQL ANSI standard published in 1992, which is called SQL-92. This book uses SQL commands that comply with the SQL-92 standard.

Most database vendors add extensions to the standard SQL commands to make their databases more powerful or easier to use. These extensions are fairly similar across different platforms, and once you become proficient with SQL on one DBMS platform, it is fairly easy to move to other platforms.

SQL commands fall into two basic categories:

- **Data definition language (DDL)** commands—Used to create new database objects (such as user accounts and tables) and modify or delete existing objects. When you execute a DDL command, the command immediately changes the database, so you do not need to save the change explicitly.

- **Data manipulation language (DML)** commands—Used to insert, update, delete, and view database data. When you execute a DML command, you must explicitly save the command to make the new data values visible to other database users.

In this chapter, you work with DDL commands to create new database tables. Usually, only database administrators execute DDL commands. Sometimes you execute DDL operations using utilities that automatically generate the underlying SQL commands. However, to create the applications in the tutorials and case projects in this book, you must first create the underlying database tables in your Oracle 10*g* database. Furthermore, you must understand the underlying structure of the database tables. To gain this understanding, you must be familiar with the syntax and use of the SQL DDL commands that create the tables. In this chapter, you use the DDL commands to create, modify, and drop Oracle 10*g* database tables. You type DDL commands into a text editor, and you execute the commands in SQL*Plus, the Oracle 10*g* command-line SQL environment.

In this book, all SQL command words, which are known as **reserved words**, appear in all uppercase letters. All user-supplied variable names appear in lowercase letters, although Oracle 10*g* displays the user-supplied values in uppercase letters when it executes the commands. Although SQL*Plus commands are not case sensitive, these conventions make SQL commands and the examples in this book easier to interpret. When the book references a table or column outside a command, the table or column name appears in all uppercase letters. When the book provides the general syntax for a command, user-supplied values appear in lowercase italic type. Optional parameters or optional phrases within commands appear in square brackets. For parameters that can have one value from a set of values, the possible values are separated by the bar (|) character.

ORACLE 10*g* USER ACCOUNTS

Before you begin working with an Oracle 10*g* database, you need to understand how Oracle 10*g* manages database user accounts. When you create a new table using a personal database such as Access, you start the database application and create a new database. Access saves the database file in your workstation's file system. You are usually the only person who uses this database file. In contrast, when a database administrator creates a new Oracle 10*g* database, the database server stores the database files, and all of the database's users share these files. To keep each user's tables and data separate and secure, the DBA creates a **user account** for each user, which he or she identifies using a unique username and password. Each user can then create database tables and other data objects that reside in his or her area of the database, which is called a **user schema**. The data objects within a user schema are called **database objects** or **schema objects**. A user's schema contains all of the objects that the user creates and stores in the database. Schema objects include all of the user's database tables. Schema objects also include views, which are logical tables that are based on a query, and stored programs.

To connect to an Oracle 10*g* database and create and manage database objects in your user schema, you must enter a username and a password.

If you are using an Oracle 10g Personal Edition database, the directions for creating the username and password are in the installation instructions. If you are using an Oracle 10g Enterprise Edition client/server database, your instructor will create your username and password.

NOTE

As a security measure, Oracle does not automatically provide privileges when a new user account is created. In fact, even with a valid user account, you must still be explicitly granted the privilege to connect to the database.

TIP

When you create a new database object in your user schema, you are the owner of this object. As the object owner, you have privileges to perform all possible actions on the object; for example, you can modify a database table's structure, delete the table, and insert, update, and view table data.

DEFINING ORACLE 10*g* DATABASE TABLES

Tables are the primary data objects in a relational database. When you create a new Oracle 10*g* database table, you specify the table name, the name of each data column, and the data type and size of each data column. You can also define **constraints**, which are restrictions on the data values that a column can store. Examples of constraints include whether the column is a primary key, whether it is a foreign key, whether a column allows NULL values, and whether the column can only contain certain values, such as FR, SO, JR, or SR for a column that specifies student class values.

Table names and column names must follow the **Oracle naming standard**, which is a series of rules that Oracle Corporation has established for naming all database objects. This Oracle naming standard states that objects must be from one to 30 characters long, can contain letters, numbers, and the special symbols ($),(_), and (#), and must begin with a character. Examples of legal Oracle 10*g* database object names are STUDENT_TABLE, PRICE$, or COURSE_ID#. Examples of illegal Oracle 10*g* database object names are STUDENT TABLE (which contains a blank space), STUDENT-TABLE (which contains a hyphen), or #COURSE_ID (which does not begin with a character). The Oracle naming standard is referenced throughout the book.

You use the CREATE TABLE SQL command to create a new table. The general syntax for the CREATE TABLE command is:

```
CREATE TABLE tablename
(columnname1 data_type,
columnname2 data_type, ...)
```

In this syntax, *tablename* represents the name of the new table and must follow the Oracle naming standard. Following the CREATE TABLE clause, you list the name of each database column, followed by its data type. The data type identifies the type of data that the column stores, such as numbers, characters, or dates. (The next section describes Oracle 10*g* data types in detail.) Database column names must also follow the Oracle naming standard. You enclose the list of columns in parentheses, and separate each column specification with a comma.

The CREATE TABLE command is an example of a SQL DDL command. (Recall that you use DDL commands to create and modify database objects.) With a CREATE TABLE command, the DBMS creates the database table as soon as the command executes, and you do not need to use any other commands to explicitly save the table.

ORACLE 10*g* DATA TYPES

Notice in the syntax for creating a table that you specify a data type for each column. The **data type** specifies the kind of data that the column stores. An Oracle 10*g* database configures data columns using specific data types for two reasons. First, assigning a data type provides a means for error checking. For example, you cannot store the character data *Chicago* in a column assigned a DATE data type. Second, data types enable the DBMS to use storage space more efficiently by internally storing different types of data in different ways. The basic Oracle 10*g* data types define character, number, date/time, and large object values.

Character Data Types

Character data columns store alphanumeric values that contain text and numbers not used in calculations, such as telephone numbers and postal codes. The Oracle 10*g* character data types are VARCHAR2, CHAR, NVARCHAR2, and NCHAR. Each data type stores data in a different way.

NOTE The Oracle 10*g* database also supports the LONG data type, which allows you to store up to 2 gigabytes of character data. Because of limitations of the LONG data type, Oracle Corporation recommends that you store large volumes of character data using one of the large object data types rather than the LONG data type. Therefore, this book does not address the LONG data type.

VARCHAR2 Data Type

The **VARCHAR2** data type stores variable-length character data up to a maximum of 4000 characters. **Variable-length character data** is character data in which values in different rows can have a different number of characters. For example, in the Northwoods University database STUDENT table (see Figure 1-25), the values in the column that stores student last names may all have a different number of characters because people's names contain different numbers of characters.

You use the following syntax to declare a VARCHAR2 data column:

```
columnname VARCHAR2(maximum_size)
```

Columnname specifies the name of the column, and must follow the Oracle naming standard. *Maximum_size* can be an integer value from 1 to 4000. If the user inserts data values that are smaller than the specified maximum size, the DBMS stores only the actual character values. The DBMS does *not* add trailing blank spaces to the end of the entry to make the entry fill the maximum column size. If a user inserts a data value that has more characters than the maximum column size, an error occurs.

You would define the S_LAST column in the Northwoods University STUDENT table as follows:

```
s_last VARCHAR2(30)
```

This column declaration states that the column that stores a student's last name is a variable-length character column with a maximum of 30 characters. Examples of data stored in this column include the values Jones and Perez.

CHAR Data Type

The **CHAR** data type stores fixed-length character data up to a maximum of 2000 characters. **Fixed-length character data** is character data in which the data values for different rows all have the same number of characters. For example, the S_CLASS column in the Northwoods University STUDENT table stores values showing the student's current class standing (senior, junior, sophomore, or freshman). The S_CLASS column stores these values using the abbreviations SR, JR, SO, or FR, which all have exactly two characters.

The general syntax for declaring a CHAR data column is as follows:

```
columnname CHAR[(maximum_size)]
```

Maximum_size, which is optional, can be an integer value from 1 to 2000. If you omit the *maximum_size* value, the default size is one character. If a user inserts a data value in a CHAR column that has more characters than the maximum column size, an error occurs. If the user inserts data values that are smaller than the specified maximum size, the DBMS adds trailing blank spaces to the end of the entry to make the entry fill the *maximum_size* value. For example, if you declare the S_LAST column in the STUDENT table using the CHAR data type as s_last CHAR(30) and insert a data value of Perez, then the actual value that the database stores is Perez, plus 25 blank spaces to the right of the last character. Therefore, you should use the CHAR data type only when every data value in a column has exactly the same number of characters.

You would use the following command to define the S_CLASS column in the STUDENT table using the CHAR data type:

```
s_class CHAR(2)
```

The CHAR data type uses data storage space more efficiently than VARCHAR2, and the DBMS processes data retrieved from CHAR columns faster. However, if there is any chance that the data values might have different numbers of characters, use the VARCHAR2 data type.

NVARCHAR2 and NCHAR Data Types

Oracle 10*g* stores character data in VARCHAR2 and CHAR columns using American Standard Code for Information Interchange (ASCII) coding, which represents each character as an 8-digit (one-byte) binary value. ASCII coding can represent a total of 256 different characters. As a result, you cannot use the VARCHAR2 or CHAR data types to represent data that is input in character sets other than standard English.

NOTE

Examples of other character sets are Kanji, which represents the Japanese language, and Cyrillic, which represents Russian and a variety of eastern European languages. To store these characters sets in an Oracle 10*g* database, you must use Unicode coding.

To address the 256-character limitation of ASCII coding, the Oracle 10*g* database provides the NVARCHAR2 and NCHAR data types. The **NVARCHAR2** and **NCHAR** data types store variable-length and fixed-length data just as their VARCHAR2 and CHAR counterparts do, except that they use Unicode coding. **Unicode** is a standardized technique that provides a way to encode data in diverse languages, and which is input using different character sets. The NVARCHAR2 data type is similar to the VARCHAR2 data type in all other ways, and the NCHAR data type is similar to the CHAR data type in all other ways. You declare NVARCHAR2 and NCHAR columns using the following general syntax:

```
columnname NVARCHAR2(maximum_size)
columnname NCHAR[(maximum_size)]
```

Number Data Types

Oracle 10*g* stores all numerical data using the NUMBER data type. The NUMBER data type stores negative, positive, fixed, and floating-point numbers between 10^{-130} and 10^{125}, with precision up to 38 decimal places. **Precision** is the total number of digits both to the left and to the right of the decimal point.

NOTE

The precision value includes only the digits, and does not include formatting characters, such as the currency symbol, commas, or the decimal point.

You use the NUMBER data type for any column that stores numerical data upon which users may perform arithmetic calculations. (Recall that you use a character data type to store numerical data that will not be involved in calculations, such as telephone numbers or postal codes.) When you declare a NUMBER column, you optionally specify the precision and the scale using the following general syntax:

```
columnname NUMBER [([precision,] [scale])]
```

The *precision* value specifies the total number of digits in the data value, and the scale value specifies the number of digits on the right side of the decimal point. The precision value includes all digits, including the digits to the right of the decimal point, so the precision value is usually larger (it cannot be smaller) than the scale value. There are three NUMBER data subtypes: integer, fixed-point, and floating-point. You specify these data subtypes using different precision and scale configurations in the column declaration.

Integer Numbers

An integer is a whole number with no digits on the right side of the decimal point. To declare an integer data column, you omit the scale value in the data declaration using the following general syntax:

```
columnname NUMBER(precision)
```

You usually use integers to declare surrogate key columns. For example, the F_ID column in the FACULTY table stores integers (student identification numbers), so the declaration for this column is as follows:

```
f_id NUMBER(5)
```

In this column declaration, NUMBER(5) specifies that the F_ID column has a maximum length of five digits. The omission of the scale value specifies that the column values have no digits to the right of the decimal point. Examples of values in this column are 100, 101, 102, and so on.

For an integer data column, if the user enters a value that is smaller than the specified precision, the DBMS stores the actual data value, and does not store or display leading zeros. If the user attempts to enter a data value that is larger than the specified precision value, an error occurs. If the user enters a value that contains digits to the right of the decimal point, the DBMS rounds the value to the nearest whole number.

Fixed-point Numbers

A **fixed-point number** contains a specific number of decimal places, so the column declaration specifies both the precision and scale values. An example of a fixed-point data column is the PRICE column in the Clearwater Traders INVENTORY table. Data values in the PRICE column can range between 0 and 999.99, and always have exactly two digits to the right of the decimal point. You would use the following column declaration to define the PRICE column:

```
price NUMBER(5,2)
```

This declaration specifies that the PRICE column can have a maximum of five digits, with exactly two digits to the right of the decimal point. Examples of data values in the PRICE column are 259.99 and 59.99. Note that the decimal points are not included in the precision value that specifies the maximum width of the data value.

If the user attempts to enter a data value that is larger than the specified precision value, an error occurs. If the user enters a value that contains more digits to the right of the decimal point than specified in the scale, the DBMS rounds the value to the specified scale. For example, suppose the user inserts the value 19.579 in the PRICE column. Because the PRICE column has a scale of 2, the DBMS rounds the value to two decimal places, and stores the value as 19.58.

Floating-point Numbers

A **floating-point number** contains a variable number of decimal places. The decimal point can appear anywhere, from before the first digit to after the last digit, or can be omitted entirely. To define a floating-point data column, you omit both the precision and the scale values in the column declaration using the following syntax:

```
columnname NUMBER
```

Although no floating-point values exist in any of the case study databases, a potential floating-point column in the Northwoods database would be student grade point average (GPA), which you would declare as:

```
s_gpa NUMBER
```

A student's GPA might include one or more decimal places, such as 2.7045, 3.25, or 4.0.

Date and Time Data Types

The Oracle 10*g* data types that store date and time values include the **datetime** data subtypes, which store actual date and time values, and the interval data subtypes, which store an elapsed time **interval** between two datetime values. The main datetime subtypes are DATE and TIMESTAMP. The interval subtypes include INTERVAL YEAR TO MONTH and INTERVAL DAY TO SECOND. The following sections describe these data types.

DATE Data Type

The **DATE** data type stores dates from December 31, 4712 BC to December 31, AD 4712. The DATE data type stores the century, year, month, day, hour, minute, and second. The default date format is DD-MON-YY, which indicates the day of the month, a hyphen, the month (abbreviated using three capital letters), another hyphen, and the last two digits of the year. The default time format is HH:MI:SS AM, which indicates the hours, minutes, and seconds using a 12-hour clock. If the user does not specify a time when he or she enters a DATE data value, the default time value is 12:00:00 AM. If the user does not specify the date when he or she enters a time value, the default date value is the first day of the current month.

To declare a DATE data column, use the following general syntax:

```
columnname DATE
```

As always, *columnname* must adhere to the Oracle naming standard. Because the Oracle 10*g* DBMS stores DATE columns in a standard internal format that does not vary, you do not need to include a length specification. An example of a data column that uses the DATE data type is the S_DOB column (student date of birth) in the Northwoods University STUDENT table, which you declare using the following command:

```
s_dob DATE
```

The S_DOB column stores a value such as 07-OCT-82 12:00:00 AM. The 12:00:00 AM time is the default time assigned to a date when you do not explicitly enter a time.

TIMESTAMP Data Type

The TIMESTAMP data type stores date values similar to the DATE data type, except it also stores fractional seconds in addition to the century, year, month, day, hour, minute, and second. An example of a TIMESTAMP data value is 15-AUG-06 09.26.01.123975 AM. You use the TIMESTAMP data type when you need to store precise time values. The following general syntax declares a TIMESTAMP data column:

```
columnname TIMESTAMP (fractional_seconds_precision)
```

In this syntax, *fractional_seconds_precision* specifies the precision (number of decimal places) that the DBMS will store for the fractional seconds. If you omit the *fractional_seconds_precision* specification, the default value is six decimal places.

For example, suppose you want to store a date value that includes the fractional seconds for when a shipment is received at Clearwater Traders, and you specify the fractional seconds using two decimal places. You would declare the SL_DATE_RECEIVED column as follows:

```
sl_date_received TIMESTAMP(2)
```

INTERVAL YEAR TO MONTH Data Type

Recall that the Oracle 10*g* interval data types store an elapsed time interval between two dates. The INTERVAL YEAR TO MONTH data type stores a time interval, expressed in years and months, using the following syntax:

```
+|- elapsed_years-elapsed_months
```

In this syntax, + | – indicates that the interval can specify either a positive or negative time interval. If you add a positive time interval to a known date, the result is a date after the known date. If you add a negative time interval to a known date, the result is a date before the known date. *Elapsed_years* specifies the year portion of the interval, *elapsed_months* specifies the month portion of the interval, and the two values are separated by a hyphen (–). An example INTERVAL YEAR TO MONTH data value is +02-11, which specifies a positive time interval of 2 years and 11 months.

You use the following command to declare an INTERVAL YEAR TO MONTH data column:

```
columnname INTERVAL YEAR[(year_precision)] TO MONTH
```

In this syntax, *year_precision* specifies the maximum allowable number of digits that the column uses to express the year portion of the interval. (If you omit the *year_precision* value in the column declaration, the default value is 6, which enables the column to display a maximum interval of 999,999 years.) The TIME_ENROLLED column in the Northwoods University STUDENT table, which displays the amount of time that a student has been enrolled at the university, is an example of a column that uses an interval data type. You would use the following command to declare the TIME_ENROLLED column using the INTERVAL YEAR TO MONTH data type, with the default value of 6:

```
time_enrolled INTERVAL YEAR TO MONTH
```

INTERVAL DAY TO SECOND Data Type

The INTERVAL DAY TO SECOND data type stores a time interval, expressed in days, hours, minutes, and seconds. An INTERVAL DAY TO SECOND data column stores data values using the following general syntax:

```
+|- elapsed_days ↵
elapsed_hours:elapsed_minutes:elapsed_seconds
```

The following data value expresses a negative time interval of 4 days, 3 hours, 20 minutes, and 32 seconds: -04 03:20:32.00. The basic syntax to define an INTERVAL DAY TO SECOND data column is as follows:

```
columnname INTERVAL DAY[(leading_precision)] ↵
   TO SECOND[(fractional_seconds_precision)]
```

In this syntax, *leading_precision* specifies the maximum allowable number of digits that the column uses to express the elapsed days, and *fractional_seconds_precision* specifies the maximum allowable number of digits that the column uses to express elapsed seconds. The default values for these expressions are 2 and 6, respectively.

You would use the following command to declare the TIME_ENROLLED column using the INTERVAL DAY TO SECOND data type and accept the default precision values:

```
time_enrolled INTERVAL DAY TO SECOND
```

Large Object (LOB) Data Types

Sometimes databases store binary data, such as digitized sounds or images, or references to binary files from a word processor or spreadsheet. In these cases, you can use one of the Oracle 10*g* large object (LOB) data types. Table 2-1 summarizes the four LOB data types.

Large Object (LOB) Data Type	Description
BLOB	Binary LOB, storing up to 4 GB of binary data in the database
BFILE	Binary file, storing a reference to a binary file located outside the database in a file maintained by the operating system
CLOB	Character LOB, storing up to 4 GB of character data in the database
NCLOB	Character LOB that supports 2-byte character codes, stored in the database—up to a maximum of 4 GB

Table 2-1   Large object (LOB) data types

NOTE   Previous versions of Oracle supported the RAW and LONG RAW data types for storing binary data. Oracle Corporation recommends storing binary data using the large object data types, so this book does not use the RAW and LONG RAW data types.

You declare an LOB data column using the following general syntax:

```
columnname LOB_data_type
```

In this syntax, *LOB_data_type* is the name of the LOB data type, and can have the value BLOB, CLOB, BFILE, or NCLOB. Note that you do not specify the object size. The database automatically allocates the correct amount of space to store an LOB object when you insert the object into the database. For example, suppose you want to store the image of the faculty members in the F_IMAGE column of the Northwoods University FACULTY database table. You could store the actual binary image data using a BLOB data type, which you declare using the following command:

```
f_image BLOB
```

Alternately, you could store a reference to the location of an external image file using the BFILE data type by making the following declaration:

```
f_image BFILE
```

CONSTRAINTS

Constraints are rules that restrict the data values that you can enter into a column in a database table. There are two types of constraints: **integrity constraints**, which define primary and foreign keys; and **value constraints**, which define specific data values or data ranges that must be inserted into columns and whether values must be unique or not NULL. There are two levels of constraints: table constraints and column constraints.

A **table constraint** restricts the data value with respect to all other values in the table. An example of a table constraint is a primary key constraint, which specifies that a column value must be unique and cannot appear in this column in more than one table row.

A **column constraint** limits the value that can be placed in a specific column, irrespective of values that exist in other table rows. Examples of column constraints are value constraints, which specify that a certain value or set of values must be used, and NOT NULL constraints, which specify that a value cannot be NULL. A value constraint might specify that the value of a GENDER column must be either M (for male) or F (for female). You might place a NOT NULL constraint on a student ADDRESS column, to ensure that users always enter address information when they create a new customer row.

You can place constraint definitions at the end of the CREATE TABLE command, after you declare all of the table columns. Or, you can place each constraint definition within the column definition, so it immediately follows the data column declaration for the column associated with the constraint.

Every constraint in a user schema must have a unique constraint name that must adhere to the Oracle naming standard. You can explicitly define the constraint name, or omit the constraint name, and allow the Oracle 10*g* DBMS to assign the constraint a system-generated name. The system-generated names are generic and use the prefix of SYS_C*n*, where *n* represents a numeric value designed to make each constraint name unique. This type of naming strategy makes it difficult to determine, solely by its name, with which table the constraint is associated. Therefore, it is a good practice to specify constraint names explicitly using descriptive names so you can easily identify the constraints that exist in your user schema.

A convention for assigning unique and descriptive constraint names is called the **constraint naming convention**. The constraint naming convention uses the following general syntax: `tablename_columnname_constraintID`. *Tablename* is the name of the table in which the constraint is defined, and *columnname* is the table column associated with the constraint. *ConstraintID* is a two-character abbreviation that describes the constraint. Table 2-2 shows commonly used *constraintID* abbreviations, and the Oracle 10*g* constraint type identifier. (You will learn how to use the Oracle 10*g* constraint type identifiers later in this chapter.)

Constraint Type	ConstraintID Abbreviation	Oracle 10*g* Constraint Type Identifier
PRIMARY KEY	pk	P
FOREIGN KEY	fk	R
CHECK CONDITION	cc	C
NOT NULL	nn	N
UNIQUE	uk	U

Table 2-2   Common constraintID abbreviations

TIP   Sometimes the constraint naming convention causes constraint names to become longer than 30 characters. When this happens, you must abbreviate either the table name or the column name. Be sure to do this in a way that enables you to identify the table and column name with which the DBMS associates the constraint.

Integrity Constraints

An integrity constraint defines primary key columns, and specifies foreign keys and their corresponding table and column references. The following paragraphs describe how to define primary keys, foreign keys, and composite primary keys.

Primary Keys

The general syntax for defining a primary key constraint within a column declaration is:

```
CONSTRAINT constraint_name PRIMARY KEY
```

The syntax for defining a primary key constraint at the end of the CREATE TABLE command, after the column declarations, is:

```
CONSTRAINT constraint_name PRIMARY KEY (columnname)
```

Note that the first syntax, which defines the primary key within the column declaration, does not specify the column name. This is because the primary key constraint is defined within the column declaration. The second syntax includes the column name within the constraint definition, because the constraint is defined independently of the column declaration.

An example of a primary key column is LOC_ID in the Northwoods LOCATION table. You would use the following command to create the LOCATION table, and specify LOC_ID as the primary key (this example uses the syntax for which the constraint definition is independent of the column declarations, and uses the constraint naming convention to define the constraint name):

```
CREATE TABLE location
(loc_id NUMBER(6),
bldg_code VARCHAR2(10),
room VARCHAR2(6),
capacity NUMBER(5),
CONSTRAINT location_loc_id_pk PRIMARY KEY (loc_id));
```

You would use the following command to define LOC_ID as the table's primary key using the syntax for which the constraint definition is created within the column definition (note that in this approach, the column name is not repeated after the constraint definition):

```
CREATE TABLE location
(loc_id NUMBER(6)
CONSTRAINT location_loc_id_pk PRIMARY KEY,
bldg_code VARCHAR2(10),
room VARCHAR2(6),
capacity NUMBER(5));
```

Foreign Keys

A foreign key constraint is a column constraint that specifies that the value a user inserts in a column must exist as a primary key in a referenced table. As with primary keys, you can define foreign key constraints independently of a column declaration, or directly within the declaration of the column that has the constraint. The general syntax for specifying a foreign key constraint at the end of the column declarations and independent of a specific column declaration is:

```
CONSTRAINT constraint_name
FOREIGN KEY (columnname)
REFERENCES primary_key_tablename (primary_key_columnname)
```

Primary_key_tablename refers to the name of the table in which the foreign key is a primary key. *Primary_key_columnname* refers to the name of the primary key column in the *primary_key_tablename* table.

The general syntax for specifying a foreign key constraint within a column declaration is:

```
CONSTRAINT constraint_name
REFERENCES primary_key_tablename
(primary_key_columnname)
```

Note that the second syntax omits the keyword FOREIGN KEY and the constraint's column name. This is because the command declares the foreign key constraint within the column declaration, and the keyword REFERENCES implicitly defines the constraint as a foreign key constraint.

An example of a foreign key is the LOC_ID column in the Northwoods FACULTY table. The command to create the LOC_ID foreign key constraint in the FACULTY table independently of the column declaration is as follows:

```
CONSTRAINT faculty_loc_id_fk FOREIGN KEY (loc_id)
REFERENCES location (loc_id)
```

The command to create the LOC_ID foreign key constraint in the FACULTY table within the LOC_ID column declaration is as follows:

```
loc_id NUMBER(6) CONSTRAINT faculty_loc_id_fk
REFERENCES location (loc_id)
```

NOTE

Before you can create a foreign key reference, the table in which the column is a primary key must already exist, and the column must be defined as a primary key. In this example, you must create the LOCATION table and define LOC_ID as the primary key of the LOCATION table before creating the FACULTY table and defining LOC_ID as a foreign key within the FACULTY table.

Composite Keys

Recall that a composite key is a primary key composed of two or more data columns. You define a composite primary key constraint by listing all of the columns that make up the composite primary key, with each column name separated by a comma, using the following syntax:

```
CONSTRAINT constraint_name
PRIMARY KEY (columnname1, columnname2, …)
```

Constraint_name should consist of the table name, the names of all columns that make up the composite key, and the identifier pk. You must place the composite key definition after the commands in the CREATE TABLE command that define the columns that make up the key.

In the Northwoods ENROLLMENT table, the primary key consists of both the S_ID and the C_SEC_ID columns. The command to create the ENROLLMENT table and specify that the S_ID and C_SEC_ID columns make up the composite primary key is:

```
CREATE TABLE enrollment
(s_id NUMBER(5) CONSTRAINT enrollment_s_id_fk
REFERENCES student(s_id),
c_sec_id NUMBER(8) CONSTRAINT enrollment_c_sec_id_fk
REFERENCES course_section(c_sec_id),
CONSTRAINT enrollment_s_id_c_sec_id_pk
PRIMARY KEY (s_id, c_sec_id));
```

TIP

The columns that make up a composite key usually have foreign key constraints as well, as seen in the preceding example.

Value Constraints

Value constraints are column-level constraints that restrict the data values that users can enter into a given column. Commonly used value constraints include:

- CHECK conditions—This enables you to specify that a column value must be a specific value or fall within a range of values.
- NOT NULL constraint—Specifies whether a column value can be NULL
- DEFAULT constraint—Specifies that a column has a default value that the DBMS automatically inserts for every row, unless the user specifies an alternate value
- UNIQUE constraint—Specifies that a column must have a unique value for every table row

Check Conditions

Check conditions specify that a column value must be a specific value (such as M), from a set of allowable values (such as M or F), or fall within a specific range (such as greater than zero but less than 1000). You should create check condition constraints only when the number of allowable values is limited and not likely to change. You must be prudent when specifying check conditions in table definitions, because once the table is populated with data, it is difficult or impossible to modify the constraint—all rows must satisfy the constraint.

The DBMS must be able to evaluate each expression in a check condition as either true or false. You can combine expressions using the logical operators AND and OR. When you join two expressions using the AND operator, both expressions must be true for the expression to be true. When you join two expressions using the OR operator, only one expression needs to be true for the expression to be true.

As an example of a check condition, consider the S_CLASS column in the Northwoods University STUDENT table, in which the values are restricted to FR, SO, JR, or SR (freshman, sophomore, junior, or senior). The syntax to define this check condition is:

```
CONSTRAINT student_s_class_cc CHECK
((s_class = 'FR') OR (s_class = 'SO')
OR (s_class = 'JR') OR (s_class ='SR'))
```

You can also use a check condition to validate a range of allowable values. An example of a range check condition is in the CREDITS column in the Northwoods COURSE table, where the allowable values must be greater than 0 and less than 12. The constraint definition is:

```
CONSTRAINT course_credits_cc
CHECK((credits > 0) AND (credits < 12))
```

Note that the DBMS can evaluate both of the expressions `(credits > 0)` and `(credits < 12)` as either true or false. The AND condition specifies that both conditions must be true to satisfy the check condition.

NOT NULL Constraints

The **NOT NULL constraint** specifies whether the user must enter a value for a specific row, or whether the value can be NULL (absent or unknown). Primary key columns automatically have a NOT NULL constraint. From a business standpoint, some columns, such as a customer's name, should not be NULL. The following syntax specifies that the S_LAST column in the Northwoods student table must contain a value or the row is rejected, rather than added to the table:

```
s_last VARCHAR2(30)
CONSTRAINT student_s_last_nn NOT NULL
```

Default Constraints

A **default constraint** specifies that a particular column has a default value that the DBMS automatically inserts for every table row. This constraint must be created in the column declaration, not as a separate command beginning with CONSTRAINT. For example, if most Northwoods University students live in Florida, you would use the following column declaration to specify that the S_STATE column has a default value of FL. If the default value is a character string, you must enclose the value in single quotation marks.

```
s_state CHAR(2) DEFAULT 'FL'
```

The DBMS will insert the default only if the user inserts a NULL value into the S_STATE column. If the user specifies an alternate =value, the alternate value replaces the default value.

UNIQUE Constraints

A **UNIQUE constraint** is a table constraint that specifies that a column must have a unique value for every table row. It is basically the same as a primary key constraint, except NULL values are allowed in the column. NULL values are not allowed in a column that is referenced by a primary key constraint. An example of a column that might require a unique constraint is the TERM_DESC column in the Northwoods University TERM table, to specify that each term has a unique description. You would use the following command to create this constraint:

```
CONSTRAINT term_term_desc_uk UNIQUE (term_desc)
```

CREATING DATABASE TABLES USING SQL*PLUS

This next section explains how to create database tables using SQL*Plus, Oracle 10*g*'s command-line SQL utility. Starting an Oracle 10*g* client application such as SQL*Plus is a two-step process: first you start the application on your client workstation, then you log onto the Oracle 10*g* database. In this first set of steps, you start SQL*Plus, log onto the database, and create some database tables.

NOTE   If you are working in a computer laboratory, your instructor will inform you of the correct procedures to start and log onto the Oracle 10*g* database. If you are using Oracle 10*g* Personal Edition, use the logon procedures in the installation instructions.

To start SQL*Plus:

1. Click **Start** on the Windows taskbar, point to **All Programs** (if you are using Windows 2000, point to **Programs**), point to **Oracle 10*g* OraHome10**, point to **Application Development**, and then click **SQL Plus**. The Log On dialog box opens and requests your username, password, and host string, as shown in Figure 2-1.

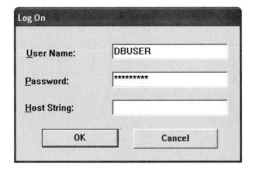

Figure 2-1   SQL*Plus Log On dialog box

NOTE

These instructions assume that you installed the Oracle 10*g* DBMS software in the OraHome10 folder. If you installed the software in a different folder, point to **All Programs**, then point to **Oracle 10*g* *foldername*** instead. If you installed the program on your home machine and have created only one database, you can leave the Host String blank. However, if there is more than one database, or if you receive a TNS error, enter the name of your database as the Host String.

2. Type your username, press **Tab**, type your password, press **Tab** again, type the host string provided by your instructor, and then click **OK**. (If you are using Oracle 10*g* Personal Edition, leave the host string blank.) The SQL*Plus program window opens, as shown in Figure 2-2. If necessary, maximize the program window.

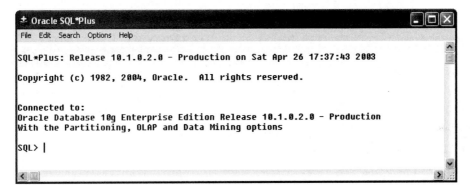

Figure 2-2   SQL*Plus window

HELP

Your release of SQL*Plus might have a different version number, depending on the client or server software you are using. The steps in the book will work the same regardless of the version number.

To use SQL*Plus, you type SQL commands at the SQL prompt. You end each command with a semicolon (;), which marks the end of the command. Then you press Enter to submit the command to the SQL*Plus interpreter. The **SQL*Plus interpreter** checks the command for syntax errors, and if the command is error free, the SQL interpreter submits the command to the database. SQL commands are not case sensitive, and the SQL interpreter ignores not only spaces between characters but line breaks as well. However, it is a good practice to format your SQL commands using uppercase letters for reserved words, and to break the command across multiple lines, so the command is easier to understand and debug.

When you create database tables that contain foreign key references to other tables, you must first create the table in which the foreign key is a primary key. For example, to create the Northwoods University STUDENT table, you must first create the FACULTY table, because the F_ID column in the STUDENT table is a foreign key that references the F_ID column in the FACULTY table. However, note that the FACULTY table contains the LOC_ID column, which is a foreign key that references the LOC_ID column in the LOCATION table. The LOCATION table has no foreign key references, so you can create the LOCATION table before you create any other tables. (In other words, you must create the LOCATION table before you can create the FACULTY table, and you must create the FACULTY table before you can create the STUDENT table.) Now you will create the LOCATION table.

To create the LOCATION table:

1. In SQL*Plus, type the command in Figure 2-3 at the SQL prompt. Do not type the line numbers. SQL*Plus adds line numbers to your command after you press Enter. Each line of the command defines a different column of the database.

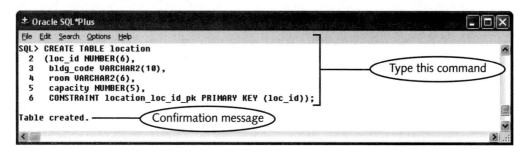

Figure 2-3 SQL command to create the LOCATION table

2. Type **;** as the last character of the last line to end the command. The semi-colon marks the end of the SQL command for the SQL interpreter.

3. Press **Enter** to execute the command. The confirmation message "Table created" appears, as shown in Figure 2-3.

 If your command had an error, the next section will describe how you can edit the command using a text editor.

Creating and Editing SQL Commands Using a Text Editor

Many SQL commands are long and complex, and it is easy to make typing errors. A good approach for entering and editing SQL*Plus commands is to type commands into a text editor such as Notepad, then copy your commands, paste the copied commands into SQL*Plus, and execute the commands. If the command has an error, you can switch back to the text editor, edit the command, copy and paste the edited text back into SQL*Plus, and then reexecute the command.

When you are creating database tables, it is a good idea to save the text of all of your CREATE TABLE commands in a single Notepad text file so you have a record of the original code. Saving all the commands in one file creates a **script**, which is a text file that contains several related SQL commands. You can run the script later to re-create the tables if you need to make changes. You can save multiple CREATE TABLE commands in such a text file. Just make sure that they are in the proper order so that foreign key references are made after their parent tables are created.

Next you learn how to use Notepad to edit SQL*Plus commands. If your SQL*Plus commands have errors, you can edit the commands, and reexecute them. You also want to save the Notepad file so you have a record of your commands. This is the beginning of the script file that creates the tables in the Northwoods University database. You continue to use this file to record and edit your SQL commands for the rest of this chapter.

To edit the command in Notepad and save the file:

1. In SQL*Plus, type the command in Figure 2-4 at the SQL prompt, exactly as shown. (This command purposely contains an error.) Press **Enter** to execute the command. An error message appears as shown in Figure 2-4 because the command that declares the ROOM column spells VARCHAR2 as VARCHR2.

2. Select the text as shown in Figure 2-4. Do not select the line numbers, the SQL prompt, confirmation, or error message.

```
± Oracle SQL*Plus
File Edit Search Options Help
SQL> CREATE TABLE location
  2 (loc_id NUMBER(6),
  3   bldg_code VARCHAR2(10),
  4   room VARCHR2(6),
  5   capacity NUMBER(5),
  6 CONSTRAINT location_loc_id_pk PRIMARY KEY (loc_id));
room VARCHR2(6),
            *                               Error location
ERROR at line 4:
ORA-00907: missing right parenthesis          Error message
```

Figure 2-4   Selecting the SQL command text

3. Click **Edit** on the menu bar, and then click **Copy**.

 Another way to copy text is to highlight the text, and then press Ctrl+C. (The C is not case sensitive.)

4. Start Notepad (or an alternate text editor) on your computer.

5. In Notepad, click **Edit** on the menu bar, and then click **Paste**. The SQL command text appears in Notepad.

Another way to paste text is to press Ctrl+V. (The *V* is not case sensitive.)

6. Correct the command text by changing the data type spelling to VARCHAR2 in the ROOM column declaration. Then copy the command, switch to SQL*Plus, paste the corrected text at the SQL prompt, and press **Enter** to run the command again. (If you successfully created the LOCATION table the first time, an error message stating "name is already used by an existing object" appears.) If a different error message appears, and you still have not yet successfully created the LOCATION table, switch back to Notepad. Debug the command, and run the command again until the message appears confirming that the LOCATION table has been successfully created.

7. Switch back to Notepad, click **File**, and then click **Save**. Navigate to the Chapter2\Tutorials folder on your Data Disk, type **Ch2Queries.sql** in the File name text box, open the Save as type list, select **All files**, and then click **Save**.

Script files normally have a .sql extension. If you do not open the Save as type list and select All Files, Notepad automatically appends a .txt extension onto the filename.

8. Minimize Notepad.

Using Oracle Online Help Resources to Debug SQL Commands

When a SQL command contains a syntax error, the SQL*Plus interpreter displays error information that includes the line number within the command that caused the error, the position of the error within the line, and an error code and description of the error. When you tried to create the LOCATION table, the error message shown in Figure 2-4 appeared. Note that the interpreter displays the command that contains the error, and shows the position of the error in the command by placing an asterisk (*) under the character that caused the error.

Oracle 10*g* error codes have a 3-character prefix (such as ORA), and a 5-digit error code. The prefix indicates which Oracle 10*g* utility generated the error. In this case, the ORA prefix indicates that the error was generated by the DBMS. The error code (00907) is a numeric code assigned by Oracle Corporation that indicates that a right (closing) parenthesis was omitted in a command. In this case, the data type for the ROOM column was misspelled as VARCHR2 instead of VARCHAR2. Although the error description ("missing right parenthesis") does not accurately describe the error, the asterisk flagging the error position makes the error easy to locate.

Sometimes the causes of SQL command errors are not readily apparent, and you need to retrieve more information about the error. For example, consider the error message in Figure 2-5.

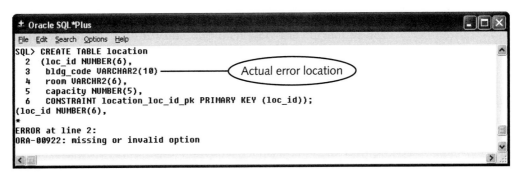

Figure 2-5 SQL command error message requiring further explanation

The error message indicates that the error occurred on Line 2, and the asterisk indicates that the error occurred at the position of the opening parenthesis before the first table column (LOC_ID) was specified. When you inspect the second line of the command, the line does not appear to have any errors. The error code explanation ("missing or invalid option") does not adequately explain the error cause, nor does it provide ideas for locating or correcting the error.

To get further information about this error, you can connect to the Oracle Technology Network (OTN) Web site and search for the error code. OTN is a Web-based resource that Oracle Corporation provides free of charge.

Although the format of the Web site tends to change, usually you can view reference material regarding Oracle error codes by selecting the Documentation option available on the home page. When the Documentation page appears, select the option to view the Documentation available for the Oracle 10*g* Database. After accessing the database documentation portion of the Web site, you can perform a search on the specific error code returned by the DBMS (error code ORA-00922 in this example). Documentation for most ORA- error codes returns the possible causes for the error and action necessary to correct the problem.

If a search for the error code does not return the desired results, you can also perform a search for "Part Number B10744-01" which provides access to Oracle® Database Error Messages 10*g* Release 1 (10.1) available on the Web site. Each chapter provides access to a different range of error codes. A list of the different chapters is provided on the Contents page.

When an error occurs that you cannot locate, a last resort debugging technique is to create the table multiple times, each time adding a column declaration repeating the process until you find the declaration causing the error. First, paste your nonworking command in a Notepad file and modify it so that it creates the table with only the first column declaration. Copy the modified command, and paste it into SQL*Plus. If SQL*Plus successfully creates the table with the first column, you now know that the error is not in the first column declaration. Delete the table using the DROP TABLE command, which has the following syntax: `DROP TABLE tablename;`. (You will learn more about the DROP TABLE command later in this chapter.) Then, modify the command in Notepad to create the table using only the first and second column declarations. If this works, you now know that the problem is not in either the first or second column declaration. Drop the table again, and modify the command to create the table using only the first, second, and third column declarations. Continue this process of adding one more column declaration to the CREATE command until you locate the column declaration that is causing the error.

NOTE

If you find the information regarding ORA error messages posted on the OTN confusing or obscure, you can also use a search engine such as Google to locate information posted by other users at different locations on the Internet.

Exiting SQL*Plus

There are three ways to exit SQL*Plus:

- Type exit at the SQL prompt.
- Click File on the menu bar, and then click Exit.
- Click the Close button on the program window title bar.

You should *never* exit SQL*Plus by simply shutting down the machine. This can lead to corrupt files and prevent you from starting the database in the future.

NOTE

When entering or editing data that is stored in a database table, you should not click the Close button to exit SQL*Plus or you risk losing any uncommitted data changes.

Your database connection disconnects automatically when you exit SQL*Plus. In the next set of steps, you exit SQL*Plus by typing exit at the SQL prompt.

To exit SQL*Plus:

1. Type **exit** at the SQL prompt.

2. Press **Enter**. Your database connection disconnects, and the SQL*Plus window closes.

Creating a Table with a Foreign Key Constraint

Recall that a foreign key creates a relationship between two database tables. In the Northwoods University FACULTY table, LOC_ID is a foreign key that references a row in the LOCATION table. Now you create the FACULTY table, and specify the LOC_ID column as a foreign key.

To create the FACULTY table and specify the foreign key:

1. Switch to Notepad, where your Ch2Queries.sql file should be open. If necessary, press **Enter** to create a new blank line, and then type the command in Figure 2-6 to create the FACULTY table.

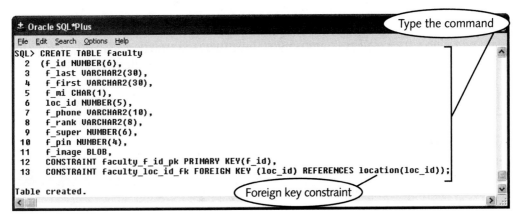

Figure 2-6 SQL command to create the FACULTY table

2. Select and copy all of the CREATE TABLE faculty command text.

3. Start SQL*Plus, log on to the database, paste the copied command at the SQL prompt, and then press **Enter**. If necessary, debug the command until you successfully create the FACULTY table.

NOTE
For the rest of the chapter, you should type SQL commands in the 2Queries.sql file, unless the instructions state to type the command at the SQL prompt. Then paste the commands into SQL*Plus, and debug the commands if necessary.

VIEWING INFORMATION ABOUT TABLES

After you create database tables, you often need to review information such as the table name, column and data type values, and constraint types. For example, when you write a command to insert a new row, you need to specify the table name, and you need to list the data values in the same order as the table column names. To view the column names and data types of an individual table, you use the DESCRIBE command, which has the following syntax:

```
DESCRIBE tablename
```

Next you use the DESCRIBE command to view information about the columns in the LOCATION and FACULTY tables that you created earlier.

NOTE The DESCRIBE command is actually a SQL*Plus command and not a SQL command so it does not require a semicolon at the end of the command to execute. Another feature of SQL*Plus commands is that they can be abbreviated. For example, you can type DESC *tablename* rather than DESCRIBE *tablename* to view the structure of a table.

To view the table columns:

1. In SQL*Plus, type the command **DESCRIBE location**, and then press **Enter**. The column names and data types for the LOCATION table should appear as shown in Figure 2-7.

2. Next, type **DESCRIBE faculty** at the SQL prompt and then press **Enter**. The column names and data types for the FACULTY table appear as shown in Figure 2-7.

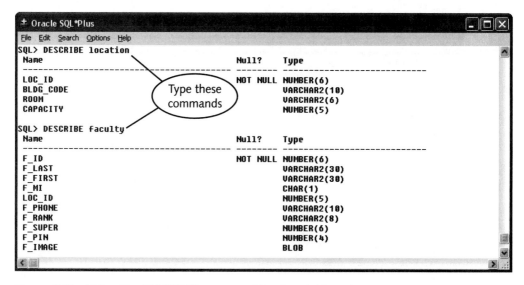

Figure 2-7 Using the DESCRIBE command to view table columns

Note that the output lists the column names and data types. It also shows the columns that have NOT NULL constraints: LOC_ID in the LOCATION table and F_ID in the FACULTY table. These columns cannot be NULL because they are the tables' primary key columns. However, note that the DESCRIBE command does not list the foreign key constraint on the LOC_ID column in the FACULTY table. To view information about constraints other than the NOT NULL constraint, you use the Oracle 10*g* data dictionary views.

The **Oracle 10*g* data dictionary** consists of tables that contain information about the structure of the database. When the DBA creates a new database, the system creates the data dictionary in a user schema named SYS. The Oracle 10*g* DBMS automatically updates the data dictionary tables as users create, update, and delete database objects. As a general rule, users and database administrators do not directly view, update, or delete values in the data dictionary tables. Rather, users interact with the data dictionary using the **data dictionary views**. A **view** is a database object that the DBMS bases on an actual database table and which enables the DBMS to present the table data in a different format based on the needs of users. A view can serve to hide some table columns in which the user has no interest or doesn't have required privileges to view.

The data dictionary views are divided into three general categories:

- USER—Shows the objects in the current user's schema

- ALL—Shows both the objects in the current user's schema and the objects that the user has privileges to manipulate. For example, another user might create a table, and then give you the privilege to update the table.

- DBA—Allows users who are database administrators to view information about all database objects

The general command to retrieve information from a data dictionary view is:

```
SELECT view_columnname1, view_columnname2, …
FROM prefix_object;
```

NOTE You learn more about using the SELECT command to retrieve database data in Chapter 3. In this chapter, you will use the SELECT command in a limited way to retrieve information about your database tables.

In this syntax, *view_columnname1, view_columnname2,* and so forth reference the names of the columns in the view that you wish to retrieve. *Prefix* specifies the view category, and can have the value USER, DBA, or ALL. *Object* is the type of database object you are examining, such as TABLES or CONSTRAINTS. For example, the following command retrieves the names of all of a user's database tables by retrieving the TABLE_NAME column from the USER_TABLES view:

```
SELECT table_name
FROM user_tables;
```

Similarly, the following command uses the ALL prefix and retrieves the names of all database tables that a user has either created or has been given object privileges to manipulate:

```
SELECT table_name
FROM all_tables;
```

In the next set of steps, you use data dictionary view commands to view information about your tables.

To view information about your tables:

1. Type **SELECT table_name FROM user_tables;** at the SQL prompt, and then press **Enter** to view information about the tables in your user schema. The output should be similar to the output in Figure 2-8. Your output will be different if you have created additional tables.

2. Type **SELECT table_name FROM all_tables;** at the SQL prompt, and then press **Enter** to retrieve the names of all tables that you have privileges to manipulate. Again, the output should appear similar to the output shown in Figure 2-8. (The output of this command will be different if other database users have created tables and given you privileges to manipulate these tables. Some of the output may scroll off the screen. The LOCATION and FACULTY tables are at the bottom of the list and are not shown in Figure 2-8.)

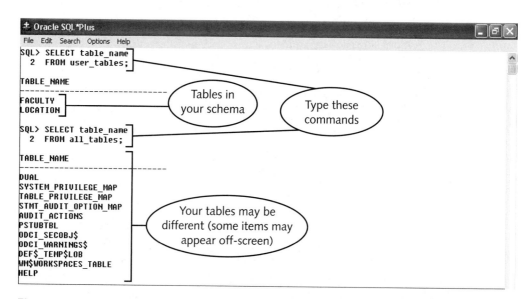

Figure 2-8   Retrieving table names from the USER_TABLES data dictionary view

You can retrieve information about a variety of database objects using different data dictionary views. Table 2-3 shows the names of database objects for which you can retrieve information from data dictionary views.

Object Name	Object Type
OBJECTS	All database objects
TABLES	Database tables
INDEXES	Table indexes created to improve query retrieval performance
VIEWS	Database views
SEQUENCES	Sequences used to generate surrogate key values automatically
USERS	Database users
CONSTRAINTS	Table constraints
CONS_COLUMNS	Table columns that have constraints
IND_COLUMNS	Table columns that have indexes
TAB_COLUMNS	All table columns

Table 2-3   Database objects with data dictionary views

You can retrieve information from these views using commands similar to the ones you used to retrieve information about your database tables. For example, the command to retrieve the names of all of your database objects is:

```
SELECT object_name
FROM user_objects;
```

All the data dictionary views have different columns. To determine the names of the columns in a specific database view, you use the DESCRIBE command along with the view name. Figure 2-9 shows a description of the column names in the USER_CONSTRAINTS data dictionary view.

Next you query the USER_CONSTRAINTS view to determine the constraints that exist in your database tables. You will select the CONSTRAINT_NAME, TABLE_NAME, and CONSTRAINT_TYPE columns from the USER_CONSTRAINTS view.

To view your table constraints:

1. In SQL*Plus, type the command in Figure 2-10 at the SQL prompt as shown.

2. Press **Enter** to execute the command. Figure 2-10 shows the output. Your output may be different if you have created additional database tables with other constraints.

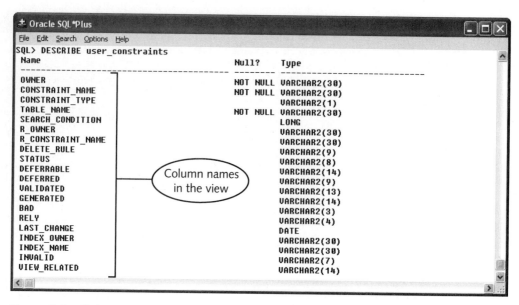

Figure 2-9   Column names in the USER_CONSTRAINTS data dictionary view

The output shows that your user schema has three constraints, two in the FACULTY table and one in the LOCATION table. The constraint names are the values you specified using the constraint naming convention when you created the tables. These values make it easy to determine a constraint's associated column name, and the constraint identifiers (**pk** and **fk**) identify the type of each constraint (primary key and foreign key). The final column in the query output also shows the constraint types. Recall that Table 2-2 shows the letter that Oracle 10*g* uses to identify different constraint types. In Figure 2-10, the CONSTRAINT_TYPE column identifies the primary key constraints using the letter *P*, and the foreign key constraints using the letter *R*. In Figure 2-10, the CONSTRAINT_TYPE column identifies the primary key constraints using the letter P and the foreign key constraints using the letter R, which stands for REFERENCE.

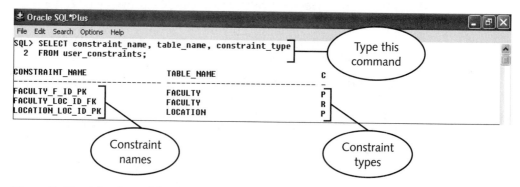

Figure 2-10   Viewing table constraint information

NOTE

The "C" displayed as the heading for the CONSTRAINT_TYPE column in Figure 2-10 is limited to the single letter C because the data in the column is declared as a VARCHAR2 of size 1. If the number of characters declared for a column is less than the number of characters in the name of the column, the column name in the heading is truncated to the number of characters in the column declaration.

To retrieve a list of all of the constraints for a specific database table in your user schema, you use the following command:

```
SELECT constraint_name, constraint_type
FROM user_constraints
WHERE table_name = 'DATABASE_TABLENAME';
```

In this syntax, the *DATABASE_TABLENAME* parameter is the name of the table for which you want to display constraints. You must specify this value in all uppercase letters because the data dictionary stores all object information in uppercase letters. You also must enclose the value in single quotation marks because it is a character string. Now you query the USER_CONSTRAINTS data dictionary view to retrieve information about the constraints in the FACULTY table.

To list the FACULTY table constraints:

1. In SQL*Plus, type the command shown in Figure 2-11 to retrieve information about the constraints in the FACULTY table.

2. Press **Enter** to execute the command. Your output should look similar to the output in Figure 2-11. (Don't worry if your constraints appear in a different order than the ones listed in Figure 2-11. You learn how to change the order of retrieved data values in Chapter 3.)

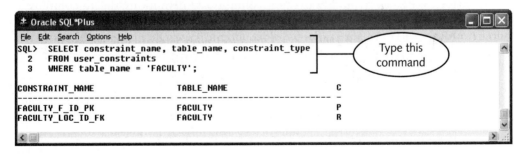

Figure 2-11   Viewing information about constraints in a specific table

MODIFYING AND DELETING DATABASE TABLES

When you design a database, you should plan your tables carefully to avoid having to change the structure of the database tables later. When you change database tables, often the applications that retrieve information from the tables no longer work correctly and have to be modified. However, Oracle 10g provides techniques for modifying existing database tables if you need to do so.

In Oracle 10g, you can modify existing database tables by performing actions such as changing the name of a table, adding new columns, deleting columns that are no longer needed, or changing the data type or maximum size of an existing column. There are some specifications of an Oracle 10g database table that you can always modify. To modify these specifications, you perform an **unrestricted action**. There are other table specifications that you can modify only in certain situations. To modify these specifications, you perform a **restricted action**. Table 2-4 lists unrestricted actions, and Table 2-5 summarizes restricted actions when modifying Oracle 10g database tables.

Unrestricted Actions
Renaming a table
Adding new fields
Deleting fields
Increasing the *maximum_size* value of a field
Deleting constraints

Table 2-4   Unrestricted actions when modifying database tables

The following sections describe how to delete and rename existing tables, add columns to existing tables, modify existing column data definitions, delete columns, and modify existing column constraints.

Deleting and Renaming Existing Tables

You should delete database tables with caution, because it is extremely difficult to rebuild a database table that you delete by mistake. And, there is always a chance that some database application somewhere is still using the table. When you delete a table from the database, you delete the table structure from the data dictionary and remove all of the table's data from the database. To delete a table, you use the DROP TABLE command, which has the following syntax:

```
DROP TABLE tablename;
```

Restricted Actions	Restriction
Deleting a table from a user schema	Allowed only if the table does not contain any columns that other tables reference as foreign keys
Changing an existing column's data type	Allowed only if existing data in the column is compatible with the new data type. For example, you could change a VARCHAR2 data type to a CHAR data type, but you could not change a VARCHAR2 data type to a NUMBER data type.
Decreasing the width of an existing column	Allowed only if existing column values are NULL
Adding a primary key constraint to an existing column	Allowed only if current column values are unique (no duplicate values) and NOT NULL
Adding a foreign key constraint	Allowed only if current column values are NULL, or exist in the referenced table
Adding a UNIQUE constraint to a column	Allowed only if current column values are all unique
Adding a CHECK constraint	Allowed, but the constraint applies only to values users insert after the constraint is added
Changing a column's default value	Allowed, but the default value is only inserted for rows that are added after the change

Table 2-5   Restricted actions when modifying database tables

Recall that deleting a table is a restricted action, so the DROP TABLE command cannot succeed if the table represented by *tablename* contains columns that other tables reference as foreign keys. For example, you cannot drop the LOCATION table in the Northwoods University database, because the FACULTY table references the LOC_ID column as a foreign key.

To drop a table that contains columns that other tables reference as foreign keys, you have two options. One option is to first drop all the tables that contain the foreign key references. For example, to delete the LOCATION table, you could first delete the FACULTY table, and then you could delete the LOCATION table. The second option is first to delete all of the foreign key constraints that reference the table to be deleted. To delete the foreign key constraints that reference a table, you use the CASCADE CONSTRAINTS option in the DROP TABLE command, which has the following syntax:

```
DROP TABLE tablename
CASCADE CONSTRAINTS;
```

When you execute the DROP TABLE command with the CASCADE CONSTRAINTS option, the system first drops all of the constraints associated with the table, and then drops the table. In the following set of steps, you try to drop the LOCATION table without the CASCADE CONSTRAINTS option, and note that an error occurs, because of the foreign key reference in the FACULTY table. You then drop the table using the CASCADE CONSTRAINTS option, which first deletes the table constraint.

To drop the LOCATION table, along with all of the constraints that reference the table, follow these steps:

1. Type **DROP TABLE location;** at the SQL prompt, and then press **Enter**. The error message "ORA-02449: unique/primary keys in table referenced by foreign keys" appears as shown in Figure 2-12, which indicates that one or more columns in the LOCATION table are referenced as foreign keys in other tables.

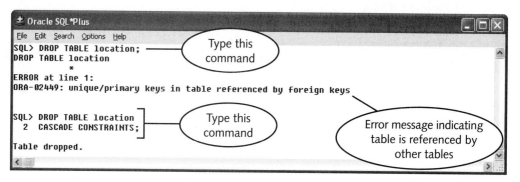

Figure 2-12   Commands to delete a database table

2. Type **DROP TABLE location CASCADE CONSTRAINTS;** at the SQL prompt, and then press **Enter**. The "Table dropped" message confirms that the DBMS successfully dropped the table.

To rename an existing table, you use the RENAME TO command, which has the following syntax:

```
RENAME old_tablename
TO new_tablename;
```

When you rename a table, the DBMS automatically transfers to the new table integrity constraints, indexes, and privileges that referenced the old table. However, when you rename a table, objects that referenced the old table, such as views and stored procedures and functions, become invalid. (You learn about views in Chapter 3, and you learn how to create stored procedures and functions in Chapter 9.) For example, you use the following command to rename the Northwoods University FACULTY table to NW_FACULTY:

```
RENAME faculty TO nw_faculty;
```

Adding Columns to Existing Tables

Sometimes an organization's data requirements change, and you need to add new columns to existing database tables. Recall from Table 2-4 that adding a new column to a table is an unrestricted action, so you can easily add new table columns as needed. The basic syntax of the command to add a new column to a table is:

```
ALTER TABLE tablename
ADD(columnname data_declaration constraints);
```

In this syntax, *columnname* is the name of the new data column. *Data_declaration* defines the new column's data type and maximum size, and *constraints* defines any constraints you want to place on the new column, such as foreign key references, check conditions, or value constraints. Northwoods University wants a column added to the FACULTY table that specifies each faculty member's employment start date. Next, you add the START_DATE column to the FACULTY table.

To add the START_DATE column to the FACULTY table:

1. Type the following command at the SQL prompt:

```
ALTER TABLE faculty
ADD (start_date DATE);
```

2. Press **Enter** to execute the command. The "Table altered" confirmation message appears and confirms that the DBMS successfully altered the table.

Modifying Existing Column Data Definitions

Sometimes you need to modify the data definition for an existing column. Recall from Table 2-5 that you can modify the data type of an existing column only if existing data values in the column rows are compatible with the new data type. As a result, you can only change a column's data type to another data type that stores the same kind of data. You can change VARCHAR2 columns to CHAR columns and vice versa, and you can change DATE columns to TIMESTAMP columns and vice versa. But, you *cannot* change VARCHAR2 columns to NUMBER columns or DATE columns, nor can you change NUMBER columns to VARCHAR2 or DATE columns.

You can also change the maximum size parameters for an existing column. Making an existing column's maximum size larger is an unrestricted action, and making an existing column's size smaller is a restricted action. You can make an existing column's size specification smaller only if the table does not contain any data. If the table contains rows, the values for the column being changed must be NULL.

The general syntax of the command to modify an existing column's data declaration is:

```
ALTER tablename
MODIFY(columnname new_data_declaration);
```

When you created the FACULTY table (see Figure 2-6), you defined the F_RANK column using the VARCHAR2 data type, with a maximum column width of eight characters. In the following set of steps, you modify the F_RANK column so it uses the CHAR data type, and has a maximum width of four characters.

To modify the data type and size of the F_RANK column:

1. At the SQL*Plus prompt, type the following command to modify the F_RANK column:

```
ALTER TABLE faculty
MODIFY (f_rank CHAR(4));
```

2. Press **Enter**. The "Table altered" message confirms that the DBMS successfully altered the table.

Deleting a Column

Sometimes you need to delete an existing column because the database applications no longer need the column's data. When the column is deleted, any data stored in that column is also removed from the database.

The general command to delete an existing column is:

```
ALTER TABLE tablename
DROP COLUMN columnname;
```

Renaming a Column

Suppose you need to change the name of the F_RANK column in the FACULTY table from F_RANK to FACULTY_RANK. To make this change, you could delete the existing F_RANK column, and then add to the table a new column named FACULTY_RANK. Of course if there were any data in the column, the data would be lost when you deleted the F_RANK column. Instead, you can issue the following ALTER TABLE command:

```
ALTER TABLE tablename
RENAME COLUMN old_columnname TO new_columnname;
```

To replace the F_RANK column with the FACULTY_RANK column:

1. At the SQL prompt, type **ALTER TABLE faculty RENAME COLUMN f_rank TO faculty_rank;**

2. Press **Enter** to execute the command. The "Table altered" message confirms that the DBMS successfully changed the name of the column.

Adding and Deleting Constraints

During the life cycle of a database system, the database applications change, and you may need to add or delete constraints in existing tables. For example, suppose after you create the Northwoods University FACULTY table, you need to add a UNIQUE constraint to the F_PIN column to ensure that each faculty member's PIN is unique. Table 2-5 shows that adding new constraints is a restricted action. It is difficult to add new constraints to tables that already contain data, because usually the data values do not conform to the new constraint. Deleting constraints is an unrestricted action, so it is easy to do.

To add a constraint to an existing table, you use the following command:

```
ALTER TABLE tablename
ADD CONSTRAINT constraint_name constraint_definition;
```

In this syntax, *constraint_name* is the name of the new constraint, based on the constraint naming convention. *Constraint_definition* specifies the constraint type (such as primary key or check condition), using the syntax you learned for creating constraints after all the column definitions are completed in the CREATE TABLE command. Next, you add the UNIQUE constraint to the FACULTY table to ensure that all F_PIN values are unique.

To add the UNIQUE constraint to the FACULTY table:

1. Type the following command at the SQL prompt:

```
ALTER TABLE faculty
ADD CONSTRAINT faculty_f_pin_uk UNIQUE (f_pin);
```

2. Press **Enter** to execute the command. The "Table altered" message confirms that the DBMS successfully added the constraint.

To remove an existing constraint, you use the following command:

```
ALTER TABLE tablename
DROP CONSTRAINT constraint_name;
```

In this syntax, *constraint_name* is the name you assigned to the constraint using the constraint naming convention. In this next set of steps you drop the UNIQUE constraint you just added to the F_PIN column in the FACULTY table.

To drop the constraint:

1. Type the following command at the SQL prompt:

```
ALTER TABLE faculty
DROP CONSTRAINT faculty_f_pin_uk;
```

2. Press **Enter** to execute the command. The "Table altered" message confirms that the DBMS successfully deleted the constraint.

Enabling and Disabling Constraints

Sometimes while you are developing new database applications, it is useful to disable constraints, then reenable the constraints when the application is finished. For example, suppose one programming team member is working on an application to add rows to the FACULTY table, while another team member is performing maintenance

operations on the LOCATION table. (Recall that the LOC_ID column in the FAC-
ULTY table references the LOC_ID column in the LOCATION table as a foreign key.)
If the team member working with the LOCATION table deletes all of the table rows,
the team member working with the FACULTY table cannot insert any new rows,
because there are no LOC_ID primary key values to reference.

When you create a new constraint using either the CREATE TABLE or ALTER TABLE
commands, the DBMS immediately enables the constraint. When a constraint is
enabled, the DBMS enforces the constraint when users attempt to add new data to the
database. You can use SQL commands to disable constraints. You use the following command to disable an existing constraint:

```
ALTER TABLE tablename
DISABLE CONSTRAINT constraint_name;
```

Similarly, you use the following command to enable a constraint that you previously disabled:

```
ALTER TABLE tablename
ENABLE CONSTRAINT constraint_name;
```

If you try to disable a constraint that is currently disabled, or enable a constraint that
is currently enabled, the DBMS executes the command anyway, and does not issue
an error. Next you execute the commands to disable the FACULTY_LOC_ID_FK
FOREIGN KEY constraint in the FACULTY table, and then reenable the constraint.

To disable and then reenable the constraint:

1. Type the following command at the SQL prompt, and then press **Enter**:

```
ALTER TABLE faculty
DISABLE CONSTRAINT faculty_loc_id_fk;
```

The "Table altered" message appears, which confirms that the DBMS successfully
disabled the constraint.

2. Type the following command at the SQL prompt, and then press **Enter**:

```
ALTER TABLE faculty
ENABLE CONSTRAINT faculty_loc_id_fk;
```

The "Table altered" message appears, which confirms that the DBMS successfully reenabled the constraint.

You are now familiar with how to create and modify Oracle 10*g* database tables using
SQL commands. Next you delete the FACULTY table, and then exit SQL*Plus.

To delete the FACULTY table and exit SQL*Plus:

1. Type the following command at the SQL prompt to delete the FACULTY
table, and then press **Enter**:

```
DROP TABLE faculty CASCADE CONSTRAINTS;
```

2. To exit SQL*Plus, type **exit** at the SQL prompt, and then press **Enter**.

3. Save the Ch2Queries.sql file and close Notepad.

CHAPTER SUMMARY

2

❐ Users interact with relational databases using high-level query languages that use English words such as SELECT, CREATE, ALTER, INSERT, and UPDATE. The standard query language for relational databases is Structured Query Language (SQL).

❐ SQL commands include data description language (DDL) commands, which create, modify, and delete database objects, and data manipulation language (DML) commands, which allow users to insert, update, delete, and view database data. Usually, only database administrators create and execute DDL commands.

❐ Each user account owns table and data objects in its own area of the database, which is called the user schema. Objects within a user schema are referred to as database objects or schema objects.

❐ When you create an Oracle 10*g* database table, you specify the table name, the name of each data column, and the data type and size of each data column. Table names and column names must follow the Oracle naming standard.

❐ When you create a table, you define constraints, which define primary and foreign keys and restrict the data values that users can store in columns.

❐ Column data types ensure that users enter correct data values. Data types also allow the DBMS to use storage space more efficiently by optimizing how specific types of data are stored.

❐ Oracle 10*g* stores character data using the VARCHAR2 and NVARCHAR2 data types, which store variable-length character data, and the CHAR and NCHAR data types, which store fixed-length character data.

❐ Oracle 10*g* stores numerical data in columns that have the NUMBER data type. The precision of a NUMBER data type declaration specifies the total number of digits both to the left and to the right of the decimal point, and the scale specifies the number of digits to the right of the decimal point. Based on the precision and scale values, NUMBER columns can store integer, fixed-point, and floating-point values.

❐ Oracle 10*g* uses the DATE and TIMESTAMP data types to store date and time data, and the INTERVAL YEAR TO MONTH and INTERVAL DAY TO SECOND data types to store time intervals.

❐ Oracle 10*g* uses the BLOB, BFILE, CLOB, and NCLOB data types to store binary data or large volumes of text data.

❐ Constraints restrict the data values that users can enter into database columns. Integrity constraints define primary key and foreign key columns; value constraints specify specific data values or ranges of values, whether the values can be NULL, whether the values have a default value, and whether a value must be unique. Table constraints restrict data values with respect to all other values in the table, and column constraints limit values for a single column, regardless of values in other rows.

❑ You can define constraints at the end of the CREATE TABLE command after all the columns are declared, or you can define each constraint within the column definition. Each constraint in a user schema must have a unique constraint name.

❑ SQL*Plus commands are not case sensitive, and the SQL*Plus interpreter ignores spaces and line breaks for formatting. You can create a SQL*Plus command using a text editor, copy the command, and then paste it into SQL*Plus. If the command has an error, you can edit the command in the text editor, and then copy, paste, and execute the command again.

❑ When SQL commands have errors, the SQL*Plus interpreter reports the line number that is causing the error, shows the position of the character that is causing the error, and returns an error code and description. When you need more information about an error code, you can look up the error code on the Oracle Technology Network Web site.

❑ You can use the DESCRIBE command to display a table's column names, data types, and NOT NULL constraints. You can use the data dictionary views to retrieve information about your database table names and constraint names.

❑ You can drop and rename tables, add new columns to tables, increase column widths, or drop existing constraints with no restrictions. You can modify a column data type or constraint only if the data that was inserted in the column is compatible with the updated data type or constraint. You can delete and rename table columns.

❑ You can disable constraints while you develop new database applications. When you disable a constraint, the DBMS does not enforce the constraint when you insert new data values.

REVIEW QUESTIONS

1. Which command is used to display the structure of a table in SQL*Plus?

2. What is the purpose of a value constraint?

3. You can use the _____ data type to make certain that if the name John is stored in a column, it is stored using a total of nine spaces, rather than just four spaces.

4. The DATE data type uses the default format of DD-MON-YY. True or False?

5. All table and column names must start with a(n) _____.

6. What command and option should be entered to delete a table that is referenced by a foreign key constraint in another table?

7. What is the default value assigned to a VARCHAR2 data type if no maximum size is stated in the column declaration?

8. All integrity constraints are displayed in the column list when you display the structure of a table. True or False?

9. Which view can be referenced to obtain a list of all objects that currently exist in the database?

10. By including a DEFAULT constraint for a column, you ensure that the column never contains a NULL value. True or False?

MULTIPLE CHOICE

1. Which of the following constraints permit NULL values?
   a. Unique
   b. primary key
   c. NOT NULL
   d. none of the above

2. When you create an object, it is stored in your _____.
   a. user table
   b. user schema
   c. data dictionary
   d. object list

3. Which of the following declares that the column can store a total of five digits, two of which are decimal values?
   a. NUMBER(3,2)
   b. NUMBER(5,2)
   c. NUMBER(float)
   d. NUMBER(p, 2)

4. Which of the following is a valid declaration for a DATE datatype?
   a. dobDATE(8)
   b. dobDATE(7)
   c. dobDATE
   d. dobDATE(DD-MON-YY)

5. Which of the following constraints must be included or the table cannot be created in Oracle 10*g*?
   a. primary key
   b. foreign key
   c. NOT NULL
   d. none of the above

6. Which of the following commands should be used to change the name of a database table?

    a. RENAME

    b. ALTER TABLE

    c. MODIFY TABLE

    d. CHANGE

7. Which of the following commands should be used to change the name of a column?

    a. RENAME

    b. ALTER TABLE

    c. MODIFY TABLE

    d. CHANGE

8. When modifying a database table, which of the following is a restricted action?

    a. renaming a table

    b. deleting a column

    c. adding a constraint

    d. deleting a constraint

9. Which view can be used to determine the name of the database tables currently stored in the database?

    a. USERS

    b. CONSTRAINTS

    c. TAB_COLUMNS

    d. TABLES

10. Which of the following ALTER TABLE clauses is used to remove a column from a database table?

    a. MODIFY

    b. DELETE

    c. REMOVE

    d. none of the above

PROBLEM-SOLVING CASES

Figures 1-24 and 1-25 display the data and relationships within the Clearwater Traders database. Reference these figures when completing the following cases.

1. Based on the description of the Clearwater Traders database, determine the appropriate constraints for each table. Make certain you consider the following constraint types for each table in the database:

   ❑ Primary key

   ❑ Foreign key

   ❑ Check condition

   ❑ NOT NULL

   ❑ Unique

   When drafting your results, list each table name, the constraint type, and the column that is referenced by the constraint.

2. Based on the sample data provided, determine the necessary column definition for each column contained in the following tables:

   ❑ CUSTOMER

   ❑ ORDERS

   ❑ ORDER_LINE

   ❑ ORDER_SOURCE

   ❑ CATEGORY

   ❑ ITEM

   ❑ INVENTORY

3. Using the results from Case 1 and Case 2, write the SQL commands necessary to create the CUSTOMER, ORDERS, ORDER_LINE, ORDER_SOURCE, CATEGORY, ITEM, and INVENTORY tables. Remember to include the appropriate constraints for these tables in your SQL statements. Enter these commands into a Notepad file named Ch2Case3.sql and save it in the Chapter02\Cases folder on your data disk. Do not execute the commands at this time.

4. Using the SQL commands created in Case 3, create the CUSTOMER, ORDERS, and ORDER_SOURCE tables using the SQL*Plus utility. After the tables are created, use the DESCRIBE command to verify that the columns have been correctly defined for each of these tables.

5. Use the SQL*Plus utility to perform the following tasks:

   a. Change the name of the ORDERS table to ORDER.

   b. Delete the C_MI column from the CUSTOMER table.

   c. Change the definition of the O_METHPMT column of the ORDER table so it is defined as a CHAR datatype that stores only two characters.

d. Use the USER_CONSTRAINTS view to display the name of all constraints for the ORDER table.

e. Delete the ORDER_SOURCE table. If you receive an error message, take the appropriate corrective action, and then remove the table.

NOTE

Recall that foreign key constraints can create problems when dropping a table that is referenced by the constraint.)

f. Delete the CUSTOMER and ORDER tables.

CHAPTER

3

USING SQL QUERIES TO INSERT, UPDATE, DELETE, AND VIEW DATA

◀ LESSON A ▶

After completing this lesson, you should be able to:

♦ Run a script to create database tables automatically
♦ Insert data into database tables
♦ Create database transactions and commit data to the database
♦ Create search conditions in SQL queries
♦ Update and delete database records and truncate tables
♦ Create and use sequences to generate surrogate key values automatically
♦ Grant and revoke database object privileges

Recall from Chapter 2 that when database developers create a database application, they must specify the underlying SQL commands that the application forwards to the DBMS for processing. Also recall that SQL commands can be data definition language (DDL) commands, which create new database objects, and data manipulation language (DML) commands, which allow users to insert, update, delete, and view database data. In Chapter 2, you became familiar with the Oracle 10g SQL DDL commands to create new database objects. The next step in developing a database is to use DML commands to insert, modify, and delete data records.

Ultimately, the people who use databases for daily tasks such as creating new customer orders or enrolling students in courses perform DML operations using forms and reports that automate data entry, modification, and summarization. To create these forms and reports, database developers must translate user inputs into SQL commands that the forms or reports submit to the database. Therefore, database developers must be very proficient with SQL DML commands. DML commands that allow users to retrieve database data are called **queries**, because the data that these commands retrieve often answers a question. DML commands that insert, update, or delete database data are called **action queries**, because the commands perform an action that changes the data values in the database.

Because of the complexity of the chapter material, this chapter is divided into three lessons. In this lesson, you learn how to create action queries to insert, update, and delete database records. In Lesson B, you learn various methods of retrieving rows from database tables. In Lesson C, methods for linking data from different tables are presented. You also learn how to issue commands to commit or discard the changes. Before you begin, you learn how to run a script to create the tables in the case study databases automatically.

USING SCRIPTS TO CREATE DATABASE TABLES

Recall from Chapter 2 that a script is a text file that contains one or more SQL commands. You can run a script in SQL*Plus to execute the SQL commands in the script automatically. Usually, database developers use script commands to create, modify, or delete database tables, or to insert or update data records, because scripts provide a way to execute a series of SQL commands quickly and easily. You created scripts in Chapter 2 when you typed a series of SQL commands, one after another, in your Notepad text file, and then saved the file using the .sql file extension.

To run a script, you type `start` at the SQL prompt, a blank space, and then the full path and filename of the script file. For example, to run the script in the c:\OraData\Chapter03\MyScript.sql file, you type the command: `START c:\OraData\Chapter3\MyScript.sql`. The folder path and script filename and extension can be any legal Windows filename, but they *cannot* contain any blank spaces.

NOTE

When you run a script, you can omit the .sql file extension because this is the default extension for script files. Also, you can type the at symbol (@) instead of START to run a script. For example, an alternate way to run the MyScript.sql file is to type @c:\OraData\Chapter3\MyScript.

Before you can complete the chapter tutorials, you need to run a script named Ch3EmptyNorthwoods.sql. This script first deletes all the Northwoods University database tables that you may have created in Chapter 2, and then it executes the commands to re-create the tables. In this next set of steps, you examine the script file in Notepad, and then run it through Sql*Plus.

To examine and run the Ch3EmptyNorthwoods.sql script file:

1. Start Notepad, and open Ch3EmptyNorthwoods.sql from the Chapter03 folder on your Data Disk. The file contains a DROP TABLE command to drop each of the tables in the Northwoods University database. It also contains a CREATE TABLE command to re-create each table. These commands are similar to the ones you learned in Chapter 2.

2. Start SQL*Plus and log onto the database.

3. To run the Ch3EmptyNorthwoods.sql script, type **START c:\OraData\Chapter03\Ch3EmptyNorthwoods.sql** at the SQL prompt, and then press **Enter**. (If your Data Disk is on a different drive letter, or has a different folder path, type that drive letter or folder path instead.)

TIP

You could also type START c:\OraData\Chapter03\Ch3EmptyNorthwoods (omitting the .sql file extension), or type @c:\OraData\Chapter03\Ch3EmptyNorthwoods (using @ instead of START).

NOTE

This book assumes that your Data Disk files are stored on your workstation's C: drive in a folder named OraData. If you store your Data Disk files on a different drive or in a different folder, type the appropriate drive letter and path specification when you are instructed to open a file on your Data Disk.

HELP

Don't worry if you receive an error message in the DROP TABLE command as shown in Figure 3-1. This error message indicates that the script is trying to drop a table that does not exist. The script commands drop all existing Northwoods University database table definitions before creating the new tables; otherwise, the script would try to create a table that already exists, and SQL*Plus would generate an error and would not re-create the table.

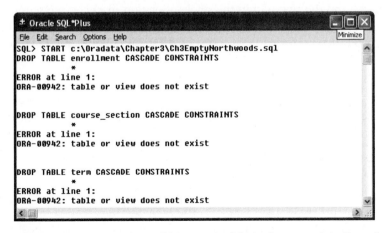

Figure 3-1   Error message that occurs while typing — script attempts to drop a nonexistent table

INSERTING DATA INTO TABLES

You use the SQL INSERT command to add new records to database tables. The following sections describe how to use the INSERT command, how to use format models to format input data, how to insert date and interval data values, and how to create transactions and commit data.

Using the INSERT Command

You can use the INSERT command either to insert a value for each column in the table or to insert values only into selected columns. The basic syntax of the INSERT statement for inserting a value for each table column is:

```
INSERT INTO tablename
VALUES (column1_value, column2_value, ...);
```

When you insert a value for every table column, the INSERT command's VALUES clause must contain a value for each column in the table. If a data value is unknown or undetermined, you insert the word NULL in place of the data value, and that column value remains undefined in that record. You must list column values in the same order that you defined the columns in the CREATE TABLE command. You can use the DESCRIBE command to determine the order of the columns in a table.

You use the following command to insert the first record in the Northwoods University LOCATION table in Figure 1-26 and specify that the CAPACITY column value is NULL:

```
INSERT INTO location
VALUES (1, 'CR', '101', NULL);
```

Recall that in the LOCATION table, the LOC_ID column has a NUMBER data type, and the BLDG_CODE and ROOM columns have the VARCHAR2 data type. When you insert data values into columns that have a character data type, you must enclose the values in single quotation marks, and the text within the single quotation marks is case sensitive. If you want to insert a character string that contains a single quotation mark, you type the single quotation mark two times. For example, you specify the address 454 St. John's Place as '454 St. John''s Place'. Note that the characters after the *n* in *John* are two single quotation marks (' '), not a double quotation mark (").

You can also use the INSERT command to insert values only in specific table columns. The basic syntax of the INSERT statement for inserting values into selected table columns is:

```
INSERT INTO tablename (columnname1, columnname2, ...)
VALUES (column1_value, column2_value, ...);
```

When using this syntax, you specify the names of the columns in which you want to insert data values in the INSERT INTO clause, then list the associated values in the VALUES clause. You can list the column names in any order in the INSERT INTO clause. However, you must list the data values in the VALUES clause in the same order as their associated columns appear in the INSERT INTO clause. If you omit one of the table's columns in the INSERT INTO clause (F_PHONE, F_RANK, etc.), the Oracle 10*g* database automatically inserts NULL as the omitted value. You use the following command to insert only the F_ID, F_LAST, and F_FIRST values for the first row in the FACULTY table:

```
INSERT INTO faculty (F_FIRST, F_LAST, F_ID)
VALUES ('Teresa', 'Marx', 1);
```

Note that in this command, the column names appear in a different order than the order they appear in the FACULTY table. However, the command succeeds, because each column's corresponding data value is listed in the VALUES clause in the same order as the column names in the INSERT INTO clause.

You must be very careful to place the data values in the correct order in the VALUES clause. Consider the INSERT command in Figure 3-2.

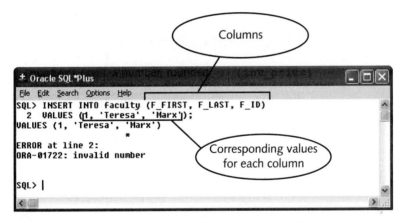

Figure 3-2 INSERT command with data values in the wrong order

Note that the column names in the INSERT INTO clause are F_FIRST, F_LAST, and F_ID, which have the VARCHAR2, VARCHAR2, and NUMBER data types, respectively. The DBMS expects the data values in the VALUES clause to have these data types. Note that the values in the VALUES clause (1, 'Teresa', 'Marx') are not in the same order as the column names in the INSERT INTO clause: The values are listed as F_ID, F_FIRST, and then F_LAST. The error occurs because in the VALUES clause, the values have the NUMBER, VARCHAR2, and VARCHAR2 data types. The DBMS automatically converts the number 1 to a character, but the DBMS cannot convert the character string 'Marx' to a number.

Before you can insert a new data row, you must also ensure that all the foreign keys that the new row references have already been added to the database. For example, suppose you want to insert the first row in the Northwoods University STUDENT table. In the first STUDENT row, Tammy Jones' F_ID value is 1. This value refers to F_ID 1 (Teresa

Marx) in the FACULTY table. Therefore, the FACULTY row for F_ID 1 must already be in the database before you can add the first STUDENT row, or a foreign key reference error occurs. Now look at Teresa Marx's FACULTY record, and note that it has a foreign key value of LOC_ID 9. Similarly, this LOCATION row must already be in the database before you can add the row to the FACULTY table. The LOC_ID 9 record in the LOCATION table has no foreign key values to reference. Therefore, you can insert LOC_ID 9 in the LOCATION table, insert the associated FACULTY record, and then add the STUDENT record. In the following steps, you insert a row into the LOCATION table.

To insert a row into the LOCATION table:

1. Switch to Notepad, create a new file, and then type the SQL action query shown in Figure 3-3. Save the Notepad file as **Ch3AQueries.sql** in the Chapter03 folder on your Data Disk. Do not close the file.

2. Copy the action query text, paste it into SQL*Plus, and then press **Enter**. The message "1 row created." confirms that the row has been added to the LOCATION table.

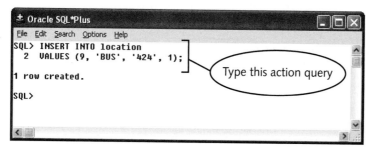

Figure 3-3   Action query to insert the first row into the LOCATION table

NOTE

For the rest of the lesson, assume that you type all commands into your Ch3AQueries.sql Notepad text file, and then copy and paste them into SQL*Plus.

Note that the LOCATION table has four columns—LOC_ID, BLDG_CODE, ROOM, and CAPACITY—and that their associated data types are NUMBER, VAR-CHAR2, VARCHAR2, and NUMBER. You enter the NUMBER column values as digits, and the VARCHAR2 columns as text strings within single quotation marks ('). Remember that you must enter the column values in the INSERT clause in the same order as the columns in the table, and you must include a value for every column in the table. When you insert a row into the LOCATION table, the DBMS expects a NUM-BER data type, then a VARCHAR2, another VARCHAR2, and then another NUM-BER. An error occurs if you try to insert the values in the wrong order or if you omit a column value.

NOTE

If you cannot remember the order of the columns or their data types, remember that you can use the DESCRIBE command to verify the table's structure.

Now you insert the row for F_ID 1 (Teresa Marx) in the FACULTY table. To practice using the INSERT action query when you insert values for specific columns, you enter values only for the F_ID, F_LAST, F_FIRST, F_MI, and LOC_ID columns, and omit the values for F_PHONE, F_RANK, F_SUPER, F_PIN, and F_IMAGE. Recall that when you omit values in an INSERT action query, the DBMS automatically specifies that the omitted values are NULL.

To insert specific columns in a row:

1. Type the following action query to insert the first row into the FACULTY table:

```
INSERT INTO faculty (F_ID, F_LAST, F_FIRST, F_MI, ↵
LOC_ID)
VALUES (1, 'Marx', 'Teresa', 'J', 9);
```

2. Execute the query. The message "1 row created." confirms that the DBMS has added the row to the table.

Format Models

Oracle stores data values in an internal binary format, and you can display data values using a variety of output formats. Suppose that a column with the DATE data type contains a value that represents 07/22/2006 9:29:00 PM. The default Oracle 10*g* output format for DATE data columns is DD-MON-YY, so by default, this value appears as 22-JUL-06 when you retrieve the value from the database. (The database stores the time portion of the date, but does not display the time in the default output format.) Instead of using the default output format, you can use an alphanumeric character string called a **format model**, also called a **format mask**, to specify a different output format, such as July 22, 2004 9:29 PM, or 21:29:00 PM (which uses 24-hour clock notation). Similarly, you could use a format model to format a NUMBER value of 1257.33 to appear as $1,257.33 or as 1257. Format models only change the way the data appears, and do not affect how Oracle 10*g* stores the data value in the database. Table 3-1 lists how you can use some common numerical data format models to display the numeric data value stored as 012345.67. Within a format model, the digit *9* is a placeholder that represents how number data values appear within the formatting characters.

Format Model	Description	Displayed Value
999999	Returns the value rounded to the number of placeholders, and suppresses leading zeroes	12346
099999	Returns the value rounded to the number of placeholders, and displays leading zeroes	012346
$99999	Returns the value rounded to the number of placeholders, and prefaces the value with a dollar sign; suppresses leading zeroes	$12346
99999MI	Prefaces negative values with – (minus sign)	–12346
99999PR	Displays negative values in angle brackets	<12346>
99,999	Displays a comma in the indicated position	12,346
99999.99	Displays the specified number of placeholders, with a decimal point in the indicated position	12345.67

Table 3-1   Common numerical format models

Table 3-2 shows common date format models using the example date of 5:45:35 PM, Sunday, February 15, 2006.

Format Model	Description	Displayed Value
YYYY	Displays all 4 digits of the year	2006
YYY or YY or Y	Displays last 3, 2, or 1 digit(s) of the year	006, 06, 6
RR	Displays dates from different centuries using two digits; year values from 0 to 49 are assumed to belong to the current century, and year values numbered from 50 to 99 are assumed to belong to the previous century	06
MM	Displays the month as digits (01-12)	02
MONTH	Displays the name of the month, spelled out, uppercase; for months with fewer than 9 characters in their names, the DBMS adds trailing blank spaces to pad the name to 9 characters	FEBRUARY
Month	Displays the name of the month, spelled out in mixed case; the DBMS adds trailing blank spaces to pad the name to 9 characters if necessary	February
DD	Displays the day of the month (01-31)	15
DDTH	Displays the day of the month as an ordinal number	15TH

Table 3-2   Common date format models

Format Model	Description	Displayed Value
DDD	Displays the day of the year (01-366)	46
DAY	Displays the day of the week, spelled out, uppercase	SUNDAY
Day	Displays the day of the week, spelled out, mixed case	Sunday
DY	Displays the name of the day as a 3-letter abbreviation	SUN
AM, PM, A.M., P.M.	Meridian indicator (without or with periods)	PM
HH	Displays the hour of the day using a 12-hour clock	05
HH24	Displays the hour of the day using a 24-hour clock	17
MI	Displays minutes (0-59)	45
SS	Displays seconds (0-59)	35

Table 3-2   Common date format models (continued)

You can specify to include front slashes (/), hyphens (-), and colons (:) as formatting characters between different date elements. For example, the format model MM/DD/YYYY appears as 02/15/2006, and the model HH:MI:SS appears as 05:45:35. You can also include additional formatting characters such as commas, periods, and blank spaces. For example, the format model DAY, MONTH DDTH, YYYY appears as SUNDAY, FEBRUARY 15TH, 2006.

Inserting Date and Interval Values

Recall that Oracle 10*g* stores date values in columns that have the DATE data type, and time interval values in columns that have the INTERVAL data type. The following sections describe how to insert data values into columns that have the DATE and INTERVAL data types.

Inserting Values into DATE Columns

To insert a value into a DATE column, you specify the date value as a character string, then convert the date character string to an internal DATE format using the TO_DATE function. The general syntax of the TO_DATE function is:

```
TO_DATE('date_string', 'date_format_model')
```

In this syntax, *date_string* represents the date value as a text string, such as '08/24/2006', and *date_format_model* is the format model that represents the *date_string* value's format, such as MM/DD/YYYY. You convert the text string '08/24/2006' to a DATE data type using the following command:

```
TO_DATE('08/24/2006', 'MM/DD/YYYY')
```

Similarly, the following command converts the character string '24–AUG–2005' to a DATE data type:

```
TO_DATE('24-AUG-2005', 'DD-MON-YYYY')
```

Recall from Chapter 2 that the DATE data type also stores time values. Time is stored in the default format HH:MI:SS, where HH represents hours, MI represents minutes, and SS represents seconds. Note that in the COURSE_SECTION table, the C_SEC_TIME column stores values for the times when course section classes begin. To convert a 10:00 AM value to a DATE format for C_SEC_ID 1, you use the following command:

```
TO_DATE('10:00 AM', 'HH:MI AM')
```

Inserting Values into INTERVAL Columns

Recall from Chapter 2 that Oracle 10g has two data types that store time intervals. The INTERVAL YEAR TO MONTH data type stores time intervals that consist of years and months, and the INTERVAL DAY TO SECOND data type stores time intervals that consist of days, hours, minutes, seconds, and fractional seconds. As with the DATE data type, you insert INTERVAL data values as character strings, then use a function to convert the character string to an internal INTERVAL data format.

To convert a character string that represents elapsed years and months to an INTERVAL YEAR TO MONTH data value, you use the TO_YMINTERVAL function. This function has the following general syntax:

```
TO_YMINTERVAL('years-months')
```

In this syntax, *years* is an integer that represents the interval years, and *months* is an integer that represents the interval months. Note that the command encloses the values in single quotation marks, and separates the *years* and *months* values with a hyphen (–). For example, you use the following command to convert an interval of four years and nine months to the INTERVAL YEAR TO MONTH data format:

```
TO_YMINTERVAL('4-9')
```

Similarly, you use the TO_DSINTERVAL function to convert a character string that represents elapsed days, minutes, hours, and seconds to an INTERVAL DAY TO SECOND data format. This function has the following general syntax:

```
TO_DSINTERVAL('days HH:MI:SS.99')
```

In this syntax, *days* is an integer that represents the interval days. *HH:MI:SS.99* represents the interval hours, minutes, seconds, and fractional seconds. As before, the command encloses the values in single quotation marks. The command separates the *days* and *HH:MI:SS.99* values with a blank space. The fractional seconds value is optional. You use the following command to convert an interval of 1 hour and 15 minutes to the INTERVAL DAYS TO SECONDS data format:

```
TO_DSINTERVAL('0 01:15:00')
```

Now you practice inserting DATE and INTERVAL data values into the database by adding the row for S_ID 1 (Tammy Jones) to the STUDENT table. The STUDENT table contains the S_DOB (student date of birth) column, which has the DATE data type. The STUDENT table also contains the TIME_ENROLLED column, which has the INTERVAL YEAR TO MONTH data type, and represents how long the student has been enrolled at Northwoods University.

To add a row to the STUDENT table

1. Type the action query shown in Figure 3-4 to insert the first row into the STUDENT table, using the functions to convert character strings to DATE and INTERVAL formats.

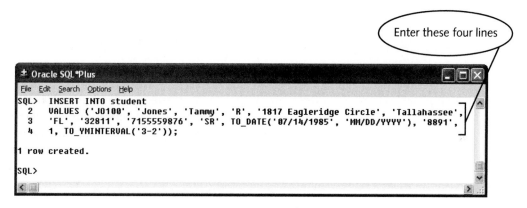

Figure 3-4   Action query to insert data values with DATE and INTERVAL data types

2. Execute the query. The message "1 row created." confirms that the DBMS has added the row to the table.

Inserting LOB Column Locators

Recall that the F_IMAGE column in the FACULTY table is a large object (LOB) data column of type BLOB, which stores the binary data for the image in the database. To make data retrievals for both LOB and non-LOB data more efficient, Oracle stores LOB data in a separate physical location from other types of data in a row. Whenever you insert data in an LOB column, you must initially insert a locator for the data. An **LOB locator** is a structure that contains information that identifies the LOB data type and points to the alternate memory location. After you insert the LOB locator, you then write a program or use a utility to insert the binary data into the database. (You will learn how to load BLOB data into the database in Chapter 10.)

To create a locator for a BLOB data column, you use the following syntax:

```
EMPTY_BLOB()
```

In the following steps you insert the row for faculty member Mark Zhulin that includes the BLOB locator. You insert data values for the F_ID, F_LAST, F_FIRST, and F_IMAGE columns only.

To insert the row for faculty member Mark Zhulin:

1. Type the following action query to insert the second row into the FACULTY table:

```
INSERT INTO faculty (F_ID, F_LAST, F_FIRST, F_IMAGE)
VALUES (2, 'Zhulin', 'Mark', EMPTY_BLOB());
```

2. Execute the query. The message "1 row created." confirms that the DBMS has added the row to the table.

CREATING TRANSACTIONS AND COMMITTING NEW DATA

When you create a new table or update the structure of an existing table, the DBMS changes the rows immediately and makes the result of the change visible to other users. This is not the case when you insert, update, or delete data rows. Recall that commands for operations that add, update, or delete data are called action queries. An Oracle 10*g* database allows users to execute a series of action queries as a **transaction**, which represents a logical unit of work. In a transaction, all of the action queries must succeed, or none of the transactions can succeed.

An example of a transaction is when a Clearwater Traders customer purchases an item. The DBMS must insert the purchase information in the ORDERS and ORDER_LINE tables, and reduce the item's quantity on hand value in the INVENTORY table. Suppose that the DBMS successfully inserts the purchase information in the ORDERS and ORDER_LINE tables, but then experiences a power failure and does not successfully reduce the quantity on hand in the INVENTORY table. The customer purchases the item, but the INVENTORY table does not contain the updated value for the quantity on hand. All three action queries need to succeed, or none of them should succeed.

After the user enters all of the action queries in a transaction, he or she can either **commit** (save) all of the changes or **roll back** (discard) all of the changes. When the Oracle 10*g* DBMS executes an action query, it updates the data in the database and also records information that enables the DBMS to undo the action query's changes. The DBMS saves this information to undo the changes until the user commits or rolls back the transaction that includes the action query.

The purpose of transaction processing is to enable every user to see a consistent view of the database. To achieve this consistency, a user cannot view or update data values that are part of another user's uncommitted transactions because these uncommitted transactions, which are called **pending transactions**, might be rolled back. The Oracle DBMS implements transaction processing by locking data rows associated with pending transactions. When the DBMS locks a row, other users cannot view or modify the row.

When the user commits the transaction, the DBMS releases the locks on the rows, and other users can view and update the rows again.

You do not explicitly create new transactions in Oracle 10g. Rather, a new transaction begins when you start SQL*Plus and execute a command, and the transaction ends when you commit the current transaction. After you commit the current transaction, another transaction begins when you type a new query. You commit a transaction by executing the COMMIT command. After you execute the COMMIT command, a new transaction begins.

TIP   You can configure SQL*Plus to commit every query automatically after you execute the query. However, when you configure SQL*Plus this way, you cannot create transactions, so this is not recommended.

Now you commit your current transaction, which contains the two INSERT action queries that you have executed so far.

To commit your changes:

1. Type **COMMIT;** at the SQL prompt.

2. Press **Enter**. The message "Commit complete." indicates that your changes are permanent and visible to other users.

TIP   The Oracle 10g DBMS automatically commits the current transaction when you exit SQL*Plus. However, it is a good idea to explicitly commit your changes often so that the rows are available to other users. This ensures your changes are saved if you do not exit normally because of a power failure or workstation malfunction.

The rollback process enables you to restore the database to its state the last time you issued the COMMIT command. To do this, the DBMS rolls back all of the changes made in subsequent action queries. Figure 3-5 shows an example of a rolled back transaction. The first INSERT action query inserts a course into the Northwoods University COURSE table. Then, the user issues the ROLLBACK command. The final SELECT query shows that the row was not inserted into the COURSE table, and that the DBMS rolled back the INSERT command.

NOTE   You will learn how to use the SELECT command to view data later in this chapter. For now, to view the current data in all of the columns in a table, you type SELECT * FROM *tablename*. The asterisk (*) specifies that the values of all table columns are displayed. However, the asterisk cannot be used with tables, such as the FACULTY table that contain the BLOB datatypes.

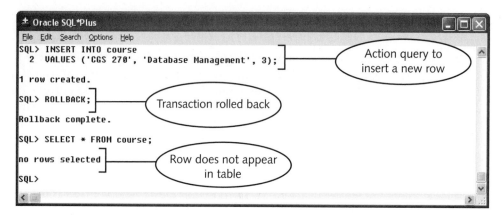

Figure 3-5   Rolling back a transaction

You can use rollbacks with savepoints. A **savepoint** is a bookmark that designates the beginning of an individual section of a transaction. You can roll back part of a transaction by rolling back the transaction only to the savepoint. Figure 3-6 shows how to create savepoints and then roll back a SQL*Plus session to an intermediate savepoint.

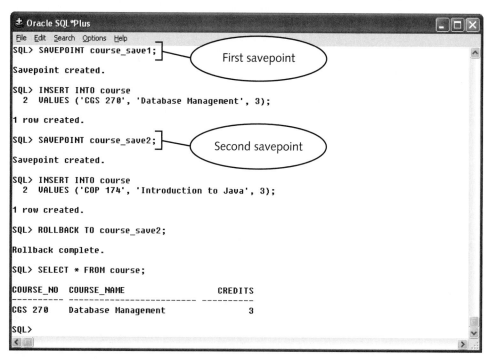

Figure 3-6   Using the ROLLBACK command with savepoints

In the first command in Figure 3-6, the user creates a savepoint named COURSE_SAVE1. Next, the user inserts course information for course CGS 270 into the COURSE table. Next, the user creates a second savepoint named COURSE_SAVE2, and then inserts a second course, COP 174, into the COURSE table. Finally, the user issues the command ROLLBACK TO course_save2;, which rolls back the transaction to the COURSE_SAVE2 savepoint. This causes the DBMS to restore the database to it's state when the user created the savepoint: the row in the COURSE table for CGS 270 is still inserted, but the row for the second course is discarded. If the user issues the command ROLLBACK TO course_save1;, the DBMS rolls back the command for inserting course COP 174.

CREATING SEARCH CONDITIONS IN SQL QUERIES

To write queries to update, delete, or retrieve database rows, you often need to specify which rows the query updates, deletes, or retrieves. For example, to delete the first row in the STUDENT table, you must specify to delete the row in which the S_ID value is JO100. To identify specific rows in a query, you use a **search condition**, which is an expression that seeks to match specific table rows. You use search conditions to compare column values to numerical, date, and character values.

The general syntax of a SQL search condition is:

```
WHERE columnname comparison_operator search_expression
```

In this syntax, *columnname* specifies the database column whose value you seek to match. The *comparison_operator* compares the *columnname* value with *search_expression*. *Search_expression* is usually a constant, such as the number 1 or the character string 'SR'. Table 3-3 lists common SQL comparison operators.

Operator	Description	Example
=	Equal to	S_CLASS = 'SR'
>	Greater than	CAPACITY > 50
<	Less than	CAPACITY < 100
>=	Greater than or equal to	S_DOB >= TO_DATE ('01-JAN-1980', 'DD-MON-YYYY')
<=	Less than or equal to	MAX_ENRL <= 30
<> != ^=	Not equal to	STATUS <> 'CLOSED' STATUS != 'CLOSED' STATUS ^= 'CLOSED'

Table 3-3   Comparison operators commonly used with search conditions

Operator	Description	Example
LIKE	Uses pattern matching in text strings; is usually used with the wildcard character (%), which indicates that part of the string can contain any characters; search string within single quotation marks is case sensitive	`term_desc LIKE 'Summer%'`
IN	Determines if a value is a member of a specific search set	`s_class IN ('FR','SO')`
NOT IN	Determines if a value is not a member of a specific search set	`s_class NOT IN ('FR','SO')`
IS NULL	Determines if a value is NULL	`s_mi IS NULL`
IS NOT NULL	Determines if a value is not NULL	`s_mi IS NOT NULL`

Table 3-3 Comparison operators commonly used with search conditions (continued)

The equal to (=), greater than (>), less than (<), and not equal to (<>, != , or ^=) comparison operators are similar to the comparison operators you use in other programming languages. (You will learn how to use the LIKE, IN, NOT IN, IS NULL, and IS NOT NULL comparison operators in Lesson B in this chapter.) Now you learn how to create search expressions, and how to use the AND, OR, and NOT logical operators to create complex search conditions.

Defining Search Expressions

Recall that in a search condition, *search_expression* is the value that the search column matches. When the search expression is a number, you simply insert the number value. An example of a search condition in which *search_expression* is a number value is WHERE f_id = 1. This search condition retrieves the row for F_ID 1.

You must enclose text strings in single quotation marks. For example, you use the following search condition to match the S_CLASS value in the STUDENT table:

```
WHERE s_class = 'SR'
```

Search_expression values within single quotation marks are case sensitive. The search condition S_CLASS = 'sr' does not retrieve rows in which the S_CLASS value is 'SR'.

When *search_expression* involves a DATE data value, you must use the TO_DATE function to convert the DATE character string representation to an internal DATE data format. The following search condition matches dates of January 1, 1980:

```
WHERE s_dob = TO_DATE('01/01/1980', 'MM/DD/YYYY')
```

NOTE
If you choose to search for a date value without using the TO_DATE function, the search condition must be in the same format as Oracle 10*g*'s internal storage format (that is, DD-MON-YY).

To search for dates before a given date, you use the less than (<) comparison operator. To search for dates after a given date, you use the greater than (>) operator. To search for dates that match a given date, you use the equal to (=) operator.

To search for data values that match an INTERVAL column, you must use the TO_YMINTERVAL and TO_DSINTERVAL functions to convert the character string representation of the interval to the INTERVAL data type. For example, you use the following search condition to match all rows in the COURSE_SECTION table for which the C_SEC_DURATION value is 1 hour and 15 minutes:

```
WHERE c_sec_duration = TO_DSINTERVAL('0 1:15:00')
```

You can also create search conditions that specify time intervals that are greater than or less than a given interval. For example, the following search condition specifies all Northwoods University students who have been enrolled for less than one year:

```
WHERE time_enrolled < TO_YMINTERVAL('1-0')
```

Creating Complex Search Conditions

A **complex search condition** combines multiple search conditions using the AND, OR, and NOT logical operators. When you use the **AND logical operator** to combine two search conditions, both conditions must be true for the complex search condition to be true. If no rows exist that match *both* conditions, then the complex search condition is false, and no matching rows are found. For example, the following complex search condition matches all rows in which BLDG_CODE is 'CR' and the capacity is greater than 50:

```
WHERE bldg_code = 'CR' AND capacity > 50
```

When you use the **OR logical operator** to create a complex search condition, only one of the conditions must be true for the complex search condition to be true. For example, the following search condition matches all course section rows that meet either on Tuesday and Thursday or on Monday, Wednesday, and Friday (at Northwoods University, R denotes courses that meet on Thursday):

```
WHERE day = 'TR' OR day = 'MWF'
```

You can use the **NOT logical operator** to match the logical opposite of a search expression, using this syntax:

```
WHERE NOT(search_expression)
```

For example, the following search condition finds all rows in the STUDENT table in which the S_CLASS column has any value other than 'FR':

```
WHERE NOT (s_class = 'FR')
```

UPDATING AND DELETING EXISTING TABLE ROWS

At Northwoods University, student addresses and telephone numbers often change, and every year students (it is hoped) move up to the next class. Existing data rows that are no longer relevant must be removed or saved in different locations. The following sections describe how to update and delete table rows.

Updating Table Rows

To update column values in one or more rows in a table, you use an **UPDATE action query**. An UPDATE action query specifies the name of the table to update, and lists the name of the column (or columns) to update, along with the new data value (or values). An UPDATE action query also usually contains a search condition to identify the row or rows to update. The general syntax of an UPDATE action query is:

```
UPDATE tablename
SET column1 = new_value1, column2 = new_value2, ...
WHERE search condition;
```

In this syntax, *tablename* specifies the table whose rows you wish to update. The SET clause lists the columns to update and their associated new values.

NOTE You can update multiple columns in the same table using a single UPDATE action query, but you can update rows in only one table at a time in a single UPDATE action query.

The WHERE clause specifies the search condition. If you omit the search condition, the query updates every table row. In this next set of steps, you create an UPDATE action query to update faculty member Teresa Marx's F_RANK value to 'FULL'. (When you originally inserted the row, you omitted her F_RANK column value, so the DBMS inserted the value as NULL.) You will use her F_ID value (1) as the *search_expression* in the search condition.

To create an UPDATE action query:

1. Type the action query shown in Figure 3-7 to update faculty member Teresa Marx's F_RANK value.

2. Execute the query. The "1 row updated." message confirms that the row was updated.

```
± Oracle SQL*Plus                                          _ □ ×
File  Edit  Search  Options  Help
SQL> UPDATE faculty           Type this action query
   2  SET f_rank = 'FULL'
   3  WHERE f_id = 1;

1 row updated.

SQL> |
```

Figure 3-7   Creating an UPDATE action query

Recall that you can use a single UPDATE action query to update multiple column values in the same table. For example, the following action query changes Teresa Marx's F_RANK value to 'ASSOCIATE' and her F_PIN value to 1181:

```
UPDATE faculty
SET f_rank = 'ASSOCIATE', f_pin = 1181
WHERE f_id = 1;
```

TIP

Because the F_RANK column has the VARCHAR2 data type, the action query encloses the value in single quotation marks.

You can update multiple rows in a table with a single UPDATE command by specifying a search condition that matches multiple rows using the greater than (>) or less than (<) operators. Be careful to always include a search condition in an UPDATE action query: If you omit the search condition in the UPDATE command, then the command updates all table rows. For example, the following command updates the value of S_CLASS to 'SR' for all rows in the STUDENT table:

```
UPDATE student
SET s_class = 'SR';
```

Deleting Table Rows

You use the SQL DELETE action query to remove specific rows from a database table, and you truncate the table to remove all of the table rows. The following sections describe these two approaches for removing table rows.

The SQL DELETE Action Query

The general syntax for a DELETE action query is:

```
DELETE FROM tablename
WHERE search condition;
```

In this syntax, *tablename* specifies the table from which to delete the row(s). A single DELETE action query can delete rows from only one table. The search condition specifies the row or rows to delete. Be careful to always include a search condition in a DELETE action query—if you accidentally omit the search condition, the DELETE action query deletes all of the table rows! In the following set of steps, you delete Tammy Jones from the STUDENT table. You specify 'Tammy' and 'Jones' in the action query's search condition.

To delete a row:

1. Type the DELETE action query shown in Figure 3-8 to delete the Tammy Jones row from the STUDENT table.

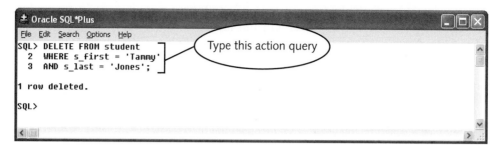

Figure 3-8   Creating a DELETE action query

2. Execute the query. The message "1 row deleted." confirms that the DBMS has deleted the row.

You can use a single DELETE action query to delete multiple rows from a table if the search condition matches several rows. If you omit the search condition, a DELETE action query deletes every table row.

When a row's value is a foreign key that references another row, it is called a **child row**. For example, recall that the row for F_ID 1 (Teresa Marx) references LOC_ID 9, which is Teresa's office location ID. The row for Teresa Marx is a child row of the row for LOC_ID 9. You cannot delete a row if it has a child row; therefore, you cannot delete the row for LOC_ID 9 (Teresa Marx's office), unless you first delete the row in which the foreign key value exists (F_ID 1, faculty member Teresa Marx). In the next set of steps, you attempt to delete a row that has a child row, and view the error message.

To try to delete a row that has a child row:

1. Type the action query shown in Figure 3-9 to attempt to delete LOC_ID 9 (Teresa Marx's office) from the LOCATION table.

2. Execute the query. The error message indicates that the row contains a value that is referenced as a foreign key. To delete the row for LOC_ID 9 successfully, you must first delete the row for F_ID 1, which is the row that references it as a foreign key.

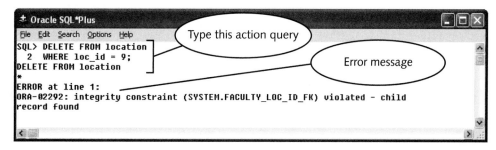

Figure 3-9   Error message that appears when you attempt to delete a row that has a child row

Truncating Tables

Recall that after a DML statement executes, you must commit the transaction before the DBMS makes the changes visible to other users. The DELETE action query is a DML statement. Whenever you delete a row using a DELETE action query, an Oracle 10*g* DBMS process records rollback information that the DBMS uses to restore the row if you roll back the transaction. For a DELETE action query, rollback information includes all of the column values for the deleted data row. If you use a DELETE action query to delete many rows, such as all of the rows in a table, the DBMS must make a copy of each deleted value. For a table with many columns and rows, this process can take a long time.

Sometimes you need to delete all of the rows in a table, and you know that you will not roll back the transaction. This may occur because you may have moved the data to a different location, or the data is outdated or incorrect. When you need to delete all of the rows in a table quickly, you can **truncate** the table, which means you remove all of the table data without saving any rollback information. When you truncate a table, the table structure and constraints remain intact. To truncate a table, you use the TRUNCATE TABLE command, which has the following general syntax:

```
TRUNCATE TABLE tablename;
```

You cannot truncate a table that has foreign key constraints as long as the foreign key constraints are enabled. (Recall from Chapter 2 that when a constraint is enabled, the DBMS enforces the constraint.) Therefore, you must disable a table's foreign key constraints before you can truncate the table. Recall that you use the following command to disable an existing constraint:

```
ALTER TABLE tablename
DISABLE CONSTRAINT constraint_name;
```

In the next set of steps, you truncate the LOCATION table to delete all of its rows. Recall that the LOC_ID column in the LOCATION table is a foreign key in both the FACULTY table and the COURSE_SECTION table, so you must first disable the LOC_ID foreign key constraints in the FACULTY and COURSE_SECTION tables. Then you truncate the LOCATION table.

To truncate the LOCATION table:

1. Type `ALTER TABLE faculty DISABLE CONSTRAINT faculty_loc_id_fk;` at the SQL> prompt, as shown in Figure 3-10, to disable the foreign key constraint for the FACULTY table.

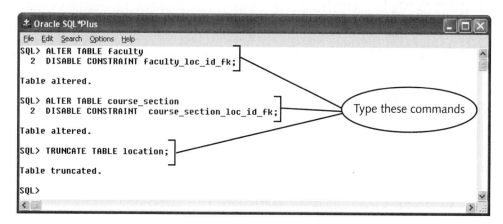

Figure 3-10   Commands that disable constraints and then truncate a table

2. Type `ALTER TABLE course_section DISABLE CONSTRAINT course_section_loc_id_fk;` to display the foreign key constraint for the COURSE_SECTION table.

3. Type `TRUNCATE TABLE location;` to remove the rows contained in the LOCATION table.

The "Table truncated." message confirms that the DBMS successfully truncated the table.

SEQUENCES

Recall from Chapter 1 that sequences are sequential lists of numbers that the Oracle 10*g* DBMS generates to create unique surrogate key values. For example, you might create a sequence to assign values for the faculty ID (F_ID) column in the Northwoods University FACULTY table. Columns that use sequence values must have the NUMBER data type. The following sections describe how to use SQL commands to create and manipulate sequences.

Creating New Sequences

A sequence is a database object, so you use the DDL (data description language) CREATE command to create a new sequence. (Recall that you used the DDL CREATE command in Chapter 2 to create new database tables.) Because the CREATE SEQUENCE command is a DDL command, the DBMS immediately creates the sequence when you execute the command, and you do not need to type **COMMIT** to save the sequence.

Figure 3-11 shows the general syntax for the command to create a new sequence. In Figure 3-11, square brackets enclose optional parameters, and a bar (|) separates parameters that include one of two values.

3

```
CREATE SEQUENCE sequence_name
   [INCREMENT BY number]
   [START WITH start_value]
   [MAXVALUE maximum_value]  |  [NOMAXVALUE]
   [MINVALUE minimum_value]  |  [NOMINVALUE]
   [CYCLE]  |  [NOCYCLE]
   [CACHE number_of_values]  |  [NOCACHE]
   [ORDER]  |  [NOORDER];
```

Figure 3-11   General syntax used to create a new sequence

In this syntax, *sequence_name* is the name of the sequence. Every sequence in your user schema must have a unique name, and the name must follow the Oracle naming standard rules. The CREATE SEQUENCE command can include the following optional parameters:

- INCREMENT BY specifies the value by which the sequence is incremented, and *number* can be any integer. By default, the DBMS increments a sequence by one.

- START WITH specifies the first sequence value. The *start_value* parameter can be any positive or negative integer value. When you omit the START WITH clause, the sequence starts with the value 1.

- MAXVALUE specifies the maximum value to which you can increment the sequence. The default maximum value is NOMAXVALUE, which allows the DBMS to increment the sequence to a maximum value of 10^{27}.

- MINVALUE specifies the minimum value for a sequence that has a negative increment value. The default minimum value is NOMINVALUE, which has a minimum value of 10^{-26}.

- CYCLE specifies that when the sequence reaches its MAXVALUE, it cycles back and starts again at its MINVALUE. For example, if you create a sequence with a maximum value of 10 and a minimum value of 5, the sequence increments up to 10 and then starts again at 5. If you omit the CYCLE parameter, or replace it with NOCYCLE, the sequence continues to generate values until it reaches its MAXVALUE, then it quits generating values.

- CACHE specifies that whenever you request a sequence value the DBMS automatically generates the specified *number_of_values* and stores them in a server memory area called a cache. This improves database performance when many users are simultaneously requesting sequence values. By default, the DBMS stores 20 sequence numbers in the cache. The NOCACHE parameter directs the DBMS not to cache any sequence values.

- ORDER ensures that the DBMS grants sequence numbers to users in the exact chronological order in which the users request the values. For example, the first user who requests a number from the sequence is granted sequence number 1, the second user who requests a value is granted sequence number 2, and so forth. This is useful for tracking the order in which users request sequence values. By default, this value is NOORDER, which specifies that the DBMS does not necessarily grant the sequence values in the same order that users request the values.

Next, you create a sequence named LOC_ID_SEQUENCE to generate values for the LOC_ID column in the LOCATION table. You specify that the sequence start with 20.

To create LOC_ID_SEQUENCE:

1. To create the sequence, type the command shown in Figure 3-12.

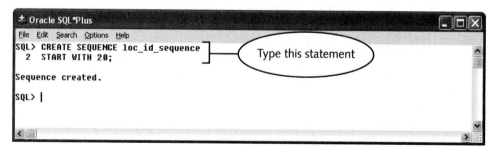

Figure 3-12   Creating a sequence

2. Execute the command. The "Sequence created." message confirms that the DBMS has created LOC_ID_SEQUENCE.

Viewing Sequence Information

A large database application may use many sequences, so sometimes you might forget the names of the sequences that you create. To review the names of your sequences, you can query the USER_SEQUENCES data dictionary view. (Recall that the data dictionary views contain information about the database structure.) The USER_SEQUENCES view has a column named SEQUENCE_NAME, which displays the names of all of the sequences in your user schema. In the next set of steps, you retrieve the names of your sequences.

To retrieve the names of your sequences:

1. Type the query shown in Figure 3-13.

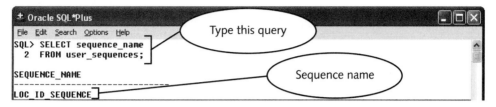

Figure 3-13 Viewing sequence names

2. Press **Enter**. The output shows the names of your sequences. (If you have created other sequences, these sequences also appear in the output.)

TIP

The USER_SEQUENCES data dictionary view contains other columns that provide additional information about sequences, such as the MINVALUE and MAXVALUE specifications, and the last value the sequence generated.

Using Sequences

To use sequences, you must first understand psuedocolumns. An Oracle 10*g* **pseudocolumn** acts like a column in a database table, but is actually a command that returns a specific value. You can use sequence pseudocolumns within SQL commands to retrieve the current or next value of a sequence. Table 3-4 summarizes the sequence pseudocolumns.

Pseudocolumn Name	Output
CURRVAL	Most recent sequence value retrieved during the current user session
NEXTVAL	Next available sequence value

Table 3-4 Oracle sequence pseudocolumns

To retrieve the next value in a sequence, you reference the NEXTVAL pseudocolumn using the following general syntax:

```
sequence_name.NEXTVAL
```

For example, you use the following INSERT action query to retrieve the next value in LOC_ID_SEQUENCE and insert the value into a row in the LOCATION table:

```
INSERT INTO location (LOC_ID)
VALUES(loc_id_sequence.NEXTVAL);
```

Next, you insert a row into the LOCATION table using LOC_ID_SEQUENCE and the NEXTVAL pseudocolumn.

To insert a new LOCATION row using the LOC_ID_SEQUENCE sequence:

    1. Type the first action query shown in Figure 3-14.

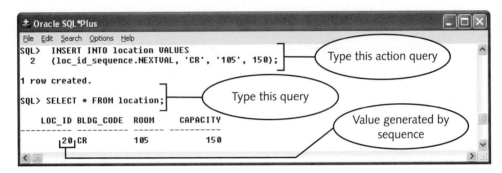

Figure 3-14   Inserting a new row using a sequence

    2. Execute the query. The confirmation message "1 row created." indicates that you inserted one row.

    3. To view the value that the DBMS inserted in this row for LOC_ID, type the second query shown in Figure 3-14, and then press **Enter** to execute the query. This query retrieves the new rows from the LOCATION table, as shown in Figure 3-14. The new row should have a LOC_ID value of 20, which was the first value in the sequence, along with the other values you specified in the INSERT action query.

Sometimes you need to access the next value of a sequence, but you don't want to insert a new row. For example, suppose you want to display a faculty ID (F_ID) value on a form, but you need to have the faculty enter more information before you actually insert the new row into the database. To do this, you create a SELECT query that uses the sequence name with the NEXTVAL pseudocolumn using the following syntax:

```
SELECT sequence_name.NEXTVAL
FROM DUAL;
```

This query's FROM clause specifies the DUAL database table. **DUAL** is a simple table that belongs to the SYSTEM user schema. It has one row, which contains a single column that contains the character string 'X'. All database users can use DUAL in SELECT queries, but they cannot modify or delete values in DUAL. Whenever you execute a SELECT query using a pseudocolumn with any database table, the query returns a value for each table row. It is more efficient to retrieve psuedocolumns from DUAL, which has only one row, than from other tables that may contain many rows.

To view the current value of the sequence, you execute a SELECT query that uses the sequence name with the CURRVAL pseudocolumn, using the following syntax:

```
SELECT sequence_name.CURRVAL
FROM DUAL;
```

In the following set of steps, you use the SELECT queries to retrieve the next LOC_ID_SEQUENCE value, and then view the current value.

To use the SELECT queries to access sequence values:

1. Type and execute the first query shown in Figure 3-15. Your query output should show 21 as the NEXTVAL value in the LOC_ID_SEQUENCE. Now that it has been accessed, 21 is the current value of the sequence.

2. Type and execute the second query shown in Figure 3-15 to display the current sequence value. The query output confirms that 21 is now the current value of the LOC_ID_SEQUENCE.

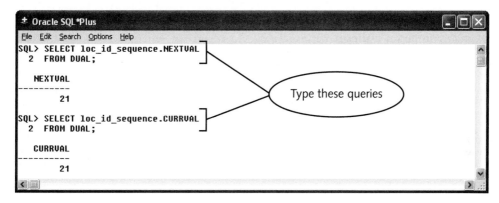

Figure 3-15   Using the NEXTVAL and CURRVAL pseudocolumns in SELECT queries

Often multiple users simultaneously use a sequence to retrieve surrogate key values. The DBMS must guarantee that each user who accesses a sequence receives a unique sequence number. When a user connects to an Oracle 10*g* database, the user creates a **user session**. A user session starts when a user starts a client application such as SQL*Plus, and logs onto the database. A user session terminates when the user exits the application. The Oracle 10*g* DBMS uses user sessions to ensure that all sequence users receive unique sequence numbers. When the DBMS issues a sequence value to your user session, no other user's session can access that sequence value. This prevents two different users from assigning the same primary key values to rows.

You can use the CURRVAL pseudocolumn to access the last value that you retrieved from a sequence only during the same database user session in which you used the NEXTVAL pseudocolumn to retrieve the value. In the following steps you confirm the CURRVAL user session restriction by retrieving the next value in LOC_ID_SEQUENCE, exiting SQL*Plus, starting it again, and then trying to use the CURRVAL command to access the current sequence value.

To examine the CURRVAL database session restriction:

1. Type and execute the query **SELECT loc_id_sequence.NEXTVAL FROM DUAL;**. The value 22 should appear as the next sequence value.

2. Exit SQL*Plus.

3. Start SQL*Plus again, and log onto the database.

4. Type and execute the query **SELECT loc_id_sequence.CURRVAL FROM DUAL;**. You receive the error message "Error at line 1: ORA-08002: sequence LOC_ ID_SEQUENCE.CURRVAL is not yet defined in this session." This indicates that no CURRVAL exists for the sequence because you exited SQL*Plus and started a new SQL*Plus session, but have not yet retrieved a sequence value in this session. A CURRVAL can be selected from a sequence only after selecting a NEXTVAL.

Deleting Sequences

To delete a sequence from the database, you use the DROP SEQUENCE DDL command. For example, to drop LOC_ID_SEQUENCE, you use the command **DROP SEQUENCE loc_id_sequence;**. Do not drop LOC_ID_SEQUENCE right now, because you will need to use it for the rest of the exercises in this chapter. As with any DDL command, you do not need to commit the action explicitly when you drop a sequence.

DATABASE OBJECT PRIVILEGES

When you create database objects such as tables or sequences, other users cannot access or modify the objects in your user schema unless you give them explicit privileges to do so. Table 3-5 summarizes some of the commonly used Oracle 10g object privileges for tables and sequences.

Object Type(s)	Privilege	Description
Table, Sequence	ALTER	Allows the user to change the object's structure using the ALTER command
Table, Sequence	DROP	Allows the user to drop (delete) the object
Table, Sequence	SELECT	Allows the user to view table records or select sequence values
Table	INSERT, UPDATE, DELETE	Allows the user to insert, update, or delete table records
Any object	ALL	Allows the user to perform all possible operations on the object

Table 3-5   Commonly used Oracle 10g object privileges

The following sections describe how to grant and revoke object privileges.

Granting Object Privileges

You grant object privileges to other users using the SQL GRANT command. The general syntax for the SQL GRANT command is:

```
GRANT privilege1, privilege2, ...
ON object_name
TO user1, user2, ...;
```

In this syntax, the GRANT clause lists each privilege to be granted, such as ALTER or DROP. *Object_name* represents the name of the object on which you are granting the privilege, such as the tablename or sequence name. The TO clause lists the name of the Oracle 10*g* user account to which you are granting the privilege, such as SCOTT or DBUSER. Note that in a single command, you can grant privileges for only one object at a time, but you can grant privileges to many users at once. If you want to grant privileges to every database user, you can use the keyword PUBLIC in the TO clause. Next, you grant object privileges on some of your database objects to other database users.

To grant object privileges:

1. Type the first command shown in Figure 3-16 to grant SELECT and ALTER privileges on your STUDENT table to the SCOTT user account. (The Oracle 10*g* DBMS automatically creates the SCOTT user account when the DBA installs the database.)

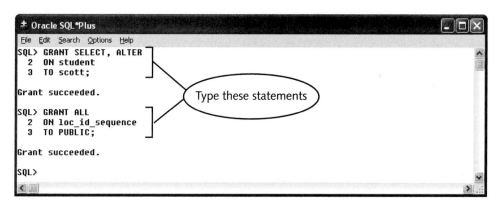

Figure 3-16   Granting object privileges

2. Execute the command. The message "Grant succeeded." confirms that the DBMS successfully granted the privileges.

3. Type the second command shown in Figure 3-16 to grant all privileges on your LOC_ID_SEQUENCE sequence to all database users, and then press **Enter** to execute the command. The message "Grant succeeded." again confirms that the DBMS successfully granted the privileges.

Revoking Table Privileges

To cancel a user's object privileges, you use the REVOKE command. The general syntax for the REVOKE command is:

```
REVOKE privilege1, privilege2, ...
ON object_name
FROM user1, user2, ...;
```

Next, you revoke the privileges on the STUDENT table that you granted in the previous set of steps.

To revoke privileges:

1. Type the command shown in Figure 3-17 to revoke the SELECT and ALTER privileges that you granted to user SCOTT on the STUDENT table, and then execute the command.

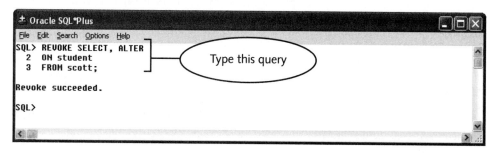

Figure 3-17   Revoking object privileges

2. Exit SQL*Plus.

3. Switch to Notepad, save the changes you made to Ch3AQueries.sql, and then exit Notepad.

NOTE

When you grant a privilege to PUBLIC, the privilege can be revoked only by a user who has been granted the DBA system privileges.

SUMMARY

❑ Database developers need to understand how to create SQL queries for inserting, updating, deleting, and viewing data because they use these commands when writing database applications.

❐ A script is a text file with a .sql extension that contains a sequence of SQL commands for creating or modifying database objects or manipulating data rows.

❐ You use an INSERT action query to insert new data rows into a table. Before you can add a new data row, you must ensure that all rows containing foreign key values referenced by the new row have already been added to the database.

❐ When you insert a new character column value, you must enclose the value in single quotation marks. Character column values are case sensitive. To add character column values that contain embedded single quotation marks, type the single quotation mark twice.

❐ To insert a new DATE column value, you insert the value as a character string, and then convert the value to the DATE format using the TO_DATE function. When you use the TO_DATE function, you specify the input format of the DATE character string using a format model, which is a text string that specifies how a data value is input or output.

❐ To insert a new INTERVAL column value, you insert the value as a character string, and then convert the value to an INTERVAL format using the TO_YMINTERVAL or TO_DSINTERVAL functions.

❐ One or more related action queries make up a transaction. A transaction is a logical unit of work. All parts of the transaction must be completed or the database might contain inconsistent data. You make the changes caused by a transaction's action queries permanent by issuing the COMMIT command. You use the ROLLBACK command to discard all of the transaction's changes.

❐ You use a SQL search condition to match one or more database rows. The DBMS must be able to evaluate a search condition as either TRUE or FALSE. You can combine search conditions using the AND, OR, and NOT logical operators.

❐ You update rows using the UPDATE action query. You can update multiple rows in a table using a single UPDATE command by specifying a search condition that matches multiple rows.

❐ You use the DELETE command to remove database rows. You can only delete rows from one table at a time. You cannot delete a row if one of its columns is referenced as a foreign key. To delete all of a table's rows quickly, you truncate the table. When you truncate a table, the DBMS does not save rollback information, so you cannot restore the rows.

❐ A sequence is a database object that automatically generates sequential numbers that are used for surrogate key values. You can access sequence values using the NEXTVAL and CURRVAL pseudocolumns within either an INSERT action query or a SELECT query.

REVIEW QUESTIONS

1. A(n) _____ represents a logical unit of work.

2. The UPDATE command is used to add new rows to an existing table. True or False?

3. What is the rationale for using a date format model?

4. The _____ command is used to update pending transactions to the database table(s).

5. What could prevent Oracle 10g from executing a TRUNCATE command that does not have a syntax error?

6. The _____ pseudocolumn is used to retrieve the next value in a sequence.

7. What is the difference between the AND and OR logical operators?

8. The DELETE command is used to remove objects from a database. True or False?

9. By default, an Oracle 10g sequence is increased by a value of _____ for each new integer generated.

10. How is a date or character string search condition different from a numeric search condition?

MULTIPLE CHOICE

1. Which of the following statements removes all rows from the ORDERS table?

   a. `DELETE * FROM orders;`

   b. `DROP * FROM orders;`

   c. `DELETE FROM orders;`

   d. `DROP FROM orders;`

2. Which of the following statements returns all orders placed on May 29, 2006?

   a. `SELECT * FROM orders WHERE o_date = 5/29/2006;`

   b. `SELECT * FROM orders WHERE o_date = '5/29/2006';`

   c. `SELECT * FROM orders WHERE o_date = '29-MAY-06';`

   d. none of the above

3. Which of the following statements correctly updates the received date for the shipment with the SHIP_ID of 2 to 11/19/2006?

   a. `UPDATE shipment_line`
      `SET sl_date_received = TO_DATE('11/19/2006',`
      `'MM/DD/YYYY')WHERE ship_id = 2;`

   b. `UPDATE shipment_line`
      `SET sl_date_received = TO_DATE('11/19/2006',`
      `'MM/DD/YYYY');`

c. `INSERT shipment_line`
   `SET sl_date_received = TO_DATE('11/19/2006',`
   `'MM/DD/YYYY');`

d. `INSERT shipment_line`
   `SET sl_date_received = TO_DATE('11/19/2006',`
   `'MM/DD/YYYY')`
   `WHERE ship_id = 2;`

4. Which of the following statements displays the next value in the ORDER_NO_SEQ sequence?

   a. `SELECT next.order_no_seq FROM DUAL;`

   b. `SELECT order_no_seq.nextvalue FROM DUAL;`

   c. `SELECT order_no_seq.currval FROM DUAL;`

   d. `SELECT order_no_seq.nextval FROM DUAL;`

5. Which of the following statements deletes all customers that reside in Florida?

   a. `DELETE FROM customer WHERE c_state = 'FL';`

   b. `DELETE FROM customer WHERE c_state = "FL";`

   c. `DELETE FROM customer WHERE c_state = FL;`

   d. `DELETE FROM customer WHERE c_state != FL;`

6. Which of the following commands is used to discard any uncommitted changes?

   a. ROLLBACK

   b. COMMIT

   c. DELETE

   d. UNDO

7. Which of the following SQL commands cannot be rolled back after it is executed?

   a. TRUNCATE

   b. DELETE

   c. INSERT

   d. UPDATE

8. Which of the following is a valid statement?

   a. You cannot truncate or drop a table that contains a constraint.

   b. You cannot drop a table that contains a foreign key constraint.

   c. You cannot drop a table that is being referenced by a foreign key constraint.

   d. all of the above

9. Which of the following is a valid statement?

   a. The CREATE SEQUENCE command is a DML, so the user must execute a COMMIT command before the sequence is permanently updated to the database.

   b. The CREATE SEQUENCE command is a DDL, so the user must execute a COMMIT command before the sequence is permanently updated to the database.

   c. The CREATE SEQUENCE command is a DML, so the sequence is permanently updated to the database when the command is executed.

   d. The CREATE SEQUENCE command is a DDL, so the sequence is permanently updated to the database when the command is executed.

10. Which of the following is a logical operator?

   a. !=

   b. OR

   c. IS NULL

   d. LIKE

PROBLEM-SOLVING CASES

Before starting any of the problem-solving cases, make certain you run the Ch3EmptyClearwater.sql script file stored in the Chapter03 folder of your Data Disk to create the necessary tables. Case 1 must be completed before attempting the remaining cases.

1. Create a script named Ch03Case1.sql to perform the actions indicated below. Store the script file in the Chapter03 folder of your Data Disk, and then execute the script in SQL*Plus.

   a. Add each customer shown in Figure 1-24 to the CUSTOMER table.

   b. Create a sequence named ORDERS_ID_SEQ to generate the O_ID values for the ORDERS table shown in Figure 1-24.

   c. Add each order shown in Figure 1-24 to the ORDERS table. Use the ORDERS_ID_SEQ sequence to enter the appropriate value for the O_ID column.

   d. Add each order source shown in Figure 1-24 to the ORDER_SOURCE table.

   e. Make certain the rows you have added are permanently updated to the database tables.

2. Execute the appropriate commands to complete the specified tasks. Save a copy of your commands in a script file named Ch03Case2.sql in the Chapter03 folder of your Data Disk.

   a. Remove the customer with the C_ID of 5 from the CUSTOMER table. If an error occurs while attempting to remove the customer's row, take the necessary corrective actions to complete the task.

   b. Update the order with the O_ID of 6 in the ORDERS table so the payment method is money order (MO) rather than credit card (CC).

   c. Add a new catalog to the ORDER_SOURCE database table. The new catalog is the Winter 2006 version and should be assigned the ID of 7.

   d. Discard the changes you made in the previous three steps.

3. Execute the appropriate commands to complete the specified tasks. Save a copy of your commands in a script file named Ch03Case3.sql in the Chapter03 folder of your Data Disk.

   a. Create a sequence named CUSTOMER_ID_SEQ that generates the ID for each customer added to the CUSTOMER table.

   b. Drop and then re-create the CUSTOMER_ID_SEQ sequence so each subsequent ID generated increases by a value of 10.

   c. Display the current contents of the CUSTOMER table to determine if the change to the sequence had any impact on the existing values in the C_ID column.

   d. Remove the CUSTOMER_ID_SEQ sequence from the database.

4. Execute the appropriate commands to complete the specified tasks. Save a copy of your commands in a script file named Ch03Case4.sql in the Chapter03 folder of your Data Disk.

   a. Disable the foreign key constraint in the Clearwater database that references the CUSTOMER table.

   b. Remove all customer rows from the CUSTOMER table.

   c. Issue the command SELECT * FROM CUSTOMER; to verify that there are no rows in the table. Use the ROLLBACK command to restore the rows. Then issue the SELECT command again to verify that the rows are now in the CUSTOMER table.

   d. Create a savepoint named SAVEPOINT1.

   e. Truncate the contents of the CUSTOMER, ORDERS, and ORDER_SOURCE tables. Use the command SELECT * FROM ORDERS; to verify removal of the table data. Attempt to roll back to the SAVEPOINT1 savepoint and then verify the contents of the ORDERS table. Were you able to roll back truncation of the table? Why or why not?

5. Execute the appropriate commands to complete the specified tasks. Save a copy of your commands in a script file named Ch03Case5.sql in the Chapter03 folder of your Data Disk.

a. Grant the user SCOTT the privilege to view and alter your CUSTOMER table.

b. Revoke from the user SCOTT the privilege to alter your CUSTOMER table.

c. Delete the CUSTOMER, ORDERS, and ORDER_SOURCE tables from your database.

d. Attempt to roll back deletion of the CUSTOMER table. Were you able to successfully complete this step? Why or why not?

◀ LESSON B ▶

After completing this lesson, you should be able to:
- ♦ Write SQL queries to retrieve data from a single database table
- ♦ Create SQL queries that perform calculations on retrieved data
- ♦ Use SQL group functions to summarize retrieved data

Now that you have learned how to insert, update, and delete data rows using SQL*Plus, the next step is to learn how to retrieve data. In this lesson, you learn how to write SQL queries to retrieve data from a single database table, sort the output, and perform calculations on data, for example, calculating a person's age from data stored in a date-of-birth column or calculating a total for an invoice. You also learn how to format retrieved data in SQL*Plus. For the rest of this chapter, you execute SELECT queries using fully populated Northwoods University database tables. You create these tables by running the Ch3Northwoods.sql script in the Chapter3 folder on your Data Disk. This script contains SQL commands to drop all of the tables in these databases, re-create the tables, and then insert all of the table rows. Follow the set of steps to run the script file.

To run the script files:

1. If necessary, start SQL*Plus and log onto the database.

2. Run the Ch3Northwoods.sql script by typing the following command at the SQL prompt:

   START c:\OraData\Chapter03\Ch3Northwoods.sql.

Recall that you might receive an error message if the script tries to drop a table that has not yet been created. (See Figure 3-1.) You can ignore this error message.

RETRIEVING DATA FROM A SINGLE DATABASE TABLE

The basic syntax for a SQL query that retrieves data from a single database table is:

```
SELECT columnname1, columnname2, ...
FROM ownername.tablename
[WHERE search_condition];
```

The SELECT clause lists the columns whose values you want to retrieve. In the FROM clause, *ownername* specifies the table's user schema, and *tablename* specifies the name of the database table. If you are retrieving data from a table in your own user schema, you can omit *ownername* (and the period) in the FROM clause. If you are retrieving data from a table in another user's schema, you must include *ownername* in the FROM clause, and the table's owner must have granted you the SELECT privilege on the table. For example, you use the following FROM clause to retrieve data values from the LOCATION table in user SCOTT's user schema:

```
FROM scott.location
```

The WHERE clause optionally specifies a search condition that instructs the query to retrieve selected rows. To retrieve every row in a table, the data values do not need to satisfy a search condition, so you can omit the WHERE clause. Next, you retrieve the first name, middle initial, and last name from every row in the Northwoods University STUDENT table.

To retrieve selected column values from every row of the STUDENT table:

1. Start Notepad, create a new file named **Ch3BQueries.sql**, and then type the query in Figure 3-18 to retrieve the student first name, middle initial, and last name from every row in the STUDENT table. Copy and paste the query into SQL*Plus.

NOTE

In this lesson, you enter all of the tutorial commands in your Ch3BQueries.sql file, and then execute them in SQL*Plus.

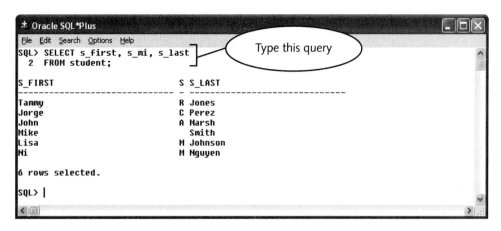

Figure 3-18   Retrieving selected column values

2. Press **Enter** to execute the query. Your output should look like the query output in Figure 3-18. Note that the column names appear as column headings, and the column values appear in the same order in which you listed them in the SELECT clause.

If you want to retrieve all of the columns in a table, you can use an asterisk (*) as a wildcard character in the SELECT clause instead of typing every column name. Now you write a query to retrieve every row and every column in the LOCATION table.

To retrieve every row and column from the LOCATION table:

1. Type and execute the following query to select all rows and columns from the LOCATION table:

```
SELECT *
FROM location;
```

NOTE

By default, the SQL*Plus environment shows query output one page at a time. You learn how to modify the display later in this lesson in the section titled "Modifying the SQL*Plus Display Environment."

Suppressing Duplicate Rows

Sometimes a query retrieves duplicate data values. For example, suppose you want to view the different ranks of faculty members at Northwoods University. Some of the faculty members have the same rank, so the query SELECT f_rank FROM faculty; retrieves duplicate values. The SQL DISTINCT qualifier examines query output before it appears on your screen and suppresses duplicate values. The DISTINCT qualifier has the following general syntax in the SELECT command:

```
SELECT DISTINCT columnname;
```

For example, to suppress duplicate faculty ranks, you use the command `SELECT DIS-TINCT f_rank FROM faculty;`. Next, you execute the query to retrieve all of the faculty ranks and display the duplicate values. Then, you execute it again using the DISTINCT qualifier to suppress the duplicates.

To retrieve and suppress duplicate rows:

1. Type and execute the first query in Figure 3-19. The Associate value appears twice because two rows in the table contain the Associate value in the F_RANK column.

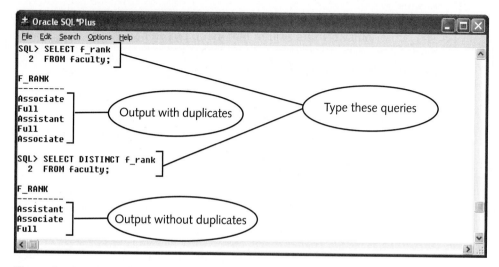

Figure 3-19   Output with duplicate rows suppressed

2. To suppress duplicate values, type and execute the second query in Figure 3-19, which uses the DISTINCT qualifier. The output now appears with the duplicate rows suppressed.

Using Search Conditions in SELECT Queries

So far, the queries that you have executed have retrieved all of the rows in the table. To retrieve rows matching specific criteria, you can use search conditions. Recall that you used search conditions in UPDATE and DELETE action queries to match specific data rows, and that search conditions use comparison operators to match a data value or range of values. The following sections describe how to use search conditions in SELECT queries.

Exact and Inexact Search Conditions

An **exact search condition** uses the equal to comparison operator (=) to match a value exactly, whereas an **inexact search condition** uses the inequality comparison operators (>, <, >=, <=) to match a range of values. Next, you use an exact search condition to retrieve the rows in the FACULTY table for which the F_RANK value is Associate.

To use an exact search condition in a SELECT query:

1. Type the following query to retrieve specific rows in the FACULTY table:

```
SELECT f_first, f_mi, f_last, f_rank
FROM faculty
WHERE f_rank = 'Associate';
```

2. Execute the query. The output lists the first name, middle initial, last name, and rank of all faculty members with the rank of Associate.

Next you create an inexact search condition that retrieves the number of every room in the Business ('BUS') building at Northwoods University that has a capacity greater than or equal to 40 seats. You use the greater than or equal to (>=) comparison operator in the inexact search condition.

To use an inexact search condition in a SELECT query:

1. Type the following query to retrieve specific rows in the LOCATION table:

```
SELECT room
FROM location
WHERE bldg_code = 'BUS'
AND capacity >= 40;
```

2. Execute the query. The output reports that rooms 105 and 211 match the search criteria.

Searching for NULL and NOT NULL Values

Sometimes you need to create a query to return rows in which the value of a particular column is NULL. For example, you might want to retrieve enrollment rows for courses in which the instructor has not yet assigned a grade, so the GRADE value is NULL. To search for NULL values, you use the following general syntax:

```
WHERE columnname IS NULL
```

Similarly, to retrieve rows in which the value of a particular column is not NULL, you use the following syntax:

```
WHERE columnname IS NOT NULL
```

Next, you create queries to retrieve rows from the ENROLLMENT table in which specific column values are NULL or NOT NULL.

To create queries using the IS NULL and IS NOT NULL search conditions:

1. Type and execute the first query in Figure 3-20 to retrieve all rows in the ENROLLMENT table for which the GRADE column value is NULL. The returned rows show all enrollment rows for which the instructor has not yet assigned a grade value.

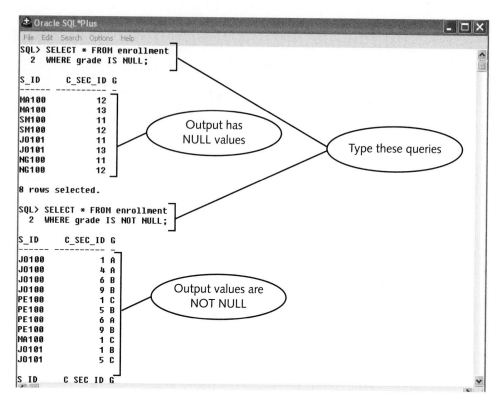

Figure 3-20   Queries using IS NULL and IS NOT NULL search conditions (partial output shown)

2. Type and execute the second query in Figure 3-20 to retrieve the enrollment rows for which the instructor has assigned a grade value.

Using the IN and NOT IN Comparison Operators

You can use the IN comparison operator to match data values that are members of a set of search values. For example, you can retrieve all enrollment rows in which the GRADE column value is a member of the set ('A', 'B'). Similarly, you can use the NOT IN comparison operator to match values that are not members of a set of search values. Now you retrieve rows using the IN and NOT IN comparison operators.

To retrieve rows using the IN and NOT IN comparison operators:

1. Type and execute the first query in Figure 3-21 to retrieve every enrollment row in which the GRADE value is either A or B.

2. Type and execute the second query in Figure 3-21 to retrieve every enrollment row in which the GRADE value is neither A nor B.

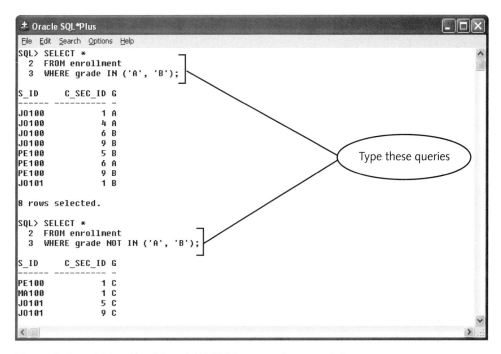

Figure 3-21   Using the IN and NOT IN comparison operators

Using the LIKE Comparison Operators

Sometimes, you need to perform searches by matching part of a character string. For example, you might want to retrieve rows for students whose last name begins with the letter M, or find all courses with the text string MIS in the COURSE_NO column. To do this, you use the LIKE operator. The general syntax of a search condition that uses the LIKE operator is:

```
WHERE columnname LIKE 'character_string'
```

Character_string represents the text string to be matched and is enclosed in single quotation marks. *Character_string* must contain either the percent sign (%) or underscore (_) wildcard characters.

The percent sign (%) wildcard character represents multiple characters. If you place (%) on the left edge of the character string to be matched, the DBMS searches for an exact match on the far-right characters and allows an inexact match for the characters represented by (%). For example, the search condition WHERE term_desc LIKE '%2006' retrieves all term rows in which the last four characters in the TERM_DESC column are 2006. The DBMS ignores the characters on the left side of the character string up to the substring 2006. Similarly, the search condition WHERE term_desc LIKE 'Fall%' retrieves all term rows in which the first four characters are Fall, regardless of the value of the rest of the string. The search condition WHERE course_name LIKE '%Systems%' retrieves every row in the COURSE table in which the COURSENAME column contains the character string Systems anywhere in the string. In the search condition LIKE '%Systems%', the (%) wildcard character does not require that characters be present before or after the character string Systems

The underscore (_) wildcard character represents a single character. For example, the search condition WHERE s_class LIKE '_R' retrieves all values for S_CLASS in which the first character can be any value, but the second character must be the letter R.

You can use the underscore (_) and percent sign (%) wildcard characters together in a single search condition. For example, the search condition WHERE c_sec_day LIKE '_T%' retrieves all course sections that meet on Tuesday, provided exactly one character precedes T in the C_SEC_DAY column. The search condition ignores all of the characters that follow T in the column value, so the query retrieves values such as MT, MTW, and MTWRF. Next, you create some queries that use the LIKE operator.

To create queries using the LIKE operator:

1. Type and execute the first query in Figure 3-22 to retrieve all rows from the TERM table in which the last four characters of the TERM_DESC column are 2006.

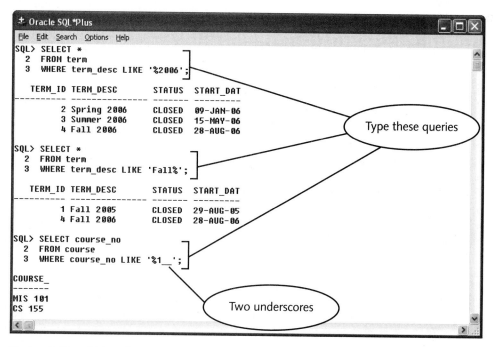

Figure 3-22   Using the LIKE comparison operator

3

2. Type and execute the second query in Figure 3-22 to retrieve all rows from the TERM table in which the first four letters are Fall. Note that the characters in single quotation marks are always case sensitive.

3. Type and execute the third query in Figure 3-22 to search for a COURSE_NO value from the COURSE table in which the third to last character is the number 1. This search expression uses the (%) wildcard operator to match any characters on the left end of the string, then specifies the number 1, followed by two underscores (__) to specify that exactly two characters follow the number 1.

Sorting Query Output

When you insert rows into an Oracle database, the DBMS does not store the rows in any particular order. When you retrieve rows using a SELECT query, the rows may appear in the same order in which you inserted them into the database, or they may appear in a different order, based on the database's storage configuration. You can sort query output by using the **ORDER BY clause** and specifying the **sort key**, which is the column the DBMS uses as a basis for ordering the data. The syntax for a SELECT query that uses the ORDER BY clause is as follows:

```
SELECT columnname1, columnname2, ...
FROM ownername.tablename
WHERE search_condition
ORDER BY sort_key_column;
```

In this syntax, *sort_key_column* can be the name of any column within the SELECT clause. If *sort_key_column* has the NUMBER data type, the DBMS by default sorts the rows in numerical ascending order. If *sort_key_column* has one of the character data types, the DBMS by default sorts the rows in alphabetical order. If *sort_key_column* has the DATE data type, the DBMS by default sorts the rows from older dates to more recent dates. To sort the rows in the reverse order from the default, use the DESC qualifier, which stands for *descending*. You place the DESC qualifier at the end of the ORDER BY clause, using the following syntax: ORDER BY sort_key_column DESC.

In the following steps, you retrieve rows from the LOCATION table and sort the rows based on the CAPACITY column.

To use the ORDER BY clause to sort data:

1. Type and execute the first query in Figure 3-23 to list the building code, room number, and capacity for every room with a capacity that is greater than or equal to 40, sorted in ascending order by capacity.

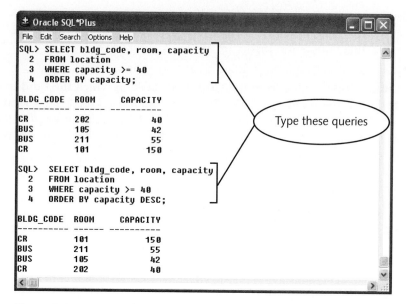

Figure 3-23   Using the ORDER BY clause to sort data

2. Type and execute the second query in Figure 3-23 to retrieve the same rows, but add the DESC qualifier at the end of the ORDER BY clause to sort the rows in descending order.

You can specify multiple sort keys to sort query output on the basis of multiple columns. You specify which column gets sorted first, second, and so forth, by the order of the sort keys in the ORDER BY clause. The next query lists all building codes, rooms, and capacities, sorted first by building code and then by room number.

To sort data on the basis of multiple columns:

1. Type the query in Figure 3-24 to sort the rows by building codes and then by room numbers.

2. Execute the query. The query output lists every row in the LOCATION table, first sorted alphabetically by building code, and then sorted within building codes by ascending room numbers.

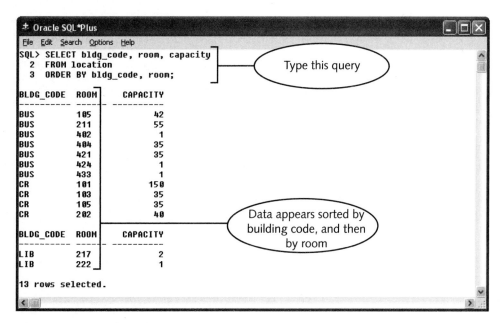

Figure 3-24   Sorting data using multiple sort keys

USING CALCULATIONS IN SQL QUERIES

Database applications often display values based on calculations on retrieved data values. For example, an application may display a student's age based on his or her date of birth, or display the total cost of a customer order. You can perform many calculations directly within SQL queries. This is a very efficient way to perform calculations, because the DBMS returns to the client workstation only the calculated value, rather than all of the data values that contribute to the calculated value. For example, to calculate the total cost of a customer order yourself, you need to retrieve the item price and the order quantity for each order item. For an order with many different items, this could involve many rows. However, if the DBMS calculates the order total and sends only the calculated value to the client, it generates less network traffic, and system performance improves.

You can create SQL queries to perform basic arithmetic calculations and to use a variety of built-in functions. The following sections describe how to create queries using calculations and functions.

Performing Arithmetic Calculations

Table 3-6 lists the SQL arithmetic operations and their associated SQL operators in order of precedence.

Operation	Operator
Multiplication, Division	*, /
Addition, Subtraction	+, −

Table 3-6   Operators used in SQL query calculations

In mathematics and in programming languages, arithmetic expressions that combine multiple operations and contain more than one operator must be evaluated in a specific order. SQL evaluates division and multiplication operations first, and addition and subtraction operations last. SQL always evaluates arithmetic calculations enclosed within parentheses first.

You can perform arithmetic calculations on columns that have the NUMBER, DATE, or INTERVAL data types. The following sections describe how to perform these calculations for these data types.

Number Calculations

You can perform any basic arithmetic operations on columns that have the NUMBER data type by placing the operator in the SELECT clause along with the column names. For example, you use the following SELECT clause to retrieve the product of the INV_PRICE times the INV_QOH columns in the INVENTORY table:

```
SELECT inv_price * inv_qoh
```

In addition, you can include constants—actual numeric values—in the calculations performed by a query. For example, suppose Northwoods University charges $86.95 per credit hour for tuition. To determine how much tuition a student is charged for a class, you can simply multiply the number of credit hours earned for a course by the credit hour tuition rate. You use the multiplication operator (*) to derive the tuition amount for each retrieved row.

To calculate values in a SQL query:

1. Type the query in Figure 3-25.

2. Execute the query. Note that the query output shows the calculated value in a column whose heading is the calculation formula.

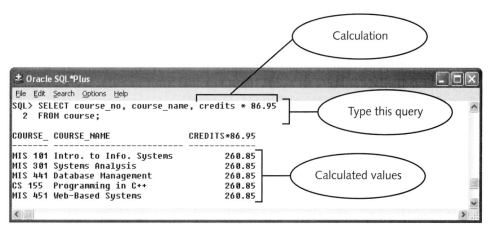

Figure 3-25 Query with calculated values

Date Calculations

Database applications often perform calculations involving dates. You may need to deter-mine a date that is a specific number of days before or after a known date. Or you may need to determine the number of days between two known dates. Often, these calcula-tions involve the current date. To retrieve the current system date from the database server, you use the **SYSDATE pseudocolumn**. (Recall from Lesson A in this chapter that a pseudocolumn acts like a column in a database table, but is actually a command that returns a specific value.) The following query retrieves the current system date:

```
SELECT SYSDATE
FROM DUAL;
```

If you retrieve the SYSDATE value along with other database columns from another table, you can omit DUAL from the FROM clause.

TIP

To add a specific number of days to a known date, you add the number of days, expressed as an integer, to the known date. To subtract a specific number of days from a known date, you subtract the number of days, expressed as an integer, from the known date. For example, the following expression specifies a date that is 10 days after the order date (O_DATE) in the Clearwater Traders ORDERS table:

```
o_date + 10
```

For the first row in the ORDERS table, this expression returns the value 6/08/2006, which is 10 days after the O_DATE value of 5/29/2006.

You cannot multiply or divide values of the DATE data type.

TIP

Sometimes you need to determine the number of days between two known dates. To return an integer value that represents the number of days between two dates, you subtract the first date from the second date. If the first date is after the second date, the result is a positive integer. If the first date is before the second date, the result is a negative integer. The following expression returns the number of days between the current date and the order date column in the ORDERS table:

```
SYSDATE - o_date
```

You usually do not store a person's age in a database because ages change from year to year. Rather, you store the person's date of birth and calculate his or her age based on the current system date. Next, you create a query that retrieves student information and calculates the student's age based on the student date of birth (S_DOB) data column and the current date.

To use the SYSDATE function to calculate student ages:

1. Type the query in Figure 3-26 to list the student ID, last name, and age for each student. The query calculates the student ages by subtracting the student dates of birth from the current date.

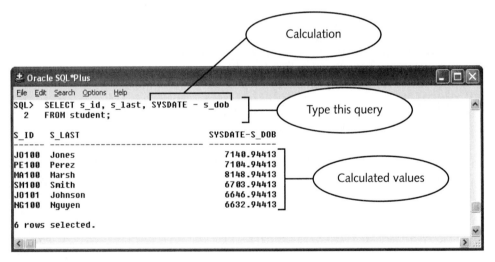

Figure 3-26 Calculating student ages based on dates of birth

2. Execute the query. The query output lists the calculated ages in days rather than in years. (Your query output will be different because the output depends on the current system date.)

To express the calculated ages in years instead of days, you must divide these values by the number of days in a year, which is approximately 365.25 (including leap years). You can do this calculation in SQL by combining multiple arithmetic operations in a single query. Recall that in SQL commands, the DBMS evaluates multiplication and division operations first, and then evaluates addition and subtraction operations. To evaluate the subtraction operation *before* the division operation, you must place the subtraction operation in parentheses using the following expression:

```
(SYSDATE - S_DOB)/365.25
```

3

To convert the days into years by performing multiple arithmetic operations in a specific order:

1. Type the query in Figure 3-27, which uses parentheses to specify the order of the arithmetic operations.

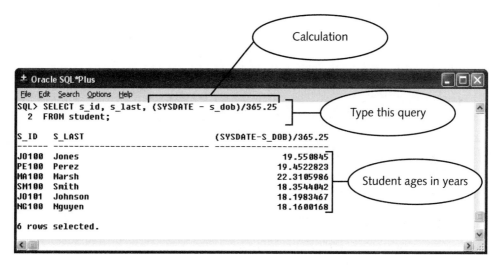

Figure 3-27   Combining multiple arithmetic operations in a single query

2. Execute the query. The query output now lists the students' ages in years. (Your output will be different, based on your current system date.)

Interval Calculations

The Oracle 10*g* DBMS can perform calculations using interval values that store elapsed time values. To specify a date that is before or after a known date, you can add to or subtract from the known date a column value that has the INTERVAL data type. Recall that in the STUDENT table, the TIME_ENROLLED column contains interval values that specify how long each student has been enrolled at Northwoods University. To determine the date on which a student enrolled in the university, you subtract the TIME_ENROLLED value from the current system date, as in the following expression:

```
SYSDATE - time_enrolled
```

Now you create a query that uses an interval calculation to determine the date each student enrolled in the university.

To use an interval calculation:

1. Type the query in Figure 3-28, which subtracts an INTERVAL value from a known date.

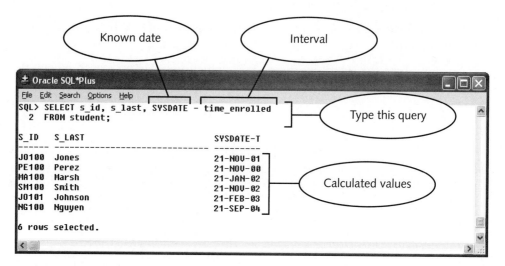

Figure 3-28 Subtracting an interval from a known date

2. Execute the query. The query output lists the dates the students enrolled at the university, based on the INTERVAL value and the known date.

You can also add or subtract intervals from one another to calculate the sum or difference of two intervals. For example, suppose you need to update the TIME_ENROLLED column every month by adding one month to the interval value. You use the following query to add an interval of one month to the current TIME_ENROLLED value:

```
SELECT s_id, time_enrolled + TO_YMINTERVAL('0-1')
FROM student;
```

Recall that you use the TO_YMINTERVAL function to convert a text string that represents a year and month interval to an INTERVAL data type.

TIP

Similarly, you use the following query to add an interval of 10 minutes to the C_SEC_DURATION column in the COURSE_SECTION table:

```
SELECT c_sec_id, c_sec_duration +
TO_DSINTERVAL('0 00:10:00')
FROM course_section;
```

Recall that you use the TO_DSINTERVAL function to convert a text string that represents days, hours, minutes, and seconds to an INTERVAL data type.

TIP

Oracle 10*g* SQL Functions

Oracle 10*g* provides a number of built-in functions to perform calculations and manipulate retrieved data values. These functions are called **single-row functions**, because they return a single result for each row of data retrieved. The following subsections describe the Oracle 10*g* SQL single-row functions for number, character, and date values.

Single-row Number Functions

Oracle 10*g* SQL has several single-row number functions that you can use to manipulate retrieved data in ways that are more complex than simple arithmetic operations. For example, Oracle 10*g* provides number functions for rounding data values or raising values to exponential powers. Table 3-7 summarizes some commonly used SQL single-row number functions.

Function	Description	Example Query	Result
ABS (*number*)	Returns the absolute value of a number	SELECT ABS(capacity) FROM location WHERE loc_id = 1;	ABS(150) = 150
CEIL (*number*)	Returns the value of a number, rounded to the next highest integer	SELECT CEIL (inv_price) FROM inventory WHERE inv_id = 1;	CEIL(259.99) = 260
FLOOR (*number*)	Returns the value of a number, rounded down to the next integer	SELECT FLOOR (inv_price) FROM inventory WHERE inv_id = 1;	FLOOR(259.99) = 259
MOD (*number, divisor*)	Returns the remainder (modulus) for a number and its divisor	SELECT MOD (inv_qoh, 10) FROM inventory WHERE inv_id = 1;	MOD(16, 10) = 6
POWER (*number, power*)	Returns the value representing a number raised to the specified power	SELECT POWER (inv_qoh, 2) FROM inventory WHERE inv_id = 2;	POWER(12, 2) = 144
ROUND (*number, precision*)	Returns a number, rounded to a specified precision	SELECT ROUND (inv_price, 0) FROM inventory WHERE inv_id = 1;	ROUND(259.99, 0) = 260
TRUNC (*number, precision*)	Removes all digits from a number beyond the specified precision	SELECT TRUNC (inv_price, 1) FROM inventory WHERE inv_id = 1;	TRUNC (259.99, 1) = 259.9

Table 3-7   Oracle 10*g* SQL single-row number functions

To use a SQL single-row number function, you list the function name in the SELECT clause, followed by the required parameter (or parameters) in parentheses. The next query demonstrates how to use a number function with a calculated database value. In Figure 3-27, you calculated each student's age based on his or her date of birth. The student ages currently appear in years, along with a fraction that represents the time since the student's last birthday. Usually, ages appear as whole numbers. You use the TRUNC function to truncate the fraction portion of the calculated student ages.

To use the TRUNC function to remove the fractional component:

1. Type the query in Figure 3-29.

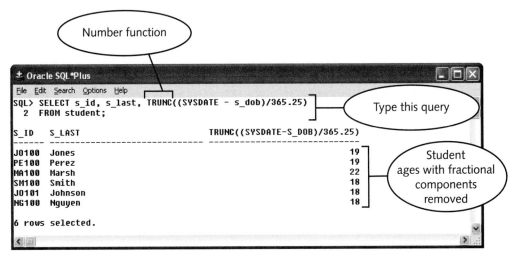

Figure 3-29 Using a SQL number function

2. Execute the query. The query output shows the student ages in years without fractional values. (Your values will be different, based on your current system date.)

Single-row Character Functions

Oracle 10*g* SQL also provides single-row character functions that you can use to format character output. Table 3-8 summarizes these character functions.

Function	Description	Example Query	String Used in Function	Function Result
CONCAT (*string1, string2*)	Concatenates (joins) two strings	`SELECT CONCAT (f_last, f_rank) FROM faculty WHERE f_id = 1;`	'Marx' and 'Associate'	'MarxAssociate'

Table 3-8 Oracle 10*g* SQL single-row character functions

3

Function	Description	Example Query	String Used in Function	Function Result
INITCAP (*string*)	Returns the string with only the first letter in uppercase text	`SELECT INITCAP (bldg_code) FROM location WHERE loc_id = 1;`	'CR'	'Cr'
LENGTH (*string*)	Returns an integer representing the string length	`SELECT LENGTH (term_desc) FROM term WHERE term_id = 1;`	'Fall 2005'	9
LPAD|RPAD (*string, number_of_ characters_ to_add, padding_ character*);	Returns the value of the string with a specified number of padding characters added to the left/right	`SELECT LPAD (term_desc, 12, '*'), RPAD (term_desc, 12, '*') FROM term WHERE term_id = 1;`	'Fall 2005'	'***Fall 2005' 'Fall 2005***'
LTRIM|RTRIM (*string, search_string*)	Returns the string with all occurrences of the search string trimmed on the left/right side	`SELECT LTRIM (course_no, 'MIS') FROM course WHERE course_name LIKE '%Intro%';`	'MIS 101'	' 101'
REPLACE (*string, search_string, replacement_ string*)	Returns the string with every occurrence of the search string replaced with the replacement string	`SELECT REPLACE (term_desc, '200', '199') FROM term WHERE term_id = 1;`	'Fall 2005'	'Fall 2005'
SUBSTR (*string, start_position, length*)	Returns a string, starting at the start position, and of the specified length	`SELECT SUBSTR (term_desc, 1, 4) FROM term WHERE term_id = 1;`	'Fall 2005'	'Fall'
UPPER|LOWER (*string*)	Returns the string with all characters converted to uppercase or lowercase letters	`SELECT UPPER (term_desc) , LOWER (term_desc) FROM term WHERE term_id = 1;`	'Fall 2005'	'FALL 2005' 'fall 2005'

Table 3-8   Oracle 10*g* SQL single-row character functions (continued)

Next, you create some queries that use the Oracle 10*g* SQL character functions. First, you use the CONCAT function to display the Northwoods University location building codes and rooms as a single text string. Then, you use the INITCAP function to display the values in the STATUS column in the TERM table in mixed-case letters, with the first letter capitalized.

To create queries using the Oracle 10*g* SQL character functions:

1. Type and execute the first query in Figure 3-30. The building codes and rooms appear as a single text string.

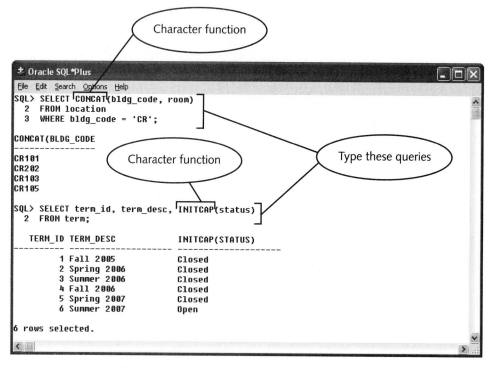

Figure 3-30   Using SQL character functions

2. Type and execute the second query in Figure 3-30. The STATUS values appear in mixed-case letters, with the first letter capitalized. (Recall that the STATUS values are stored in the database in all capital letters.)

Single-row Date Functions

Earlier you learned how to use date arithmetic within SQL queries to display a date that occurs a specific number of days before or after a known date. Oracle 10*g* SQL provides a variety of single-row date functions to support additional date operations. Table 3-9 summarizes these date functions.

Function	Description	Example Query	Date(s) Used in Function	Function Result
`ADD_MONTHS` `(date,` `months_` `to_add)`	Returns a date that is the specified number of months after the input date	`SELECT ADD_MONTHS` `(ship_date_` `expected, 2) FROM` `shipment WHERE` `ship_id = 1;`	9/15/2006	11/15/2006
`LAST_DAY` `(date)`	Returns the date that is the last day of the month specified in the input date	`SELECT LAST_DAY` `(ship_date_` `expected) FROM` `shipment WHERE` `ship_id = 1;`	9/15/2006	9/30/2006
`MONTHS_` `BETWEEN` `(date1,` `date2)`	Returns the number of months, including decimal fractions, between two dates	`SELECT MONTHS_` `BETWEEN (ship_` `date_expected,` `TO_DATE('10-AUG-` `2006', 'DD-MON-` `YYYY')) FROM` `shipment WHERE` `ship_id = 1;`	9/15/2006	1.1612903

Table 3-9   Oracle10*g* SQL single-row date functions

Now you execute some queries that use the Oracle 10*g* SQL date functions. First, you use the ADD_MONTHS function to display a date that is two months after the date that the Spring 2007 term started. Then, you use the MONTHS_BETWEEN function to display the number of months between the current system date and the date on which the Summer 2007 term started.

To execute queries using the Oracle 10*g* SQL date functions:

1. Type and execute the first query in Figure 3-31. The query output shows 08-MAR-07, which is two months after the retrieved date value.

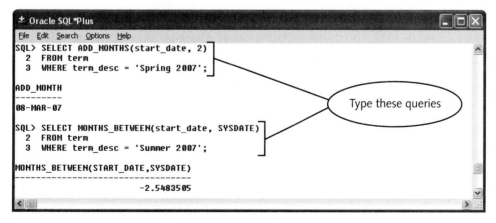

Figure 3-31   Using SQL date functions

2. Type and execute the second query in Figure 3-31. The output shows the number of months between the current system date and the term's starting date. In this example, the term started approximately two and a half months before the computer's current system date. (Your value will be different, based on your current system date.)

ORACLE 10g SQL GROUP FUNCTIONS

So far, you have displayed and manipulated individual database rows. However, database applications often need to display information about a group of rows. For example, a query might display the total number of students who enroll in a specific Northwoods University course section, or the total revenue generated from all orders that customers place using the Clearwater Traders Web site. To display data that summarizes multiple rows, you use one of the Oracle 10g group functions. An Oracle 10g SQL **group function** performs an operation on a group of queried rows and returns a single result, such as a column sum. Table 3-10 describes commonly used Oracle SQL group functions.

Function	Description	Example Query	Result
AVG (*fieldname*)	Returns the average value of a numeric field's returned values	SELECT AVG(capacity) FROM location;	33.30769
COUNT(*)	Returns an integer representing a count of the number of returned rows	SELECT COUNT(*) FROM enrollment;	20
COUNT (*fieldname*)	Returns an integer representing a count of the number of returned rows for which the value of *fieldname* is NOT NULL	SELECT COUNT(grade) FROM enrollment;	12
MAX (*fieldname*)	Returns the maximum value of a numeric field's returned values	SELECT MAX(max_enrl) FROM course_section;	140
MIN (*fieldname*)	Returns the minimum value of a numeric field's returned values	SELECT MIN(max_enrl) FROM course_section;	30
SUM (*fieldname*)	Sums a numeric field's returned values	SELECT SUM(capacity) FROM location;	432

Table 3-10   Oracle 10g SQL group functions

To use a group function in a SQL query, you list the function name, followed by the column name on which to perform the calculation, in parentheses. In the following steps, you execute a query that uses group functions to sum the maximum enrollment for each course section in the Summer 2007 term and calculate the average, maximum, and minimum current enrollments.

To use group functions in a query:

    1. Type the query in Figure 3-32 to sum the maximum enrollment for all course sections and calculate the average, maximum, and minimum current enrollment for each course section for the Summer 2007 term (TERM_ID = 6).

3

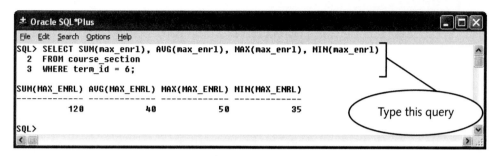

Figure 3-32   Using group functions in a SQL query

    2. Execute the query. The output shows the summary data for the Summer 2007 term.

The following section describes how to use the COUNT group function, which works differently than the other group functions. You also learn how to use the GROUP BY clause to group similar data in group functions, and how to use the HAVING clause to add search conditions to group functions.

Using the COUNT Group Function

The COUNT group function returns an integer that represents the number of rows that a query returns. The COUNT(*) version of this function calculates the total number of rows in a table that satisfy a given search condition. The COUNT(*) function is the only group function in Table 3-10 that includes NULL values. The other functions ignore NULL values. The COUNT(*columnname*) version calculates the number of rows in a table that satisfy a given search condition and also contain a non–null value for the given column. Next, you use both versions of the COUNT function to count the total number of courses in which student Lisa Johnson (S_ID = JO101) has enrolled and the total number of courses in which Johnson has received a grade (GRADE is NOT NULL).

To use the COUNT group function:

    1. Type and execute the first query in Figure 3-33, which uses the COUNT(*) group function to count the total number of courses in which student Lisa Johnson (S_ID = JO101) has enrolled.

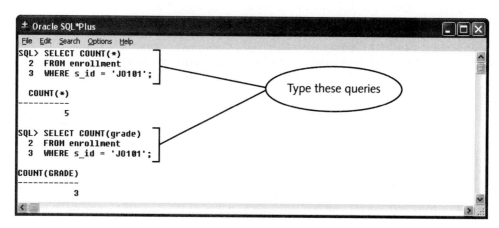

Figure 3-33   Using the COUNT group function

2. Type and execute the second query in Figure 3–33, which uses the COUNT(columnname) function to count the total number of courses for which student Lisa Johnson has received a grade.

Notice the difference between the output for the first and second queries. Lisa has enrolled in a total of five courses, and has been assigned a GRADE value in only three of those courses.

Using the GROUP BY Clause to Group Data

If a query retrieves multiple rows and the rows in one of the retrieved columns have duplicate values, you can group the output by the column with duplicate values and apply group functions to the grouped data. For example, you might want to retrieve the names of the different building codes at Northwoods University and calculate the sum of the capacity of each building. To do this, you add the GROUP BY clause after the query's FROM clause. The GROUP BY clause has the following syntax:

```
GROUP BY group_columnname;
```

Now you create a query that uses a group function with the GROUP BY clause to list the Northwoods University building codes and sum each building's capacity.

To use the GROUP BY clause to group rows:

1. Type the query in Figure 3–34 to list the building code and the total capacity of each building in the LOCATION table.

2. Press **Enter** to execute the query. The output lists the building codes and their associated capacities.

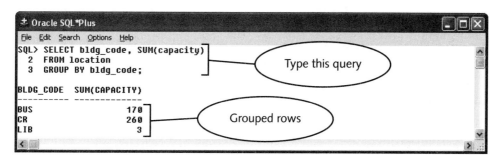

Figure 3-34   SQL query that uses the GROUP BY clause to group rows

When you create a query containing a GROUP BY clause, all columns listed in the SELECT clause must be included in the GROUP BY clause. In other words, if ungrouped columns are included in the SELECT clause, Oracle 10*g* will return an error message. This error occurs because SQL cannot display single-row results and group function results in the same query output. In the next query, you view the error that occurs when you attempt to mix single rows and grouped rows in the same query. You repeat the query to return building codes and the sums of their capacities, but omit the GROUP BY clause.

To repeat the query without the GROUP BY clause:

1. Type the query in Figure 3-35 to list the building code name and the total capacity of each building in the LOCATION table, omitting the GROUP BY clause.

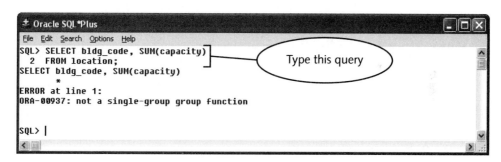

Figure 3-35   Error that occurs when you omit the GROUP BY clause

2. Execute the query. The error message "ORA-00937: not a single-group group function" appears.

This error message indicates that a SELECT clause cannot contain both a group function and an individual column expression unless the individual column expression is in a GROUP BY clause. To solve this problem, you must include BLDG_CODE in the GROUP BY clause, as shown in the query in Figure 3-34.

Using the HAVING Clause to Filter Grouped Data

You can use the HAVING clause to place a search condition on the results of queries that display group function calculations. The HAVING clause has the following syntax:

```
HAVING group_function comparison_operator value
```

In this syntax, *group_function* is the group function expression, *comparison_operator* is one of the comparison operators in Table 3-3, and *value* is the value that the search condition matches. For example, suppose you want to retrieve the total capacity of each building at Northwoods University, but you are not interested in the data for buildings that have a capacity of less than 100. You use the following HAVING clause to filter the output:

```
HAVING SUM(capacity) >= 100
```

Next, you execute the query to retrieve this data.

To filter grouped data using the HAVING clause:

1. Type the query in Figure 3-36 to retrieve building codes and total capacities where the total capacity is greater than or equal to 100.

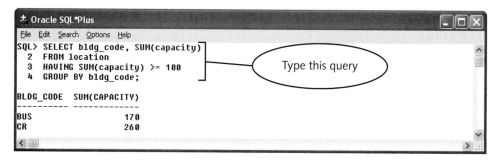

Figure 3-36 Using the HAVING clause with a group function

2. Execute the query. Compare the query output with the output in Figure 3-34, and note that the output omits the LIB building code because the sum of this building's capacity does not meet the search criteria in the HAVING clause.

FORMATTING OUTPUT IN SQL*PLUS

So far, you have accepted the default output formats in SQL*Plus—output column headings are the same as the database column names. You have also accepted the default screen widths and lengths. Now you learn how to modify the format of output data. You learn how to change the column headings, the SQL*Plus display width and length, and the format of retrieved data values.

Creating Alternate Column Headings

In SQL*Plus query output, column headings for retrieved columns are the names of the database table columns. When you create retrieve values that perform arithmetic calculations, or that use single-row functions or group functions, the output headings appear as the formula or function. For example, in Figure 3-36, the heading for the values that calculate the sum of the capacity for each building appears as SUM(CAPACITY). Similarly, when you calculated the student ages (see Figure 3-29), the heading appeared as the full formula for the calculation. To display a different heading in the query output, you can create alternate output heading text, or you can create an alias.

Alternate Output Heading Text

To specify alternate output heading text, you use the following syntax in the SELECT clause:

```
SELECT columnname1 "heading1_text",
columnname2 "heading2_text", ...
```

In this syntax, *heading1_text* specifies the alternate heading that appears for *columnname1*, *heading2_text* specifies the alternate heading that appears for *columnname2*, and so forth. You enclose the alternate heading text in double quotation marks, and it can contain characters, including blank spaces. In the following steps, you create a query that specifies alternate output headings.

To specify alternate output headings:

1. Type the query in Figure 3-37 to create the alternate output headings for the query columns.

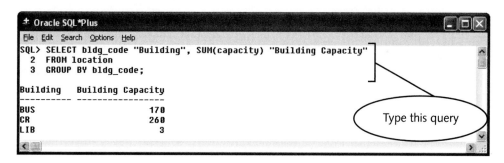

Figure 3-37   Creating alternate output headings

2. Execute the query. The output appears with the alternate headings.

Changing the output heading changes only the SQL*Plus output display—you cannot use the alternate heading to reference the column in a GROUP BY or ORDER BY clause in a query. In the query in Figure 3-37, if you wanted to order the query output by the building capacity value, which is stored in the CAPACITY column, you need to specify the ORDER BY clause as `ORDER BY SUM(capacity);`. You could not order the query output by specifying the heading "Building Capacity" in the ORDER BY clause.

Aliases

An **alias** is an alternate name for a query column. After you create an alias, you can reference the alias in other parts of the query, such as in the GROUP BY or ORDER BY clause. The general syntax for creating an alias is:

```
SELECT columnname1 AS alias_name1...
```

The *alias_name1* value must follow the Oracle naming standard and cannot contain blank spaces. You now repeat the query to sum the capacity of all of the buildings but modify it to create an alias for the summed capacity values. Then, you use the alias as the sort key to specify the sort order of the output.

To create an alias:

    1. Type the query in Figure 3-38.

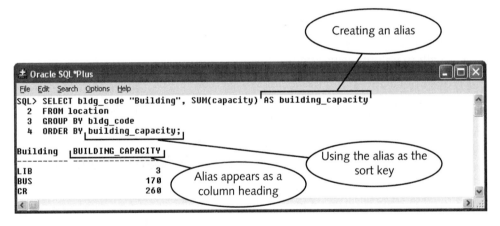

Figure 3-38   Creating an alias

    2. Execute the query. The alias appears as the new output heading, and the output values appear sorted by capacity.

Note that the query uses the alias as the sort key. Also note that the output heading ("BUILDING_CAPACITY") appears in all uppercase letters. The DBMS automatically converts alias names to uppercase characters.

Modifying the SQL*Plus Display Environment

Recall that SQL*Plus displays query output one page at a time. A SQL*Plus page consists of a specific number of characters per line and a specific number of lines per page.

If the query output contains more characters than the current SQL*Plus environment's page width, then the output wraps to the next line. If the query output contains more lines than the current SQL*Plus page length, then the column headings repeat at the top of each page. Next, you type a query that spans multiple SQL*Plus lines and pages, and examine the output.

To type a query that spans multiple lines and pages:

    1. Type the query in Figure 3-39.

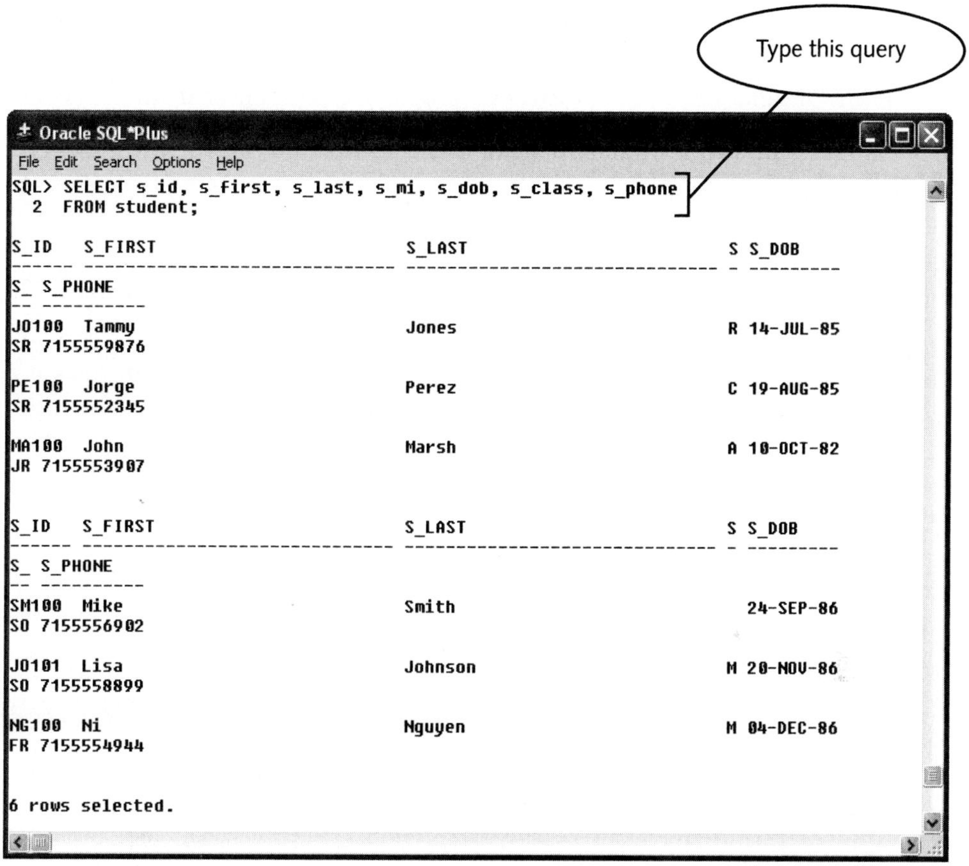

Figure 3-39   Query output that spans multiple lines and pages

    2. Execute the query. The query output should look like Figure 3-39, spanning multiple lines and pages.

If your output does not look like Figure 3-39, it is because your SQL*Plus display environment settings have been changed from the default settings. You adjust your settings in the next set of steps.

The output spans multiple lines, and the data values for the S_CLASS and S_PHONE columns wrap to the next line. The output spans multiple pages, and the column headings repeat after the values for the first three rows. You can configure the SQL*Plus page and line size by clicking Options on the menu bar, clicking Environment, and then configuring the environment in the Environment dialog box. SQL*Plus saves these changes on your workstation, and the configuration values remain the same until you or someone else change them again.

The SQL*Plus environment **linesize** property specifies how many characters appear on a display line, and the **pagesize** property specifies how many lines appear on a SQL*Plus page. Now you modify your SQL*Plus display settings, and make the line and page sizes larger.

To modify the SQL*Plus display settings:

1. Click **Options** on the menu bar, and then click **Environment**. The Environment dialog box opens.

2. Select **linesize** in the Set Options list. Select the **Custom** option button, and type **120** in the Value column to display 120 characters per line.

3. Select **pagesize** in the Set Options list. Select the **Custom** option button, and then type **40** in the Value column to display 40 lines per page.

4. Click **OK** to save your settings.

5. Type the query in Figure 3-40 to test your new settings. The query output should appear as shown, with each row's value on the same line, and all of the values on the same display page.

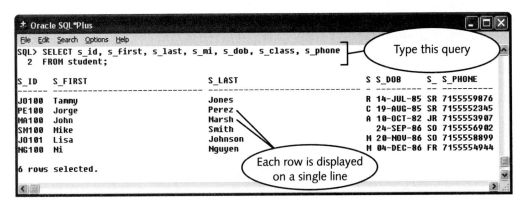

Figure 3-40 Query output with modified display settings

Formatting Data Using Format Models

When you retrieve data from NUMBER or DATE data columns, the data values appear using the SQL*Plus default date and number formats. Sometimes, you might want to display number columns using a currency format model that shows a currency symbol, or you might want to display DATE columns using an alternate format model. To use an alternate format model, you can use the TO_CHAR function to convert the column to a character string, and then apply the desired format model to the value.

The TO_CHAR function has the following syntax:

```
TO_CHAR(column_name, 'format_model')
```

In this syntax, *column_name* is the column value you wish to format, and *format_model* is the format model you wish to apply to the column value. Note that the format model appears in single quotation marks.

It is necessary to convert output columns to characters and apply a specific format model when you store time values in a column that has the DATE data type. Recall that DATE data columns store time as well as date information; however, the default DATE output format model is DD-MON-YY, which does not have a time component. Therefore, you need to format time output using an alternate format model. In the next set of steps, you retrieve the values from the C_SEC_TIME column in the COURSE_SECTION table to examine the output. When the Ch3Northwoods.sql script originally inserts the values for the C_SEC_TIME column into the COURSE_SECTION table, it specifies the time values in the default format. However, the script does not specify a date value, so the DBMS inserts the default date value, which is the first day of the current month. You realize that you first need to retrieve the values, then convert them to characters using the TO_CHAR function, and finally display the values using a time format model.

To examine and format the C_SEC_TIME column values:

1. Type and execute the first query in Figure 3-41 to retrieve the C_SEC_TIME column from the COURSE_SECTION table. Note that the values appear in the default DATE format and do not display the time values. (Your values will be different, because you ran your script during a different month.)

2. Type and execute the second query in Figure 3-41 to convert the course section times to characters, then display the values using a format model that has a time (rather than a date) component.

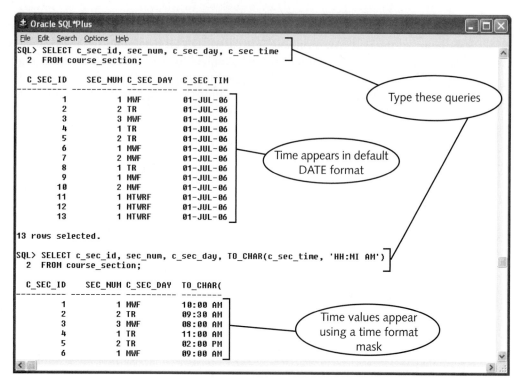

Figure 3-41 Displaying time values using the TO_CHAR function (partial output shown)

You can also use the TO_CHAR function to apply format models to NUMBER columns. For example, the credit hours multiplied by the tuition rate for a course at Northwoods University represents a currency value. Next, you create a query that formats output values using a currency format model.

To create a query that formats number values as currency:

1. Type and execute the following command to format the INV_PRICE values using a currency format model:

```
SELECT course_name, TO_CHAR(credits*86.95, '$99,999.99')
FROM course
WHERE course_no = 'MIS 101';
```

2. Close SQL*Plus. Switch to Notepad, save Ch3BQueries.sql, then close Notepad and all other open applications.

SUMMARY

- In a single-table SELECT query, the SELECT clause lists the columns that you want to display in your query, the FROM clause lists the name of the table involved in the query, and the WHERE clause specifies a search condition.

- If you are querying a table that you did not create, you must preface the table name with the owner's username, and the owner must have given you the privilege to retrieve data from that table.

- To retrieve every row in a table, the data values do not need to satisfy a search condition, so you omit the WHERE clause.

- If you want to retrieve all of the columns in a table, you can use an asterisk (*) in place of the individual column names in the SELECT clause.

- You can use the DISTINCT qualifier to suppress duplicate rows in query output, and you can sort query outputs using the ORDER BY clause.

- To retrieve specific rows, you can use exact search conditions, which exactly match a value, or inexact search conditions, which match a range of values.

- To retrieve rows in which the value of a particular column value is NULL, you use the search condition `WHERE columnname IS NULL`. To retrieve rows in which the value of a particular column is NOT NULL, you use the search condition `WHERE columnname IS NOT NULL`.

- You use the IN comparison operator to match data values that are members of a set of search values, and you use the NOT IN comparison operator to match data values that are not members of a set of search values.

- You use the LIKE comparison operator to retrieve values that match partial text strings. To use the LIKE operator, you specify the search string using the percent (%) wildcard character to specify multiple characters and the underscore (_) wildcard character to specify a single character.

- You can create SQL queries that perform addition, subtraction, multiplication, and division operations on retrieved data values.

- The SYSDATE pseudocolumn returns the current system date and time. You can create queries that use date arithmetic to add or subtract a specific number of days from a known date or determine the number of days between two known dates. You can perform calculations using interval data values by adding or subtracting an interval to a known date. You can also calculate the sum or difference of two interval values.

- Oracle 10*g* SQL provides single-row number, character, and date functions that allow you to manipulate retrieved values for each row of data that a query retrieves.

- Oracle 10*g* SQL provides group functions to calculate the sum, average, or maximum or minimum value of a group of rows that a query retrieves. Oracle 10*g* SQL also provides a function to count the number of rows that a query retrieves. You can use the GROUP BY clause to group the output on a specific output value, and you can use the HAVING clause to add a search condition to a query that contains a group function.

❑ By default, SQL*Plus displays query output using database column names or formulas for queries that perform arithmetic calculations. You can create alternate output headings by specifying the alternate heading text for each output column. Or, you can create an alias, which is an alternate name for the column that can be used in the GROUP BY or ORDER BY clause.

❑ You can change the appearance of the SQL*Plus environment by changing the pagesize and linesize properties. The linesize property specifies the number of characters that appear on the SQL*Plus screen, and the pagesize property specifies the number of lines that appear on the screen.

❑ To format values stored in DATE and NUMBER columns in a SQL query, you can convert the column to a character column using the TO_CHAR function and format the column using a format model.

REVIEW QUESTIONS

1. The _____ wildcard character is used to specify exactly one character.

2. The IN operator is used to identify that the search condition includes a pattern. True or False?

3. Which clause is used to restrict the number of rows returned by a query that is based on a stated condition?

4. The _____ keyword is included in the query to suppress duplicate results.

5. The COUNT(*) function can be used to include NULL values in its results. True or False?

6. The _____ clause is used to restrict the groups that are included in the output of a query.

7. Which logical operator requires that both conditions be met for a row to be included in the results?

8. Which function is used to format the display of a date value in the results of a query?

9. The _____ keyword is used to denote a column alias.

10. Which SQL*Plus property determines how many lines are displayed for column headings that are repeated in the results?

MULTIPLE CHOICE

1. Which of the following clauses present query results in a sorted order?

   a. GROUP BY

   b. ORDERED BY

   c. SORT BY

   d. none of the above

2. The _____ wildcard character is used to represent any number of characters.

   a. *

   b. %

   c. _

   d. ^

3. Based on its syntax, which of the following is a valid query?

   a. `SELECT acolumn FROM atable WHERE acolumn LIKE NULL;`

   b. `SELECT acolumn FROM atable WHERE acolumn = NULL;`

   c. `SELECT acolumn FROM atable WHERE acolumn IS NOT NULL;`

   d. `SELECT acolumn FROM atable WHERE acolumn <> NULL;`

4. Which function is used to determine the total credit hours currently being taken by a student?

   a. TOTAL

   b. SUM

   c. CALC

   d. ADD

5. Which function is used to determine how many professors currently hold the rank of assistant professor?

   a. TOTAL

   b. ADD

   c. SUM

   d. COUNT

6. Which keyword is used to retrieve the computer's current date and time?

   a. DATE

   b. SYSTEMTIME

   c. TIME

   d. none of the above

7. Which of the following is a single-row function?

   a. COUNT

   b. SUM

   c. INITIALCAP

   d. none of the above

8. When searching for nonnumeric data, the search condition must be enclosed in:

   a. double-quotation marks (")

   b. single-quotation marks (')

   c. parentheses ( )

   d. none of the above

9. If you use the character string Faculty as a column alias, the string must be enclosed in _____ or it is converted to uppercase characters in the results.

   a. double-quotation marks (")

   b. single-quotation marks (')

   c. parentheses ( )

   d. none of the above

10. By default, Oracle 10*g* sorts data in _____.

    a. ascending order

    b. descending order

    c. natural order

    d. the order in which it is stored in the database table

PROBLEM-SOLVING CASES

For all cases, use Notepad or another text editor to write a script using the specified filename. Always use the search condition text exactly as the case specifies. Place the queries in the order listed, and save the script files in your Chapter03\Cases folder on your Data Disk. If you haven't done so already, run the Ch3Clearwater.sql script in the Chapter03 folder on your Data Disk to create and populate the case study database tables. All the following cases are based on the Clearwater Traders database.

1. Determine the inventory price of inventory item (INV_ID) #1.

2. Determine which customers were born in the 1970s.

3. Determine how many different categories of inventory are carried by Clearwater Traders.

4. Determine how many shipments have not yet been received.

5. Calculate the total quantity on hand for each inventory item in the INVENTORY table—ignore their different sizes and colors.

6. Determine the total number of orders received on May 29, 2006.

7. Determine how many orders placed on May 31, 2006, were paid by credit card.

8. Create a list of all Florida and Georgia customers.

9. List all inventory items that do not have an associated size.

10. Identify which shipments are expected by September 1, 2006.

11. Determine how many inventory items have an inventory price greater than $60.00 and are available in a size of L or XL.

12. In the INVENTORY table, ITEM_ID 5 is available in how many different colors?

13. In the INVENTORY table, ITEM_ID 5 is available in how many different sizes?

14. Determine the current age of each customer. Display the customer's first and last names and their age in years.

15. Display the SHIP_DATE_EXPECTED of each shipment in the SHIPMENT table using the format MONTH DD,YYYY.

16. Determine how many orders were received from each of the available order sources (OS_ID).

17. Identify which items in stock (that is INV_QOH > 0) are available in sizes Medium or Large and are available in the colors Royal, Bright Pink, or Spruce.

18. Create a listing that identifies the different items (ITEM_ID) in the INVEN-TORY table and the number of colors available for each item.

19. Determine how many items are not in categories 2 or 4.

20. List the unique ITEM_IDs along with their inventory prices from the INVENTORY table. Format the inventory prices so they are displayed in the format $999.99.

◀ LESSON C ▶

After completing this lesson, you should be able to:
- ◆ Create SQL queries that join multiple tables
- ◆ Create nested SQL queries
- ◆ Combine query results using set operators
- ◆ Create and use database views

So far, all of your SQL queries have retrieved data from a single database table. In this lesson, you learn how to create queries that retrieve data from multiple tables. You also learn how to create nested queries, in which the output from one query serves as a search condition in a second query. And, you learn how to create advanced SQL queries that select rows for updating; you learn how to create and query database views; finally, you learn how to create database indexes. For this lesson, you execute SQL commands and queries using the fully populated Northwoods University database tables. Your first task is to create or refresh these tables by running the Ch3Northwoods.sql script in the Chapter03 folder on your Data Disk. You also change the page and line sizes in your SQL*Plus environment.

To run the script files and change the page and line sizes:

1. If necessary, start SQL*Plus and log onto the database. Run the Ch3Northwoods.sql script by typing the following command at the SQL prompt:

   `START c:\OraData\Chapter03\Ch3Northwoods.sql`

2. Click **Options** on the menu bar, click **Environment**, and then select **linesize** in the Set Options list. Select the **Custom** option button, and type **120** in the Value column to display 120 characters per line.

3. Select **pagesize** in the Set Options list. Select the **Custom** option button, type **40** in the Value column to display 40 lines per page, and then click **OK** to save your settings.

NOTE

As an alternative to Steps 2 and 3, you can issue the commands SET LINESIZE 120 and SET PAGESIZE 40 on separate lines at the SQL*Plus prompt.

JOINING MULTIPLE TABLES

One of the strengths of SQL is its ability to **join**, or combine, data from multiple database tables using foreign key references. The general syntax of a SELECT query that joins two tables is:

```
SELECT column1, column2, ...
FROM table1, table2
WHERE table1.joincolumn = table2.joincolumn
AND search_condition(s);
```

The SELECT clause lists the names of the columns to display in the query output. The FROM clause lists the names of all of the tables involved in the join operation. If you display a column that exists in more than one of the tables in the FROM clause, you must qualify the column name in the SELECT clause. To **qualify** a column name in the SELECT clause, you specify the name of the table that contains the column, followed by a period, and then specify the column name. Because the column exists in more than one table, you can qualify the column name using the name of any table listed in the FROM clause that contains the column. For example, suppose you write a query that lists the F_ID column in the SELECT clause, and the STUDENT and FACULTY tables in the FROM clause. Because F_ID exists in both the STUDENT and FACULTY tables, you must qualify the F_ID column in the SELECT clause. To do so, you can use the following syntax: **STUDENT.F_ID**. You could also qualify the F_ID column as **FACULTY.F_ID**.

The WHERE clause contains the **join condition**, which specifies the table names to be joined and column names on which to join the tables. The join condition contains the foreign key reference in one table and the primary key in the other table. You can add search conditions using the AND and OR operators.

SQL supports multiple types of join queries. In this book, you learn to create inner joins, outer joins, and self-joins.

Inner Joins

The simplest type of join occurs when you join two tables based on values in one table being equal to values in another table. This type of join is called an **inner join**, **equality join**, **equijoin**, or **natural join**. For example, suppose you want to retrieve student last and first names, along with each student's advisor's ID and last name. This query retrieves data from two database tables; the student last and first names are in the STUDENT table, and the advisor last names are in the FACULTY table. Each student's advisor ID appears in the F_ID column in the STUDENT table, and each F_ID value in the STUDENT table corresponds to an F_ID value in the FACULTY table, as shown in Figure 3-42.

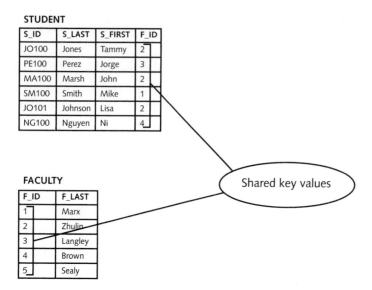

STUDENT

S_ID	S_LAST	S_FIRST	F_ID
JO100	Jones	Tammy	2
PE100	Perez	Jorge	3
MA100	Marsh	John	2
SM100	Smith	Mike	1
JO101	Johnson	Lisa	2
NG100	Nguyen	Ni	4

FACULTY

F_ID	F_LAST
1	Marx
2	Zhulin
3	Langley
4	Brown
5	Sealy

Shared key values

Figure 3-42   Joining two tables based on shared key values

Figure 3-42 shows a partial listing of the columns in the FACULTY and STUDENT tables, and shows how you join the tables on the F_ID column, which is the primary key in the FACULTY table and a foreign key in the STUDENT table. Note that because the F_ID column exists in both the STUDENT and FACULTY tables, you must qualify the F_ID column in the SELECT clause by prefacing the column name with the name of one of the tables. In the next set of steps, you create a SELECT query to retrieve student IDs and last and first names, along with each student's advisor's ID and last name, by joining the STUDENT and FACULTY tables.

To retrieve rows by joining two tables based on shared key values:

1. Start Notepad, create a new file named **Ch3CQueries.sql**, and then type the first query in Figure 3-43 to retrieve the student ID, student last and first names, advisor ID, and advisor last name. Note that you must qualify the F_ID column in the SELECT clause, because the F_ID column exists in both the STUDENT and FACULTY tables. You could qualify F_ID using either the STUDENT table or the FACULTY table.

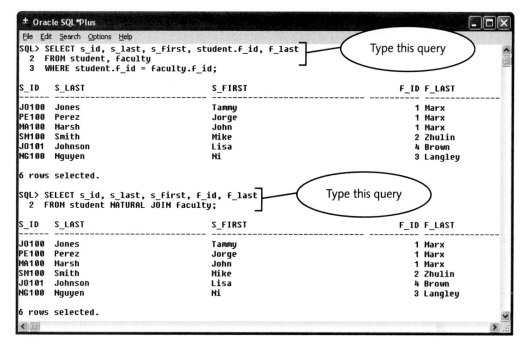

Figure 3-43   Queries joining two tables

2. Copy the command, paste the command into SQL*Plus, and then press **Enter** to execute the query. The output shows the values from both the STUDENT table and the FACULTY table.

NOTE

For the rest of this lesson, you type all queries into your Ch3CQueries.sql file, and then copy, paste, and execute the queries in SQL*Plus.

In cases where the tables have a single commonly named and defined column, you can use the NATURAL JOIN keywords to join the tables in the FROM clause of the SELECT statement, as shown by the second query in Figure 3-43. Notice the syntax of the FROM clause when the NATURAL JOIN keywords are included; there is no comma separating the table names, simply the keywords. In addition, notice that the qualifier is no longer included for the F_ID column in the SELECT clause of the statement. Why? For the tables to be joined correctly, the value of the F_ID column in each table must be equivalent so the same data is displayed regardless of which table is referenced. In fact, if you do include a qualifier, Oracle 10*g* returns an error message.

In the preceding examples, you joined two tables. In SQL queries, you can join any number of tables in a SELECT command. When you join tables, the name of each table in the query must appear in the FROM clause. This includes tables whose columns are

display columns, which are columns that appear in the SELECT clause, and whose columns are **search columns**, which appear in search conditions. The primary key and foreign key columns on which you join the tables are called **join columns**.

When you join multiple tables, sometimes you must join the tables using an intermediary table whose columns are not display or search columns, but whose columns are join columns that serve to join the two tables. For example, suppose you want to create a query that lists the last names of all faculty members who teach during the Summer 2007 term. Figure 3-44 shows that this query involves three tables: FACULTY, COURSE_SECTION, and TERM.

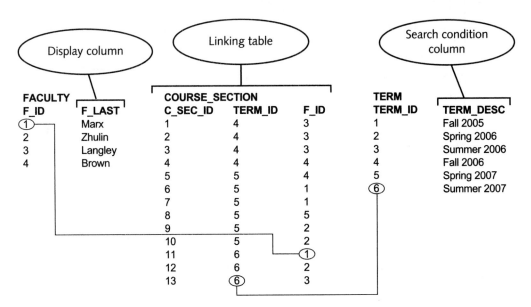

Figure 3-44   Joining three tables

Figure 3-44 shows a partial listing of the columns in the FACULTY, COURSE_SECTION, and TERM tables. The FACULTY table contains the F_LAST column, which is the display column. The TERM table contains the TERM_DESC search column, which specifies the search condition as the Summer 2007 term. The query joins the FACULTY and TERM tables using the COURSE_SECTION table as in intermediary, or linking table. A **linking table** does not contribute any columns as display columns or search condition columns, but contains join columns that link the other tables through shared foreign key values. Even though the query does not display columns from the COURSE_SECTION table or include them in search conditions, you must include the COURSE_SECTION table in the query's FROM clause, and include join conditions to not only specify the links between the FACULTY and COURSE_SECTION tables, but also the links between the COURSE_SECTION and TERM tables. Now you execute this query in SQL*Plus.

To execute a query that joins three tables:

    1. Type the query in Figure 3-45 to join the FACULTY, COURSE_SECTION, and TERM tables.

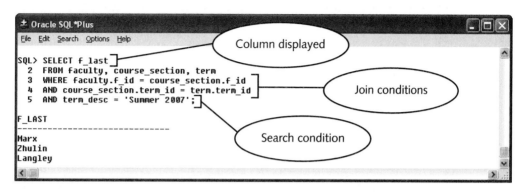

Figure 3-45   Query joining three tables

    2. Execute the query. The query output shows the display column from the FACULTY table, based on the search column in the TERM table.

NOTE

The query in Figure 3-45 cannot be modified to join the three tables with the NATURAL JOIN keywords because the FACULTY and COURSE_SECTION have two commonly named and defined columns: F_ID and LOC_ID. If F_ID had been the only common column between the tables, the FROM clause could have read: FROM faculty NATURAL JOIN course_section NATURAL JOIN term, leaving the WHERE clause free to only restrict the rows retrieved to the course offered in the Summer term of 2007.

Sometimes queries that join multiple tables can become complex. For example, suppose that you want to create a query to display the COURSE_NO and GRADE values for each of student Tammy Jones' courses. This query requires you to join four tables: STU-DENT (to search for S_FIRST and S_LAST), ENROLLMENT (to display GRADE), COURSE (to display COURSE_NO), and COURSE_SECTION (to join ENROLL-MENT to COURSE using the C_SEC_ID join column).

You join the STUDENT and ENROLLMENT tables using the S_ID column as the join column, because S_ID is the primary key in the STUDENT table and a foreign key in the ENROLLMENT table. You need to include the COURSE_SECTION table to join the ENROLLMENT table to the COURSE table, using the COURSE_NO foreign key link between COURSE and COURSE_SECTION and the C_SEC_ID foreign key link between COURSE_SECTION and ENROLLMENT. At this point, you are probably hopelessly confused. For complex queries such as this one, it is help-ful to draw a query design diagram such as the one in Figure 3-46.

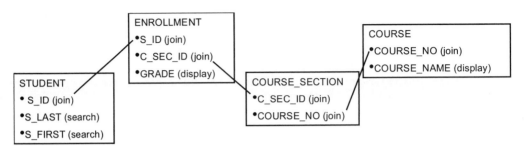

Figure 3-46 Query design diagram

A **query design diagram** shows the display and search columns in a SQL query that joins multiple tables, as well as the required join columns and their links. To create a query design diagram, you draw a rectangle to represent each table in the query. Within each table's rectangle, you list the names of each display column, search column, and join column. Then, you draw the links between the join columns as shown.

Table 3-11 describes the process for deriving the SQL query based on a query design diagram. Note that Figure 3-46 shows four tables and three links between the tables. Because there are three links, the query must have three join conditions. You must always have one fewer join condition than the total number of tables that the query joins. In this query, you are joining four tables, so you have three join conditions. Now you execute the query derived in Table 3-11 to join the four database tables.

Step	Process	Result
1	Create the SELECT clause by listing the display fields	`SELECT course_no, grade`
2	Create the FROM clause by listing the table names	`FROM student, enrollment,` `course_section, course`

3	Create a join condition for every link between the tables	`WHERE student.s_id =`
		`enrollment.s_id`
		`AND enrollment.c_sec_id`
		`= course_section.c_sec_id`
		`AND course_section.course_no =`
		`course.course_no`

4	Add additional search conditions for remaining search fields	`AND s_last = 'Jones'` `AND s_first = 'Tammy'`

Table 3-11 Deriving a SQL query from a query design diagram

To join four tables in a single query:

1. Type the query in Figure 3-47.

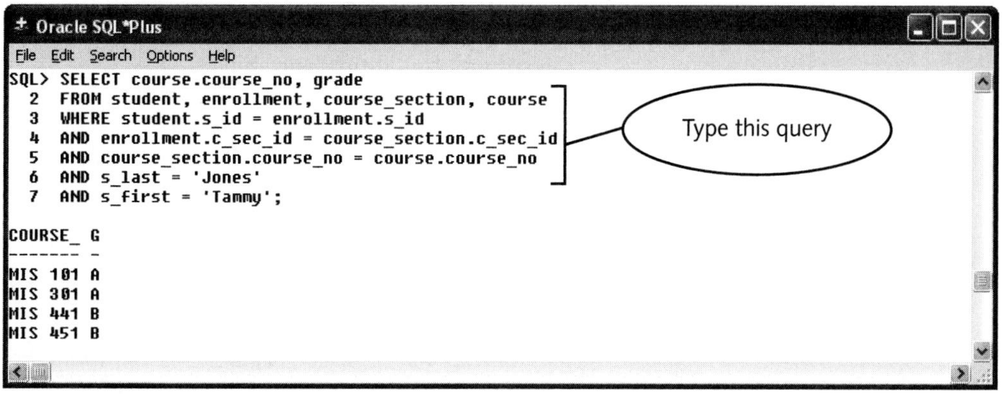

```
± Oracle SQL*Plus                                                    _ □ X
File  Edit  Search  Options  Help
SQL> SELECT course.course_no, grade
  2  FROM student, enrollment, course_section, course
  3  WHERE student.s_id = enrollment.s_id
  4  AND enrollment.c_sec_id = course_section.c_sec_id        Type this query
  5  AND course_section.course_no = course.course_no
  6  AND s_last = 'Jones'
  7  AND s_first = 'Tammy';

COURSE_ G
------- -
MIS 101 A
MIS 301 A
MIS 441 B
MIS 451 B
```

Figure 3-47   Query joining four tables

2. Execute the query. The output shows the values based on joining four tables in a single query.

If you accidentally omit a join condition in a multiple-table query, the output retrieves more rows than you expect. When you omit a join condition, the query creates a **Cartesian product**, whereby every row in one table is joined with every row in the other table. For example, suppose you repeat the query to show each student row, along with each student's advisor (see Figure 3-43), but you omit the join condition. Every row in the STUDENT table (six rows) is joined with every row in the FACULTY table (five rows). The result is 6 times 5 rows, or 30 rows. You create this query next.

To create a Cartesian product by omitting a join condition:

1. Type the query in Figure 3-48.

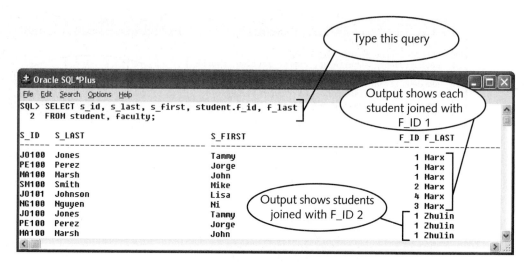

Figure 3-48   Query that creates a Cartesian product (partial output shown)

2. Execute the query. The output appears as shown in Figure 3-48. (Some of the output appears off the screen.)

The query first joins each row in the STUDENT table with the first row in the FACULTY table (F_LAST 'Marx'). Next, each row in the STUDENT table is joined with the second row in the FACULTY table (F_LAST 'Zhulin'). This continues until each STUDENT row is joined with each FACULTY row, for a total of 30 rows returned. When a multiple-table query returns more rows than you expect, look for missing join conditions.

Outer Joins

An inner join returns rows only if values exist in all tables that are joined. If no values exist for a row in one of the joined tables, the inner join does not retrieve the row. For example, suppose you want to retrieve the different locations of the courses included in the COURSE_SECTION table. This query requires joining rows in the LOCATION and COURSE_SECTION tables. However, not every location in the LOCATION table has an associated COURSE_SECTION row, so the query retrieves rows only for locations that have associated COURSE_SECTION rows. Now you retrieve this data using an inner join, and see how the inner join query omits some of the rows.

To retrieve location and course section information using an inner join:

    1. Type the query in Figure 3-49.

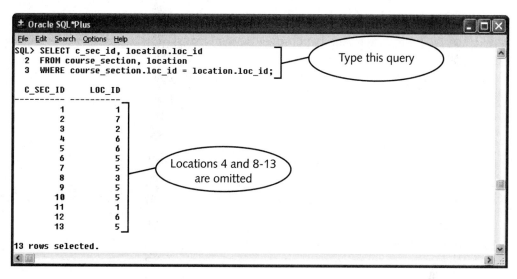

Figure 3-49   Inner join query that omits rows

    2. Execute the query. The output displays 13 rows.

The output shows values for 13 course sections. However, notice that not all the locations from the LOCATION table are included in the output. The query retrieved information only for locations that have rows in the COURSE_SECTION table, and omitted LOC_ID values that did not have rows in the COURSE_SECTION table. To retrieve information for all locations, regardless of whether they have associated course sections or not, you must use an outer join.

An **outer join** returns all rows from one table, which is called the **inner table**. An outer join also retrieves matching rows from a second table, which is called the **outer table**. The query designer specifies which table is the inner table and which table is the outer table. In this case, because you want to retrieve all of the rows in the LOCATION table, you specify LOCATION as the inner table. Because you want to display the course section rows if they exist, you specify COURSE_SECTION as the outer table.

To create an outer join in Oracle 10*g* SQL, you label the outer table in the join condition using the following syntax:

    inner_table.join_column = outer_table.join_column(+)

The outer join operator (+) signals the DBMS to insert a NULL value for the columns in the outer table that do not have matching rows in the inner table. In the following set of steps, you modify the query in Figure 3-49 so that it retrieves all locations, even if those values that do not have associated rows in the COURSE_SECTION table. In this query, the LOCATION table is the inner table, and the COURSE_SECTION table is the outer table.

NOTE

The outer join operator (+) can be placed on either side of the equal sign; the key is that it must be adjacent to the outer table.

To retrieve location and course_section information using an outer join:

1. Type the query in Figure 3-50, which includes the outer join operator (+) in the WHERE clause to specify that the COURSE_SECTION table is the outer table.

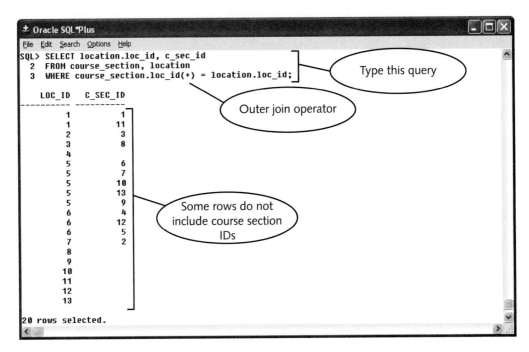

Figure 3-50   Query using an outer join

NOTE

Your output in Figure 3-50 may display the rows in a different order.

2. Execute the query. The output now shows values for every row in the LOCA-TION table, as well as the corresponding COURSE_SECTION rows, when they exist.

Note that the locations that do not have associated course sections scheduled in that location contain NULL values in the C_SEC_ID column. In Oracle 10*g*, NULL (undefined) values appear as blank spaces rather than as the word NULL or as an alternate symbol.

Self-joins

3

Sometimes a relational database table contains a foreign key that references a column in the same table. For example, at Northwoods University each assistant and associate professor is assigned to a full professor who serves as the junior professor's supervisor. The faculty ID of the supervisor for each professor is listed in the F_SUPER column of the FACULTY table. The ID stored in the F_SUPER column can be linked back to the F_ID column in the table to identify the name of the individual's supervisor. In essence, the F_SUPER column is a foreign key in the FACULTY table that references that same table's primary key. To create a query that lists the names of each junior faculty member and the names of their supervisor, you must join the FACULTY table to itself. When you create a query that joins a table to itself, you create a **self-join**.

To create a self-join, you must create a table alias and structure the query as if you are joining the table to a copy of itself. A **table alias** is an alternate name that you assign to the table in the query's FROM clause. The syntax to create a table alias in the FROM clause is:

```
FROM table1 alias1, ...
```

When you create a table alias, you must then use the table alias, rather than the table name, to qualify column names in the SELECT clause and in join conditions. Next, you create the query that lists the names of all junior faculty members and their supervisors. To make the process easier to understand, you create two table aliases—FAC for the faculty version of the table and SUPER for the supervisor version of the same table, as shown in Figure 3-51.

FACULTY (Table name: FAC)

F_ID	F_LAST	F_FIRST	F_MI	LOC_ID	F_PHONE	F_RANK	F_SUPER
1	Marx	Teresa	J	9	4075921695	Associate	4
2	Zhulin	Mark	M	10	4073875682	Full	
3	Langley	Colin	A	12	4075928719	Assistant	4
4	Brown	Jonnel	D	11	4078101155	Full	
5	Sealy	James	L	13	4079817153	Associate	2

SUPERVISOR
(Table name: SUPER)

Figure 3-51   Tables referenced in self-join query

The query names the table that displays the original faculty member information as FAC and the table that displays supervisor information as SUPER. Because the F_SUPER column represents the faculty ID of the supervisor, the join condition is:

```
WHERE fac.f_super = super.f_id
```

Next, you type the query to list junior professors and their supervisor's names by creating a self-join on the FACULTY table.

To create a self-join on the FACULTY table:

1. Type the query in Figure 3-52. The SELECT clause specifies alternate column headings for the output columns to clarify the output.

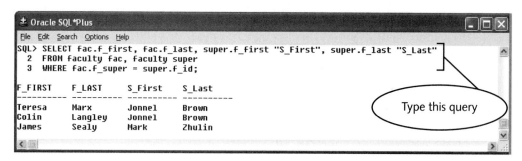

Figure 3-52　Self-join of the FACULTY table

2. Press **Enter** to execute the query. The output shows the names of all faculty members that have supervisors.

CREATING NESTED QUERIES

Sometimes you need to create a query that is based on the results of another query. For example, suppose you need to retrieve the first and last names of all students who have the same S_CLASS value (freshman, sophomore, and so forth) as student Jorge Perez. You first create a query that retrieves Jorge's S_CLASS value. Then, you create a second query to retrieve the names of the students with the same S_CLASS value. Rather than retrieving this data using two separate queries, you can retrieve the same data using a single nested query.

A **nested query** consists of a main query and one or more subqueries. The **main query** is the first query that appears in the SELECT command. A **subquery** retrieves values that the main query's search condition must match. In a nested query, the DBMS executes the subquery first, and the main query second. You create nested queries to retrieve intermediate results that are then used within the main query's search conditions. The following sections not only describe how to create nested queries containing subqueries that return a single value, but also how to create queries containing subqueries that might return multiple values. You also learn how to create nested queries with multiple subqueries, and subqueries that are also nested queries.

Creating Nested Queries with Subqueries that Return a Single Value

3

Figure 3-53 shows the general syntax for creating a nested query.

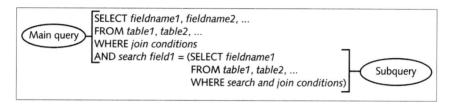

Figure 3-53   Syntax for nested query

In this syntax, the data values that the *subquery* retrieves specify the search condition *search_expression* value for the main query. The search column in the main query (*search_field1*) must reference the same column as the SELECT column (*fieldname1*) in the subquery. When the DBMS executes a nested query, it first evaluates the subquery. Then, the DBMS substitutes the subquery output into the main query, and executes the main query. If the subquery retrieves no rows or multiple rows, an error message appears.

NOTE

You can also use subqueries in search conditions in UPDATE and DELETE commands.

In the next set of steps, you create a nested query that retrieves the first and last names of all students who have the same S_CLASS value as student Jorge Perez. The subquery retrieves a single value, which is Jorge's S_CLASS value. The main query retrieves the student first and last names.

To create a nested query:

1. Type the query in Figure 3-54 to retrieve the names of all students who have the same S_CLASS value as student Jorge Perez. To make the query easier to read and understand, indent the subquery as shown. Note that the subquery must retrieve the S_CLASS column, and the main query's search condition must also specify the S_CLASS column as the search column.

2. Execute the query. The output shows the students who have the S_CLASS value 'SR', which includes Tammy Jones and Jorge Perez.

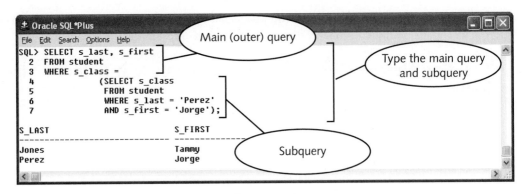

Figure 3-54   Creating a nested query that returns a single value

TIP

To debug a nested query, run the subquery separately to ensure that it is retrieving the correct results before you combine it with the main query.

Creating Subqueries that Return Multiple Values

Because the subquery in Figure 3-54 returns a single value, which is Jorge Perez's S_CLASS value, the main query's search condition uses an equal to comparison operator (=). What if the subquery might return multiple results, each of which satisfies the search condition? For example, suppose you want to retrieve the names of all students who have ever been enrolled in the same course section as Jorge Perez. The subquery retrieves all C_SEC_ID values in the ENROLLMENT table for Jorge, which probably includes multiple values. The main query then retrieves the names of all students who have enrolled in any of these same course sections. To structure this query, you must use the IN comparison operator instead of the equal to comparison operator (=). (In Lesson B, you learned how to use the IN comparison operator to create a search condition to match a set of values.) You can also use the NOT IN operator to specify that the main query should retrieve all rows except those that satisfy the search condition.

Next, you create a nested query in which the subquery retrieves multiple values. The subquery retrieves the C_SEC_ID value of all courses in which Jorge Perez has enrolled, and the main query uses the IN operator to match the names of all students who have ever been enrolled in one of these courses. You use the DISTINCT qualifier in the main query's SELECT clause to suppress duplicate names.

To write a nested query with a subquery that retrieves multiple values:

1. Type the query in Figure 3-55 to retrieve the names of all students who have enrolled in the same course sections as Jorge Perez. Note that the main query's search condition uses the IN comparison operator, because the subquery returns multiple values.

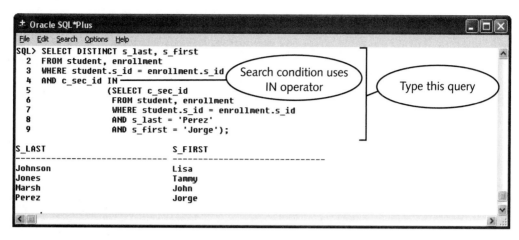

Figure 3-55   Creating a nested query whose subquery returns multiple values

2. Execute the query. The output values appear as shown.

Using Multiple Subqueries Within a Nested Query

You can specify multiple subqueries within a single nested query by using the AND and OR operators to join the search conditions associated with the subqueries. For example, suppose you want to retrieve the names of all students who have the same S_CLASS value as Jorge Perez and have also been enrolled in a course section with him. You use a query that joins the two search conditions using the AND operator, and each search condition is specified using a separate subquery.

To create a nested query that uses multiple subqueries:

1. Type the query in Figure 3-56 to retrieve the names of all students who have the same S_CLASS value as Jorge Perez and who have also been enrolled in a course section with him.

2. Execute the query. The output shows that only one student other than Jorge himself, Tammy Jones, satisfies both these conditions.

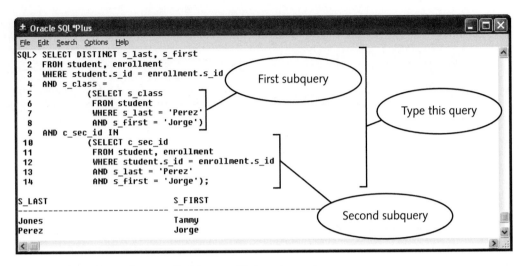

Figure 3-56   Nested query that has multiple subqueries

Creating Nested Subqueries

A **nested subquery** is a subquery that contains a second subquery that specifies its search expression. Now you use a nested subquery to create a query to retrieve the names of students who have taken courses with Jorge Perez in the CR building. The innermost subquery retrieves the course section ID values for all course sections located in the CR building. This subquery's main query retrieves the course sections from this subset taken by Jorge Perez. The outermost main query retrieves the names of all students who enrolled in the course sections that satisfy both of these conditions.

To create a query with a nested subquery:

1. Type the query in Figure 3-57.

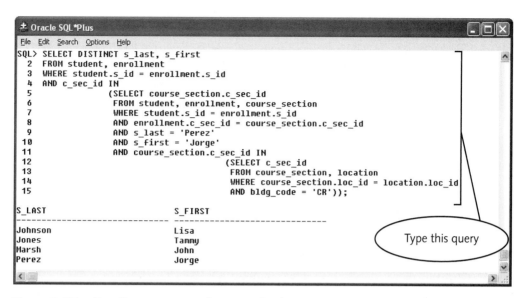

Figure 3-57   Creating a query with a nested subquery

2. Execute the query. Note that the output excludes any student that may have taken the same course section as Jorge Perez, but the course was located in a different building.

You can create subqueries that nest to multiple levels. However, as a general rule, nested queries or queries with nested subqueries execute slower than queries that join multiple tables, so you should probably not use nested queries unless you cannot retrieve the desired result using a nonnested query.

USING SET OPERATORS TO COMBINE QUERY RESULTS

Sometimes, you need to combine the results of two separate queries in ways that are not specified by foreign key relationships. For example, you might need to create a telephone directory that shows the names and telephone numbers of all students and faculty members at Northwoods University. Or, you might want to retrieve the first and last names of faculty members who have offices in the BUS building and have taught a class in BUS. These rows do not have any foreign key relationships, so the only way to retrieve them is to create multiple queries. Alternately, you can use common set operators to combine the results of separate queries into a single result. Table 3-12 describes the Oracle 10*g* SQL set operators, and the following sections describe how to use them.

Set Operator	Description
UNION	Returns all rows from both queries, and suppresses duplicate rows
UNION ALL	Returns all rows from both queries, and displays duplicate rows
INTERSECT	Returns only rows returned by both queries
MINUS	Returns the rows returned by the first query minus the matching rows returned by the second query

Table 3-12   Oracle 10*g* SQL set operators

UNION and UNION ALL

A query that uses the UNION set operator joins the output of two unrelated queries into a single output result. The general syntax of a query that uses the UNION operator is:

```
query1 UNION query2;
```

In this syntax, *query1* represents the first query, and *query2* represents the second query. Both queries must have the same number of display columns in their SELECT clauses, and each display column in the first query must have the same data type as the corresponding column in the second query. For example, if the display columns returned by *query1* are a NUMBER data column and then a VARCHAR2 data column, then the display columns returned by *query2* must also be a NUMBER data column followed by a VARCHAR2 data column.

To create a telephone directory of every student and faculty member at Northwoods University, you need to create a query to list the last name, first name, and telephone number of every student and every faculty member in the Northwoods University database. You can create two separate queries to retrieve these rows, but you cannot retrieve the results using a single query. (The foreign key student/faculty advising relationship that joins these tables is not relevant for this query, because you want to display all students and all faculty members, not just students and their faculty advisors.) Because a single query cannot return data from two unrelated queries, you must use a UNION set operator to join the query outputs. Next, you create a query that uses the UNION set operator.

To create a query that uses the UNION set operator:

    1. Type the query in Figure 3-58.

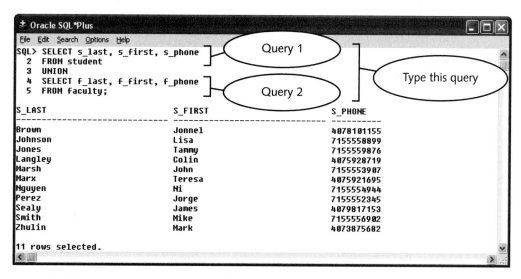

Figure 3-58   Query using the set UNION operator

2. Execute the query. The output shows the student and faculty names and telephone numbers in a single list. The results are sorted by the first display column.

In the query output in Figure 3-58, the default output column titles are the column names in the first query's SELECT statement. The output results are sorted based on the first column of the first SELECT statement.

If one row in the first query exactly matches a row in the second query, the UNION output suppresses duplicates, and shows the duplicate row only once. To display duplicate rows, you must use the UNION ALL operator. For illustrative purposes, suppose you need to create a query that displays the name of every course that was taken by freshmen, sophomores, and juniors (i.e., not seniors) and a list of all courses offered in term 6. Because some of the students may have taken more than one course in a term, this query may retrieve duplicate names. If you use the UNION operator to combine the results of these two queries, the duplicate rows are suppressed. If you use the UNION ALL operator, then duplicate rows appear. In the next set of steps, you create a query with the UNION operator and then with the UNION ALL operator, and compare the results.

To create UNION and UNION ALL queries:

1. Type and execute the query in Figure 3-59, which uses the UNION operator. Note that duplicate rows do not appear.

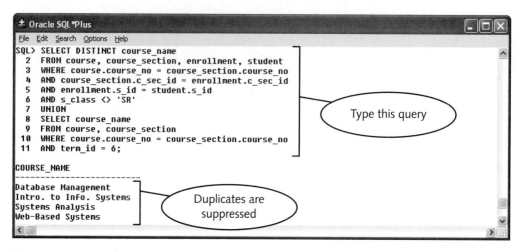

Figure 3-59   Duplicate course names suppressed by UNION set operator

2. Now type and execute the query in Figure 3-60, which uses the UNION ALL operator. This time the duplicate rows appear, and show that three courses satisfy the search requirements in both queries.

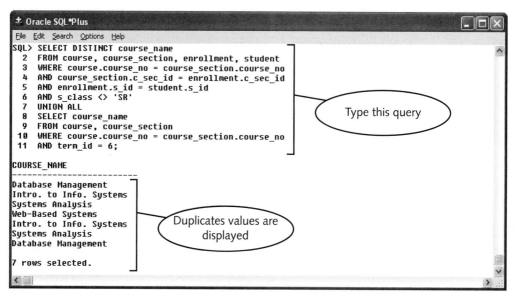

Figure 3-60   Using the UNION ALL set operator

INTERSECT

Some queries require an output that finds the intersection, or matching rows, in two queries. Like a UNION, an INTERSECT query requires that both queries have the same number of display columns in the SELECT statement and that each column in the first query has the same data type as the corresponding column in the second query. An INTERSECT query automatically suppresses duplicate rows.

In the previous section, you created a UNION query that displayed the names of courses taken only by students who were not seniors, in addition to courses that were offered in term 6. Suppose you want to create a query that retrieves only the courses that satisfy both of these requirements. One approach is to re-create the query using the INTERSECT set operator. You do this next.

To create a query that uses the INTERSECT set operator:

1. Type the query in Figure 3-61.

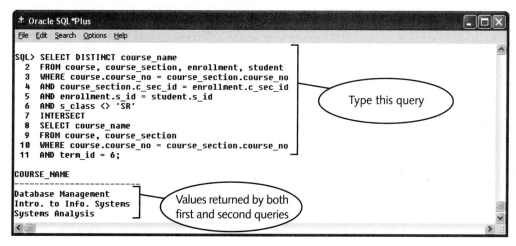

Figure 3-61   Query using the INTERSECT set operator

2. Execute the query. The query output shows that three courses satisfy both search conditions.

MINUS

The MINUS operator allows you to find the difference between two unrelated query result lists. As with the UNION and INTERSECT operators, the MINUS operator requires that both queries have the same number of display columns in the SELECT statement, and that each column in the first query has the same data type as the corresponding column in the second query. And as with the other set operators, the MINUS operator automatically suppresses duplicate rows.

Suppose you want to retrieve the courses that were taken by freshmen, sophomores, and juniors, but were not offered in term 6. Because the first portion of the previous query determined the courses that were not taken by seniors, and the latter portion identified the courses offered in term 6, you can simply modify the previous query by using the MINUS set operator. The results of the modified query display courses taken by freshmen, sophomores, and juniors with any courses not offered in term 6 removed from the results.

To create a query using the MINUS operator:

1. Type the query in Figure 3-62, which uses the MINUS operator to display the difference between two queries.

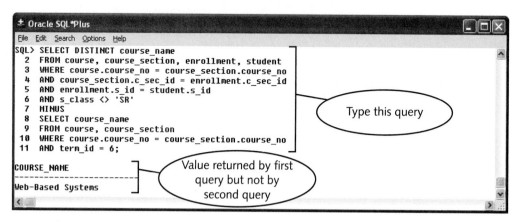

Figure 3-62   Query using the MINUS set operator

2. Execute the query. The query output shows only the course that meets the conditions specified by the first portion of the query, but did not meet the condition established by the second portion of the query.

CREATING AND USING DATABASE VIEWS

A database view presents data in a different format than the one in which the DBMS stores the data in the underlying database tables. A view is similar to storing the result of a query in the database. You create a view using a SQL query called the **source query**. The source query can specify a subset of a single table's columns or rows, or the query can join multiple tables. You can then retrieve data from the view the same way you retrieve data from a database table. You can use special views called **updatable views**, also called simple views, to insert, update, and delete data in the underlying database tables.

A view derives its data from database tables called **source tables**, or underlying base tables. When users insert, update, or delete data values in a view's source tables, the view reflects the updates as well. If a DBA alters the structure of a view's source tables, or if a DBA drops a view's source table, then the view becomes invalid and can no longer be used.

Once you create a view, you can use the view to retrieve rows just as with a table. This saves you from having to reenter complex query commands. For example, you can create a view to list the item description, size, color, price, and quantity ordered for all items associated with order ID (O_ID) 1 in the Clearwater Traders database. This view is based on a query that joins four tables: ORDERS, ORDER_LINE, INVENTORY, and ITEM. Once you create the view, you can easily query it, and you do not need to type the join or search conditions to display the rows.

DBAs use views to enforce database security by allowing certain users to view only selected table columns or rows. A specific user might be given privileges to view and edit data in a view based on a table, but she or he may not be given privileges to manipulate all of the data in the table. For example, suppose a data entry person is needed to insert, update, and view data in the FACULTY table. For security reasons, the DBA does not want this person to have access to the F_PIN column. The DBA creates a view based on the FACULTY table that contains all table columns except F_PIN. The following sections describe how to create and query views.

Creating Views

The syntax to create a view is:

```
CREATE VIEW view_name
AS source_query;
```

In this syntax, *view_name* must follow the Oracle naming standard. *View_name* cannot already exist in the user's database schema. If there is a possibility that you have already created a view using a specific name, you can use the following command to create or replace the existing view:

```
CREATE OR REPLACE VIEW view_name
AS source_query;
```

After you create a view, you can grant or revoke privileges to other users to perform operations such as SELECT, INSERT, or UPDATE. When you replace an existing view, you replace only the view column definitions. All of the existing object privileges that you granted on the view remain intact.

Source_query can retrieve columns from a single table or can join multiple tables. *Source_query*'s SELECT clause can contain arithmetic and single-row functions, as long as you create an alias for each column that performs a calculation or uses a function. *Source_query* can also contain a search condition.

Recall that you can create updatable views that users can use to insert, update, or delete values in the view's source tables. When you create an updatable view, the SQL query has specific restrictions: The SELECT clause can contain only column names, and it cannot specify arithmetic calculations or single-row functions. The query also cannot contain the ORDER BY, DISTINCT, or GROUP BY clauses, group functions, or set operators. And, the query's search condition cannot contain a nested query. Next, you create an updatable view named

FACULTY_VIEW, which is based on columns in the FACULTY table. This view contains all of the FACULTY columns except F_PIN and F_IMAGE.

To create an updatable view:

1. Type the command in Figure 3-63 to create the FACULTY_VIEW. Because the query does not contain any of the restricted items, this view is an update-able view.

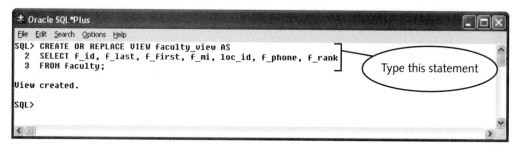

Figure 3-63   Creating an updateable view

2. Execute the command. The "View created." confirmation message confirms that the DBMS created the view.

Next you use a complex source query to create a view that contains a listing of the courses being taught by faculty members for term 6. This source query contains columns from multiple tables so it restricts the DML operations that can be performed. The view lists the faculty member's name and the name of the course he or she is teaching. In addition, the data is sorted in alphabetical order by the last names of the faculty members.

To create a complex view that cannot be updated:

1. Type the command in Figure 3-64 to create the view that cannot be updated.

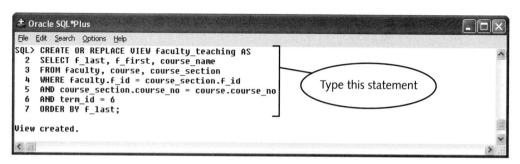

Figure 3-64   Creating a view that cannot be updated

2. Execute the command. The message "View created." confirms that the DBMS successfully created the view.

Executing Action Queries Using Views

After you create an updatable view, you can use the view to execute action queries that insert, update, or delete data in the underlying source tables. In the following steps, you use FACULTY_VIEW to insert a row into the FACULTY source table.

To insert a row using the FACULTY_VIEW:

1. Type the following action query to insert a new row into FACULTY_VIEW:

   ```
   INSERT INTO faculty_view VALUES
   (6, 'May', 'Lisa', 'I', 11, '7155552508', 'INST');
   ```

2. Press **Enter** to execute the query. The message "1 row created." confirms that the DBMS successfully inserted the row. Although you inserted the row using the view, the DBMS actually inserted the row into the FACULTY table.

3. Type **SELECT * FROM faculty_view;** to determine whether the new faculty member is included in the table.

4. Type **DELETE FROM faculty_view WHERE f_last = 'MAY';** to delete the new faculty member.

You can also execute update action queries and delete action queries using the view just as with a database table, provided that you do not violate any constraints that exist for the underlying database table, and provided that you have sufficient object privileges for updating or deleting rows in the view.

Retrieving Rows from Views

You can query a view using a SELECT statement, just as with a database table, and use the view in complex queries that involve join operations and subqueries. Next, you create queries using your database views. You create a query that joins FACULTY_VIEW with the LOCATION table to list the names of each faculty member, along with the building code and room number of the faculty member's office.

To create queries using database views:

1. Type the query in Figure 3-65 to retrieve the faculty information by joining FACULTY_VIEW with the LOCATION table.

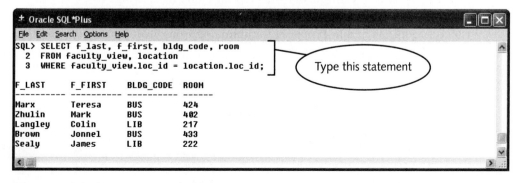

Figure 3-65   Query using a database view

2. Execute the query. Note that the query to retrieve data from the view is much simpler than the query to retrieve the data from the underlying database tables.

Removing Views

You use the DROP VIEW command to remove a view from your user schema. This command has the following syntax:

```
DROP VIEW view_name;
```

Recall that the DBMS bases a view on a source query. When you drop a view, you do not delete the data values that appear in the view—you drop only the view definition. Now you drop the FACULTY_VIEW view that you created earlier.

To drop FACULTY_VIEW:

1. Type and execute the following command at the SQL prompt:

```
DROP VIEW faculty_view;
```

The "View dropped." confirmation message indicates that the DBMS successfully dropped the view.

2. Close SQL*Plus. Switch to Notepad, save **3CQueries.sql**, and then close Notepad.

SUMMARY

❑ To join multiple tables in a SELECT query, you must list in the FROM clause every table that contains a display column, search column, or join column. You include a join condition to specify every link between those tables in the WHERE clause.

❑ If two tables only have one commonly defined column, the NATURAL JOIN keyword can be used to join the tables in the FROM clause.

❑ If you display a column in a multiple-table query that exists in more than one of the tables, you must qualify the column name by writing the table name, followed by a period, and then the column name.

❑ If you accidentally omit a join condition in a multiple-table query, the result is a Cartesian product, which joins every row in one table with every row in the other table.

❑ An inner join retrieves all matching columns in two or more tables.

❑ An outer join retrieves all rows in the first, or inner, table, even if no corresponding values exist in the second, or outer, table.

❐ A self-join joins a table to itself. To create a self-join, you must create a table alias and structure the query as if you are joining the table to a copy of itself.

❐ A nested query consists of a main query and a subquery that retrieves an intermediate result that serves as the main query's search expression. You can nest subqueries to multiple levels.

❐ A query that uses the UNION set operator combines the results of two unrelated queries and suppresses duplicate rows. A query that uses the UNION ALL set operator combines the results of two unrelated queries and displays duplicate rows. Queries that use either one of the UNION set operators must have the same number of display columns in the SELECT clause, and each column in the first query must have the same data type as the corresponding column in the second query.

❐ A query that uses the INTERSECT set operator returns the intersection, or matching rows, in two unrelated queries.

❐ A query that uses the MINUS set operator retrieves the difference between two unrelated query result outputs.

❐ A database view is a logical table that you create using a query. You can use a view to enforce security measures within a table or to simplify the process of displaying data that a complex query retrieves.

❐ An updatable view is a view that allows users to insert, update, or delete values in the view's source tables. When you create an updatable view, the SQL query's SELECT clause can contain only column names, and cannot specify arithmetic calculations or single-row functions. The query also cannot contain the ORDER BY, DISTINCT, or GROUP BY clauses, group functions, set operators, or a nested query.

REVIEW QUESTIONS

1. If you are creating a query that joins four tables, how many join conditions should you create?

2. The _____ operator is used to return multiple values to an outer query.

3. In a query that includes the equal to (=) operator, to return values to an outer query, duplicate values are suppressed. True or False?

4. How are the underlying tables affected by removing a view?

5. When should you create a subquery?

6. The _____ set operator does *not* suppress duplicate values.

7. The NATURAL JOIN keywords can be used in the FROM clause of a query to join tables that have a single commonly named and defined column. True or False?

8. A(n) _____ view does not have an ORDER BY or GROUP BY clause and does not include any computed values.

9. A(n) _____ join requires the use of a table alias to mimic the referencing of two different tables, when in reality, they are the same table.

10. An outer join operator is used to indicate that unmatched rows should be included in the query results. True or False?

MULTIPLE CHOICE

1. Which of the following types of joins include a row from table A if it has an equivalent value in the common column in table B?

    a. inner join

    b. natural join

    c. equijoin

    d. all of the above

2. Table A contains eight rows and table B contains four rows; the output of their Cartesian product includes _____ rows.

    a. 32

    b. 12

    c. 8

    d. 4

3. Which of the following is a set operator?

    a. MINUS

    b. INTERSEPT

    c. SUM

    d. all of the above

4. Which line in the following query causes an error to occur?
```
SELECT f_last, f_first
FROM faculty
WHERE f_id =
      (SELECT f_id, super_id
        FROM faculty
        WHERE rank = 'Assistant');
```

    a. WHERE f_id =

    b. (SELECT f_id, super_id

    c. WHERE rank = 'Assistant');

    d. none of the above

5. The following statement is an example of what type of query?
```
SELECT s_first, s_last
FROM student
WHERE s_id IN
      (SELECT s_id
       FROM enrollment
       WHERE grade = 'A');
```
a. inner query

b. outer query

c. self-join query

d. nested query

6. The following statement is an example of what type of query?
```
SELECT s_first, s_last, grade, course_sec_id
FROM student s, enrollment e
WHERE s_s_id = e_sid;
```
a. inner query

b. outer query

c. self-join query

d. nested query

7. Which of the following commands removes the view named FAC from the database?

a. `DROP Fac;`

b. `DROP VIEW Fac;`

c. `DELETE VIEW Fac;`

d. `DELETE Fac;`

8. Which of the following is a valid statement?

a. A view is a logical table that prevents a user from accessing the underlying tables.

b. An updateable view can be used to delete data from its underlying table.

c. A view based on only one table can be used to update calculated columns.

d. all of the above

9. If you create an inner join between table A that contains six rows and table B that contains four rows, what is the maximum number of rows that can be included in the results?

a. 0

b. 4

c. 6

d. 24

10. Which of the following operators is used to indicate the inner table in an outer join query?

a. (*)

b. (+)

c. (-)

d. (^)

PROBLEM-SOLVING CASES

For all cases, use Notepad or another text editor to write a script using the specified filename. Always use the search condition text exactly as the case specifies. Place the queries in the order listed, and save the script files in your Chapter03\Cases folder on your Data Disk. If you haven't done so already, run the Ch3Clearwater.sql script in the Chapter03 folder on your Data Disk to create and populate the case study database tables. All of the following cases are based on the Clearwater Traders database.

1. List the name of each item included on order #1.

2. Determine which customer placed order #1 and the total amount of the order.

3. List the customers who have ordered Boy's Surf Shorts and identify how many were purchased.

4. Identify which shipments have not yet been received by Clearwater Traders and the items on each of those shipments.

5. Create a view named ITEM that displays the name of each item in inventory, its retail price, and the quantity on hand.

6. Using the ITEM view, determine the total amount of investment Clearwater Traders currently has in its inventory. (*Hint:* You should only receive one value from this query.)

7. Create a query that displays each shipment currently listed in the SHIPMENT table with its expected shipment date. For the shipments that have already been received, include actual received dates in the results.

8. Determine which customers have purchased the same items as customer #3.

9. Determine the total amount received for all sales of outdoor gear.

10. Determine the total amount of sales generated by the company's Web site.

4

INTRODUCTION TO PL/SQL

◀ LESSON A ▶

Objectives
After completing this lesson, you should be able to:
- ♦ Describe the fundamentals of the PL/SQL programming language
- ♦ Write and execute PL/SQL programs in SQL*Plus
- ♦ Execute PL/SQL data type conversion functions
- ♦ Display output through PL/SQL programs
- ♦ Manipulate character strings in PL/SQL programs
- ♦ Debug PL/SQL programs

You have learned how to create Oracle database tables and write queries to insert, update, delete, and view records. However, most database users don't use SQL commands to interact with a database. Database users use database applications with graphical user interfaces in which they can enter inputs into text boxes, and then point to and click buttons to execute programs that perform database functions. Oracle10*g* provides several different utilities for developing user applications and for writing programs that help DBAs manage the database. In this book, you use Forms Builder to create database applications that display data on the screen and allow users to insert, update, and delete data values. You use Reports Builder to create database reports, which provide a summary view of data at a specific point in time. To use these utilities effectively, database developers need to understand PL/SQL, which is a procedural programming language that Oracle utilities use to manipulate database data. This chapter presents an introduction to the PL/SQL programming language. It assumes you have already used a programming or scripting language.

A **procedural programming language** is a programming language that uses detailed, sequential instructions to process data. A PL/SQL program combines SQL queries with procedural commands for tasks such as manipulating variable values, evaluating IF/THEN decision control structures, and creating loop structures that repeat instructions multiple times until the loop reaches an exit condition. Although other procedural programming languages can contain SQL commands and interact with an Oracle 10*g* database, Oracle Corporation designed PL/SQL expressly for interacting with the Oracle 10*g* database and previous versions of Oracle databases. As a result, PL/SQL interfaces directly with the Oracle 10*g* database and is available within Oracle development environments. When you use other procedural programming languages to interact with a database, you pass the SQL queries as text strings, and it becomes tricky to create complex queries that contain values that are represented by program variables.

Another advantage of using PL/SQL to create programs that interface with the Oracle 10*g* database is that database developers can store PL/SQL programs directly in the database, which makes the programs available to all database users. This feature also makes it easier to manage database applications.

Fundamentals of PL/SQL

PL/SQL is a full-featured programming language. You write PL/SQL programs using a text editor, then execute the program commands in Oracle 10*g* utilities such as SQL*Plus or Forms Builder. PL/SQL is an interpreted language, which means that a program called the PL/SQL interpreter checks each program command for syntax errors, translates each command into machine language, and then executes each program command, one command at a time. PL/SQL commands are not case sensitive, except for character strings, which you must enclose in single quotation marks. The PL/SQL interpreter ignores blank spaces and line breaks. A semicolon (;) marks the end of each PL/SQL command.

As with other programming languages, the PL/SQL programming language contains reserved words that you use within command instructions and built-in functions to perform common tasks such as manipulating numbers or character strings. You declare variables to reference data using predefined data types, which are similar to the data types you used to define Oracle 10*g* database columns. To make PL/SQL programs easier to understand, you generally structure parts of commands in uppercase letters, and other parts of commands in lowercase letters. Table 4-1 summarizes capitalization styles in PL/SQL commands.

Item Type	Capitalization	Example
Reserved word	Uppercase	BEGIN, DECLARE
Built-in function	Uppercase	COUNT, TO_DATE
Predefined data type	Uppercase	VARCHAR2, NUMBER
SQL command	Uppercase	SELECT, INSERT
Database object	Lowercase	student, f_id
Variable name	Lowercase	current_s_id, current_f_last

4

Table 4-1   PL/SQL command capitalization styles

To introduce the PL/SQL programming language, the following sections discuss PL/SQL variables and data types and the structure of PL/SQL program blocks. They also address how to create comment lines to document programs, how to use arithmetic operators and assignment statements, and how to enable interactive output from PL/SQL programs in the SQL*Plus environment.

PL/SQL Variables and Data Types

Programs use variables to store and reference values such as numbers, character strings, dates, and other data values. In PL/SQL, variable names must follow the Oracle naming standard for naming database objects. Although PL/SQL variables can be reserved SQL words, such as NUMBER or VALUES, it is not a good practice to use these names because they can cause unpredictable results. PL/SQL variable names can also use the same names as such database objects as tables and columns, but again, to avoid unpredictable results, you should not give PL/SQL variables the same names as database objects. PL/SQL variable names should be as descriptive as possible; for example, you should create a variable name such as current_s_id, rather than a variable name such as x. Oracle Developers normally type PL/SQL variable names in lowercase letters, and use underscores to separate words within phrases.

PL/SQL is a **strongly typed language**, which means that you must write a command that explicitly declares each variable and specifies its data type before you use the variable. When you declare a variable, the program creates an area in your workstation's main memory that stores information about the variable and associates the variable name with that memory location. With a strongly typed language such as PL/SQL, you can write commands that assign values to variables and compare variable values only for variables with the same data type. You use the following syntax to declare a PL/SQL program variable:

```
variable_name data_type_declaration;
```

When you declare a variable, the variable's default value is always NULL. PL/SQL supports scalar, composite, reference, and large binary object (LOB) variables. The following sections discuss these different variable types.

Scalar Variables

Scalar variables reference a single value, such as a number, date, or character string. Scalar variables can have data types that correspond to the Oracle 10*g* database data types, and you usually use scalar variables to reference data values that programs retrieve from the database. Table 4-2 summarizes the PL/SQL scalar database data types.

Data Type	Description	Sample Declaration
VARCHAR2	Variable-length character string	`current_s_last VARCHAR2(30);`
CHAR	Fixed-length character string	`student_gender CHAR(1);`
DATE	Date and time	`todays_date DATE;`
INTERVAL	Time interval	`curr_time_enrolled INTERVAL YEAR TO MONTH;` `curr_elapsed_time INTERVAL DAY TO SECOND;`
NUMBER	Floating-point, fixed-point, or integer number	`current_price NUMBER(5,2);`

Table 4-2   Scalar database data types

The PL/SQL VARCHAR2 data type references variable-length character data up to a maximum length of 32,767 characters. This is different from its database counterpart, which can hold a maximum of 4000 characters in an Oracle 10*g* database column. When you declare a VARCHAR2 variable in PL/SQL, you must specify the maximum number of characters that the variable can reference.

The PL/SQL CHAR data type references fixed-length character strings, also up to a maximum length of 32,767 characters. When you declare a CHAR variable, if you do not specify the maximum column width, the default width is 1. When a command assigns a data value to a CHAR variable and a data value has fewer characters than the maximum width specification, the system pads the rest of the width with blank spaces so the value completely fills the maximum width. For example, if you declare a CHAR variable with a maximum width of 20 and then enter the data value John into the column, 16 blank spaces are added to the right side of the data to extend its width to 20 spaces. If the assigned value is larger than the maximum column width, an error occurs.

The PL/SQL DATE data type is the same as its Oracle 10*g* database counterpart and stores both date and time values. The INTERVAL data type stores time interval values. The command to declare an INTERVAL variable must specify whether the variable references INTERVAL YEAR TO MONTH or INTERVAL DAY TO SECOND values, as shown in the code examples in Table 4-2.

The PL/SQL NUMBER data type is identical to the Oracle 10*g* database NUMBER data type. You declare a NUMBER variable using the syntax *variable_name* NUMBER [(*precision*, [*scale*])]. As with database field declarations, the precision specifies the total length of the number, including decimal places, and the scale specifies the number of digits to the right of the decimal point. When you declare a NUMBER variable, you specify only the precision for integer values and both the precision and scale for fixed-point values. For floating-point values, you omit both the precision and scale specifications.

PL/SQL also has general scalar data types that do not correspond to database data types. Table 4-3 summarizes the PL/SQL general scalar data types.

4

Data Type	Description	Sample Declaration
Integer number subtypes (BINARY_INTEGER, INTEGER, INT, SMALLINT)	Integer	`counter BINARY_INTEGER;`
Decimal number subtypes (DEC, DECIMAL, DOUBLE PRECISION, NUMERIC, REAL)	Numeric value with varying precision and scale	`student_gpa REAL;`
BOOLEAN	True/False value	`order_flag BOOLEAN;`

Table 4-3   General scalar data types

The integer number subtypes reference integer values. The subtypes differ with regard to the sizes of the values that each can reference and how the system internally represents the values. You usually use the BINARY_INTEGER data type for loop counters. Oracle10g stores BINARY_INTEGER data values internally in a binary format, which takes slightly less storage space than the NUMBER data type, so the system can perform calculations more quickly on BINARY_INTEGER data values than on integer NUMBER values.

You can use the decimal number subtypes with varying precision and scale values to reference different types of numeric values. PL/SQL supports these data types to provide ANSI compatibility and compatibility with other database systems such as IBM DB/2.

The BOOLEAN data type references values that are TRUE, FALSE, or NULL. When you declare a BOOLEAN variable, it has a value of NULL until a command assigns to the variable a value of TRUE or FALSE.

Composite Variables

A **data structure** is a data object made up of multiple individual data elements. A **composite variable** references a data structure that contains multiple scalar variables, such as a record or a table. In PL/SQL, the composite variable data types include RECORD, which specifies a structure that contains multiple scalar values, and is similar to a table record; TABLE, which specifies a tabular structure with multiple columns and rows; and VARRAY, which specifies a variable-sized array. A variable-sized array is a tabular structure that can expand or contract based on the data values it contains.

Reference Variables

Reference variables directly reference a specific database column or row and assume the data type of the associated column or row. Table 4-4 summarizes the reference data types.

Reference Data Type	Description	Sample Declaration
%TYPE	Assumes the data type of a database field	`cust_address` `customer.c_address%TYPE;`
%ROWTYPE	Assumes the data type of a database record	`order_row orders%ROWTYPE;`

Table 4-4   Reference data types

The %TYPE reference data type specifies a variable that references a single database field. The syntax for a %TYPE data declaration is *variable_name tablename.fieldname%TYPE;*. For example, you would use the following command to declare a variable named current_f_last that references values retrieved from the F_LAST column in the Northwoods University FACULTY table: `current_f_last FACULTY.F_LAST%TYPE;`. The current_f_last variable assumes a data type of VARCHAR2(30), because this is the data type of the F_LAST column in the FACULTY table.

The %ROWTYPE reference data type creates composite variables that reference an entire data record. The general format for a %ROWTYPE data declaration is *variable_name tablename%ROWTYPE;*. The following command declares a variable that references an entire row of data in the FACULTY table: `faculty_row FACULTY%ROWTYPE;`. This variable references all of the columns in the FACULTY table, and each column has the same data type as its associated database column.

LOB Data Types

PL/SQL **LOB data types** declare variables that reference binary data objects, such as images or sounds. The LOB data types can store either binary or character data up to four gigabytes in size. LOB values in PL/SQL programs must be manipulated using a special set of programs, called the DBMS_LOB package.

PL/SQL Program Blocks

Figure 4-1 shows the general structure of a PL/SQL program block.

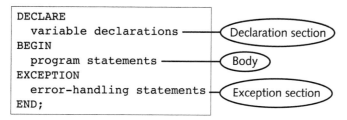

Figure 4-1   Structure of a PL/SQL program block

The program's **declaration section** begins with the reserved word DECLARE, and contains the commands that declare the program's variables. You can declare multiple variables in the declaration section and separate each individual variable declaration command with a semicolon (**;**).

The **execution section** of a PL/SQL block starts with the BEGIN reserved word and consists of program statements, such as assignment statements, conditional structures, and looping structures. The **exception section** begins with the EXCEPTION reserved word, and contains program statements for error handling. PL/SQL program blocks end with the END reserved word.

NOTE   Recall from earlier in the chapter that a semicolon (**;**) marks the end of each PL/SQL command. You also terminate program blocks with a semicolon placed after the END reserved word. You do not terminate keywords that mark the beginning of program block sections, DECLARE, BEGIN, and EXCEPTION, with semicolons.

In a PL/SQL program block, the DECLARE and EXCEPTION sections are optional. If there are no variables to declare, you can omit the DECLARE section and start the program with the BEGIN command. If there are no error-handling statements, then you can omit the EXCEPTION section.

The PL/SQL program includes commands such as comment statements, commands to perform arithmetic operations, assignment statements that assign values to variables, conditional (IF/THEN) structures to perform decisions, looping structures to repeat instructions multiple times, SQL commands that retrieve database data, and other types of commands. The following sections describe how to create comment statements, use arithmetic operators, and create assignment statements in the PL/SQL program body.

Comment Statements

Comment statements are lines of text that explain or document a program step or series of steps. You must mark, or delimit, comment statements, so that the interpreter does not try to interpret them as program commands. You delimit PL/SQL comment statements in two ways. To create a block of comments that spans several lines, you enclose the block in the symbols /* and */. For example, the following commands create a comment block:

```
/* Script: student_register
Purpose: To enroll students in classes
Revisions: 3/30/2006 LM Script */
```

To create a comment statement that appears on a single line, you delimit the statement by typing two hyphens at the beginning of the line, as follows:

```
DECLARE
--variable to reference current S_ID value
current_S_ID NUMBER;
```

Arithmetic Operators

PL/SQL commands often use arithmetic operators to manipulate numerical data. The PL/SQL arithmetic operators are similar to those used in most programming languages. Table 4-5 summarizes the PL/SQL arithmetic operators.

Operator	Description	Example	Result
**	Exponentiation	2 ** 3	8
*	Multiplication	2 * 3	6
/	Division	9/2	4.5
+	Addition	3 + 2	5
–	Subtraction	3 – 2	1
–	Negation	–5	–5

Table 4-5   PL/SQL arithmetic operators in describing order of precedence

Table 4-5 displays the operators in the order in which the PL/SQL interpreter evaluates each operator in an expression. For example, in the following code to calculate an employee's overtime pay, 40 is multiplied by over_time_rate before the overtime hours are determined because multiplication (*) has higher precedence than subtraction (–).

```
total_hours_worked - 40 * over_time_rate
```

As with most programming languages, you can force the PL/SQL interpreter to evaluate arithmetic operators in a specific order by placing an operation in parentheses. To correct the logic error, the code is changed by using parentheses to have the number of overtime hours calculated before the hours are multiplied by the overtime rate.

```
(total_hours_worked - 40) * over_time_rate
```

Assignment Statements

In programming, an **assignment statement** assigns a value to a variable. In PL/SQL, the assignment operator is a colon followed by an equal sign (:=). An assignment statement has the following syntax:

variable_name := *value*;

Note that you place *variable_name*, which is the name of the variable to which you are assigning the new value, on the left side of the assignment operator. You place *value* on the right side of the assignment operator. *Value* can be a literal (constant) data value, such as 3.14159 for the value of *pi* or "Tammy Jones" for a student name, or it can be a variable. In a PL/SQL command, you must always enclose a **string literal**, which is a literal value that is a character string, within single quotation marks. A string literal can contain any combination of valid PL/SQL characters. If a single quotation mark appears within a string literal, you must type two single quotation marks (for example, 'Tammy''s Computer'). An example of an assignment statement that assigns a literal value to a variable is `current_s_first_name := 'Tammy';`.

You can also create assignment statements that assign a variable value to another variable. An example of an assignment statement that assigns the value of one variable to another is `current_s_first_name := s_first_name;`.

Recall that when you declare a new PL/SQL variable, the default value is NULL. You can place an assignment statement within a variable declaration to assign an initial value to a new variable. For example, you use the following combined declaration/assignment statement to declare a variable named current_student_ID with data type NUMBER and assign to it an initial value of 100: `current_student_ID NUMBER := 100;`.

NOTE

> When declaring a variable with an initial value in the declaration section of the PL/SQL block, you can use the keyword DEFAULT rather than the assignment operator (:=). For example, you can assign the initial value of 100 to a variable with the following syntax:
> `current_student_ID NUMBER DEFAULT 100;`.

If you use a NULL value in an assignment statement that performs an arithmetic operation, the resulting value is always NULL. For example, consider the following program code:

```
DECLARE
     variable1 NUMBER;
     variable2 NUMBER := 0;
BEGIN
     variable2 := variable1 + 1;
END;
```

This program does not assign an initial value to variable1 before the command in the program body adds the value 1 to variable1. The result of adding a value to a NULL value is another NULL value, so this command assigns a NULL value to variable2. To avoid making this error, always assign initial values to variables that you use in arithmetic operations.

EXECUTING A PL/SQL PROGRAM IN SQL*PLUS

Now that you are familiar with the fundamentals of the PL/SQL programming language, you are ready to write a simple PL/SQL program. You can create and execute PL/SQL program blocks within a variety of Oracle 10*g* development environments. For example, later in the book, you add PL/SQL programs to your Forms Builder and Reports Builder forms and reports. In this chapter, however, while you are being introduced to PL/SQL, you execute PL/SQL programs in SQL*Plus.

Displaying PL/SQL Program Output in SQL*Plus

Whenever you start a new SQL*Plus session and plan to display PL/SQL program output, you should first increase the size of the **PL/SQL output buffer**, which is a memory area on the database server that stores the program's output values before they are displayed to the user. You use the following SQL*Plus command to configure the PL/SQL output buffer:

```
SET SERVEROUTPUT ON SIZE buffer_size
```

The *buffer_size* value specifies the size of the output buffer, in bytes. For example, the following command sets the output buffer size as 4000 bytes:

```
SET SERVEROUTPUT ON SIZE 4000
```

If you do not specify the buffer size, the default buffer size is 2000 bytes. If you write a PL/SQL program that displays output containing more than 2000 bytes, an error message appears stating that a buffer overflow error has occurred. Therefore, if you are displaying more than just a few lines of output, it is a good idea to make the output buffer larger than the default size.

To display program output in the SQL*Plus environment, you use the DBMS_OUTPUT.PUT_LINE procedure, which has the following syntax:

```
DBMS_OUTPUT.PUT_LINE('display_text');
```

DBMS_OUTPUT is an Oracle built-in package, consisting of a set of related programs that users can access. PUT_LINE is the name of the DBMS_OUTPUT procedure that displays output in SQL*Plus. The *display_text* value can contain literal character strings, such as 'My name is Jorge', or variable values, such as current_s_first. An example command to display the current value stored in a variable named current_s_first is:

```
DBMS_OUTPUT.PUT_LINE(current_s_first);
```

The PUT_LINE procedure can display a maximum of 255 characters of text data. If you try to display more than 255 characters, an error occurs. Follow these steps to start SQL*Plus and set up the DBMS_OUTPUT buffer.

To start SQL*Plus and set up the output buffer:

1. Start SQL*Plus and log onto the database.

2. Type **SET SERVEROUTPUT ON SIZE 4000** at the SQL prompt to set up the PL/SQL output buffer, then press **Enter** to execute the command. Note that in Figure 4-2, a semicolon is not required at the end of this command. Why? Because SET is a SQL*Plus command and not a SQL command.

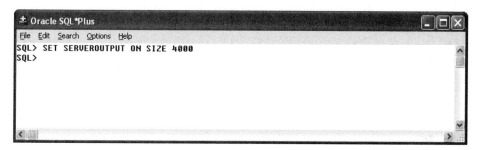

Figure 4-2 Setting up the output buffer in SQL*Plus

Writing a PL/SQL Program

In this chapter, you write a PL/SQL program in Notepad or another text editor, copy and paste the program commands into SQL*Plus, then execute the program. To execute a PL/SQL program in SQL*Plus, you press Enter after the last program command, then type a front slash (/), then press Enter again. The slash instructs the PL/SQL interpreter to execute the program code.

Next you write a PL/SQL program to declare a variable named todays_date that has the DATE data type. The program then assigns the current system date to todays_date and displays the variable value as output. It is a good programming practice to place the DECLARE, BEGIN, and END commands flush with the left edge of the text editor window, and then indent the commands within each section. The blank spaces do not affect the program's functionality, but they make the program easier to read and understand. You can indent the lines by pressing the Tab key or by pressing the spacebar to add blank spaces.

To write and execute a PL/SQL program:

1. Start Notepad, or the text editor of your choice, and create a new file named **Ch4APrograms.sql** in the Chapter04\Tutorials folder on your Solution Disk.

2. To create the PL/SQL program, type the commands shown in Figure 4-3 in the new Notepad file. (You paste the commands into SQL*Plus in the next step.) Indent the program lines as shown to format the different program blocks.

```
--PL/SQL program to display the current date
DECLARE
    todays_date DATE;
BEGIN
    todays_date := SYSDATE;
    DBMS_OUTPUT.PUT_LINE('Today''s date is ');
    DBMS_OUTPUT.PUT_LINE(todays_date);
END;
```

Figure 4-3   PL/SQL program commands

3. Copy the text, and then paste the copied text into SQL*Plus. Press **Enter** after the last program command to move to a new program line, type /, and then press **Enter** again to execute the program. The program output appears as shown in Figure 4-4. (Your output will be different, based on your current system date.)

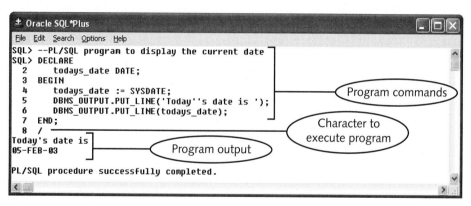

Figure 4-4   Executing a PL/SQL program in SQL*Plus

If your program does not run correctly, examine your code carefully to make sure you typed it correctly. If you think you typed it correctly, read the section titled "Debugging PL/SQL Programs" later in this lesson to learn how to debug PL/SQL programs.

PL/SQL DATA CONVERSION FUNCTIONS

Sometimes a program needs to convert a value from one data type to another data type. For example, if a user inputs the value 2 in a text input on a form, the program must convert the character 2 to a number before it can assign the value to a NUMBER variable and use the value to perform arithmetic calculations. PL/SQL sometimes performs **implicit data conversions**, in which the interpreter automatically converts a value from one data type to another. For example, in the PL/SQL program that you

just wrote, when the PL/SQL interpreter encountered the command DBMS_
OUTPUT.PUT_LINE(todays_date), it automatically converted the todays_date
DATE variable to a text string for output. Sometimes the PL/SQL interpreter is unable
to implicitly convert a value and an error occurs. Other times the PL/SQL interpreter
performs an implicit conversion, but not to the data type expected, and the results of
operations on the data are invalid. For these reasons, it is a good idea to perform **explicit
data conversions**, which convert variables to different data types using data conversion
functions that are built into PL/SQL. (A **function** is a program that receives one or
more inputs, called parameters, and returns a single output value.) Table 4-6 lists the main
PL/SQL data conversion functions.

4

Data Conversion Function	Description	Example

TO_CHAR	Converts either a number or a date value to a string using a specific format model	`TO_CHAR(2.98, '$999.99');`
		`TO_CHAR(SYSDATE,`
		`'MM/DD/YYYY');`

TO_DATE	Converts a string to a date using a specific format model	`TO_DATE('07/14/2003',` `'MM/DD/YYYY');`
TO_NUMBER	Converts a string to a number	`TO_NUMBER('2');`

Table 4-6   PL/SQL data conversion functions of PL/SQL

From Chapter 3, you are already familiar with how to use the TO_CHAR and
TO_DATE functions to convert data values in SQL commands to alternate data types.
As you can see from Table 4-6, PL/SQL also provides the TO_NUMBER function to
convert a character value to a NUMBER data type.

MANIPULATING CHARACTER STRINGS WITH PL/SQL

A **string** is a character data value that consists of one or more characters. Manipulating
strings is an important topic in any programming language. For example, you might want
to create a single string by **concatenating**, or joining, two separate strings. Or, you
might want to **parse**, or separate, a single string consisting of two data items separated
by commas or spaces into two separate strings. SQL*Plus provides a variety of built-in
string-handling functions to perform concatenation and parsing operations.

Concatenating Character Strings

To concatenate two strings in PL/SQL, you use the double bar (||) operator with the
following syntax:

```
new_string := string1 || string2;
```

Suppose you are working on a PL/SQL program that uses the following variable names and values:

Variable Name	Value
s_first_name	Tammy
s_last_name	Jones

You would like to concatenate these two values into a single variable named s_full_name that has the value "Tammy Jones". The command to concatenate the example name values is:

```
s_full_name := s_first_name || s_last_name;
```

This command looks pretty simple, but there is a catch: this command puts the two strings together with no spaces between them, so the value of s_full_name is now "TammyJones", which is not what you wanted. You need to insert a blank space between the first and last name. To insert a blank space, you use the same command, but concatenate a string consisting of a blank space between the two variable names, as in the following command:

```
s_full_name := s_first_name || ' ' || s_last_name;
```

Concatenation operations can become even more complex. Suppose you saved the following values for a building code, room, and capacity in three variables:

Variable Name	Data Type	Value
bldg_code	VARCHAR2	CR
room_num	VARCHAR2	101
room_capacity	NUMBER	150

Now suppose you need to save the string "CR Room 101 has 150 seats" in a variable named room_message. Note that the room_capacity value is a numeric data type and not one of the character data types. Before you can use the number in a concatenation operation, you must use the TO_CHAR function to convert it to a string. The command to create this string and assign its value to the room_message variable is:

```
room_message := bldg_code ||' Room '|| room_num
|| ' has '|| TO_CHAR(room_capacity) ||' seats.';
```

Always remember to add spaces before and after the variable values so that the strings don't run together.

Now you modify the PL/SQL program you wrote earlier so that it displays the lead-in message "Today's date is" and the date as a single concatenated string. You will use the TO_CHAR function to convert the todays_date variable to a string.

In the following exercises, you are modifying the code that you have already written. To keep a record of all of your PL/SQL programs, copy the existing program in your text editor, paste the copy below the original program, and then modify the copy. Save all program code in the Ch4APrograms.sql file in your Chapter04\Tutorial folder on your Solution Disk.

NOTE

To modify the PL/SQL program to display the date as a single string:

1. In Notepad, copy your current program commands, and paste the copied code below the existing commands. Then, modify the copied commands so they look like the commands in Figure 4-5.

```
± Oracle SQL*Plus
File  Edit  Search  Options  Help
SQL> --PL/SQL program to display the current date
SQL> DECLARE
  2      todays_date DATE;
  3  BEGIN
  4      todays_date := SYSDATE;
  5      DBMS_OUTPUT.PUT_LINE('Today''s date is ' || TO_CHAR(todays_date));
  6  END;
  7  /
Today's date is 05-FEB-06

PL/SQL procedure successfully completed.          Modified command
```

Figure 4-5   Displaying program output using a concatenated character string

2. Copy the modified code, and paste it into SQL*Plus. Press **Enter** after the last line of the program, type /, and then press **Enter** again to execute the program. The output should appear as shown in Figure 4-5.

The displayed results will vary based on the system date of your computer.

NOTE

Removing Blank Leading and Trailing Spaces from Strings

Sometimes when you retrieve data values from the database, or convert numeric or date values to characters, the data values contain blank leading or trailing spaces. To remove blank leading spaces, you use the LTRIM function. To remove blank trailing spaces, you use the RTRIM function. These functions have the following syntax:

```
string := LTRIM(string_variable_name);
string := RTRIM(string_variable_name);
```

Suppose a variable named student_address has the CHAR data type of size 20. Its current value is '951 Rainbow Dr' (14 characters followed by six spaces). To remove all spaces from the right side of the variable named student_address and assign the result back to the student_address variable, you use the RTRIM function as follows:

```
student_address := RTRIM(student_address);
```

Finding the Length of Character Strings

In parsing operations, you often need to find the number of characters in a character string. To find the number of characters in a character string, you use the PL/SQL LENGTH function, which returns an integer that represents the length of a character string. The LENGTH function has the following syntax:

```
string_length := LENGTH(string_variable_name);
```

In this syntax, *string_length* is the name of an integer variable that represents the string length. Suppose you declare a character variable named bldg_code, and its current value is 'CR'. You use the following command to declare an integer variable named code_length, and assign to it the string length:

```
code_length as NUMBER(3) := LENGTH(bldg_code);
```

Because the value of bldg_code is 'CR', the LENGTH function returns the number 2 to the code_length variable. If the bldg_code variable contains additional leading or trailing blank spaces, the LENGTH function includes these spaces in the length.

Character String Case Functions

Sometimes PL/SQL programs need to modify the case of character strings. For example, the TERM table in the Northwoods University database stores the STATUS data values in all uppercase letters, as "OPEN" and "CLOSED." You might want to display the STATUS value in program output using mixed-case letters with the initial letter of each word capitalized, as "Open" and "Closed," or in all lowercase letters, as "open" and "closed." You can use the UPPER function to convert lowercase or mixed-case characters to all uppercase characters, the LOWER function to convert uppercase and mixed-case characters to all lowercase letters, and the INITCAP function to convert uppercase, lowercase, or mixed-case letters to a string in which only the first letter is uppercase. These functions have the following syntax:

```
string := UPPER(string_variable_name);
string := LOWER(string_variable_name);
string := INITCAP(string_variable_name);
```

For example, suppose that a variable named s_full_name stores the value "Tammy Jones". The following command converts the s_full_name value to all uppercase characters: s_full_name := UPPER(s_full_name);.

Next, you add the INITCAP and LENGTH functions to your PL/SQL program so that it displays the current day of the week and the number of characters in the name of the current day. You declare a VARCHAR2 variable named current_day and assign to it the current day of the week. You declare another VARCHAR2 variable named current_day_length and assign to it the length of the current day's character string. The program then converts the day of the week to mixed-case characters with the first letter capitalized, trims the blank trailing spaces from the day using the RTRIM function, and displays the day and the length of the name of the day.

To modify your PL/SQL program to use the INITCAP, RTRIM, and LENGTH functions:

4

1. In Notepad, copy your program commands, and then modify the copied code so it looks like the code in Figure 4-6.

```
Oracle SQL*Plus
File  Edit  Search  Options  Help
SQL> --PL/SQL program to display the current date
SQL> DECLARE
  2      todays_date         DATE;
  3      current_day         VARCHAR2(9);
  4      current_day_length  BINARY_INTEGER;
  5  BEGIN
  6      todays_date := SYSDATE;
  7          -- extract day portion from current date, and trim trailing blank spaces
  8      current_day := TO_CHAR(todays_date, 'DAY');
  9      current_day := RTRIM(current_day);
 10          -- convert day to mixed case letters with initial letter capitalized
 11      current_day := INITCAP(current_day);
 12          -- determine length of day's character string
 13      current_day_length := LENGTH(current_day);
 14      DBMS_OUTPUT.PUT_LINE('Today''s date is ' || current_day || ', ' || TO_CHAR(todays_date));
 15      DBMS_OUTPUT.PUT_LINE('The length of the word ' || current_day || ' is ' ||
 16          TO_CHAR(current_day_length) || ' characters.');
 17  END;
 18  /
Today's date is Sunday, 05-FEB-06
The length of the word Sunday is 6 characters.
```

Your output will differ based on your computer's system date

Figure 4-6 PL/SQL program using character string functions

2. Copy the text, then paste it into SQL*Plus and execute the program. The program output should be similar to the output in Figure 4-6. (Your output will be different depending on your system date.)

Parsing Character Strings

Sometimes you need to parse an existing character string to extract one or more substrings. For example, the COURSE_NO column in the Northwoods University COURSE table combines department abbreviations and course numbers into a single character string, such as "MIS 101" and "CS 163." A program might search through the COURSE_NO values and parse out the characters before the blank space to determine the department abbreviation within each COURSE_NO value. To parse a substring within a string, you use the INSTR and SUBSTR functions.

The INSTR Function

The INSTR function searches a string for a specific substring. If the function finds the substring, it returns an integer that represents the starting position of the substring within the original string. If the function does not find the substring, the function returns the value 0. The general syntax for the INSTR function is:

```
start_position := INSTR(original_string, substring);
```

The following code example uses the INSTR function to return the starting position of the single blank space in a variable named curr_course_no that currently contains the value "MIS 101". The function returns the starting position value to an integer number variable named blank_position:

```
blank_position := INSTR(curr_course_no, ' ');
```

Note that in this command, the substring is a blank string literal (' '). If the curr_course_no string variable contains the value "MIS 101", the function returns the value 4, because the blank space is the fourth character in the string.

The SUBSTR Function

The SUBSTR function extracts a specific number of characters from a character string, starting at a given point. The general syntax for the SUBSTR function is:

```
extracted_string := SUBSTR(string_variable, ↵
starting_point, number_of_characters);
```

In this syntax, *starting_point* represents the position in the original string where the SUBSTR function begins extracting characters, and *number_of_characters* represents the number of characters that the function extracts. For example, you use the following command to extract the string "MIS" from the curr_course_no variable and set it equal to a character variable named curr_dept:

```
curr_dept := SUBSTR(curr_course_no, 1, 3);
```

In this command, 1 represents the starting place (the first character in the string), and 3 represents the number of characters to extract (three, one for every letter in "MIS").

To parse a string, you use the INSTR function with the SUBSTR function. You use the INSTR function to find the **delimiter**, which is the character that separates the target substring from the rest of the string. To parse out all of the characters before the delimiter, you specify that the starting point is 1, and specify that the *number_of_characters* value is the delimiter position minus 1. For example, you use the following command to return the department abbreviation in a single command:

```
blank_space := INSTR(curr_course_no,' ');
curr_dept := SUBSTR(curr_course_no, 1, (blank_space - 1));
```

In this command, the last parameter (the number of characters to extract) is specified by the INSTR function, which returns the position of the blank space. You must subtract 1 from the result of the INSTR function, because you want to return only the characters up to, but not including, the blank space.

To parse out all of the characters after the delimiter, you specify that the starting point is the delimiter position plus 1, and specify that the *number_of_characters* value is the string length minus the delimiter position. For example, use the following command to return the course number (such as "101" in "MIS 101") in a single command:

```
blank_space := INSTR(curr_course_no,' ');
curr_number := SUBSTR(curr_course_no, blank_space + 1),
               (LENGTH(curr_course_no) - blank_space));
```

Now you write a PL/SQL program that uses the INSTR and SUBSTR functions to parse a character string that contains the COURSE_NO value "MIS 101". The program output displays the course department ("MIS") and the course number ("101") as individual character strings.

To write a program that uses the INSTR and SUBSTR functions to parse a character string:

1. In Notepad, type the commands in Figure 4-7.

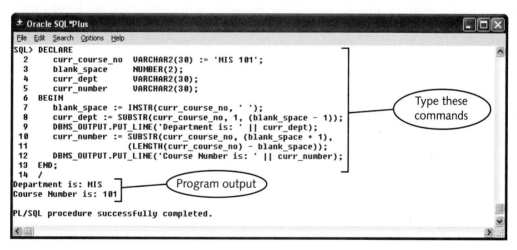

Figure 4-7 PL/SQL program that parses a character string

2. Execute the program in SQL*Plus. The output appears as shown in Figure 4-7. If your program does not execute correctly, find the error using the debugging strategies that the next section describes.

3. Exit SQL*Plus.

4. Save your file in Notepad, and then close Notepad.

DEBUGGING PL/SQL PROGRAMS

Program errors usually fall into one of two categories: syntax errors and logic errors. A **syntax error** occurs when a command does not follow the guidelines of the programming language. Syntax errors generate compiler or interpreter error messages. Figure 4-8 shows an example of a syntax error in the program you just wrote to parse a character string.

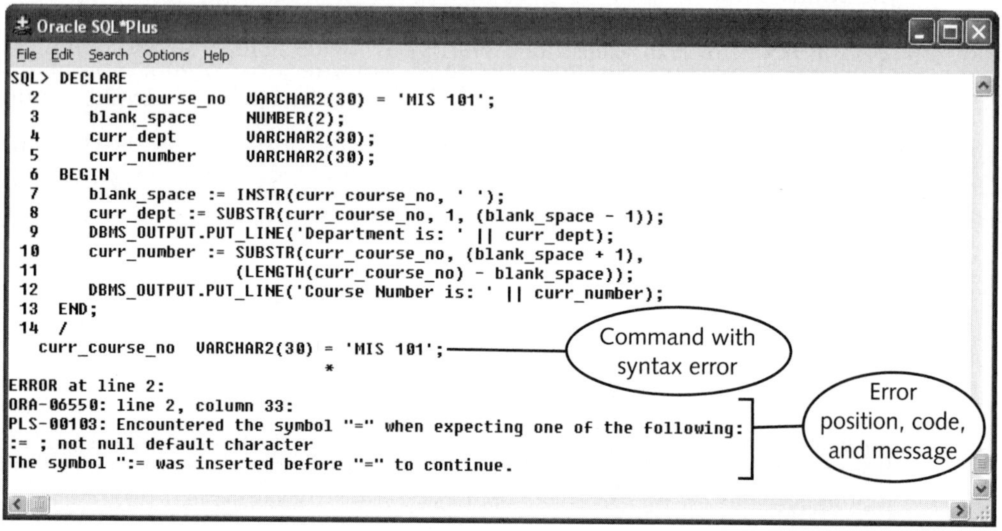

Figure 4-8   Program with a syntax error

The first command in the declaration section uses the equal sign (=) instead of the assignment operator (:=) to assign the initial value of the curr_course_no variable. The PL/SQL interpreter flags the error, displays an error code with a PLS prefix, and shows the line number and position of the error.

A **logic error** does not stop a program from running, but results in an incorrect result. Figure 4-9 shows an example of a program with a logic error.

```
Oracle SQL*Plus
File  Edit  Search  Options  Help
SQL> DECLARE
  2      curr_course_no   VARCHAR2(30) := 'MIS 101';
  3      blank_space      NUMBER(2);
  4      curr_dept        VARCHAR2(30);
  5      curr_number      VARCHAR2(30);
  6  BEGIN
  7      blank_space := INSTR(curr_course_no, ' ');
  8      curr_dept := SUBSTR(curr_course_no, 1, (blank_space - 1));
  9      DBMS_OUTPUT.PUT_LINE('Department is: ' || curr_dept);
 10      curr_number := SUBSTR(curr_course_no, blank_space,            Command with
 11                       (LENGTH(curr_course_no) - blank_space));      logic error
 12      DBMS_OUTPUT.PUT_LINE('Course Number is: ' || curr_number);
 13  END;
 14  /
Department is: MIS
Course Number is:  10                      Incorrect output

PL/SQL procedure successfully completed.
```

Figure 4-9   Program with a logic error

The program successfully executes, but displays incorrect output, showing the course number value as " 10" rather than as "101". This error results from a parsing error in the command that uses the SUBSTR function to extract the course number: the function should extract the substring starting *after* the blank space, but it extracts the substring starting *at* the blank space. As a result, the function parses the string " 10" (a blank space, followed by the string "10") rather than the string "101". The following sections present examples of common syntax and logic errors, and tips for debugging both types of errors.

Finding Syntax Errors

Syntax errors often involve misspelling a reserved word, omitting a required character in a command, such as a parenthesis, single quotation mark, or semicolon, or using a built-in function improperly. In SQL*Plus, the PL/SQL interpreter flags the line number and character location of syntax errors, and displays an error code and message. In Figure 4-8, the ORA-06550 error code indicates that the error is a PL/SQL compile error. The PLS-00103 error indicates that the PL/SQL interpreter encountered the equal sign (=) when it was expecting a different symbol, namely the assignment operator (:=). You can find explanations and solution strategies for PLS error codes on the Oracle Technology Network Web site, which you used in Chapter 2 to interpret ORA error codes.

NOTE

Remember, to look up error codes online, use your Web browser to go to *http://otn.oracle.com*. In the documentation section for the Oracle 10g Database, search for PLS-00103 (or whatever PLS error number you are currently dealing with).

Sometimes the PL/SQL interpreter's syntax error location does not correspond to the command in which the error actually occurs. For example, Figure 4-10 shows a syntax error in which the ending semicolon is omitted after the curr_course_no variable declaration.

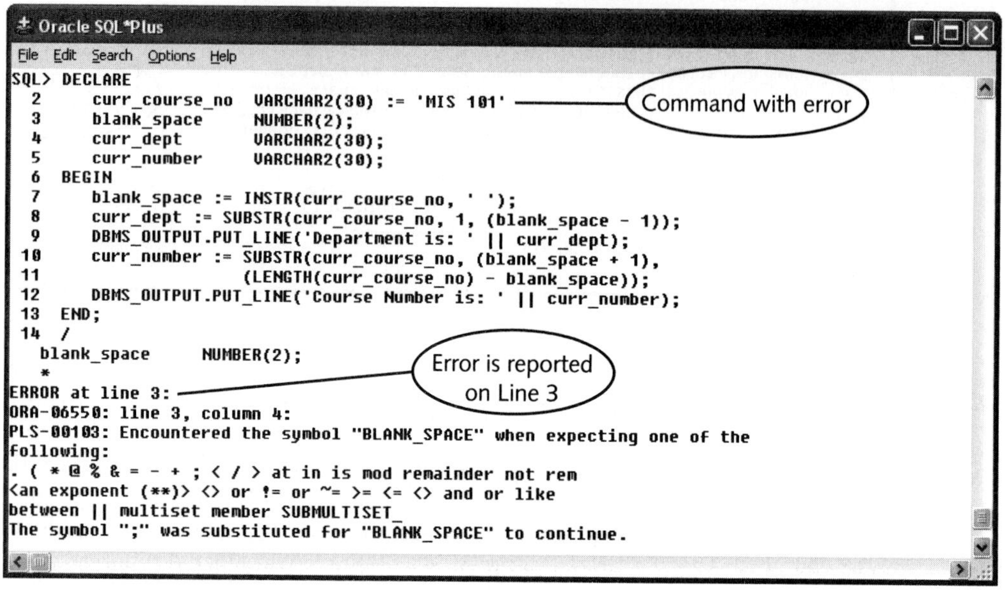

Figure 4-10   Syntax error in which the reported line number is incorrect

The PL/SQL interpreter reports on Line 3, but the error actually occurs on Line 2. When an error message appears and the flagged line appears correct, the error usually occurs on program lines *preceding* the flagged line.

Sometimes it is difficult to spot syntax errors visually, especially when the interpreter reports the incorrect line number or does not specify the error line location. If you cannot find a syntax error visually, you need to determine systematically which program line is generating the error. A useful debugging technique for isolating program errors is to **comment out** program lines, which means to turn program code statements into comment statements so the interpreter does not attempt to interpret the commands. Then you run the program. If the program runs successfully, then you know that the lines you commented out were causing the error.

Sometimes one syntax error can generate many more errors. These types of errors are called **cascading errors**. A good debugging strategy is to locate and fix the first error and then rerun the program. Don't try to fix all of the syntax errors before rerunning the program, because fixing that one first error may correct the others as well.

In summary, keep the following points in mind when resolving syntax errors:

- Error locations in error messages might not correspond to the line of the actual error.

- Try to isolate the error location systematically by commenting out suspect lines, modifying suspect lines, or both.

- One error might result in several other cascading errors.

4

Finding Logic Errors

Logic errors that result in incorrect program output can be caused by not using the proper order of operations in arithmetic functions, passing incorrect parameter values to built-in functions, creating loops that do not terminate properly, using data values that are out of range or not of the right data type, and so forth. In complex commercial software applications, legions of professional testers locate logic errors during rigorous beta-testing processes. For the programs you will write while you are learning PL/SQL and developing Oracle 10*g* database applications, you will usually be able to locate logic errors by determining what the output should be and comparing it to your actual program output.

The best way to locate logic errors is to view variable values during program execution using a **debugger**, which is a program that enables software developers to pause program execution and examine current variable values. Unfortunately, the SQL*Plus environment does not provide a PL/SQL debugger. However, you can track variable values using DBMS_OUTPUT statements. Recall that Figure 4-9 shows the program that is supposed to display the course number string "101". However, the program is displaying the string as " 10" instead.

The variable value with the error is curr_number. Therefore, the error is most likely occurring in the assignment statement on Lines 10 and 11, which assigns curr_number the value returned by the SUBSTR function. To debug the program, you can place a DBMS_OUTPUT statement before the line that uses the SUBSTR function in order to display the values of the variables that the command passes to the SUBSTR function. When you place debugging statements in your code, always include strings that label the variables, so you do not become confused about which variable values you are viewing. Figure 4-11 shows the DBMS_OUTPUT debugging statements and their resulting output.

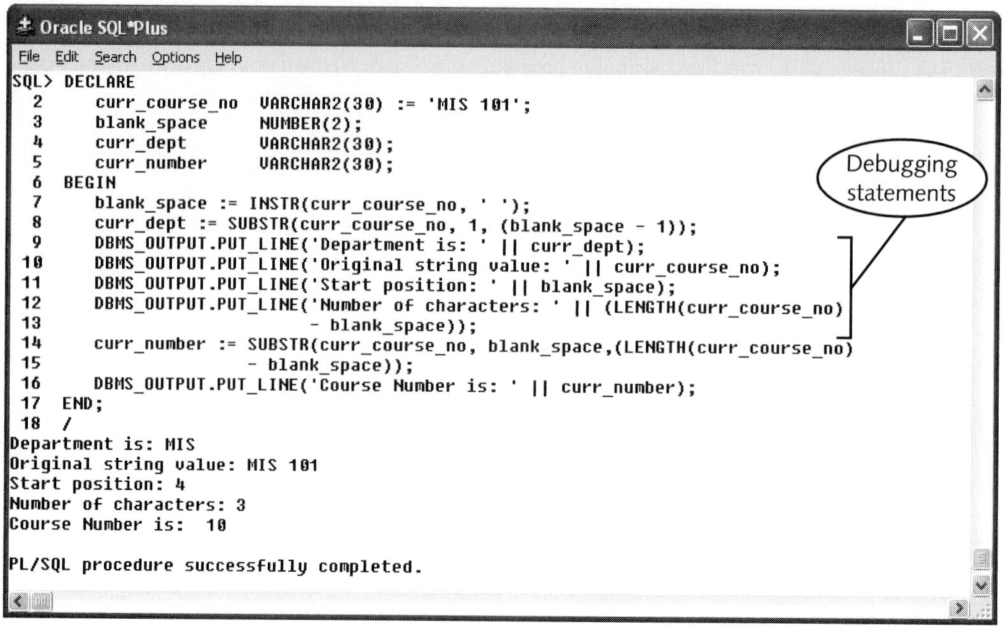


```
SQL> DECLARE
  2     curr_course_no  VARCHAR2(30) := 'MIS 101';
  3     blank_space     NUMBER(2);
  4     curr_dept       VARCHAR2(30);
  5     curr_number     VARCHAR2(30);
  6  BEGIN
  7     blank_space := INSTR(curr_course_no, ' ');
  8     curr_dept := SUBSTR(curr_course_no, 1, (blank_space - 1));
  9     DBMS_OUTPUT.PUT_LINE('Department is: ' || curr_dept);
 10     DBMS_OUTPUT.PUT_LINE('Original string value: ' || curr_course_no);
 11     DBMS_OUTPUT.PUT_LINE('Start position: ' || blank_space);
 12     DBMS_OUTPUT.PUT_LINE('Number of characters: ' || (LENGTH(curr_course_no)
 13                        - blank_space));
 14     curr_number := SUBSTR(curr_course_no, blank_space,(LENGTH(curr_course_no)
 15                  - blank_space));
 16     DBMS_OUTPUT.PUT_LINE('Course Number is: ' || curr_number);
 17  END;
 18  /
Department is: MIS
Original string value: MIS 101
Start position: 4
Number of characters: 3
Course Number is:  10

PL/SQL procedure successfully completed.
```

Figure 4-11 Program with debugging statements

The debugging statements show that the SUBSTR function is extracting a substring from the "MIS 101" string, starting at position 4, and extracting 3 characters. The blank space is at position 4, so the SUBSTR function extracts the blank space, followed by the characters "10". Instead, the SUBSTR function should start at position 5 and extract 3 characters. This indicates that the error is in the SUBSTR function's *starting_point* parameter, which should have the value 5 instead of 4.

Here is a checklist to use when you debug a logic error:

1. Identify the incorrect output variable(s).
2. Identify the inputs and calculations that contribute to the invalid output.
3. Use DBMS_OUTPUT debugging statements to find the values of the inputs that are contributing to the invalid output.
4. If you still can't locate the problem, take a break and try to find the error later.
5. If you still can't locate the problem, ask a fellow student for help.
6. If you still can't locate the problem, ask your instructor for help.

SUMMARY

- ❐ PL/SQL is a procedural programming language that Oracle 10*g* applications use to manipulate database data. PL/SQL is an interpreted language in which the PL/SQL interpreter translates each command into machine language and then executes each program command, one command at a time.

- ❐ PL/SQL commands are not case sensitive, except for character strings, which you must enclose in single quotation marks. The PL/SQL interpreter ignores blank spaces and line breaks, and a semicolon (;) marks the end of each PL/SQL command.

- ❐ PL/SQL variable names adhere to the Oracle naming standard. PL/SQL is a strongly typed language, so you must declare all variables before you can reference them in program commands.

- ❐ PL/SQL has scalar data types for variables that reference a single value; composite data types for variables that reference multiple scalar values; reference data types for variables that reference retrieved data values; and LOB data types for variables that reference large binary objects. Reference data types assume the same data type as a referenced database column or row.

- ❐ A PL/SQL program block consists of the declaration section, which contains variable declarations; the execution section, which contains program commands; and the exception section, which contains error-handling commands. The declaration and exception sections are optional.

- ❐ You assign values to PL/SQL variables using the assignment operator (:=). You place the variable name on the left side of the operator and the value on the right side of the operator.

- ❐ In PL/SQL programs, a variable's value is NULL until a command explicitly assigns to it a value. Numerical operations involving NULL values (such as adding a value to a NULL variable value) result in NULL values.

- ❐ PL/SQL performs implicit data conversions, but the output is unpredictable, so you should always use the explicit data conversion functions.

- ❐ PL/SQL has operators and built-in functions that allow you to concatenate and parse character strings. To concatenate two character strings, you use the double bar (||) operator. To parse strings, you use the INSTR, SUBSTR, and LENGTH functions.

- ❐ A syntax error occurs when the program code does not follow the guidelines of the programming language. A logic error does not stop a program from compiling, but results in incorrect output.

- ❐ To correct syntax errors in PL/SQL programs in the SQL*Plus environment, you must systematically identify the line(s) generating the error. You can do this by commenting out any program line that you suspect might be causing the problem.

❏ To correct logic errors in PL/SQL programs in the SQL*Plus environment, you must identify the program line that is generating the incorrect output, and then examine the values that are being used by that line to create the incorrect output. You can do this by temporarily displaying intermediate output values.

REVIEW QUESTIONS

1. In a PL/SQL block, which are the mandatory section(s)?

2. Write PL/SQL commands to declare the following variables. Select an appropriate data type for each variable.

Variable Name	Data Description
artist_name	Up to 4000 characters of variable-length text
apartment_no	An integer with a maximum value of 999
return_status	TRUE or FALSE
purchase_price	A fixed-point number with two decimal places, with a maximum value of 999.99
withdrawal_date	a date
state	a text column that always contains exactly two characters

3. To terminate a PL/SQL block, include _____ at the end of the block.

4. The PUT_LINE package of the DBMS_OUTPUT procedure is used to display output from a PL/SQL block. True or False?

5. Use _____ to override order of precedence for arithmetic operators in an expression.

6. You can add two hyphens to the beginning of a line so that the line is not interpreted as program commands. True or False?

7. What is the value of the second variable after execution of the following PL/SQL block?

```
DECLARE
   first_variable NUMBER;
   second_variable NUMBER;
BEGIN
   first_variable := first_variable * 3;
END;
```

8. The _____ symbol is used to end each PL/SQL command within a PL/SQL block.

9. How do you know if a program has a logic error?

10. You can use the LTRIM function to remove blank spaces from the end of a character string. True or False?

MULTIPLE CHOICE

1. When attempting to debug a logic error, the first step is to:

   a. Use the DBMS_OUTPUT debugging statements to determine the values of variables.

   b. Ask a fellow student for help.

   c. Ask your instructor for help.

   d. Identify the incorrect output variable(s).

2. Which of the following functions converts a number or date to a character string?

   a. CHAR_STRING

   b. TO_CHAR

   c. TO_DATE

   d. TO_NUMBER

3. Based on the order of precedence, what is the result of the following expression:

   14 + 8 * 4 / 2 - 6

   a. 24

   b. 18

   c. -48

   d. none of the above

4. Which of the following assigns a variable a reference data type that is the same structure as the record structure from a specified database table?

   a. %TYPE

   b. %COLUMNTYPE

   c. %ROWTYPE

   d. %REF

5. Which of the following is *not* classified as a scalar data type?

   a. BOOLEAN

   b. BINARY_INTEGER

   c. VARCHAR2

   d. VARRAY

6. Which of the following terms describes the joining of two separate strings?

   a. parsing

   b. concatenating

   c. data conversion

   d. procedural

7. Which of the following will *not* stop a PL/SQL program from running?

    a. logic error

    b. cascading error

    c. syntax error

    d. equivalency error

8. Which of the following symbols is used to assign the value of one variable to another variable?

    a. =

    b. :=

    c. ^

    d. %

9. Identify the line of code containing the error in the following PL/SQL block.

```
1 DECLARE
2   first_variable NUMBER :=2;
3 BEGIN
4   first_variable := first_variable ^ 3;
5 DBMS_OUTPUT.PUT_LINE('The value is ' || first_variable);
6 END;
```

    a. 2

    b. 4

    c. 6

    d. There is no error.

10. Identify the line of code containing the error in the following PL/SQL block.

```
1 DECLARE
2   second_variable NUMBER DEFAULT 2;
3 BEGIN
4   second_variable := second_variable + 3;
5   DBMS_OUTPUT.PUT_LINE(second_variable);
6 END;
```

    a. 2

    b. 4

    c. 6

    d. There is no error.

PROBLEM-SOLVING CASES

Execute the Ch4Clearwater.sql script file in the Chapter4 folder on your Data Disk before attempting the following cases. Use Notepad or another text editor to create the following PL/SQL programs. Save all solution files in the Chapter04\Cases folder on your Solution Disk.

1. A PL/SQL program contains the following variables with the associated values:

Variable	Value
curr_customer_first	Neal
curr_customer_last	Graham
curr_order_number	1
curr_order_date	5/29/06

a. Write a PL/SQL program command that concatenates the variables to create a string with the text "Neal Graham" and assigns the result to a variable named curr_customer_name. Make certain to display the value of the curr_customer_name to test the logic of the PL/SQL program. Save the file as Ch4ACase1A.sql.

b. Write a PL/SQL program command that concatenates the variables to create a string with the text "Order# 1 was placed on 5/29/06 by Neal Graham" and assigns the result to a variable named order_message. Make certain to display the value of the order_message to test the logic of the PL/SQL program. Save the file as Ch4ACase1B.sql.

2. Create a PL/SQL program to print a mailing label for the following customer information, and save the file as Ch4ACase2.sql.

curr_customer_first	Neal
curr_customer_last	Graham
curr_address	9815 Circle Dr.
curr_city	Tallahassee
curr_state	FL
curr_zip	32308

3. The following PL/SQL block extracts and displays the area code and phone number for a customer. However, it contains an error that must be corrected before the correct results are displayed.

```
DECLARE
    curr_phone            VARCHAR2(30) := '9045551897';
    curr_areacode    VARCHAR2(30);
    curr_number      VARCHAR2(30);
BEGIN
    curr_areacode := SUBSTR(curr_phone, 1, 3);
    DBMS_OUTPUT.PUT_LINE('The area code is: ' ||
    curr_areacode);
    curr_number := SUBSTR(curr_phone, 3, -7);
    DBMS_OUTPUT.PUT_LINE('The phone number is: ' ||
    curr_number);
END;
```

Enter and execute the corrected version of the PL/SQL block. Save the modified version as Ch4ACase3.sql.

4. Create a PL/SQL block that assigns the date 1/31/06 to the variable EXAMPLE_DATE and then displays the variable's value in the format of "January 31, 2006". Make certain you include the appropriate documentation in the program to identify its purpose and its author (your name). Save the program as Ch4ACase4.sql.

5. Write program commands to declare the following variables and assign to them the given values:

Variable Name	Data Type	Value
inventory_ID	Numeric	5
inventory_color	Character	Sky Blue
inventory_price	Numeric	259.99
inventory_QOH	Numeric	23

Write commands that use the variables to display the output in the following format:

Inventory ID: 5

Color: Sky Blue

Price: $259.99

Quantity on Hand: 23

Save the completed file as Ch4ACase5.sql. Do not insert, or hard-code, the actual data values in the output procedure.

◀ LESSON B ▶

Objectives

After completing this lesson, you should be able to:

♦ Create PL/SQL decision control structures
♦ Use SQL queries in PL/SQL programs
♦ Create loops in PL/SQL programs
♦ Create PL/SQL tables and tables of records
♦ Use cursors to retrieve database data into PL/SQL programs
♦ Use the exception section to handle errors in PL/SQL programs

You have learned how to create simple PL/SQL programs that declare variables, make arithmetic calculations, and process character strings. In this lesson, you learn how to create decision control structures and loops. You also learn how to write PL/SQL commands that insert, display, and manipulate database values, and how to create exception handlers to handle runtime errors.

PL/SQL DECISION CONTROL STRUCTURES

The PL/SQL programs you have written so far use **sequential processing** that processes statements one after another. However, most programs require decision control structures that alter the order in which statements execute, based on the values of certain variables. In PL/SQL, you can create IF/THEN, IF/THEN/ELSE, and IF/ELSIF decision control structures.

IF/THEN

The PL/SQL IF/THEN decision control structure has the following syntax:

```
IF condition THEN
  commands that execute if condition is TRUE;
END IF;
```

The *condition* is an expression that PL/SQL has to be able to evaluate as either TRUE or FALSE. The *condition* can compare two values, such as a variable and a literal, or the *condition* can be a Boolean variable. Table 4-7 shows the PL/SQL comparison operators and usage examples.

Operator	Description	Example
=	Equal to	count = 5
<>	Not equal to	count <> 5
!=	Not equal to	count != 5
>	Greater than	count > 5
<	Less than	count < 5
>=	Greater than or equal to	count >= 5
<=	Less than or equal to	count <= 5

Table 4-7   PL/SQL comparison operators

NOTE

Remember that the operator that compares two values is (=), whereas the assignment operator that assigns a value to a variable is (:=).

If the *condition* evaluates as TRUE, one or more program statements execute. If the *condition* evaluates as FALSE or NULL, the program skips the statement(s). It is good programming practice to format IF/THEN structures by indenting the program statements that execute if the condition is TRUE so that the structure is easier to read and understand. You can indent these program statements by pressing the Tab key or by pressing the spacebar to add blank spaces.

If a *condition* evaluates as NULL, then the IF/THEN structure behaves the same as if the condition evaluates as FALSE. A *condition* can evaluate to NULL under two circumstances: if you specify the *condition* using a Boolean variable that has not been assigned an initial value, or if any variables used in the *condition* currently have a value of NULL.

Next you create a program that displays the current day of the week. You use an IF/THEN decision control structure in the PL/SQL program so that it displays the current day of the week only if the current day happens to be Friday. Note that the condition comparison is case sensitive.

To add an IF/THEN structure to a PL/SQL program:

1. Start Notepad, and create a new file named **Ch4BPrograms.sql** and save it in the Chapter04\ folder on your Solution Disk. Type the commands shown in Figure 4-12.

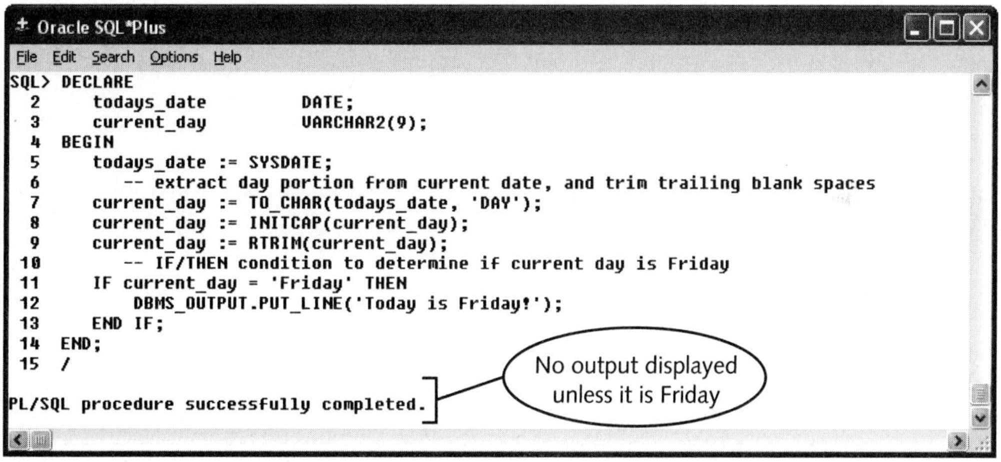

Figure 4-12   PL/SQL program that uses an IF/THEN decision control structure with outcome of FALSE

2. Start SQL*Plus and log onto the database. To allow output to be displayed, type **SET SERVEROUTPUT ON SIZE 4000** and press **Enter.**

3. Copy the program commands from the program you entered from Figure 4-12, and paste the copied commands into SQL*Plus. Execute the program (and debug it if necessary). If it happens to be Friday (which it is not in the example), the condition evaluates as TRUE, and the program generates the specified output.

4. In Notepad, copy your program code and paste it back into Ch4BPrograms.sql below the code for your previous program. Modify the copied code as shown in Figure 4-13, so that the condition evaluates to TRUE if it is not Friday.

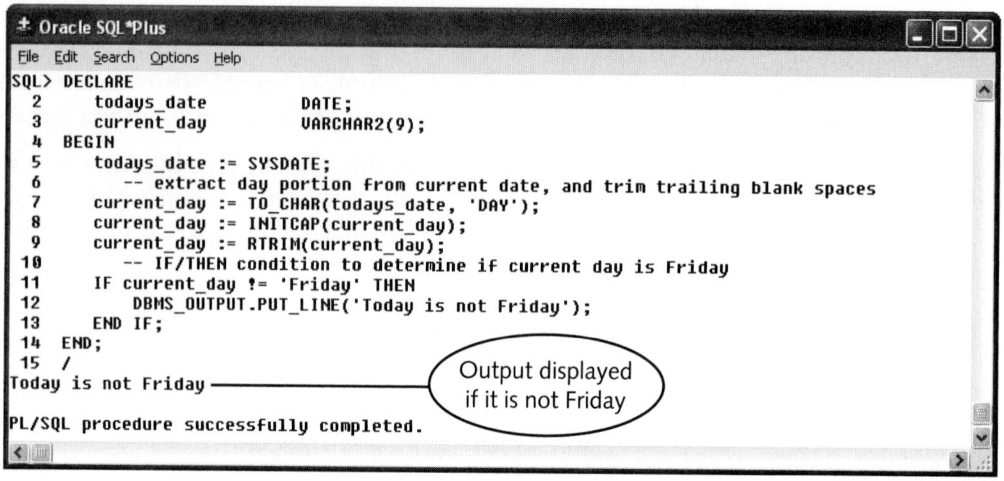

```
Oracle SQL*Plus
File  Edit  Search  Options  Help
SQL> DECLARE
  2      todays_date          DATE;
  3      current_day          VARCHAR2(9);
  4  BEGIN
  5      todays_date := SYSDATE;
  6      -- extract day portion from current date, and trim trailing blank spaces
  7      current_day := TO_CHAR(todays_date, 'DAY');
  8      current_day := INITCAP(current_day);
  9      current_day := RTRIM(current_day);
 10      -- IF/THEN condition to determine if current day is Friday
 11      IF current_day != 'Friday' THEN
 12          DBMS_OUTPUT.PUT_LINE('Today is not Friday');
 13      END IF;
 14  END;
 15  /
Today is not Friday ──────────  Output displayed
                                 if it is not Friday
PL/SQL procedure successfully completed.
```

Figure 4-13   PL/SQL program that uses an IF/THEN decision control structure with outcome of TRUE

     5. Copy the modified commands, paste the commands into SQL*Plus, then execute the program. Unless you were one of the lucky ones whose condition evaluated as TRUE in Step 3, your output should now look like the output in Figure 4-13.

IF/THEN/ELSE

The previous example suggests the need for a decision control structure that executes alternate program statements when a condition evaluates as FALSE. This is an IF/THEN/ELSE structure. In PL/SQL, the IF/THEN/ELSE structure has the following syntax:

```
IF condition THEN
 commands that execute if condition is TRUE;
ELSE
 commands that execute if condition is FALSE;
END IF;
```

Now you modify the program with the IF/THEN structure so that it uses an IF/THEN/ELSE structure. The program displays one output if the current day is Friday, and a different output if the current day is not Friday.

To modify your PL/SQL program to use an IF/THEN/ELSE structure:

     1. In Notepad, copy your second program's code and paste it below the second program. Modify the code as shown in Figure 4-14 to use an IF/THEN/ELSE structure.

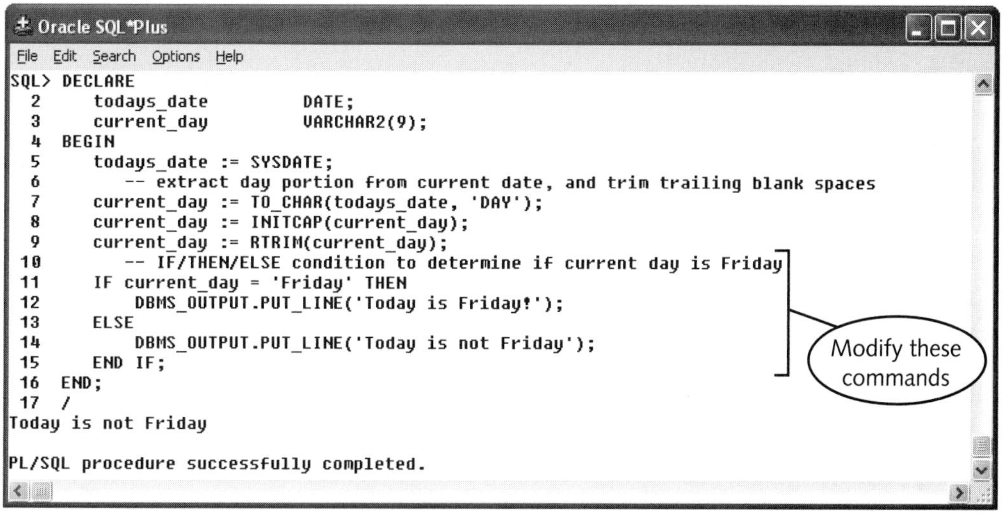

```
± Oracle SQL*Plus
File  Edit  Search  Options  Help
SQL> DECLARE
  2      todays_date          DATE;
  3      current_day          VARCHAR2(9);
  4  BEGIN
  5      todays_date := SYSDATE;
  6      -- extract day portion from current date, and trim trailing blank spaces
  7      current_day := TO_CHAR(todays_date, 'DAY');
  8      current_day := INITCAP(current_day);
  9      current_day := RTRIM(current_day);
 10      -- IF/THEN/ELSE condition to determine if current day is Friday
 11      IF current_day = 'Friday' THEN
 12          DBMS_OUTPUT.PUT_LINE('Today is Friday!');
 13      ELSE
 14          DBMS_OUTPUT.PUT_LINE('Today is not Friday');
 15      END IF;
 16  END;
 17  /
Today is not Friday

PL/SQL procedure successfully completed.
```

Modify these commands

Figure 4-14   PL/SQL program that uses an IF/THEN/ELSE structure

2. Copy the code into SQL*Plus, and execute the program. Depending on the current day, the appropriate output appears.

Nested IF/THEN/ELSE

You can nest IF/THEN/ELSE structures by placing one or more IF/THEN/ELSE statements within the program statements that execute after the IF or ELSE command. When you create nested IF/THEN/ELSE structures, it is especially important to properly indent the program lines following the THEN and ELSE commands. Correct formatting enables you to better understand the program logic and spot syntax errors, such as missing END IF commands. Next you modify your PL/SQL program so that it uses a nested IF/THEN/ELSE structure if the current day is not Friday; the program also tests to see if the current day is Saturday.

To modify your PL/SQL program to use a nested IF/THEN/ELSE structure:

1. In Notepad, copy your third PL/SQL program and paste it below your third program. Modify it as shown in Figure 4-15 to create a nested IF/THEN/ELSE structure.

```
± Oracle SQL*Plus                                                    [_][□][X]
File  Edit  Search  Options  Help
SQL> DECLARE
  2       todays_date           DATE;
  3       current_day           VARCHAR2(9);
  4  BEGIN
  5       todays_date := SYSDATE;
  6        -- extract day portion from current date, and trim trailing blank spaces
  7       current_day := TO_CHAR(todays_date, 'DAY');
  8       current_day := INITCAP(current_day);
  9       current_day := RTRIM(current_day);
 10        -- NESTED IF/THEN/ELSE condition to determine if current day is Friday or Saturday
 11       IF current_day = 'Friday' THEN
 12          DBMS_OUTPUT.PUT_LINE('Today is Friday!');
 13       ELSE
 14          IF current_day = 'Saturday' THEN
 15             DBMS_OUTPUT.PUT_LINE('Today is Saturday!');
 16          ELSE
 17             DBMS_OUTPUT.PUT_LINE('Today is not Friday or Saturday.');
 18          END IF;                                                  Add/modify
 19       END IF;                                                    these commands
 20  END;
 21  /
Today is not Friday or Saturday.

PL/SQL procedure successfully completed.
```

Figure 4-15   Nested IF/THEN/ELSE structure

2. Copy and paste the code into SQL*Plus, and execute the program. Depending on the current day, the appropriate output appears.

IF/ELSIF

The IF/ELSIF decision control structure allows you to test for many different conditions. The general syntax for this structure is:

```
IF condition1 THEN
     commands that execute if condition1 is TRUE;
ELSIF condition2 THEN
     commands that execute if condition2 is TRUE;
ELSIF condition3 THEN
     commands that execute if condition3 is TRUE;
...
ELSE
     commands that execute if none of the
     conditions are TRUE;
END IF;
```

NOTE

In the ELSIF command, there is no second E in ELS and no space between ELS and IF.

In an IF/ELSIF decision control structure, the interpreter first evaluates *condition1*. If *condition1* is true, the interpreter executes the associated program statement(s), and then exits the IF/ELSIF structure. If *condition1* is false, the interpreter evaluates *condition2*. If *condition2* is true, the interpreter executes the associated program statements(s), then exits the IF/ELSIF structure. If *condition2* is false, the interpreter evaluates the next condition. This process continues until the interpreter finds a true condition. If none of the conditions are true, the interpreter executes the commands following the ELSE keyword.

4

Next you modify your PL/SQL program so that it uses an IF/ELSIF structure to test for each day of the week and display the appropriate output. Note that the program statements following all of the ELSIF commands and the program statement following the ELSE command are all indented to make the structure easier to read and interpret.

To modify your PL/SQL program to use the IF/ELSIF structure:

1. In Notepad, copy your fourth PL/SQL program and paste it below your fourth program. Modify it as shown in Figure 4-16 to use the IF/ELSIF structure.

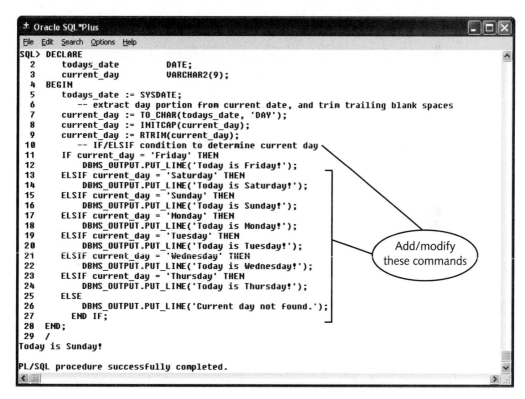

Figure 4-16 Using an IF/ELSIF structure

2. Copy the code, paste it into SQL*Plus, and execute the program. Depending on the current day, the appropriate output appears.

Logical Operators AND, OR, and NOT

You used the AND, OR, and NOT logical operators in Chapter 3 SQL queries. You can also use these operators to create complex expressions for a decision control structure condition. This is a powerful technique; however, it can also be tricky to use. Keep in mind that each individual expression's TRUE and FALSE values are combined into a single TRUE or FALSE result for the entire condition. When you use AND in a complex expression, the expressions on both sides of the AND operator must be true for the combined expression to evaluate to TRUE. When you use OR in a complex expression, the combined expression evaluates to TRUE if the expression on either side of the OR operator is true (both expressions do not need to be true). Finally, it is important to note that PL/SQL evaluates NOT conditions first, then AND conditions, and finally OR conditions are evaluated last.

Next, you learn about one of the problems you can encounter when combining AND and OR operators in an expression. The next program appears to work correctly until the current day is Sunday. Then the program fails. This exercise illustrates why you must be especially careful when using AND and OR. You can easily miss the error in the following program, until Sunday arrives.

To modify your PL/SQL program to use AND and OR:

1. In Notepad, copy your PL/SQL program from Figure 4-16 and paste it below your last program. Modify it as shown in Figure 4-17 to use AND and OR.

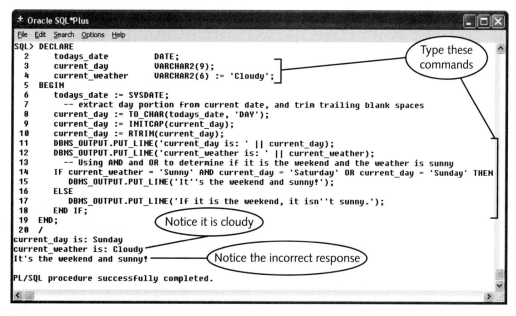

```
± Oracle SQL*Plus                                                    _ □ X
File  Edit  Search  Options  Help
SQL> DECLARE
  2      todays_date          DATE;
  3      current_day          VARCHAR2(9);
  4      current_weather      VARCHAR2(6) := 'Cloudy';
  5  BEGIN
  6      todays_date := SYSDATE;
  7          -- extract day portion from current date, and trim trailing blank spaces
  8      current_day := TO_CHAR(todays_date, 'DAY');
  9      current_day := INITCAP(current_day);
 10      current_day := RTRIM(current_day);
 11      DBMS_OUTPUT.PUT_LINE('current_day is: ' || current_day);
 12      DBMS_OUTPUT.PUT_LINE('current_weather is: ' || current_weather);
 13          -- Using AND and OR to determine if it is the weekend and the weather is sunny
 14      IF current_weather = 'Sunny' AND current_day = 'Saturday' OR current_day = 'Sunday' THEN
 15          DBMS_OUTPUT.PUT_LINE('It''s the weekend and sunny!');
 16      ELSE
 17          DBMS_OUTPUT.PUT_LINE('If it is the weekend, it isn''t sunny.');
 18      END IF;
 19  END;
 20  /
current_day is: Sunday
current_weather is: Cloudy
It's the weekend and sunny!

PL/SQL procedure successfully completed.
```

Type these commands

Notice it is cloudy

Notice the incorrect response

Figure 4-17   Incorrectly using AND and OR

2. Copy the code, paste it into SQL*Plus, and execute the program. If it is not Sunday, the appropriate output appears. If you were running this program on Sunday, you would see the same incorrect output as shown in Figure 4-17.

Why is the output from the program shown in Figure 4-17 incorrect on Sundays? Remember that the AND operator is evaluated before the OR operator. Because current_weather = 'Sunny' evaluates to FALSE, it does not matter whether current_day = 'Saturday' evaluates to TRUE. The expressions on both sides of an AND operator must evaluate to TRUE for the complex AND expression to evaluate to TRUE; so in this example, the AND expression evaluates to FALSE. The program then combines this FALSE result with current_day = 'Sunday' using an OR operation. Recall that if the expression on either side of an OR operator is true, the expression evaluates to TRUE. If current_day is Sunday, it doesn't matter whether current_weather is Cloudy; the expression still evaluates to TRUE. Figure 4-18 illustrates how PL/SQL evaluates this complex expression. The numbers show the order in which the parts of the condition are evaluated.

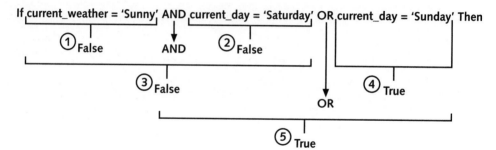

Figure 4-18   Evaluating AND and OR in an expression

Although PL/SQL evaluates AND before OR in a complex expression, you can use parentheses to override this precedence and force the program to evaluate the OR side of the expression first. In the next exercise, you use parentheses to correct the logical error in the expression in the last program.

To correct the AND and OR program:

1. In Notepad, copy your program from Figure 4-17 and paste a copy beneath the original program. Add parentheses as shown in Figure 4-19.

```
Oracle SQL*Plus
File  Edit  Search  Options  Help
SQL> DECLARE
  2       todays_date          DATE;
  3       current_day          VARCHAR2(9);
  4       current_weather      VARCHAR2(6) := 'Cloudy';
  5  BEGIN
  6       todays_date := SYSDATE;
  7          -- extract day portion from current date, and trim trailing blank spaces
  8       current_day := TO_CHAR(todays_date, 'DAY');
  9       current_day := INITCAP(current_day);
 10       current_day := RTRIM(current_day);
 11       DBMS_OUTPUT.PUT_LINE('current_day is: ' || current_day);
 12       DBMS_OUTPUT.PUT_LINE('current_weather is: ' || current_weather);
 13          -- Using AND and OR to determine if it is the weekend and the weather is sunny
 14       IF current_weather = 'Sunny' AND
 15          (current_day = 'Saturday' OR current_day = 'Sunday') THEN
 16          DBMS_OUTPUT.PUT_LINE('It''s the weekend and sunny');
 17       ELSE
 18          DBMS_OUTPUT.PUT_LINE('If it is the weekend, it isn''t sunny.');
 19       END IF;
 20  END;
 21  /
current_day is: Sunday
current_weather is: Cloudy
If it is the weekend, it isn't sunny.

PL/SQL procedure successfully completed.
```

Add these parentheses

Figure 4-19   Correcting the AND and OR program

2. Copy the code, paste it into SQL*Plus, and execute the program. Now the correct output always appears, regardless of the day.

USING SQL QUERIES IN PL/SQL PROGRAMS

Recall that SQL data definition language (DDL) commands, such as CREATE, ALTER, and DROP, change the structure of database objects. Data manipulation language (DML) commands, such as SELECT and INSERT, query or manipulate data in database tables. Transaction control commands, such as COMMIT and ROLLBACK, organize DML commands into logical transactions and commit them to the database or roll them back. Table 4-8 summarizes the SQL command categories and shows which commands you can and cannot use in PL/SQL programs.

4

Category	Purpose	Examples	Can Be Used in PL/SQL Programs
DDL	Creates and modifies database objects	CREATE, ALTER, DROP	No
DML	Manipulates data values in tables	SELECT, INSERT, UPDATE, DELETE	Yes
Transaction Control	Organizes DML commands into logical transactions	COMMIT, ROLLBACK, SAVEPOINT	Yes

Table 4-8   Using SQL commands in PL/SQL programs

To use a SQL action query (INSERT, UPDATE, or DELETE) or transaction control command in a PL/SQL command, you simply put the query or command in the PL/SQL program, using the same syntax you use to execute the query or command in SQL*Plus. In PL/SQL action queries, you can use variables instead of literal values to specify data values. For example, suppose you have a PL/SQL program containing a variable named curr_first_name that has the value "Tammy". You could use the following command to insert the value "Tammy" into the S_FIRST column in the Northwoods University STUDENT table:

```
INSERT INTO student (s_first)
VALUES (curr_first_name);
```

You can also use variables to specify search expressions in SQL queries. For example, you use the following search condition to match every S_FIRST value that has the same value as the curr_first_name variable:

```
WHERE s_first = curr_first_name;
```

NOTE

You can create PL/SQL commands to retrieve data values, assign the retrieved values to variables, and then manipulate the values in program commands. To do this, you must create cursors, which are described later in this lesson.

Next you write a PL/SQL program that uses DML and transaction control commands to insert the first three records into the Northwoods University TERM table. First, you run the EmptyNorthwoods.sql script to delete your current Northwoods University database tables and re-create the Northwoods University database tables without any data in them. Then, you confirm that the TERM table is empty.

To run the script to create the database tables and check the TERM table:

1. If you stopped SQL*Plus, restart it and type **SET SERVEROUTPUT ON SIZE 4000** to enable interactive output and set up the output buffer.

2. Type **START a:\Chapter04\Ch4EmptyNorthwoods.sql** at the SQL prompt to run the script to drop your existing Northwoods University tables and create the empty Northwoods University database tables. If your data files are stored in a different location, you will need to substitute the appropriate path in the START command.

3. Type the following query at the SQL prompt: **SELECT * FROM term;**. The message "no rows selected" appears, indicating that the table is empty.

Next you write a PL/SQL program to insert the first three records into the TERM table. You declare variables corresponding to each of the table columns, assign the variables to the appropriate values, insert the records, and then commit the records.

To write the program to insert the TERM records:

1. Return to the Ch4BPrograms.sql file in Notepad, and type the PL/SQL program commands in Figure 4-20.

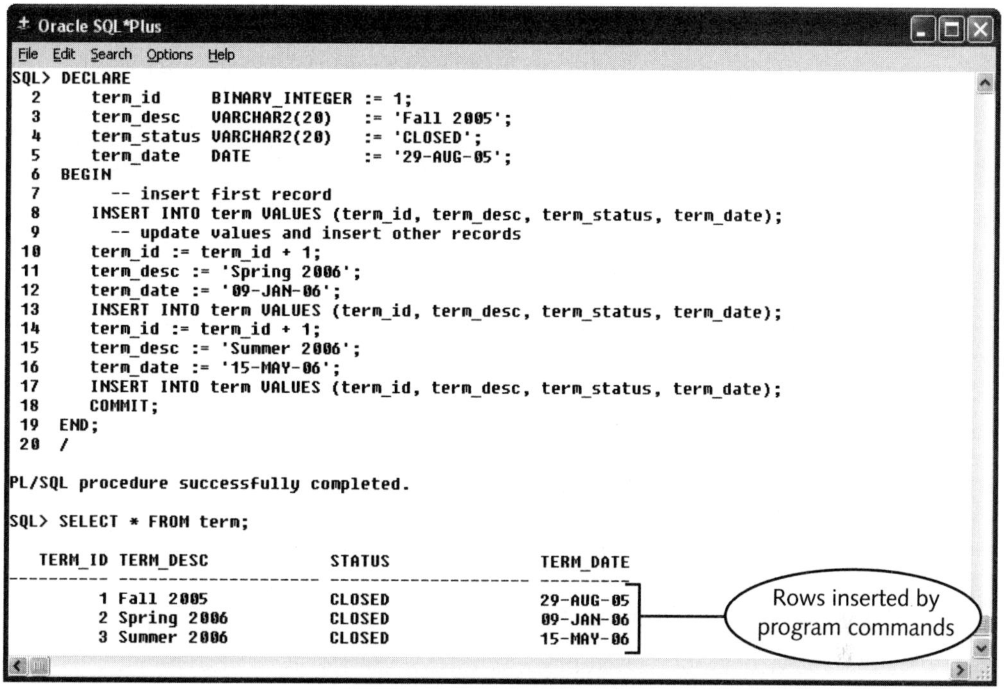

Figure 4-20   PL/SQL program that uses an INSERT action query

2. Copy the commands, paste them into SQL*Plus, and run the program. The message "PL/SQL procedure successfully completed." appears when your program runs correctly. If a syntax error appears, debug and rerun your program until it successfully executes.

3. At the SQL prompt, type the SELECT command at the bottom of Figure 4-20 to view the TERM table records. The output shows that the program correctly inserted the records.

LOOPS

A **loop** is a program structure that systematically executes a series of program statements, and periodically evaluates an exit condition to determine if the loop should repeat or exit. A **pretest loop** evaluates the exit condition before any program commands execute, and a **posttest loop** executes one or more program commands before the loop evaluates the exit condition for the first time. You use a pretest loop if there is a case in which the looping program statements should never execute. You use a posttest loop when you are sure that the looping program statements always need to execute at least once. PL/SQL has five loop structures: LOOP...EXIT, LOOP...EXIT WHEN,

WHILE...LOOP, numeric FOR loops, and cursor FOR loops. The following sections describe the first four loop structures. You learn about cursor FOR loops in the section on cursors later in this lesson.

To illustrate the different types of loop, you create a database table named COUNT_TABLE that has one numerical data column named COUNTER. You create PL/SQL programs that use the different loop structures to insert the numbers 1, 2, 3, 4, and 5 automatically into the COUNTER column in COUNT_TABLE, shown in Table 4-9.

COUNTER
1
2
3
4
5

Table 4-9   COUNT_TABLE

Next you create COUNT_TABLE in SQL*Plus.

To create COUNT_TABLE:

1. In SQL*Plus, type the following command at the SQL prompt:

```
CREATE TABLE count_table
(counter NUMBER(2));
```

2. Press **Enter** to execute the command: The message "Table created." confirms that the DBMS successfully created the table.

The LOOP...EXIT Loop

The LOOP...EXIT loop can be either a pretest or a posttest loop. The basic syntax of a LOOP...EXIT loop is:

```
LOOP
   [program statements]
   IF condition THEN
     EXIT;
   END IF;
   [additional program statements]
END LOOP;
```

In this syntax, the LOOP keyword signals the beginning of the loop. A set of optional program statements follow the LOOP keyword. An IF/THEN decision structure evaluates the loop exit condition. When the condition is true, the EXIT command exits the loop and resumes execution of program statements after the END LOOP command. If the IF/THEN decision structure is the first code in the loop, it is a pretest loop. If it is the last code in the loop, it is a posttest loop. It is good programming practice to indent the program lines between the LOOP and END LOOP commands to make the loop structure easier to read and understand.

Next you write a PL/SQL program that uses a LOOP...EXIT loop to insert records 1 through 5 into COUNT_TABLE. The program statements in the loop execute five times, once for each record. Because the looping program statements always execute at least once, you use a posttest loop. The test condition evaluates the value of a variable named loop_count. The program exits the loop prior to inserting a record when loop_count is equal to 6.

To use a posttest LOOP...EXIT loop to insert the COUNT_TABLE records:

  1. In Notepad, type the commands in Figure 4-21 to insert the table records.

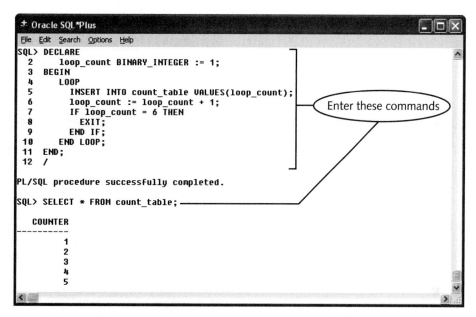

Figure 4-21 PL/SQL program that uses a LOOP...EXIT loop to insert records

  2. Copy the commands, paste them into SQL*Plus, execute the program, and debug it if necessary.

  3. Type and execute the SELECT query in Figure 4-21 to view the COUNT_TABLE records and confirm that the program correctly inserted the records.

The LOOP...EXIT WHEN Loop

The LOOP...EXIT WHEN loop can also be either a pretest or a posttest loop. The general syntax of the LOOP...EXIT WHEN loop is:

```
LOOP
     program statements
     EXIT WHEN condition;
END LOOP;
```

This loop executes the program statements, and then uses the EXIT WHEN command to test the exit condition. If the condition is true, the loop exits. As with the LOOP...EXIT loop, you should indent the loop commands as shown to make the loop easier to understand. Next, you delete the records you inserted into COUNT_TABLE, and then write a PL/SQL program to use a LOOP...EXIT WHEN loop to insert the records again.

To delete the records and insert them again using a LOOP...EXIT WHEN loop:

1. In Notepad, type the commands in Figure 4-22 to use a LOOP...EXIT WHEN loop to insert the records again.

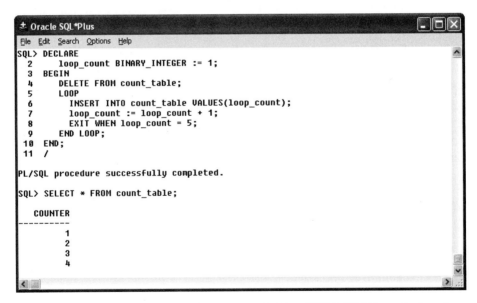

Figure 4-22   PL/SQL program that uses a LOOP...EXIT WHEN loop to insert records

2. Copy the commands, paste them into SQL*Plus, execute the program, and debug it if necessary.

3. Type and execute the SELECT query in Figure 4-22 to view the COUNT_TABLE records and confirm that the program successfully inserted the records.

The WHILE...LOOP

The WHILE...LOOP is a pretest loop that evaluates the exit condition before it executes any program statements. The general syntax of the WHILE...LOOP is:

```
WHILE condition LOOP
    program statements
END LOOP;
```

You again delete the records in COUNT_TABLE, and then write a PL/SQL program to insert the records using the WHILE...LOOP structure.

To delete the records and use the WHILE...LOOP to insert the records again:

1. In Notepad, type the commands in Figure 4-23 to use a WHILE...LOOP to insert the records again.

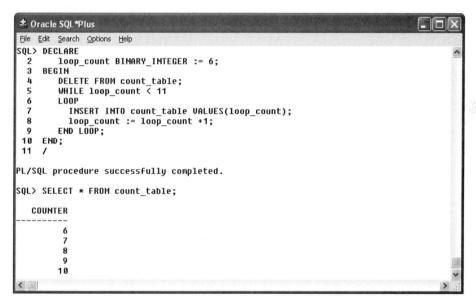

Figure 4-23   PL/SQL program that uses a WHILE...LOOP to insert records

2. Copy the commands, paste them into SQL*Plus, run the program, and debug it if necessary.

3. Type and execute the SELECT query in Figure 4-23 to confirm that the program successfully inserted the records.

The Numeric FOR Loop

In the loop exercises you have completed, you have had to declare and increment a counter variable within your loop commands. The numeric FOR loop does not require that you explicitly increment a counter variable. You declare the loop counter variable and its start and end values in the loop's FOR statement. The loop automatically increments the counter variable until it reaches the end value, and then exits. The general syntax of the numeric FOR loop is:

```
FOR counter_variable IN start_value .. end_value
LOOP
      program statements
END LOOP;
```

Start_value and *end_value* must both be integers. The loop always increments *counter_variable* by one, until it reaches *end_value*. Next, you delete the COUNT_TABLE records and use a numeric FOR loop to reinsert the records.

To delete the records and insert the records again using a numeric FOR loop:

1. In Notepad, type the commands in Figure 4-24 to use a numeric FOR loop to insert the records again. (Because this program does not require you to declare any variables, there is no declare section.)

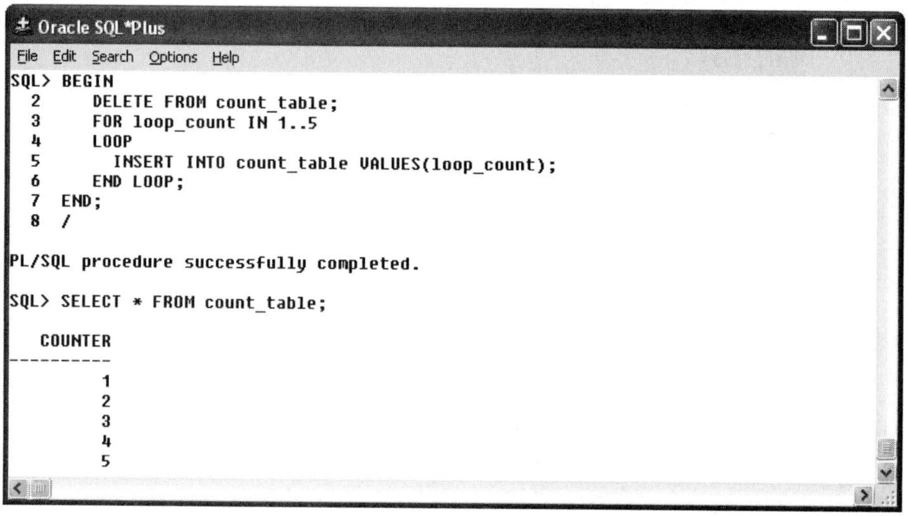

Figure 4-24   PL/SQL program that uses a numeric FOR loop to insert records

2. Copy the commands, paste them into SQL*Plus, run the program, and debug it if necessary.

3. Type and execute the SELECT query in Figure 4-24 to confirm that the program successfully inserted the records.

CURSORS

In Chapter 1, you learned that a pointer is a link to a physical location that stores data. In Oracle 10*g* (and previous Oracle versions), a **cursor** is a pointer to a memory location on the database server that the DBMS uses to process a SQL query. You use cursors to retrieve and manipulate database data in PL/SQL programs. There are two kinds of cursors: implicit and explicit. The following sections describe how to use cursors to retrieve database data values from the tables into PL/SQL programs.

Implicit Cursors

Whenever you execute an INSERT, UPDATE, DELETE, or SELECT query, the DBMS allocates a memory location on the database server called the **context area**. The context area contains information about the query, such as the number of rows that the query processes, and a parsed (machine language) representation of the query. For SELECT queries, the context area also stores the **active set**, which is the set of data rows that the query retrieves. An **implicit cursor** is a pointer to the context area. This cursor is called an implicit cursor because you do not need to write commands explicitly to create it or retrieve its values. Figure 4-25 illustrates an implicit cursor that points to the context area for a query that retrieves the first record from the Northwoods University COURSE table.

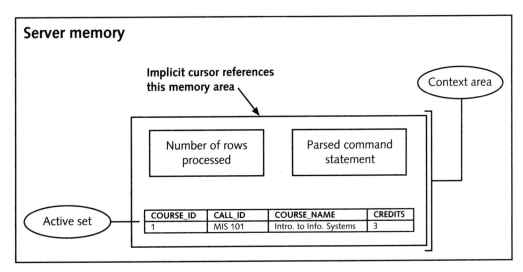

Figure 4-25   Implicit cursor

You can use an implicit cursor to assign the output of a SELECT query to PL/SQL program variables when you are sure that the query will return one—and only one— record. If the query returns more than one record, or does not return any records, an error occurs.

To retrieve data values from a query's implicit cursor into PL/SQL program variables, you add an INTO clause to the SELECT query, using the following syntax:

```
SELECT field1, field2, ...
INTO variable1, variable2, ...
FROM table1, table2, ...
WHERE join_ conditions
AND search_condition_to_retrieve_1_record;
```

The SELECT clause specifies the columns to retrieve, and the INTO clause specifies the names of corresponding PL/SQL variables that reference the column values. The query assigns the value of *field1* to *variable1*, the value of *field2* to *variable2*, and so on. The variables must have been declared in the program's declaration section, and the variables must have the same data types as the associated database columns.

It is useful to use the %TYPE reference data type to declare variables that you use with implicit cursors. Recall that the %TYPE data type creates a variable that has the same data type as a table column, and that you use the following syntax to declare a variable using the %TYPE data type: *variable_name tablename.fieldname*%TYPE. For example, to declare a variable named current_f_last that has the same data type as the F_LAST column in the FACULTY table, you use the command `current_f_last faculty.f_last%TYPE`.

Next you run the scripts to refresh the Northwoods University databases. Then, you write a PL/SQL program that uses an implicit cursor to display the last and first names of the faculty member with the faculty ID (F_ID) value of 1 from the Northwoods University FACULTY table. You assign the last and first name values to variables named current_f_last and current_f_first, which you declare using the %TYPE reference data type.

To refresh your database tables and create a program with an implicit cursor:

1. In SQL*Plus, run the Ch4Northwoods.sql script in the Chapter04 folder of your Data Disk.

2. In Notepad, type the commands in Figure 4-26 to create the program that uses an implicit cursor. Note that you insert a blank space enclosed in single quotation marks before the faculty member's first name, and between the faculty member's first and last names.

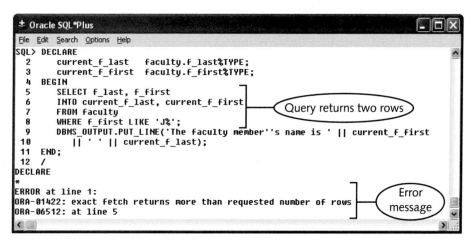

Figure 4-26   PL/SQL program that displays database data using an implicit cursor

    3. Copy the commands, paste them into SQL*Plus, run the program, and debug it if necessary. The output displays the retrieved data values.

Always remember that you can use an implicit cursor only when the SQL query retrieves one and only one record. Figure 4-27 shows the error message that appears when the query in the preceding program is modified to retrieve all of the FACULTY table records.

```
Oracle SQL*Plus
File  Edit  Search  Options  Help
SQL> DECLARE
  2      current_f_last    faculty.f_last%TYPE;
  3      current_f_first   faculty.f_first%TYPE;
  4  BEGIN
  5      SELECT f_last, f_first
  6      INTO current_f_last, current_f_first          Query returns two rows
  7      FROM faculty
  8      WHERE f_first LIKE 'J%';
  9      DBMS_OUTPUT.PUT_LINE('The faculty member''s name is ' || current_f_first
 10          || ' ' || current_f_last);
 11  END;
 12  /
DECLARE
*
ERROR at line 1:
ORA-01422: exact fetch returns more than requested number of rows     Error
ORA-06512: at line 5                                                   message
```

Figure 4-27   Error that appears when an implicit cursor query retrieves multiple records

The error message "ORA-01422: exact fetch returns more than requested number of rows" indicates that the implicit cursor query tried to retrieve multiple records. If you write a program that uses an implicit cursor, and this message appears, you need to modify the implicit cursor query by adding a search condition so that the query retrieves only one record.

If the query for an implicit cursor retrieves no records, then another error occurs. Figure 4-28 shows the error that appears when an implicit cursor query returns no records.

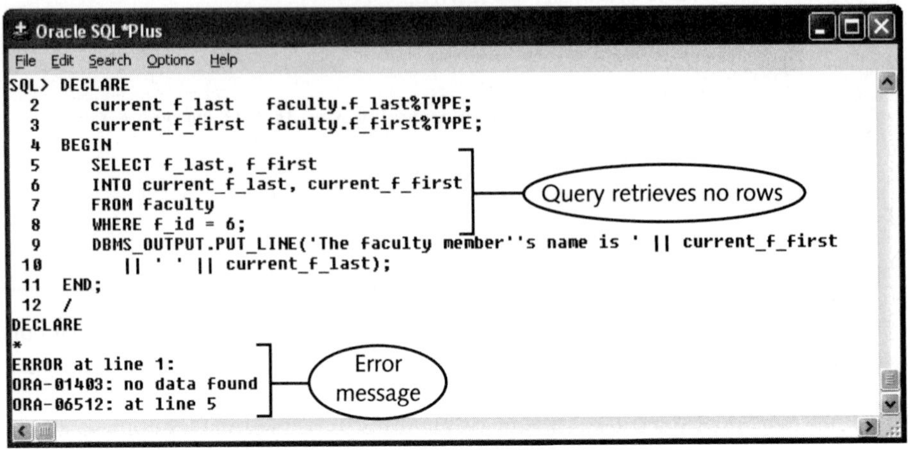

Figure 4-28   Error that appears when an implicit cursor query retrieves no records

In Figure 4-28, the query search condition tries to match F_ID 6. However, the FACULTY table does not contain a record in which F_ID = 6. As a result, the error message stating "no data found" appears.

Implicit cursors provide a fast and easy way to retrieve data into a PL/SQL program from a single record. However, if you want to retrieve data from multiple records, or if there is a chance that a cursor query might retrieve no records, you need to create an explicit cursor.

Explicit Cursors

You create an **explicit cursor** to retrieve and display data in PL/SQL programs for a query that might retrieve multiple records, or that might return no records at all. It is called an explicit cursor because you must explicitly declare the cursor in the program's declaration section, and you must write explicit commands to process the cursor. The steps for creating and using an explicit cursor are:

1. Declare the cursor.

2. Open the cursor.

3. Fetch the data rows.

4. Close the cursor.

Declaring an Explicit Cursor

When you declare an explicit cursor, you create a memory location on the database server that processes a query and stores the records that the query retrieves. You use the following syntax to declare an explicit cursor:

```
CURSOR cursor_name IS select_query;
```

Cursor_name can be any valid PL/SQL variable name. *Select_query* is the query that retrieves the desired data values. You can use any valid SQL SELECT query, including queries that involve set operators and queries with subqueries or nested subqueries. The query's search condition can contain PL/SQL variables, as long as the variables are declared before the cursor is declared and are assigned values before the program opens and processes the cursor.

Opening an Explicit Cursor

The syntax to open an explicit cursor is:

```
OPEN cursor_name;
```

When you execute the command to open a cursor, the PL/SQL interpreter examines, or parses, the cursor's SQL query, confirms that the query contains no syntax errors, and translates the query into a machine-language format. The system then stores the parsed query in the cursor's context area, and creates the memory structure that stores the active set. (Recall that the active set contains the data values that the query retrieves.) However, the cursor does not retrieve the data values yet.

Fetching the Data Rows

You use the FETCH command to retrieve the query data from the database into the active set, one row at a time. The FETCH command associates each column value with a program variable. Because a query might return several rows, you execute the FETCH command within a loop. The following syntax shows how to use the FETCH command within a LOOP...EXIT WHEN loop:

```
LOOP
     FETCH cursor_name INTO variable_name(s);
     EXIT WHEN cursor_name%NOTFOUND;
```

In this syntax, *cursor_name* represents the name of the cursor, and *variable_name(s)* represents either a single variable or a list of variables that receives data from the cursor's SELECT query. Usually, you declare the cursor variable or variables using either the %TYPE or %ROWTYPE reference data types.

When you process an explicit cursor, a pointer called the **active set pointer** indicates the memory location of the next record that is retrieved from the database. When the FETCH command executes past the last record of the query, the active set pointer points to an empty record. When you use the LOOP...EXIT WHEN structure to process a cursor, you use the exit condition WHEN *cursor_name*%NOTFOUND to determine if the last cursor record has been fetched and if the active set pointer is now pointing to an empty record. When the active set pointer points to an empty record, the loop exits.

Closing the Cursor

You should always close a cursor after it processes all of the records in the active set so that its memory area and resources are available to the system for other tasks. The general syntax for the cursor CLOSE command is:

```
CLOSE cursor_name;
```

If you forget to close a cursor, the system automatically closes the cursor when the program that processes the cursor ends.

Processing Explicit Cursors Using a LOOP...EXIT WHEN Loop

The LOOP...EXIT WHEN loop structure is often used to process explicit cursors that retrieve and display database records. The first program in this section uses LOOP...EXIT WHEN and uses a variable that has the %TYPE data type to display a single column value. The second program uses the same loop structure and a variable that has the %ROWTYPE data type to display multiple column values.

Using a %TYPE Variable to Display Explicit Cursor Values

In the next exercise, you write a program that uses an explicit cursor to retrieve and display the ROOM value for every record in the LOCATION table in which the BLDG_CODE value is "LIB". The program stores retrieved values in a variable named current_room that you declare using the %TYPE data type. You structure the search condition to reference the string "LIB" as a variable value. Note that when you use a variable in the cursor query's search condition, the command to declare the variable must appear in the declaration section before the command that declares the cursor. Also note that the EXIT WHEN command in the loop shown in Figure 4-29 is neither a pretest nor a posttest. It is in the middle of two other lines of code. If you move the EXIT WHEN command before or after the other lines in the loop, the LIB 222 output appears twice instead of once, as shown in Figure 4-29. Although nothing is stopping you from placing the test wherever you need to within the loop, you do not get the desired result in this program if you place the test in another location.

To create a program that processes an explicit cursor using a LOOP...EXIT WHEN loop:

1. In Notepad, type the commands in Figure 4-29.

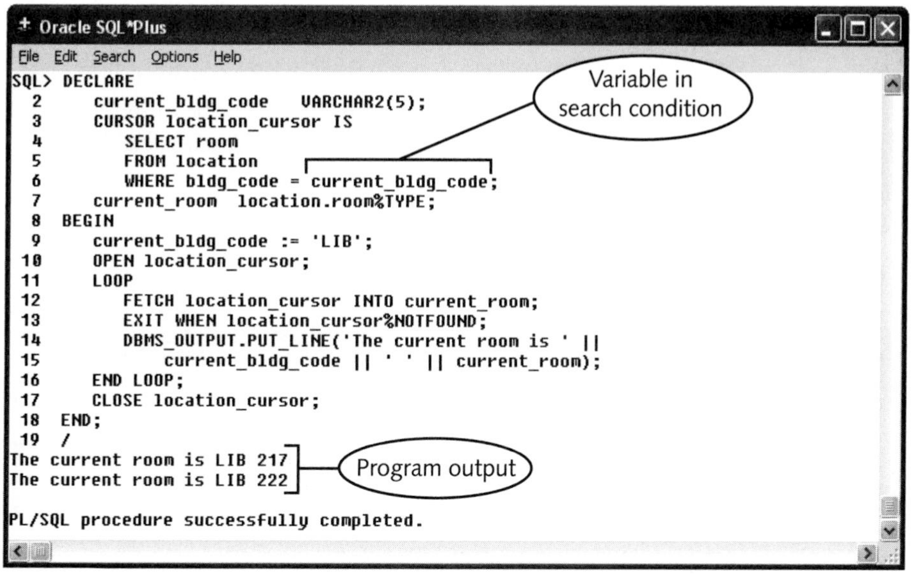

Figure 4-29   Processing an explicit cursor using a %TYPE variable

> 2. Copy the commands, paste them into SQL*Plus, run the program, and debug it if necessary. The formatted output should appear as shown in Figure 4-29.

Using a %ROWTYPE Variable to Display Explicit Cursor Values

In the preceding program, the cursor query retrieved only one column, ROOM. When a cursor retrieves multiple data fields, you can declare multiple %TYPE variables and fetch the output into each individual variable. Or, you can fetch the output into a single variable that has the %ROWTYPE reference data type, and assumes the same data type as the row that the cursor retrieves. By using the %ROWTYPE data type, you have to declare only a single variable to reference all of the cursor fields.

You use the following command to declare a cursor %ROWTYPE variable:

> *row_variable_name cursor_name*%ROWTYPE;

For example, you would use the following command to declare a row variable named location_row that references all of the columns in the SELECT clause for the location_cursor's query:

> `location_row location_cursor%ROWTYPE;`

To reference individual data columns within a %ROWTYPE variable, you use the following syntax: *row_variable_name.fieldname*. For example, you use the expression `location_row.room` to reference the ROOM column in the location_row variable.

Next you modify the program you just wrote so that the query retrieves both the ROOM and CAPACITY columns from the LOCATION table. You modify the FETCH command to fetch the retrieved values into a %ROWTYPE variable named location_row. The commands that display the program output individually reference the ROOM and CAPACITY columns within the %ROWTYPE variable.

To process an explicit cursor using a %ROWTYPE variable:

1. In Notepad, make a copy of the explicit cursor program code, and then modify the commands as shown in Figure 4-30.

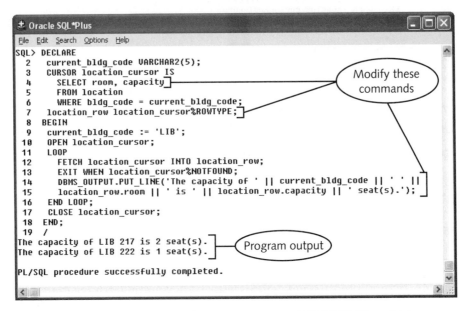

Figure 4-30   Processing an explicit cursor using a %ROWTYPE variable

2. Copy the commands, paste them into SQL*Plus, execute the program, and debug it if necessary. The output should appear as shown in Figure 4-30.

Processing Explicit Cursors Using a Cursor FOR Loop

When you use a LOOP...EXIT WHEN loop to process an explicit cursor, you must write commands explicitly to open the cursor, fetch the records, and then close the cursor. To make it easier to process explicit cursors, PL/SQL has a cursor FOR loop that automatically opens the cursor, fetches the records, and then closes the cursor. The general syntax of a cursor FOR loop is:

```
FOR variable_name(s) IN cursor_name LOOP
processing commands
END LOOP;
```

In this syntax, *variable_name(s)* can include previously declared %TYPE and %ROW-TYPE variables that reference the columns that the cursor query retrieves. *Processing commands* are additional commands that process the cursor values, such as DBMS_OUT-PUT procedure calls to display the cursor output.

NOTE When you process a cursor using a LOOP...EXIT WHEN loop, you can reference the cursor variables outside the loop. With a cursor FOR loop, you cannot reference the value of the current cursor variables outside the cursor FOR loop.

Next you write a program that uses a cursor FOR loop to process location_cursor. You do not need to explicitly open, fetch into, or close the cursor, so the processing requires less code.

To process a cursor using a cursor FOR loop:

1. In Notepad, copy your explicit cursor program, and modify it as shown in Figure 4-31.

```
± Oracle SQL*Plus                                                    [_][□][X]
File  Edit  Search  Options  Help
SQL> DECLARE
  2     current_bldg_code VARCHAR2(5);
  3     CURSOR location_cursor IS
  4        SELECT room, capacity
  5        FROM location
  6        WHERE bldg_code = current_bldg_code;
  7     location_row location_cursor%ROWTYPE;
  8  BEGIN
  9     current_bldg_code := 'LIB';
 10     FOR location_row IN location_cursor LOOP
 11        DBMS_OUTPUT.PUT_LINE('The capacity of ' || current_bldg_code || ' ' ||
 12           location_row.room || ' is ' || location_row.capacity || ' seat(s).');
 13     END LOOP;
 14  END;
 15  /
The capacity of LIB 217 is 2 seat(s).                    Add/modify
The capacity of LIB 222 is 1 seat(s).                   these commands

PL/SQL procedure successfully completed.
```

Figure 4-31   Processing an explicit cursor using a cursor FOR loop

2. Copy the commands, paste them into SQL*Plus, run the program, and debug it if necessary. The LOCATION records appear as shown.

HANDLING RUNTIME ERRORS IN PL/SQL PROGRAMS

Programmers prefer to concentrate on the positive aspects of their programs. Creating programs that support business processes such as generating invoices, processing paychecks, and tracking time usage is difficult and time consuming. As a result, programmers tend to breathe a sigh of relief once a program works and overlook or ignore the consequences of a user entering an incorrect data value or pressing the wrong key. In reality, however, users enter incorrect data and press wrong keys. Networks fail, computers crash, and anything else that can go wrong (eventually) goes wrong. Programmers can't do much if a user's computer fails, but programmers should do everything possible to prevent users from entering incorrect data and to keep incorrect keystrokes from damaging the system.

Well-written programs help users avoid errors, inform users when an error occurs, and provide advice for correcting errors. PL/SQL supports **exception handling**, whereby programmers place commands for displaying error messages and giving users options for fixing errors in the program's exception section. (Recall from Figure 4-1 that the exception section is optional, and appears after the execution section and before the final END; command.)

You can classify program errors as syntax errors, runtime errors, and logic errors. Syntax errors occur when the interpreter or compiler checks the program syntax, and usually involve spelling or syntax problems. You must correct syntax errors before a program executes. Logic errors occur when the program runs, yet does not display the correct information. This section focuses on runtime errors, which are errors that cause the program to fail during execution (the program having made it past the interpreter or compiler syntax checks). Logic errors often contribute to runtime errors. These errors usually are reclassified as runtime errors. Such errors usually involve problems with data values, such as trying to retrieve no rows or several rows using an implicit cursor, trying to divide by zero, or trying to insert into a database table a value that violates the table's constraints. You use exception handling to handle runtime errors. Figure 4-32 shows a runtime error that occurs when an action query to insert a record into the TERM table does not succeed.

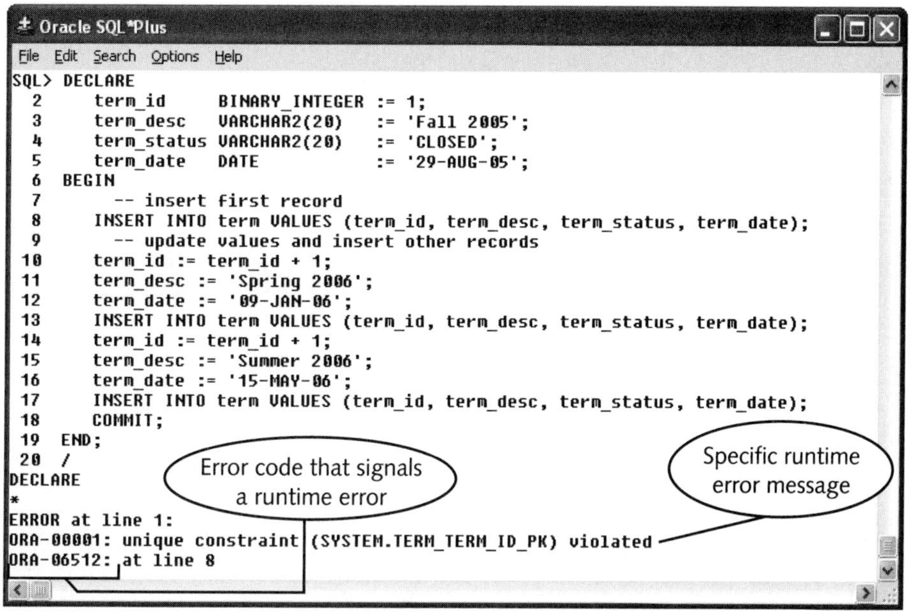

```
± Oracle SQL*Plus                                                    [_][□][X]
File  Edit  Search  Options  Help
SQL> DECLARE
  2       term_id      BINARY_INTEGER := 1;
  3       term_desc    VARCHAR2(20)   := 'Fall 2005';
  4       term_status VARCHAR2(20)   := 'CLOSED';
  5       term_date    DATE           := '29-AUG-05';
  6  BEGIN
  7       -- insert first record
  8       INSERT INTO term VALUES (term_id, term_desc, term_status, term_date);
  9       -- update values and insert other records
 10       term_id := term_id + 1;
 11       term_desc := 'Spring 2006';
 12       term_date := '09-JAN-06';
 13       INSERT INTO term VALUES (term_id, term_desc, term_status, term_date);
 14       term_id := term_id + 1;
 15       term_desc := 'Summer 2006';
 16       term_date := '15-MAY-06';
 17       INSERT INTO term VALUES (term_id, term_desc, term_status, term_date);
 18       COMMIT;
 19  END;
 20  /
DECLARE
*
ERROR at line 1:
ORA-00001: unique constraint (SYSTEM.TERM_TERM_ID_PK) violated
ORA-06512: at line 8
```

Error code that signals a runtime error

Specific runtime error message

Figure 4-32   Example of a runtime error

The INSERT action query does not succeed because a record already exists in the table that has the same primary key value (TERM_ID 1) as the new record. This violates the TERM table constraint that ensures that the primary key for each record is unique. The error code ORA-00001 provides specific details about the error. The error code ORA-06512 indicates that this is a runtime error rather than a compile error, and shows the line number that generates the error.

When a runtime error occurs, an **exception**, or unwanted event, is **raised**. Program control immediately transfers to the program's exception section, where exception handlers exist to deal with, or **handle**, different error situations. An exception handler contains one or more commands that provide operation instructions when a specific exception is raised. An exception handler might correct the error without notifying the user of the problem, or it might inform the user of the error without taking corrective action. An exception handler could also correct the error and inform the user of the error, or it could inform the user of the error and allow the user to decide what action to take. After an exception handler executes, the program ends.

There are three kinds of exceptions: predefined, undefined, and user defined. The following sections describe these exceptions and how to write their associated exception handlers.

Predefined Exceptions

Predefined exceptions are the most common errors that occur in programs. The PL/SQL language assigns an exception name and provides a built-in exception handler for each predefined exception. Table 4-10 summarizes the most common predefined exceptions.

Oracle Error Code	Exception Name	Description
ORA-00001	DUP_VAL_ON_INDEX	Command violates primary key unique constraint
ORA-01403	NO_DATA_FOUND	Query retrieves no records
ORA-01422	TOO_MANY_ROWS	Query returns more rows than anticipated
ORA-01476	ZERO_DIVIDE	Division by zero
ORA-01722	INVALID_NUMBER	Invalid number conversion (such as trying to convert "2B" to a number)
ORA-06502	VALUE_ERROR	Error in truncation, arithmetic, or data conversion operation

Table 4-10   Common PL/SQL predefined exceptions

With predefined exceptions, the system automatically displays an error message informing the user of the nature of the problem. For example, the program in Figure 4-32 raised the ORA-00001 exception (this is the predefined DUP_VAL_ON_INDEX exception), and the message "unique constraint (SYSTEM.TERM_TERM_ID_PK) violated" appeared. The expression in parentheses is the name of the constraint that was violated, which consists of the user schema name (SYSTEM) and the constraint name (TERM_TERM_ID_PK).

You can create exception handlers to display alternate error messages for predefined exceptions. This allows you to make the error messages easier to understand and allows you to provide instructions to the user for correcting the error. Figure 4-33 shows the general syntax of an exception handler that traps predefined errors and replaces the system error messages with alternate error messages.

```
EXCEPTION
  WHEN exception1_name THEN
     exception1 handler commands;
  WHEN exception2_name THEN
     exception2 handler commands;
  ...
  WHEN OTHERS THEN
     other handler commands;
END;
```

Figure 4-33   Exception handler syntax

In this syntax, *exception1_name, exception2_name*, and so forth refer to the name of the predefined exception, as shown in Table 4-10. *Exception handler commands* represent the program commands that display the appropriate error message for each exception. The WHEN OTHERS THEN clause is a catchall exception handler that allows you to present a general message to describe errors that do not have specific exception handlers. After an exception handler finishes processing, the program exits. You indent the exception handlers as shown in Figure 4-33 to make the program's exception section easier to read and understand.

Next you write a program that displays an alternate message for a predefined exception. In this program, an implicit cursor tries to select a course name from the COURSE_NAME column in the COURSE table using a search condition that returns no records. When an implicit cursor query returns no records, the NO_DATA_FOUND predefined exception is raised. First, you write the program that generates the predefined exception.

To write a program that generates a predefined exception:

1. In Notepad, type the commands in Figure 4-34, which generate a predefined exception.

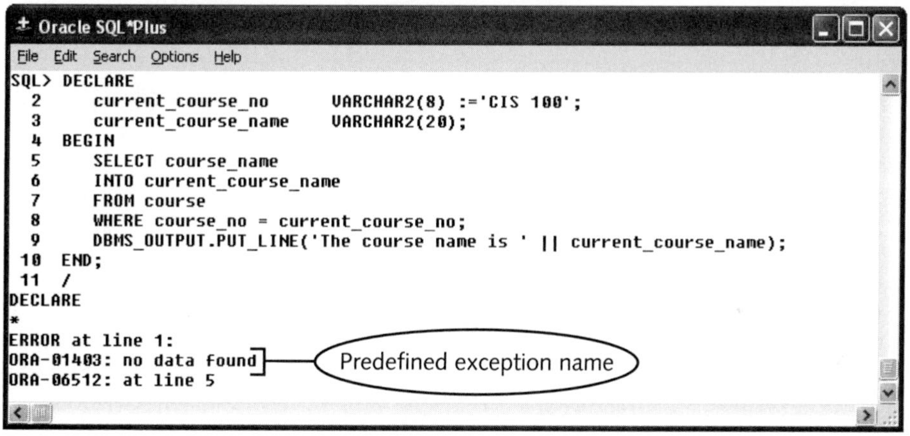

Figure 4-34   PL/SQL program that generates a predefined exception

2. Copy the commands, paste them into SQL*Plus, and then run the program. The Oracle error code ORA-06512 appears, indicating that there is a runtime error. The specific Oracle error code ORA-01403 and associated predefined error message "no data found" also appear. (If another error appears, debug the program until your output matches the output in Figure 4-34.)

Table 4-10 shows that this error corresponds to the NO_DATA_FOUND predefined exception. Next you modify the program by adding an exception handler that handles the NO_DATA_FOUND error. This exception handler displays an alternate error message that informs the user of the nature of the error and suggests a corrective action.

To add the exception handler that displays the alternate error message:

1. In Notepad, make a copy of your previous program, then add the commands in Figure 4-35 to create the exception handler.

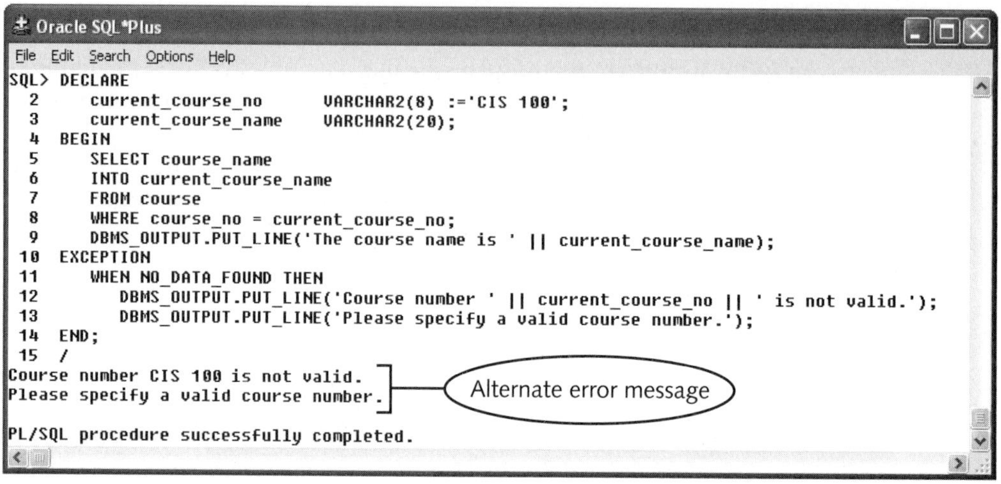

Figure 4-35 Creating an exception handler that displays an alternate message for a predefined exception

2. Copy the commands, paste them into SQL*Plus, run the program, and debug it if necessary. The more informative and instructive error messages specified by the exception handler now appear.

During program development, it is helpful to use the WHEN OTHERS exception handler to display the associated Oracle error code number and error message for unanticipated errors. To do this, you use the SQLERRM built-in function. The SQLERRM function returns a character string that contains the most recent error's error code and error message text. To use this function, you must first declare a VARCHAR2 character variable to which the text of the error message is assigned. The maximum length of the character variable is 512, because the maximum length of an Oracle error message is 512 characters.

Now you modify your program so that the implicit cursor query returns all of the rows from the COURSE table. Because an implicit cursor query cannot process more than one row, the query generates an exception. The WHEN OTHERS exception handler is used to display the associated Oracle error code and message when this or any other unhandled error occurs.

To create a WHEN OTHERS exception handler to handle unanticipated errors:

1. In Notepad, make a copy of your previous program. In the copied program, modify the cursor SELECT command, and add the commands for the WHEN OTHERS exception handler, as shown in Figure 4-36.

2. Copy the commands, paste the commands into SQL*Plus, run the program, and debug it if necessary. The error code and message appear as shown in Figure 4-36.

4

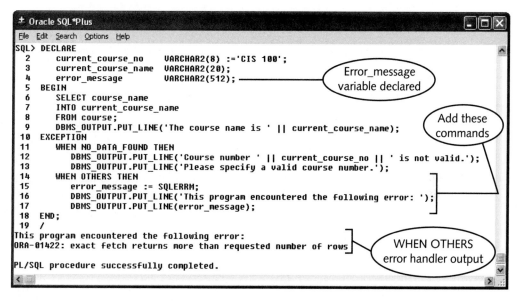

Figure 4-36   Creating a WHEN OTHERS error handler

Undefined Exceptions

You have learned that predefined exceptions are the common errors that have predefined names, as listed in Table 4-10. **Undefined exceptions** are less common errors that do not have predefined names. Figure 4-37 shows an example of a program that generates an undefined exception.

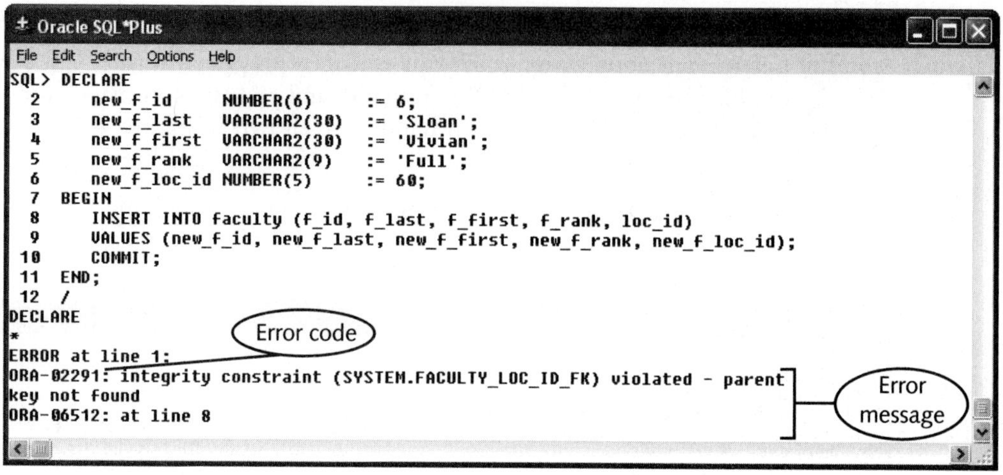

```
Oracle SQL*Plus
File  Edit  Search  Options  Help
SQL> DECLARE
  2      new_f_id     NUMBER(6)      := 6;
  3      new_f_last   VARCHAR2(30)   := 'Sloan';
  4      new_f_first  VARCHAR2(30)   := 'Vivian';
  5      new_f_rank   VARCHAR2(9)    := 'Full';
  6      new_f_loc_id NUMBER(5)      := 60;
  7  BEGIN
  8      INSERT INTO faculty (f_id, f_last, f_first, f_rank, loc_id)
  9      VALUES (new_f_id, new_f_last, new_f_first, new_f_rank, new_f_loc_id);
 10      COMMIT;
 11  END;
 12  /
DECLARE
*
ERROR at line 1:                                     Error code
ORA-02291: integrity constraint (SYSTEM.FACULTY_LOC_ID_FK) violated - parent          Error
key not found                                                                        message
ORA-06512: at line 8
```

Figure 4-37   Example of an undefined exception

In this program, an INSERT action query tries to insert into the FACULTY table a record for which the LOC_ID data value is 60. Recall that LOC_ID is a foreign key column in the FACULTY table, and values that action queries insert as foreign keys must already exist in the parent table. If you examine the current values in the LOCATION table in the Northwoods University database in Figure 1-26, you see that there is no value for LOC_ID 60. As a result, the action query violates the foreign key constraint. Because this is an undefined error, the error message does not display an exception name, but instead displays the Oracle error code and associated error message.

To handle an undefined exception such as the -02291 error shown in Figure 4-37, you must explicitly declare the exception in the program's declaration section, and associate the new exception with a specific Oracle error code. Then, you create an exception handler in the exception section using the same syntax as for predefined exceptions. The general syntax for declaring an exception in the program's declaration section is:

```
DECLARE
    e_exception_name EXCEPTION;
    PRAGMA EXCEPTION_INIT(e_exception_name,
    -Oracle_error_code);
```

In this syntax, e_*exception_name* represents the name you assign to the exception and can be any legal PL/SQL variable name. Usually, you preface user-declared exceptions with the characters e_ to mark them as being handled by exception handlers. The PRAGMA EXCEPTION_INIT command associates your exception name with a specific Oracle error code. The *Oracle_error_code* parameter is the 5-digit numeric error code that Oracle assigns to runtime errors. For the error in Figure 4-37, the *Oracle_error_code* value is -02291. You must preface the error code number with a hyphen (-), and you can omit leading zeros, so you could express the *Oracle_error_code* parameter value for the error in Figure 4-37 as **-2291**.

Recall that you can find a complete listing of ORA- error codes on the Oracle Technology Network Web site at *http://otn.oracle.com*. Usually, you are familiar with the error codes that your programs generate, because you see them so often during development! Next, you write a program that declares an exception named e_foreign_key_error that you associate with the code ORA-02291. When the user tries to insert a value that has a foreign key constraint, and the value does not exist in its parent table, the exception handler executes and displays a message.

To write a program to handle an undefined exception:

1. In Notepad, type the commands in Figure 4-38 to create the exception handler for the undefined exception.

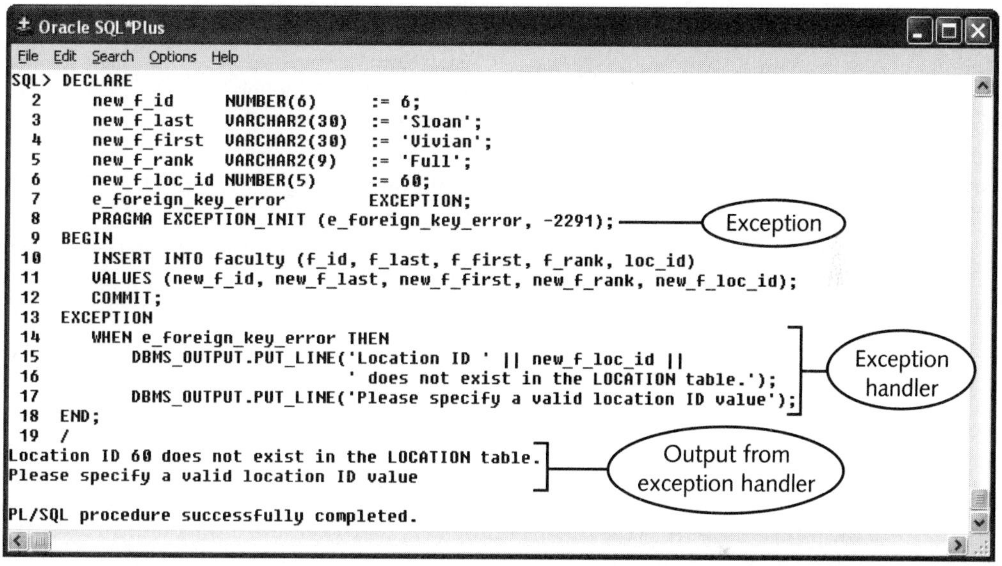

Figure 4-38   Creating an exception handler for an undefined exception

2. Copy the commands, paste them into SQL*Plus, run the program, and debug it if necessary. The exception handler error message appears instead of the Oracle error codes and messages as shown in Figure 4-38.

User-defined Exceptions

You create user-defined exceptions and associated exception handlers to handle errors that do not raise an Oracle runtime error, but that require exception handling to enforce business rules or to ensure the integrity of the database. For example, suppose that Northwoods University has a business rule that states that the only records that users can delete from the ENROLLMENT table are records in which the GRADE value is NULL. Records for which instructors have assigned grades cannot be deleted. If a user tries to delete an ENROLLMENT record in which the GRADE value is not NULL, the program should raise an exception, and display a message advising the user that the grade column is not NULL and the record cannot be deleted.

Figure 4-39 shows the general syntax for declaring, raising, and handling a user-defined exception.

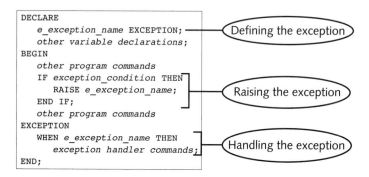

```
DECLARE
    e_exception_name EXCEPTION;          Defining the exception
    other variable declarations;
BEGIN
    other program commands
    IF exception_condition THEN
        RAISE e_exception_name;          Raising the exception
    END IF;
    other program commands
EXCEPTION
    WHEN e_exception_name THEN
        exception handler commands;      Handling the exception
END;
```

Figure 4-39   General syntax for declaring, raising, and handling a user-defined exception

In this syntax, you define the exception in the declaration section, using the same syntax as you used to define undefined exceptions in the program in Figure 4-38. To raise the exception, you use an IF/THEN decision structure in which the condition evaluates the exception condition. In the example for deleting GRADE values from the ENROLLMENT table, *exception_condition* evaluates whether the GRADE value is not NULL. If *exception_condition* is TRUE, the RAISE command raises the exception and transfers processing to the exception section. In the exception section, the exception handler executes and displays the error-handling messages.

In the next exercise, you write a program that contains commands to create, raise, and handle a user-defined exception handler to avoid deleting an ENROLLMENT record with an assigned grade. Then the program attempts to delete the first ENROLLMENT record, which has a GRADE value, so that the program raises the user-defined exception.

To create a program with a user-defined exception handler:

1. Type the commands in Figure 4-40 to define, raise, and handle the user-defined exception.

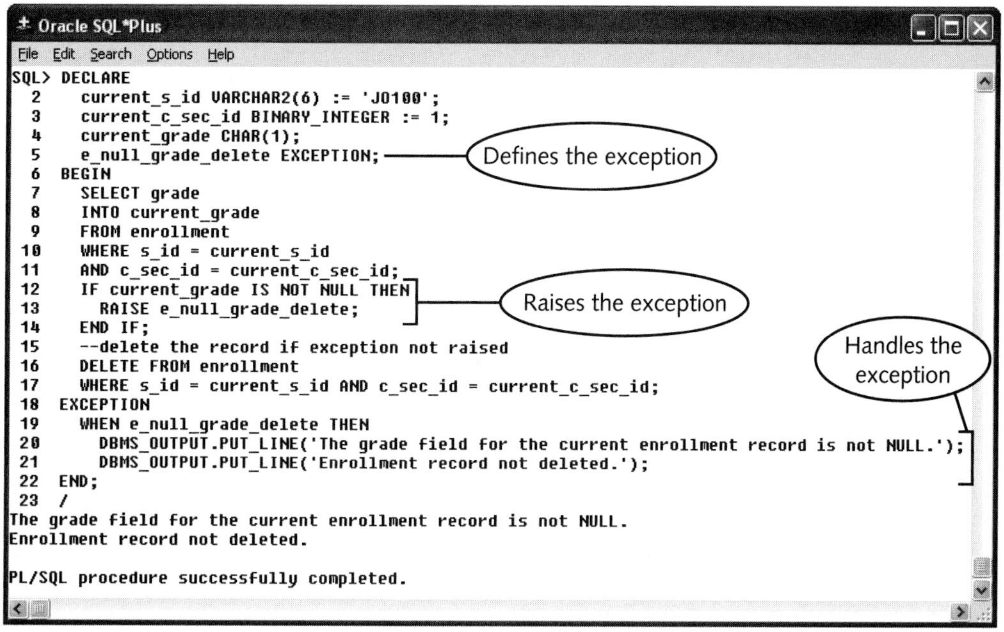

Figure 4-40 PL/SQL program to define, raise, and handle a user-defined exception

2. Copy the commands, paste them into SQL*Plus, run the program, and debug it if necessary. The user-defined exception message appears.

3. Exit SQL*Plus.

4. Save your Notepad file, and then close Notepad.

You now understand the basic fundamentals of the PL/SQL programming language. In later chapters, you learn how to use PL/SQL commands to create database applications and to create programs that you store in the database and make available to other users.

SUMMARY

- The IF/THEN decision control structure evaluates a condition and executes a series of program commands if the condition is true. The IF/THEN/ELSE structure evaluates a condition, executes one series of program commands if the condition is true, and executes an alternate set of commands if the condition is false. The IF/ELSIF structure allows you to test for many different conditions and execute associated commands.

- You can execute DML and transaction control commands in PL/SQL programs. You cannot execute DDL commands in PL/SQL programs. To use a SQL action query or transaction control command in a PL/SQL command, you insert the query or command in the PL/SQL program using the same syntax you use to execute the query or command in SQL*Plus. You can use PL/SQL variables to represent values in action queries and search expressions in search conditions.

❐ A loop repeats an action multiple times until it reaches an exit condition. PL/SQL has five different types of loop: LOOP...EXIT, LOOP...EXIT WHEN, WHILE...LOOP, numeric FOR loops, and cursor FOR loops.

❐ A cursor is a pointer to a memory location that the DBMS uses to process a SQL query. You use cursors to retrieve and manipulate database data in PL/SQL programs.

❐ Whenever you execute an INSERT, UPDATE, DELETE, or SELECT query, the DBMS creates an implicit cursor. You can use implicit cursors to retrieve values into PL/SQL programs from SELECT queries that return one and only one record.

❐ You create an explicit cursor to retrieve and display data in PL/SQL programs for a query that might retrieve multiple records or might return no records at all. To use an explicit cursor, you must declare the cursor, open it, fetch the records, and then close the cursor. Usually, an explicit cursor fetches records into a variable that is declared using either the %TYPE or %ROWTYPE reference variable data type. When you process an explicit cursor using a cursor FOR loop, you do not need to explicitly open the cursor, fetch the rows, or close the cursor.

❐ PL/SQL uses exception handling, whereby all code for displaying error messages and giving users options for fixing errors is placed in the program's exception section. When a runtime error occurs, an exception (or unwanted event) is raised, and program execution immediately transfers to the exception section. The exception section contains exception handlers, which are commands that display error messages for specific errors.

❐ Predefined exceptions are common errors that have predefined exception names. You can create exception handlers that display alternate error messages for predefined exceptions.

❐ Undefined exceptions are less common errors. To handle an undefined exception, you must define the exception and associate it with an Oracle error code, then create an exception handler.

❐ User-defined exceptions handle exceptions that do not cause an Oracle runtime error, but require exception handling to enforce business rules or ensure the integrity of the database.

REVIEW QUESTIONS

1. What is the purpose of including exception handling in a PL/SQL block?

2. An explicit cursor is not required if a SELECT statement in a PL/SQL block returns _____ row(s) of results.

3. An explicit cursor must be declared if a SELECT statement is included in a Cursor FOR loop. True or False?

4. Which types of commands cannot be executed in a PL/SQL program block?

5. What happens to a LOOP...EXIT WHEN loop if the exit condition is never TRUE?

6. In a complex condition that uses both AND and OR logical operators, the _____ operator is evaluated first.

7. In an IF/ELSIF decision control structure, the commands following the ELSIF keyword are executed if no other conditions are true. True or False?

8. If a condition in an IF/THEN/ELSE decision control structure is evaluated as NULL, the program statements are executed. True or False?

9. What command is used to declare a user-defined exception?

10. Predefined errors are handled in the _____ section of a PL/SQL program block.

4

MULTIPLE CHOICE

1. Which of the following symbols is used for comparisons in a PL/SQL block?
   a. :=
   b. =
   c. ^
   d. =>

2. An explicit cursor may be required for which type of operation in a PL/SQL program?
   a. INSERT
   b. SELECT
   c. DELETE
   d. UPDATE

3. Which of the following types of loops never executes if the initial condition is FALSE?
   a. WHILE...LOOP
   b. LOOP...EXIT WHEN
   c. LOOP...EXIT
   d. all of the above

4. Which of the following commands cannot be used in a PL/SQL program?
   a. ROLLBACK
   b. SAVEPOINT
   c. DROP
   d. INSERT

5. Given that first_variable = 12 and second_variable = 48, which of the following conditions is evaluated as TRUE?

    a. first_variable < 33 OR second_variable < first_variable

    b. first_variable > 33 AND second_variable < first_variable

    c. first_variable > 33 OR second_variable < first_variable

    d. first_variable > 33 AND second_variable > first_variable

6. Which line in the following IF/THEN/ELSE structure contains an error?

```
1     IF current_day = 'Wednesday' THEN
2         DBMS_OUTPUT.PUT_LINE('Today is Wednesday!');
3     ELSE
4      IF current_day = 'Friday' THEN
5         DBMS_OUTPUT.PUT_LINE('Today is the last day of ↵
          the workweek!');
6       ELSIF
7         DBMS_OUTPUT.PUT_LINE('Today is not Wednesday or ↵
          Friday.');
8       END IF;
9     END IF;
```

    a. 1

    b. 4

    c. 6

    d. 9

7. Given that first_variable = 12 and second_variable = 48, which of the following conditions is evaluated as TRUE?

    a. first_variable = 12 AND second_variable < first_variable OR second_variable > 48

    b. (first_variable = 12 AND second_variable < first_variable) OR second_variable = 48

    c. first_variable = 12 OR (second_variable < first_variable AND second_variable = 48)

    d. (first_variable = 12 OR second_variable < first_variable) AND second_variable = 48

8. When a SELECT statement is used with an explicit cursor in a PL/SQL block, which of the following clauses is *not* required?

    a. SELECT

    b. FROM

    c. INTO

    d. None—all of the above are required.

9. Which cursor attribute can be used to determine when a loop should be exited?

   a. %ROWFOUND

   b. %NOTFOUND

   c. %OPEN

   d. %ROWTYPE

10. The ORA-01422 error code is an example of a(n) _____ exception.

   a. predefined

   b. unnamed

   c. undefined

   d. user-defined

4

PROBLEM-SOLVING CASES

Execute the Ch4Clearwater.sql script file in the Chapter4 folder on your Data Disk before attempting the following cases. Use Notepad or another text editor to create the following PL/SQL programs. Save all solution files in the Chapter4\Cases folder on your Solution Disk.

1. Create a PL/SQL program block that retrieves each customer from the CUSTOMER table and creates a list that displays each customer's name, address, and daytime phone number on a single line. Print a heading at the beginning of the list identifying it as the "Clearwater Traders Mailing List". Save the file as Ch4BCase1.sql.

2. Create a PL/SQL program block that identifies each customer's name and the total amount owed by each customer. Also display the total amount owed by all customers as the final line of output. Save the file as Ch4BCase2.sql.

3. Create a PL/SQL program block that displays each order and the items on that order. Save the file as Ch4BCase3.sql.

4. The management of Clearwater Traders is considering implementing a new discount policy. If the total order is more than $100, the customer receives a 10% discount; if it totals more than $200, a 20% discount is applied. Create a PL/SQL block that calculates the total amount due for each order before any discount is applied and then the revised amount due with the appropriate discount applied. Include only the order ID, original amount, and discounted amount in the results. Save the file as Ch4BCase4.sql.

5. Create a PL/SQL block that can be used to retrieve the description and price of an inventory item based on its item ID. Include the appropriate exception handling to display a message if the item is not found. Save the file as Ch4BCase5.sql.

5

INTRODUCTION TO FORMS BUILDER

◀ LESSON A ▶

After completing this lesson, you should be able to:
- ♦ Display Forms Builder forms in a Web browser
- ♦ Use a data block form to view, insert, update, and delete database data
- ♦ Create a data block form that displays a single record at a time
- ♦ Use the Object Navigator to change form object names
- ♦ Use the Data Block and Layout Wizards to modify form properties
- ♦ Create a tabular-style data block form that displays multiple records

You have learned to use SQL commands to insert, update, delete, and view database data. It is not practical, however, to expect users to regularly interact with a database by creating SQL queries. Instead, users use database applications called forms to interact with a database. A **form** looks like a paper form and provides a graphical interface that allows users to easily insert new database records and to modify, delete, or view existing records. Programmers use the Forms Builder utility to create forms. Forms Builder is part of the Oracle Developer Suite 10*g*. This lesson describes how to create and use data block forms. A **data block form** is a form that is associated with one or more specific database tables.

DISPLAYING FORMS IN A WEB BROWSER

Figure 5-1 shows an example form that displays data values from the Northwoods University LOCATION table. You can use the form's text fields to enter new data values, and to display, update, and delete existing data values. Developer10*g* displays forms as Web pages in a Web browser. The form appears in the Forms Services window within the browser window. The Forms Services application displays menus and a toolbar that provide general functions for all form applications.

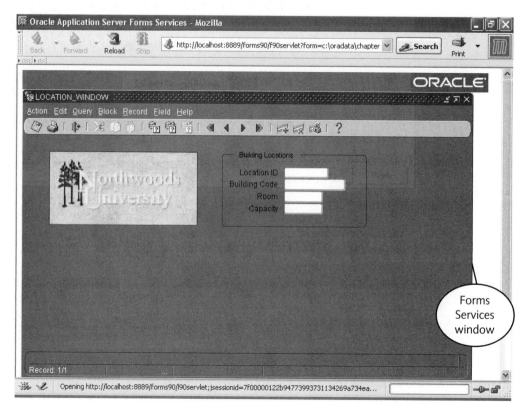

Figure 5-1 Forms Builder Location form displayed using a Web browser

Figure 5-1 shows the form in Mozilla. Although you can display forms in other Web browsers, this book uses Mozilla and sometimes the Internet Explorer Web browsers.

Displaying forms in a Web browser makes it easy to distribute data to users. A brief introduction to the architecture of the World Wide Web will help you understand how Forms Builder displays forms as Web pages.

Architecture of the World Wide Web

The World Wide Web consists of networked computers on the Internet. Some of the computers act as servers, and other computers act as clients. (Recall that a server shares its resources, such as files, hardware, or programs, with other computers, and a client requests and uses server resources.) Today, almost any computer can be both a client and a server at the same time. For example, your workstation can share files with other workstations, which makes your workstation a server, and it can also request files from other servers.

On the Web, users at home or in offices work on client-side computers that are connected to the Internet, and use programs called **Web browsers**, or simply **browsers**, to access information on the Internet. Some popular browsers are Mozilla, Netscape Navigator, Foxfire, and Microsoft Internet Explorer. **Web servers** are computers that are connected to the Internet and run special Web server software. Web servers store the files that people can access via the Internet using a browser. A Web server listens for messages that are sent to the server from client browsers. When a Web server receives a message from a browser, it reads and sends, or downloads, the requested Web page back across the Internet to the user's browser. Figure 5-2 illustrates the Web architecture.

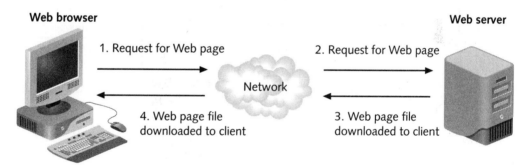

Figure 5-2   Web architecture

Web pages can be static or dynamic. In a **static Web page**, the developer determines the content when creating the page, and the page always displays the same information. In a **dynamic Web page**, the content varies based on user inputs or data retrieved from external sources. A Forms Builder form that appears in a Web browser is a dynamic Web page that derives its content from an Oracle 10*g* database.

How Forms Builder Displays a Form in a Browser

To create a dynamic Web page that displays database data, database application developers specify the form's contents, appearance, and functionality in Forms Builder. When they run the form, the Forms Builder system translates the design specification into a Java applet. A **Java applet** is a self-contained Java program that runs in a Web browser's generic Java runtime environment. Java is a full-featured, object-oriented programming language that programmers use to create many different kinds of applications. Most popular Web browsers, including Internet Explorer and Netscape Navigator, support a generic Java run-time environment that executes Java commands and displays Java objects on Web pages.

How does Forms Builder translate the form design specification into a Java applet? It uses a Web server process called an **OC4J Instance**. When you are developing forms, your workstation runs both the Web server and the Web browser in its main memory. Then, when you run a form, the Forms Builder development environment compiles the form design file, which has an .fmb extension, into a file with an .fmx extension, and sends the compiled .fmx file to the OC4J Instance. The OC4J Instance translates the compiled .fmx file into a Java applet, and then downloads the Java applet to your Web browser, as shown in Figure 5-3.

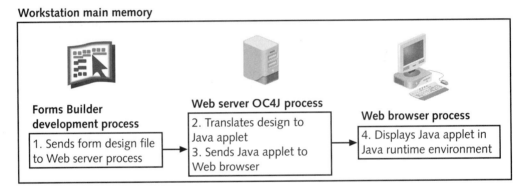

Figure 5-3   Displaying a Forms Builder form using a Web browser

USING A DATA BLOCK FORM

A **block** is a group of related form items, such as text fields and option buttons. A **data block** corresponds to a specific database table, and contains objects, such as text fields or option buttons, that display values from the table's data fields. A data block form contains one or more data blocks. The most common block items are **text items**, which display text data values in text fields. The Northwoods University LOCATION form contains a single data block with text items that represent all of the fields in the LOCATION table. Because this form displays data from only one table, it is a single-block form. Later, you learn how to create a multiple-block form, which displays data from multiple database tables.

Next you start Forms Builder, open the Northwoods University LOCATION form shown in Figure 5-1, and run the form to become familiar with its appearance and functionality. Before you open and run the form, you run the SQL scripts to refresh your database tables, and you start the OC4J Instance.

NOTE

The currently released version of Oracle 10*g* Developer tools do not support the INTERVAL data type. Therefore, the scripts that you use in this chapter to generate the case study databases store the interval data values as character strings.

To refresh your database tables:

1. Start SQL*Plus and log onto the database.

2. Use the **START** command to run the Ch5Northwoods.sql script in the Chapter05 folder on your Data Disk, and then close SQL*Plus.

3. To start the OC4J Instance, click **Start**, point to **All Programs**, point to **Oracle Developer Suite – OraHome**, point to **Forms Developer**, and click **Start OC4J Instance**. A command window with the title Start OC4J Instance opens.

HELP

If you are using Windows 2000, you click Programs rather than All Programs, point to Oracle Developer Suite — OraHome, point to Forms Developer, and click Start OC4J Instance.

4. Minimize the Start OC4J Instance window.

Now you are going to start the Forms Builder development environment, open the LOCATION form, save the form on your Solution Disk, and run the form.

To start Forms Builder, open the form, and run the form:

1. Click **Start**, point to **All Programs**, point to **Oracle Developer Suite – OraHome**, point to **Forms Developer**, and click **Forms Builder**. The Forms Builder development environment opens. (You learn how to use the development environment later in this lesson.)

2. To open the LOCATION form, click **File** on the menu bar, click **Open**, navigate to the Chapter05 folder on your Data Disk, select **Location.fmb**, and click **Open**. The form appears in the Object Navigator window, which displays the form and its components as a hierarchical tree.

TIP

Another way to open a form is to click the Open button 📂 on the toolbar.

3. To save the form on your Solution Disk, click **File** on the menu bar, click **Save As**, navigate to the Chapter05\Tutorials folder on your Solution Disk, and click **Save**.

When you run a form in Forms Builder, the form filename and folder path cannot contain any blank spaces.

CAUTION

4. To run the form, click **Program** on the menu bar, and then click **Run Form**. A dialog box opens, and asks if you want to log on before compiling the form. Click **Yes**.

Another way to run a form is to click the Run Form button on the toolbar.

TIP

5. Type your username, password, and connect string in the Connect dialog box, and then click **Connect**. After a few moments, the Northwoods University LOCATION form appears in your Web browser, as shown in Figure 5-4.

If a Security Warning dialog box opens asking if you want to install and run Oracle JInitiator, click Yes. The JInitiator application initializes the Java applet in your Web browser. Accept the defaults on the Choose Destination Location page, and click Next. When the Installation Complete dialog box appears, click OK.

HELP

If an error message with the text "FRM-90928: Positional parameter after key on command line" appears, it means that your form filename or folder path contains blank spaces. Save the file using a filename and folder path that do not contain blank spaces.

HELP

If the Web page does not appear and your Web browser displays the characters http://%%20 (or something similar) in the Address field, you need to configure Forms Builder so it correctly accesses your Web browser. Close the browser window, click Edit on the menu bar, click Preferences, select the Runtime tab, click the insertion point in the Web Browser Location field, click Browse, and then navigate to the location of your Web browser executable file. If you use Internet Explorer, the browser executable file is C:\Program Files\Internet Explorer\iexplore.exe. Click Open, and then click OK.

HELP

You should log onto the Oracle database using the same username and password, and connect string values you used to log on to SQL*Plus in Chapters 2, 3, and 4.

NOTE

Within the browser window, the Java applet displays the Forms Services window. Figure 5-4 shows a full-screen view of the Forms Services window in Internet Explorer.

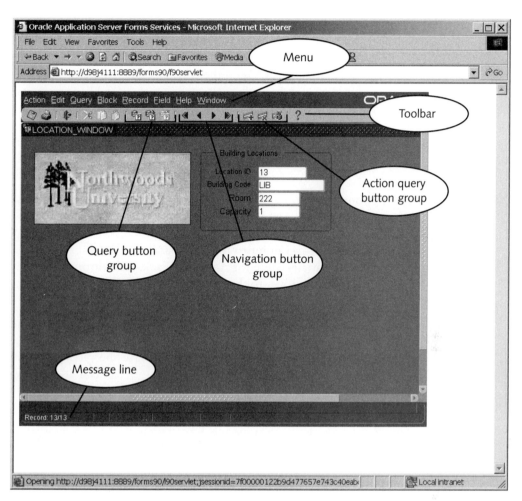

Figure 5-4   Forms Services window

The Forms Services window has a form menu that contains selections that allow you to manipulate the form. The Forms Services window also has a toolbar with buttons that you can use to insert, view, modify, and delete records. In this book, you primarily use the toolbar buttons to control the form. At the bottom of the window are lines that display information about the form's status.

From the top left, the first toolbar button is the Save button, which you use to save a new record or to save changes to an existing record. The second button is the Print button, which prints the current form. The third button is the Exit button, which allows you to exit the form without exiting the Web browser.

The Query button group allows you to query data. When a form is running, it can be in one of two modes: Normal or Enter Query. When a form is in **Normal mode**, you can view data records, sequentially step through the records, and change data values. When a form is in **Enter Query mode**, you can enter a search expression in one of the form fields, and then retrieve the associated records. To place the form in Enter Query mode, you click the Enter Query button ⬚ on the toolbar. Clicking ⬚ clears the form fields, and allows you to enter a search expression in one or more fields. To return the form to Normal mode, you must either execute the query or cancel the query. To execute the query, you click the Execute Query button ⬚. Clicking ⬚ retrieves the records associated with the search expression, and returns the form to Normal mode. To cancel the query and return the form to Normal mode without retrieving any records, you click the Cancel Query button ⬚.

The Navigation button group allows you to navigate among different records and different blocks. In the Navigation button group, the Previous Block button ◀ moves the insertion point to the previous data block in a multiple-block form, and the Next Block button ▶ moves the insertion point to the next data block. When the results of a query appear, you can sequentially step forward and backward through the records. The Previous Record ◀ button moves back to the previous record in the table, and the Next Record button ▶ moves forward to the next record in the table.

The Action query button group inserts, deletes, and locks records. The Insert Record button ⬚ clears the form fields and creates a blank record into which you may enter new data. You must enter all required fields and then save the record or else delete the new record by clicking the Remove Record button ⬚. You can also use ⬚ to delete an existing record. The Lock Record button ⬚ locks the current record, which prevents other users from updating or deleting it.

The bottom of the window has a **message line**, which displays error or status messages, and a status line, which reports how many records the current query has retrieved.

Using a Form to View Table Records

You can use a form to view existing table records. To retrieve specific table records, you click the Enter Query button ⬚ to place the form in Enter Query mode, type a search expression in one or more of the form text items, and then click the Execute Query button ⬚ to execute the query. Alternatively, you can retrieve all of the records in a table by placing the form in Enter Query mode, and then executing the query without typing a search expression. This is equivalent to creating a SQL query that does not contain a search condition. When a query retrieves multiple records, you can use the buttons in the Navigation button group to step through the records sequentially. Now you retrieve all of the records in the LOCATION table, and use the Navigation buttons to step through the records sequentially.

To retrieve all of the LOCATION records, and then step through the records:

1. Click the **Enter Query** button 🖳. The form fields are cleared, and the insertion point appears in the Location ID field. Notice that the message "Enter a query; press CTRL+F11 to execute, F4 to cancel" appears in the message line at the bottom of the form. Pressing CTRL+F11 is equivalent to clicking the Execute Query button 🖳 on the toolbar, and pressing F4 is equivalent to clicking the Cancel Query button 🖳. Also, notice that the mode indicator in the status line at the bottom of the form indicates that the form is in Enter Query mode.

2. Because you want to retrieve all of the table records, do not type a search condition, and click 🖳. The first LOCATION record, room 101 in building CR, appears on the form.

3. Click the **Next Record** button ▶ on the toolbar. The record displaying data for the next location in the database, room 202 in building CR, appears.

4. Click the **Previous Record** button ◀ on the toolbar. Room 101 in building CR appears again because it is the record preceding room 202 in building CR.

5. Click ▶ again to display the record for Location ID 2. Continue clicking ▶ until you scroll through all the LOCATION records. The last record is for Location ID 13, as shown in Figure 5-4.

Viewing table records sequentially works well for the small sample databases in this book, but is not a feasible way to find a specific record in a database that contains thousands of records. When you are working with large databases, you usually retrieve specific records by entering a search expression in a form text item. Entering a search expression in a form is similar to creating a search condition in SQL*Plus. As with SQL*Plus, searches involving text strings are always case sensitive, so you must enter the search expression exactly as the value appears in the database. Now you retrieve all LOCATION records for which the Building Code value is LIB.

To use a form to retrieve specific records:

1. Click the **Enter Query** button 🖳 on the toolbar.

2. Place the insertion point in the Building Code field, and type **LIB**. Note that the search expression must contain uppercase letters, because the database stores building abbreviations using uppercase letters, as shown in Figure 5-5.

5

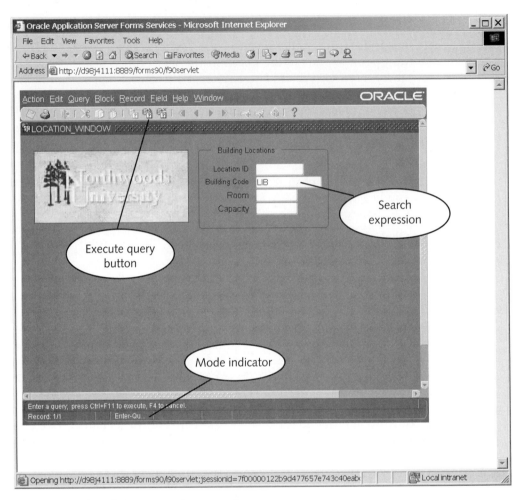

Figure 5-5   Specifying a search expression

      3. Click the **Execute Query** button ![icon]. Scroll through the retrieved records, and note that the query retrieved the records for Location ID 12 and 13, whose Building Code values are LIB.

This is an exact search. In forms, you can also perform inexact searches, which retrieve records containing data values that fall within a range of values or contain partial text strings. In a form, you structure inexact search expressions the same way you structured inexact search expressions in SQL queries. You can search for number values that are greater than, less than, greater than or equal to, less than or equal to, or not equal to search values by typing the comparison operator and the search value directly in the search text item. For example, to retrieve the records for all locations whose Location ID field value is greater than 4, you would type >4 in the form's ID field.

You can also use the underscore (_) and percent sign (%) character wildcard operators to search for partial text strings, as you did with the LIKE operator in SQL queries. For example, you could perform an inexact search that looks for all first-floor rooms. Because the room numbers are assigned a 100 value if they are on the first floor and a 200 value if they are on the second floor, you can use the expression 1% in the form's Room field to retrieve the desired records. Next you perform an inexact search to retrieve all locations whose Location ID is greater than 4. Then, you perform an inexact search to retrieve all rooms on the first floor.

To perform inexact searches:

1. Click the **Enter Query** button , type **>4** in the Location ID field as the search condition, and then click the **Execute Query** button . The record for Location ID 5 appears.

2. Click the **Next Record** button . The record for Location ID 6 appears. Continue clicking until a blank record appears, indicating that there are no more records.

3. Click again, type **1%** in the Location ID field as the search condition, and then click . The first record the query retrieves is for Location ID 1, which is the first room in the LOCATION table that is located on the first floor.

Using a Form to Insert, Update, and Delete Records

You can use a form to perform the DML operations of inserting, updating, and deleting records. Using a form to insert a new record is equivalent to inserting a record using the SQL INSERT action query. When you first open a form, a new blank record appears. Whenever you run a query, a new blank record appears as the last record. Now you navigate to the blank record at the end of the previous query, and insert a new record into the LOCATION table.

To insert a new record:

1. Click the **Next Record** button to move to the last record in the previous query. If necessary, place the insertion point in the Location ID field, and type **14**.

2. Pressing **Tab** to navigate among the text items, enter the data values in Figure 5-6.

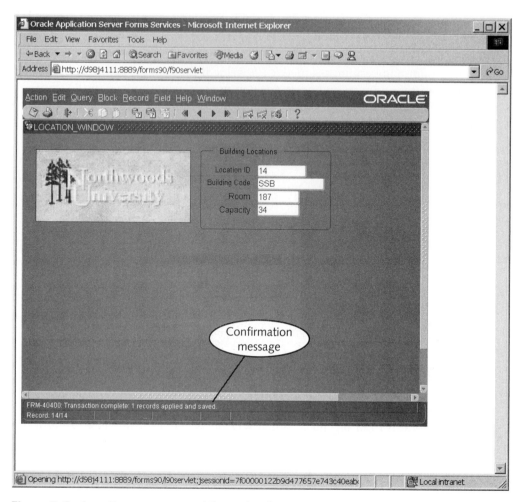

Figure 5-6   Inserting a new record through a form

3. Click the **Save** button on the toolbar. The confirmation message "FRM-40400: Transaction complete: 1 records applied and saved" appears on the message line in the lower-left corner of the screen. (You may need to scroll down in the browser window to view the message. If you moved your mouse pointer off and then back onto the Save button, move it off the Save button now, or the Save message appears instead.)

Next you modify an existing LOCATION record. Using a form to modify records is equivalent to using the SQL UPDATE action query to modify records. You retrieve the record for Location ID 14 and then change the room capacity from 34 to 40.

To retrieve and update an existing record:

1. Click the **Enter Query** button ![icon], type **14** in the Location ID field, which is the location you previously added to the LOCATION table, and then click the **Execute Query** button ![icon] to execute the query. The complete data record appears in the form.

2. Place the insertion point in the Capacity field, and then change the capacity to **40**.

3. Click the **Save** button. The confirmation message "FRM-40400: Transaction complete: 1 records applied and saved" appears on the message line.

5

Next you remove a record, which is equivalent to deleting a record using the SQL DELETE action query. You delete the new record for Location ID, which you inserted earlier.

To delete a record:

1. First retrieve the desired record by clicking the **Enter Query** button ![icon], typing **14** in the Location ID field, and then clicking the **Execute Query** button ![icon]. The record for the room in the SSB building appears.

2. Click the **Remove Record** button ![icon] on the toolbar to delete the record. The data values no longer appear. Unlike updating and inserting records, a confirmation message does not appear on the message line when you delete a record.

3. Click the **Save** button. The confirmation message "FRM-40400: Transaction complete: 1 records applied and saved" appears on the message line. Although the record appeared to be deleted after Step 2, at this point the deletion was not yet committed to the database. Clicking the **Save** button committed the deletion and removed the record from the database.

Viewing and Interpreting Form Errors

When a user makes an error while entering form data, an error code and message appear in the message line. Some errors are detected, or **trapped**, within the form, and the form issues the error message. Other errors are not detected until the form sends the data values to the database, so the Oracle DBMS issues the error message. Now you enter some incorrect data values, view error messages that both the form and the DBMS generate, and learn how to get more information about user errors. First you try to enter a record that uses the incorrect data type in the ID field. The form traps this error. Then you try to delete a record that is referenced by a foreign key constraint. The DBMS traps this error.

To create errors and view error information:

1. Click the **Insert Record** button ![icon], make sure the insertion point is in the Location ID field, type **aaa**, which is the wrong data type for the Location ID field, and then press **Tab**. The error message "FRM-50016: Legal characters are

0 – 9 – + E" appears in the message line, indicating that the entered value must be a number. (You may need to scroll down in the browser window to view the error message.) The form traps this error, and the form does not send the data to the database until you correct the error.

2. Before you can execute a query to retrieve the record for Location ID 12, you must remove the current record. Press the **Remove** button. Press the Enter Query button and enter **12** in the Location ID field. Press the **Execute Query** button. To remove this record from the form, press the **Remove** button. Delete this record from the database by pressing the **Save** button. An error message appears indicating that the record cannot be deleted.

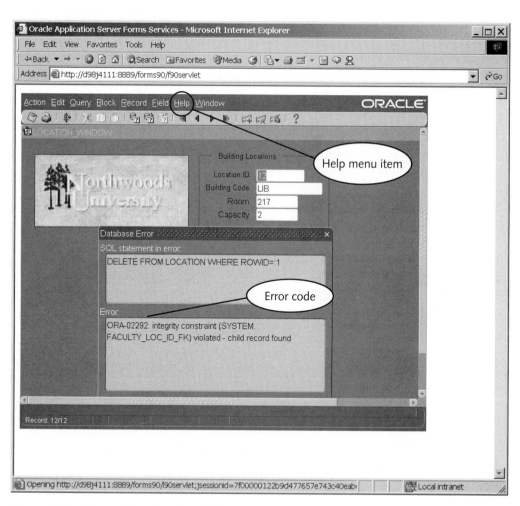

Figure 5-7   Information about a database error

To view more information about the database error:

1. To view information about the error, click **Help** on the form's menu bar, and then click **Display Error**. (Be sure to click Help on the form menu rather than on the browser window's menu. The form menu items appear as white text on a dark gray background, and the word "ORACLE" is on the right edge of the menu, as shown in Figure 5-7.)

2. The Database Error dialog box opens and indicates that the error is caused by a foreign key violation, as shown in Figure 5-7. Click **OK** to close the Database Error dialog box.

5

Closing a Form and Committing Changes

You can close a form by clicking the Exit button ▐▶ on the toolbar, by clicking Action on the form menu and then clicking Exit, or by closing the browser window. When you close a form, if you have made any changes that you have not yet saved, a dialog box opens and prompts you to either commit or roll back your changes. Now you update a record, close the form, and then commit your changes.

To update a record, and then close the form and commit the changes:

1. Click the **Enter Query** button ▒▒. Click **No** in the message box that appears, type **1** in the Location ID field, and then click the **Execute Query** button ▒▒. The first record in the table is displayed.

2. Change the Capacity value from 150 to **100**.

3. Click the **Exit** button ▐▶ on the toolbar. A message box appears asking if you want to save your changes. You can click Yes to commit your changes, No to roll back your changes, or Cancel to return to the form.

4. Click **Yes** to commit your changes, and then click **OK** to save your changes and close the form.

5. Close the browser window.

CREATING A DATA BLOCK FORM

Recall that a data block is associated with a specific database table. When you create a data block form, Forms Builder automatically generates the text items and labels for data fields in that table and then provides the code for inserting, modifying, deleting, and viewing data records. Before beginning to create the FACULTY data block form, you need to become familiar with the Forms Builder Object Navigator.

The Object Navigator

The Object Navigator window displays Forms Builder objects and their underlying components as a hierarchical tree. Figure 5-8 shows the Object Navigator window.

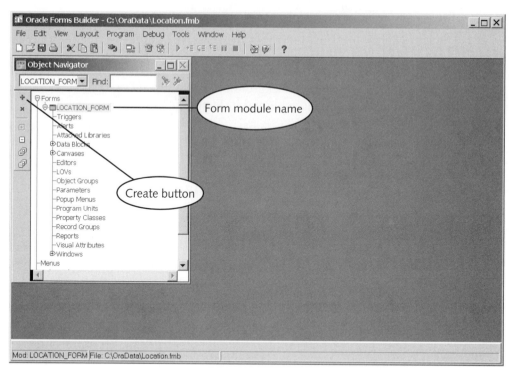

Figure 5-8   Object Navigator window

The Forms Builder toolbar is available with the Object Navigator window as it is with all Forms Builder windows. The Forms Builder toolbar displays buttons for creating a new form, opening and saving files, printing the current form, and cutting, copying, and pasting text. The Forms Builder toolbar also displays the Connect button ⬛, which allows you to connect to the database; the Compile Module button ⬛, which compiles the current form; the Run Form button ⬛, which runs the current form; and the Run Form Debug button ⬛, which runs the current form in a debugging environment. The debugging control buttons control execution during debugging, and appear enabled only during debugging sessions. The Layout Wizard button ⬛ and Data Block Wizard button ⬛ start the Layout Wizard and Data Block Wizard visual configuration tools, respectively.

The Object Navigator tree represents objects as nodes. The top level Forms Builder objects are Forms, Menus, PL/SQL Libraries, Object Libraries, Built-in Packages, and Database Objects. In this chapter, you create and manipulate forms. When you open a form, its node appears under the Forms node, and is called a **form module**, or just a form. Figure 5-8 shows that the Location form module is named LOCATION_FORM.

In the Object Navigator tree, if an object node is a plus sign (+), it means that the object has lower-level objects that you can view by expanding the node. Object nodes that appear as a minus sign (-) are currently expanded. If neither a plus sign nor a minus sign is beside a node, the node is empty and does not currently contain lower-level objects.

In Figure 5-8, the LOCATION_FORM form module node is currently expanded, and shows that the form module contains data blocks, canvases, and windows. These particular node types are the basic building blocks of every form—a form must have a window, canvas, and block.

A **window** is the familiar rectangular area on a computer screen that has a title bar at the top. Windows usually have horizontal and vertical scroll bars, and they usually can be resized, maximized, and minimized. In Forms Builder forms, you can specify window properties, such as title, size, and position on the screen. A **canvas** is the area in a window in which you place graphical user interface (GUI) objects, such as buttons and text fields. Recall that a block is a structure that contains a group of objects such as text fields or command buttons. A data block is a block that is associated with a database table. When you create a data block, the system automatically generates the text items and labels for data fields in that table and provides the code for inserting, modifying, deleting, and viewing data records.

A form can contain one or more windows. Simple applications usually have one window; more complex applications might have several windows. A window can have multiple canvases. A canvas can have multiple blocks, and individual items within a block can appear on different canvases. It is useful to think of the form as a painting, with the window as the painting's frame, the canvas as the canvas of the painting, and a block as a particular area of the painting.

Creating a New Data Block Form

Now you create the FACULTY form, which is a data block form that displays fields from the Northwoods University FACULTY table, as shown in Figure 5-9.

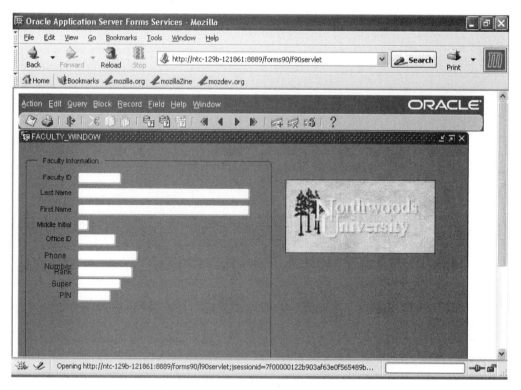

Figure 5-9 FACULTY form

To create a new form module object, you select File on the menu bar, point to New, and then click Form. Or, you can select the Forms node in the Object Navigator tree, and click the Create button ✚.

To create the Faculty form:

1. Select the **Forms** node in the Object Navigator tree, and then click the **Create** button ✚ on the toolbar.

2. A new module appears in the Object Navigator tree. The module has a default name, such as MODULE2.

NOTE

If you continue adding forms, their default names will be MODULE3, MODULE4, and so on. We use the name MODULE2 in the following text. If you have been experimenting with creating new forms, your default module might have a different number.

After you create a new form module, you need to create the data block that associates the form with a database table, and you need to specify the form layout, which specifies how the form appears to the user. You create the data block first, and then specify the layout properties.

Creating a Data Block

To create a new data block, you select the Data Blocks node under the form module, and then click the Create button ✚. When you create a new data block, you have the option of creating the data block manually or using the Data Block Wizard, which is a tool that allows you to specify the data block's properties. The Data Block Wizard has the following pages:

- **Welcome page**—Introduces the Data Block Wizard
- **Type page**—Allows you to specify whether the data block's data comes from a database table, a database view, or a stored procedure

NOTE
A stored procedure is a PL/SQL program that the database stores. You can create a stored procedure to retrieve and manipulate data values, and then base a data block on the stored procedure's data. You do not use stored procedures as data block sources in this book.

- **Table page**—Allows you to select the database table that provides the data for the data block. The Table page also allows you to select the specific table fields that you include in the block, and whether you want to enforce integrity constraints in the form application. The database itself always enforces integrity constraints, such as unique primary keys and foreign key references. However, you can also specify to have the form enforce integrity constraints, so data values that violate integrity constraints are flagged before they are submitted to the database.
- **Name page**—Allows you to specify the name of the data block as it appears in the Object Navigator tree
- **Finish page**—Presents the options of either creating the data block and then immediately starting the Layout Wizard to specify the form layout properties, or just creating the data block

Now you are going to create a new data block and use the Data Block Wizard to specify the data block configuration.

To create and configure a new data block:

1. Select the **Data Blocks** node under the MODULE2 form node, and then click the **Create** button ✚. The New Data Block dialog box shown in Figure 5-10 opens and asks if you want to use the Data Block Wizard to create the new block, or if you want to create the new data block manually.

Figure 5-10   New Data Block dialog box

2. Make sure that the Use the Data Block Wizard option button is selected, and then click **OK**. If the Data Block Wizard Welcome page appears, click **Next**.

3. The Data Block Wizard Type page, shown in Figure 5-11, appears, allowing you to associate the data block with either a table or view, or with a stored procedure.

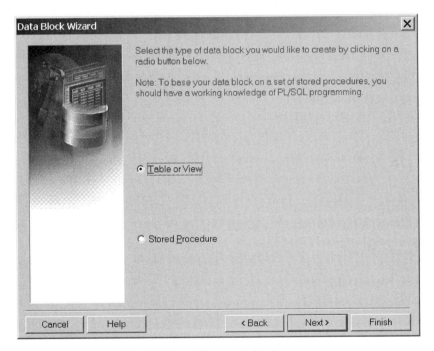

Figure 5-11   Data block Wizard Type window

4. Because you are going to associate this data block with the FACULTY table, confirm that the Table or View option button is selected, and then click **Next**. The Table page appears, which allows you to select the data block's database table and to specify the table fields that the data block will contain.

5. Click **Browse**. The Tables dialog box opens, as shown in Figure 5-12. (You might have to connect to the database again if you exited Forms Builder and restarted it.)

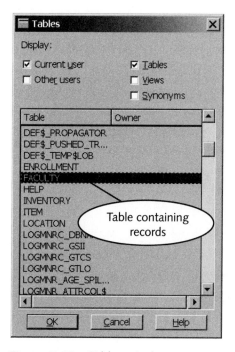

Figure 5-12 Tables window

The Tables dialog box displays a list of database tables from which you can select the table on which to base the data block. If the Current user check box is checked, every table in your user schema appears in the Tables list. If the Other users check box is checked, every table in the database appears. You would check the Other users check box only when you want to base the data block on tables in another user's database schema.

Before you can use a table in another user's database schema, that user must first grant you the privileges for inserting, updating, deleting, or viewing data.

NOTE

If the Tables check box is checked, database tables appear in the list. If the Views check box is checked, database views appear in the list, and if the Synonyms check box is checked, synonyms appear in the list.

A synonym is an alternate name for a table. You do not use synonyms in this book.

NOTE

Next you select the FACULTY table in the Tables list, and specify the table fields that the data block is going to include. A data block usually contains all the fields in a table, but that isn't necessary. For example, you might exclude a field that you do not want users to view or change, such as a password or PIN value.

To select the table and specify the table fields:

1. In the Tables dialog box, make sure that the **Current user** and **Tables** check boxes are checked and that the other check boxes are cleared. Note that the list displays every database table in your user schema. Select **FACULTY** in the Table list.

2. Click **OK** to close the Tables dialog box and return to the Data Block Wizard Table page, which appears as shown in Figure 5-13.

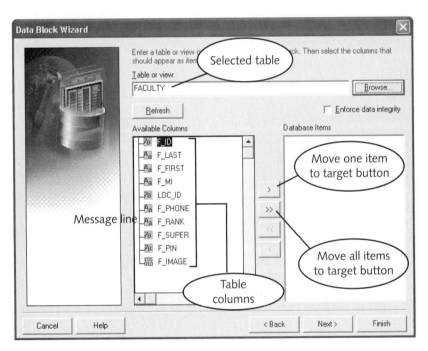

Figure 5-13   Data Block Wizard Table page

On the Table page, FACULTY appears in the Table or view field, and the FACULTY table fields appear in the Available Columns list. The icon beside each field identifies the field's data type. The Number icon ⁷⁸⁹ identifies F_ID, LOC_ID, F_SUPER and F_PIN as numeric data fields; the Character icon **A** identifies F_LAST, F_FIRST, F_MI, F_PHONE, and F_RANK as character data fields; and the Binary icon 101/010 identifies F_IMAGE as a binary field.

To include a specific field in the data block, you select the field in the Available Columns list, and then click the **Move one item to target** button ⊡ to move the field to the Database Items list. To remove a field from the data block, you select the field in the Database Items list, and click the **Move one item to source** button ⊡. To select all the table fields, click the **Move all items to target** button ⊡. To remove all of the selected fields, click the **Move all items to source** button ⊡. To select several adjacent fields, select the first field, press and hold the Shift key, and then click the last field. To select several nonadjacent fields, select the first field, press and hold the CTRL key, and then click each of the other fields.

If you check the Enforce data integrity check box on the Table page, the form flags integrity constraint violations, such as nonunique primary keys or foreign key values that do not exist in the referenced parent table. This means that when an error occurs, you are going to see FRM-error codes that Forms Builder generates, rather than ORA-error codes that the Oracle 10*g* DBMS generates. Remember, the DBMS enforces the integrity constraints even if you do not check the Enforce data integrity check box.

Now finish creating the data block. The data block includes all of the FACULTY fields except the F_IMAGE field. You do not want to enforce data integrity in the form, so you do not check the Enforce data integrity check box. And you accept the default data block name, which is the same as the name of the database table.

To finish creating the data block:

1. Click the **Move all items to target** button ⊡ to include all the FACULTY table fields in the data block. The selected fields appear in the Database Items list.

2. Select the **F_IMAGE** field and click the **Move one item to source** button ⊡. Make sure that the Enforce data integrity check box is cleared, as shown in Figure 5-14, and then click **Next**.

5

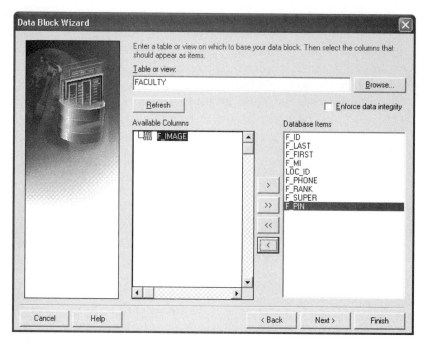

Figure 5-14   Desired columns selected

3. The Name page shown in Figure 5-15 appears, and displays the default data block name, which is FACULTY. Accept the default name, and click **Next**.

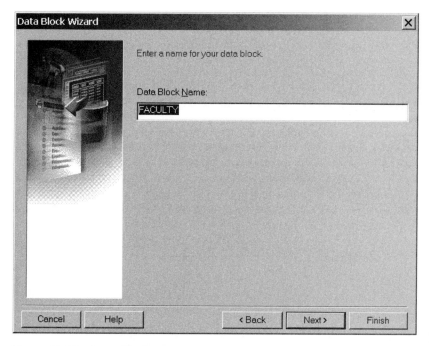

Figure 5-15   Data Block Name page

4. The Data Block Wizard Finish page appears, which allows you either to use the Layout Wizard to specify the form layout, or to create the form layout manually.

Creating the Form Layout

A database application should display descriptive field labels. The fields should be wide enough to display each data value, so no text is clipped. The fields should appear aligned, and neatly stacked on top of one another. Related fields should be enclosed within a frame, which is a rectangle on the form that encloses like objects.

The **form layout** specifies how a given form looks to the user. The basic components of a form layout include the specific data block fields that appear, the field labels, the number of records that appear at one time, and the form frame title. You can use the Layout Wizard visual configuration tool to specify the form layout properties. The Layout Wizard pages include:

- **Welcome page**—Introduces the Layout Wizard

- **Canvas page**—Specifies the canvas on which the form objects appear. Recall that a canvas is the area on a form in which you place the form objects, such as command buttons and text items. A canvas can be a content canvas, stacked canvas, toolbar canvas, or tab canvas. A **content canvas** fills the entire window. A **stacked canvas** contains multiple canvases that are stacked on top of one another and that you can display or hide to change the appearance of the canvas as needed. A **toolbar canvas** has horizontal or vertical toolbars that program commands can display or hide. A **tab canvas** enables different related canvases to appear on a single tab page; users can access each individual canvas by clicking a labeled tab.

- **Data Block page**—Allows you to select the fields that will be displayed in the form

- **Items page**—Allows you to specify the prompts, widths, and heights for the form text items being displayed. A **prompt** is the label that describes the data value that appears in the associated text field, such as Faculty ID or Middle Initial. By default, the prompt values are the same as the database field names, with blank spaces inserted in place of underscores. You specify the widths and heights using **points**, which correspond to font sizes, as the default measurement unit. The default widths correspond to the maximum field data widths and the default form font.

- **Style page**—Allows you to specify the layout style and properties. In a **form-style layout**, only one record appears on the form at a time. In a **tabular-style layout**, multiple records appear in a table on the form. In a tabular layout, if more records exist than can appear on the form at one time, the developer can optionally display a scroll bar to allow the user to view records.

- **Rows page**—Allows you to specify the title that appears on the frame that encloses the data block items, the number of rows that appear in a tabular-style layout, and whether a tabular-style layout displays a scroll bar

- **Finish page**—Signals that the layout is complete

Now you start the Layout Wizard directly from the Data Block Wizard's Finish page. Then you use the Layout Wizard to specify the form layout.

To use the Layout Wizard to specify the form layout:

1. On the Data Block Wizard Finish page, be sure that the Create the data block, then call the Layout Wizard option button is selected, as shown in Figure 5-16, and then click **Finish**. The Layout Wizard Welcome page appears.

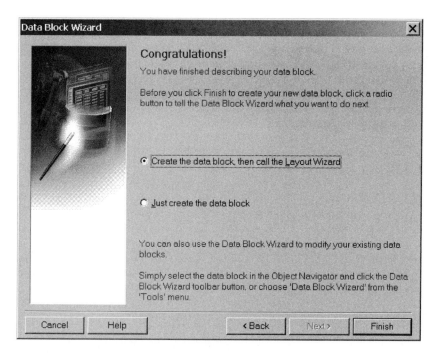

Figure 5-16   Data Block Wizard Finish page

2. Click **Next**. The Layout Wizard Canvas page appears. Because you have not yet created any canvases, you accept the default (New Canvas) selection in the Canvas list. Because you want the FACULTY data to appear on a single canvas that fills the entire window, you use a content canvas, so make sure that Content is selected in the Type list, as shown in Figure 5-17.

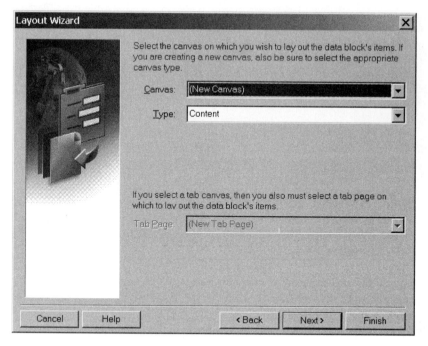

Figure 5-17   Layout Wizard window

3. Click **Next**. The Data Block page appears, which allows you to select the display fields. Because you want to display all of the data block fields on the layout, click the **Move all items to target** button ⟩⟩ to select all block fields for display. The block field names appear in the Displayed Items list, as shown in Figure 5-18.

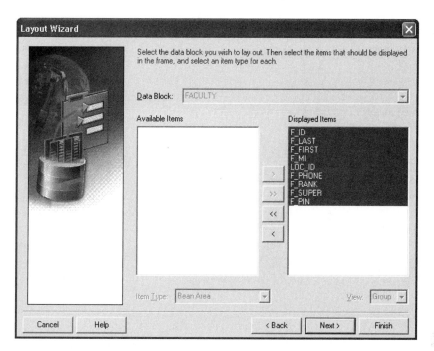

Figure 5-18   Data Block page

4. Click **Next**. The Items page appears. To modify the prompts so that they are more descriptive than the table field names, place the insertion point in the first Prompt row if necessary, delete the current text, and type **Faculty ID**.

5. Modify the rest of the prompts as shown in Figure 5-19, and accept the default width values, even if they are different from the ones shown in the figure.

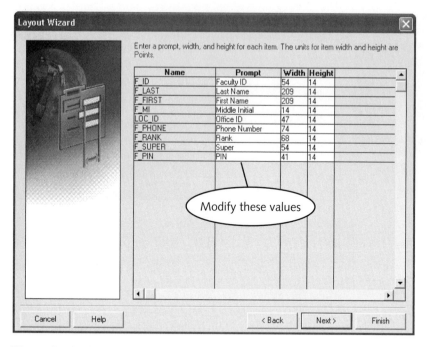

Figure 5-19   Items page

6. Click **Next**. The Style page appears. Because you want to display only one order record on the form at a time, you specify a form–style layout, so make sure that the Form option button is selected.

7. Click **Next**. The Rows page appears. Type **Faculty Information** in the Frame Title text box, as shown in Figure 5-20. Because you specified a form layout style, only one record should appear at a time. Be sure that the Records Displayed field value is *1*.

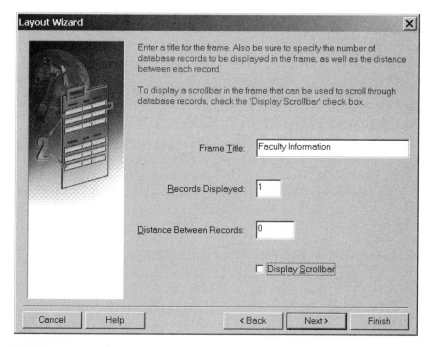

Figure 5-20   Rows page

NOTE

The Distance Between Records field specifies the distance between each individual record for a tabular-style layout, and the Display Scrollbar check box specifies whether to display a scroll bar on a tabular-style layout. These values do not change the appearance of a form-style layout.

8. Click **Next**, and then click **Finish** to finish the layout. The Layout Editor window opens, with all of the form items selected. If necessary, maximize the Layout Editor window.

9. Click anywhere on the canvas to deselect the selected form items. The Layout Editor window appears, as shown in Figure 5-21.

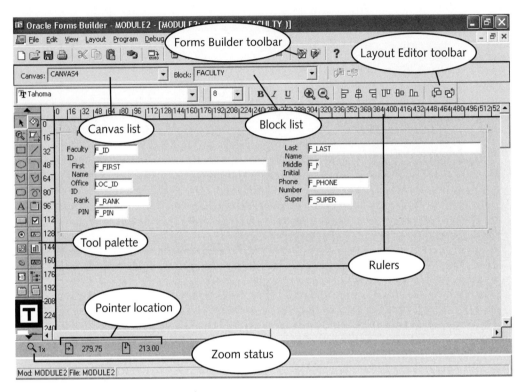

Figure 5-21   Layout Editor window

The Layout Editor

The **Layout Editor** provides a graphical display of the form canvas. You use Layout Editor to draw and position form items, and to add boilerplate objects such as labels, titles, and graphic images. **Boilerplate objects** do not contribute directly to the form's functionality, but enhance its appearance. The Canvas list at the top of the Layout Editor window shows the name of the current canvas. When you use the Layout Wizard to create a new canvas, the canvas receives a default name. In Figure 5-21, the default canvas name is CANVAS7, but your default canvas name might be different. The Block list shows the name of the current data block, which is FACULTY.

The Layout Editor toolbar allows you to edit the form and modify the form's text properties. (Note that the Forms Builder toolbar appears at the top of the window, and the Layout Editor toolbar appears under it.) The tool palette, which is at the far left side of the editor, provides tools for creating boilerplate objects and other objects. Rulers appear along the top and left edges of the canvas, and the zoom status and pointer location indicators are on the bottom-left edge of the window. Now you save the form in the Layout Editor. Recall that files that contain the design specifications for a form have an .fmb extension.

To save the form:

1. Click the **Save** button on the Forms Builder toolbar.

2. Navigate to the Chapter5\Tutorials folder on your Solution Disk, and save the file as **Ch5AFaculty.fmb**.

Now you use the Layout Editor to modify the form's visual appearance so it looks like the finished form in Figure 5-9, in which the form's text items appear on the left side of the form, stacked over one another. The finished form also displays the Northwoods University logo, which is a boilerplate image. First you resize and move the form frame and items. When you resize a form's frame, Forms Builder automatically repositions the items within the frame so that they fit into the new frame size. When you move a frame, the items within the frame move also.

To resize and move the form frame:

1. In the Layout Editor, click on the borderline to select the frame that encloses the form text items. Selection handles appear on the frame's perimeter.

2. Drag the lower-right selection handle to make the form frame longer and narrower, so that the form text items appear stacked over one another. Adjust the frame size so that it just encloses the items, as shown in Figure 5-22.

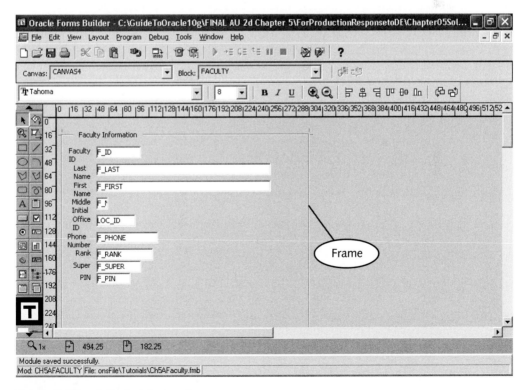

Figure 5-22   Adjusted frame size

NOTE

If the message "FRM-10855: Not enough horizontal space in frame *frame_name* for object *object_name*." appears, you made the frame too narrow to enclose one or more of the text items. Click OK, and then make the frame wider so that the frame fully encloses all text items.

Next you add the Northwoods University logo to the form. The form displays the logo as a static imported image. A **static imported image** is a graphic enhancement that appears the same regardless of the data that the form displays. To create a static imported image, you import a file into the form that contains a graphic image. Forms Builder incorporates the image data into the form design (.fmb) file. Static imported images make the form's .fmb file larger and make the Web page image of the form load more slowly, so you should always select small (less than 100 KB) files for static imported images.

To import a graphic image as a static imported image, you click Edit on the Forms Builder menu bar, point to Import, and then click Image to open the Import Image dialog box, which is shown in Figure 5-23.

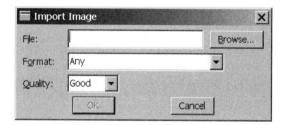

Figure 5-23 Import Image dialog box

The Import Image dialog box enables you to import an image that is stored either in the file system or in the database, by specifying the image file and file format. In this book, you import images from the file system. The Format list specifies the image file format. Popular image file formats include bitmap (.bmp), GIF (.gif), TIFF (.tif), and JPEG (.jpg) files. Bitmap files are usually uncompressed, whereas .gif, .tif, and .jpg files use different compression methods. Most graphics applications can create files using one or more of these file types. If you select Any in the Format list, Forms Builder automatically imports the image using the image type that the image file's extension specifies.

The Quality list specifies how the form stores the image in terms of image resolution and number of colors. The Quality choices include Excellent, Very Good, Good, Fair, and Poor. If you import an image using the Excellent quality selection, Forms Builder saves the image using the maximum number of colors and highest possible resolution, which uses the maximum amount of file space and takes the longest time to load. The default Quality value is Good, which is satisfactory for most graphic images. You can experiment with the different quality levels to determine the minimum acceptable quality for each image.

Now you add the Northwoods University logo to the form as a static imported image. The image is stored in a file named Nwlogo.jpg in the Chapter5 folder on your Data Disk. You import the image using the default (Good) image quality selection.

To add the logo as a static imported image:

1. Click **Edit** on the menu bar, point to **Import**, and then click **Image**. The Import Image dialog box opens, as shown in Figure 5-23.

2. Click **Browse** beside the File field, navigate to the Chapter5 folder on your Data Disk, select **Nwlogo.jpg**, and then click **Open**. Make sure that Good is selected in the Quality list, and then click **OK**. The image appears in the upper-left corner of the form.

3. In the Layout Editor, select the image, resize it if necessary, and position it as shown in Figure 5-24. Then save the form.

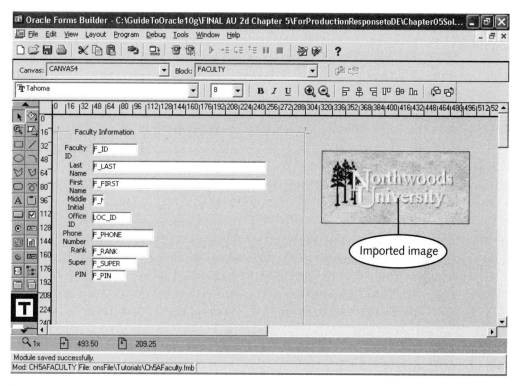

Figure 5-24   Inserted image

Next you need to format the labels for each of the field values displayed on the form. Currently the Faculty ID and several other labels are displayed across two lines. The following steps have you reformat the labels so each is displayed on one line.

To display the labels on one line:

1. Click the **Faculty ID** label in the Northwoods University form.

2. Click the **Faculty ID** label in the Northwoods University form again. When the Faculty ID label box is in Edit mode, remove the return at the end of the first line of the label so the second line of text is moved to the first line. Press the spacebar to add a space between the words of the label. Click outside the label box to exit Edit mode.

3. Repeat this procedure for each label, as shown in Figure 5-25.

4. Move the text boxes to make room for the modified label by "lassoing" the text boxes. To lasso the text boxes, place the pointer just left and above the F_ID text box, hold down the left mouse button, and drag the pointer to the right and down so that the selection frame encloses all the text boxes. When you release the mouse button, all the text boxes will be selected. Use the right arrow key on the keyboard to move the text boxes and the labels to the right.

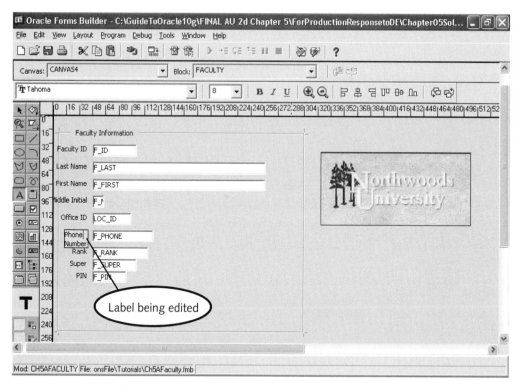

Figure 5-25   Modifying label appearance

Now you run the form. When you run a form, Forms Builder automatically creates two files in the folder that contains the form's .fmb file. One file has the same filename as the .fmb file, but has an .fmx file extension. Recall that the OC4J process translates this file into a Java applet that appears in the browser window. Forms Builder also sometimes creates a text file that has the same name as the form's .fmb file, but has an .err file extension. The .err file records and displays messages about errors that occur when you run the form. If compile errors occur when you run a form, the error messages automatically appear on the screen, so you probably will not need to use the .err file.

To run the form:

1. Click the **Run Form** button 🔘 on the Forms Builder toolbar. After a few moments, your form should appear in your browser window. The form appears as shown in Figure 5-9. (Your form does not yet display any data values.)

2. Click the **Enter Query** button 🔲, and then click the **Execute Query** button 🔲. The data values for the first record in the FACULTY form appear in the form.

3. Close the browser window.

Form Components and the Object Navigator

Recall that the Object Navigator provides a hierarchical display of all the form components. The following sections show how to use the Object Navigator to view form components and describe the different views of form objects that the Object Navigator provides.

Viewing Form Components

Recall that in the Object Navigator, nodes containing underlying objects appear as a plus sign, and you can view the underlying objects by clicking the plus sign to open the node. You can select an object and expand all of its underlying nodes by clicking the Expand All button 📁 on the Object Navigator toolbar. And, you can select an object and close all of its underlying nodes by clicking the Collapse All button 📁. Now you open the Object Navigator, close all of the nodes in the FACULTY form, and then view the new form's components.

To close the FACULTY form nodes and view the new form components:

1. In the Layout Editor, click **Window** on the menu bar, and then click **Object Navigator**. The Object Navigator opens and displays the FACULTY form under the Forms node.

2. Select the **FACULTY** node, and then click the **Collapse All** button 📁 on the Object Navigation toolbar to close the form node and hide its underlying objects. (The form node still appears in the Object Navigator tree.)

The Data Blocks, Canvases, and Windows nodes have plus signs to indicate that they contain additional objects, or minus signs if they are already expanded. (Recall that data blocks, canvases, and windows are the basic components of a form.) Now you expand these form objects and view their underlying components.

To expand the form objects:

1. Open the **Canvases** node, and then open the **Windows** node. The expanded form objects are displayed when the node is opened.

   The form has one block (FACULTY), one canvas (CANVAS7), and one window (WINDOW1). (CANVAS7 and WINDOW1 are the default names that the Layout Wizard gave to the canvas and window when you made the form using the Layout Wizard. Your form's default names might be different.)

Each object in the Object Navigator displays an icon that indicates its object type. Double-clicking the icon allows you to edit the object's properties. For example, the Canvas icon ▨ appears beside a canvas, and double-clicking this icon opens the selected canvas in the Layout Editor. The plus sign nodes to the left of the FACULTY block and CANVAS7 canvas indicate that these objects contain underlying objects that you can view by expanding the object node. Now you expand the block objects further.

To expand the block objects:

1. Open the **FACULTY** node under Data Blocks. The underlying data block objects are Triggers, Items, and Relations. The empty node symbols beside Triggers and Relations indicate that the data block does not contain any triggers or relations, and the plus sign next to the Items node indicates that the data block contains objects.

2. Open the **Items** node under FACULTY. The objects within the form data block appear as shown in Figure 5-26.

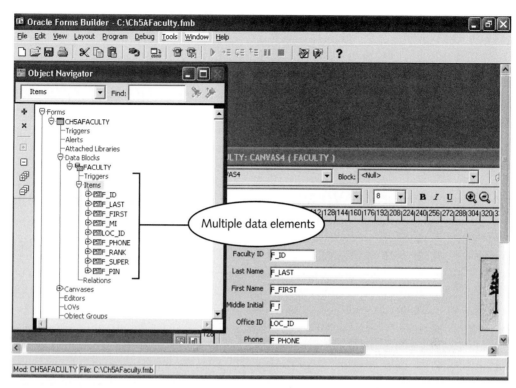

Figure 5-26   Object Navigator window

The items listed in the FACULTY form's Items node are the fields that users see on the form. **Items** are form objects that a user sees and interacts with on the canvas. All of the items in the customer form are **text items**, which display values in text fields. But what are the triggers and relation objects shown in Figure 5-26? A trigger is a PL/SQL program that starts in response to an event. An event is a response to a user or system action, such as clicking a button or loading a form, that occurs in a Windows application. In Forms Builder, events start triggers. You can create triggers that you associate with a specific data block. Currently, the FACULTY data block does not have any trigger objects. You learn how to create triggers when you create custom forms in Chapter 6.

A **relation** is a form object that Forms Builder creates for a form that displays two data blocks whose underlying database tables have a foreign key relationship. To create a relation, you must specify that the value of a primary key field in one block is equal to the corresponding foreign key field in the second block. Because the FACULTY form involves only one data block, it has no relations.

Now you expand the form canvas and form window.

To expand the form canvas:

1. Open the **CANVAS7** node. (Your canvas node may have a different name.) The canvas contains a Graphics node, which represents boilerplate objects, such as frames, lines, and graphic images.

2. Open the **Graphics** node. You see the FRAME8 (your default frame name might be different) object, which represents the frame that encloses the data block items, and the IMAGE9 (your default image name might be different) object appears, which represents the static imported image of the Northwood University logo.

Object Navigator Views

So far you have viewed the form objects in the Object Navigator in **Ownership View**, which presents the form as the top-level object, and then lists all form object nodes on the next level. You can also view the form objects in **Visual View**, which shows how form objects "contain" other objects: a form contains windows, a window contains canvases, and a canvas contains data blocks. Figure 5-27 shows the form components in Visual View.

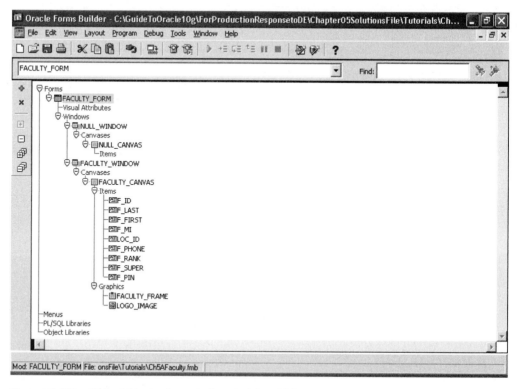

Figure 5-27   Object Navigator window in Visual View

Recall that in Ownership View, data blocks, canvases, and windows all appear as objects directly below the form module, and the Object Navigator tree does not show hierarchical relationships among windows, canvases, and blocks. (Ownership View displays one hierarchical relationship: it shows which items are in a specific data block.) Conversely, Visual View shows that WINDOW1 contains CANVAS7, and that CANVAS7 contains the form items and graphics. Visual View does not show data blocks.

Ownership View is useful for quickly accessing specific objects without having to open all of the higher-level objects in which the target object resides. Visual View is useful for viewing and understanding form object relationships. Now you examine your form's objects in Visual View.

5

To examine the form objects in Visual View:

1. Select **CH5AFACULTY**, and then click the **Collapse All** button 🗗 to close all of the form's nodes.

2. Click **View** on the menu bar, and then click **Visual View**. The top-level object—the form module—still appears.

3. Open the **CH5AFACULTY** node to display the next-level object. (Your form module object may have a different name.) The Windows node appears because it is the highest-level object within a form. The plus sign in the Windows node indicates that the form contains Windows objects.

4. Open the **Windows** node. The form's objects appear in the Object Navigator tree.

In Visual View, the form's Windows node contains the form's windows, which contain all of the other form objects. Two window nodes appear: NULL_WINDOW, which Forms Builder creates when it creates the form, and WINDOW1 (your window name might be different), which the Layout Wizard creates, and which contains the rest of the form objects. Now you expand WINDOW1 to examine the rest of the form objects.

To expand WINDOW1:

1. Open the **WINDOW1** node. (Your default window name may be different.) The Canvases node appears.

2. Open the **Canvases** node. The default canvas (CANVAS5) appears. (Your canvas name might be different.)

3. Open the **CANVAS5** node. The Items and Graphics nodes appear.

4. Open the **Items** node. The nodes that represent the form text items appear.

5. Open the **Graphics** node. The nodes that represent the form's frame and image appear.

The canvas node contains the final level of Visual View, which consists of the form items and graphics. Note that data blocks are not shown in Visual View.

Changing Object Names in the Object Navigator

After you create a data block form using the Data Block and Layout Wizards, it is a good idea to change the default object names to more descriptive names. When you have multiple forms open in the Object Navigator and when you start creating forms with multiple windows, canvases, and frames, it is hard to distinguish between different objects in the Object Navigator unless they have descriptive names. Now you change the form module name to FACULTY_FORM, the window name to FACULTY_WINDOW, the canvas name to FACULTY_CANVAS, the frame name to FACULTY_FRAME, and the image name to LOGO_IMAGE.

To change the object names:

1. Select **CH5AFACULTY** to select the form module. Its background color changes to yellow. Click the yellow highlighted **CH5AFACULTY** node again to change its background color to blue, type **FACULTY_FORM**, and then press **Enter** to save the change. (Another way to save your changes is to select another item.)

2. Select **WINDOW1** (or your form's default window), click it again so its background color changes to blue, and type **FACULTY_WINDOW**. Press **Enter** to save the change.

3. Select **CANVAS7** (or your form's default canvas), click it again so its background color changes to blue, type **FACULTY_CANVAS**, and then press **Enter**.

4. Select **FRAME4** (or your form's default frame), and change its name to **FACULTY_FRAME**.

5. Select **IMAGE8** (or your form's default image), and change its name to **LOGO_IMAGE**.

6. Click **View** on the menu bar, and then click **Ownership View** to redisplay the form objects in Ownership View.

7. Save the form.

MODIFYING FORMS USING THE DATA BLOCK WIZARD AND LAYOUT WIZARD

A powerful characteristic of the Data Block Wizard and Layout Wizard is that they are **reentrant**, which means that you can use them to modify the properties of an existing data block or layout. To use a wizard to modify an existing data block or layout, you must start the wizard in reentrant mode. To start a wizard in reentrant mode, you first select the data block or layout frame that you want to modify. If you do not want to start the Data Block Wizard or Layout Wizard in reentrant mode, you must be sure that no data block or layout frame is currently selected when you start the wizard.

You can visually tell when a wizard is in reentrant mode because all of the wizard pages appear at once, rather than one at a time with a Next button to move through them in the wizard window, and tabs appear at the top of the window. You select an individual wizard page from tabs on the top of the reentrant Wizard page. Now you modify the FACULTY data block so that the form enforces database integrity constraints. You make this change by selecting the block, and then starting the Data Block Wizard in reentrant mode.

To modify a block using the Data Block Wizard in reentrant mode:

1. In the Object Navigator, open the Data Blocks node if necessary, and then select **FACULTY** under the Data Blocks node.

2. Click **Tools** on the menu bar, and then click **Data Block Wizard** to open the Data Block Wizard in reentrant mode. The Type, Table, and Name tabs appear at the top of the Data Block Wizard window, indicating that the wizard is in reentrant mode, as shown in Figure 5-28.

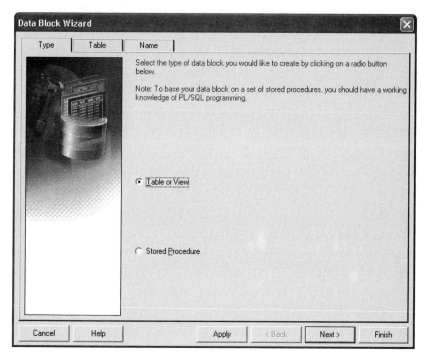

Figure 5-28   Data Block Wizard in reentrant mode

NOTE

Another way to start the Data Block Wizard in reentrant mode is to select one of the block items in the Layout Editor, and then click the Data Block Wizard button 🗔 on the Forms Builder toolbar. Or, you can select the data block in the Object Navigator, right-click, and then select Data Block Wizard.

3. Select the **Table** tab, and then check the **Enforce data integrity** check box. Click **Finish** to save the change and close the Data Block Wizard.

NOTE

Clicking Apply in the Data Block Wizard saves the change, but it does not close the Data Block Wizard. Clicking Finish saves the change and also closes the Data Block Wizard. Clicking Cancel closes the Data Block Wizard without saving any changes.

Next, you modify the form layout so that the prompt for the F_SUPER field appears as Supervisor ID instead of as Super. To do this, you select the form layout, and then open the Layout Wizard in reentrant mode. To select the layout, you must open the form in the Layout Editor, select the layout frame, and then start the Layout Wizard.

NOTE

If you delete the frame around a layout, you cannot revise the layout using the Layout Wizard in reentrant mode.

To modify the form layout using the Layout Wizard in reentrant mode:

1. With FACULTY_FORM selected in the Object Navigator window, click **Tools** on the menu bar, and then click **Layout Editor** to open the Layout Editor window.

2. Select the frame that encloses the form text items. Selection handles appear on the frame's perimeter.

3. Click **Tools** on the menu bar, and then click **Layout Wizard**. The Layout Wizard opens with the Data Block, Items, Style, and Rows tabs displayed at the top of the page, indicating that the wizard is in reentrant mode.

NOTE

Another way to start the Layout Wizard in reentrant mode is to select the frame in the Layout Editor, and then click the Layout Wizard button on the toolbar. Or, you can select the frame, right-click, and then click Layout Wizard.

4. Click the **Items** tab to move to the Items page, and then change the prompt for the F_SUPER field to **Supervisor ID**, as shown in Figure 5-29.

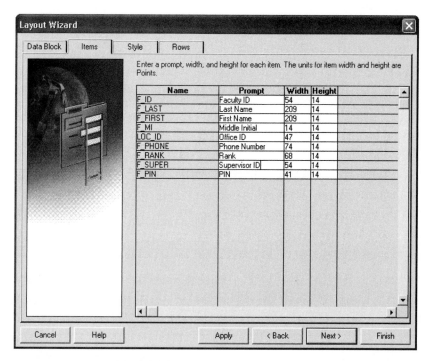

Figure 5-29 Modified field prompt

5. Click **Finish**. The modified prompt appears in the Layout Editor.

6. Save the form.

Because you are finished working with the FACULTY_FORM form and the LOCATION_FORM form, you close these forms in Forms Builder. To close a form, you select the form in the Object Navigator tree, click File on the menu bar, and then click Close.

To close the forms in Forms Builder:

1. Click **Window** on the menu bar, and then click **Object Navigator** to switch to the Object Navigator window.

2. Select **FACULTY_FORM** in the Object Navigator tree, click **File** on the menu bar, and then click **Close**. The form module no longer appears in the Object Navigator tree.

TIP

Another way to close a form in the Object Navigator is to select the form module node, and then click the Delete button ✖ on the Object Navigator toolbar. This action closes the form in Forms Builder, but does not delete the form's .fmb file.

3. Select the **LOCATION_FORM** form module, click **File** on the menu bar, save if necessary, and then click **Close** to close the form module.

CREATING A FORM TO DISPLAY MULTIPLE RECORDS

The FACULTY_FORM form has a form-style layout, which displays a single record at a time. Now you create the Northwoods University Faculty Listing form in Figure 5-30. This form has a tabular-style layout, which displays multiple records on the same form. The purpose of the form is to provide a listing of the faculty members, their office locations, and phone numbers. However, users referencing the list have no reason to view the faculty member's PIN, rank, etc., so those fields have been omitted.

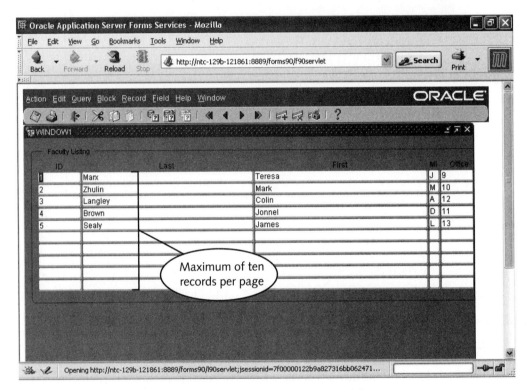

Figure 5-30   Northwoods University Faculty multiple record form

Now you create a new form module, and then create a new data block as follows:

1. Select the **Forms** node in the Object Navigator, and then click the **Create** button.

2. Make sure that the new form module is selected in the Object Navigator tree, and then click the **Data Block Wizard** button. If the Welcome to the Data Block Wizard page appears, click **Next**.

3. Make sure that the Table or View option button is selected, and then click **Next**.

4. On the Type page, click **Browse**. Make sure that the Current user and Tables check boxes are checked, select **FACULTY** on the table list, and then click **OK**.

5. On the Table page, use the **Move one item to target** button [>] to select the F_ID, F_LAST, F_FIRST, F_MI, LOC_ID, and F_PHONE fields, as shown in Figure 5-31. Do not check the Enforce data integrity box. Click **Next**.

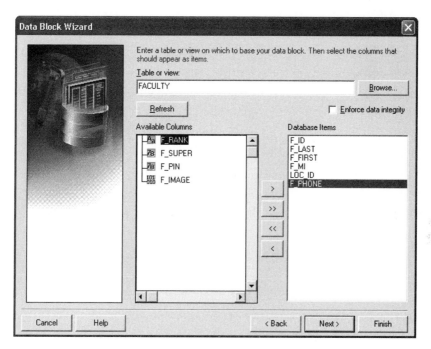

Figure 5-31   Selected fields for the FACULTY_LISTING form

6. On the Name page, type FACULTY_LISTING in the Data Block Name textbox, then click **Next**.

7. On the Finish page, make sure that the Create the data block, and then call the Layout Wizard option button is selected, and then click **Finish**.

Now you use the Layout Wizard to specify the layout properties. This form layout will be different from the one you made before, because you specify a tabular-style layout, and display ten records on the form at one time. You now create the form layout.

To create the form layout:

1. On the Layout Wizard Welcome page, click **Next**.

2. On the Canvas page, accept **(New Canvas)** as the layout canvas and **Content** as the canvas type, and then click **Next**.

3. Click the **Move all items to target** button $\boxed{\gg}$ to select all of the FACULTY_LIST block items to appear in the layout, as shown in Figure 5-32, and then click **Next**.

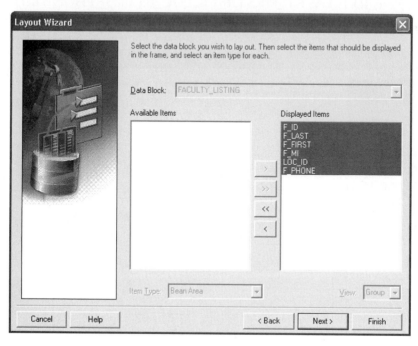

Figure 5-32   Selecting block items

4. Enter the item prompts shown in Figure 5-33, and then click **Next**:

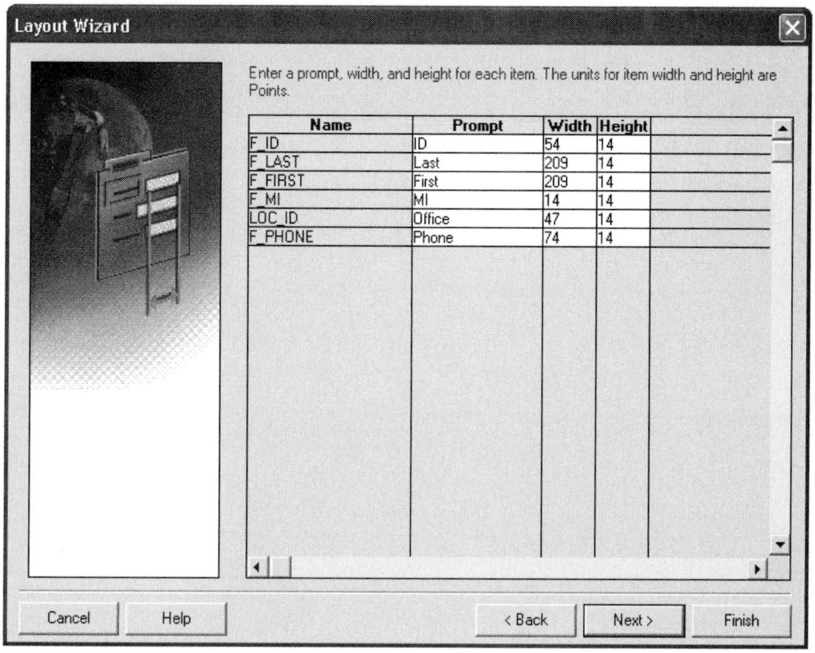
Layout Wizard
Enter a prompt, width, and height for each item. The units for item width and height are Points.
Name
Prompt
Width
Height
F_ID
ID
54
14
F_LAST
Last
209
14
F_FIRST
First
209
14
F_MI
MI
14
14
LOC_ID
Office
47
14
F_PHONE
Phone
74
14
Cancel
Help
< Back
Next >
Finish

Figure 5-33 Prompts for new data form

5. On the Style page, select the **Tabular** option button to specify that multiple records appear on the form, and then click **Next**.

6. On the Rows page, type **Faculty_Listing** in the Frame Title field, and change the Records Displayed field value to 10. Leave the Distance Between Records field value as 0 and the Display Scrollbar check box unchecked, as shown in Figure 5-34.

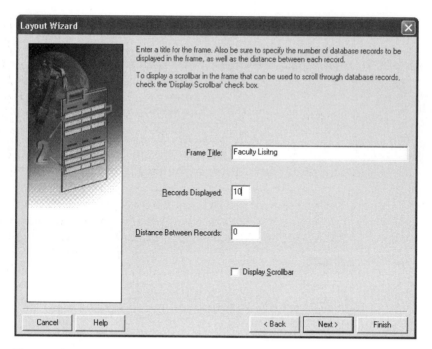

Figure 5-34 Rows page

7. Click **Next**, and then click **Finish**. The form layout appears in the Layout Editor. Click the mouse pointer anywhere on the form to deselect the selected form frame.

8. Click the **Save** button, and save the file as **Ch5AFaculty_Listing.fmb** in the Chapter05\Tutorials folder on your Solution Disk. The form layout appears as shown in Figure 5-30. (Your layout does not display the data that is shown in this figure.)

The tabular-style form layout in Figure 5-30 displays information about the faculty members contained in the FACULTY table on the form all at one time. Because you specified the distance between records as 0 in the layout properties, the records are stacked directly on top of one another.

You can perform insert, update, delete, and query operations in a tabular-style form just as you did in a form-style form that displayed a single record. Now you will run the form, and perform some query operations.

To run the form and perform query operations:

1. Click the **Run Form** button to run the form. The form appears in your browser window, as shown in Figure 5-35. The form does not yet display any data values.

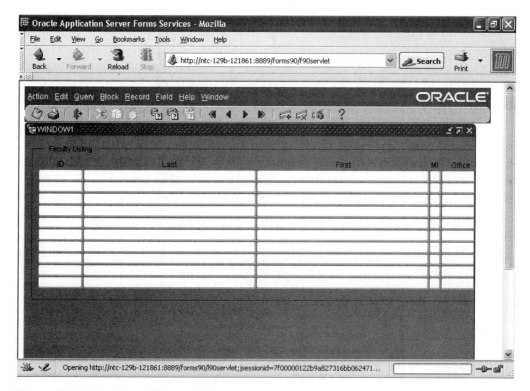

Figure 5-35   Empty multi-record form

2. Click the **Enter Query** button 🔳, and then click the **Execute Query** button 🔳. The form displays the data values of the FACULTY records, as shown in Figure 5-30.

3. Close the browser window, and then close Forms Builder. If you are asked to save changes to any of your forms as Forms Builder exits, click **Yes**.

4. To shut down the OC4J Instance, select **Start** on the taskbar, point to **All Programs**, point to **Oracle Developer Suite – OraHome**, point to **Forms Developer**, and then click **Shutdown OC4J Instance**.

As with the single-record form, you can use a tabular-style form to search for specific records, as well as insert, edit, and delete records.

SUMMARY

❑ A form is a database application that looks like a paper form and provides a graphical interface that allows users to add new database records and to modify, delete, or view existing records. A data block form is a form that is associated with a specific database table.

❑ All Developer10*g* forms appear as Web pages in a Web browser. To display forms as Web pages, a Web server process named OC4J translates the form into a Java applet, which is a self-contained Java program that runs in a Web browser's Java runtime environment.

❑ You can use a form to view, insert, update, and delete records. You can perform exact searches in forms to retrieve records that contain specific data values. You can perform restricted searches to retrieve records containing data values that fall within a range of values or contain partial text strings.

❑ A data block is a group of related form items that are associated with a single database table. When you create a data block form using the Data Block Wizard and Layout Wizard, the system automatically generates the text items and prompts for data fields in that table and supplies the underlying code for inserting, modifying, deleting, and viewing data records.

❑ The form layout specifies how the form appears to the user. A canvas is the area on a form in which you place layout objects such as buttons and text items. In a form-style layout, only one record appears at a time, whereas in a tabular-style layout, multiple records appear on the form at one time.

❑ The Layout Editor provides a graphical display of the form canvas. You use Layout Editor to draw and position form items, and to add boilerplate objects, such as labels, titles, and graphic images.

❑ The Object Navigator enables you to access all form objects either by object type in Ownership View, or hierarchically by opening objects that contain other objects in Visual View.

❑ Form modules contain windows, windows contain canvases, and canvases contain items such as text items and buttons.

❑ You can use the Data Block Wizard and Layout Wizard to modify existing data blocks and form layouts by selecting the data block or layout frame and then starting the wizard in reentrant mode.

REVIEW QUESTIONS

1. A Web browser is used to display a _____ Web page that displays data from an Oracle 10*g* database.

2. If an error occurs while performing data manipulation operations through a form, the database disconnects from the form. True or False?

3. If you do not check the Enforce data integrity check box when you are creating a form, the DBMS still enforces any integrity constraints that exist in the database. True or False?

4. A form can be used to display data from a table or _____ contained in the database.

5. What actions are permitted when a form is running in Normal mode?

6. When a form is run in a browser, it is translated into a Java _____.

7. What is a static imported image?

8. What is a data block form?

9. The _____ Navigator provides access to all form objects.

10. Before you access Forms Builder, you need to start the _____ instance.

5

MULTIPLE CHOICE

1. Canvases contain:
   a. form modules
   b. windows
   c. text items
   d. all of the above

2. You can use the _____ to place an image on a form.
   a. Layout Editor
   b. Layout Wizard
   c. Data Block Wizard
   d. Object Navigator

3. Programs collectively called a _____ can be used to access information on the Internet.
   a. Web editor
   b. Web browser
   c. Web server
   d. none of the above

4. The _____ button group inserts, deletes, and locks records.
   a. Navigation
   b. Action query
   c. Execute
   d. Query

5. Objects that contribute to the form's appearance, but not its functionality, are called _____ objects.
   a. static
   b. boilerplate
   c. image
   d. embedded

6. If the Data Block Wizard displays all the wizard pages at one time, the wizard is in _____ mode.

    a. multiple-tabular

    b. reentrant

    c. design

    d. edit

7. When a form has a _____ -style layout, only one record can be displayed at a time.

    a. form

    b. tabular

    c. multiple

    d. singular

8. Which of the following is *not* a valid quality option when importing an image?

    a. Excellent

    b. Poor

    c. Draft

    d. Fair

9. In _____ View, the form is presented as the top-level object, and then all form object nodes are listed on the next level.

    a. Ownership

    b. Visual

    c. Inheritance

    d. Form

10. To include a specific field on a form, select the field from the Available Columns list, and then click the _____ button.

    a. Move one item to target

    b. Move one item to source

    c. Move all items to target

    d. Move all items to source

PROBLEM-SOLVING CASES

Before starting any of the problem-solving cases, make certain you run the Ch5Clearwater.sql script file stored in the Chapter05 folder of your Data Disk to set up the necessary data. The case problems reference the Clearwater Traders sample database. For all cases, save the solution files in the Chapter05\Cases folder on your Solution Disk.

1. Use Forms Builder to create a form that displays all the data contained in the ITEM table of the Clearwater Traders database, except images in the ITEM_IMAGE field.

Assign prompts that are descriptive of the fields' contents. Make certain that at least five records are displayed on the form at the same time. In addition, include scroll bars in the form so users can view other records that may exist. Save the form as Ch05Case1A in the Chapter05\Cases folder of your data disk.

2. Using the form you created in the previous case, perform the following:

   a. Display all items containing the word "Shorts" in the ITEM_DESC field.

   b. Edit ITEM_ID 2 and change the ITEM_DESC to "All Season Tent".

   c. Add a new item to the table using the following data values:

   ITEM_ID: 8
   ITEM_DESC: Sleeping Bag
   CAT_ID: 4

   d. Attempt to delete the item that contains the CAT_ID value of 1. Explain why you were unable to delete the record.

   e. Save the changes you made to the ITEM table and close the form.

3. Use Forms Builder to create the form shown in Figure 5-36 to display the contents of the CUSTOMER table from the Clearwater Traders database. Save your completed form as Ch05Case3A in the Chapter05\Cases folder of your data disk.

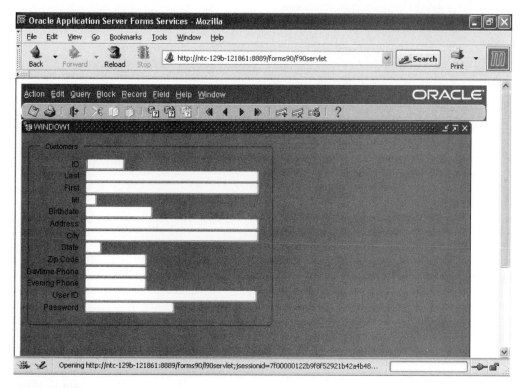

Figure 5-36

4. Modify the form created in Case 3 to include the image shown in Figure 5-37. The Clearlogo.jpg image is stored in the Chapter05 folder on your data disk. Save the modified form as Ch05Case4A in the Chapter05\Cases folder on your Data Disk.

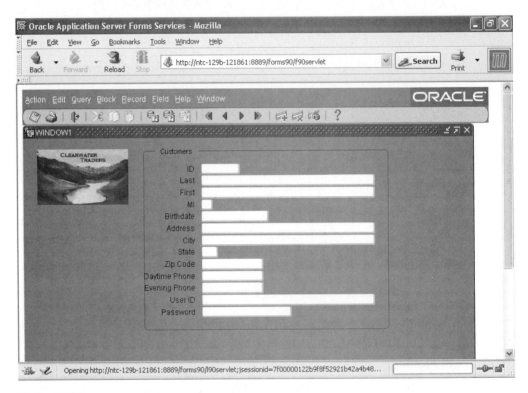

Figure 5-37

5. Use the modified form from Case 4 to perform the following:

   a. Display all customers that live in Florida.

   b. Change the daytime phone number for the customer with the last name of Lewis to 352-555-7881, but don't save these or any other changes to customer records.

   c. Display all customers who have an area code of 482 in their evening phone number.

   d. Close the form.

◄ LESSON B ►

After completing this lesson, you should be able to:
- ♦ Create a data block form that is based on a database view
- ♦ Modify form properties to improve form appearance and function
- ♦ Create a master-detail form that contains multiple data blocks
- ♦ Format form text items using format masks

CREATING A FORM BASED ON A DATABASE VIEW

Recall that a database view looks and acts like a database table but is based on a query. A view can contain a subset of a table's fields, or it can contain fields from multiple tables. Database administrators often use views to enforce security in databases, or to make it easier to retrieve related data that is in multiple tables. It is useful to create views that retrieve data values for records that are related by foreign keys, and then display the actual data values in a form, rather than the foreign key values. For example, modification of the multiple-record FACULTY form from Figure 5-35 to include the building name and office number of the faculty member is more informative than simply including the Office ID.

You can create data block forms based on views. When you create a form based on a view, *you can only view* the data values. You cannot insert new values into the underlying database tables, and you cannot update or delete the values. Now you are going to create a form that includes the faculty member's office information. The form will include data from a view that links the FACULTY and LOCATION tables.

First, you need to create the view from which this form derives its data. To do so, you start SQL*Plus, and create a view that retrieves all of the FACULTY records, along with their associated office locations.

To create the view in SQL*Plus:

1. Start SQL*Plus, log onto the database, and type the following command to create the view:

```
CREATE OR REPLACE VIEW faculty_location AS
SELECT f_id, f_last, f_first, f_mi, f_phone, bldg_code, room
FROM faculty NATURAL JOIN location;
```

2. Press **Enter** to execute the query. The message "View created" appears. Close SQL*Plus.

Now you create a form that uses FACULTY_LOCATION as its data source and displays five records at a time. You select the view from your list of database objects when you create the data block.

To create a form based on a view:

1. Start the OC4J Instance, minimize the window, and then start Forms Builder. The Object Navigator opens, and displays a new form module.

Make sure that the form module is selected, then click the **Data Block Wizard** button 📝 on the toolbar to create a new data block. If the Data Block Wizard Welcome page appears, click **Next**.

2. Make sure the Table or View option button is selected as the data block type, and then click **Next**.

3. When the Table page appears, click **Browse**, and then log onto the database, if necessary. Make sure that the Current User check box is checked. Clear the **Tables** check box, and then check the **Views** check box so that only your database views appear, as shown in Figure 5-38. Select **FACULTY_LOCATION**, and then click **OK**.

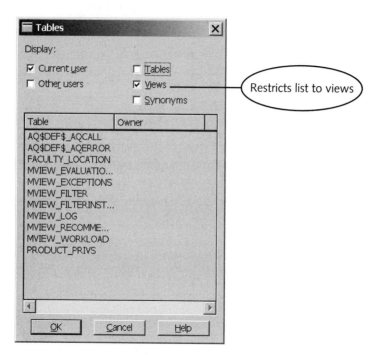

Figure 5-38 Available views

4. Select all of the view columns to be included in the data block, make sure that the Enforce data integrity check box is cleared, and then click **Next**.

5. Accept FACULTY_LOCATION as the Data Block Name, and click **Next**.

6. Make sure the Create the data block, then call the Layout Wizard option button is selected, and then click **Finish**. The Layout Wizard Welcome page appears.

The form layout should display five records and include scroll bars. Now you create the layout using the Layout Wizard.

To create the form layout:

1. On the Layout Wizard Welcome page, click **Next**.

2. On the Canvas page, accept the **(New Canvas)** and **Content** default selections, and then click **Next**.

3. On the Data Block page, select all of the block items for display in the layout, and then click **Next**.

4. Modify the item prompts and field widths as shown in Figure 5-39, and then click **Next**.

5

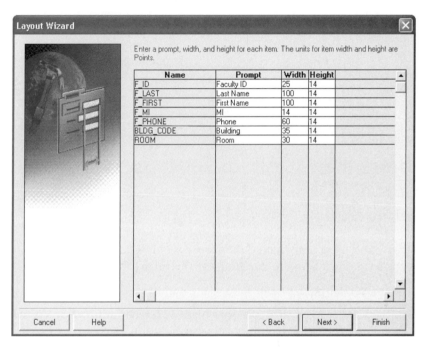

Enter a prompt, width, and height for each item. The units for item width and height are Points.

Name	Prompt	Width	Height		
F_ID	Faculty ID	25	14		
F_LAST	Last Name	100	14		
F_FIRST	First Name	100	14		
F_MI	MI	14	14		
F_PHONE	Phone	60	14		
BLDG_CODE	Building	35	14		
ROOM	Room	30	14		

Cancel   Help   < Back   Next >   Finish

Figure 5-39   Modified item prompts and field widths

5. Select the **Tabular** layout style option button, and then click **Next**.

6. Type **Faculty Locations** in the Frame Title field, change the Records Displayed field value to **5**, leave the Distance Between Records field value as 0, and check the **Display Scrollbar** check box. Click **Next**, and then click **Finish**. The form appears as shown in Figure 5-40.

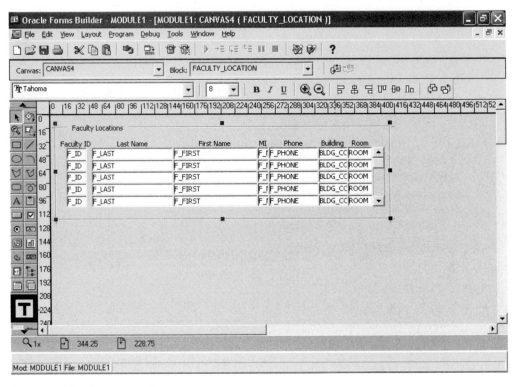

Figure 5-40   Form layout

Now you save the form and change the form's window, canvas, and frame names in the Object Navigator. Then you run and test the form.

To save the form, change the form item names, and run the form:

1. In the Object Navigator, save the form as **Ch5BFacultyLocView.fmb** in the Chapter5\Tutorials folder on your Solution Disk.

2. Click the **Run Form** button 🔯 to run the form. When the form opens in the browser window, click the **Enter Query** button 🔐, and then click the **Execute Query** button 🔐 to retrieve all of the view records into the form. The form should look like Figure 5-41, although it might look slightly different, depending on your display settings.

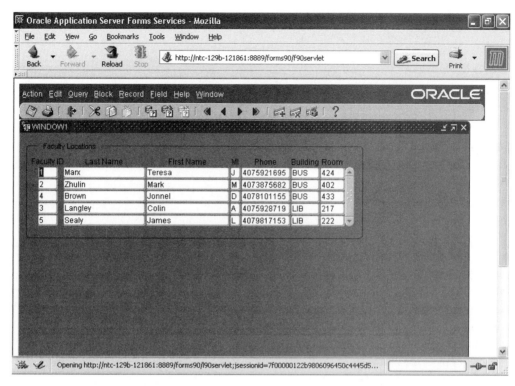

Figure 5-41   Faculty Locations form

3. Close the browser window.

4. Close the form in Forms Builder. If a dialog box opens asking if you want to save the form, click Yes.

Recall that when you create a data block using a view, you can only examine the data. You cannot modify the data by inserting new records or updating or deleting existing records.

MODIFYING FORM PROPERTIES

Form applications should be attractive, easy to use, and configurable. Often, you need to modify some of the properties of data block forms you create using the Data Block and Layout Wizard to meet these standards. For example, the data block form based on the Northwoods University FACULTY table shown in Figure 5-42 requires some formatting modifications.

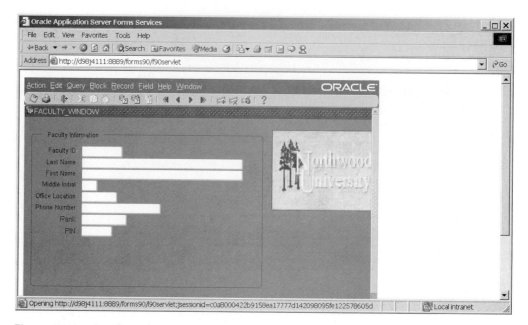

Figure 5-42 Faculty Information form

The form window title should be changed from the window object's name (FACULTY_WINDOW) to a more descriptive name that uses mixed-case letters. In addition, the color of the text boxes should be changed to provide a more professional appearance to the form.

Next, you modify these properties of the Northwoods University COURSE form. First you change the window title. To do this, you modify the window Title property in the window's Property Palette. Every form object has a **Property Palette**, which is a window that displays a list of object properties and their associated values. To open a form object's Property Palette, you select the object in the Object Navigator or on the Layout Editor, right-click, then click Property Palette. Now you open the form and examine the window's Property Palette.

To open the form and examine the window Property Palette:

1. In Forms Builder, open the **Faculty_Information.fmb** file from the Chapter05 folder on your Data Disk, and save the file as **Ch5BFaculty.fmb** in the Chapter05\Tutorials folder on your Solution Disk.

2. In the Object Navigator, open the **Windows** node, select **FACULTY_ WINDOW**, right-click, and then click **Property Palette**. The Property Palette for the window opens, as shown in Figure 5-43.

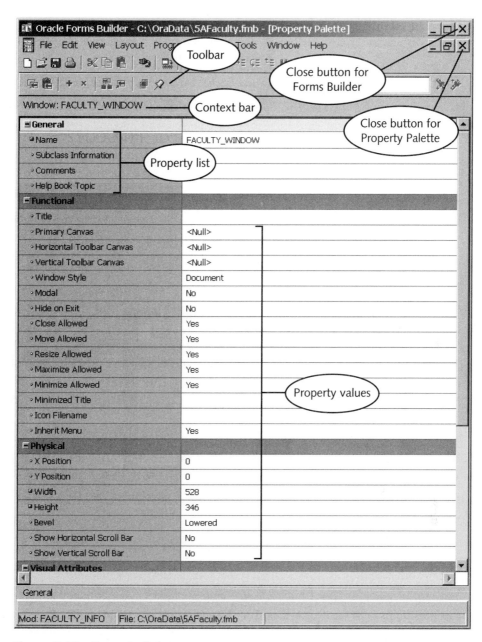

Figure 5-43   Property Palette

You can also open the Property Palette for most objects by double-clicking the object icon in the Object Navigator window, or by selecting the object, clicking Tools on the menu bar, and then clicking Property Palette.

The Property Palette has a toolbar that provides functions for working with the Property Palette. It also has a context bar that describes the object associated with the Property Palette and a property list that lists object properties and their associated values. Because no property is currently selected, the only active button on the Property Palette toolbar is the Pin/Unpin button ✄. Normally, whenever you open the Property Palette window, it displays the property list for the object that is currently selected in the form. When you click ✄, the current Property Palette window remains open, and you can open and work in multiple Property Palette windows at the same time. To close the Property Palette, you click the Close button on the Property Palette window.

Different form object types have different properties. For example, a text item has different properties than a window or a canvas.

The Property Palette groups related properties within the property list. For example, the Property Palette groups window properties such as X Position, Y Position, Width, and Height together, and groups properties related to the window's fonts such as Font Name, Font Size, and Font Weight together. Each group of related properties is called a **property node**. You can expand and collapse property nodes. By default, all property nodes are expanded when the Property Palette opens.

To specify a property value, you can type the desired value or select the value from a list. For some properties, you specify a value using a dialog box that opens when you click the More button [...] that appears in the middle of the property value box, or on the right side when you select the property.

Changing the Window Title

When you create a new form, you should always change the form window title to a descriptive title. Now you change the Course form window title in the Property Palette, then run the form and view the new title.

To change the window title and run the form:

1. In the FACULTY_WINDOW Property Palette, place the insertion point in the space to the right of the Title property under the Functional node. The property background turns yellow to indicate that the property is selected.

2. Type **Northwoods University** as the new window title, then select another property to apply the change. When you change a property value, the node marker to the left of the property name changes to a green color, as shown in Figure 5-44.

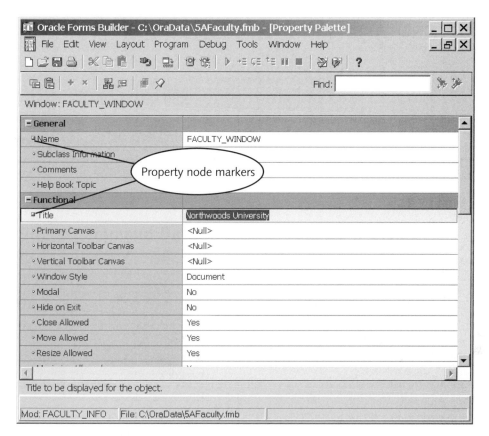

Figure 5-44   Modifying the window title

3. Close the Property Palette window, which is the inner window within the Forms Builder window.

If the message "Save changes to FACULTY_FORM?" appears when you try to close the Property Palette, you closed the outer Forms Builder window instead of the inner Property Palette window. Click Cancel, and then close the inner Property Palette window.

4. Run the form. The new window title appears on the form window within the Forms Services window. Close the browser.

Modifying Frame Properties

Recall that a frame encloses a data block's items. When you create a form layout in the Layout Wizard, you specify the frame title and the distance between the records. Usually, the text items appear stacked on top of each other in a single column. In the Faculty form, the frame is a wider size and displays two columns of text items. Now you resize the frame and change the frame properties. First, you examine the frame properties in the frame Property Palette.

To examine the frame Property Palette:

1. Open the Layout Editor by clicking **Tools** on the menu bar, and then clicking **Layout Editor**.

2. Select the **FACULTY_FRAME** frame, right-click, and then click **Property Palette**. The frame Property Palette opens, as shown in Figure 5-45.

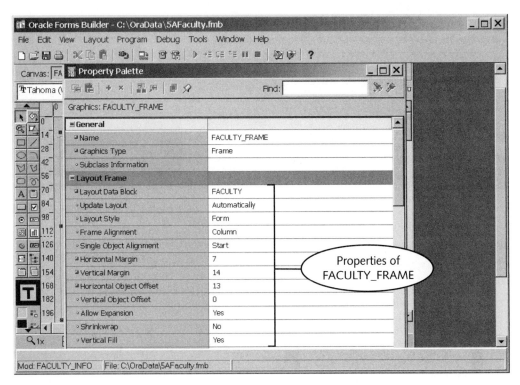

Figure 5-45   Frame properties

You configured many of the frame properties, such as the frame data block, the frame title, the frame layout style, the number of records the frame displays, and the space between the records, when you created the form layout using the Layout Wizard. For example, the Layout Data Block property value specifies the name of the data block whose items the frame encloses, and the Layout Style specifies that the frame encloses a form-style layout.

An important frame property is the Update Layout property, which specifies how the frame behaves. Currently, the Update Layout property value is set to Automatically, which means that when you resize or move the frame, Forms Builder automatically moves and repositions the items within the frame. The frame displays the items in the order they appear in the Data Blocks list in the Object Navigator. If the frame is resized to where it is not long enough to accommodate all of the items in a single vertical column, then the items appear in multiple columns. Other possible Update Layout property values are Manually, which enables you to resize the frame and then position the items manually, and Locked, which behaves the same as Manually, except that the item positions do not update when you change the layout properties using the Layout Wizard in reentrant mode.

When you are developing simple data block forms, you usually leave the Update Layout property value set to Automatically, so you do not have to position and align the form items manually. Now you resize the frame and observe how Forms Builder automatically repositions the frame items. Then, you move the frame to the center of the canvas.

To resize and reposition the form frame:

1. Close the frame Property Palette.

2. Make sure that the frame is selected, and then drag the lower-right selection handle all the way to the right edge of the screen display to make the frame wider.

3. Select **Undo** from the **Edit** menu to remove the changes made in the last step.

You can modify frame properties using the Layout Wizard in reentrant mode, or by changing the property values in the frame Property Palette. Be careful if you modify the frame properties using the Layout Wizard in reentrant mode, because this restores all frame properties to their default values. Any manual formatting you have done, such as resizing the frame, modifying the prompts directly, or modifying frame properties in the Property Palette, will be lost. If you want to preserve custom formatting in a frame layout, first open the Layout Wizard in reentrant mode and apply changes, change the frame Update Layout value to Locked, and then you can proceed.

Modifying Form Prompts and Prompt Properties Using the Layout Editor

In the Layout Editor, you can change the text that appears in item prompts by selecting the prompt, and then clicking the selected prompt. This opens the prompt for editing. Now you change the text of the LOC_ID prompt to "Office Location."

To modify the prompt text:

1. In the Layout Editor, select the **LOC_ID** item prompt, and then click it again to open the prompt text for editing. The insertion point appears in the prompt text.

2. Use the backspace or delete key as needed to remove the current prompt, and then type **Office Location** as the new prompt.

HELP

If, after you edit a prompt, the prompt extends outside the frame, resize the frame slightly. This causes Forms Builder to realign the frame objects within the frame.

3. Save the form.

You can also use the Layout Editor to modify the font and font size of item prompts. To modify a prompt's font, select the prompt, and then select the desired font and font size from the font list and font size list on the Layout Editor toolbar. You can modify all of the prompts in one step by selecting all of the prompts at the same time, and then changing their font type and size. To select multiple items, you select an item, press and hold the Shift key, and then select the remaining items.

TIP

Another way to select form objects as an object group is to click the mouse pointer on the canvas at a point below the lower-left corner of the objects, and then drag the mouse pointer to draw a rectangle around the objects to select them.

Modifying Text Item Properties Using the Property Palette

The text item Property Palette enables you to control the appearance of the data value that appears in a text item. Table 5-1 summarizes some of the important text item appearance properties.

Property Node	Property Name	Description
Physical	Visible	Determines whether the item appears when the form is running
Color	Foreground Color, Background Color, Fill Pattern	Specifies the text color, background color, and fill pattern
Font	Font Name, Font Size, Font Weight, Font Style, Font Spacing	Specifies the appearance of the text item's font
Prompt	Prompt, Prompt Display Style, Prompt Justification, Prompt Alignment	Specifies the text and appearance of the item's prompt (label)

Table 5-1 Text item appearance properties

Now you modify the appearance of the form text items by changing the background color to a light shade of gray. To modify this property for all of the items in one step, you select all of the items and format them by using the Property Palette in intersection mode. You use a Property Palette in **intersection mode** to modify one or more properties of a group of objects so that all objects in the group have the same value or values. To open a Property Palette in intersection mode, you select multiple form objects as an object group, and then open the Property Palette. When you open a Property Palette in intersection mode, the context bar displays the text "Multiple selection." When all group objects have the same value for a property, the common property value appears. When the objects have different values, the property value appears as **** in the intersection Property Palette, and the property node appears as a question mark (?).

To modify the text item background color using the Property Palette in intersection mode:

1. In the Layout Editor, select the **F_ID** text item (not the item prompt), press and hold the **Shift** key, and then select the **F_LAST**, **F_FIRST**, and all the remaining text items. All the text items should display selection handles, as shown in Figure 5-46 (but the prompts are not yet aligned as shown in this figure).

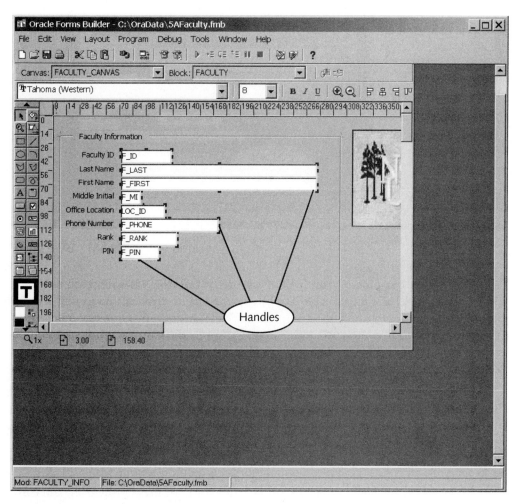

Figure 5-46 Selected text items

2. Right-click, and then click **Property Palette**. The intersection Property Palette opens, as shown in Figure 5-47. Note that the context bar displays the message "Multiple selection." The Name property appears as ****, which indicates that the selected objects have different names. Properties that display a value, such as the Enabled property, which displays the value Yes, indicate that all selected objects share this value.

3. Scroll down the Property Palette to the Color property node, select the **Background Color** property, and then click the **More** button. The Background Color dialog box opens, showing color choices for text item background colors.

4. Select a **light gray** square. The Background Color dialog box closes, and the color name, such as gray16, appears as the property value, as shown in Figure 5-47.

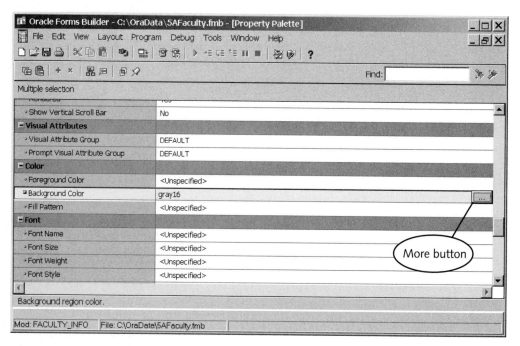

Figure 5-47   Changing background color of text items

5. Close the Property Palette window. The text items now appear with a light gray background color. Save the form.

Text items have properties that control how users enter data values and how the form displays data values. Table 5-2 summarizes some of these properties. Later in the chapter, you will format the phone numbers of the faculty members to be displayed in the format of (999) 999-9999 using a format mask.

Property Node	Property Name	Possible Values	Description
Functional	Multi-Line	Yes or No	Specifies whether the text item allows multiple-line input
Functional	Case Restriction	Yes or No	Converts entered value to uppercase or lowercase letters
Functional	Enabled	Yes or No	Specifies whether the user can navigate to the item using the Tab key or mouse pointer
Functional	Conceal Data	Yes or No	Specifies that characters entered in the field are hidden and replaced by "*"
Data	Data Type	Oracle or PL/SQL data types	Specifies the text item data type
Data	Initial Value	Value or variable	Specifies a default value that is inserted into the text item each time the user creates a new record
Data	Maximum Length	Integer value	Specifies the maximum number of characters that the user can enter into the text item
Data	Required	Yes or No	When set to Yes, specifies that the user must enter a value in the text item in order to save a new record
Data	Format Mask	Specific format mask	Specifies the format mask of the item value

Table 5-2   Text item data value properties

Text items also have properties that specify how an item interacts with the database. Table 5-3 summarizes these properties, which appear under the Database property node.

Property Name	Possible Values	Description
Primary Key	Yes or No	Specifies that the item corresponds to a database table's primary key
Query Only	Yes or No	Specifies that the item can appear in the form, but cannot be used for an insert or update operation
Query Allowed	Yes or No	Determines whether or not the user can perform queries using this item
Query Length	Integer value	Specifies the maximum length allowed in the text item for a restricted query search condition; a value of 0 means there is no length limit
Insert Allowed, Update Allowed, Delete Allowed	Yes or No	Determines whether the user can manipulate the item value when inserting, updating, or deleting a record

Table 5-3   Text item database properties

The form automatically sets these properties to default values that allow the user to insert, update, delete, and view database records. A developer changes these values only when the form needs to restrict user operations.

The Help text item properties add information that makes the form easier to use as the user navigates among form text items. You can add two different types of information: hints and ToolTips. A **hint** appears in the form message line when the insertion point is in the text item. A **ToolTip** appears beside the mouse pointer when the user moves the mouse pointer over the text item. Table 5-4 summarizes the text item hint and ToolTip properties in the Help property node.

Property Name	Possible Values	Description
Hint	Hint text	Specifies the text that appears in the message line when the insertion point is in the text item
Display Hint Automatically	Yes or No	Specifies whether the hint appears automatically when the user places the insertion point in the text item, or whether the hint appears only when the insertion point is in the text item and the user presses the F1 function key or selects the Help command on the menu bar
ToolTip	ToolTip text	Specifies the text that appears in the ToolTip

Table 5-4 Text item help properties

Now you add hints and a ToolTip to the Faculty_Information form. The hints describe the contents of the F_ID text item, and the ToolTip describes the allowable range of values for the text item. Then, you run the form and view the hints and ToolTip.

To add and view the text item hints and ToolTip:

1. Open the Property Palette of the F_ID text item, and change its Hint, Display Hint Automatically, and Tooltip properties to the values shown in Figure 5-48.

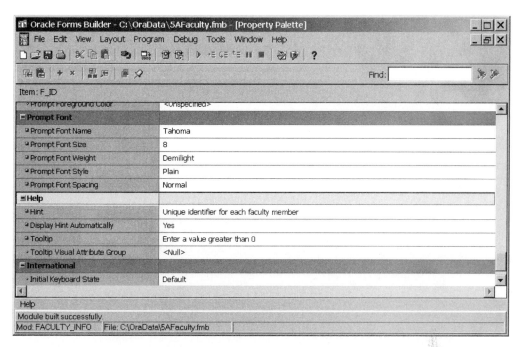

Figure 5-48   Properties of the F_ID text item

2. Save the form, and then run the form. When the form appears in the browser window, the insertion point is in the F_ID field, and the hint appears in the form message line, as shown in Figure 5-49. (You may need to scroll down in the browser window to view the hint. In order to view the ToolTip, you may have to point to the F_ID field.)

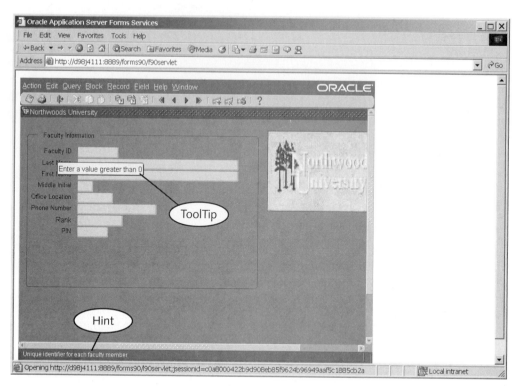

Figure 5-49 A form's ToolTip and hint

3. Close the browser window, then close the form in Forms Builder. If a dialog box opens asking if you want to save the form, click **Yes**.

CREATING A DATA BLOCK FORM THAT DISPLAYS DATA FROM MULTIPLE TABLES

You can create data block forms that display data from multiple database tables that have master-detail relationships. In a **master-detail relationship**, one database record (the master record) can have multiple related (detail) records through foreign key relationships. For example, a master-detail relationship exists between the Northwoods University FACULTY and STUDENT tables. One FACULTY record can have multiple associated STUDENT records. For instance, Teresa Marc (F_ID 1) has three different students she advises (S_ID JO100, S_ID PE100, and S_ID MA100). The FACULTY record is the master record, because a faculty member can have many student records, but a student can have only one associated faculty record.

When you create a form that contains a master-detail relationship, you always use the Data Block Wizard to create the **master block**, which is the block that displays the master records first. Then, you start the Data Block Wizard a second time, and create the **detail block**, which is the block that displays the detail records. When you create the detail block, you specify the master-detail relationship on a special page in the Data Block Wizard called the Master-Detail page. This page appears only when you create a data block in a form that already contains one or more data blocks.

Now you create the Northwoods University Faculty Advisor master-detail form in Figure 5-50.

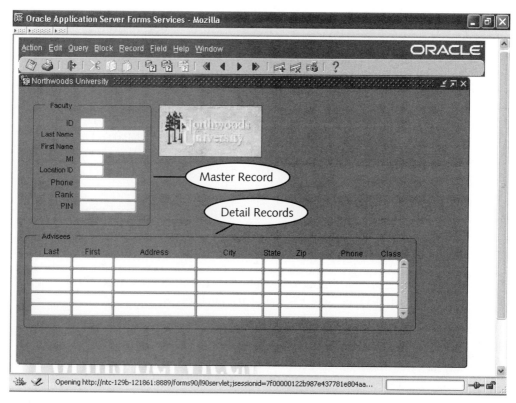

Figure 5-50   Layout of Faculty Advisor form

This form allows employees at Northwoods University to select a master faculty record, and then view and edit the associated student records. For the master block, you use a file in the Chapter5 folder on your Data Disk named Faculty.fmb. This form contains a master block that is associated with the FACULTY table, and uses a form-style layout to display a single faculty member's record one at a time. You modify this file by creating the detail block, which is the list of advisees for the selected faculty member. You specify the master-detail relationship when you create the detail block. First, you open the Faculty.fmb file, save it using a different filename, and change the form module name.

To open Faculty.fmb, save it using a different filename, and change the form module name:

1. In the Object Navigator, open **Faculty.fmb** from the Chapter05 folder on your Data Disk, and then save the file as **Ch5BFacultyAdvisor.fmb** in the Chapter05\Tutorials folder on your Solution Disk.

2. Change the form module name to **FACULTY_STUDENT**.

3. Save the form.

Creating the Detail Data Block

When the user selects a faculty member in the FACULTY block, the form shows the selected advisee information. Now you create the detail data block that displays the student information from the Northwoods University STUDENT table. Recall that to create a new data block, you select the Data Blocks node in the Object Navigator, and then click the Create button ✚. This block is going to be the detail block, so you specify the master-detail relationship on the Data Block Wizard's Master-Detail page. The wizard stores the master-detail relationship information in a relation object within the master block. Now you create the new detail data block, specify its data source, and display the Master-Detail page.

To create the detail data block and display the Master-Detail page:

1. In the Object Navigator, select the **Data Blocks** node under the FACULTY_STUDENT form module, and then click the **Create** button ✚. In the New Data Block dialog box, make sure that the Use the Data Block Wizard option button is selected, and then click **OK**. If the Data Block Wizard Welcome page appears, click **Next**.

2. On the Type page, make sure that the Table or View option button is selected, and then click **Next**.

3. On the Table page, click **Browse**, make sure that the Tables check box is checked, select **STUDENT** as the database table, and then click **OK**.

NOTE

If this action generates error number FRM-10095, then execute the Ch5Northwoods.sql script to create the database for this chapter

4. Using Figure 5-50 as a guide, select all of the block fields to be included in the data block and the F_ID field. The F_ID field will not be displayed in the final form but is needed to create the master-detail relationship. Make sure that the Enforce data integrity check box is cleared, and then click **Next**. The Wizard Master-Detail page appears.

The Master-Detail page allows you to specify master-detail relationships among data blocks. This page appears when you create a new data block in a form that already contains at least one other data block. The Create Relationship button allows you to create a new relationship, and the Delete Relationship button allows you to delete a relationship. The Master Data Blocks list shows the data blocks that are available to be selected as master blocks. When you create a master-detail relationship, you create the relationship by selecting the master block and then specifying the primary key/foreign key link that establishes the relationship. After you create the relationship, Forms Builder automatically adds triggers and program units to the form. These triggers and program units update the form's detail data values when you select a different master record value. (A program unit is a self-contained PL/SQL program that you can write to process data in a form. You learn how to create program units in forms in Chapter 6.)

There are two ways to create a master-detail relationship: you can use the Auto-join feature to allow the Data Block Wizard to create the relationship automatically, or you can specify the relationship manually. To use the Auto-join feature, you check the Auto-join data blocks check box, and then click Create Relationship. To specify the relationship manually, you make sure that the Auto-join data blocks check box is cleared, and then click Create Relationship. When you create a relationship manually, you explicitly select the text item in the master block and the text item in the detail block that will join the two blocks.

First, you use the Auto-join feature to create the relationship. To do this, you select the name of the master block from the Master Data Blocks list. The Auto-join feature automatically creates a join condition between the text item in the selected master block that has a foreign key relationship with a text item in the current detail block.

To create the relationship using the Auto-join feature:

1. Make sure that the Auto-join data blocks check box is checked, and then click **Create Relationship**. The FACULTY_STUDENT: Data Blocks dialog box opens, which shows all of the other blocks in the form. Currently, the only other form block is FACULTY.

HELP

Sometimes when you click Create Relationship when the Auto-join data blocks check box is checked, the message "FRM-10757: No master blocks are available" appears. If this message appears, click OK, clear the Auto-join data blocks check box, and then create the relationship manually, as described in the next set of steps.

2. Make sure that FACULTY is selected, and then click **OK**. FACULTY appears in the Master Data Blocks list, and the join condition for the two blocks appears in the Join Condition box, as shown in Figure 5-51.

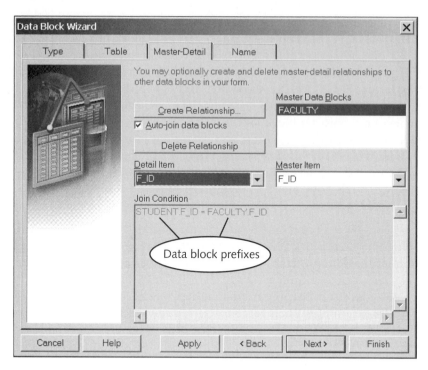

Figure 5-51   Master-detail relationship

The Join Condition field shows the join condition, which has the following syntax: *detail_block.join_item = master_block.join_item*. The join condition specifies the block names and text item names on which you are joining the two blocks. For this form, the join condition is STUDENT.F_ID = FACULTY.F_ID. Note that F_ID is the primary key in the FACULTY table, and a foreign key in the STUDENT table.

Now you create a master-detail relationship manually. You need to create the join condition manually if the form contains multiple blocks, or if the blocks contain multiple primary key/foreign key relationships that the wizard cannot resolve. First, you delete the relationship you just created and clear the Auto-join data blocks check box. Then, you create the join condition manually.

To delete the relationship and create the join condition manually:

1. On the Data Block Wizard Master-Detail page, click **Delete Relationship**. The master data block and join condition no longer appear.

2. Clear the **Auto-join data blocks** check box.

3. Click **Create Relationship**. The Relation Type dialog box opens. You can create the relation based on a join condition or on a REF item. (A REF item is a data type that contains pointers, or physical addresses, of related data

objects. REF items are an advanced Oracle 10*g* topic that this book does not address.) You base the relation on a join condition.

4. Confirm that the Based on a join condition option button is selected, and then click **OK**. The FACULTY_STUDENT: Data Blocks dialog box opens, which allows you to select the master block.

5. Click **FACULTY**, and then click **OK**. Notice that the join condition does not appear in the Join Condition field on the Data Block Wizard Master-Detail page. When you manually create a relationship, you must manually create the join condition.

6. Open the **Detail Item** list, which lists all the text items in the detail (STUDENT) block. Click **F_ID**.

7. Open the **Master Item** list, which lists all the text items in the master (FACULTY) block. Click **F_ID**. (You might need to scroll up to display F_ID.) The join condition appears in the Join Condition box.

8. Click **Next**, accept STUDENT as the Data Block Name, and click **Next** again.

9. Make sure the Create the data block, then call the Layout Wizard option button is selected, and then click **Finish**.

Now you create the layout for the STUDENT detail block. You will use a tabular layout, display five student records, and use a scroll bar.

To create the detail block layout:

1. Click **Next** on the Layout Wizard Welcome page. Because you want the detail block fields to appear on the same canvas as the FACULTY block fields, accept **FACULTY_CANVAS** as the layout canvas, and then click **Next**.

2. Click the **Move all items to target** button ⟫ , and then click **Next**.

3. Change the field prompts and widths as indicated below, and then click **Next**.

Name	Prompt	Width
S_LAST	Last Name	50
S_FIRST	First Name	50
S_ADDRESS	Address	100
S_CITY	City	80
S_STATE	State	20
S_ZIP	Zip	50
S_PHONE	Phone	74
S_CLASS	Class	20
F_ID	<blank>	0

4. On the Style page, select the **Tabular** option button, and then click **Next**.

5. On the Rows page, type **Advisees** for the frame title, change the Records Displayed field value to **5**, and check the **Display Scrollbar** check box. Click **Next**, and then click **Finish**. The Advisees frame and text items appear on the canvas. (You may need to scroll to the bottom of the canvas to view the frame.) If the frame is at the bottom of the form, select and drag it so that it is right under the Faculty frame.

6. Open the **Object Navigator** window, and if necessary, open the **FACULTY** node under the Data Blocks node. Note that a relation object named FACULTY_STUDENT has been created. The new FACULTY_STUDENT relation specifies the join condition created on the Master-Detail page. Also note that Forms Builder has added some trigger objects and some program unit objects to the form to update the data in the master-detail text items when you run the form.

7. Save the form.

Running the Master-detail Form

Now you run the form to see how a master-detail form works. First, you run the form, and retrieve all of the faculty records along with their associated advisees. When you run a form with multiple data blocks, you must always put the insertion point in the master block before you execute a query. Otherwise, the form retrieves only the records for the detail block.

To run the form and retrieve the faculty and student records:

1. Run the form. When you run the form, a Compile dialog box appears briefly, which indicates that Forms Builder is compiling the triggers and program units for running the master-detail form.

2. When the form appears in the browser window, place the insertion point in the Faculty ID field in the Faculty frame, click the **Enter Query** button 📇, and then click the **Execute Query** button 📇. Teresa Marx's data values appear in the Faculty frame text items, and her associated advisees appear in the Advisee frame, as shown in Figure 5-52.

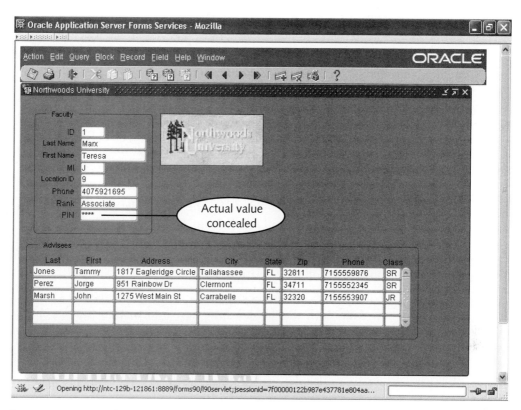

Figure 5-52   Faculty and student information

HELP

If the insertion point is in a text field in the Advisee frame when you execute the query, the query will retrieve only student records.

3. Click the **Next Record** button ▶. The record for the next faculty member appears, along with the appropriate student information if the faculty member has been assigned any advisees.

4. Continue to scroll through the rest of the faculty records, noting how the advisee information changes to reflect the current faculty member.

5. Close the browser window.

You can add, update, delete, and query database records in a master-detail form just as you did in the single-table form. The only difference is that when a form displays multiple data blocks, you can only insert, update, delete, or query records in the block that is currently selected and displays the insertion point. For example, to enter a new STUDENT record for the current faculty member, you must place the form insertion point in a text item that is in the STUDENT block. (Currently, the insertion point is in the FACULTY block.) There are two ways to move the form insertion point: click the Next Block button ▶ on the toolbar to move from the FACULTY block to the STUDENT block, or click the mouse pointer in any field in the STUDENT block.

Adding Another Detail Data Block to the Form

A data block can be the master block in multiple master-detail relationships. For example, a student has an advisor and also enrolls in courses at the university. In this case, the student block is the detail block in the student-advisor relationship, and it is also the master block in the student-enrollment relationship.

A data block can also be the detail block in one master-detail relationship, and the master block in a second master-detail relationship. For example, a faculty member can have many advisees, and an advisee (student) can be enrolled in many classes. In this scenario, the student is the detail block in the faculty-student relationship, and it is the master block in the student-enrollment relationship.

To create a second master-detail block within a form, simply use the Data Block Wizard to create a new data block that contains the new information (that is, enrollment). On the Master-Detail page of the wizard, create a new relationship that identifies the master block (STUDENT), and select the linking field (S_ID). Then you simply format the layout of the form as before. In the Problem-solving Cases at the end of this lesson, you will practice creating a second master-detail block.

USING FORMAT MASKS TO FORMAT CHARACTER STRINGS

To improve the appearance of the Faculty_Student form, you can modify the Format Mask property for the Phone Number text item. A **format mask** is the same as a format model. Recall that a format model specifies how an application formats a data item, such as displaying a date value using the format MM/DD/YYYY, or displaying a number value as currency. In this section, you modify the format mask of the Phone Number text item so that the output appears with parentheses and a dash. Then, you run the form and confirm that the data value appears correctly.

Recall that in a text item's Property Palette, you can change the Format Mask property to format how the text item's data values appear. In Figure 5-52, the telephone number for Teresa Marx appears as 4075921695, but users could read it more easily if the digits were formatted as (407) 592-1695. Recall that the Northwoods University database stores telephone numbers as character strings because users do not perform arithmetic calculations on telephone numbers. You can embed formatting characters in character string data fields by preceding the format mask with the characters "FM", and placing embedded formatting characters in quotation marks. For the telephone number, you would embed an opening parenthesis (( ), list the first three numbers, embed a closing parenthesis and a blank space ( )), list the next three numbers, embed a hyphen (–), and then list the last four numbers. The complete format mask appears as follows: FM"("999") "999"–"9999. Figure 5-53 illustrates this format mask.

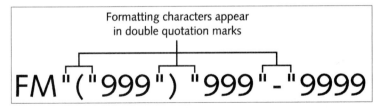

Figure 5-53 Creating a format mask that embeds formatting characters in a character string

Note that the format mask encloses the embedded formatting characters in double quotation marks, and uses instances of the number 9 as markers to represent the number values. Now you change the format masks for the faculty and student telephone numbers.

NOTE

For numeric columns, simply enter the desired format of the display, such as 99.99 to display two decimal places. The double quotation marks are not required if there are no character strings included in the display. To format date values, simply enter the desired format (that is, MM/DD/YYYY) without double quotation marks.

To specify the format masks and run the form:

1. If necessary, open the FACULTY_STUDENT form in the Layout Editor by clicking Tools on the menu bar, and then clicking Layout Editor.

TIP

You can also open a form in the Layout Editor by double-clicking the form's Canvas icon ▨ in the Object Navigator, or right-clicking the form and selecting Layout Editor from the pop-up menu.

2. In the Faculty frame, select the **F_PHONE** text item, right-click, and then click **Property Palette**.

3. Scroll down to the Data property node, and type the following format mask for the telephone number: FM"("999") "999"–"9999, as shown in Figure 5-54. (There are three embedded character strings in the format mask: an opening parenthesis, a closing parenthesis followed by a blank space, and a hyphen.)

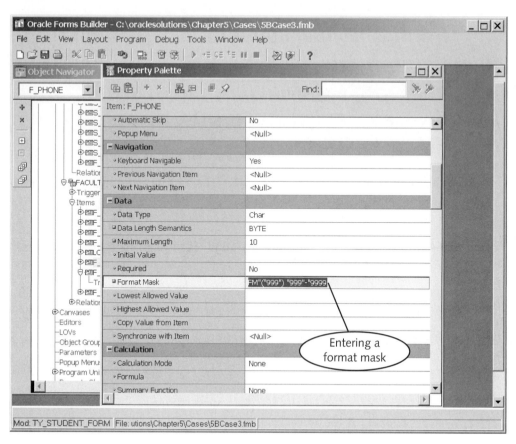

Figure 5-54 Inserting a property mask

4. Close the Property Palette.

5. In the Advisees frame of the Object Navigator, repeat the same process for the **S_PHONE** text item.

6. Save the form, and then run the form. When the form opens in the browser window, place the insertion point in the ID field in the FACULTY frame, click the **Enter Query button** 🔍, and then click the **Execute Query button** 🔍 to retrieve all of the database records. The results should look like Figure 5–55 and show the formatted telephone number values.

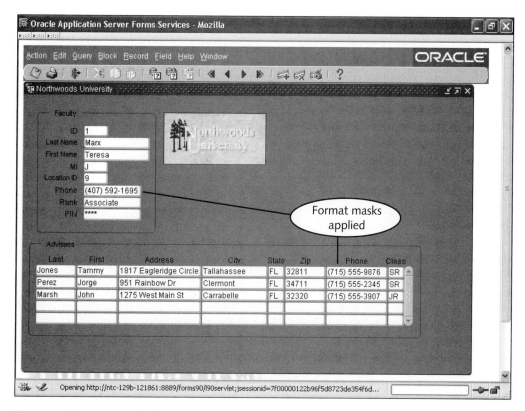

Figure 5-55   Form with formatted data

7. Close the browser window.

8. Close Forms Builder. If a dialog box opens asking if you want to save the form, click **Yes**.

9. Shut down the OC4J Instance.

SUMMARY

- ❑ You can create data block forms that derive their data from database views. This allows you to display data from multiple tables in the same data block. When you create a data block form based on a view, you can view data, but you cannot modify the database by inserting new records or updating or deleting existing records.

- ❑ Every form object has a Property Palette that allows you to modify the object's properties. Different objects have different property lists, which you can modify by typing new values, selecting from predefined lists, or specifying values using dialog boxes.

5

❏ You can configure frames three ways: to automatically position the items enclosed in the frame, to allow the developer to manually position the items, or to lock the items and preserve the custom formatting when the developer updates the layout in the Layout Wizard in reentrant mode.

❏ You can open a Property Palette in intersection mode and use the intersection Property Palette to modify the properties of multiple objects to the same value.

❏ You can modify text item prompts in the Layout Editor by opening the prompt for editing, and then typing the new value. You can also change the prompt's font style and size.

❏ You can use the Property Palette to modify the appearance of form text items by specifying alternate values for colors, font styles and sizes, and background colors. You can also modify the behavior of the form by specifying text item properties such as whether the value is required in the form, or whether the user can update or delete the value.

❏ A text item can have a hint, which is informative text that appears in the message line when you place the insertion point in the text item. A text item can also have a ToolTip, which is informative text that appears directly beside the text item when you place the mouse pointer over the item.

❏ In a master-detail relationship, one database record can have multiple related records through foreign key relationships. A master record can have many detail records, but a detail record always has only one associated master record.

❏ You can create data block forms that display master-detail relationships. A block can be the master block in one or more master-detail relationships. A block can also be the detail block in one master-detail relationship and the master block in a second master-detail relationship with a different block.

❏ When you create a master-detail relationship, you always create the master block first and then specify the master-detail relationship when you create the detail block. The form stores the relationship information in a relation object within the master block, and creates triggers and program units to display the correct detail record when the user changes the master record value.

❏ When you create a master-detail relationship, checking the Auto-join data blocks check box instructs the system to automatically create the join condition between the master block and detail block. You also can create the join condition manually.

❏ You can add, update, delete, and query database records in master-detail blocks just as in single-table data blocks, except that you need to first place the insertion point in the appropriate data block before performing the insert, update, or query.

❏ You can specify format masks for a form text item by entering the desired format mask in the item Property Palette. You can add formatting characters to character data by embedding the characters within the format mask.

REVIEW QUESTIONS

1. What is the name of the window that displays the values of an object's properties?

2. You can update database records using a data block form that is based on a database view. True or False?

3. Use the Property Palette in _____ mode to modify several properties to make them the same value at the same time.

4. To display a date as MM/DD/YYYY, you need to set the _____ property.

5. When displaying data from two tables that represent a one-to-many relationship in a form, the many side of the relationship is the master block. True or False?

6. What is the purpose of a ToolTip?

7. When do you use double quotation marks in a format mask?

8. How does the Auto-join feature of the Data Block Wizard work?

9. You can change the properties of a frame using the Layout Wizard in _____ mode, or by changing the values in the Property Palette.

10. A group of related properties is called a property node. True or False?

5

MULTIPLE CHOICE

1. If a property node appears as a(n) _____, it means that the property value has been changed.

   a. question mark (?)

   b. black square or dot

   c. green square or dot

   d. asterisk (*)

2. If a frame's Update Layout property value is _____, it means that when you resize the frame, the items within the frame automatically move.

   a. automatically

   b. locked

   c. manually

   d. required

3. In an intersection Property Palette, what does it mean when a Property Palette displays **** for a property value?

   a. Multiple items with different property values have been selected.

   b. The selected items are unavailable.

   c. Several properties are available for the selected item.

   d. The item is locked and a password is required.

4. Which of the following is a valid format mask?
   a. 99/99/9999
   b. 9.99
   c. $9,999.99
   d. "MM"/"DD"/"YYYY"

5. If Tables A and B have a master-detail relationship, Table A is the master block, and Table B is the detail block, the foreign key is most likely stored in:
   a. the TABLEA_B join block
   b. Table A
   c. Table B
   d. TABLEA master block

6. You should add a(n) _____ to a form if you want a message to be displayed in the message line when a user is accessing a text item.
   a. ScreenTip
   b. ToolTip
   c. exception message
   d. hint

7. When you create a form based on a _____, you cannot edit any data displayed in the form.
   a. view
   b. merged table
   c. master-detail relationship
   d. multiple tables

8. To change the window title for a form, enter the desired title in the Title property under the _____ node.
   a. Group
   b. Display
   c. General
   d. Functional

9. In a form, a frame:
   a. makes the form look more professional
   b. encloses items from the same data block
   c. creates groups of data that can be edited at the same time
   d. none of the above

10. A scroll bar is displayed in the detail block of a form only if:

    a. There are more than five records.

    b. There are more than ten records.

    c. The option is selected in the Layout Wizard.

    d. The property is assigned the value of Automatic.

PROBLEM-SOLVING CASES

The case problems reference the Clearwater Traders sample database. For all cases, save the solution files in the Chapter5\Cases folder on your Solution Disk. For all forms, rename the window and canvas using descriptive object names. Specify descriptive item prompts, and if necessary, format the item prompts so that they appear on a single line.

1. In this case, you create a form based on a view that displays information about Clearwater Traders customer orders.

    a. In SQL*Plus, create a view named **ORDER_VIEW** that retrieves the order ID, order date, payment method, associated customer first and last names, and order source description for every record in the Clearwater Traders ORDERS table. Save the command to create the view in a file named **Ch05Case1B.sql**.

    b. Create a form named **ORDER_VIEW_FORM** based on ORDER_VIEW. Title the frame **Customer Orders**, use a tabular layout, display five records on the form at a time, and display a scrollbar. Resize the layout text items if necessary so that all items appear in the layout without scrolling horizontally. Format the order date so that the date for Order ID 1 appears as "05/29/2006." Save the form design file as **Ch05Case1B.fmb**.

2. In this case, you create a form based on a view that displays the items ordered by Clearwater Traders customers.

    a. In SQL*Plus, create a view named **ORDER_LINE_VIEW** that retrieves the customer's first and last names, the order ID, the item IDs contained on the order, and a description of each item. Save the command to create the view in a file named **Ch05Case2B.sql**.

    b. Modify the **ORDER_VIEW_FORM** from Case 1 by adding a data block to display the information regarding the items purchased by each customer. Save the modified form design file as **Ch05Case2B.fmb**.

3. Create a master-detail form named **CUSTOMER_ORDERS_FORM** that displays records from the CUSTOMER and ORDERS tables. Format the form to resemble the sample form shown in Figure 5-56a. Remember to assign the appropriate values to the form's Title property and the text items prompts. Retrieve the customer and order data to verify that the form functions correctly. Save the form design file as **Ch05Case3B.fmb**.

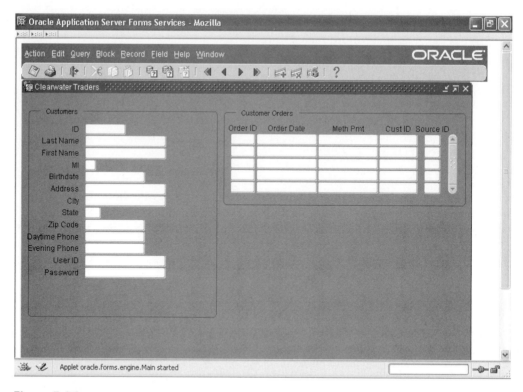

Figure 5-56a

4. Modify the **CUSTOMER_ORDERS_FORM** from Case 3 to also display data from the ORDER_LINE table, as shown in Figure 5-56b. (I: The ORDERS table is the master block in the ORDERS, ORDER_LINE relationship.) Run the form to verify the data is displayed correctly. Save the modified form design file as **Ch05Case4B.fmb**.

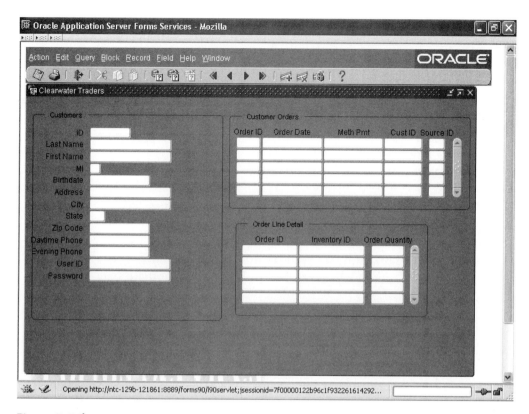

Figure 5-56b

5. Use the modified form created in Case 4 to perform the following tasks:

   a. Attempt to delete the first customer from the CUSTOMER table using the Customer block in the form. Why were you unable to perform this task?

   b. Add yourself as a customer to the CUSTOMER table using the CUSTOMER block on the form.

   c. Create a new order for yourself in the Customers Orders frame of the form.

   d. In the Order Line Detail frame of the form, add two items for your order— remember to enter items that actually exist in the ITEMS table of the Clearwater Traders database or the new data will be rejected.

   e. Save the changes made to the database tables.

   f. Explain why you would not have been able to add the new data in Steps b-d if the C_ID and O_ID fields were not included in the form.

◀ LESSON C ▶

After completing this lesson, you should be able to:

♦ Use sequences to automatically generate primary key values in a form

♦ Create lists of values (LOVs) to provide lists for foreign key values

♦ Describe the different form items that you can use to enter and modify data values

In this lesson, you learn how to write your own PL/SQL programs to make your forms more functional and easier to use. You use triggers to automatically retrieve sequence values into forms for primary key values. You also use triggers to display a pick list of values from which the user can select a value for a form text item. And, you explore how to use alternate form items, such as radio buttons and check boxes, to display and manipulate data.

USING SEQUENCES TO GENERATE PRIMARY KEY VALUES

With the forms you have created so far, whenever you insert a new record into a database, you have to type the value for the record's primary key. This could be a source of errors if two users type the same primary key value for different records. It is inconvenient for the user always to have to query the database or look up the next available primary key value using a printed report.

In Chapter 3, you learned how to create sequences to generate surrogate key values for primary key fields automatically. Now you are going to learn how to create a form that automatically retrieves the next value in a sequence and displays it in the form text item that is associated with the database table's primary key. To learn how to do this, you work with the Northwoods University Location form in Figure 5-57.

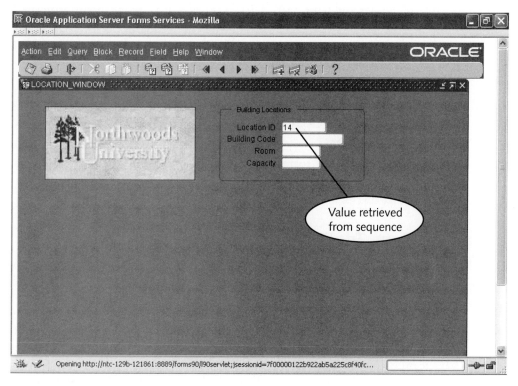

Figure 5-57   Northwoods University Location form

In this form, when the user or the system creates a new blank record, the form automatically retrieves the next value from a sequence, and displays the value in the Location ID text item. (The Location ID text item corresponds to the LOC_ID database field, which is the primary key of the LOCATION table.) Before you create the form, you run the script that refreshes the database tables. Then, you create the sequence to generate the

LOC_ID values that the form will retrieve. Because the highest current LOC_ID value is 13, you set the sequence start value at 14.

To run the scripts and create the sequence:

1. Start SQL*Plus and log onto the database.

2. To refresh your database table, run the **Ch5Northwoods.sql** script in the Chapter05 folder on your Data Disk.

3. Type the following command to create LOC_ID_SEQUENCE (do not close SQL*Plus; you will use it later in this lesson):

```
CREATE SEQUENCE loc_id_sequence
START WITH 14;
```

Now you start Forms Builder and open a form named Location.fmb that is in the Chapter05 folder on your Data Disk. This is a data block form associated with the Northwoods University LOCATION table that displays one location at a time. First, you start Forms Builder, open the form, and save the form with a different filename.

To open the form and save it with a different filename:

1. Start the OC4J Instance, and then minimize the window. Start Forms Builder, and open **Location.fmb** from the Chapter05 folder on your Data Disk. Save the file as **Ch5CLocation.fmb** in the Chapter05\Tutorials folder on your Solution Disk.

2. Click **File** on the menu bar, click **Connect**, and connect to the database.

To insert the next LOC_ID_SEQUENCE value into the Location ID form field automatically, you create a form trigger. The next section describes form triggers.

Creating Form Triggers

A **form trigger** is a program unit in a form that runs in response to an **event**, which is a user action such as clicking a button, or a system action, such as loading the form or exiting the form. (Recall that a program unit is a self-contained PL/SQL program in a form.) You associate form triggers with specific form objects, such as text items, command buttons, data blocks, or the form itself. For example, when you click the Save button on the Forms Services toolbar, a trigger containing a SQL INSERT command executes. The trigger's form object is the Save button, and the action is clicking the button. To write a trigger, you must specify the trigger's object, the event that starts the trigger, and the code that executes when the trigger fires. Every form object has specific events that execute, or **fire**, triggers. For example, buttons have a WHEN-BUTTON-PRESSED event that fires a trigger when the user clicks the button. Forms have a PRE-FORM event that fires a trigger just before the form first opens.

Now you create a trigger that automatically inserts the next sequence value into the Location ID text item on the Location Items form whenever the user creates a new block record. You associate this trigger with the LOCATION block, and with a block event called WHEN-CREATE-RECORD. This block event fires whenever a user creates a new blank record in a data block.

To create the trigger:

1. In the Object Navigator window in Ownership View, open the **Data Blocks** node, open the **LOCATION** node, and then select the **Triggers** node under the LOCATION data block.

2. Click the **Create** button ✚ to create a new trigger. The LOCATION_FORM: Triggers dialog box opens.

The Triggers dialog box shows all the block events that can have associated triggers. An object can have several associated events, so the Find field allows you to enter a search string, which helps you to quickly find the event that you want to associate with the trigger. You want to attach the trigger code to the WHEN-CREATE-RECORD event, so you perform a search using the string "WHEN%" to retrieve all events that start with the text string "WHEN". (Recall that the % wildcard character matches a variable number of characters.)

To find and select the trigger event:

1. Place the insertion point just before the % in the Find field, type **WHEN** and then press **Enter**. The list displays the events that start with the word "WHEN", as shown in Figure 5-58.

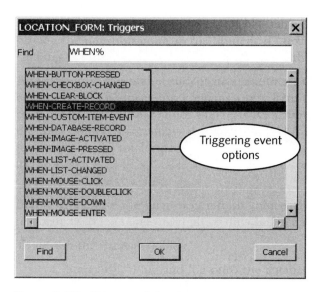

Figure 5-58   Triggers dialog box

NOTE

In the Triggers dialog box, searches for event names are not case sensitive.

2. Select the **WHEN-CREATE-RECORD** event, and then click **OK**. The PL/SQL Editor opens, as shown in Figure 5-59. If necessary, maximize the window.

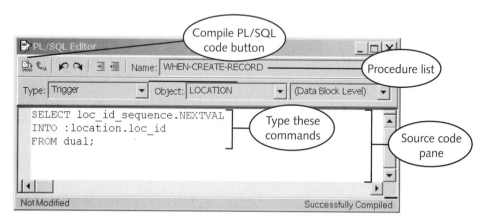

Figure 5-59 PL/SQL Editor

The PL/SQL Editor provides an environment for writing, compiling, and editing PL/SQL programs. The Name list shows the name of the current program unit, and allows you to access other program units. The Type list shows the program unit type, which is Trigger in Figure 5-59. The Object list shows the object to which the program unit is attached and the object level in the Object Navigator. In Figure 5-59, the WHEN-CREATE-RECORD trigger is attached to the LOCATION object, which is under the Data Blocks node.

The **source code pane** provides a text editor in which you can type PL/SQL program statements. The **status line**, which is also called the status bar, displays the program unit's current modification status (Modified or Not Modified) and compile status (Not Compiled, Successfully Compiled, or Compiled with Errors). The PL/SQL Editor toolbar provides the following buttons: Compile PL/SQL code, which compiles the current program unit; Revert PL/SQL code, which reverts the code to the state it was in the last time you saved or compiled, and discards subsequent changes; Undo, which discards the most recent change; Redo, which restores a change after you click the Undo button; Indent, which indents a line or block of code toward the right edge of the source code pane; and Outdent, which moves a line or block of code back to the left edge of the source code pane.

The source code pane uses colored text to define different command elements. Reserved words (such as BEGIN and UPDATE) appear blue. Command operators for arithmetic, assignment, and comparison appear red. User-defined variables are black, literal values (such as the number 11) are blue-green, and comment statements are light green.

Now you type the code for the WHEN-CREATE-RECORD trigger. This trigger needs to retrieve the next value in the LOC_ID_SEQUENCE and insert the value into the LOC_ID item on the form. To do this, you use an implicit cursor. Recall from Chapter 4 that you learned how to use an implicit cursor to retrieve values in a single record and assign the result to a program variable, using the following general syntax:

```
SELECT field_name
INTO variable_name
FROM table_name
WHERE search_condition;
```

You use the NEXTVAL pseudocolumn to select the next value of the sequence, and you display the next value in the form's Location ID text item. To reference a form text item in a PL/SQL program, you use the following syntax:

```
:block_name.item_name
```

In this syntax, *block_name* is the name of the block that contains the text item, and *item_name* is the name of the text item as it appears in the Object Navigator. Note that in the expression, a colon (:) precedes the block name. In the Location form, the block name is LOCATION, and the text item name is LOC_ID. Therefore, you use the following expression to reference the LOC_ID text item in a PL/SQL program: `:location.loc_id`.

After you create a trigger, you must compile the trigger to translate it into executable code. When you compile a trigger, the compiler checks the code for syntax errors and for incorrect references to database or form objects. Forms Builder automatically compiles all form triggers when it runs the form, but it is a good practice to compile each trigger just after you enter the source code so that you can find and correct errors immediately. When the compiler detects an error, it displays the line number of the statement causing the error, along with an error description. Now you enter the PL/SQL code for the WHEN-CREATE-RECORD trigger, and then compile the trigger.

To create and compile the WHEN-CREATE-RECORD trigger:

1. Place the insertion point on the first line in the source code pane in the PL/SQL Editor if necessary, and then type the code in Figure 5-59 to create the trigger. Note that PL/SQL and SQL commands appear in blue text, and user-entered variables appear in black text.

5

2. Click the **Compile PL/SQL code** button on the Editor toolbar to compile the trigger. If the trigger compiles successfully, the "Successfully Compiled" message appears on the status line.

> If your trigger does not compile successfully, you need to debug it. Debugging form triggers is similar to debugging other PL/SQL programs, except that there are some different kinds of errors that you might encounter. The next section discusses common form trigger errors.
>
> **NOTE**

3. If your trigger compiles successfully, close the PL/SQL Editor window by closing the inner window in Forms Builder. (Do not close the outer window, because this closes Forms Builder.)

4. Save the form.

Correcting Common Trigger Syntax Errors

When you create simple form triggers such as the one in Figure 5-59, errors usually occur as a result of syntax errors such as referencing database objects or form items incorrectly, or as a result of not being connected to the database. This section describes the error messages you often see as a result of these common syntax errors.

> In Chapter 6, you learn how to use the Forms Debugger to debug complex triggers that contain many lines of code and contain logic errors.
>
> **NOTE**

Figure 5-60 shows an example of a syntax error that appears when compiling the WHEN-CREATE-RECORD trigger.

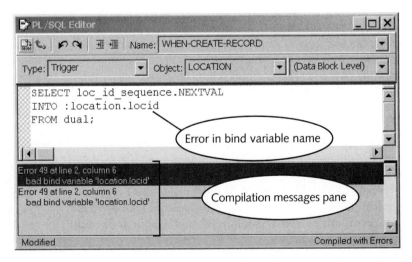

Figure 5-60 Trigger syntax error caused by referencing a form object incorrectly

When an error occurs in a program unit, the PL/SQL Editor's **compilation messages pane** displays the line number of the error, the error message, and correction suggestions. The insertion point shows the location for the selected error in the source code pane, or at least the line on which the compiler thinks the error occurs. Sometimes multiple errors appear in the compilation messages pane. The selected error is the error message that appears shaded. If you select a different error, the insertion point moves to the location of the different error.

In Figure 5-60, the syntax error occurs because the code references the form text item as `locid` rather than as `loc_id`, so the command is referencing a form object that does not exist. The error message "bad bind variable '*variable_name*'" usually indicates that you are referencing a form object incorrectly. In this example, to correct the error, you change the variable name to `location.loc_id`, and then recompile the trigger.

Figure 5-61 shows a common syntax error with the error message "identifier '*identifier_name*' must be declared."

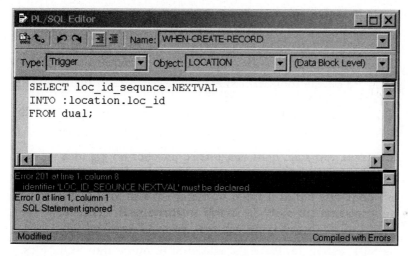

Figure 5-61   Trigger syntax error caused by referencing a database object incorrectly

This error usually indicates that the command references a database object, such as a table, field, or sequence, incorrectly. This figure shows that the command references the sequence as loc_id_sequence, which is misspelled. To correct the error, you correct the sequence name to loc_id_sequence, and then recompile the trigger.

Figure 5-62 shows another common syntax error, with the message "identifier 'DUAL' (or some other table name used in the trigger) must be declared."

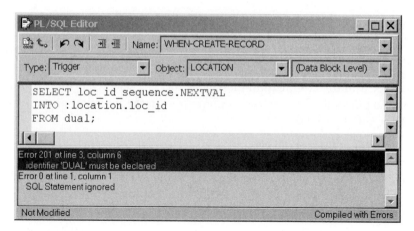

Figure 5-62   Trigger syntax error caused by not being connected to the database

This error (or one naming a database table other than DUAL) occurs when you are not connected to the database at the time you compile the trigger. This can happen when you start Forms Builder and open an existing form, and then revise a trigger and try to compile it. Because Forms Builder does not explicitly prompt you to connect to the database, you might forget to connect before compiling. The solution is to click the Connect button 📇 on the toolbar, connect to the database in the usual way, and then recompile the trigger.

NOTE

The error message shown in Figure 5-62 appears if you are not connected to the database when you compile a trigger with a SELECT command for any database table, not just DUAL.

Errors often occur when a SQL command within the PL/SQL trigger code has an error. If you suspect that an error is occurring as a result of an error in a SQL command, copy the command, paste the command into a text editor, and remove any PL/SQL commands in the SQL statement. Paste the edited SQL query into SQL*Plus, run it, and debug the query in SQL*Plus.

There are many more possible syntax errors, but these are some of the most common ones. The best way to locate errors is to isolate the line of source code generating the error, make sure that the form objects and database objects are specified correctly, and make sure that the SQL query or PL/SQL syntax is correct.

Now you run the form to test the trigger that displays the next sequence value in the LOC_ID text item. Then you insert a new record using the sequence value.

To test the trigger:

1. If necessary, correct any syntax errors in your trigger, recompile the code until it compiles successfully, close the PL/SQL Editor, and then save the form. The WHEN-CREATE-RECORD trigger appears under the Triggers node in the LOCATION data block.

2. Run the form. The form appears with the next sequence number (14) inserted into the Location ID field, as shown previously in Figure 5-57. Enter a new location into the form, and then save the record. The confirmation message appears, confirming that the record was successfully inserted.

3. Click the **Insert Record** button 🔲 to insert another record. A new blank record appears, with the next sequence value (15) inserted into the Location ID field.

4. Click the **Remove Record** button 🔲 to remove the blank record, and then close the browser window.

5. Close the form in Forms Builder.

CREATING A LIST OF VALUES (LOV)

Currently, when you use the Northwoods University Student form to insert a new record into the STUDENT table, you must type data values for foreign keys, such as F_ID. To make data entry easier and avoid errors for data fields that contain foreign key values or restricted data values, it is a good practice to allow the user to select from a list of allowable values. For example, allowable values for Advisor ID are integer values 1-5, which are values already stored in the F_ID field of the FACULTY table.

A **list of values (LOV)** (pronounced ell-oh-vee) displays a list of possible data values for a text item. Figure 5-63 shows an LOV for the Advisor ID field in the Students form.

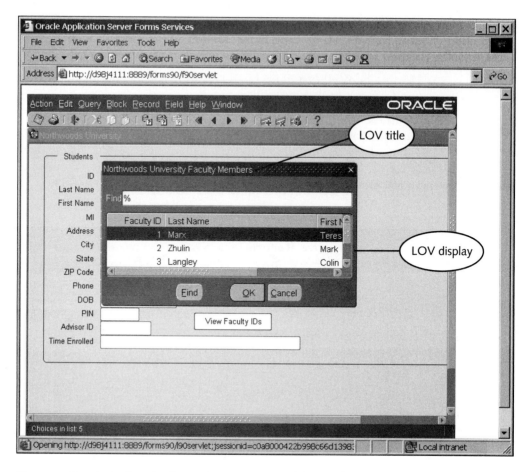

Figure 5-63   LOV that displays values for the Advisor ID field

The **LOV display** is the dialog box that displays possible choices for a form text item. The user opens the LOV display by placing the insertion point in the text item associated with the LOV and pressing CTRL+L. Or, the user can open the LOV display by clicking Edit on the Forms Services menu bar, and then clicking Display List. Or, the user can open the LOV by clicking a command button that has a trigger containing commands to open the LOV. In the LOV display, the user can perform exact or inexact searches by typing a search string in the Find field, and then clicking Find. The user can select a value in the list and click OK to display the selection's value in one or more form text items, or click Cancel to close the LOV display without selecting a value.

The LOV display derives its data values from a **record group**, which is a form object that represents data in a tabular format. However, unlike database tables, record groups are separate objects that belong to the form module and exist only as long as the form is running. A record group object can have a maximum of 255 fields of the following data types: CHAR, LONG, NUMBER, or DATE. Record group fieldnames must follow the Oracle naming standard.

You can create a new LOV using the LOV Wizard, which automates the process. You can also create an LOV manually by creating an LOV object and a new record group in the Object Navigator, and then defining their properties. Creating an LOV using the LOV Wizard is a five-step process:

1. **Specify the LOV display values**—When you create a new LOV, you create a SQL query that returns the data values that appear in the LOV display. This query creates a new record group. You can also specify for the LOV to use an existing record group.

2. **Format the LOV display**—To format an LOV display, you change the field titles and widths, specify which display fields the LOV returns to form text items, modify the LOV display title, and change the position of the LOV display on the form. It is a good practice to resize the LOV so that the user does not have to scroll to view the field values.

3. **Attach the LOV to a text item**—You must associate an LOV with a specific form text item. When the form insertion point is in this text item and the user presses CTRL+L, the LOV display opens.

4. **Change the new LOV and record group default names to descriptive names in the Object Navigator**—When you create an LOV using the LOV Wizard, the wizard automatically creates LOV and record group objects, and assigns default names to these objects. It is a good practice to change these to descriptive names, especially in forms that contain multiple LOVs.

5. **Create an optional command button to open the LOV display**—The LOV Wizard automates the first three steps in the process. The following sections describe how to use the LOV Wizard to create and modify an LOV.

Creating an LOV

The LOV Wizard contains the following pages to automate the LOV creation process:

- **LOV Source page**—Allows you to select whether to create a new record group or use an existing record group

- **SQL Query page**—Allows you to enter the SQL query that retrieves the records that appear in the LOV display, build a SQL query button using a visual tool called Query Builder, or import a query from a text file (you do not use the Query Builder tool in this book)

- **Column Selection page**—Specifies which record group columns appear in the LOV display

- **Column Display page**—Enables you to format the LOV fields by specifying the display title, field headings, column widths, and form text items to which the fields return values

- **LOV Display page**—Allows you to specify the LOV display title, size, and position

- **Advanced Options page**—Allows you to specify how many records the LOV retrieves; the default value is 20 (For an LOV that might return hundreds or thousands of records, it is desirable to limit the number of records to improve system response time.)

- **Items page**—Specifies the text item to which the LOV is attached; when the form insertion point is in this text item and the user presses CTRL+L, clicks Edit on the Forms Services menu bar, and then clicks Display List, or clicks the LOV command button, the LOV display opens.

- **Finish page**—Signals that the LOV Wizard has successfully created the new LOV

Now you use the LOV Wizard to create the LOV in Figure 5-63, which allows users to select from a list of faculty IDs and names in the Student form. Then you use the LOV Wizard to specify the LOV display values, format the display, and attach the LOV to a text item.

To use the LOV Wizard to create the LOV:

1. In Forms Builder, open **Student.fmb** from the Chapter5 folder on your Data Disk, and save the file as **Ch5CStudent_LOV.fmb** in the Chapter05\Tutorials folder on your Solution Disk.

2. To start the LOV Wizard, click **Tools** on the menu bar, and then click **LOV Wizard**. The LOV Source page, shown in Figure 5-64 appears, which allows you to create a new record group or use an existing record group. Because there are no existing record groups in the form, you cannot use an existing record group, so this option is disabled. Click **Next**. The SQL Query page opens.

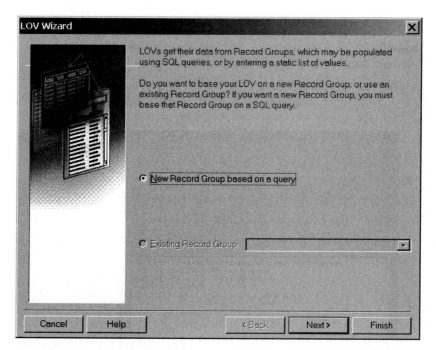

Figure 5-64   LOV source page

3. To retrieve the item ID and item description fields for the LOV display, type the following query shown in the SQL Query Statement field in Figure 5-65.

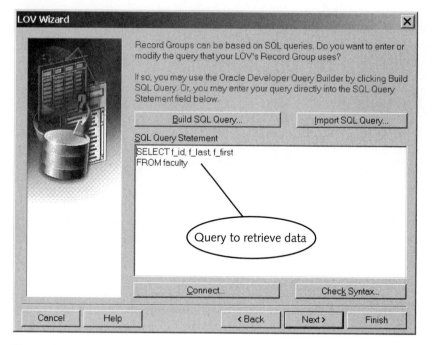

Figure 5-65   SQL query page

Do not type a semicolon (**;**) at the end of the command or an error will occur.

NOTE

    4. Click **Next**. The Column Selection page appears.

If an error message appears when you click Next on the SQL Query page, it means that you did not enter the SQL query correctly. If you cannot find the error visually, copy the query text and paste the text into a text editor. To copy the command text, select the text, and then press CTRL+C. Then execute and debug the query in SQL*Plus.

NOTE

    5. On the Column Selection page, click the **Move all items to target** button ⟫ to display all of the record group fields in the LOV display, and then click **Next**.

    6. The Column Display page appears, as shown in Figure 5-66.

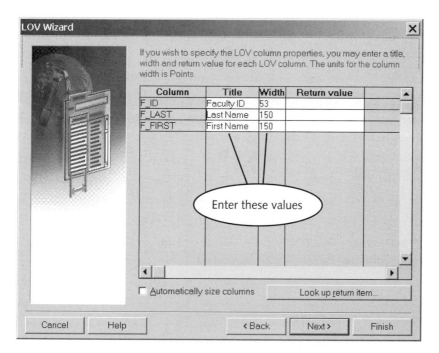

Figure 5-66   Column display page

The Column Display page enables you to format the fields in the LOV display. The Title column shows the title that appears for each LOV display field, and the Width column shows the width of each LOV display field. By default, the LOV display fields have the same titles and widths as the corresponding database fields, which might result in column titles that are hard to understand and column widths that are wider than required for the LOV display.

The Return value column specifies the form text item (preceded by its block name and a period) to which the LOV transfers the user's LOV selection for the current field. For example, if the user selects Teresa Marx in the LOV display and clicks OK, the ID value for the faculty member, which is 1, appears in the form's F_ID text item.

Now you specify the Column Display page values. You select the form block and text item name (FACULTY.F_ID) as the return value for the F_ID column. In this LOV, you do not specify a return value for the faculty member's first or last name, because there is no form text item that displays these items.

To specify the LOV display column titles, widths, and return value:

1. Modify the column titles and widths as shown in Figure 5-66.

2. Click the insertion point in the Return value column for F_ID, and then click **Look up return item**. The Items and Parameters dialog box shown in Figure 5-67 opens, and displays a list of all the items in the STUDENT

block. Each item is preceded by the block name and a period (STUDENT.). These are the form text items to which the selected item ID value can be returned.

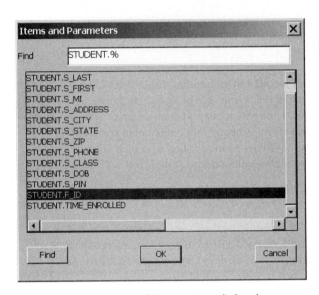

Figure 5-67   Items and Parameters dialog box

3. Select **STUDENT.F_ID**, which is the text item in which the selected item ID will appear after the user makes a selection from the LOV display.

4. Click **OK**. STUDENT.F_ID appears as the Return value for the F_ID text item on the Column Display page. Click **Next**. The LOV Display page appears.

5. To specify the LOV display properties, type **Northwoods University Faculty Members** in the Title field, as shown in Figure 5-68, and accept the default Width and Height field values. Make sure that the Yes, let Forms position my LOV automatically option button is selected, then click **Next**. The Advanced Options page shown in Figure 5-69 appears.

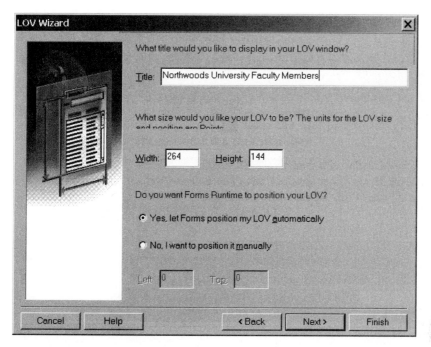

Figure 5-68   LOV Display page

As shown in Figure 5-69, the LOV Wizard Advanced Options page specifies how many records the LOV retrieves. For an LOV that might return hundreds or thousands of records, it is desirable to use the options on the Advanced Options page to improve system response time.

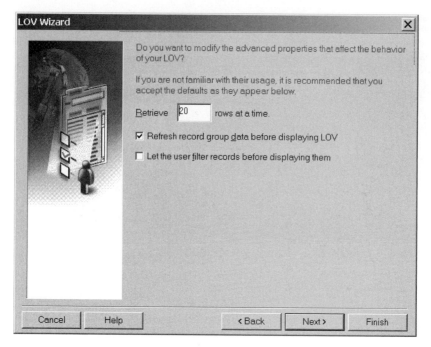

Figure 5-69   Advanced Options page

On the Advanced Options page, the Refresh record group data before displaying LOV check box allows you to specify whether the LOV queries the database and refreshes the record group values each time the LOV display opens. If this check box is cleared, the LOV retrieves the data values from the database once, when the user first opens the LOV display, and stores the data values. When the user opens the LOV display subsequent times, the stored data values appear. This improves system response time for an LOV that retrieves many data values and whose values don't change very often.

The Let the user filter records before displaying them check box allows the user to type a search condition in the Find field on the LOV display before any records appear. This reduces the number of records that the LOV retrieves, and improves system performance for an LOV that retrieves many records. Because the LOV display in the Student form retrieves only the five records in the FACULTY table, you accept the LOV defaults (refresh the data each time the LOV is displayed, and do not let the user filter the data before displaying values).

To finish the LOV, you must specify the text item to which the LOV is attached. Recall that you attach an LOV to a form text item, and when the form insertion point is in this text item and the user either presses CTRL+L, clicks Edit, and then clicks Display List, or clicks the LOV command button, the LOV display opens. The field to which the LOV is attached must be one of the text items to which the LOV returns values. Now you complete the LOV Advanced Options page, and attach the LOV to the FAC-ULTY_ID text item.

To complete the LOV Advanced Options page and attach the LOV to the FACULTY_ID text item:

1. Accept the default selections on the Advanced Options page, and then click **Next**. The Items page shown in Figure 5-70 appears. Because the STUDENT.F_ID text item is the only text item to which the LOV returns a value, it is the only item that appears in the Return Items list.

5

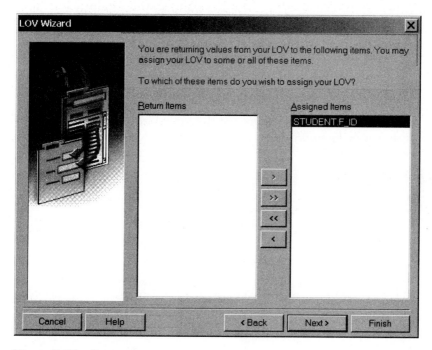

Figure 5-70   Items page

2. Make sure that STUDENT.F_ID is selected in the Return Items list, and then click the **Move one item to target** button ⏵ to move this text item to the Assigned Items list.

3. Click **Next** to display the LOV Finish page, and then click **Finish** to finish creating the LOV. The Object Navigator window displays the new LOV and record group.

LOVs appear only in the Object Navigator Ownership View. They do not appear in the Object Navigator Visual View.

NOTE

4. Save the form.

Renaming the LOV and Record Group Objects

After you create an LOV using the LOV Wizard, a new LOV object appears under the LOVs node in the Object Navigator, and a new record group appears under the Record Groups node. The LOV object specifies the LOV display properties, and the record group object specifies the data that the LOV displays. It is a good practice to replace the default names with descriptive names, especially in forms that contain multiple LOVS. Descriptive names allow you to distinguish among different LOVs and record groups if you want to modify their properties.

Now you change the names of the new LOV and record group objects to descriptive names. (Usually, you name the LOV and record group the same name, so you can visually see the association between each LOV and record group.) Then you run the form, and test the LOV.

To change the names of the LOV and record group objects, then test the LOV:

1. In the Object Navigator, change the name of the LOV object to **FACULTY_ID**, and the name of the record group to **FACULTY_ID**, as shown in Figure 5-71.

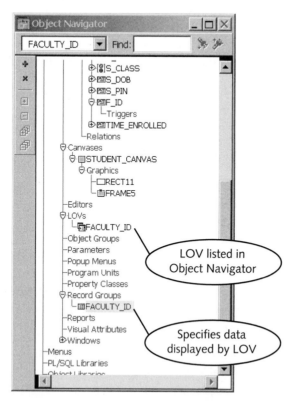

Figure 5-71   Object Navigator containing LOV object

2. Save the form, then run the form. The form appears in the browser window.

3. Type **MR100** in the Student's ID field, and then click the Advisor ID text box, the field to which the LOV is attached. Note that the message "List of Values" appears in a box on the status line. (The message may be truncated.) This indicates that the current field has an LOV.

4. Press **CTRL+L**. (The letter L can be either uppercase or lowercase.) The LOV display appears in the middle of the form.

5. Select the faculty member named **Mark Zhulin**, and then click **OK**. The Advisor ID value for the selection, which is 2, appears in the form's advisor field.

6. Close the browser window.

Adding a Command Button to Open the LOV Display

Recall that users can open the LOV display by placing the insertion point in the text item associated with the LOV, and then pressing CTRL+L. You can also create a command button to allow users to open the LOV display. You create a command button by drawing the button on the canvas using the Button tool ▣ on the tool palette. To configure a form button item, you modify the Name property, which is the name that Forms Builder uses to reference the button. You also modify the Label property, which specifies the text that appears on the button.

To make the button operational, you must create an associated button trigger that contains the commands first to place the insertion point in the LOV text item, and then to open the LOV display. To create a button trigger, you can select the button in the Layout Editor, right-click, and then click PL/SQL Editor. You then select the trigger event, and specify the trigger code. To create a button whose trigger code executes when the user clicks the button, you select the WHEN-BUTTON-PRESSED button event.

The commands to place the insertion point in a form text item, and then open an LOV display are:

```
GO_ITEM('block_name.item_name');
LIST_VALUES;
```

In the Student form, the LOV is associated with the F_ID text item in the STUDENT block, so the GO_ITEM command would appear as `GO_ITEM('student.f_id');`. Now you create a command button to open the LOV display, and then run the form and test the button.

To create and test a command button to open the LOV display:

1. In the Object Navigator, click **Tools** on the menu bar, and then click **Layout Editor** to open the Layout Editor.

2. Select the **Button** tool ▣ on the tool palette, and draw a command button on the right side of the Student frame.

3. Make sure the button is selected, right-click, and then click **Property Palette**. Change the button Name property value to **FACULTY_ID_BUTTON** and the Label property value to **View Faculty IDs**, as shown in Figure 5-72. Close the Property Palette. The LOV command button appears.

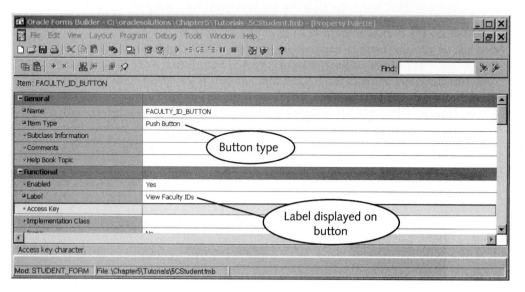

Figure 5-72 Property Palette

4. Make sure that the View Faculty IDs button is selected, right-click, and then click **PL/SQL Editor**. The STUDENT_FORM: Triggers dialog box opens, as shown in Figure 5-73. Select **WHEN-BUTTON-PRESSED** from the list, and then click **OK**. The trigger opens in the PL/SQL Editor.

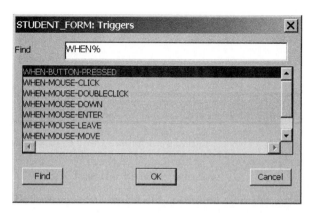

Figure 5-73 STUDENT_FORM: Triggers dialog box

5. Type the commands in the PL/SQL Editor, as shown in Figure 5-74.

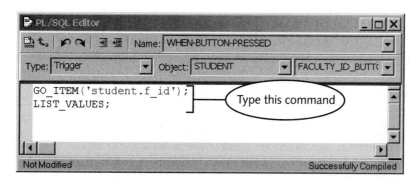

Figure 5-74   PL/SQL Editor

6. Click the **Compile PL/SQL Code** button ![button] to compile the trigger, debug the trigger if necessary, and then close the PL/SQL Editor.

7. Save the form, then run the form. The form appears in the browser window.

8. Type **MR100** in the Student's ID field, and then click **View Faculty IDs**. The LOV display opens, as shown in Figure 5-75. (You may need to resize the LOV to see all three rows.)

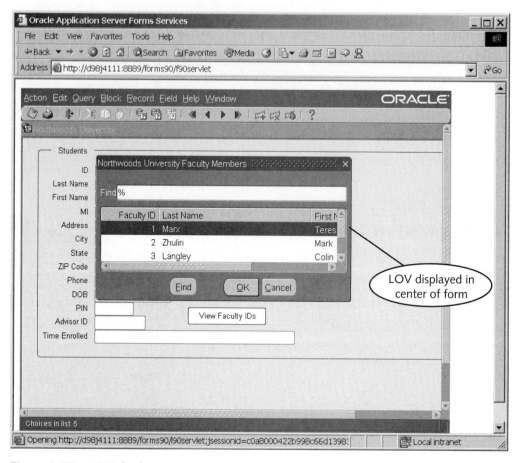

Figure 5-75   LOV display

9. Select Faculty ID **2** (Zhulin, Mark), and then click **OK**. The Item ID value for the selection, which is 2, appears in the form's ID field.

10. Close the browser window.

Using the LOV Wizard in Reentrant Mode

As with the other Forms Builder wizards, you can open the LOV Wizard in reentrant mode to modify the properties of an existing LOV. To modify an LOV using the LOV Wizard in reentrant mode, you select the LOV in the Object Navigator, and then start the LOV Wizard.

Currently, the LOV display in the Student form appears in the upper-left corner of the form. The form would be more attractive if the LOV display appeared toward the right side of the form, so the item prompts are still visible, as shown in Figure 5-76. The LOV also needs to be longer, so that the user can view all of the records without scrolling.

And, the display needs to be narrower, so that the display window does not appear clipped at the edge of the Forms Services window.

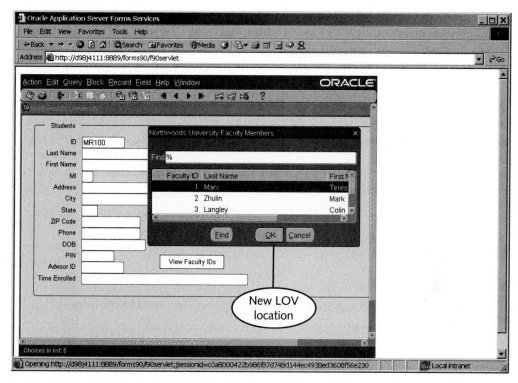

Figure 5-76   Modified LOV display

First you determine the position where the LOV display should appear on the form. To do this, you open the form in the Layout Editor, place the mouse pointer on the position where you want the upper-left corner of the LOV display to appear, and note the pointer location position. (Recall from Lesson A that the pointer location appears at the bottom of the Layout Editor window.) Then you open the LOV Wizard in reentrant mode, and change the LOV display position to the desired location. Finally, you run the form and test your changes.

To modify the LOV display position and then run the form:

1. If necessary, click **Tools** on the menu bar, and then click **Layout Editor** to open the form in the Layout Editor.

2. Place the mouse pointer along the frame border and slightly to the left of the center of the screen. Note the pointer position. (In Figure 5-76, the horizontal position is 168, and the vertical position is 25.)

3. Click **Window** on the menu bar, and then click **Object Navigator** to open the Object Navigator. Select the LOV object named **FACULTY_ID**, click **Tools** on the menu bar, and then click **LOV Wizard**. The LOV Wizard opens in reentrant mode, with tabs across the top of the page to allow you to access each of the Wizard pages.

4. Select the **LOV Display** tab, and select the **No, I want to position it manually** option button. Type your horizontal mouse position value in the Left field, and your vertical mouse position value in the Top field, as shown in Figure 5-77.

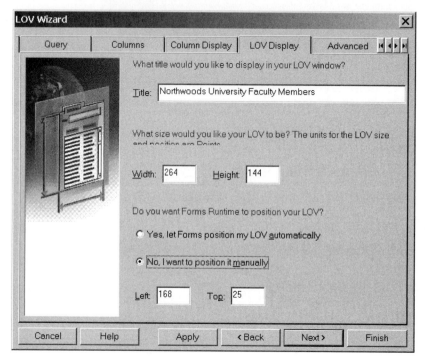

Figure 5-77 LOV Display page

5. Click **Finish** to save your changes, and close the LOV Wizard. Then save the form.

6. To test your changes, run the form, type **MR100** for the Student's ID field, click the Advisor ID text item, and then press **CTRL+L** to open the LOV display. The LOV display appears on the right side of the form.

NOTE

If your LOV display does not appear in the correct position, click Cancel to close the LOV display, close the browser window, open the LOV in the LOV Wizard in reentrant mode, and then modify the LOV display position as necessary.

7. Click **Cancel** to close the LOV display, and then close the browser window.

Next, you modify the size of the LOV display. You open the LOV Wizard in reentrant mode, and modify the display size and the column widths.

To modify the size of the LOV display:

1. If necessary, select the FACULTY_ID node in the Object Navigator, click **Tools** on the menu bar, and then click **LOV Wizard** to open the wizard in reentrant mode.

2. To change the width of the columns, select the **Column Display** tab, and change the width to the values displayed in Figure 5-78.

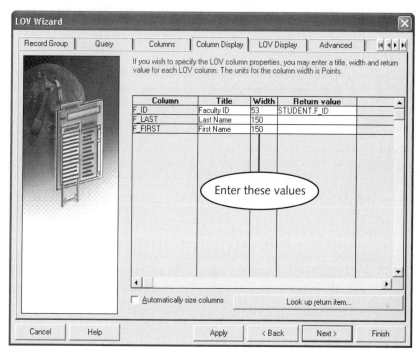

Figure 5-78   Modified column display

3. To change the overall size of the LOV display, select the **LOV Display** tab, and change the Width field value to **210**, and the Height field value to **220**, as shown in Figure 5-79. Then click **Finish** to close the LOV Wizard.

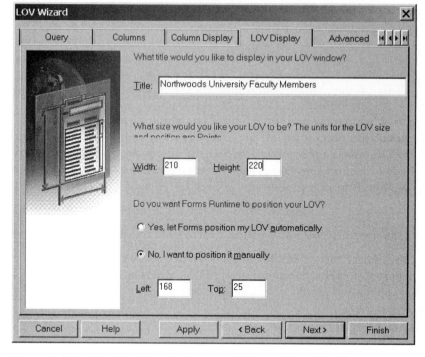

Figure 5-79   Modifications to the overall size of the LOV display

4. Save the form, and then run the form.

5. Type **MR100** in the Student's ID field, click in the Advisor ID text item, and then press **CTRL+L**. The LOV display appears, as shown in Figure 5-80. Click **Cancel** to close the LOV display, and then close the browser window.

6. Close **Ch5CStudent_LOV.fmb** in Forms Builder.

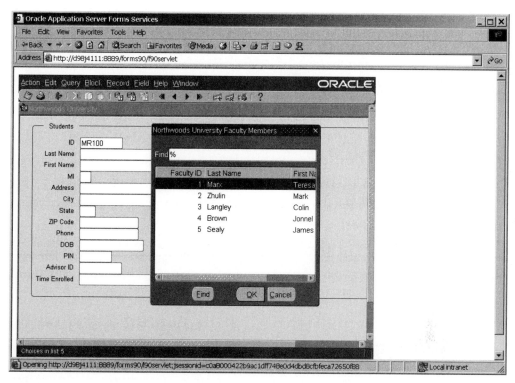

Figure 5-80   Modified LOV display

NOTE

If your LOV display does not display all of the field values or all of the records, click Cancel, close the browser window, open the LOV Wizard in reentrant mode, and adjust the LOV display size and the column widths as necessary.

You should always format the LOV display so that the user does not have to scroll to view the records or field values. Formatting the LOV display is a trial-and-error process of changing the configuration, running the form, and then adjusting the configuration as necessary.

REPRESENTING DATA VALUES USING OTHER ITEM TYPES

Up to this point, you have used text items to display data values in forms. You can also use radio button and check box form items to represent data values. The following sections describe how to create and use these other types of form items.

Creating Radio Buttons

You can use **radio buttons**, which are also called option buttons, to represent data fields whose values must be one of a small set of mutually exclusive selections. In the Northwoods University database, you might use radio buttons to represent the faculty rank (F_RANK) field in the FACULTY table, which can have values of ASSISTANT, ASSOCIATE, FULL, or INSTRUCTOR. You could also use radio buttons to represent GRADE values in the ENROLLMENT table, which can have values of A, B, C, D, or F. You should use radio buttons only when a data field has a limited number of possible values (usually five choices or less), and the possible choices do not change.

In Forms Builder, individual radio buttons exist within a **radio group**. A user can select only one button in a radio group at a time. Each radio button has an associated data value, and the radio group has the data value of the currently selected radio button.

When you use the Layout Wizard to create a form layout, you can select the type of form item that represents each data field. So far, you have used the default item type, which is a text item, to display data fields as text items. Now you create a form that displays a data field using a radio group, as shown in the Northwoods University Student form in Figure 5-81.

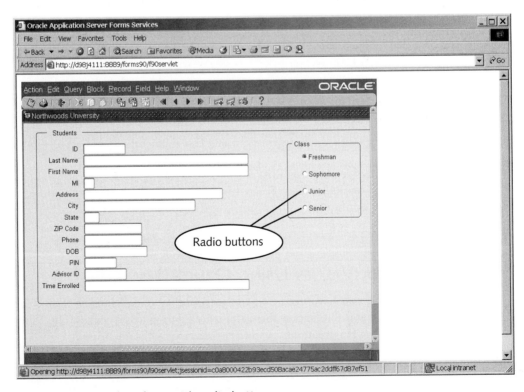

Figure 5-81   Student form with radio buttons

This data block form allows users to view, insert, update, and delete records in the Northwoods University STUDENT table. In the STUDENT table, the S_CLASS field currently is limited to four values: FR, SO, JR, and SR, so the form displays the S_CLASS field values using a radio group.

Now you create the form, and use the Data Block Wizard to create a data block associated with the STUDENT table. On the Layout Wizard Data Block page, you specify that the S_CLASS field values appear in a radio group. Then, you use the Layout Wizard to create the form layout.

To create the new form, data block, and layout:

1. Create a new form, and save the form as **Ch5CStudent.fmb** in the Chapter5\Tutorials folder on your Solution Disk.

2. Use the Data Block Wizard to create a new data block associated with the STUDENT table. Include all STUDENT data fields in the data block, and make sure that the Enforce data integrity check box is cleared.

3. Use the Layout Wizard to create a form layout. Accept the (New Canvas) canvas and Content type default options, and then click **Next**.

4. On the Data Block page, select all data block items for display.

5. Select **S_CLASS** in the Displayed Items list. The Item Type list box is activated. Open the **Item Type** list, as shown in Figure 5-82. The default item type is Text Item.

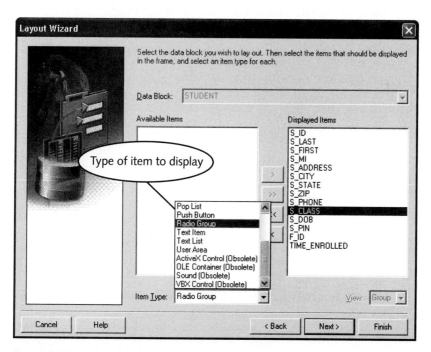

Figure 5-82   Selecting an alternative item type

6. Select **Radio Group** as the item type, and then click **Next**.

7. Accept the default widths, set the prompts as shown in Figure 5-81, and then click **Next**.

8. Make sure that the Form option button is selected for the layout style, and then click **Next**.

9. Type **Students** for the frame title, do not change any of the other values, click **Next**, and then click **Finish**. The form layout appears in the Layout Editor.

10. Click anywhere on the canvas to deselect the selected form items, and then save the form.

In the form layout, no radio buttons appear, and nothing appears for the S_CLASS data field. When you specify to use a radio group for a data block field, the Layout Wizard creates the radio group in the Object Navigator. However, you need to create the individual radio buttons on the canvas. Now you open the Object Navigator, rename the form objects, and view the new radio group.

To rename the form objects and view the new radio group:

1. Click **Window** on the menu bar, and then click **Object Navigator** to open the Object Navigator. Change the form module name to **STUDENT_FORM**, change the canvas name to **STUDENT_CANVAS**, and change the window name to **STUDENT_WINDOW**.

2. Open the STUDENT_WINDOW Property Palette, change the Title property value to **Northwoods University**, select another property to apply the change, and then close the Property Palette.

3. If necessary, open the Data Blocks node, and then open the STUDENT node.

4. Open the Items node, and then open the S_CLASS node, which represents the new radio group. The data block items appear as shown in Figure 5-83.

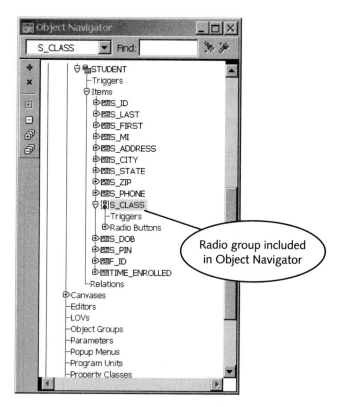

Figure 5-83   Data block items showing new radio group

In the data block item list, the Text Item icon ▦ appears beside all of the objects except S_CLASS, and indicates that the form represents these fields as text items. The Radio Group icon [▨] beside S_CLASS indicates that the form represents this field as a radio group. The S_CLASS radio group has two child nodes: Triggers and Radio Buttons. There are no objects in the Radio Buttons node yet. You need to manually create the individual radio buttons to represent the four data values that can appear in the S_CLASS data field. To create radio buttons, you draw the radio buttons on the canvas using the Radio Button tool ⊙ on the Layout Editor tool palette.

Now you format the text items on the canvas and make the data block frame larger to accommodate the radio buttons. You also change the frame's Update Layout property. Currently, the frame automatically positions the block items within the frame. You change the Update Layout property to Locked, so that when you update the form using the Layout Wizard later, you do not lose your custom formatting. Then, you create the first radio button. When you create a new radio button, you must specify the radio button group in which to place the radio button.

To resize and format the frame, and create a radio button:

1. If necessary, click **Tools** on the menu bar, and then click **Layout Editor** to open the Layout Editor. Edit the item prompts in the Layout Editor so that the prompts appear as shown in Figure 5-81.

2. Select the **Students** frame, and, if necessary, make the frame narrower so that all of the text items appear in a single column and are stacked on top of one another.

3. Verify that the **Students** frame is selected, open its Property Palette, select the **Update Layout** property, and select **Locked** from the property list. Close the Property Palette.

 TIP

If you do not change the Update Layout frame property to Locked, you lose all custom formatting when you open the Layout Wizard in reentrant mode.

4. If necessary, select the **Students** frame, and drag the lower-right selection handle to the right to make the frame wide enough to accommodate the radio button group.

5. Select the **Radio Button** tool ⊚ on the tool palette, and draw a rectangle on the canvas to correspond with the first radio button, as shown in Figure 5-84. After you release the mouse button, the Radio Groups dialog box opens, which allows you to specify the radio button's radio group.

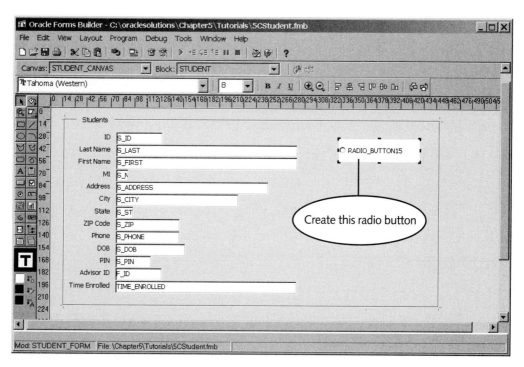

Figure 5-84   Creating a radio button

6. Make sure that the *S_CLASS* radio group is selected, and then click **OK**.
The radio button and its corresponding label appear, as shown in Figure 5-84.

Now you change some of the radio button's properties. You specify values for the button's Name property, which is how the form references the button. You also change the radio button's Label property, which is the description that appears beside the button on the canvas. And, you specify the button's Radio Button Value, which corresponds to the actual data value in the database that the radio button represents. Recall that when a user selects a radio button, its data value becomes the data value for the radio group.

To change the radio button properties:

1. Select the new radio button, right-click, click **Property Palette** to open its Property Palette, and then change the following property values. When you are finished, close the Property Palette.

Property	Value
Name	**FR_RADIO_BUTTON**
Label	**Freshman**
Radio Button Value	**FR**

The Radio Button Value property is the database data value that the radio button represents, and the Label property is the label that appears beside the radio button on the form.

TIP

Now you change the background color of the radio button so that it is the same color as the canvas. To do this, you use the Fill Color tool . This tool allows you to change the fill color of any form item. This tool is displayed at the bottom of the tool palette. Depending on your screen display setting, it might not currently be visible. If it is not visible, scroll down on the tool palette to find it.

To change the radio button fill color:

1. If the Fill Color tool is not visible, click the down arrow on the tool palette to scroll down.

2. In the Layout Editor, click the **radio button** if necessary, and then click. The Fill Color palette opens. Select a color to match the form background.

You cannot change the Fill Color property of a radio button to No Fill. Instead, you must match the radio button's fill color to the background color of the canvas if you want the rectangle to disappear.

TIP

Now that you have created and formatted the Freshman radio button, you can easily create the rest of the radio buttons in the radio group by copying this button, pasting the copy onto the canvas, and changing the individual properties of each button to correspond to the other S_CLASS data values. First, you copy the radio button and paste it three times, so you have a total of four radio buttons.

To copy and paste the radio button:

1. Select the **Freshman** radio button if necessary, and then click the **Copy** button on the Layout Editor toolbar to make a copy of the radio button.

You also can press CTRL+C to copy and CTRL+V to paste, or click Copy or Paste on the Edit menu on the menu bar.

2. Click the **Paste** button on the Layout Editor toolbar, make sure that S_CLASS is the selected radio group, and then click **OK** to paste the copied radio button onto the canvas. The new radio button appears directly on top of the existing radio button.

3. Select the newly pasted radio button, and position it below the first radio button. (Do not worry about aligning the new radio button with the other button right now.)

4. To create the next radio button, click 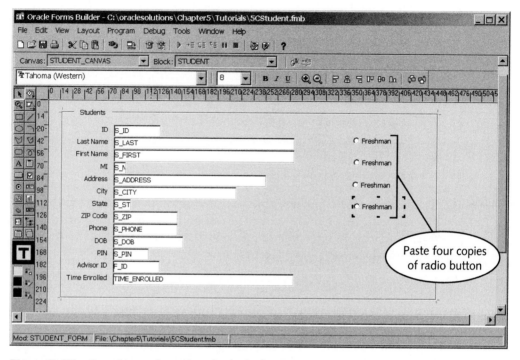 again, make sure that the S_CLASS radio group is selected, and then click **OK**. Reposition the new button so that it is below the current radio buttons. Repeat this process one more time to create one more radio button. Your pasted radio buttons should look something like Figure 5-85.

Figure 5-85 Copying and pasting the radio buttons

You probably need to align the radio buttons and space them evenly. To do this, you select all of the radio buttons as an object group, and then align the left edges and stack them vertically using the Align Objects dialog box.

To align and format the radio buttons:

1. Select the radio buttons as an object group by selecting the first radio button, pressing and holding the **Shift** key, and then selecting the remaining radio buttons. Selection handles appear around the object group.

2. With the buttons selected as an object group, click **Layout** on the menu bar, and then click **Align Components**. The Align Objects dialog box opens.

3. In the Horizontally frame, select the **Align Left** option button. In the Vertically frame, select the **Stack** option button. Then click **OK**. The four radio buttons should now appear with their left edges aligned and stacked on top of each other.

4. Click anywhere on the canvas to deselect the selected group object.

Next, you change the properties of the pasted radio buttons. You specify each button's Name, Label, and Value property.

To change the properties of the pasted radio buttons:

1. Open the Property Palettes for the individual radio buttons, and change the properties of the second, third, and fourth buttons in the radio group as follows:

Property	Button 2	Button 3	Button 4
Name	SO_RADIO_ BUTTON	JR_RADIO_ BUTTON	SR_RADIO_ BUTTON
Label	Sophomore	Junior	Senior
Radio Button Value	SO	JR	SR

2. Save the form.

Next you specify the radio group's initial value. The **initial value** of a radio group is the Radio Button Value property of the radio button that is selected when the form first opens, or when you click the Insert Record button [icon] on the Forms Services toolbar to create a new blank record. If you do not specify the initial value of a radio group, an error occurs when you run the form.

To specify the radio group's initial value, you must change the Initial Value property on the radio group's Property Palette so that it has the same Radio Button Value as the radio button that is selected when the form first opens. This is usually the first button in the group. (Recall that the Radio Button Value is the database value that corresponds to the radio button.) For example, in the S_CLASS radio group, the Radio Button Value for the first radio button is "FR." If you want the first radio button to be selected when the form first appears, you specify the Initial Value property of the S_CLASS radio group to be "FR" also.

To change the radio group's initial value:

1. Open the Object Navigator window, right-click the **S_CLASS** radio group, and then click **Property Palette**.

2. Scroll down to the Data property node, type **FR** for the Initial Value property, and then close the Property Palette and save the form.

Next you format the radio group to enhance its appearance. You create a frame around the radio group and reposition the radio buttons on the form. You also format the frame by changing the appearance of the bevel of the frame border so that it appears as an inset line. Lastly, you use the Fill Color tool [icon] to specify that the frame does not have a fill color.

To create and format a frame around the radio group and reposition the radio buttons:

1. Open the Layout Editor, select the **Frame** tool on the tool palette, and then draw a frame around the radio buttons, as shown previously in Figure 5-81.

HELP
If the radio buttons overlap the frame, resize the radio buttons so that the frame encloses the buttons.

2. Select the frame if necessary, click **Layout** on the menu bar, point to **Bevel**, and then click **Inset**.

3. Right-click the frame, click **Property Palette**, and change the frame's Name property to **S_CLASS_FRAME**. Scroll down to the Frame Title node, and change the Frame Title property value to **Class**. Close the Property Palette.

4. Make sure that the frame is still selected, then select the **Fill Color** tool. Click **No Fill** to remove the frame fill color.

5. Resize the radio buttons, adjust the frame size, and position the frame so that your form looks like Figure 5-81, and then save the form.

TIP

To resize all of the buttons in one operation, select the buttons as an object group, and drag the group selection handle to the desired size.

Now you test the form to confirm that the radio buttons display the student data correctly. First, you run the form and step through all of the table records. Then, you insert a new record and use the radio buttons to specify the student class value.

To test the form:

1. Run the form. The form appears in the browser window.

HELP

If the radio button labels have a different background color than the form, close the form, then adjust the radio button fill color to match the form background color.

2. To retrieve all of the records, click the **Enter Query** button , and then click the **Execute Query** button . The first STUDENT record (Tammy Jones) appears, with the Senior radio button selected. Click the **Next Record** button ▶.

3. Scroll through all the STUDENT records, and confirm that the radio group displays the correct data value.

4. Click the **Insert Record** button to insert a new blank record. Note that the Freshman radio button is selected, because it is the radio group's initial value.

5. Enter the following data values, and then save the new record. The save confirmation message appears, confirming that the record was inserted.

Field	Value
ID	**10**
Last Name	**Andersen**
First Name	**Ashley**
Class	**Sophomore**

Next you use the radio group to create a query. You enter a query with a search condition that returns all items for which the S_CLASS value is "JR."

To create a query using the radio group:

1. Click the **Enter Query** button 🖥️, and then select the **Junior** radio button.

2. Click the **Execute Query** button 🖥️. The record for John Marsh, who is the only student who is a junior, appears.

3. Close the browser window.

4. Close the form in Forms Builder.

Creating Check Boxes

Database applications use check boxes to represent data values that can have only one of two opposing values, such as True or False, or Yes or No. Normally, the label that appears beside the check box is one of the two opposing values. If the check box is checked, the check box represents the data value shown on the label. If the check box is cleared, the data value represented by the label is false, and the check box represents the opposing data value. For example, the Northwoods University Term form in Figure 5-86 uses check boxes to represent the STATUS field in the TERM table.

Note that the check boxes have the label "Closed." If a check box is checked, the associated term status value is CLOSED. If a check box is cleared, the user would understand that the term status value is the opposite of closed, which is OPEN.

To configure a form check box item, you use the Label property to specify the label that appears beside the check box. A check box also has a Value when Checked property, which specifies the data value the check box represents when the box is checked, and a Value when Unchecked property, which specifies the data value the check box represents when it is cleared. In addition, a check box has a Check Box Mapping of Other Values property. This property allows you to specify how the check box appears when the form first opens, when a user creates a new blank record, or when the value for the check box field is a different value than the ones for when the check box is checked or unchecked. For example, if the value for a STATUS field in the TERM table is NULL (rather than OPEN or CLOSED), the check box appears in this state. The Check Box

Mapping of Other Values property can have values of Not Allowed, Checked, or Unchecked.

Now you configure the check boxes in the Northwoods University Term form in Figure 5-86. Currently, the form represents the STATUS field as a text item. You change the STATUS field's item type to a check box, and change the check box label to "Closed." You specify that the check boxes represent the data value CLOSED when they are checked, and the data value OPEN when they are unchecked (cleared). When a user adds a new term to the table, the STATUS value of the new term is usually OPEN, so the Check Box Mapping of Other Values property value for the check box is Unchecked.

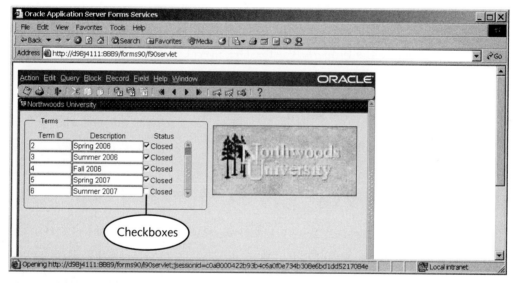

Figure 5-86   Northwoods University Term form

To modify the data block form to use a check box:

1. In Forms Builder, open **Term.fmb** from the Chapter05 folder on your Data Disk, and save the file as **Ch5CTerm.fmb** in the Chapter05\Tutorials folder on your Solution Disk.

2. Open the **Layout Editor**. The records appear in a tabular layout style, with five records on the form at one time. You change the text item to a check box item using the item Property Palette.

3. Select any **STATUS** text item. Selection handles appear around all five items. Press the **F4** function key on the keyboard to open the item Property Palette. In the Property Palette, confirm that the item name is STATUS, and that the Item Type value is Text Item.

4. Select the **Item Type** property, open the list, and select **Check Box** to change the item type to a check box. (You have to scroll up to find Check Box.) The Property Palette properties change to display properties for a Check Box item.

5. Change the Label property value to **Closed**.

6. Change the Value when Checked property value to **CLOSED**. This represents the data value of the STATUS item when the check box is checked.

7. Change the Value when Unchecked property value to **OPEN**. This represents the data value of the STATUS item when the check box is cleared.

8. Select the **Check Box Mapping of Other Values** property, open the list, and select **Unchecked**. This specifies that the check box is cleared when the form first opens or when the user adds a new blank record to the form.

9. Close the Property Palette. The STATUS items now appear as check boxes on the canvas.

10. Select the STATUS check boxes, and adjust the column width so that it appears as shown in Figure 5-86. Adjust the frame width as necessary so that your form appears as shown in Figure 5-86. Then save the form.

Now you run the form and confirm that the check boxes display the correct values for the STATUS field in the term table.

To run the form and test the check boxes:

1. Run the form. Click the **Enter Query** button, and then click the **Execute Query** button to retrieve all of the TERM records.

2. Scroll down to display the record for Term ID 6, as shown in Figure 5-86. Note that the check boxes are checked for all terms for which the STATUS value is CLOSED, and the check box is cleared for Term ID 6, in which the STATUS value is OPEN.

3. Clear the **Closed** check box for Term ID 5 (Spring 2007), and then save the record. The confirmation message appears, indicating that the change was successfully made.

HELP

If an error occurs when you try to save the record, open the Property Palette for the STATUS check box, and confirm that the Value when Unchecked property is set to OPEN. The property must be specified in all uppercase letters and cannot have any blank spaces at the end of the value.

4. Select the record for Term ID **6**, and then click the **Insert Record** button. A new blank record appears under the record for Term ID 6.

5. In the blank record, type **7** in the Term ID field and **Fall 2007** in the Description field, and leave the Status check box cleared to indicate that the

status of the new term is OPEN. Save the record. The confirmation message appears, indicating that the record was saved.

6. Close the browser window.

7. Close the form in Forms Builder, then close Forms Builder and all other open Oracle applications. If a dialog box opens asking if you want to save the form, click Yes.

You are now familiar with creating data block forms, creating master-detail relationships in data block forms, and displaying form data in different ways. In the next chapter, you learn how to create custom forms, which display data from multiple database tables in a more flexible format.

5

SUMMARY

◻ Triggers are program units that execute when a user performs an action, such as clicking a button, or when the system performs an action, such as loading a form. You associate triggers with specific form objects, such as text items, buttons, data blocks, or the form itself.

◻ To write a trigger, you specify the form object with which the trigger is associated; the action, or event, that starts the trigger; and the PL/SQL code that executes when the trigger fires.

◻ A list of values (LOV) displays a list of selections for a form text item. After the user selects a value, the LOV enters the selection in a form text item. You usually use an LOV to provide selections for fields that represent foreign key fields or fields with restricted data values.

◻ To create an LOV, you use a SQL query to create a record group. The LOV display shows the record group records. Users open the LOV display by placing the form insertion point in the text item associated with the LOV and pressing CTRL+L, by clicking Edit on the Forms Services menu bar, and then clicking Display List, or by clicking the LOV command button.

◻ You can use the LOV Wizard to create the LOV display and record group, specify the display properties, and associate the LOV with a form text item.

◻ You can use radio buttons to display, enter, and modify data values when a table field has a limited number of related choices. In a form, individual radio buttons exist within a radio group. Each individual radio button has an associated data value, and the radio group has the data value of the currently selected radio button.

◻ You can use check boxes to represent data values that can have only one of two opposing values. If the check box is checked, the check box represents the data value that the check box label represents. If the check box is cleared, the check box represents a data value that is the opposite of the label value.

REVIEW QUESTIONS

1. A(n) _____ causes a form trigger to execute.
2. List the steps to perform when creating an LOV.
3. The Column Display page of the LOV Wizard allows you to format the display title and field headings of the LOV fields. True or False?
4. By default, LOV objects are assigned names that end with _LOV. True or False?
5. List the steps for creating a trigger.
6. When a data field can only assume one value from a small set of possible values, _____ can be used as the field's item type.
7. Identify two different methods of accessing an LOV display.
8. When you specify that a data block item is a radio group, two child nodes, _____, and Radio Buttons, will appear in the Object Navigator.
9. By default, LOV display fields have the same titles and widths as their corresponding data items in the form. True or False?
10. When an error occurs during the compilation of a program unit, where is the error code identifying the type of error displayed?

MULTIPLE CHOICE

1. A program unit in a form that runs when a user clicks a button is called a form _____.
   a. status monitor
   b. compiler
   c. trigger
   d. LOV source
2. A record group can have a maximum of _____ fields.
   a. 30
   b. 52
   c. 255
   d. 1000
3. When you create an LOV, what is used to create a record group?
   a. a canvas object
   b. a PL/SQL block
   c. a SQL query
   d. a data block

4. Which item type is appropriate only when a data field can contain only two possible values?

   a. radio button

   b. command button

   c. text box

   d. check box

5. Which of the following is *not* a valid page in the LOV Wizard?

   a. PL/SQL page

   b. SQL Query page

   c. Column Display page

   d. Items page

6. You can specify the initial value for a radio group using the _____ property on the radio group's Property Palette.

   a. Default

   b. Initial Value

   c. View

   d. Format Mask

7. Which of the following is *not* a valid data type for a record group object in an Oracle form?

   a. VARCHAR2

   b. NUMBER

   c. DATE

   d. CHAR

8. When you create a trigger that automatically displays the next value in a sequence in a form text item when the form first opens, you associate the trigger with the following form object:

   a. form module

   b. window

   c. data block

   d. text item that displays the sequence value

9. You enter the trigger code, the code executed when an event occurs, in the _____ Editor.

   a. Data Block

   b. PL/SQL

   c. LOV

   d. Object

10. A checkmark in a check box on a form means that the data value represented by the label is:

   a. NOT NULL

   b. NULL

   c. false

   d. true

PROBLEM-SOLVING CASES

The case problems reference the Clearwater Traders sample database. For all cases, save the solution files in the Chapter05\Cases folder on your Solution Disk. For all forms, rename the window and canvas using descriptive names, and change the window titles to descriptive values. Use descriptive item prompts, and if necessary, format item prompts so that they appear on a single line. Execute the Ch5Clearwater.sql script file before attempting these cases.

1. Create a form named INVENTORY using the format shown in Figure 5-87. Create a sequence named **INVENTORY_SEQUENCE** and perform the steps necessary so the Inventory ID is automatically filled with the next available sequence number when a new record is created. Have the sequence value begin with 41. After creating the form, run the form and add a new inventory item to verify that there are no errors. Save the form as **Ch05Case1C.fmb**. Save the code to create the sequence as Ch05Case1C.sql.

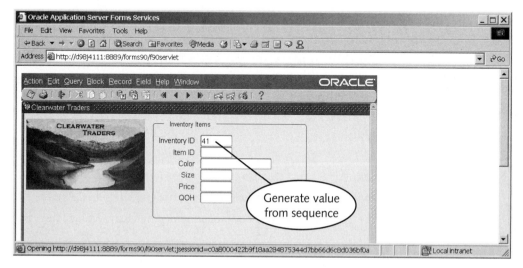

Figure 5-87

2. Modify the INVENTORY form created in Case 1 to include an LOV for the Item ID item. Format the LOV so when it is displayed it resembles the example shown in Figure 5-88. Run the form to ensure there are no errors and add a sample record to the INVENTORY table. Save the form as **Ch05Case2C.fmb**.

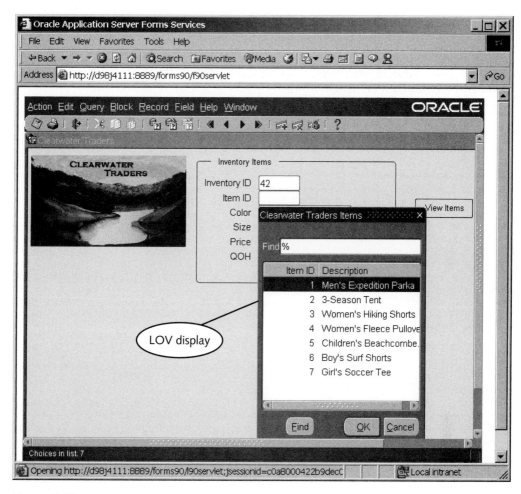

Figure 5-88

3. Create a form named **ITEM_FORM** that displays all of the fields except ITEM_IMAGE in the Clearwater Traders ITEM table. Save the form as **Ch05Case3C.fmb**.

   a. Create a new data block, and use a tabular-style layout that displays five records at one time and has a scrollbar. Change the frame title to **Items**. Show the **Clearwater Traders logo** on the form. (The Clearwater Traders logo is stored in the Clearlogo.jpg file in the Chapter5 folder on your Data Disk.)

    b. Create an LOV named **CATEGORY_LOV** to display the allowable values for the CAT_ID foreign key field. Attach the LOV to the **CAT_ID** text item. Display the CAT_ID and CAT_DESC values in the LOV display. Format the LOV display so that it appears under the CAT_ID text item, and resize the display so that the user does not have to scroll to view the records or field values.

4. Create an ORDER form, based on the ORDERS table, which allows the user to select the Customer ID (C_ID) and Order Source (OS_ID) from LOVs. Also include a sequence named O_ID_SEQUENCE that enters the Order ID of any new orders. Start the O_ID_SEQUENCE at 100. Save the form as **Ch05Case4C.fmb** and the SQL code as **Ch05Case4C.sql**.

5. Create a master-detail form named **CUSTOMER_ORDER** that allows you to enter orders for a customer. Use the O_ID_SEQUENCE created in the previous case to assign the Order ID for each order. After creating the form, run the form, and enter a new order for Customer ID 5. Save the form as **Ch05Case5C.fmb**.

CHAPTER 6

CREATING CUSTOM FORMS

◀ LESSON A ▶

After completing this lesson, you should be able to:

♦ Create and use custom forms
♦ Create command buttons that use form triggers to manipulate data
♦ Use the Forms Debugger to find form logic and runtime errors
♦ Work with form triggers
♦ Create form navigation triggers

The data block forms that you created in Chapter 5 displayed values from a single database table or values from multiple tables related by foreign key values. The appearance of a data block form reflects the structure of the underlying database tables. However, sometimes you need to create forms to insert, update, delete, and display data from many related tables in a way that reflects the organization's processes rather than the structure of the database tables. For example, when a new student registers for a class at Northwoods University, the database application may need to update the STUDENT table with the new student's information, and update the ENROLLMENT table with this student's registration in a course. In this chapter, you learn how to create custom forms that display and update data in many database tables. These forms use form triggers that contain SQL commands to process data.

INTRODUCTION TO CUSTOM FORMS

A **custom form** displays the data fields from a variety of database tables, and contains programs that support organizational processes. You do not associate a custom form with a specific database table. Rather, you identify the processes the form needs to support, and then you create the form. Custom forms use lists of values (LOVs) to retrieve data values. Custom forms manipulate data using form triggers that contain SQL commands to insert, update, and delete data values. (Recall from Chapter 5 that a form trigger is a program that you associate with a form object and that executes in response to an event such as clicking a button.) To create a custom form, you identify the business processes and related database operations that the form supports, design the interface, and then create the form.

Identifying the Business Processes and Database Operations

To create a custom form, first you must identify the processes that the form is intended to support, and then identify the associated database tables. The best way to start is to describe the process. Suppose you need to create a custom form to support the registration process at Northwoods University. When students enroll in a new course, the information is tracked in the ENROLLMENT table of the Northwoods University database (see Figure 1-26).

Student information includes the student's ID, first and last names, and classification (for example, freshman, sophomore, and so on). When the student enters the course number, the course name is displayed on the form. To ensure that there is sufficient room in the course, the remaining capacity for the course is also displayed. After the student verifies that the displayed course information is correct, the student is ready to enroll in the course. To simplify the process, a command button is added to the Registration form to update the ENROLLMENT table by adding the student ID and course number to the ENROLLMENT table.

Designing the Interface

The next step is to visualize how the form is going to look. Figure 6-1 shows a design sketch of the interface design for the Northwoods University registration form.

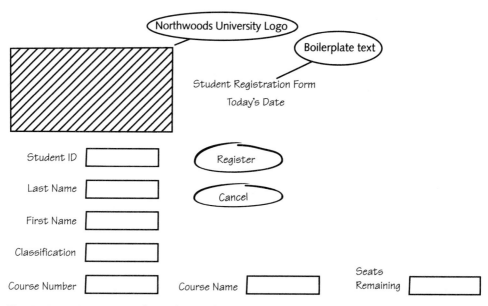

Figure 6-1 Registration form design sketch

The form displays the Northwoods University logo and the boilerplate text "Student Registration Form." The form also has fields to display values for the student's ID, last and first names, classification, course number, name, seats remaining, and command buttons to control the form actions. The user selects the student information using an LOV that displays records from the STUDENT table. (Users can open the LOV display by placing the insertion point in the Student ID text item and then pressing CTRL+L, or clicking Edit on the Forms Services menu and then clicking Display List.)

After the user selects the student data, another LOV retrieves available courses for the current term. The form also contains a text item to display the current capacity for the course. The value displayed for the Seats Remaining text item is calculated by counting the total number of students already registered for the course and subtracting this total from the course's total capacity. The form has a Register button to add registration data to the ENROLLMENT table and a Cancel button to cancel the current operation and clear the form text items.

Before you create the form, you refresh your database tables by running the scripts that re-create the tables and insert all of the sample data records.

To refresh the database tables:

1. Start SQL*Plus, and log onto the database.

2. Run the **Ch6Northwoods.sql** script from the Chapter06 folder on your Data Disk. Do not close SQL*Plus because you use it again later in the chapter.

CREATING A CUSTOM FORM

To create a custom form, you manually create the form canvas in the Object Navigator. Then, you create the form items by "painting" the items on the canvas, using tools on the Layout Editor tool palette. Finally, you write the code that controls the form functions. The following sections describe these steps.

Creating the Form Canvas

When you created data block forms in Chapter 5, the Data Block Wizard and the Layout Wizard automatically created the form canvas. When you create a custom form, you manually create the form canvas in the Object Navigator. Then you start Forms Builder, create and rename the form canvas, rename the form module and the form window, and change the Title property of the form window. You perform these operations in Visual View in the Object Navigator, so you can see the relationships among the form objects.

To create the form canvas and rename the form objects:

1. Start the OC4J Instance, and then minimize the window.

2. Start Forms Builder. The Object Navigator window opens and displays a new form module. Save the form as **Ch6ARegistration.fmb** in the Chapter06\Tutorials folder on your Solution Disk. Maximize the Object Navigator window if necessary, and change the form module name to **REGISTRATION_FORM**.

3. Click the **Connect** button 🖳 on the Forms Builder toolbar, and connect to the database in the usual way.

4. To open the Object Navigator in Visual View, click **View** on the menu bar, and then click **Visual View**. The form objects appear in Visual View.

5. Open the Windows node. Two windows appear: the NULL_WINDOW and the default form window, which is named WINDOW1. (Your default window name may be different.) Recall that Forms Builder automatically creates the NULL_WINDOW object.

6. Change the name of WINDOW1 to **REGISTRATION_WINDOW**.

7. Open the REGISTRATION_WINDOW node, select the Canvases node, and then click the **Create** button ➕ to create a new canvas. Change the name of the new canvas to **REGISTRATION_CANVAS**, as shown in Figure 6-2.

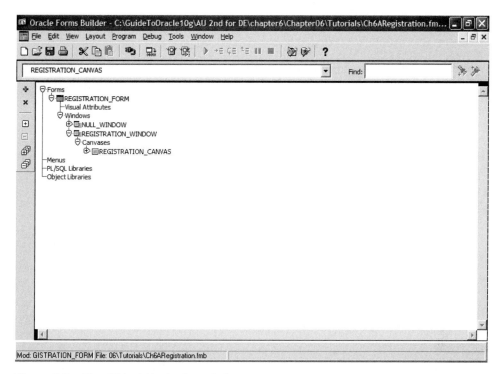

Figure 6-2   The Object Navigator window

8. Select the REGISTRATION_WINDOW node, right-click, click **Property Palette**, change the window Title property value to **Northwoods University**, then close the Property Palette.

9. Save the form.

Creating a Control Block

The next step in creating a custom form is to create a control data block that contains the form items. A **control data block**, which is also called a **control block**, is a data block that you do not associate with a particular database table. A control block contains text items, command buttons, radio buttons, and other form items that you manually draw on the form canvas and then control through form triggers. To create a control block, you create a new data block in the Object Navigator, and specify that the data block is to be created manually, rather than by using the Data Block Wizard. You create the control block in Ownership View, because the Data Blocks node does not appear in Visual View.

To create the control block:

1. View the form in Ownership View by clicking **View** on the menu bar, and then click **Ownership View**.

2. Under the REGISTRATION_FORM node, select the Data Blocks node, and then click the **Create** button ✚ to create a new data block. The New Data Block dialog box opens. Select the **Build a new data block manually** option button, and then click **OK**. The new data block appears in the Data Blocks list in the Object Navigator window. Because you did not create the data block using the Data Block Wizard and you have not associated the block with a database table, it is a control block rather than a data block.

3. Change the block name to **REGISTRATION_BLOCK**, and then save the form.

Creating the Form Items

The next step is to create the boilerplate logo image and text, the form text items, and the command button items that appear in the form design sketch in Figure 6-1. To create these items, you display the form in the Layout Editor, and then draw the items on the form canvas using tools in the tool palette. First you create the boilerplate logo image and the boilerplate text. (Recall from Chapter 5 that boilerplate items are items that do not directly contribute to the functionality of a form, but make the form more attractive or easier to understand.) You learned how to create boilerplate images such as the Northwoods University logo in Chapter 5. To create boilerplate text on a form, you use the Text tool on the tool palette to draw a field to represent the text. Then you configure the text's properties. Now you add the logo image and boilerplate text to the form.

To create the logo and boilerplate text:

1. Open the form in the Layout Editor by clicking **Tools** on the menu bar and then clicking **Layout Editor**. The Layout Editor displays a blank canvas and shows the current canvas and block in the Canvas and Block lists at the top of the Layout Editor window. Make sure that REGISTRATION_CANVAS appears as the current canvas and REGISTRATION_BLOCK appears as the current block.

HELP

If you are not working in REGISTRATION_BLOCK, or if REGISTRATION_BLOCK is a data block rather than a control block, your form will not work correctly. To move to a different block or canvas in the Layout Editor window, open the Block or Canvas list, and select the desired block or canvas.

2. To create the logo image, click **Edit** on the menu bar, point to **Import**, click **Image**, and then click **Browse**. Then navigate to the Chapter06 folder on your Data Disk, select **NWlogo.jpg**, click **Open**, and then click **OK**. The Northwoods University logo appears on the form. If necessary, move the image so it appears on the upper-left side of the form, as shown in Figure 6-3.

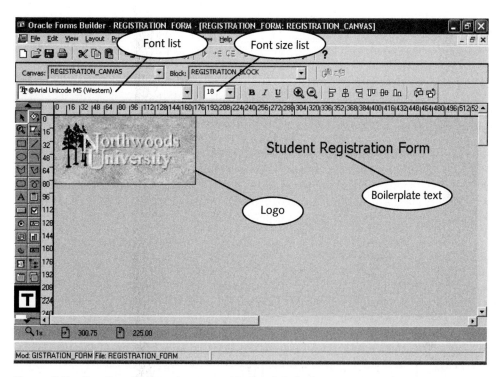

Figure 6-3   Creating the logo image and boilerplate text

3. To create the "Student Registration Form" boilerplate text, select the **Text** tool **A** on the tool palette, and then click the **Text tool pointer** on the canvas, to the right of the logo image. A text entry field appears on the form.

4. Make sure that the insertion point is in the text entry field, then type **Student Registration Form.** To close the text entry field, click anywhere on the form. The new text appears selected.

5. With the text selected, open the **Font** list, as shown in Figure 6-3, and select **Arial Unicode MS (Western).** Open the **Font Size** list, and select **18.** Your image and boilerplate text should look similar to Figure 6-3.

6. Because you changed the font and font size, all new items that you create on the canvas will have the new font size. To change the form font size back to the default size, which is 8 points, click anywhere on the canvas to cancel the boilerplate text selection, make sure that no item is selected, and then open the **Size** list and select **8.**

7. Save the form.

Next, you create the text items to represent the fields that display database values. To create a new text item, you select the Text Item tool on the Layout Editor tool palette, and then draw the text item on the form. Then you configure the text item by modifying the following properties:

- **Name**—Represents how Forms Builder internally references the item. For items that represent database fields, this is usually the same name as the database field.

- **Data Type**—Represents the type of data the text item displays. When you create a text item to represent a database field, the text item must have the same data type as the database field.

- **Maximum Length**—Represents the maximum width of the data that the text item can display. When you create a text item to represent a database field, the text item should have the same maximum width as the associated database field, plus any formatting characters that you include in the text item's Format Mask property.

- **Prompt**—Represents the label that appears beside the text item.

Now you create the Student ID text item. The Student ID text item is associated with the S_ID field in the STUDENT and ENROLLMENT database tables. These fields have the VARCHAR2 data type, and a maximum width of six characters.

TIP

To find the data type and maximum width of a database field, recall that you can execute the DESCRIBE command in SQL*Plus, which has the syntax `DESCRIBE tablename;`.

To create the Student ID text item:

1. Select the **Text Item** tool on the tool palette, and then draw a rectangular box for the Student ID text item below the boilerplate image and text. (Don't worry about the text item's exact position, because you reposition it later.) The system assigns a default name to your text item, such as TEXT_ITEM2, and the prompt does not appear yet.

2. Right-click the new text item, and then click **Property Palette**. Confirm that the Item Type property value is Text Item, and then change the Name property value to **Student_ID**.

3. Scroll down to the Data property node, click in the space next to the Data Type property to open the value list, and, if necessary, select **Char**.

4. Change the Maximum Length property to **6**.

5. Scroll down to the Prompt property node, and change the Prompt property value to **Student ID**, followed by two blank spaces. You include two blank spaces after the final character in Student ID to provide space on the form between the prompt and the text item. Reposition the COURSE_NAME text item if any overlapping occurs.

6. Close the Property Palette. The new prompt and text item name appear, as shown in Figure 6-4. Save the form.

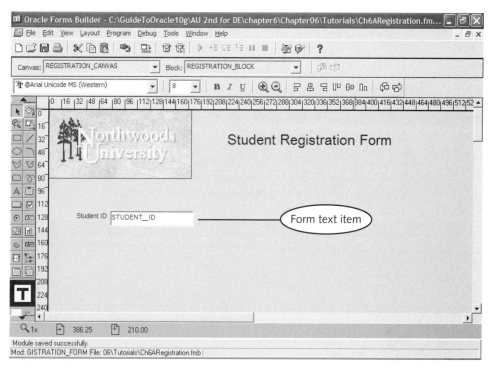

Figure 6-4   Student_ID text item

Now you create the rest of the form text items. After you create each item, you change the item name and data type to correspond with the item's associated database field.

To create the rest of the form text items:

1. Create text items with the following properties, and position the text items as shown in Figure 6-5. Do not worry about the exact item sizes and positions, because you specify the sizes and positions using an object group in a future step. Include two blank spaces after the final character in the prompt values, so that a space appears between the prompt and the text item.

Name	Data Type	Maximum Length	Prompt
S_LAST	Char	30	Last Name
S_FIRST	Char	30	First Name
S_CLASS	Char	2	Classification
COURSE_NO	Char	7	Course Number
COURSE_NAME	Char	25	Course Name
SEATS_REMAINING	Number	5	Seats Remaining

NOTE

When entering the SEATS_REMAINING prompt, press Shift+Enter, or click the Ellipsis button [...] that appears when you click in the Prompt property, to open the Prompt window. After typing SEATS, press Enter to place the words on separate lines within the prompt.

2. To resize the text items, select all of the items as an object group, right-click, and then click **Property Palette**. Scroll down to the Physical property node, then change the Width property value to **80**, and the Height property value to **14**. Close the Property Palette. The text items now all appear as the same size.

3. To reposition the text items, select the first five of the text items only as an object group, click **Layout** on the menu bar, and then click **Align Components**. Select the **Align Left** option button to specify the horizontal alignment, select the **Stack** option button to specify the vertical alignment, and then click **OK**.

4. With the objects still selected as an object group, drag the objects to the position shown in the left side of the form in Figure 6-5.

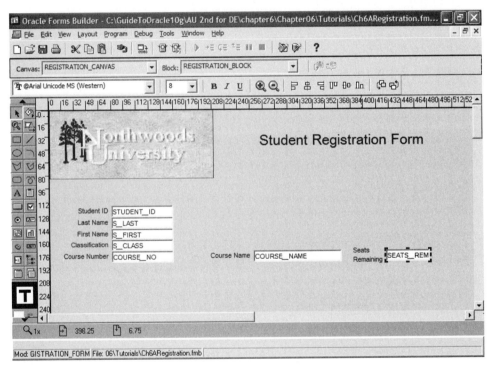

Figure 6-5   Creating the form text items

5. To reposition the **COURSE_NAME** and **SEATS_REMAINING** text items, select them as an object group with the **COURSE_NO** text item, click **Layout** on the menu bar, and then click **Align Components**. Select the **None** option button to specify the horizontal alignment, select the **Align Top** option button to specify the vertical alignment, and then click **OK**. Change the Width property of the COURSE_NAME text item to **112** and the SEATS_REMAINING text item to **55** and save the form.

Creating the LOVs

Next, you create the LOV to retrieve student data from the STUDENT table.

To create the LOV:

1. Click **Tools** on the menu bar, and then click **LOV Wizard**. The LOV Source page appears. Accept the default values, and then click **Next**.

2. On the Query page, type the SQL query shown in Figure 6-6 to retrieve the LOV display fields, and then click **Next**.

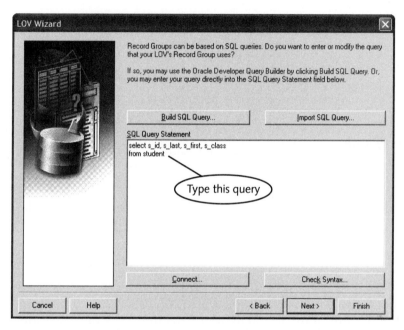

Figure 6-6   SQL query to retrieve student data

3. When the Column Selection page appears, select and move all of the query fields from the Group Columns list to the LOV Columns list for the LOV display, and then click **Next**.

NOTE

In a real-world situation, you would not display both the student's ID and name because of issues of possible identity theft. They are included in the example for illustrative purposes. An alternative is to require the student to enter his or her ID and password before retrieving the student's information.

4. When the Column Display page appears, enter the widths shown in Figure 6-7 if necessary. Enter the appropriate Return value by clicking the **Look up return item** button and double-clicking the appropriate item for each column.

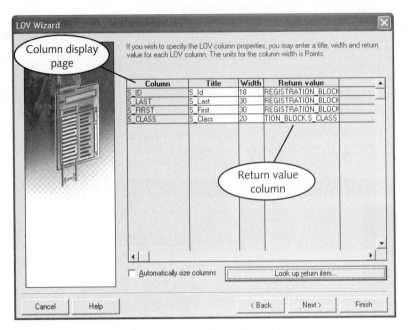

Figure 6-7 LOV column properties for student data

5. Click **Next**. The LOV Display page appears. Type **Students** for the LOV title, and then click **Next**. The Advanced Options page appears.

6. Accept the default options, and then click **Next**. The Items page appears. This page enables you to specify the form text item to which the LOV is assigned. You assign the LOV to the S_ID text item.

7. Select **REGISTRATION_BLOCK.STUDENT_ID** in the Return Items list (this is the first item in the list), then click the **Move one item to target** button [>] to move the selection to the Assigned Items list. Click **Next**, and then click **Finish** to finish the LOV.

8. Open the Object Navigator window, and change the name of the new LOV object and record group object to **STUDENTS_LOV**. Then save the form.

To display the courses offered during the current term and the number of seats still available in each course, you create a SQL query referencing the ENROLLMENT, COURSE, COURSE_SECTION, and TERM tables. The TERM table is referenced to identify the current term based on a status value of OPEN. The COURSE_SECTION table is included in the query to determine which courses are offered in the current (open) term. The name of the course is retrieved from the COURSE table, while the ENROLLMENT table is included to calculate the number of students already enrolled in the course, and therefore, the number of seats still available in the course. Now you use the LOV Wizard to create the LOV for the course data.

To create the LOV:

1. Click **Tools** on the menu bar, and then click **LOV Wizard**. The LOV Source page appears. Accept the default values, and then click **Next**.

2. On the Query page, type the SQL query shown in Figure 6-8 to retrieve the LOV display fields, and then click **Next**.

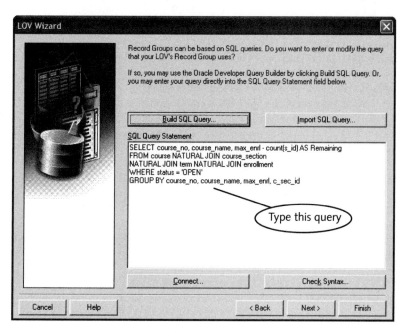

Figure 6-8   SQL query to retrieve course data

3. When the Column Selection page appears, select all of the query fields for the LOV display, and then click **Next**.

4. When the Column Display page appears, enter the width values shown in Figure 6-9. Then assign the appropriate return value for each column.

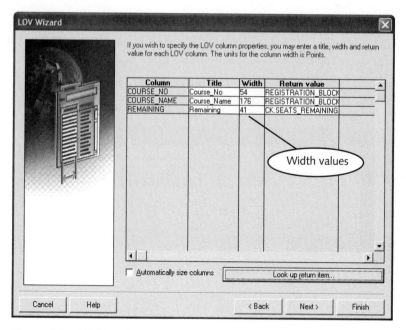

Figure 6-9   Width values

5. Click **Next**. The LOV Display page appears. Type **Course Information** for the LOV title, then click **Next**. The Advanced Options page appears.

6. Accept the default options, and then click **Next**. The Items page appears. This page enables you to specify the form text item to which the LOV is assigned. You assign the LOV to the COURSE_NO text item.

7. Select **REGISTRATION_BLOCK.COURSE_NO** in the Return Items list, then click the **Move one item to target** button [>] to move the selection to the Assigned Items list. Click **Next**, and then click **Finish** to finish the LOV.

8. Open the Object Navigator window, and change the name of the new LOV object and record group object to **COURSE_LOV**. Then save the form.

Now you run the form, and test the LOV. You open the LOV display, confirm that the columns appear correctly, and confirm that the LOV returns the selected values to the form text items.

To test the LOV:

1. Click the **Run Form** button 🔲 to run the form. When the form appears in the browser window, make sure that the insertion point is in the Student ID text item, and then press **CTRL+L**. The LOV display opens.

NOTE

If you run a form multiple times, the message "FRM-30087 Unable to create form file *folder_path\filename*.fmx" may appear. If this happens, click OK. This error occurs when an Oracle process named ifweb90.exe opens the .fmx file, but does not close the file. To force ifweb90.exe to release (close) the file, press CTRL+ALT+Delete to open the Windows Task Manager, select the Processes tab, select the Image Name column to sort the processes, select the ifweb90.exe process, click End Process, click OK, and then close the Task Manager window. This action will not harm Forms Builder or your computer.

HELP

You may need to close the browser window, then adjust the LOV display width and height in Forms Builder so all of the fields appear.

6

2. Select the record for Student ID **MA100** and then click **OK**. The selected values appear in the Student ID, Last Name, First Name, and Classification text items, as shown in Figure 6–10.

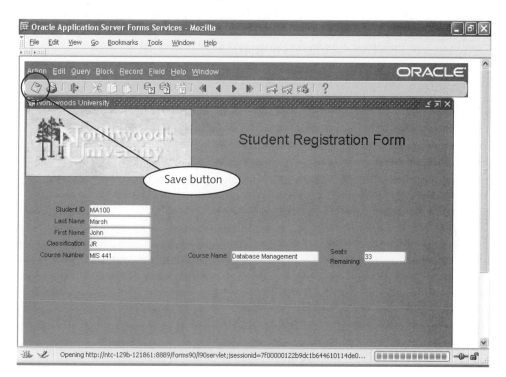

Figure 6-10   Data displayed in Student Registration form

HELP If the returned values are not displayed correctly or if only some of the values appear, it is probably because you did not specify the LOV return values correctly. Close the browser window, open the Object Navigator in Forms Builder, select the STUDENT_LOV object under the LOV's node, open the LOV Wizard in reentrant mode, and then examine the return values on the Column Display page to confirm that the column names are listed correctly and that each column corresponds with the correct return value.

HELP An LOV selection's values may not appear in the form text items if the form text item Data Type and Maximum Length properties are not exactly the same as the database field properties. If your LOV's values did not appear in the form text items, open the Property Palette for each text item, and make sure that the data type and maximum width for each form text item are the same as for the text item's corresponding database field.

3. Place the insertion point in the Course Number text item, and then press **CTRL+L**. The LOV display opens.

4. Select the record for Course Number **MIS 441** and then click **OK**. The selected values appear in the Course Number, Course Name, and Seats Remaining text items, as shown in Figure 6-10.

5. Close the browser window.

Displaying System Date and Time Values in Form Text Items

Recall that on the Northwoods University registration form design sketch in Figure 6-1, the current date is displayed beneath the boilerplate text. Forms Builder provides system variables that you can use to display system date and time values in form text items. A **system variable** is a variable representing a value that is always available to any form. Table 6-1 summarizes the Forms Builder date and time system variables.

System Variable	Return Value
$$DATE$$	Current operating system date
$$TIME$$	Current operating system time
$$DATETIME$$	Current operating system date and time
$$DBDATE$$	Current database server date
$$DBTIME$$	Current database server time
$$DBDATETIME$$	Current database server date and time

Table 6-1 Forms Builder date and time system variables

To display a value in a text item automatically when a form first opens, you set the text item's Initial Value property to the desired value. Now you create the TODAY_DATE text item and change its Initial Value property to $$DATE$$ so it displays the current operating system date. Then you run the form and confirm that the system date appears in the text item when the form first opens.

To display the current operating system date in the text item:

1. Click **Tools** on the menu bar, and then click **Layout Editor** (if the Layout Editor is not already showing).

2. Click the **Text Item** tool ⌨ from the tool palette and click beneath the boilerplate text to insert the text item box.

3. Right-click the inserted text item box and then click **Property Palette** or press **F4**. Change the Name in the General node to **TODAY_DATE**. Scroll down to the Data property node, and change the Initial Value property to **$$DATE$$** as shown in Figure 6-11.

6

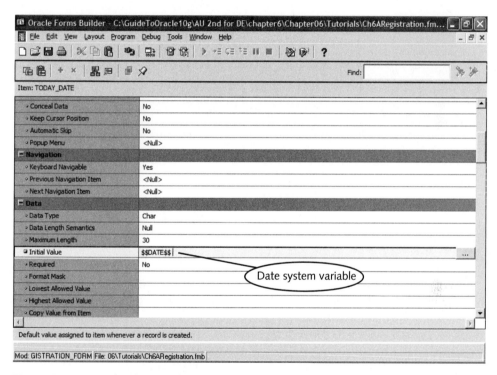

Figure 6-11 Initial Value property

4. Close the Property Palette, then save the form.

5. Run the form. The current system date appears in the Date Received text item, as shown in Figure 6-12.

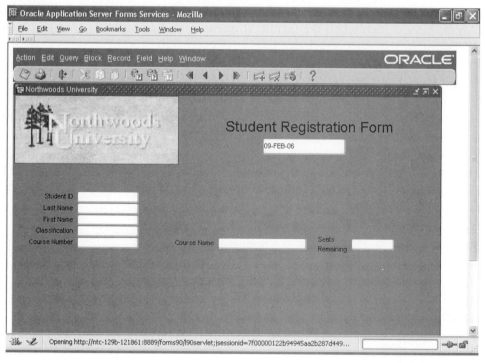

Figure 6-12   Form displaying system date

> 6. Close the browser window.

CREATING COMMAND BUTTONS

Recall that on the registration form design sketch in Figure 6-1, the form displays Register and Cancel command buttons that allow the user to insert a new row into the ENROLLMENT table or cancel the current operation. The following sections describe how to create and configure the buttons, and how to create the form triggers associated with the buttons.

Creating and Configuring Command Buttons

You create a command button by drawing the button on the canvas using the Button tool 🔲 on the tool palette. Usually, a form contains multiple command buttons that appear in a button group. Figure 6-1 shows that the receiving form displays two buttons: Register and Cancel. Whenever you create a button group, the buttons should all be the same size and should be wide enough to accommodate the longest button's label, which is the text that appears on the button. To make all the command buttons the same size, draw the button with the longest label first, and configure its properties and size. Then copy and paste the button to create the other buttons in the button group.

To configure a form button item, you modify the Name property, which is the name that Forms Builder uses to reference the button. You also modify the Label property, which specifies the text that appears on the button. And, you modify the Width and Height properties to specify the button size. Now you create the Register button, and configure its properties. Then you copy and paste the Register button to create the Cancel button.

To create the form buttons:

1. In the Layout Editor, select the **Button** tool 🔲 on the tool palette, and then draw the command button that is going to be the Register button, as shown in Figure 6-13.

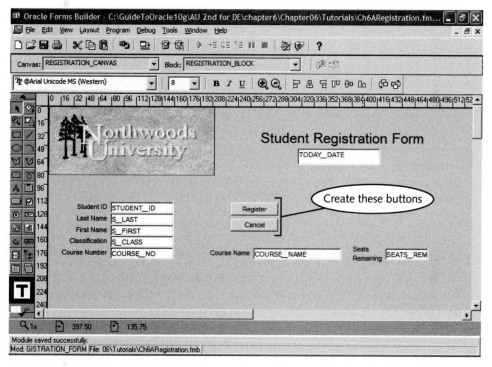

Figure 6-13   Creating the form buttons

2. Open the new button's Property Palette, and change its Name property to **REGISTER_BUTTON**, Label property to **Register**, Width property to **60**, and Height property to **16**. Then close the Property Palette.

3. In the Layout Editor, select the **Register** button if it is not already selected. Then click the **Copy** button 📋 on the toolbar to copy the button, and click

the **Paste** button 📋 to paste the copied button onto the form. Forms Builder pastes the new button directly on top of the first button.

4. Select the top button, and drag it below the original Register button.

5. Select both buttons as an object group, click **Layout** on the menu bar, and then select **Align Components**. Select the **Align Left** option button to align the buttons horizontally, and select the **Distribute** option button to evenly space the buttons vertically, and then click **OK**. Position the button group on the form as shown in Figure 6-13.

6. Select the second button, open its Property Palette, change its Name property to **CANCEL_BUTTON** and its Label property to **Cancel**, and then close the Property Palette.

7. Save the form.

Creating the Button Triggers

In the data block forms you created in Chapter 5, the Forms Services window automatically provided triggers for retrieving, inserting, updating, and deleting data. For the custom forms in this chapter, you need to create these triggers manually. To create a button trigger, you select the Triggers node under the button in the Object Navigator, click the Create button ➕ to create the trigger, select the trigger event, and then specify the trigger code. To create a button whose trigger code executes when the user clicks the button, you select the WHEN-BUTTON-PRESSED button event. In the following sections, you create the Register and Cancel button triggers.

Creating the Register Button Trigger

After the user has retrieved his or her student information and selected a course, the user then needs to click the Register button to enroll in that course. The Register button trigger must insert values in the S_ID and C_SEC_ID columns in the ENROLLMENT table.

To insert a new row with the trigger, you can use the same SQL commands that you would use in SQL*Plus. However, you replace the values with references to the form text item values. Recall that to reference form text item values, you use the syntax `:block_name.item_name`. For example, you reference the current value of the STUDENT_ID text item as `:REGISTRATION_BLOCK.STUDENT_ID`. However, recall that the ENROLLMENT table references a course by its section number. Because the course section number (C_SEC_ID) is not included in the registration form, a PL/SQL block must be included in the trigger to retrieve the value and then insert that value into the ENROLLMENT table. Next, you create the Register button trigger.

To create the Register button trigger:

1. Click **Window** on the menu bar, and then click **Object Navigator** to open the Object Navigator. Open the REGISTER_BUTTON node, select the Triggers node, and then click the **Create** button ➕ to create a new trigger. Scroll down the list of events, and double-click **WHEN-BUTTON-PRESSED**. The PL/SQL Editor opens.

2. In the PL/SQL Editor Source code pane, type the commands shown in Figure 6-14, click the **Compile PL/SQL code** button 🖳 to compile the trigger, and then close the PL/SQL Editor.

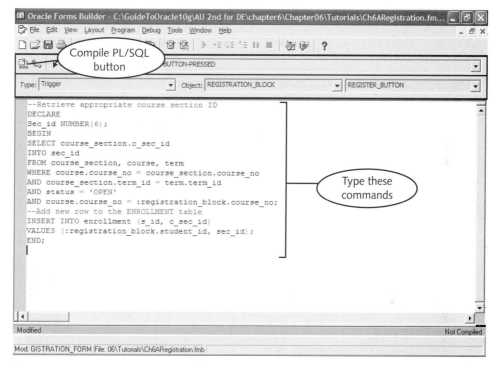

Figure 6-14   Register trigger command

 If your trigger has a syntax error, refer to the section titled "Correcting Common Trigger Syntax Errors" in Chapter 5.

3. Save the form.

Now you test the trigger by retrieving the records for student ID SM100 and course number MIS 441, and clicking the Register button to enroll the student in the course.

To test the trigger:

1. Run the form, press **CTRL+L** to open the LOV display, and select Student ID **SM100**. Move the Insertion point to the Course Number text box and press **Ctrl+L** to open the LOV display, and then select course **MIS 441**. The appropriate student and course information are displayed in the form.

2. Click the **Register** button, and then click the **Save** button 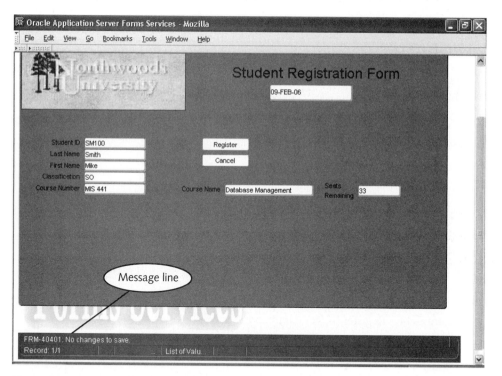 on the form's toolbar. The message "FRM-40401: No changes to save" appears in the message line, as shown in Figure 6-15. (You may need to scroll down in the browser window to view the message line.) This message confirms that the Register button trigger executed correctly, and updated the database. Although the confirmation message ("FRM-40401: No changes to save") seems counterintuitive, Forms Services issues this confirmation message when it successfully executes a COMMIT command in a form trigger in a control block.

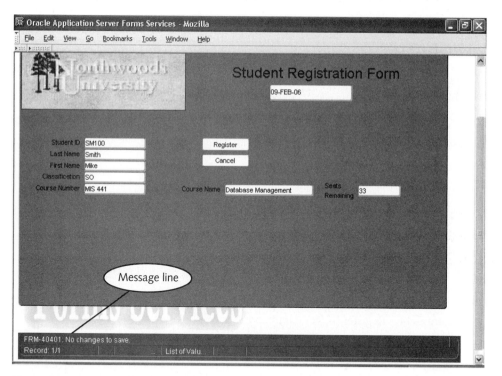

Figure 6-15   Registering for a course

3. Close the browser window.

Before you can be totally confident that this operation succeeded, you should double-check the value in the ENROLLMENT table. Now you use SQL*Plus to confirm that the form inserted the row in the ENROLLMENT table.

To use SQL*Plus to confirm that the form updated the table:

1. Switch to SQL*Plus, and type and execute the following query to confirm that the form added the record in the ENROLLMENT table. The student number should be included in the ENROLLMENT table along with the appropriate section number of the course.

```
SELECT s_id, c_sec_id
FROM enrollment NATURAL JOIN course_section
     NATURAL JOIN course NATURAL JOIN term
WHERE s_id = 'SM100' AND course_no = 'MIS 441';
```

2. Verify that one row is retrieved by the query. If there is no row retrieved, the form did not correctly insert the data into the ENROLLMENT table.

6

Creating a Program Unit to Clear the Form Fields in the Register Button Trigger

After you processed the student registration, but before you closed the browser window, values still appeared in the form text items. If you had clicked the Register button again, the Register button trigger would have attempted to modify the ENROLLMENT table again. However, this second insertion would have violated the table's primary key constraint and the message "FRM-40735: WHEN-BUTTON-PRESSED trigger raised unhandled exception ORA-00001" would have appeared on the message line. To avoid this potential error, the Register button needs to clear the form text items after it processes an incoming shipment.

There are two ways to clear the form text items in a form trigger: use the CLEAR_FORM built-in procedure, which clears all of the form text items; or you can create a program unit to set the value of the text items to a blank text string. You use the CLEAR_FORM procedure when you want to clear all form text items, and you create a program unit when you want to clear selected text items and retain the current values in other items. (Recall that a program unit is a self-contained block of PL/SQL code that you can use within a larger program.) Because you would like to retain the student information in case the student wants to register for a second course, you create a program unit to clear the values in course-related text items only.

You could place the commands to clear selected text items directly in the Register button trigger. But remember, the Cancel button is also going to clear the form items. If you place these commands in a program unit, the Cancel button can call the program unit, and you do not need to repeat the commands in the Cancel button's trigger. When multiple form triggers execute the same set of program commands, you should create a program unit to specify those commands.

To create a new program unit, you open the Object Navigator, and select the Program Units node under the form module and then click the create button. Then you specify the program unit name, which must follow the Oracle naming standard, and the program unit type. A program unit can be a **procedure**, which is a code block that executes commands to change one or more values. A program unit can also be a **function**, which is a code block that returns a single value. Finally, you type the program unit's PL/SQL commands in the PL/SQL Editor.

Now you create a program unit in Forms Builder named CLEAR_COURSE. This program unit is a procedure that clears the current values in all course-related text items by assigning a blank text string as the text item values.

To create the program unit:

1. Open the Object Navigator if necessary, select the Program Units node, and then click the **Create** button ➕ to create a new program unit. The New Program Unit dialog box opens.

2. Type **CLEAR_COURSE** in the Name field, make sure the Procedure option button is selected, and then click **OK**. The PL/SQL Editor opens and displays a heading template for the new program unit, as shown in Figure 6-16.

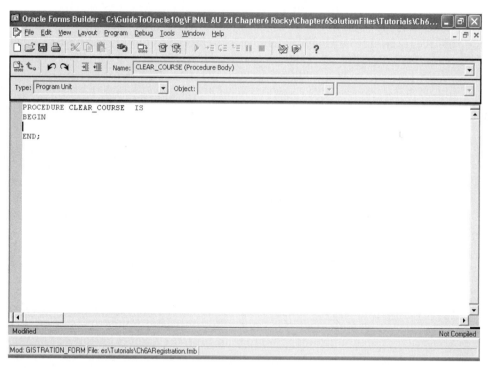

Figure 6-16   Program unit template

The program unit template specifies the program unit type (PROCEDURE or FUNCTION), followed by the program unit name, and the keyword IS. The BEGIN command signals the beginning of the program unit body, and the END command signals the end of the program unit. Recall that a PL/SQL program block consists of a declaration section, in which you declare program variables; a body section, in which you place the program commands; and an exception section, in which you place exception handlers. A program unit has the same sections. You place variable declaration commands after the IS keyword. Note that unlike other PL/SQL programs, a program unit does not use the DECLARE keyword. You place the program body commands after the BEGIN keyword. You can add the EXCEPTION keyword to add an optional exception section.

Now you type the commands for the CLEAR_COURSE program unit. The program unit does not declare any variables, so you type the commands to clear the form text items directly in the program unit body. The program unit contains commands to assign an empty string, specified by two single quotation marks (''), as the value of each of the course text items. Recall that you use the (:=) assignment operator to assign values in PL/SQL commands. The program unit does not contain any exception handlers.

To add the program unit commands:

1. Make sure that the insertion point is on the line after the BEGIN keyword, and then type the following commands to clear the selected form text items:

```
:registration_block.course_no := '';
:registration_block.course_name := '';
:registration_block.seats_remaining := '';
```

2. Compile the program unit, correct any syntax errors, and then close the PL/SQL Editor.

To call a program unit, you list the name of the program unit, along with the parameter values enclosed in parentheses, in the code of the calling program. Now you modify the Register button trigger to call the program unit.

To modify the trigger to call the program unit:

1. In the Object Navigator, double-click the **Trigger** icon under the REGISTER_BUTTON node. The PL/SQL Editor opens, and displays the Register button trigger commands.

2. Modify the trigger by adding the following command as the last line of the trigger code immediately above the END statement:

```
CLEAR_COURSE;
```

3. Compile the trigger, and then close the PL/SQL Editor and save the form.

Now you test the program unit. You run the form, and then retrieve the record for student ID NG100 and course number MIS 441. Then, you confirm that the trigger clears the form text items when you click the Register button.

To test the program unit:

1. Run the form, open the Students LOV display, select Student ID **NG100**, and then click **OK**.

2. Open the Course LOV display, select Course Number **MIS 441**, and then click **OK**.

3. Click **Register**. The trigger clears the last three fields shown on the form.

4. Close the browser window.

Creating the Cancel Button Trigger

Now you create and test the trigger for the Cancel button. The Cancel button calls the CLEAR_COURSE program unit to clear the form text items. Recall that to create a button trigger, you can select the Triggers node under the button in the Object Navigator, click the Create button ➕ to create the trigger, select the trigger event, and then specify the trigger code. An easier way to create a button trigger is to right-click the button in the Layout Editor, point to SmartTriggers, and then select the WHEN-BUTTON-PRESSED event. Now you use the SmartTriggers feature to create the Cancel button trigger.

To create and test the Cancel button trigger:

1. Open the Layout Editor, select the **Cancel** button, right-click, point to **SmartTriggers**, and then click **WHEN-BUTTON-PRESSED**. The PL/SQL Editor opens for the new trigger.

2. Type the following command in the PL/SQL Editor:

   CLEAR_COURSE;

3. Compile the trigger, and then close the PL/SQL Editor.

4. Save the form, and then run the form.

5. Open the LOV display, select the record for Student ID **MA100**, move the Insertion point to the Course Number text box, press **Ctrl+L** to open the LOV display, select Course Number **MIS 101**, and then click **OK**.

6. Click **Cancel**. The button's trigger clears the course text items.

7. Close the browser window, and then close the form in Forms Builder.

Often when you close a form in Forms Builder, a dialog box opens asking if you want to save the form. Always click Yes.

USING THE FORMS DEBUGGER TO FIND RUNTIME ERRORS

In Chapter 5, you learned how to identify and correct some of the more common syntax errors that occur in form triggers. As you start writing more complex triggers and program units to process custom forms, you might need to use the Forms Debugger to find runtime errors that occur while a form is running. A **runtime error** is an error that does not keep a program from compiling, but that generates an error while the program is running. Runtime errors are often the result of a user error, such as entering an incorrect value or clicking a button at an inappropriate time. Runtime errors may also result from incorrect program commands that attempt to assign a value to a variable that is the wrong data type or that is too large for the variable's maximum data size.

To find and correct runtime errors, you must retrieve the error messages associated with errors so you have an idea of the nature of the error. To determine the error cause, you need to identify the program line that is causing the error, and examine the variable values used within the command that has the error. The Forms Debugger allows you to step through triggers and other PL/SQL programs one line at a time to examine variable values during program execution.

Now you open a form named Registration_ERR.fmb from the Chapter06 folder on your Data Disk. This file contains the Northwoods University Registration form, except that the code in this form's Register button trigger generates a runtime error. Before working with the Registration_ERR.fmb file, you refresh your Northwoods University database tables by running the Ch6Northwoods.sql script in SQL*Plus. Then, you open the form, save the form using a different name, run it the usual way, and observe the error message.

To run the script and then open, save, and run the form:

1. In SQL*Plus, run the **Ch6Northwoods.sql** script.

2. In Forms Builder, open **Registration_ERR.fmb** from the Chapter06 folder on your Data Disk, and save the file as **Ch6ARegistration_CORR.fmb** in the Chapter06\Tutorials folder on your Solution Disk.

3. Run the form. (A Compile dialog box showing the percentage that the compilation process is complete might appear. This dialog box always appears the first time you compile a new form that has multiple triggers.)

4. Open the LOV display, select the record associated with Student ID **MA100** and Course Number **MIS 441**, and then click **OK**.

5. Click **Register**. The error message "FRM-40735: WHEN-BUTTON-PRESSED trigger raised unhandled exception ORA-01722" appears on the message line.

6. Close the browser window.

6

The following sections describe how to retrieve FRM- error messages to identify the error cause, use the Forms Debugger to identify the command that is causing the error, and examine the variable values used within the command that has the error to determine the error cause.

Retrieving FRM- Error Messages

The first step toward finding and fixing a runtime error such as this one is to investigate the nature of the error by looking up the error code explanation. Recall that error codes with the FRM- prefix are Forms Builder error codes, and error codes with the ORA- prefix are generated by the DBMS. First, you connect to the Oracle Technology Network (OTN) Web site and look up the description of the FRM- error code.

To find the description of the FRM- error message:

1. Click **Help** on the Forms Builder menu bar, and then click **Online Help**.

2. Click the **Search** tab, type **FRM-40735** in the Search field, and then click the **Search** icon. The search presents a list of documents that reference this error.

3. Select the link associated with the document titled "FRM-40735." Click the **Open** button to move to the Help Topic Window for FRM-40735.

The error message for FRM-40735 appears as follows: "FRM-40735: <trigger name> trigger raised unhandled exception <exception name>. Cause: Application Design error. The current trigger raised an exception (other than FORM_TRIGGER_FAILURE), but it did not handle the exception." This error message indicates that the generated error needed to be handled in the EXCEPTION section of the trigger. However, you still need to learn what caused the error so you can write the code to handle it. Next, you look up the description of the ORA-01722 error code that appeared on the form.

1. Close the Help Topic Window before proceeding.

To look up the ORA- error description:

1. Start your Web browser, and type **http://otn.oracle.com** in the Address field. The Oracle Technology Network Web site appears.

2. Select the **Documentation** link under the **Services** header in the left-hand column. Click the **Oracle Database 10g** link. Click the **View Library** link under the **Oracle Database 10g Documentation Library** heading. Click the **Search** tab at the top of the page. Type ORA-0172, not ORA-01722, in the text box of the **Error Search** portal. Click the **Search for this error message** button. You see a list of current error codes and messages.

NOTE

At the time of this writing, the Oracle 10g error documentation does not contain the error code ORA-01722. Previous editions of this error documenation provide the following information.

The explanation for the ORA-01722 error message is:

ORA-01722 invalid number

Cause: The attempted conversion of a character string to a number failed because the character string was not a valid numeric literal. Only numeric fields or character fields containing numeric data may be used in arithmetic functions or expressions. Only numeric fields may be added to or subtracted from dates.

Action: Check the character strings in the function or expression. Check that they contain only numbers, a sign, a decimal point, and the character "E" or "e" and retry the operation.

Now you have an idea that you are looking for an error that involves a character data type that is used incorrectly, but you don't know which command in the Register button trigger is causing the error. The following section describes how to run the form in the Forms Debugger to locate the command that generates the error.

Using the Forms Debugger

To run a form in the Forms Debugger, you click the Run Form Debug button on the Forms Builder toolbar. The form appears in the browser window as usual, and you can enter data values and click form command buttons. To debug the form, you must set a breakpoint, which pauses execution on a specific program command. You can then examine the current values of all program variables, step through the program commands to observe the execution path, and examine variable values to see how the values change. During a debugging session, you multitask between the browser window, in which you enter inputs and click buttons that execute triggers, and the **Forms Debug console**, which allows you to control form execution and examine variable values.

Setting a Breakpoint

A **breakpoint** pauses execution on a specific command. You can set a breakpoint before you run the form in the Forms Debugger or while the form is running in the Forms Debugger. You can set breakpoints only on program lines that contain executable program commands, and on SQL queries. You cannot set breakpoints on comment lines or variable declarations. You usually set a breakpoint on the first command of the trigger that is generating the error.

To set a breakpoint, you open the trigger or program unit in the PL/SQL Editor, and then double-click the mouse pointer in the gray shaded area on the left side of the PL/SQL Editor window. After you set a breakpoint, the Breakpoint icon appears, as shown in Figure 6-17, to mark the program commands that have breakpoints.

After you set a breakpoint, the breakpoint exists until you explicitly remove the breakpoint, or close the form. To remove a breakpoint, you double-click the Breakpoint icon.

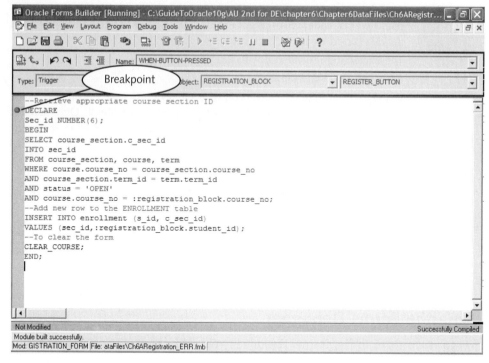

Figure 6-17 Setting a breakpoint

In the Ch6ARegistration_CORR.fmb form, the error occurs when you click the Register button, so you set a breakpoint on the first line of the REGISTER_BUTTON trigger. Then you run the form in the Forms Debugger.

To create a breakpoint, then run the form in the Forms Debugger:

1. In the Object Navigator, open the Data Blocks node, open the REGISTRATION _BLOCK node, open the Items node, open the REGISTER_BUTTON node, open the Triggers node, and then double-click the **Trigger** icon beside WHEN-BUTTON-PRESSED. The trigger opens in the PL/SQL Editor.

2. Move the mouse pointer onto the gray area on the left edge of the PL/SQL Editor window. Double-click on the left side of the first command, which is **DECLARE**. The Breakpoint icon ● appears, as shown in Figure 6-17.

TIP

Another way to set a breakpoint is to place the insertion point on the command, click Debug on the menu bar, then click Insert/Remove Breakpoint. Or, you can place the insertion point on the command, and press F5. If the command does not yet have a breakpoint, Forms Builder creates a breakpoint. If the command already has a breakpoint, Forms Builder removes the breakpoint.

3. Click the **Run Form Debug** button 🔲 to run the form in the Forms Debugger.

TIP

You can also run a form in the Forms Debugger by clicking Debug on the menu bar, and then clicking Debug Module.

Note that when you click , the form opens in the browser window as usual. You do not start debugging until you click the Register button, which is the button whose trigger contains the breakpoint. Before you start debugging, you learn how to use the Forms Debug Console.

Using the Forms Debug Console

Recall that to debug a form, you multitask between the browser window, which displays the form in the Forms Services window, and the Forms Debug console, which allows you to control form execution and examine form values. Figure 6-18 shows the Forms Debug Console during a debugging session. (This window is not visible until you click the Register button in the browser window.)

6

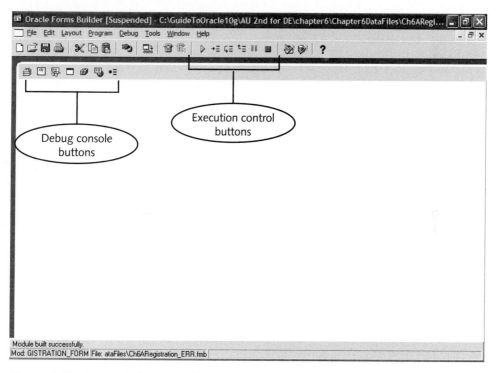

Figure 6-18   Forms Debug Console

In Figure 6-18, the Forms Builder window title bar displays the text "Oracle Forms Builder [Suspended]," which indicates that the form is running in the Forms Debugger, and execution is currently paused on a breakpoint. When form execution pauses on a breakpoint, the execution control buttons on the Forms Builder toolbar are enabled. The Go button ▶ resumes execution after pausing on a breakpoint. The Step Into button ⊕ allows you to step through the program one line at a time. The Step Over button ⊏⊒ allows you to bypass a call to a program unit. The Step Out button ⊏⊒ executes all program lines to the end of the current trigger. The Pause button ⏸ temporarily pauses execution, and the Stop button stops execution, and exits the Forms Debugger.

The Forms Debug Console has windows that allow you to examine program variable values and retrieve other debugging information. Table 6-2 summarizes the Debug Console windows, and shows each window's associated toolbar button.

Window Name	Window Description	Toolbar Button
Stack	Displays stack frames; each time you call a subprogram, such as a program unit, from a trigger, Forms Builder creates a **stack frame**, which stores information about the subprogram	
Variables	Lists the current values of all **stack variables**, which are variables that you declare in the declaration sections of individual triggers and procedures	
Watch	Allows you to create a **watch**, which monitors the value of a specific variable or form item value during program execution	
Form Values	Lists the current values of all form objects, such as text items and radio groups	
PL/SQL Packages	Lists all PL/SQL packages that are in use in the form (a PL/SQL package is a collection of related programs; you will learn how to create and use packages in Chapter 9)	
Global/System Variables	Lists the values of all global and system variables; global variables are variables that are visible to all forms that are running, and system variables are values that are available to all system users	
Breakpoints	Lists all current breakpoints	

Table 6-2   Debug console windows

You can open the Debug Console windows by clicking the associated button on the Debug Console toolbar, or by clicking Debug on the menu bar, pointing to Debug Windows, and then selecting the window so a check mark appears beside the window name. To close a window, you click Debug on the menu bar, point to Debug Windows, and then click the window to clear the check mark beside the window name.

You can display the Debug Console windows either docked inside the Debug Console, or undocked and floating outside the Debug Console. When a window is docked, it appears inside the Debug Console, and an upward-pointing arrow 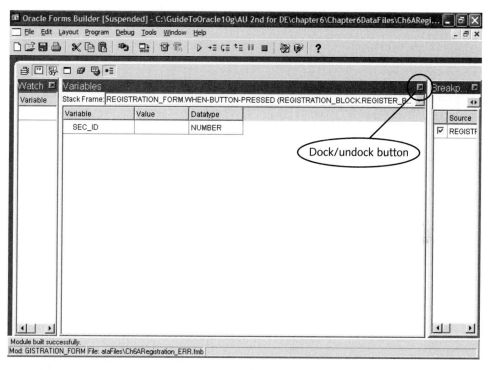 appears in the top-right corner of the window. Figure 6-19 shows the Debug Console, with the Watch, Form Values, and Breakpoints windows docked within the console.

Figure 6-19   Debug Console with docked windows

To undock a window, you click the Undock button in the window's top-right corner. This removes the window from the Debug Console, and displays it as a separate window within Forms Builder.

Now you select student and course records, and click the Register button to begin the debugging session. Then you switch to Forms Builder, and open the PL/SQL Editor to examine the debugging environment.

To insert an ENROLLMENT record and examine the PL/SQL Editor in the debugging environment:

1. In the browser window, make sure that the insertion point is in the Student ID text item, open the LOV display, select Student ID **MA100** and click **OK**.

2. Move the insertion point to the Course Number text item, open the LOV display and select Course Number **MIS 441**, and then click **OK**. Then click **Register**.

3. If you are not already in Oracle Forms Builder, click the **Oracle Forms Builder** button on the taskbar. Click **Window** on the menu bar, and then click **PL/SQL Editor**. The PL/SQL Editor appears, as shown in Figure 6-20.

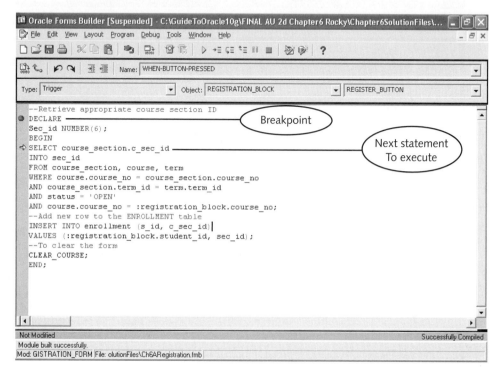

Figure 6-20   PL/SQL Editor window with execution paused on a breakpoint

Notice that the execution control buttons are enabled on the Forms Builder toolbar. Also notice the position of the execution arrow on Line 5. The **execution arrow** shows the program line that the Forms Debugger is going to execute next. As you step through the program, the execution arrow stops on SQL and PL/SQL statements, and skips comment lines. Now you open the Debug Console, and open the Form Values window to view the values of the form text items.

To open the Debug Console and view the values of the form text items:

1. Click **Window** on the menu bar, and then click **Debug Console**. The Debug Console opens, and automatically displays the Stack window when you execute a trigger.

If the Debug Console does not appear on your Window list, click Debug on the menu bar, and then click Debug Console.

2. To close the Stack window, click **Debug** on the menu bar, point to **Debug Windows**, and then click **Stack**. Your Debug Console should now appear as shown in Figure 6-18, and no windows should be open.

If other windows are open in your Debug Console, click Debug on the menu bar, and point to Debug Windows. If one or more windows appear checked, click the checked windows to close the windows.

6

3. To open the Form Values window, click **Debug** on the menu bar, point to **Debug Windows**, and then click **Form Values**. The Form Values window appears docked in the Debug Console.

If the Form Values window is floating in the Forms Builder window, click the downward-pointing arrow in the top-right corner of the window to dock the window.

4. The Form Values window shows the values of all form data block items. Currently, the Form Values window shows the REGISTRATION_BLOCK node. Open the REGISTRATION_BLOCK node to view the individual block item values. The form text item values appear as shown in Figure 6-21. (You may need to adjust the widths of the Items and Value columns to view the values.)

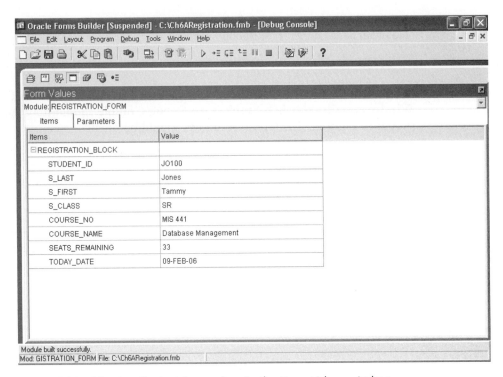

Figure 6-21   Viewing the text item values in the Form Values window

The Form Values window shows the current values of the form text items. When the value of a form text item changes based on user input or a program command, the new value appears in red text. At that point, you switch to the PL/SQL Editor, and step through the trigger commands. You identify the program line that is causing the ORA-01722 error and determine the cause of the error by examining the values of the form variables.

To step through the trigger commands:

1. Click **Window** on the menu bar, and then click **PL/SQL Editor** to switch to the PL/SQL Editor window.

2. Click the **Step Into** button. The execution arrow appears on the INSERT command, indicating the query was successfully executed.

3. Click again. This time the execution arrow does not appear on the next command. This indicates that the error occurred on the program line that was just executed, which was the line containing the INSERT command.

4. Switch to the browser window. The error message appears on the message line, which confirms that the INSERT command is the command that caused the error.

When you step through a program using the Forms Debugger, you can identify the program line that contains the runtime error because when the program line with the error executes, execution immediately halts, and an error message appears in the browser window. Before you execute the program line that causes the error, you should use the Debug Console to examine the variable values used in the program line. Then you can identify the cause of the error.

At this point, you can reconstruct the INSERT statement that the form submits to the database by substituting all of the text item values that are used in the command. The form's INSERT statement has the following syntax:

```
INSERT INTO enrollment (s_id, c_sec_id)
VALUES (sec_id,:registration_block.student_id);
```

6

One approach to resolving the error is to substitute values from the Form Values window into the INSERT statement; however, the **c_sec_id** is not included in the form. Because the error message previously indicated by the application server indicated that the problem stemmed from attempting to convert a character string to a numeric value, you can double-check that the value being assigned to each column has the correct data type. In this particular example, the **sec_id** variable is assigned a NUMBER data type, and is being assigned to the **S_ID** column, which has a VARCHAR2 data type assigned in the ENROLLMENT table. However, this is not the source of the problem, because the ORA-01722 error message indicates the attempted conversion of a character string to a number. Notice that the INSERT command attempts to assign the value of **:registration_block.student_id**, which has a CHAR data type in the Registration Form to a NUMBER column (C_SEC_ID) in the ENROLLMENT form. Apparently, the arguments in the VALUES clause are reversed, so the statement attempts to assign the data to the wrong columns.

To correct the error, remove the breakpoint, and run the form:

1. Switch to the Forms Builder, and click the **Stop** button ■ on the Forms Builder toolbar to stop execution.

2. Switch to your browser, and close the browser window.

3. Switch back to Forms Builder, which displays the PL/SQL Editor for the Register button. Change the INSERT statement in the trigger so it appears as follows:

```
INSERT INTO enrollment (s_id, c_sec_id)
VALUES (:registration_block.student_id, sec_id);
```

4. Double-click the **Breakpoint** icon ● beside the DECLARE statement to clear the breakpoint.

5. Compile the trigger, correct any syntax errors, and then close the PL/SQL Editor.

6. Save the form, click the **Run Form** button 🗇 to run the form, open the LOV display, select the third row in the LOV display, and click **OK**.

7. Move the insertion point to the Course Number text item, select the last row in the LOV display, and click **OK**.

8. Accept the retrieved values, click **Register**, and then click **Save**. The "No changes to save" confirmation message confirms that the trigger worked correctly.

If an error message appears instead of the confirmation message, repeat the debugging steps to locate the error and determine its cause. Fix your code, and then test it again.

9. Close the browser window, and then close the form in Forms Builder.

In summary, use the following process for debugging runtime errors in form triggers:

1. Look up explanations for FRM- and ORA- error messages to determine the nature of the error.

2. Run the form in the Forms Debugger, set a breakpoint on the first line of the trigger that generates the error, and step through the trigger to identify the program line causing the error. Execution immediately halts when the line causing the error executes.

3. Run the form in the Forms Debugger again. Before the line that causes the error executes, examine the variable values used in the command to determine the cause of the error.

FORM TRIGGER PROPERTIES

Recall that a form trigger is a block of PL/SQL code attached to a form object, such as a form data block, a block item, or the form itself. The trigger is activated, or fires, in response to an event, such as clicking a button. The trigger name defines the event that activates it. For example, a button's WHEN-BUTTON-PRESSED trigger activates the trigger's code.

Form triggers are a very important part of forms, and as you create more complex forms, it becomes important to understand the underlying properties of form triggers. This section explores trigger categories, timing, scope, and execution hierarchy in detail.

Trigger Categories

Table 6-3 summarizes the different categories of form triggers.

Trigger Category	Fires in Response to	Example Trigger Event
Block processing	Events in form data blocks, such as inserting new records or deleting existing records	ON-DELETE
Interface event	User actions, such as clicking a button	WHEN-BUTTON-PRESSED
Master-detail	Master-detail relationship processing, such as updating a detail block when the user selects a new master record	ON-POPULATE-DETAILS
Message handling	Events that display messages, such as inserting a record or reporting an error	ON-ERROR
Navigational	Actions that change the form focus or the location of the insertion point, such as moving to a new block or new form record, or internal form actions, such as inserting a new blank record	WHEN-NEW-RECORD-INSTANCE
Query time	Data block form query processing involving retrieving records	POST-QUERY
Transactional	Database transaction events, such as inserting, updating, or deleting records	PRE-INSERT
Validation	Database transaction events that require form data validation, such as checking to make sure required primary key values have been entered	WHEN-VALIDATE-ITEM

Table 6-3 Form trigger categories

Most of the triggers you create in this book are interface triggers. These triggers generally have the following format for the trigger name: WHEN-object-action. Interface triggers can fire in response to user actions. For example, the WHEN-BUTTON-PRESSED trigger fires when the user clicks the associated button. Interface triggers can also involve mouse events, such as WHEN-MOUSE-CLICK.

Trigger Timing

Trigger timing specifies when a trigger fires—just before, during, or after its triggering event. PRE- triggers, which are triggers with PRE as the first three characters of their name, fire just before an event successfully completes. For example, a PRE-FORM trigger fires just before a form appears in the Forms Services window, and a PRE-BLOCK trigger fires just before the user successfully navigates to a new form block. POST- triggers fire just after an event successfully completes. For example, a POST-QUERY trigger fires just after a query in a data block form retrieves records. In contrast, the ON-, WHEN-, and KEY- triggers fire in response to actions. For example, the ON-DELETE trigger fires in response to deleting a record in a data block form, and the WHEN-BUTTON-PRESSED trigger fires in response to a user clicking a button.

Trigger Scope

Trigger scope defines where an event must occur in order for the associated trigger to fire. The scope of a trigger includes the object to which the trigger is attached, as well as any objects within the trigger object. For example, when a trigger is attached to a block, the trigger's scope extends to all items within the block. If the triggering event occurs within any block item, the block trigger fires. For example, suppose you create a KEY-F1 trigger associated with a data block. This trigger fires whenever the user presses the F1 key, as long as the form insertion point is in any item in the block. Similarly, if you create a KEY-F1 trigger associated with a form, the trigger fires whenever the user presses the F1 key anywhere in the form.

Trigger Execution Hierarchy

Trigger execution hierarchy defines which trigger fires when an object within a form object contains the same trigger that the form object contains. For example, suppose a data block has a KEY-F1 trigger, and a text item within the data block also has a KEY-F1 trigger. By default, the trigger in the higher-level object, which is the block, overrides the trigger in the lower-level object, which is the text item. If you anticipate a conflict and want both triggers to execute, you can specify a custom execution hierarchy in the Property Palettes of both triggers and have one trigger execute immediately before or after the other.

DIRECTING FORM EXTERNAL NAVIGATION

In a form, two types of navigation, or movement between form objects, occur: external navigation and internal navigation. **External navigation** occurs when the user causes the form focus, which specifies the location of the insertion point, to change by making a different form item active. A form item has the **form focus** when it is the item that is currently selected on the form. For example, a text item has the form focus when the insertion point is in the text item. A command button has the form focus when the user presses Tab and causes the button to be selected. External navigation occurs when, for example, the user presses Tab, and moves the insertion point to a different text item.

TIP

When a form item, such as a command button, has the form focus, it is highlighted. If the user presses the Enter key when a button has the form focus, it is equivalent to clicking the button.

Internal navigation occurs as a result of internal form code that responds to external navigation operations or trigger commands. For example, when a user opens a data block form, a series of triggers fire that cause the form to open and a new blank record to appear.

In a custom form, you may need to adjust and control external navigation. For example, you may want the insertion point to appear in a specific text item when the form opens, and you may want to allow the user to be able to press Tab to navigate among the form text fields and other form items in a top-to-bottom, left-to-right order. The following sections describe how to set the tab order of form items, and how to use built-in programs to control external navigation.

Setting the Form Tab Order

To set the tab order of items in a custom form, you place the items in the correct order under the Items node in the Object Navigator window. The text item in which the insertion point is to appear at form start-up should be listed first, the item where the insertion point goes when the user subsequently presses Tab should be listed next, and so on. To move an object in the Object Navigator, you select the item, and place it in the desired location. To move an item so it appears directly under an existing object, you place the item on the existing object.

Now you check the tab order of the items in a form named Registration_NAV.fmb. You open the form, save it using a different filename, and then run the form and navigate through the text items.

To open and run the form:

1. Open **Registration_NAV.fmb** from the Chapter06 folder on your Data Disk, and save the file as **Ch6ARegistration_NAV.fmb** in the Chapter06\ Tutorials folder on your Solution Disk.

2. Run the form. Note the insertion point is currently in the Student ID text item.

3. Press **Tab**. The insertion point moves to the Last Name text item.

4. Press **Tab** again. The insertion point moves to the First Name text item.

5. Continue to press **Tab**, and note how the insertion point moves through the form items.

6. Close the browser window.

As you can see, currently when the user presses the Tab key, the insertion point moves down the column of text items on the left side of the form and then across the bottom of the form. After the user has moved through all of the text items, the form focus moves to the first command button (Register) and then moves to the Cancel button. Now, you adjust the tab order of the block items by moving the items into the correct order in the Object Navigator.

To adjust the tab order of the form items:

1. In the Object Navigator, open the Data Blocks node, open the REGISTRATION _BLOCK node, and then open the Items node to display the block items.

2. Select the COURSE_NO node, drag it so the mouse pointer is immediately below the STUDENT_ID node, and drop it so it is the second item listed under the Items node in REGISTRATION_BLOCK.

3. Select the REGISTER_BUTTON node, drag and drop it beneath the COURSE_NO node so that REGISTER_BUTTON appears as the third item under the Items node.

4. Select the CANCEL_BUTTON node, drag and drop it beneath the REGISTER_BUTTON node so that CANCEL_BUTTON appears as the fourth item under the Items node.

5. Save the form, and then run the form to check the tab order. The insertion point should move from the Student ID text item to the Course Number text item when you first press the Tab key. When you press the Tab key a second time, the Register button is selected. Pressing the Tab key a third time results in selecting the Cancel button.

6. Close the browser window.

Directing External Navigation Using Built-in Subprograms

Forms Builder provides several built-in subprograms called **built-ins** that you can use to direct external form navigation. You can use these built-ins to place the insertion point or form focus on a specific item, move to a different block or record, and so forth. Table 6-4 summarizes the built-in subprograms to direct external navigation.

Built-in Name	Description	Example
GO_ITEM	Moves focus to a specific form item	`GO_ITEM('receiving_block.shipment_id_text');`
GO_BLOCK	Moves focus to a specific block	`GO_BLOCK('receiving_block');`
GO_FORM	In a multiple-form application, moves focus to a specific form	`GO_FORM('receiving_form');`

Table 6-4   Built-in subprograms to control external navigation

Note that in the GO_ITEM, GO_BLOCK, and GO_FORM built-ins, you place the name of the target item, block, or form module within single quotation marks. Forms Builder does not verify this target item when it compiles the form, but verifies the item name at runtime. If the item does not exist or is not specified correctly, a runtime error occurs. Also note that in the GO_ITEM and GO_BLOCK built-ins, you do not preface the block name with a colon (:). You cannot use navigational built-in subprograms in navigational triggers, because navigational triggers fire in response to navigation events, and the navigational built-ins cause navigation events to occur.

Now you use the GO_ITEM built-in to place the form insertion point automatically in the COURSE_NO item after the user clicks the Cancel button.

To use the GO_ITEM built-in in the Cancel button trigger:

1. In the Object Navigator, open the CANCEL_BUTTON node, open the Triggers node, and then double-click the **Trigger** icon 🐟 beside WHEN-BUTTON-PRESSED to open the Cancel button trigger code. Add the following command as the last line of the trigger:

```
GO_ITEM('registration_block.course_no');
```

2. Compile the trigger, close the PL/SQL Editor, and save the form.

3. Run the form, open the LOV display, and select the first record. The values appear on the form.

4. Press **Tab** to move the insertion point. The insertion point moves to the COURSE_NO item. Open the LOV display, and select the first record. The values appear on the form.

5. Press **Tab** to move the insertion point. Click **Cancel**. Note that as a result of the GO_ITEM navigational built-in, the insertion point now appears in the Course Number text item.

6. Close the browser window.

7. Shut down the OC4J Instance, and close SQL*Plus.

SUMMARY

- A custom form displays the data fields from a variety of database tables, and processes data using triggers that contain SQL commands to retrieve, insert, update, and delete data values.

- To create a custom form, you manually create a canvas and a control data block, "paint" the form items on the canvas, and write the code that controls the form functions. A control data block contains form items, and is not associated with a specific database table.

- When you create a text item to display a database field value, the text item must have the same data type as the corresponding data field, and its maximum length must be large enough to accommodate the maximum length of the data, plus any characters included in a format mask.

- You create a command button on a custom form by drawing the button on the canvas, and then writing the button trigger. You associate the trigger with the button's WHEN-BUTTON-PRESSED event.

- To clear the text items in a form, you can use the CLEAR_FORM built-in procedure, which clears all the form text items; or you can clear selected text items, while retaining the current values in others, by setting the value of the text items to a blank text string.

❑ Runtime errors are errors that occur while a form is running. Runtime errors are usually the result of incorrect user inputs or actions, or program commands that assign illegal values to variables.

❑ To find and correct a runtime error, you must identify the nature of the error by looking up the FRM- or ORA- error codes. Then you must identify the program command that is causing the error.

❑ The Forms Debugger allows you to create breakpoints that pause execution. Then, you can step through program commands one line at a time to identify commands that cause runtime errors and examine variable values during execution.

❑ You can create form triggers to support block processing, interface events, master-detail processing, message handling, navigational events, query processing, transaction processing, and validation.

❑ Trigger scope defines where an event must occur for a trigger to fire. The scope of a trigger includes the object to which the trigger is attached, as well as any objects within the trigger object.

❑ External navigation occurs when the user causes the form focus, or item that is currently selected on the form, to change. Internal navigation is performed by internal form code within the form in response to external navigation operations.

❑ Navigational triggers fire when internal navigation occurs. Navigational triggers are associated with forms, data blocks, and form items, and they fire at different times, depending on internal and external navigation events.

❑ To set the tab order of form items, you place the items in the correct order under the Items node in the Object Navigator window. Forms Builder provides several built-ins to direct external form navigation.

Review Questions

1. What is the purpose of the GO_ITEM subprogram?

2. When you create a custom form, the control data block must be associated with at least one database table. True or False?

3. In a custom form, how do you specify which value returned by an LOV display is displayed in which text item box?

4. A(n) _____ is used to specify the action that occurs when a command button is pressed.

5. When clearing the value displayed in a text item box, what operator assigns the blank space, or empty string, to the text item?

6. To insert the system date from the database server into a form, use
_____ as the system variable.

7. When including several commands in a trigger, make certain only the last statement is followed by a semicolon. True or False?

8. What is the purpose of a breakpoint?

9. A syntax error prevents a program from compiling although a runtime error does not prevent a program from compiling. True or False?

10. The _____ window of the Debug Console allows you to view the current values of all stack variables.

6

MULTIPLE CHOICE

1. Which type of navigation is a response to a trigger command?

   a. external navigation

   b. focus navigation

   c. internal navigation

   d. hierarchy navigation

2. Which of the following system variables returns the system date and time of the database server?

   a. $$DBDATETIME$$

   b. $$DBDATE+TIME$$

   c. $$DBDATE&TIME$$

   d. none of the above

3. When you create a control block for a form, you are actually building a _____ manually.

   a. canvas

   b. trigger

   c. data block

   d. palette

4. The associated _____ determines the action taken when a button is clicked.

   a. control block

   b. data item

   c. trigger

   d. system variable

5. A block of code that returns only a single value is a _____.

    a. procedure

    b. function

    c. trigger

    d. control block

6. The _____ indicates which statement is going to be executed next by the Form Debugger.

    a. breakpoint indicator

    b. Execution window

    c. Statement window

    d. execution arrow

7. What is the correct syntax for referencing form text item values?

    a. block_name:item_name

    b. block_name.item_name

    c. .block_name:item_name

    d. none of the above

8. A breakpoint is set in the _____.

    a. Debug Editor

    b. PL/SQL Editor

    c. Forms Editor

    d. Layout Editor

9. Which built-in procedure clears all form text items?

    a. GO_ITEM

    b. BLANK_STRING

    c. SET_NULL

    d. CLEAR_FORM

10. The tab order of form items is specified by _____.

    a. the order in which you paint the items on the form canvas

    b. the order the items appear in the data block Items list

    c. the Tab Order property values of the items

    d. the order the items appear on the form canvas

PROBLEM-SOLVING CASES

The following cases use the Clearwater Traders sample database. When creating forms, rename all form components using descriptive names. Make all form items have a top-to-bottom, left-to-right external navigation order. Save all form files in the Chapter06\Cases folder on your Solution Disk. Save all SQL commands in a file named Ch6AQueries.sql. Before starting any of the cases, you must run the Ch6Clearwater.sql script file stored in the Chapter06 folder of your Data Disk to set up the necessary data.

1. Create the custom form shown in Figure 6-22 that displays values for the Shipment ID, Inventory ID, Inventory Description, Size, Color, Quantity, and the Date the shipment is Received. The purpose of the form is to allow the receiving clerk to confirm the shipment quantity, change the quantity if the shipment contains a different quantity than expected, and update the SL_DATE_RECEIVED column of the SHIPMENT_LINE table. The date the shipment is received should be assigned the default value of the operating system's date. Allow the user to select the Shipment ID and retrieve data from the SHIPMENT_LINE table using an LOV display. Save the file as Ch6ACase1.fmb. (The Clearwater Traders logo is stored as the CWlogo.tif file in the Chapter06 folder on your Data Disk.)

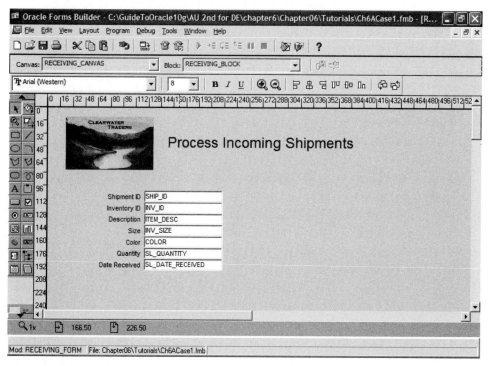

Figure 6-22   Receiving form

2. Modify the form created in Case 1 to include an Update button that updates the SHIPMENT_LINE table with the actual quantity received and date the shipment

is received. In addition, clicking the Update button should also update the Quantity on Hand (QOH) field of the INVENTORY table with the revised quantity of the inventory item. Also include a Cancel button to cancel the current operation and clear all the form text items. Place the Update and Cancel buttons to the right of the form text items. Save the modified form as Ch6ACase2.fmb.

3. Modify the form from Case 2 and change the form tab order so the insertion point is in the Shipment ID text item when the form is opened. If the user presses the Tab key, the insertion point should move to the Quantity text item. When the Tab key is pressed a third time, the insertion point should move to the Date Received text item. Modify the Update trigger so that all text items are cleared after the SHIPMENT_LINE and INVENTORY tables are updated. Save the modified form as Ch6ACase3.fmb.

4. Create the custom form shown in Figure 6-23 to insert, update, and delete records in the Clearwater Traders SHIPMENT table. When the user clicks the Create button, a new Shipment ID value appears. The user enters the expected date value and clicks the Save New button, and the form saves the new record in the database. The Shipment ID field has an associated LOV that lists existing shipment records. When the user selects an existing shipment, the user can modify the data values and then click the Update button to save the changes. Perform the following steps and save the file as Ch6ACase4.fmb.

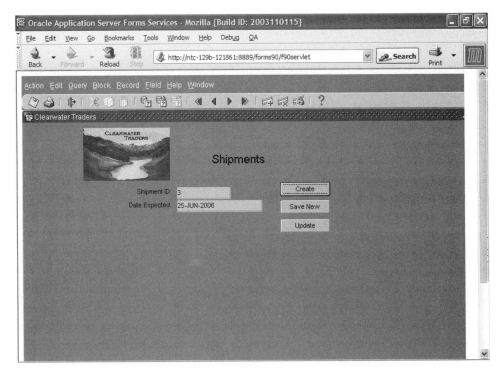

Figure 6-23 Clearwater Traders Shipments form

a. In SQL*Plus, create a new sequence named SHIP_ID_SEQUENCE that starts with 30 and has no maximum value. Use the sequence to generate the Shipment ID values. (If you already created this sequence in a case in a previous chapter, drop the existing sequence and then create a new sequence.) Then, create a trigger for the Create button that clears the form text items, automatically displays the next sequence value in the Shipment ID text item, and automatically displays the system date in the Date Expected text item. (*Hint:* Use the command `SELECT sequence_name INTO :block_name .text_item FROM DUAL;` in the trigger.)

b. Create an LOV associated with the Shipment ID text item that displays the shipment ID and date expected for all shipments. The LOV should return the selected record's values to the form text items.

c. Set the Initial Value property of the Date Expected text item so the current operating system date appears when the form first opens.

d. When the user clicks the Save New button, insert the form text item values into the database, commit the changes, and then clear the form text items. Redisplay the current system date in the Date Expected text item. (*Hint:* Use the SYSDATE pseudocolumn to retrieve the system date.)

e. When the user clicks the Update button, update the SHIPMENT table with the modified date expected value for the selected shipment ID, commit the change, and then clear the form text items. Redisplay the current system date in the Date Expected text item.

5. **Debugging Problem:** A custom form saved in a file named Case5_ERR.fmb in the Chapter06 folder on your Data Disk updates records in the Clearwater Traders SHIPMENT_LINE and INVENTORY tables. Debug the form and correct all of the syntax and runtime errors. Save the corrected form as Ch6ACase5_CORR.fmb in the Chapter06\Cases folder on your Solution Disk.

◀ LESSON B ▶

After completing this lesson, you should be able to:

♦ Suppress default system messages
♦ Create alerts and messages to provide system feedback to users
♦ Create applications that avoid user errors
♦ Trap common runtime errors

CONTROLLING SYSTEM MESSAGES

Forms display default system messages with FRM- and ORA- prefixes on the Forms Services message line. These messages are useful for determining if the DBMS successfully inserts, updates, or deletes a record, as well as for determining the nature of errors that occur while you are running a form. Oracle Corporation classifies messages according to their severity, and whether or not they require user intervention. Table 6-5 summarizes the levels of form system messages.

Message Severity Level	Description	Example
5	Informative message that does not require user intervention	FRM-40400: Transaction complete: 1 records applied and saved.
10	Informative message that identifies a procedural mistake made by the user	FRM-40201: Field is full. Can't insert character.
15	Informative message that identifies a data-entry error, such as entering an incorrect data value in a text item	FRM-50016: Legal characters are 0-9 - + E.
20	Error message that identifies a condition that keeps a form trigger from working correctly	FRM-40602: Cannot insert into or update data in a view.
25	Error message that identifies a condition that causes the form to operate incorrectly	FRM-40919: Internal SQL statement execution error: %d.
>25	Error message that identifies a condition that must be corrected immediately for the form to continue running	FRM-40024: Out of memory.

Table 6-5   System message severity levels

Level 5 error messages provide information about what is happening and usually do not require user action. Error messages that appear as a result of user errors have message severity levels of 10, 15, or 20 and require user intervention to correct the condition. Error messages caused by errors within a form trigger have levels of 25 or greater and require intervention by the form developer or DBA.

Oracle Corporation assigns the error severity levels, and the scale simply compares relative severity, with 5 being low severity, 15 somewhat higher, 20 somewhat higher still, and >25 the highest severity. Message severity levels do not correspond to error code numbers—Oracle Corporation assigns the error code numbers sequentially as it documents new errors.

Sometimes form developers suppress the default system messages and replace them with custom messages. Forms Builder determines which messages to display on the message line using the :SYSTEM.MESSAGE_LEVEL variable. This is a system variable that stores a value corresponding to one of the message severity levels (0, 5, 10, 15, 20, or 25). While a form is running, Forms Services suppresses all messages with a severity level that is lower (less severe) than the current :SYSTEM.MESSAGE_LEVEL value. The default :SYSTEM.MESSAGE_LEVEL value is 0, so by default, all system messages appear. If you assign the value 20 to :SYSTEM.MESSAGE_LEVEL, only messages with severity level 20 or higher appear on the message line.

To assign a value to :SYSTEM.MESSAGE_LEVEL, you usually create a PRE-FORM trigger, which executes before the form opens. This is a form-level trigger, so you create it by selecting the Triggers node under the form module and then clicking the Create button on the toolbar. Or, you can right-click the Triggers node under the form module, point to SmartTriggers, and then select PRE-FORM.

The PRE-FORM trigger to suppress system messages contains the following command:

```
:SYSTEM.MESSAGE_LEVEL := message_level;
```

In this command, *message_level* is the number that represents the severity level of system messages that you want to suppress. For example, you use the following command to suppress all messages except those of level 25 or greater:

```
:SYSTEM.MESSAGE_LEVEL := 25;
```

To gain practice with suppressing system messages, you use the Faculty Information Form shown in Figure 6-24.

6

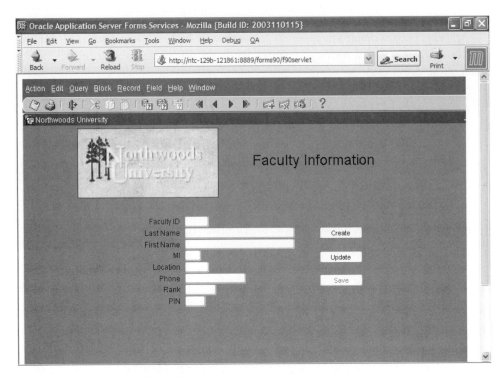

Figure 6-24 Faculty Information form

The Faculty Information form is a custom form that allows new faculty members to be added to the FACULTY table or modify existing data in the table. When entering a new faculty member, the user can click the Create button 🕂 to generate the new faculty ID, which will appear in the Faculty ID text item box. After the remaining values have been added to the form, the user can click the Save button to add the record to the FACULTY table. To update existing faculty records, an LOV display can retrieve a record. Then the user clicks the Update button to save the modified record.

Now you refresh your database tables, and then open the Faculty Information form shown in Figure 6-24, which is stored in a file named Ch6BFaculty.fmb in the Chapter06 folder on your Data Disk. You create a PRE-FORM trigger to set the :SYSTEM.MESSAGE_LEVEL value to 25, so that only the most severe messages (severity level greater than or equal to 25) appear.

To refresh your database tables, open the form, and create the PRE-FORM trigger:

1. Start SQL*Plus, log onto the database, and run the **Ch6Northwoods.sql** script in the Chapter06 folder on your Data Disk.

2. In SQL*Plus, type the following command to create a sequence to generate project ID values. (If you already created this sequence in a previous chapter

or case, drop the existing sequence and create it again. Do not close SQL*Plus, because you use it again later in this lesson.)

CREATE SEQUENCE f_id_sequence START WITH 6;

3. Start the OC4J Instance.

4. Start Forms Builder, click the **Connect** button 🖐 on the toolbar, and then connect to the database.

5. Open **Ch6BFaculty.fmb** from the Chapter06 folder on your Data Disk, and save the file as **Ch6BFaculty_Form.fmb** in the Chapter06\Tutorials folder on your Solution Disk.

6. If necessary, maximize the Object Navigator window. To create the PRE-FORM trigger, select the Triggers node under the FACULTY_FORM node, right-click, point to **SmartTriggers**, and then click **PRE-FORM**. The PL/SQL Editor opens.

7. Type the following command in the PL/SQL Editor:

:SYSTEM.MESSAGE_LEVEL := 25;

8. Compile the trigger, correct any syntax errors, close the PL/SQL Editor, and save the form.

Now you run the form to see the effect of modifying the PRE-FORM trigger. You change the rank for faculty ID 3 (Colin Langley) to Associate and confirm that the PRE-FORM trigger suppresses the Level 5 "FRM-40401: No changes to save" message.

To test the trigger:

1. Run the form, make sure that the insertion point is in the F_ID field, open the LOV display, select the record for F_ID **3**, and then click **OK**. The data for the faculty member appear in the form text items.

2. Place the insertion point in the **Rank** field, type **Associate'**. Click **Update** to update the record. Note that the standard "No changes to save" confirmation message does not appear on the message line.

3. Place the insertion point in the **F_ID** field again, open the LOV display, select Faculty ID **3** again, and then click **OK**. Notice that the updated value (Associate) for the faculty member's rank appears, which confirms that the Update button trigger worked correctly, and the PRE-FORM trigger successfully suppressed the system message.

4. Close the browser window.

The update was successful, and the modified PRE-FORM trigger suppressed the confirmation message. However, the form still needs to provide users with explicit system feedback. You can provide explicit feedback by replacing the system messages with custom messages or by displaying dialog boxes, called alerts, which enable users to respond in different ways.

PROVIDING SYSTEM FEEDBACK

An important application design principle is to provide users with feedback about what is happening in an application. You should caution users when they are about to make a change that could potentially cause harm. And, you should make applications forgiving, and allow users to undo unintended operations. Right now, the Faculty Information form does not provide any confirmation that an update operation was successful, nor does it provide an opportunity for canceling the update, or rolling back the update. To provide this feedback in forms, you create custom messages and alerts.

Custom Messages

A **custom message** is a short (up to 200 characters) text string that the form developer displays on the form message line. Figure 6-25 shows an example of a custom message.

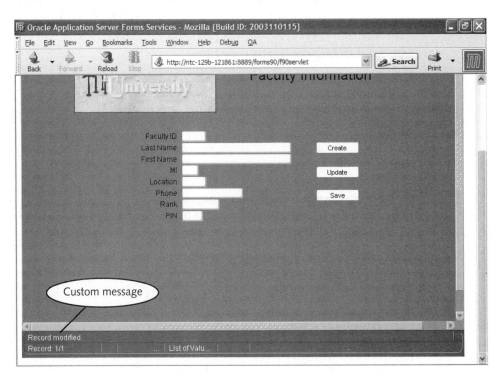

Figure 6-25   Example of a custom message

You should create a custom message when the form needs to provide a short, simple message that does not require an acknowledgement from the user. To create a custom message, you use the MESSAGE built-in subprogram, which has the following general syntax:

```
MESSAGE('message_string');
```

In the MESSAGE built-in, *message_string* is a text string of up to 200 characters that specifies the message text. Now you add some custom messages to the Faculty Information form to provide user feedback when the form inserts a new record or updates an existing record.

To add custom messages to the Faculty Information form:

1. In Forms Builder, save the Faculty Information form from the previous exercise as **Ch6BFaculty_Form_MESSAGE.fmb** in the Chapter6\Tutorials folder on your Solution Disk.

2. Open the Layout Editor, right-click the **Save** button on the form layout, and then click **PL/SQL Editor** to open the button trigger.

3. To display a custom message when the user saves a new record, add the following command as the last command in the trigger:

   MESSAGE('Record added.');

4. Compile the trigger, correct any syntax errors, and then close the PL/SQL Editor.

5. To display a custom message when the user updates an existing record, right-click the **Update** button, click **PL/SQL Editor**, and then add the following command as the last command in the trigger:

   MESSAGE('Record modified.');

6. Compile the trigger, correct any syntax errors, close the PL/SQL Editor, and then save the form.

Now you confirm that the messages appear. You add a new record to the form, and then you update the new record.

To test the messages:

1. Run the form. To create a new record, click **Create**. The next sequence value appears in the Faculty ID field.

2. Type **Moore** in the Last Name field and **Cecil** in the First Name field. Move the insertion point to the Rank field and type **Full**.

3. Click **Save** to save the record. The message "Record added." appears on the message line. (You may need to scroll down in the browser window to view the message.)

4. Make sure that the insertion point is in the Faculty ID field, open the LOV display, select the faculty member you just added (Cecil Moore), and click **OK**.

5. Place the insertion point in the Rank field, and type **ASSOCIATE**.

6. Click **Update** to update the record. The "Record modified." message appears in the message line.

7. Close the browser window, and then close the form in Forms Builder.

Alerts

An **alert** is a dialog box that can display a text message longer than 200 characters, and displays one or more buttons that allow the user to select between alternatives that execute associated program statements. Figure 6-26 shows an example of an alert.

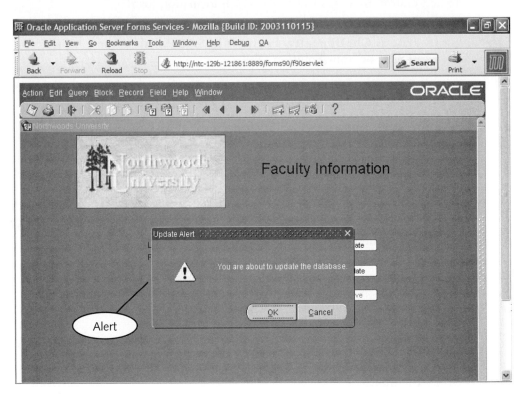

Figure 6-26   Example alert

You should create an alert for these situations: the feedback requires a longer message than will fit on the message line, the user needs to select between alternate ways to proceed, or the user needs to acknowledge important messages. The following sections describe how to create and display an alert.

Creating an Alert

An alert is a top-level form object. To create a new alert, you select the Alerts node in the Object Navigator, and then click the Create button ✚. Then you specify the alert properties. Figure 6-26 shows the properties that define the appearance of an alert.

The Title property determines the title that appears in the alert window title bar. The Message property defines the text that appears in the alert. The Style property value specifies the icon that appears on the alert. Possible alert styles are Note, Caution, and

Stop. Note alerts display an "i" for information. A Caution alert, which you can see in Figure 6-26, displays an exclamation point (!), and Stop alerts display a red "X" or a red stoplight. An alert that has the Note style conveys information to the user, such as confirming that the form has inserted a record. Caution alerts inform the user that he or she is about to make a choice that cannot be undone and could lead to a potentially damaging situation, such as deleting a record. Stop alerts inform the user that he or she has instructed the system to perform an action that is not possible, such as trying to delete a record that is referenced as a foreign key in another table.

The Button Label property determines how many buttons appear on the alert and the labels that appear on the buttons. An alert can have a maximum of three buttons, so there are three Button Label properties: Button 1 Label, Button 2 Label, and Button 3 Label. If you delete the label for a given button, that button no longer appears on the alert.

Now you open the Ch6BFaculty_Form.fmb file on your Solution Disk, and save the file as Ch6BFaculty_Form_ALERT.fmb. Then you create an alert named UPDATE_ALERT that informs users when the FACULTY table is about to be updated and gives them the option of continuing or canceling the operation. You configure the alert properties so the alert appears as shown in Figure 6-26.

To create the alert:

1. Open the **Ch6BFaculty_Form.fmb** form in the Chapter06\Tutorials folder on your Solution Disk, and save the form as **Ch6BFaculty_Form_ALERT.fmb** in the Chapter06\Tutorials folder on your Solution Disk. (If you did not create the **Ch6BFaculty_Form.fmb** form earlier in the lesson, a copy of this file is in the Chapter06 folder on your Data Disk.)

2. Make sure that the Object Navigator window is open in Ownership View, and then select the Alerts node under the FACULTY_FORM node.

3. Click the **Create** button ╬ on the Object Navigator toolbar to create a new alert object.

4. Double-click the **Alert** icon ⚠ beside the new alert to open its Property Palette, and then change its properties as follows:

Property	Value
Name	**UPDATE_ALERT**
Title	**Update Alert**
Message	**You are about to update the database.**
Alert Style	**Caution**
Button 1 Label	**OK**
Button 2 Label	**Cancel**
Button 3 Label	(deleted)

5. Close the Property Palette, and then save the form.

Displaying an Alert

To display an alert in a form, you use the SHOW_ALERT built-in function. In programming, a function always returns a value. The SHOW_ALERT function returns a numeric value that corresponds to the button that the user clicks on the alert. To display an alert during the execution of a trigger, you need to declare a numeric variable and then assign to this variable the value that the SHOW_ALERT function returns, using the following general syntax:

```
DECLARE
     alert_button NUMBER;
BEGIN
     alert_button := SHOW_ALERT('alert_name');
END;
```

In this syntax, the alert_name is the name of the alert, as it appears in the Object Navigator. Note that in the command, the alert_name value is enclosed in single quotation marks. Alert_button is a variable that stores a number representing the alert button that the user clicks. If the user clicks the first button on the alert, the SHOW_ALERT function assigns to alert_button the value stored in a numeric variable named ALERT_BUTTON1. If the user clicks the second button on the alert, the function assigns this value to a numeric variable named ALERT_BUTTON2 to alert_button, and if the user clicks the third button, the function assigns the value of a numeric variable named ALERT_BUTTON3 to alert_button.

To execute alternate program commands depending on the alert button that the user clicks, you create an IF/ELSIF decision control structure. This structure evaluates the value that the SHOW_ALERT function returns, and then executes the appropriate program commands. Figure 6-27 shows the general syntax for a PL/SQL program that first displays an alert, then executes different program commands depending on the alert button that the user clicks.

```
DECLARE
   alert_button NUMBER;
BEGIN
   alert_button := SHOW_ALERT('alert_name');
   IF alert_button = ALERT_BUTTON1 THEN
     commands to execute for first alert button
   ELSIF alert_button = ALERT_BUTTON2 THEN
     commands to execute for second alert button
   ELSIF alert_button = ALERT_BUTTON3 THEN
     commands to execute for third alert button
   END IF;
END;
```

Figure 6-27   Syntax to display an alert and execute alternate commands depending on the button the user clicked

Now you create a new program unit named DISPLAY_ALERT that uses the SHOW_ALERT function to display UPDATE_ALERT, which has two buttons: OK and Cancel. The program unit uses an IF/ELSIF structure to specify the correct action, depending on which alert button the user clicks. If the user clicks the OK button, the program unit commits the transaction and displays a custom message confirming that the record was updated. If the user clicks the Cancel button, the program unit rolls back the transaction and displays a message stating that the transaction was rolled back.

To create the program unit to display the alert:

1. In the Object Navigator window, select the Program Units node, and then click the **Create** button ➕ to create a new program unit.

2. Type **DISPLAY_ALERT** for the new program unit name, make sure that the Procedure option button is selected, and then click **OK**.

3. Modify the procedure so it appears as shown in Figure 6-28. Then compile the code, correct any syntax errors if necessary, close the PL/SQL Editor, and save the form.

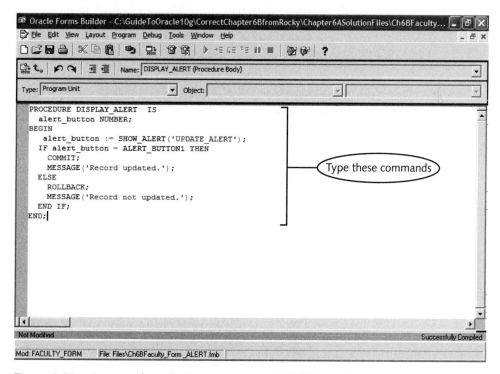

Figure 6-28   Commands to display and process UPDATE_ALERT

Next, you need to modify the trigger for the Update button so that it calls the DISPLAY_ALERT program unit instead of committing the transaction to the database. Then you run the form, select and update a record, and display the alert.

To modify the Update button trigger and then run the form:

1. Open the Layout Editor, right-click the **Update** button, and then click **PL/SQL Editor**.

2. Delete the **COMMIT;** command, and replace it with the following command that calls the DISPLAY_ALERT program unit:

   DISPLAY_ALERT;

3. Compile the trigger, correct any syntax errors, close the PL/SQL Editor, and save the form.

4. Run the form, make sure that the insertion point is in the Faculty ID field, open the LOV display, select the record for Faculty ID **1**, and then click **OK**. The data values for the project appear in the form text items.

5. Place the insertion point in the Phone field, and change the faculty member's phone number to 4075921659, and then click **Update**. UPDATE_ALERT appears, as shown in Figure 6-26.

6. Click **OK**. The confirmation message "Record updated." appears on the message line.

7. Make sure that the insertion point is in the F_ID field, open the LOV display, select the record for F_ID **1** again, and then click **OK**. The data values for the faculty member appear in the form text items, showing the updated phone number.

8. Place the insertion point in the F_ID field, open the LOV display, select the record for F_ID **3** (Colin Langley), click **OK**, and then click **Update**. The alert appears again. This time, you cancel your changes.

9. Click **Cancel**. The "Record not updated." message appears, and the form fields are cleared.

10. Close the browser window, and then close the form in Forms Builder.

AVOIDING USER ERRORS

Forms should help users avoid errors such as entering an incorrect data value, or clicking a button at the wrong time. The following sections describe how to help users avoid these errors by configuring forms that validate input values, programmatically disable form command buttons, and disable navigation for form text items containing values that users should not change.

Validating Form Input Values

Users must enter valid date values in date fields, and valid number values in number fields. To ensure that users enter correct values, form text items **validate** input values to ensure that the values meet specific preset requirements. You can validate form input values using text item validation properties or form validation triggers.

Text Item Validation Properties

A form can validate a text item's value using specific text item validation properties. Table 6-6 summarizes the Forms Builder text item validation properties.

Property Node	Property	Allowable Values	Description
Data	Data Type	Char, Number, Date, Alpha, Integer, Datetime, Long, Rnumber, Jdate, Edate, Time	Ensures that input values are of the correct data type
Data	Maximum Length	Integer value	Defines the maximum number of characters the item will accept
Data	Required	Yes or No	Specifies whether or not the value can be NULL
Data	Lowest Allowed Value	Integer value	For numerical fields, specifies the lowest acceptable value
Data	Highest Allowed Value	Integer value	For numerical fields, specifies the highest acceptable value
Data	Format Mask	Legal format masks	Ensures that user input is in the correct format
List of Values	Validate From List	Yes or No	Specifies that the value entered by the user should be validated against the item's LOV

Table 6-6   Text item validation properties

You can use these properties to validate values that users enter in either data block forms or in custom forms. The Data Type property specifies that the value that the user enters in the item must be of a specific data type. So far, you have used this property in custom forms to match the text item property with the associated database field data type. You can use this property to restrict input values further. For example, you can specify the Alpha data type if the field should allow only nonnumerical characters, the Datetime data type if the value must include both a date and time component, or the Time data type if the value must include only a time component. Other choices specify the data value format: Edate specifies a European date format (DD/MM/YY); Jdate specifies a

Julian calendar date that is stored in a NUMBER data field and is included for compatibility with prior Oracle versions; and Rnumber specifies a right-justified number.

The Validate From List property causes the form to compare the value entered by the user with the first column of the LOV associated with the item. If the LOV does not contain a matching value, the LOV display opens, and the LOV uses the user's input value as a search condition. For example, if the user enters the character M in a field that contains an LOV with state abbreviations, all state abbreviations that begin with M appear in the LOV display. The Validate From List property is useful to validate foreign key values.

Now you open the Students form shown in Figure 6-29. This form allows users to insert and update records in the Northwoods University STUDENT table.

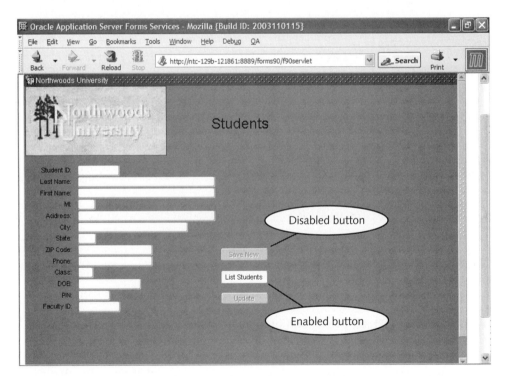

Figure 6-29   Northwoods University Students form

The user enters values in the form fields and clicks Save New to save the record. The user can select the Student ID field value from an LOV that displays values from the STUDENT table. The user can click the List Students button to open an LOV display that displays the names of existing students, update the existing values, and then click the Update button to save the changes.

You modify properties of some of the text items to provide additional validation. You specify that the State and Class fields accept two characters of character (Alpha) data. You also specify that the form validates the Faculty ID field value using the text item's LOV. Now you open the form and modify its validation properties.

To open and modify the form:

1. In Forms Builder, open **Student.fmb** from the Chapter06 folder on your Data Disk, and save the file as **Ch6BStudent.fmb** in the Chapter06\Tutorials folder on your Solution Disk.

2. Open the form in the Layout Editor, select the **State** text item, press and hold down the **Shift** key, and then click the **Class** text item to select the text items as an object group. Right-click, and then click **Property Palette** to open the intersection Property Palette.

3. To specify that the State and Class text items contain a maximum of 2 alphabetical characters, scroll down to the Data property node, and change the Data Type to **Alpha**. Confirm that the Maximum Length property value is 2, and then close the Property Palette.

4. Open the Property Palette for the Faculty ID text item, scroll down to the List of Values (LOV) property node, change the Validate from List property to **Yes**, close the Property Palette, and then save the form.

A form can be set to validate input values each time the user moves the insertion point out of an item, record, or block, or when the user exits a form. The level at which validation occurs is called the **validation unit**, which represents the largest chunk of data that a user can enter before the form validates all input values. By default, the validation unit is at the item level, so each time the user moves the insertion point out of a specific item, the form validates the item's current value, based on the item's validation properties. To change the validation unit, you open the form's Property Palette, and modify the Validation Unit property. This property has possible values of Form, Data Block, Record, or Item. With a data block form, you can adjust the validation unit so that validation occurs when the user moves the insertion point to a different record within a data block, or to a different data block within the form. With a custom form, validation should always occur at the item level.

With the Students form, which is a custom form, validation occurs when the user moves the insertion point to a different item. Now you run the Students form, create a new record, and test the validation properties that you have just set.

To test the validation properties:

1. Run the form. Place the insertion point in the State text item, type **22**, which is not a legal value, and then press **Tab** to move the insertion point out of the item, and validate the item value. The message "FRM-50001: Acceptable characters are a-z, A-Z, and space" appears on the message line, indicating that numerical characters cannot be used in this text item.

2. Delete the current State value, type **WI**, and then press **Tab**. No message appears, indicating the value was validated.

3. Place the insertion point in the Faculty ID field, type **10** (which is not a legal F_ID value), and then press **Tab**. The Faculty LOV display opens. Select ID **1** (Teresa Marx) and then click **OK**. The new value is accepted.

4. Close the browser window.

Form Validation Triggers

You can use text item validation properties to validate that a value has a specific data type or maximum or minimum size. However, you cannot use the text item validation properties to specify validation conditions such as allowing users to enter only numeric values in character fields that contain numeric values such as postal codes or telephone numbers, or to specify that the user can enter only certain characters. You can perform complex validation operations such as these using validation triggers. An **item validation trigger** is an item-level trigger that you associate with the item's WHEN-VALIDATE-ITEM event. The trigger fires when the item is validated, as determined by the form validation unit. The trigger code tests the current item value to determine if it satisfies the validation condition or conditions. If the current item value does not satisfy the validation condition(s), the form displays a message and then raises a built-in exception named FORM_TRIGGER_FAILURE, which automatically aborts the current trigger. Now you create an item validation trigger for the Class text item on the Students form to confirm that the input value is one of the legal class codes of FR, SO, JR, or SR. Then, you test the trigger.

To create and test an item validation trigger:

1. In the Layout Editor, select the **Class** text item, right-click, point to **SmartTriggers**, and then click **WHEN-VALIDATE-ITEM**. The PL/SQL Editor opens.

2. Type the following command in the PL/SQL Editor to create the validation trigger:

```
IF NOT (:student.s_class) IN
('FR', 'SO', 'JR', 'SR') THEN
  MESSAGE('Legal values are FR, SO, JR, SR');
  RAISE FORM_TRIGGER_FAILURE;
END IF;
```

3. Compile the trigger, correct any syntax errors, close the PL/SQL Editor, and save the form.

4. Run the form. Place the insertion point in the Class field, type **AA**, and then press **Tab**. The message "Legal values are FR, SO, JR, SR" appears on the message line, indicating that the item validation trigger fired correctly.

5. Close the browser window.

Disabling Form Command Buttons to Avoid User Errors

To create a new record, the user must enter values into the form text items, and then click the Save New button. If the user clicks the Save New button without entering a student's last name, an incomplete record is saved. Therefore, when the form first opens, the Save New button should be disabled (grayed-out). The Save New button should not be enabled until after the user has made entries into the Last Name text item.

To disable a command button when a form first opens, you open the command button's Property Palette and set the Enabled property to No. To enable or disable a button while a form is running, you use the SET-ITEM-PROPERTY built-in, which has the following general syntax:

```
SET_ITEM_PROPERTY('item_name', property_name,
property_value);
```

In this syntax, *item_name* is the form item name, as it appears in the Object Navigator. Note that the command encloses *item_name* in single quotation marks. *Property_name* is the name of the property to change, and *property_value* is the new property value. For example, you would use the following command to disable UPDATE_BUTTON by setting its Enabled property value to False:

```
SET_ITEM_PROPERTY('UPDATE_BUTTON', ENABLED,
PROPERTY_FALSE);
```

TIP

In the SET_ITEM_PROPERTY built-in, *property_value* is usually not the same value as the value that appears in the item's Property Palette. For a complete list of property values to use in the SET_ITEM_PROPERTY built-in, click Help on the Forms Builder menu bar, click Online Help, select the Index tab, and search for SET_ITEM_PROPERTY.

Now you modify the form so the Save New button is disabled when the form first opens. You also add a command to the Last Name text item trigger to enable the Save New button after an entry is made. Then you run the form to test the button properties.

To change and test the button properties:

1. In the Layout Editor, right-click the **Save New** button, and then click **Property Palette**. To disable the button when the form first opens, change the Enabled property value to **No**, and then close the Property Palette.

2. Right-click the **Last Name** text item, and then click **PL/SQL Editor**. Select the WHEN-VALIDATE-ITEM trigger. Click the **OK** button. The PL/SQL editor appears. To enable the Save New button, enter the following command in the trigger:

```
SET_ITEM_PROPERTY('SAVE_BUTTON', ENABLED, PROPERTY_TRUE);
```

3. Compile the trigger, debug it if necessary, and then close the PL/SQL Editor.

6

4. Save the form, and then run the form. The Save New button appears disabled when the form first opens.

5. Enter the letter **L** in the Last Name text item, and then click the **First Name** text item. The Save New button is now enabled.

6. Close the browser window.

Disabling Text Item Navigation

A primary key of one table can be referenced by the foreign keys in other tables. This can present a problem if a user accidentally changes a record's primary key. The error can cause problems throughout the database if the primary key is referenced by multiple tables. Therefore, you should prevent users from accidentally changing a record's primary key. One strategy for not allowing users to directly update text items that contain primary key values is to make these text items nonnavigable. When a text item is **nonnavigable**, the user cannot press the Tab key to place the insertion point in the text item. To make a text item nonnavigable, you set the item's Keyboard Navigable property to No, so the only way a user can place the insertion point in the item is by clicking the mouse pointer.

When a text item is nonnavigable, the user can still click the mouse pointer in the text item to enter a value. To overcome this problem, you create a trigger that moves the insertion point, or form focus, to another form item whenever the user clicks the text item using the mouse. To move the insertion point to an alternate form item when a user clicks the mouse, you create a trigger associated with the WHEN-MOUSE-UP event. The WHEN-MOUSE-UP event occurs whenever the user clicks the mouse button in an item, and then releases the mouse button. (You could alternately use the WHEN-MOUSE-CLICK event, because users usually click the mouse button to place the form insertion point in a text item. However, you would have to create an additional trigger for the WHEN-MOUSE-DOUBLECLICK event, because some users might double-click the field to place the form insertion point in a text item.)

Now you make the Student ID text item nonnavigable, and create a WHEN-MOUSE-UP trigger to switch the form focus to the Create command button whenever the user clicks the Student ID text item. You use the GO_ITEM built-in subprogram to switch the form focus to the command button. Then you run the form and confirm that you cannot directly type the Student ID text item value.

To make the text item nonnavigable and switch the form focus when the user clicks the mouse in the text item, and then run the form:

1. In the Layout Editor, double-click the **S_ID** text item to open its Property Palette. Scroll down to the Navigation node, select the **Keyboard Navigable** property, open the list, select **No**, and then close the Property Palette.

2. To create the trigger to change the form focus to the List Students button when the user clicks the mouse pointer in the S_ID text item, right-click the **S_ID** text item, and then click **PL/SQL Editor**. Select **WHEN-MOUSE-UP** from the list, and then click **OK**.

3. Type the following command in the Source code pane, compile the trigger, correct any syntax errors, and then close the PL/SQL Editor and save the form.

```
GO_ITEM('STUDENT.LIST_STUDENTS_BUTTON');
```

4. Run the form. Note that because the Student ID text item is now nonnavigable, the insertion point initially appears in the Last Name text item instead of in the Student ID text item.

5. Click the insertion point in the Student ID text item. Note that the form focus switches to the List Students button.

6. Press **Tab**. Note that the insertion point does not appear in the Student ID text item, but moves directly to the Last Name text item.

7. Close the browser window, and then close the form in Forms Builder.

Some errors are common to many forms, such as trying to delete a record that is referenced as a foreign key by other records, duplicating the value of an existing primary key, or trying to insert a new record when fields with NOT NULL constraints have been left blank. It is a good practice to **trap** these errors, which means to intercept the default system error message and replace it with a custom error message. The custom error message gives more detailed information to the user about how to correct the error. To investigate how to trap and handle common errors, you again use the Student form shown in Figure 6-29.

In the following sections, you run the form and generate runtime errors, and then learn how to trap these errors.

Generating Runtime Errors

In Figure 6-30, the Student form has been modified to include a Delete button. Recall that the List Students button is used to open an LOV display for student information. The Update button saves modifications made to a student record, and the Save New button, when available, inserts new records into the table. The new button, the Delete button, allows you to remove rows from the STUDENT table. In this section you deliberately generate errors while updating and deleting records, and view the error messages. First, you try to delete a Student ID that has associated rows in the ENROLLMENT table.

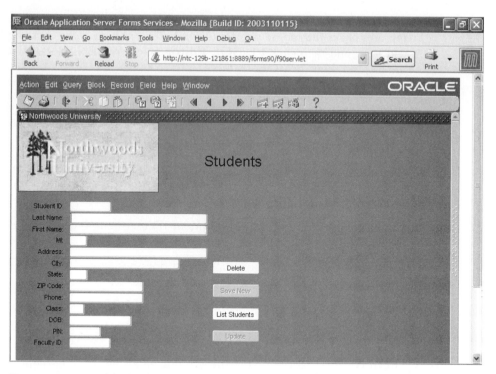

Figure 6-30 Modified Students form

To try to delete a student row:

1. Open the **Ch06Student_ERR.fmb** form. Run the form. Click the **List Students** button, select the first student, and then click **OK**. The values appear in the form text items.

2. Click **Delete** in the Students form. The error message "FRM-40735: WHEN-BUTTON-PRESSED trigger raised unhandled exception ORA-02292" appears. This ORA- (database) error occurred because you tried to delete a record that had associated child records as foreign key references.

3. Click the **List Students** button to open the LOV display, select Student ID **PE100**, and then click **OK**.

4. Delete the Faculty ID value at the bottom of the form, and type the letter **A** as the new Faculty ID value. This should generate a data type error, since the F_ID form field has a NUMBER data type.

5. Click **Update** in the Students form. The Form error message "FRM-50016: Legal Characters are 0-9-+E." appears.

6. Close the browser window.

Trapping Form Runtime Errors

Whenever an ORA- or FRM- error occurs while a form is running, the ON-ERROR event occurs. To trap form runtime errors and display alternate messages, you create a form-level trigger that corresponds to the ON-ERROR event. If the form has an ON-ERROR trigger, execution immediately transfers to the ON-ERROR trigger. Forms Builder has built-in procedures that you use in this trigger to trap errors and report error conditions. Table 6-7 summarizes these built-in error-handling procedures.

Procedure Name	Data Returned
DBMS_ERROR_CODE	Error number of the most recent database (ORA) error, represented as a negative integer
DBMS_ERROR_TEXT	Error number and message text of the most recent ORA- error
ERROR_CODE	Error number of the most recent Forms Services (FRM) error, represented as a positive integer
ERROR_TEXT	Error number and text of the most recent FRM- error or message

Table 6-7   Forms Builder built-in procedures for handling errors

If an FRM- error occurs, the ERROR_CODE procedure returns the corresponding FRM- error code. If an ORA- error occurs, the DBMS_ERROR_CODE procedure returns the corresponding ORA- error code. You can create an IF/ELSIF decision structure in the ON-ERROR trigger that traps errors based on the values of ERROR_CODE and DBMS_ERROR_CODE, and then displays custom messages or alerts to provide users with informative messages and alternatives.

Figure 6-31 shows the general syntax for an ON-ERROR trigger.

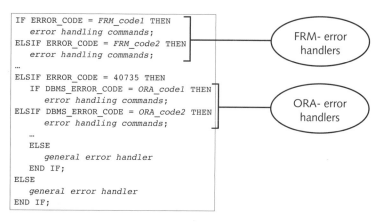

```
IF ERROR_CODE = FRM_code1 THEN
    error handling commands;
ELSIF ERROR_CODE = FRM_code2 THEN
    error handling commands;
...
ELSIF ERROR_CODE = 40735 THEN
    IF DBMS_ERROR_CODE = ORA_code1 THEN
        error handling commands;
ELSIF DBMS_ERROR_CODE = ORA_code2 THEN
        error handling commands;
    ...
    ELSE
        general error handler
    END IF;
ELSE
    general error handler
END IF;
```

FRM- error handlers

ORA- error handlers

Figure 6-31   General syntax for an ON-ERROR trigger

The trigger code uses a nested IF/ELSIF decision structure. The outer IF/ELSIF structure tests for specific FRM- error codes, and executes associated error handlers. (Recall that an error handler is one or more program commands that display an error message.) The inner IF/ELSIF structure tests for specific ORA- error codes. Recall that FRM-40735 indicates that an error is an ORA- (DBMS) error. If an FRM-40735 error occurs, the inner IF/ELSIF structure, which contains error handlers to handle common ORA- errors, executes.

When a form contains an ON-ERROR trigger and a runtime error occurs, form execution immediately transfers to the ON-ERROR trigger. If the error does not have an associated error handler in the ON-ERROR trigger, then no error handler executes, and no error message appears on the form message line. Therefore, you need to create general error handlers in the ELSE sections of the IF/ELSIF structures to handle all other errors. Usually the general error handler in the outer IF/ELSIF structure displays the result of the ERROR_TEXT procedure, which displays the error number and text of the most recent FRM- error. The general error handler in the inner IF/ELSIF structure displays the result of the DBMS_ERROR_TEXT procedure, which displays the error number and text of the most recent ORA- error.

Next, you create and test an ON-ERROR trigger that traps FRM- error 50016 (entering an illegal data type in a form field) and ORA- error 02292 (trying to delete a parent record that has child records as foreign key references in another table). When an error occurs, the trigger first tests if the value of the ERROR_CODE variable value is 50016. If it is, a custom message appears. Next, the trigger tests if the ERROR_CODE value is 40735. If it is, then an ORA- error has occurred. The trigger then tests for the value of DBMS_ERROR_CODE, which corresponds to an ORA- error code. If DBMS_ERROR_CODE is -02292, then a different custom message appears. Note that FRM- error code values are positive integers, whereas ORA- error code values are negative integers. The trigger also contains general error handlers to handle all other FRM- and ORA- errors.

To create and test the ON-ERROR trigger:

1. In Forms Builder, make sure that the Object Navigator window is open. To create a form-level trigger, select the Triggers node directly below the form module, and then click the **Create** ➕ button.

2. In the STUDENT_FORM: Triggers dialog box, select **ON-ERROR** in the list, and then click **OK**. The PL/SQL Editor opens.

3. Type the commands in Figure 6-32, compile the code, correct any syntax errors, and then close the PL/SQL Editor.

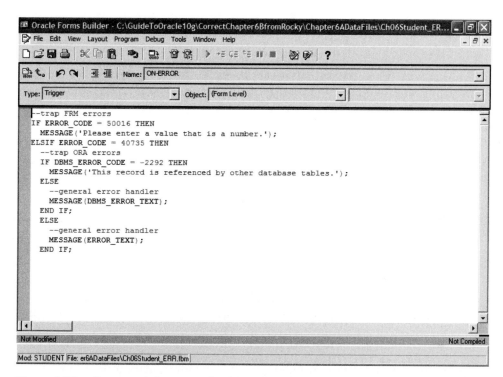

Figure 6-32   Commands to create the ON-ERROR trigger

4. Save the form and then run the form. Open the LOV display, select Student ID **PE100**, click **OK**, and then click the **Delete** button. The event handler message for the ORA-02292 error ("This record is referenced by other database tables.") appears on the message line.

5. Place the insertion point in the Faculty ID field, change the value to the letter **A**, and then click **Update**. The event handler message for the FRM-50016 error ("Please enter a value that is a number.") appears.

6. To test the general error handler, change the Faculty ID field value to **10**, then click **Update**. Click **Cancel** to close the Faculty LOV dialog box. Because 10 is not a valid Faculty ID value in the FACULTY table, the ORA-02291 error code and message appear in the message line.

7. Close the browser window, close the form in Forms Builder, and then close Forms Builder, the OC4J Instance, and SQL*Plus.

SUMMARY

- ❑ Oracle Corporation categorizes default system messages by severity level and message type. Messages with a severity level of 5 provide information about what is happening, and usually do not require any user action. Messages with severity levels of 10, 15, 25, and >25 require user, developer, or DBA action.

- ❑ While a form is running, Forms Services suppresses all messages with a severity level that is lower than (less severe) the current :SYSTEM.MESSAGE_LEVEL value. You can assign different values to this variable in the PRE-FORM trigger, which executes before the form appears.

- ❑ In a form, you can create custom messages and alerts to provide feedback to users. Custom messages display short text strings on the message line that do not require an acknowledgement. Alerts are dialog boxes that provide information to users and allow them to choose different options for proceeding.

- ❑ You create a custom message using the MESSAGE built-in. A custom message can display a maximum of 200 text characters.

- ❑ An alert is a top-level form object that you create and configure in the Object Navigator. To display an alert in a form, you use the SHOW_ALERT function, which returns a numeric value corresponding to the alert button that the user clicks.

- ❑ A form can validate user inputs using text item validation properties, or using a form validation trigger that verifies input values. The form validation unit represents the largest chunk of data that a user can enter before the form validates input values. Form validation occurs when the user moves the insertion point outside the form validation unit.

- ❑ When clicking a form command button would cause a runtime error, it is a good practice to disable the form command button. To do so, you use the SET_ITEM_PROPERTY built-in.

- ❑ If a form has an ON-ERROR trigger, execution transfers to the ON-ERROR trigger when a runtime error occurs. You can create error handlers in the ON-ERROR trigger to handle common runtime errors.

REVIEW QUESTIONS

1. You use the _____ system variable to control the default system messages that appear on a form.

2. A custom message can contain up to 200 characters. True or False?

3. Which trigger automatically executes before the form opens?

4. Which type of dialog box can be used to display an error message and requires the user to click a button before returning to the form?

5. By default, validation units are at the form level. True or False?

6. Which procedure is used to return the error number of the most recent Form Services (FRM) error?

7. An alert can display a maximum of _____ buttons.

8. When a text item is nonnavigable, the user can still click the mouse pointer to access the text item. True or False?

9. A level _____ message is an informative message that does not require user intervention.

10. A message requires an acknowledgement, such as clicking the OK button, from the user. True or False?

6

MULTIPLE CHOICE

1. By default, error messages with a severity level of _____ are displayed.

   a. 25 or above

   b. 15 or below

   c. 10 or above

   d. zero

2. The _____ property of an alert specifies the icon displayed in the dialog box.

   a. Icon

   b. Prompt

   c. Style

   d. Label

3. Which of the following built-in functions is used to display an alert?

   a. ON_ERROR

   b. ERROR_RESULT

   c. SHOW_ALERT

   d. MESSAGE

4. Which of the following procedures returns the error code generated by the most recent ORA- error?

   a. DBMS_ERROR_CODE

   b. DBMS_ERROR

   c. ERROR_CODE

   d. ERROR_TEXT

5. Which of the following is *not* a validation unit supported by Oracle Forms?

   a. Form

   b. Record

   c. Data Block

   d. Column

6. Use the _____ built-in function to disable a form command button.

   a. ON_ERROR

   b. SET_ITEM_PROPERTY

   c. PRE_FORM

   d. SHOW_DISABLE

7. A(n) _____ error code value is a positive integer.

   a. ORA–

   b. FRM–

   c. SRV–

   d. MSG–

8. Of the following error severity levels, which is considered the most severe?

   a. 0

   b. 10

   c. 15

   d. 20

9. With a custom form, validation should always occur at the _____ level.

   a. control block

   b. item

   c. form

   d. data block

10. If a value does not meet specified validation requirements, the _____ built-in exception is raised.

   a. WHEN_INVALID

   b. WHEN-VALIDATE-ITEM

   c. FORM_TRIGGER_FAILURE

   d. none of the above

PROBLEM-SOLVING CASES

For the following cases, use the Clearwater Traders sample database. The form files for all cases are in the Chapter06 folder on your Data Disk. Save all solutions files in the Chapter06\Cases folder on your Solution Disk. Save all SQL commands in a file named Ch6BQueries.sql.

1. The customer form in Figure 6-33 allows users to modify and delete items from the ITEM and INVENTORY tables. In the following tasks, you trap and handle common errors.

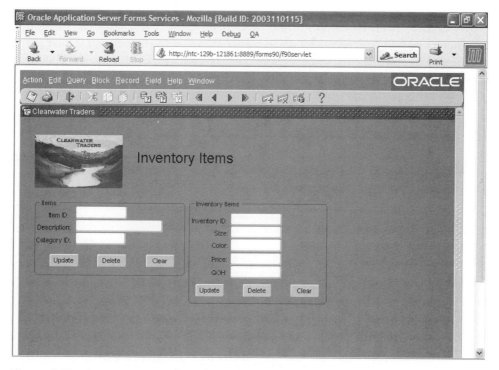

Figure 6-33 Inventory Items form for Clearwater Traders

The custom form is saved as INVENTORY_FORM.fmb in the Chapter06 folder of your Data Disk. Save the modified form as Ch6BCase1.fmb in the Chapter06\Cases folder after performing the following tasks:

a. Create an event handler that displays the message "This record is referenced by other database tables" if the user attempts to delete or modify the Item ID of a referenced item in the ITEM table.

b. Create an event handler that displays the message, "Enter a numeric value" if the user attempts to enter a nonnumeric value in the Item ID, Category ID, Inventory ID, Price, or QOH text items.

c. If the user attempts to enter a Category ID that is not contained in the CATEGORY table, generate a list of valid category IDs and descriptions from which the user can select a valid category ID.

2. Revise the Clearwater Traders Inventory Items from Case 1 (Ch6BCase1.fmb), so that the event handlers created for Items a and b display the specified error messages in alert dialog boxes rather than as messages at the bottom of the screen. Save the modified form as Ch6BCase2.fmb in the Chapter06\Cases folder.

The following information relates to Cases 3 through 5:

The custom form in Figure 6-34 allows users to insert new records, update existing records, and delete records from the Clearwater Traders CUSTOMER, ORDERS, and ORDER_LINE tables. When the user clicks Create New Customer in the Customers frame, the form retrieves a new Customer ID value from a sequence. The user enters the data values for the new customer and clicks Save New Customer, and the form saves the new customer record in the database. The user can open an LOV display associated with the Customer ID text item, select an existing CUSTOMER record, edit the data fields, and then click Update Customer to update the customer record, or click Delete Customer to delete the customer record.

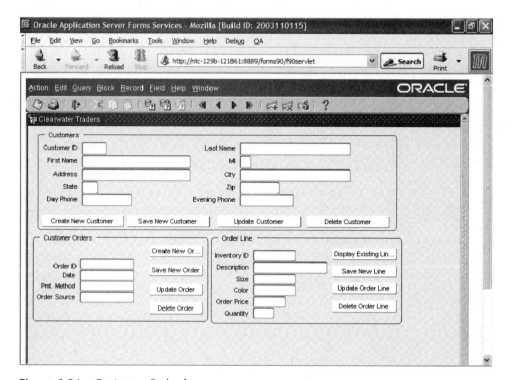

Figure 6-34   Customer Order form

After the user creates a new customer record or selects an existing customer record, he or she can click Create New Order in the Customer Orders frame to display a new customer order ID. The user can enter the order data in the data fields, and then click Save New Order to save the order. If the user wants to edit an existing order, he or she opens the LOV display associated with the Order ID text item to view the selected customer's orders. The user can select an existing order, edit the order information, and then click Update Order to update the order, or Delete Order to delete the order.

After the user creates or selects a customer order, he or she can create new order lines for the order by opening the LOV display associated with the Inventory ID text item. The user can select an Inventory ID, and enter an order quantity, and then click the Save New Line button. (The primary key of the ORDER_LINE table is the combination of ORDER_ID and INV_ID, so the form does not need to generate a primary key.) If the user clicks the Display Existing Lines button, the form displays an LOV that displays order line information for the current Customer Order ID. The user can select an existing order line and edit or delete it as needed.

3. Open **CUSTOMER_ORDER.fmb**, make the following modifications, and save the modified file as **Ch6BCase3.fmb**.

   a. In SQL*Plus, create a new sequence named **C_ID_SEQUENCE** that starts with 10 and has no maximum value.

   b. In SQL*Plus, create a new sequence named **O_ID_SEQUENCE** that starts with 100 and has no maximum value. (If you created this sequence previously, delete the existing sequence and create a new sequence.)

   c. Add a Caution alert to notify the user that a record is about to be inserted. The alert should appear before the INSERT command is committed and show the message, "Add the new record to the database?" If the user clicks the OK button, commit the insert, and show a message stating, "Record successfully inserted." If the user clicks the Cancel button, do not commit the change, and show a message stating, "Changes not saved." Display the alert when the user clicks the Save New Customer, Save New Order, or Save New Line button.

4. Open the Ch6Case3.fmb from Case 3 and make the following modifications. Save the modified file as Ch6BCase4.fmb.

   a. Add a Caution alert to notify the user when a record is going to be updated. The alert should appear before the UPDATE command is committed and display the message, "Update the current record?" If the user clicks OK, commit the update, and show a message stating. "Record successfully updated." If the user clicks Cancel, do not commit the change, and show a message stating, "Changes not saved." Display the alert when the user clicks the Update Customer, Update Order, or Update Order Line button.

   b. Add a Caution alert to notify the user when a record is about to be deleted. The alert should appear before the DELETE command is committed and show the message, "Delete the current record?" If the user clicks the OK button, commit the delete, and show a message stating, "Record successfully deleted."

If the user clicks the Cancel button, do not commit the change, and show a message stating, "Changes not saved." Display the alert when the user clicks the Delete Customer, Delete Order, or Delete Order Line button. After the user deletes an order record, the text items in the Customer Orders and Order Line frames should be cleared, but the text items in the Customers frame should still appear. After the user deletes an order line record, the text items in the Order Line frame should be cleared, but the text items in the Customers and Customer Orders frames should still appear. (*Hint*: Program units named CLEAR_CUSTORDER_FIELDS and CLEAR_ORDERLINE_FIELDS are included in the file to clear these fields correctly.)

c. Validate the Customer ID and Order ID values using the LOVs associated with the text items.

5. Open the **Ch6BCase4.fmb** created in Case 4 and make the following modifications. Save the modified file as **Ch6BCase5.fmb**.

a. Create a Stop alert with an OK button that appears only if the user clicks the Create New Order button in the Customer Orders frame before selecting a value for Customer ID. The alert should remind the user that he or she must select a customer before he or she can create or select a customer order.

b. Create a Stop alert with an OK button that only appears if the user clicks the Display Existing Lines button or the LOV command button beside Inventory ID in the Order Line frame before selecting a value for Order ID. The alert should remind the user that he or she must select an order ID before he or she can create or select an order line.

◀ LESSON C ▶

After completing this lesson, you should be able to:

♦ Convert data blocks to control blocks
♦ Link data blocks to control blocks
♦ Create a form that has multiple canvases
♦ Create tab canvases
♦ Create stacked canvases

CONVERTING A DATA BLOCK TO A CONTROL BLOCK

When you create a custom form, you manually create the control block and the form canvas, draw the text items on the form, and then configure the text item properties to match the corresponding database field data types and sizes. This process becomes tedious for control blocks that contain several text items. Often, control blocks contain multiple text items from a single database table. For example, the Northwoods University Faculty custom form in Figure 6-35 displays multiple fields from the FACULTY, TERM, and ENROLLMENT tables, and allows users to update information in the FACULTY and ENROLLMENT tables.

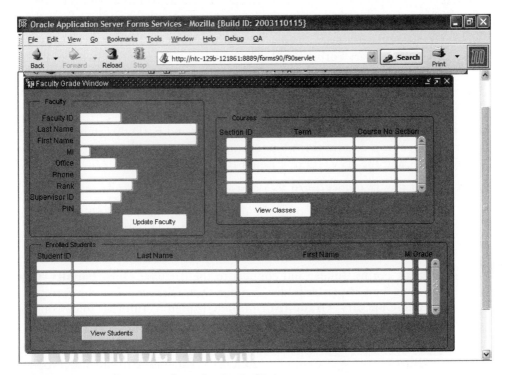

Figure 6-35 Final version of Faculty Grade form

To make it easier to create custom forms that contain many text items from a single database table, you can use the Data Block and Layout Wizards to create a data block form and a form layout. Then you convert the data block to a control block, and add command buttons to control the form processing. You can add additional text items as necessary to represent fields in other database tables.

You perform the following steps to convert a data block to a control block:

1. Change the data block's Database Data Block property value to No.

2. Change the Required property value of the text item that represents the data block table's primary key to No.

When you create a data block using the Data Block Wizard, the Database Data Block property value of the block is Yes. This specifies that the data block retrieves its values from the database. To convert a data block to a control block, you change the Database Data Block property value to No. This allows the control block to retrieve data values using LOVs.

Recall that a text item's Required property validates a text item and ensures that the text item contains a data value. Because a record's primary key value cannot be NULL, the Required property value for the text item that corresponds to the form's primary key is always Yes. When a text item's Required property value is Yes, text item validation does not allow the user to move the insertion point out of the text item until he or she enters a value. In a custom form, you must change this value to No or else the user cannot use an LOV associated with the primary key text item to retrieve values from the form.

Before you start working with this form, you start SQL*Plus, and refresh your database tables. Then you start Forms Builder and open the Faculty Grade form. Currently, this form is a data block form created using the Data Block and Layout Wizards. The form has an LOV associated with the F_ID text item that displays values for existing records and returns the selection to the form fields.

In the following steps, you modify the data block and F_ID text item properties to convert the form to a custom form. You change the data block's Database Data Block property to No, and the INV_ID text item's Required property to No.

To convert the data block form to a custom form:

1. Start SQL*Plus, log onto the database, and run the **Ch6Northwoods.sql** script in the Chapter06 folder on your Data Disk.

2. Start the OC4J Instance.

3. Start Forms Builder and connect to the database. Open **Faculty_Grade.fmb** from the Chapter06 folder on your Data Disk, and save the form as **Ch6CFaculty_Grade.fmb** in the Chapter06\Tutorials folder on your Solution Disk.

4. In the Object Navigator, open the Data Blocks node, double-click the **Data Block** icon ▦ beside the FACULTY data block to open its Property Palette, scroll down to the Database property node, change the Database Data Block property to **No**, and then close the Property Palette.

5. Open the Layout Editor, select the **F_ID** text item, right-click, and then click **Property Palette**. Scroll down to the Data property node, change the Required property value to **No**, and then close the Property Palette.

Now you add the Update Faculty button to the form. You change the Update Layout property of the layout frame to Manually, so you can add the button to the form layout. (Recall that if you do not change the frame Update Layout property to Manually, whenever you resize or move the frame, the frame items automatically appear stacked on top of each other in the order they appear in the Object Navigator.) Then you add the View Classes button to the form, create the button trigger, and test the form.

To add the button to the form and test the form:

1. In the Layout Editor, select the **Faculty** frame, right-click, and then click **Property Palette**. Change the Update Layout property to **Manually**, and then close the Property Palette.

2. Resize the frame, as shown in Figure 6-35.

3. Select the **Button** tool ▢ on the tool palette, and draw a new command button in the frame. Select the new button, right-click, click **Property Palette**, change the button's Name property value to **UPDATE_FACULTY_BUTTON** and the Label property value to **Update Faculty**, and then close the Property Palette.

4. To create the button trigger, right-click the button, point to **SmartTriggers**, and then click **WHEN-BUTTON-PRESSED**. Type the following commands in the PL/SQL Editor:

```
UPDATE faculty
SET f_last = :faculty.f_last, f_first = :faculty.f_first,
f_mi = :faculty.f_mi,
    loc_id = :faculty.loc_id, f_phone = :faculty.f_phone,
f_rank = :faculty.f_rank,
    f_super = :faculty.f_super, f_pin = :faculty.f_pin
WHERE f_id = :faculty.f_id;
COMMIT;
MESSAGE('Record successfully updated.');
CLEAR_FORM;
```

5. Compile the trigger, correct any syntax errors, and then close the PL/SQL Editor.

6. Save the form, and then run the form. Make sure that the insertion point is in the Faculty ID field, open the LOV display, select Faculty ID **1** (Teresa Marx), and then click **OK**.

7. Change the MI field value to **P**, and then click the **Update Faculty** button. The confirmation message, "Record successfully updated." appears on the message line.

6

8. Place the insertion point in the **Faculty ID** field, open the LOV display, and confirm that the new MI value for Faculty ID 1 is P.

9. Close the browser window, and then close the form in Forms Builder.

LINKING A DATA BLOCK TO A CONTROL BLOCK

Recall that a data block form contains one or more data blocks that contain text items or other form items corresponding to the fields in a specific database table. You can use the Data Block and Layout Wizards to make data block forms quickly and easily. The Forms Services application has built-in programs for viewing, inserting, updating, and deleting data in data block forms. Conversely, a custom form contains control blocks that display text items or other form items from multiple database tables. You write form triggers to process the text items and data values on custom forms. Data block forms are not as flexible as custom forms, because data block forms reflect the structure of the database tables rather than the organizational processes that the form supports. But, data block forms are much easier to create and don't require you to write form triggers to perform form processing.

Sometimes it is useful to create forms that link data blocks and control blocks to work together. These forms take advantage of the flexibility of custom forms, as well as the ease of creating data block forms. Usually, these linked control/data blocks represent a master-detail relationship, in which the control block represents the master records, and the data block represents the detail records. (Recall that in a master-detail relationship, a master record can have several related detail records.) To learn how to create a form that links control blocks and data blocks, you work with the Northwoods University Faculty Grades form shown in Figure 6-35.

This form allows users to view faculty and course data, update faculty information, and post student grades. The form has three frames:

- Faculty, which corresponds to a control block named FACULTY, and allows the user to change faculty data

- Courses, which corresponds to a control block named COURSES and allows the user to view courses taught by the current faculty member in various terms

- Enrolled Students, which corresponds to a data block named ENROLLED _STUDENTS and displays the students enrolled in the course currently selected in the COURSES frame

In this form, the COURSES control block is the master block, and the ENROLLED_STUDENTS data block is the detail block.

Now you create views to display course and student information.

1. In SQL*Plus, type the following command to create a view displaying course and term information:

```
CREATE OR REPLACE VIEW course_view
AS (SELECT f_id, c_sec_id, term_desc, course_no, sec_num
FROM term NATURAL JOIN course_section);
```

NOTE

If you have already created this view, drop the existing view using the command DROP VIEW course_view;, then repeat Step 1.

6

2. In Forms Builder, open **Ch6CFaculty_Grade.fmb** from the Chapter06\Tutorials folder on your Solution Disk.

3. Run the form. The form appears in the browser window. Currently, the form does not contain the Enrolled Students frame that shows the detail data block records. You will add the data block and create the relationship between the control block and the data block later in this lesson.

4. Click the insertion point in the Faculty ID field, open the LOV display, select Faculty ID **2** (Zhulin Ma), and then click **OK**.

5. Click the **View Classes** button in the Courses frame. The courses taught by Zhulin are displayed. Close the browser window.

To allow the user to enter student grades, you need to add the Enrolled Students frame shown in Figure 6-35 to the bottom of the Faculty Grades form. The Enrolled Students frame displays a data block based on a database view. This database view retrieves the student's information, which consists of the student's ID and name. The data block in the Courses frame displays course information, whereas the Enrolled Students frame displays the students who are enrolled in the selected course. Because the Section ID text item is in a control block, and the Class List information is in a data block, you need to create a link between the control block and the data block.

To link a control block to a data block, you complete the following steps:

1. Create the control block.

2. Create the data block.

3. Specify the link between the control block and the data block.

4. Modify the form triggers to refresh the data block when the underlying data values change.

The Faculty Grade form already contains the control block that displays the Section ID value. Now you create the data block, create the link between the control block and the data block, and then modify the form triggers to refresh the data block when the form order line values change.

Creating the Data Block

You use the Data Block Wizard and Layout Wizards to create and configure a data block that you plan to link to a control block. The data block must contain the field that links it to the control block. In this form, you link the CLASS_LIST data block to the COURSES control block using the C_SEC_ID field. This linkage causes the ENROLLED_STUDENTS data block to display information for only the current course section ID value.

Before you can create the ENROLLED_STUDENTS data block, you must create the database view that provides the information that appears in the data block. The view must contain the C_SEC_ID field, because that is the field that links the data block to the control block. The view also contains the student IDs, names, and grades from the ENROLLMENT and STUDENT tables. Now you create the view in SQL*Plus, and then create the associated data block in Forms Builder.

To create the view and data block:

1. In SQL*Plus, type the following command to create a view displaying student information:

   ```
   CREATE OR REPLACE VIEW class_list
   AS (SELECT c_sec_id, student.s_id, s_last, s_first, ⏎
   s_mi, grade
   FROM enrollment, student
   WHERE enrollment.s_id = student.s_id);
   ```

2. In Forms Builder, make sure that the FACULTY_GRADE_FORM node is selected in the Object Navigator. Because you are going to create a new data block using the Data Block Wizard, you do not want to have any data block currently selected. If a data block is selected, the Wizard opens the selected data block in reentrant mode for editing, rather than creating a new data block.

3. Click the **Data Block Wizard** button 📝, and then click **Next**. When the Type page appears, make sure that the Table or View option button is selected, and then click **Next**.

4. On the Table page, click **Browse**, check the **Views** check box, clear the **Tables** check box, select **CLASS_LIST**, and then click **OK**.

5. Click the **Move all items to target** button ⟩⟩ to select all of the view data fields for the block, and then click **Next**. The Master-Detail page appears.

When you link a control block to a data block, you cannot use the Data Block Wizard to create a master-detail relationship between the data block and the control block. Instead, you must create the data block without a master-detail relationship and then link the two blocks later. For now, you omit the master-detail relationship, finish creating the data block, and then create the block layout. You use a tabular layout, and display five order line records at a time. You do not display the Course Section ID (C_SEC_ID) value in the layout, because the value already appears in the Section ID field in the Courses frame.

To omit the master/detail relationship, finish the data block, and create the block layout:

1. On the Master-Detail page, click **Next**. Type ENROLLED_STUDENTS as the data block name, click **Next**, and then click **Finish**. The Layout Wizard Welcome page appears. Click **Next**.

2. On the Canvas page, make sure that FACULTY_CANVAS is the selected canvas, and then click **Next**.

3. On the Data Block page, click the **Move all items to target** button [»] to select all of the block items for the layout. Then, select **C_SEC_ID** in the Displayed Items list, and click the **Move one item to source** button [<] to remove **C_SEC_ID** from the list. If necessary, drag the list items and then drop them so they appear in the following order: S_ID, S_LAST, S_FIRST, S_MI, GRADE. Click **Next**.

4. On the Items page, modify the prompts as follows, and then click Next.

Name	Prompt
S_ID	Student ID
S_LAST	Last Name
S_FIRST	First Name
S_MI	MI
GRADE	Grade

5. On the Style page, select the **Tabular** option button, and then click **Next**.

6. On the Rows page, type **Enrolled Students** in the Frame Title field. Type **5** in the Records Displayed field, and check the **Display Scrollbar** check box. Click **Next**, and then click **Finish**. The Enrolled Students frame appears on the canvas.

7. If necessary, move the Enrolled Students frame so it appears beneath the Faculty and Courses frames and create the View Students button with the properties below, as shown in Figure 6-35, and then save the form.

Name	VIEW_STUDENTS_BUTTON
Label	View Students
Number of Items Displayed	1

Linking the Control Block and the Data Block

To link a master control block and a detail data block, you open the detail data block's Property Palette, and modify the data block's WHERE Clause property. A data block's WHERE Clause property appends a search condition to the SQL query that specifies the data block records. This allows you to change the data block's contents at runtime, using a search condition based on current user inputs.

The general syntax of the WHERE Clause property that creates a link between a master control block and a detail data block is as follows:

```
fieldname = :control_block.text_item_name
```

In this syntax, *fieldname* is the name of the field in the data block's SQL query on which you want to place the search condition. In the ENROLLED_STUDENTS data block, this is the C_SEC_ID field, because you want to display the students for the currently selected course. The *:control_block.text_item_name* expression specifies the name of the text item within the control block that displays the master data value. For the COURSES control block, this is the C_SEC_ID text item, so this expression is `:courses.c_sec_id`.

Now you create the relationship between the COURSES master control block and the ENROLLED_STUDENTS detail data block.

To create the relationship between the data block and control block:

1. Open the Object Navigator window, and double-click the **Data Block** icon beside the ENROLLED_STUDENTS node to open the data block Property Palette.

2. Scroll down to the Database property node, type the following value for the WHERE Clause property, and then close the Property Palette and save the form.

   ```
   c_sec_id = :courses.c_sec_id
   ```

Refreshing the Data Block Values

After you create a data block and link it to a control block, you need to add commands to initially populate the data block display, and to periodically refresh the data block records when the data in the master block or the data in the underlying table records change. To populate and refresh the data block display, you use the GO_BLOCK built-in to move the form focus to the data block. (Recall that the GO_BLOCK built-in has the following syntax: `GO_BLOCK('block_name');`.) Then, you execute the EXECUTE_QUERY built-in. The EXECUTE_QUERY built-in **flushes** the block, which makes its information consistent with the corresponding database data. To populate and refresh the ENROLLED_STUDENTS data block display, you use the following commands:

```
GO_BLOCK('ENROLLED_STUDENTS');
EXECUTE_QUERY;
```

In the Faculty Grade form, you need to refresh the ENROLLED_STUDENTS data block when the user changes the course selected in the COURSES frame or if the referenced faculty member changes. Now you add the commands to refresh the ENROLLED_STUDENTS display to the form triggers for these events. Then you run the form and confirm that the student values appear correctly.

To add the code to refresh the Enrolled Students display and run the form:

1. Open the Layout Editor.

2. Right-click the **View Students** button in the Enrolled Students frame, and click **PL/SQL Editor**. Select the When-Button-Pressed trigger. The

PL/SQL Editor opens. To refresh the class list when the user selects another course, add the following commands:

```
GO_BLOCK('ENROLLED_STUDENTS');
EXECUTE_QUERY;
```

3. Compile the trigger, correct any syntax errors, and then close the PL/SQL Editor and save the form.

4. Run the form, make sure that the insertion point is in the Faculty ID field, open the LOV display, select Faculty ID **2** (Zhulin Ma), and click **OK**.

5. In the Courses frame, click the **View Classes** button to view the faculty member's courses.

6. Select Section ID **1** in the Courses frame, then click the **View Students** button in the Enrolled Students frame to display all students enrolled in the selected course.

7. Select Section ID **12** in the Courses frame, and then click the **View Students** button in the Enrolled Students frame to display all students enrolled in the other course.

8. Close the browser window, and then close the form in Forms Builder.

CREATING FORMS WITH MULTIPLE CANVASES

As you create complex forms, it is a good practice not to show too much information on the user's screen display. It is best to place only the amount of information on the screen that the user can view without scrolling. Otherwise, the user can become disoriented about her or his position on the form, and can suffer from information overload. For complex forms that display a lot of information, you sometimes need to use multiple screens to limit the amount of information that is displayed on the user's screen at one time.

An example of an application that requires multiple screens is the Northwoods University Student Services application system. When a student accesses the system, the user can change his or her personal information, view grades, or display the student's current class schedule. The student uses the first screen to log into the system, the second screen acts as a menu and allows the student to access personal information on a separate screen, or retrieve grades or a current class schedule on alternative screens.

Forms Builder provides two approaches for creating applications with multiple screens. The **single-form approach** involves creating one form with multiple canvases. This approach enables the form to share data among the different canvases, but makes it impossible for multiple programmers to work simultaneously on different canvases of the same application, because only one programmer can open and modify the form design (.fmb) file at a time.

The **multiple-form approach** involves creating multiple forms with a different .fmb file for each application canvas. This approach works well when multiple programmers collaborate to create a complex application. It also enables programmers to use a form in many different applications, which reduces redundant programming efforts. However, it is more difficult for related forms to share data.

In general, you use the single-form approach to create closely related application screens that share many variable values. You use the multiple-form approach to integrate several loosely related form applications into a single database system. For example, you might create individual form applications to support the Northwoods University Student Services system, and then integrate these forms into a single database application.

In this section, you learn how to create a single form with multiple canvases to support the Northwoods University Student Services system. (You learn how to create multiple-form database applications in Chapter 8.)

The Northwoods University Student Services Process

When you design a form with multiple canvases, you need to identify the sequence of actions that the user will employ to interact with the canvases. Figure 6–36 uses a flow-chart to illustrate the sequence of canvases in the Northwoods University Student Services system. The application starts on the Student Log On canvas. The student enters his or her Student ID and PIN so the database can verify the identity of the student accessing the system. If the student is authenticated, a menu canvas is displayed and allows the student to select from a series of options: display or update personal information, view grades, or view current class schedule. For each option, the student can return directly to the menu canvas and select a different option. The flowchart also identifies each database table accessed by each canvas. For a student to update personal information, the STUDENT table is accessed. However, if the student displays his or her current schedule, the ENROLLMENT, COURSE_SECTION, FACULTY, and LOCATION tables must be referenced to retrieve all the required data. The student is free to move among the Student Information, Course Grades, and Enrollment canvases until exiting the system.

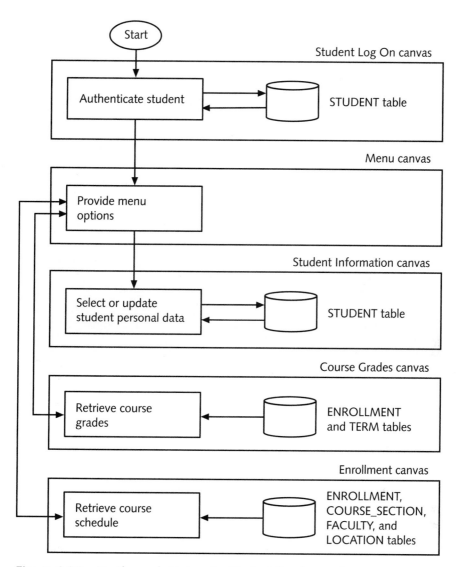

Figure 6-36   Northwoods University Student Services system

Interface Design

Figure 6-37 shows the Student Log On canvas. The student is required to enter a valid student ID and the corresponding PIN, which are stored in the STUDENT table.

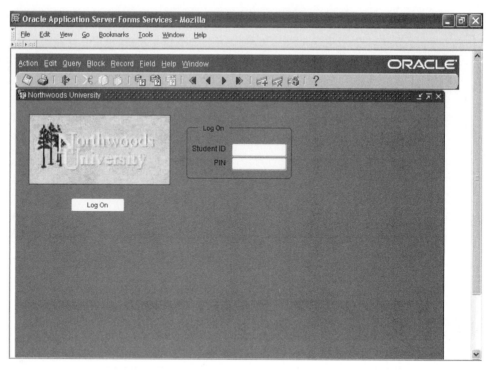

Figure 6-37 Student Log On canvas

If the data entered is contained in the table, the Menu canvas shown in Figure 6-38 appears. This screen provides access to the student's information, grades, and current schedule available in the Northwoods University Student Services system.

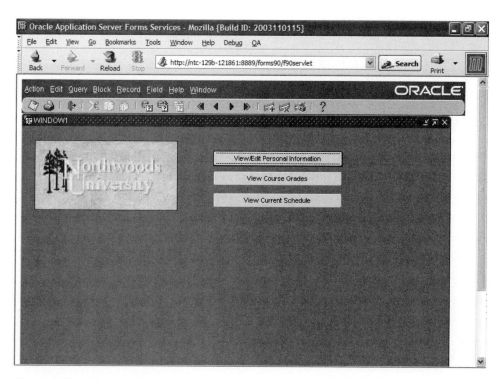

Figure 6-38   Menu canvas

Figure 6–39 shows the layout of the Student Information canvas. After students review their information, they click the Return button to return to the Menu canvas where they have the option of accessing other information.

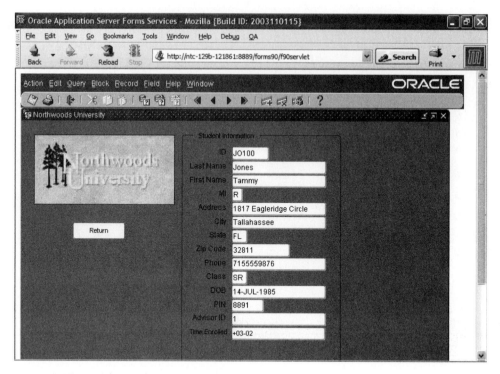

Figure 6-39 Student Information canvas

Figure 6-40 shows the layout of the Course Grades canvas. All courses taken by the student are displayed, including the term the course was taken and grade received. A Return button is included, allowing the student to access the Menu canvas.

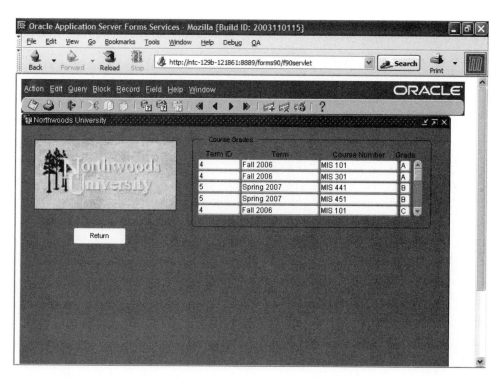

Figure 6-40   Course Grades canvas

Figure 6-41 shows the layout of the Enrollment canvas, which displays the student's schedule for the current term. Again, a Return button is included to provide the student with a convenient method for accessing the Menu canvas.

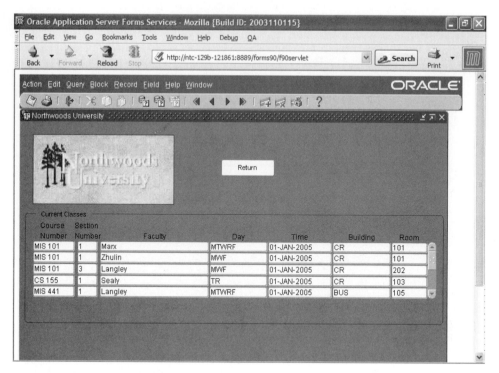

Figure 6-41   Enrollment canvas

Working with Multiple Canvases and Data Blocks

So far, all the forms you have created have contained a single canvas, so all of the block items appeared on that canvas. When you create a form that contains multiple canvases, you create the individual canvases, and then specify to display the specific block items on a specific canvas. When you create a new data block, you can select the canvas name on which the block items are to appear on the Layout Wizard Canvas page. To specify on which canvas an item appears, you open the item's Property Palette and select the target canvas name from the Canvas property list.

Often forms that have multiple canvases also have multiple data blocks. Usually you group items that appear on the same canvas in a single data block to keep the blocks small and manageable. Furthermore, you cannot create two items that have the same name in the same data block. For example, you might want to display a text item named S_ID on two different canvases. To do this, you must place the items that have the same name in separate data blocks.

When you create a new text item or command button on a canvas, you must ensure that Forms Builder places the new item in the correct block. To specify the block in which Forms Builder places a new item, you select the target block in the Block item list in the Layout Editor before you create the item on the canvas. If you accidentally create an item in the wrong data block, you can move it to the correct data block by dragging the object in the Object Navigator and then dropping it into the correct block.

At this point, to become familiar with working in a form that has multiple canvases and multiple data blocks, you open the Student_Information.fmb form file in the Chapter06 folder on your Data Disk. This form contains the layouts for the canvases in the Student Information system. You modify this form by adding the trigger code to the form buttons so the canvases work together to create a finished application.

To open and examine the form:

1. Open **Student_Information.fmb** from the Chapter06 folder of your Data Disk and save it as **Ch6CStudent_Information.fmb** in the Chapter06\ Tutorials folder on your Solution Disk.

2. In the Object Navigator, open the Data Blocks node. Note that the form contains five data blocks: LOGON_BLOCK, MENU_BLOCK, STUDENT, GRADE_VIEW, CURRENT_ENROLLMENT_VIEW. When you create a form with multiple canvases, you should place the items that appear on each canvas in a separate data block to keep the block item lists small and manageable.

3. To examine the form canvases, click **Tools** on the menu bar, and then click **Layout Editor**. The STUDENT_FORM: Canvases dialog box opens, and lists the form canvases. When a form contains multiple canvases, you must select the canvas you want to appear in the Layout Editor.

4. Select **STUDENT_CANVAS**, and then click **OK**. The canvas appears in the Layout Editor.

5. Select the **S_ID** text item, and note the values that appear in the Canvas and Block item lists at the top of the Layout Editor window, as shown in Figure 6-42. The Canvas list displays STUDENT_CANVAS, which is the name of the canvas that currently appears in the Layout Editor. The Block list displays STUDENT, which is the name of the block that contains the current selection.

6

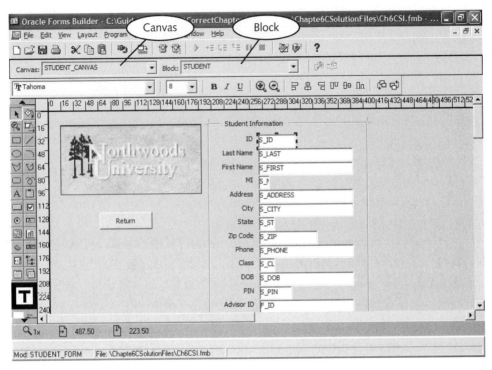

Figure 6-42 Viewing a text item's associated canvas and block

6. Open the **Canvas** list, and select **GRADE_CANVAS**. The canvas displaying the previous courses and grades appears.

7. Select the **TERM_ID** text item. Note that the Block list value changes to GRADE_VIEW, to indicate the block of the selection.

To complete the MENU_CANVAS layout, you need to create the command buttons shown in Figure 6-38 for viewing and editing a student record, viewing course grades, and viewing the current schedule. You draw the command buttons, and then modify the button properties. You must be careful to place the command buttons in the MENU_BLOCK data block.

To create the command buttons:

1. Click **Tools** on the menu bar, and then click **Layout Editor**. If necessary, select **MENU_CANVAS** in the canvas list, and then click **OK**.

2. To place the new command buttons in the MENU_BLOCK data block, select **MENU_BLOCK** in the Block list if necessary. Then draw the three command buttons on the right side of the logo, as shown in the MENU_ CANVAS canvas layout in Figure 6-38.

3. Open the Property Palette for each button, modify the button properties as follows, and then close the Property Palette:

Name	Label	Width	Height
VIEW_EDIT_STUDENT	**View/Edit Personal Information**	119	14
VIEW_GRADES	**View Course Grades**	119	14
VIEW_SCHEDULE	**View Current Schedule**	119	14

4. Select all of the buttons as an object group, click **Layout** on the menu bar, click **Align Components**, select the Horizontally **Align Left** and Vertically **Distribute** option buttons, and then click **OK**. Save the form.

Specifying the Block Navigation Order

6

When a form contains multiple data blocks and multiple canvases, you must specify which canvas initially appears when the user runs the form. When a form contains multiple canvases, the canvas that appears first is the canvas whose block items appear first under the Data Blocks node in the Object Navigator window. The order of the canvases in the Object Navigator Canvases list does not matter, only the order of the data blocks.

Recall from Figure 6-37 that when the Student Services system runs, the Student Log On canvas is the first canvas that appears. Now you open the Object Navigator, examine the order of the blocks, and if necessary, move the LOGON_BLOCK block so it is the first block that appears in the Data Blocks list. Then, you run the form to confirm that the canvas containing the LOGON_BLOCK data block items appears first.

To modify the order of the form blocks and run the form:

1. Open the Object Navigator window, select the Data Blocks node, and then click the **Collapse All** button 🔁 on the toolbar to close all of the data block nodes.

2. Open the Data Blocks node to examine the block order. If LOGON_BLOCK is not the first block in the list, select the **LOGON_BLOCK** data block node, and drag the node toward the top edge of the screen display. When the pointer is on the Data Blocks node and the Data Blocks node appears selected, drop the LOGON_BLOCK data block to make it the first block in the list. Open the Canvases node. Your form Data Blocks list should appear as shown in Figure 6-43.

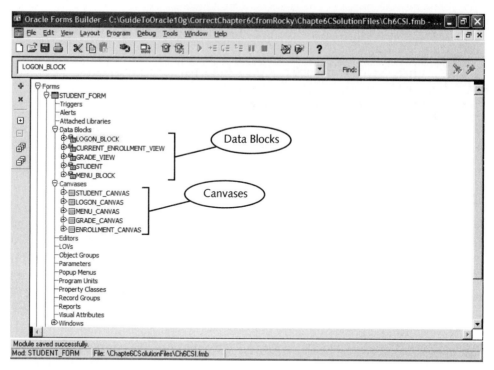

Figure 6-43   Object Navigator

3. Save the form, and then run the form. The canvas with the Logon frame appears in the browser window. Close the browser window.

Referencing Different Canvases, Form Blocks, and Block Items

All form items have a Canvas property that specifies the name of the canvas on which the item appears. When an item has the form focus, the form always displays the canvas on which that item appears. To display a different canvas within a form that has multiple canvases, you execute the GO_ITEM built-in, and move the form focus to an item on the target canvas.

The Student Log On canvas initially appears when the user first runs the Student Services form. To display the Menu canvas after the Student ID and PIN are verified, you execute the GO_ITEM built-in to move the form focus to an item in the Menu block. Usually, you move the form focus to the block item in which you want the insertion point to appear when the canvas first opens. For example, you use the following command to move the form focus to the VIEW_UPDATE_BUTTON command button in the MENU_BLOCK data block:

```
GO_ITEM('MENU_BLOCK.view_update_button');
```

In the GO_ITEM command, you do not expressly need to preface the item name with the block name, so you could use the following command to move the form focus to the VIEW_UPDATE_BUTTON: `GO_ITEM('VIEW_UPDATE_BUTTON');`. However, forms that contain multiple blocks often contain text items in different blocks that have the same name. All block item values are visible to all other blocks, including text items, command buttons, and so on. If you execute the GO_ITEM command using only the item name, without the block name preface, and multiple items with that name exist in the form, a runtime error occurs. Therefore, it is good practice to include the block name when identifying a particular item.

Now you create and test triggers for the command buttons displayed on the MENU_CANVAS canvas. To create and test the buttons:

6

1. Open **MENU_CANVAS** in the Layout Editor, select the **View/Edit Personal Information** button, right-click, point to **SmartTriggers**, and then click **WHEN-BUTTON-PRESSED**. Type the following commands to move the form focus to STUDENT_CANVAS:

   ```
   GO_ITEM('STUDENT.S_ID');
   ```

   Compile the trigger, correct any syntax errors, and then close the PL/SQL Editor.

2. Select the **View Course Grades** button, right-click, point to **SmartTriggers**, and then click **WHEN-BUTTON-PRESSED**. Type the following commands to move the form focus to GRADE_CANVAS:

   ```
   GO_ITEM('GRADE_VIEW.TERM_ID');
   ```

   Compile the trigger, correct any syntax errors, and then close the PL/SQL Editor.

3. Select the **View Current Schedule** button, right-click, point to **Smart Triggers**, and then click **WHEN-BUTTON-PRESSED**. Type the following commands to move the form focus to ENROLLMENT_CANVAS:

   ```
   GO_ITEM('CURRENT_ENROLLMENT_VIEW.COURSE_NO');
   ```

4. Compile the trigger, correct any syntax errors, and then close the PL/SQL Editor.

5. Save the form, and then run the form. Type **PE100** as the student ID, and then click the **PIN** text box. Type **1230** as the PIN and click the **Log On** command button. The MENU_CANVAS canvas appears.

6. Click the **View/Edit Personal Information** command button. After the STUDENT_CANVAS appears, click **Return**. The focus changes back to the Menu canvas. Use the same procedure to test the View Course Grades and View Current Schedule command buttons.

7. Close the browser and close the form in Forms Builder.

You now know the basic steps for creating an application that contains multiple canvases and multiple data blocks. The main points to remember are:

- The first canvas that appears when you run the form is the canvas whose block items appear first in the Object Navigator Data Blocks list.

- When you create a form that contains multiple canvases and multiple data blocks, be careful to place new form items within the correct data block. Otherwise, errors will occur when you try to reference the items.

- You can reference items in any trigger using the syntax: `block_name.item_name`.

CREATING AND CONFIGURING TAB CANVASES IN FORMS

So far, all the forms you have created use content canvases, which provide the basic background for a form window. Forms Builder also supports **tab canvases**, which are multiple-page canvases that allow users to move between multiple canvas surfaces by selecting tabs at the top of the canvas. You can use a tab canvas to display a large number of related items in a modular way, or to direct a user through a sequence of steps for performing a task. Now you create a form that uses a tab canvas to support the Northwoods University Course List system.

Creating a Tab Canvas

A tab canvas lies on top of a content canvas within the form window. The tab canvas contains two or more tab pages. A **tab page** is an object representing a surface that displays form items, and has a tab label identifier at the top. Figure 6-44 shows the components of the tab canvas for the Northwoods University Course List system. The system has two tab pages that the user can access to retrieve information for various sections and courses.

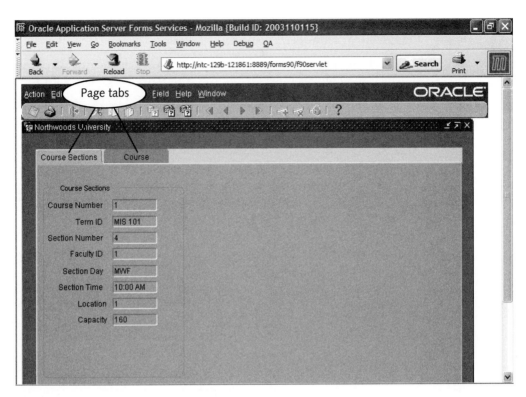

Figure 6-44   Tab canvas components

To create a form that contains a tab canvas, you create the form, and then create a new content canvas in the form. Then, you create a tab canvas on the content canvas, and configure the tab pages.

Creating a Tab Canvas

To create a tab canvas, you use the Tab Canvas tool 🗒 on the Layout Editor tool palette to draw the tab canvas area on the content canvas. Now you create the form module for the Course Listing system that uses a tab canvas. You create a content canvas in the form, and then draw a tab canvas on the content canvas.

To create the form, content canvas, and tab canvas:

1. In the Forms Builder Object Navigator, select the Forms node, and then click the **Create** button ➕ to create a new form module. Change the form module name to **COURSE_SEC_FORM**. Save the form as **Ch6CCourse _Sec_Form.fmb** in the Chapter06\Tutorials folder on your Solution Disk.

2. Open the Windows node, and change the default window name to **COURSE_SEC_WINDOW**. Open the COURSE_SEC_WINDOW

Property Palette, and change the Title property to **Northwoods University**. Then close the Property Palette.

3. Select the Canvases node, and then click ✚ to create a new content canvas. Change the content canvas name to **COURSE_SEC_CANVAS**.

4. Open the Layout Editor. Currently, the canvas is blank. Select the **Tab Canvas** tool 📋 on the tool palette, and draw a large rectangle on the canvas to represent the tab canvas, as shown in Figure 6-45. Note that by default, a new tab canvas contains two pages.

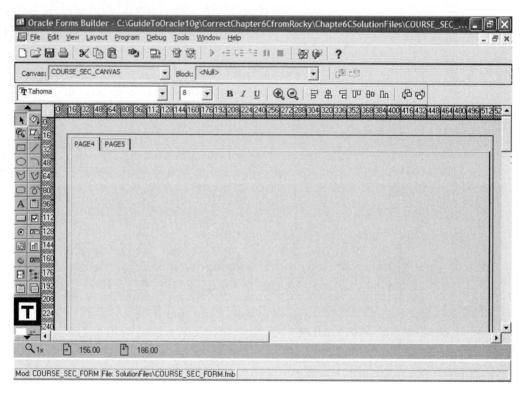

Figure 6-45   Creating a new tab canvas

Creating and Configuring Tab Pages

When you use the Tab Canvas tool to create a new tab canvas, Forms Builder creates a new tab canvas object in the Object Navigator. The tab canvas object contains a Tab Pages node. The Tab Pages node contains a tab page object to represent each tab page on the canvas. To create a new tab page, you select the Tab Pages node, and click the Create button. Figure 6-46 shows the Object Navigator tab canvas components.

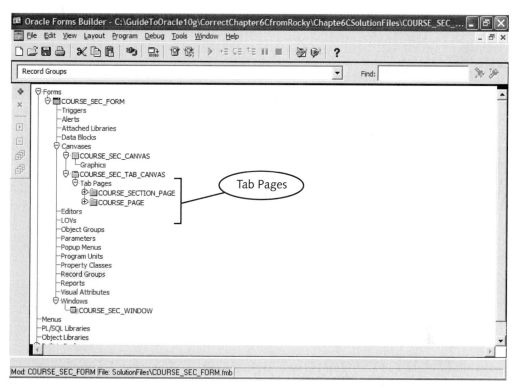

Figure 6-46   Tab canvas components in the Object Navigator

Now you examine the tab canvas and tab pages in the Object Navigator, and change the names of the tab canvas and tab pages. Currently, the tab canvas contains two tab pages.

To work with the tab canvas and tab pages in the Object Navigator:

1. Open the Object Navigator, and note that a new tab canvas appears under the Canvases node. (Note that in the Object Navigator tree, the Canvas icon [ ] identifies content canvases, and the Tab Canvas icon [ ] identifies tab canvases.) Change the name of the new tab canvas to **COURSE_SEC_TAB_CANVAS**.

2. Open the COURSE_SEC_TAB_CANVAS node, and then open the Tab Pages node. The Tab Pages node appears, and contains two page objects.

3. Change the name of the first tab page to **COURSE_SECTION_PAGE** and the second page to **COURSE_PAGE**. When you finish, your Object Navigator window should look like Figure 6-46.

4. Save the form.

The order in which the tab pages appear in the Object Navigator specifies the order in which the tab pages appear on the tab canvas. The first tab page appears on the left edge of the tab canvas, the second tab page is next, and so forth.

After you create tab pages, you modify the properties of the tab page. You change the labels that appear on the tab pages by specifying the desired label text in the Label property on the tab page's Property Palette. You can use the Background Color property to change the background color of the tab pages. Now you open the tab page Property Palettes, and change the labels and background colors.

To change the tab page properties:

1. Select the COURSE_SECTION_PAGE node, right-click, and then click **Property Palette**. Change the Label property value to **Course Sections**.

2. Select the **Background Color** property, click the **More** button [...], and select a light gray square. The color name (such as "gray12") appears as the property value. Close the Property Palette.

3. Open the Property Palette of COURSE_PAGE, and change the Label property value to **Courses**. Select the **Background Color** property, click [...], select the same light gray square you selected in Step 2, and then close the Property Palette and save the form.

It is a good design practice to use the same background color for all tab canvas pages.

Creating Form Items on a Tab Canvas

Now you can use the tab pages just as you would use any content canvas. You can create items on the tab pages using the Data Block Wizard, or by creating a control block and drawing text items directly on the tab canvases. You must be sure to specify placement of the items on the tab canvas, and select the tab page within the tab canvas that displays the item. Now you create a data block on the Course Section tab page that corresponds to the COURSE_SECTION database table. Then you run the form and test the tab canvas.

To create the data block on the tab page, and then run the form:

1. In the Object Navigator, click the **Data Block Wizard** button [icon] on the toolbar. The Welcome page appears. Click **Next**.

2. Make sure that the Table or View option button is selected, and then click **Next**. On the Table page, click **Browse**, select the **COURSE_SECTION** database table, and click **OK**.

3. Select all the columns to include in the data block, return the **C_SEC_ID** column to the Available Columns list, and then click Next.

4. Accept COURSE_SECTION as the block name, click **Next**, and then click **Finish**. The Layout Wizard Welcome page appears. Click **Next**.

5. On the Canvas page, open the **Canvas** list, and note that both the content canvas and the tab canvas appear. Because you want the items to appear on the tab canvas, select **COURSE_SEC_TAB_CANVAS**. Because the tab canvas has multiple pages, you must also specify the tab page, so make sure that **COURSE_SECTION_PAGE** is selected in the Tab Page list. Then click **Next**.

6. Click the **Move all items to target** button [>>] to select all of the data block items for the layout, and then click **Next**.

7. Change the prompts to match the prompts displayed in Figure 6-44, and then click **Next**.

8. Accept Form as the layout style, click **Next**, type **Course Sections** for the frame title, accept the other default values, click **Next**, and then click **Finish**. The canvas layout appears in the Layout Editor. Resize the frame and align the items so they appear as shown in Figure 6-44.

9. Save the form, and then run the form. The tab canvas appears and displays the Course Section tab page, as shown in Figure 6-44.

10. Click the **Courses** tab. The Courses tab page appears. It does not display any items, because you haven't created any items on it yet.

11. Close the browser window, and then close the form in Forms Builder.

To complete the Courses tab page, you simply repeat the steps required to select items for the Course Sections tab page, save, and then run the form again.

In summary, use the following steps to create a form that uses a tab canvas:

1. Create a content canvas in the form.

2. Create a tab canvas on the content canvas.

3. Create and configure the individual tab pages.

4. Add block and layout items to the individual tab pages.

SUMMARY

- You can convert a data block to a control block by changing the data block's Database Data Block property value to No, and changing the Required property value of the text item that represents the data block table's primary key to No.

- To create a relationship between a control block and a data block, create the data block without a master-detail relationship, and then modify the data block's WHERE Clause property so the matching key field on the data block is equal to the value of the master key field on the control block.

❐ To update a data block in a form, you must first move the form focus to the block using the GO_BLOCK built-in, and then issue the EXECUTE_QUERY command, which automatically flushes the block and makes its information consistent with the block's corresponding database data.

❐ It is a good practice to display information on different display screens to avoid placing more information on a single screen than a user can view without scrolling.

❐ To create a form that displays information on multiple canvases, you can use either a single-form approach or a multiple-form approach. With a single-form approach, you create a single form with multiple canvases. With a multiple-form approach, you create multiple forms with a single canvas and then integrate the forms into a single application.

❐ When you create a single form that has multiple canvases, the canvas that appears first is the canvas whose block items appear first in the Object Navigator Data Blocks node.

❐ To display a different canvas in a multiple-canvas form, use the GO_ITEM built-in to move the insertion point to a text item on the canvas to be displayed.

❐ When you create a form that contains multiple canvases and multiple data blocks, be careful to place new form items within the correct data block. In a form with multiple blocks, every item within a data block must have a unique name so the combination of the block name and the item name uniquely identifies each form item.

❐ A tab canvas is a multiple-page canvas that allows users to move between multiple canvas surfaces by clicking tabs at the top of the canvas. You use a tab canvas to display a large number of related items in a modular way, or to direct a user through a sequence of steps for performing a task.

REVIEW QUESTIONS

1. By default, when you create a data block using the Data Block Wizard, the Database Data Block property value is set to _____.

2. To create a control block, you must use the Control Block Wizard or set the Control Block property value to Yes. True or False?

3. List the steps necessary to link a control block to a data block.

4. You can use the Data Block Wizard to create a master-detail relationship between a data block and a control block. True or False?

5. Modify the data block's _____ Clause property to link a master control block and a detail data block.

6. What is the difference between using the single-form approach and multiple-form approach to create forms?

7. How do you specify which canvas appears first when a form contains multiple canvases?

8. Items specified within a block cannot be referenced by another block in the same form. True or False?

9. Provide two rationales for using tab canvases in a form.

10. By default _____ tab pages are created for each tab canvas object, but you can add more if necessary.

MULTIPLE CHOICE

1. Which of the following built-in procedures is used to refresh the data displayed in a data block when the contents of the control block change?

   a. GO_BLOCK

   b. REFRESH_BLOCK

   c. UPDATE_DATA

   d. EXECUTE_QUERY

2. Which of the following properties is used to link a control block and a detail data block?

   a. the data block's WHERE Clause property

   b. the control block's WHERE Clause property

   c. the data block's GROUP property

   d. none of the above

3. Which of the following is a valid statement?

   a. A form cannot contain two text items with the same name.

   b. A data block can contain two text items with the same name as long as they have different prompts.

   c. A data block can contain two text items with the same name as long as they have different widths.

   d. A form can contain two text items with the same name if they are placed in different data blocks.

4. The sequence of the _____ objects determines which canvas appears first in the form.

   a. trigger

   b. data block

   c. canvas

   d. window

5. If you have multiple canvases within a form, use the _____ built-in procedure to display a different canvas.

   a. GO_TO

   b. GO_ITEM

   c. GO_CANVAS

   d. EXECUTE_QUERY

6. When you are creating a form that contains tab pages, which of the following must be created first?

   a. tab page

   b. tab canvas

   c. content canvas

   d. text item

7. You convert a data block to a control block by changing its _____ and Required properties to No.

   a. Automatically Update

   b. QUERY Link

   c. Match Data

   d. Database Data Block

8. To display a tab name in the tab at the top of a tab page, specify the name in the page's _____ property.

   a. Label

   b. Title

   c. Prompt

   d. Name

9. To allow a control block to retrieve data values from an LOV, set the control block's _____ property to No.

   a. Database Data Block

   b. LOV Retrieve

   c. Update from List

   d. Layout Source

10. To identify a specific text item within a form, you should preface the text item name with the appropriate:

    a. form name

    b. canvas name

    c. window name

    d. block name

PROBLEM-SOLVING CASES

All cases refer to the Clearwater Traders sample database. Run the Ch6Clearwater.sql script in the Chapter06 folder on your Data Disk to refresh the tables in the sample database. Rename all form components using descriptive names. Make all form items have a top-to-bottom, left-to-right external navigation order. Data files for all cases are in the Chapter06 folder on your Data Disk. Save the text for all SQL commands in a file named Ch6CQueries.sql in the Chapter06\Cases folder on your Solution Disk. Save all solutions in the Chapter06\Cases folder on your Solution Disk.

1. Create the Customer Order form shown in Figure 6-47. For the Create New button in the Orders block, have the button generate a new order sequence number from the O_ID_SEQUENCE. To drop, if necessary, an existing sequence and recreate the sequence, enter the following command in SQL*Plus:

   ```
   Drop SEQUENCE o_id_sequence; CREATE SEQUENCE o_id_sequence
   START WITH 100;
   ```

   The Save button should be designed to save the completed order information to the ORDER table of the Clearwater Traders database. Save the completed form as Ch6CCase1.fmb.

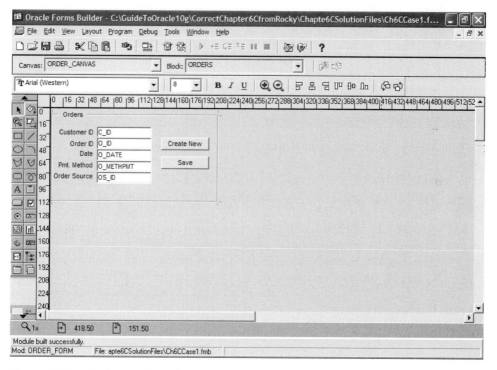

Figure 6-47   Customer Order form

2. Modify the Customer Order form created in the previous case to include an Order Line detail block as shown in Figure 6-48. Make certain that the Orders block is the control block and that the Order Line block is the detail data block. Include the necessary commands to make each command button functional. Save the modified form as Ch6CCase2.fmb.

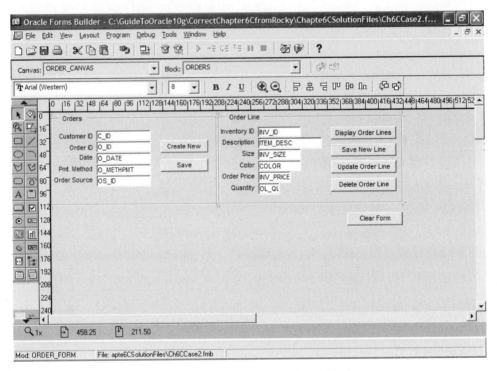

Figure 6-48   Customer Order form with Order Line detail block

3. Modify the Customer Order form from Case 2 to include an Order Summary detail data block, as shown in Figure 6-49. The purpose of the Order Summary data block is to display all items contained on a particular order. Therefore, if any changes are made through the Order Line data block, the changes should automatically be reflected in the Order Summary detail data block. Make certain you perform the necessary steps so the Order Line block is the control block for the Order Summary data block. Add a trigger to the Clear Form button that resets the data values displayed in the form to blank spaces. Save the completed form as Ch6CCase3.fmb.

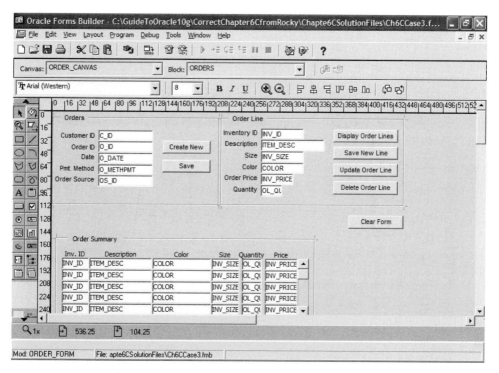

Figure 6-49 Modified Customer Order form with an Order Summary section

4. Modify the Customer Order form from Case 3 to ensure that if a change is made in one data block, the data displayed in any remaining detail block is automatically updated. (*Hint*: Add the appropriate built-in procedures to the form's command buttons.) Save the modified form as Ch6CCase4.fmb.

5. Create the form shown in Figure 6-50. Make certain the tab canvas contains three pages which are labeled as shown in the figure. Only include text items for the first tab page. Save the form as Ch6CCase5.fmb and run the form. Test the form by clicking each tab to make the various pages visible, even if they do not display any data.

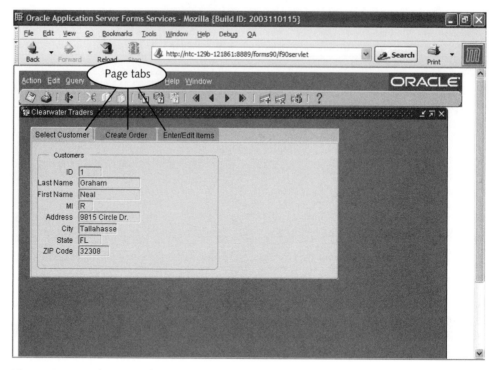

Figure 6-50   Tab canvas form

7

CREATING DATABASE REPORTS

◀ LESSON A ▶

After completing this lesson, you should be able to:
♦ Use the Reports Builder report styles
♦ Use the Report Wizard to create a report
♦ Configure the appearance of a report
♦ View a report in a Web browser
♦ Create a master-detail report
♦ Create a custom template
♦ Apply a custom template to a report

You have learned how to create forms that allow users to insert, update, delete, and view data. In this chapter, you learn how to use Reports Builder, which is the Oracle 10*g* Developer utility for creating reports. A **report** is a summary view of database data that users can view on a screen or print on paper. Reports retrieve database data using SQL queries, perform mathematical or summary calculations on the data, and format the output to create documents such as tabular reports, invoices, or form letters.

INTRODUCTION TO REPORTS BUILDER DATABASE REPORTS

Reports Builder allows application developers to create reports that display data from an Oracle 10*g* database. In Developer10*g*, Reports Builder allows developers to preview and distribute reports in a variety of different formats, including Web pages and portable document format (.pdf) files. (A .pdf file appears the same on the screen and in print, regardless of what kind of computer or printer the user has.)

Reports Builder reports appear in the following layout styles:

- **Tabular**—Presents data in a tabular format with columns and rows. Figure 7-1 shows a tabular report that appears in a Web browser. This report displays a listing of the terms stored in the Northwoods University database, including the term ID, description, status, and start date of each term.

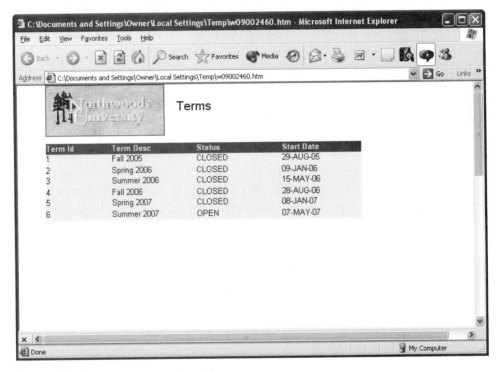

Figure 7-1   Tabular report in a Web browser

- **Form**—Resembles a Forms Builder form-style layout. This report style displays one record per page and shows data values to the right of field labels.
- **Mailing label**—Prints mailing labels in multiple columns on each page
- **Form letter**—Includes the recipient's name and address (which are stored in the database), as well as other database values embedded in the text of a letter

- **Group left** and **group above**—Display master-detail relationships. Recall that in a master-detail relationship, one master record has one or more related detail records, and the DBMS establishes the relationship using foreign keys. The Northwoods University TERM, COURSE, and COURSE_SECTION tables have a master-detail relationship, because each TERM record can have many associated COURSE records, linked by sections in the COURSE_SECTION table. Figure 7-2 shows a group left report for this relationship.

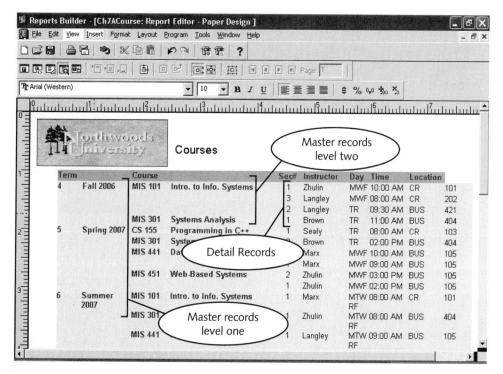

Figure 7-2   Group left report

In this group left report, the master items, which are the term ID and term description, appear on the left side of the report. A detailed listing of the corresponding detail COURSE records appears to the right of the master values, along with the relevant section information from the COURSE_SECTION table. Figure 7-3 shows the same data in a report that has the group above style, in which each master item appears above the detail lines.

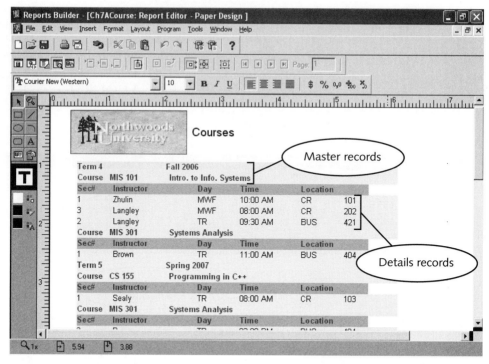

Figure 7-3 Group above report

- **Matrix**—Displays field headings across the top and down the left side of the page. A matrix layout displays data at the intersection point of two data values. Figure 7-4 shows an example of a matrix report in which Northwoods University student names appear in the row headings, course call IDs appear in the column headings, and the specific student's grade for the associated course appears at the intersection.

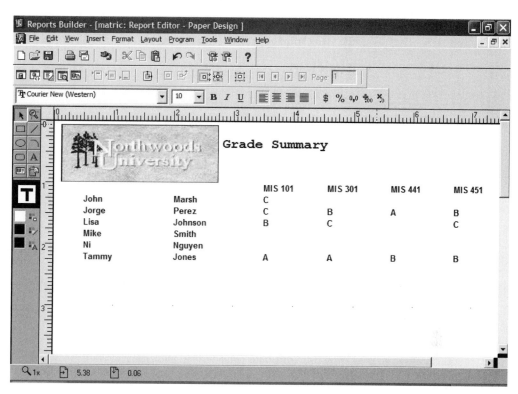

Figure 7-4   Matrix report

- **Matrix with group**—Displays a detail matrix for a master-detail relationship. Figure 7-5 shows an example of a matrix with group report, in which the master record is the Northwoods University student name, and the detail matrix shows course call IDs in the rows, term descriptions in the columns, and the student's grade at the intersection.

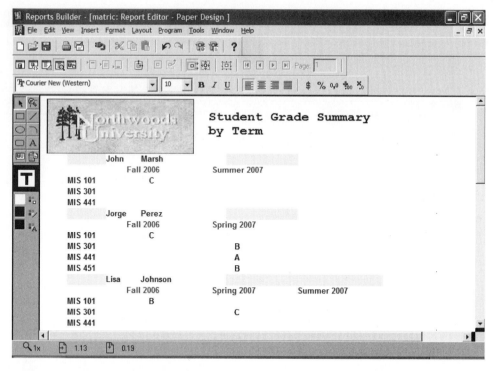

Figure 7-5   Matrix with group report

In Developer10*g*, you can create reports that can appear in a browser window and print on paper, or that can only print on paper. The tabular, group left, group above, matrix, and matrix with group report styles can either appear in a Web browser or print on paper. The form, form letter, and mailing label report styles can only print on paper. You can view a screen display that shows how the form, form letter, and mailing label report styles will appear when they print on paper, but you cannot view these styles in a browser window.

You are now familiar with the different kinds of reports that Reports Builder creates, from simple tabular reports to more complex master-detail and matrix reports. In the following section, you learn how to use Reports Builder to create a tabular report.

USING THE REPORT WIZARD TO CREATE A REPORT

Creating a report is a three-step process:

1. Specify the data that the report displays.

2. Select the report style.

3. Configure the report properties and layout.

Before you create a report, you need to identify the data the report is going to display, make a design sketch for the report, and then select the appropriate report style. If the report displays data that has a master-detail relationship, you must select one of the master-detail report styles (group above or group left). Then you create the report.

To automate the report creation process, you use the Report Wizard. The Report Wizard is similar to the Data Block and Layout Wizards you used in Chapters 5 and 6 to create forms. The Report Wizard has the following pages:

- **Welcome page**—Welcomes you to the Report Wizard

- **Report type page**—Allows you to specify how you want to display the report output within the Reports Builder development environment. Normally, you develop reports based on how you plan to display the output. You can specify to display the report as a Web page, as a paper document, or as either a Web page or a paper document.

- **Style page**—Allows you to specify the title that appears on the report, and select the report style, such as tabular or group above. If you specify on the Report type page to display the report output only as a Web page, then the form, form letter, and mailing label report styles are disabled, because these report styles can print only on paper.

- **Data source page**—Lists all the data sources for report data that are available on your workstation. Reports Builder allows you to create reports using a variety of data sources, such as data from databases other than Oracle 10*g* or data from XML files, which are text files that store data using a standard structure. In this book, you retrieve report data using SQL queries that retrieve data from an Oracle 10*g* database.

- **Data page**—Allows you to specify the SQL query that retrieves the report data. You can enter SQL queries using one of three methods: by directly typing a SQL query, by importing a query from a script file, or by building a SQL query using Query Builder. Query Builder is a high-level tool that has a graphical user interface for creating SQL queries. In this book, you directly type SQL queries for all report data sources.

- **Fields page**—Allows you to select the query fields that appear in the report

- **Totals page**—Allows you to select fields on which to calculate a group function, such as calculating the average of the field values

- **Labels page**—Allows you to specify the report field labels and widths

- **Template page**—Allows you to select a report template that defines the characteristics of the report appearance, such as fonts, graphics, and color highlights. You can use one of the predefined templates, select a template that you have created, or opt to format the report manually by not selecting a template. When you select one of the predefined templates, a representative example of the template appears on the Template page.

- **Finish page**—Signals that the Report Wizard has finished creating the report

7

Next you create the Student report in Figure 7-6 that displays information about Northwoods University students.

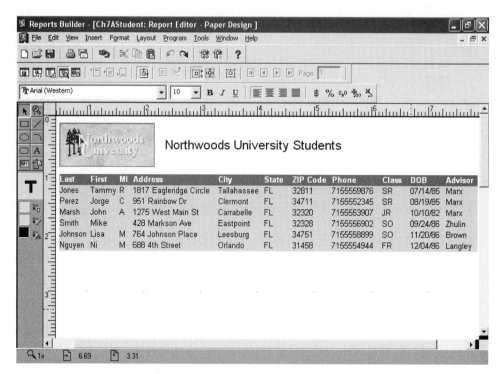

Figure 7-6   Student report

To ensure that you are starting with a complete database, you first run the script file in the Chapter07 folder on your Data Disk. Then you start Reports Builder and create the report.

To run the scripts, start Reports Builder, and create the report:

1. Start SQL*Plus and log onto the database. Run the **Ch7Northwoods.sql** scripts in the Chapter07 folder on your Data Disk to refresh your database tables. Then exit SQL*Plus.

2. To start Reports Builder, click **Start** on the taskbar, point to **All Programs**, point to **Oracle 10g Developer Suite—ORADEV**, point to **Reports Developer**, and then click **Reports Builder**. The Welcome to Reports Builder dialog box opens. This dialog box enables you to build a new report using the Report Wizard, build a new report manually, or open an existing report. You use the Report Wizard to create the new report.

3. Make sure that the **Use the Report Wizard** option button is selected, and then click **OK**. When the Report Wizard Welcome page appears, click **Next**.

4. The Report type page appears, which allows you to specify how the report output appears in the Reports Builder environment. This report appears both as a Web page and as a paper document, so be sure that the **Create both Web and Paper Layout** option button is selected, and then click **Next**.

5. The Style page appears, which allows you to specify the report title and style. Type **Northwoods University Students** in the Title field, make sure that the **Tabular** option button is selected, and then click **Next**.

6. The Data source page appears, which allows you to specify the source of the report data. Make sure that **SQL Query** is selected, and then click **Next**.

7. The Data page appears, which allows you to type a SQL query, import a query from a script file, or build a SQL query using Query Builder. Type the following query to retrieve the report data fields, and then click **Next**. Note that you do not need to type a semicolon at the end of the command:

```
SELECT s_last, s_first, s_mi, s_address, s_city,
s_state, s_zip, s_phone, s_class, s_dob, f_last
FROM student NATURAL JOIN faculty
```

8. Because you have not yet connected to the database, the Connect dialog box opens. Log onto the database in the usual way, and then click **Connect**. If your SQL query is correct, the Fields page appears, showing the fields the query returns. If the Fields page does not appear, debug your query until it works correctly.

The Report Wizard Fields page shows the data fields that the SQL query returns. You can select one or more fields in the Available Fields list to display in the report. Note that the icons beside the fieldnames indicate the field data types. Now you select the report fields, and specify the rest of the report properties.

To select the report fields and specify the other report properties:

1. Click the **Move all items to target** button ⟩⟩ to select all query fields for the report. The Available Fields list clears, and all of the query fields now appear in the Displayed Fields list. Click **Next**.

2. The Totals page appears, allowing you to specify one or more fields for which you might want to calculate a total. None of the fields in this report require totals, so don't select any fields. Click **Next**.

3. The Labels page appears, which allows you to specify the report labels and field widths. Modify the field labels and widths as follows, and then click **Next**.

Field	Label	Width
s_last	Last	10
s_first	First	10
s_mi	MI	1
s_address	Address	10
s_city	City	10
s_state	State	2
s_zip	ZIP Code	9
s_phone	Phone	10
s_class	Class	2
s_dob	DOB	9
f_last	Advisor	10

4. The Template page appears, which allows you to select a report template to define the characteristics of the report appearance. To select a predefined template, make sure the **Predefined template** option button is selected. Note that a representative example of the current Predefined template selection, which is Beige, appears on the Template page. Select **Peach** in the Predefined template list, and note that the color scheme for the example template changes. Click **Next**, and then click **Finish**. The report appears in the Report Editor – Paper Design window.

5. Maximize the Paper Design window, and maximize the Reports Builder window. Your screen should look similar to Figure 7-7.

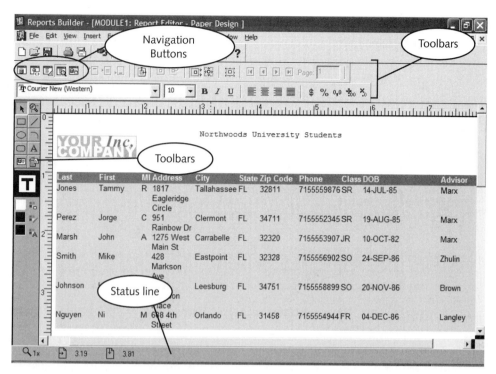

Figure 7-7   Paper Design window

CONFIGURING THE REPORT APPEARANCE

After you use the Report Wizard to create a new report, you use the Reports Builder environment to modify the report appearance and customize the report features. Recall that Reports Builder allows you to design reports that ultimately appear on paper or as Web pages. The **Paper Design window**, which is the default editing environment, provides an environment for refining the appearance of reports, and shows how the report will appear on paper. If a report is to appear as a Web page, you preview the report as a Web page, and then alter the design in the Paper Design window as necessary. The Paper Design window has toolbars for working in the Reports Builder environment, a tool palette for altering the report's appearance, and a status line for showing the current zoom status and pointer position.

Figure 7-7 shows the navigation buttons, which allow you to move to different windows within the Reports Builder environment. The Data Model button allows you to open the Data Model window, which you use to modify the report data. The Web Source button allows you to open the Web Source window, which displays the underlying HTML (Hypertext Markup Language) code that defines how the report will appear in a Web browser. (HTML is a document-layout language that defines the content and

524 Chapter 7 Lesson A   Creating Database Reports

appearance of Web pages.) The Paper Layout button opens the report in Paper Layout view, which displays the report components as symbols, and shows the relationships among the report components. The Paper Design button reopens the Paper Design window. The Paper Parameter Form button allows you to view the parameter form in the Paper Parameter Form window. On a parameter form the user selects input parameter values to customize the form appearance and functionality at runtime. You learn more about the Data Model, Web Source, Paper Layout, and Paper Parameter Form windows later in the chapter.

Next, you save the report. You can save a report design specification using a variety of formats. In this book, you save report design files as Reports Builder design files, which have an .rdf extension.

To save the report:

1. Click the **Save** button on the toolbar. The Save dialog box opens.

2. Navigate to the Chapter07\Tutorials folder on your Solution Disk, then type **Ch7AStudent** in the File name field.

3. Open the **Save as type** list, select **Reports Binary (*.rdf)**, and then click **Save**.

You can use the Paper Design window to modify report characteristics such as the position of report objects, font sizes and styles, the width of the report columns, and the format of the column data. You can also open the Report Wizard in reentrant mode and modify properties such as what data the report displays, and the order in which the fields and records appear. However, be careful when you modify the report by opening the Report Wizard in reentrant mode: if you modify the report in the Paper Design window, then open the Report Wizard in reentrant mode and change the report properties, you lose the changes you made in the Paper Design window. It is a good practice first to make sure that the report displays the correct data values before you format the report in the Paper Design window.

Using the Report Wizard in Reentrant Mode

Currently, the report does not display the student data records in any particular order. Now you open the Report Wizard in reentrant mode, and modify the SQL query so it uses the ORDER BY clause to sort the student records by the S_LAST field.

To modify the report using the Report Wizard in reentrant mode:

1. Click **Tools** on the menu bar, and then click **Report Wizard**. The Report Wizard window opens and displays tabs at the top of the page display, which indicates the Wizard is in reentrant mode.

2. Select the **Data** tab, then add the following ORDER BY clause as the last line of the SQL query, as shown in Figure 7-8:

```
ORDER BY s_last
```

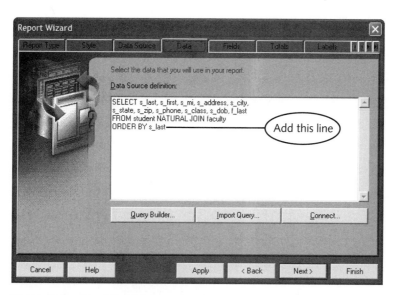

Figure 7-8   Modified SQL query

    3. Click **Finish** to save the change and close the Report Wizard. The report data records now appear sorted by student last names.

The Reports Builder Object Navigator Window

As with the Object Navigator window in Forms Builder, you use the Reports Builder Object Navigator to view report components in a hierarchical tree structure, to access different components in the Reports Builder environment, and to access the components of an individual report. To view a report in the Object Navigator, you click Window on the menu bar, and then click Object Navigator. Now you open the Student report in the Object Navigator, and view its components.

To view the report in the Object Navigator:

    1. Click **Window** on the menu bar, and then click **Object Navigator**.

    2. The Object Navigator window opens, as shown in Figure 7-9.

Different nodes may be open in your window, depending on the items you have selected in your report.

Figure 7-9   Reports Builder Object Navigator

The top-level node is Reports. Currently, the CH7ASTUDENT report is open. Other environment objects include Templates, PL/SQL Libraries, Debug Actions, Stack, Built-in Packages, and Database Objects. Templates specify report formatting; PL/SQL Libraries are collections of related PL/SQL functions or procedures; and Debug Actions are actions you can enable or disable while using the Forms Debugger. The Stack node shows current values of local variables during a debugging session. Built-in Packages are code libraries provided by Oracle Corporation to simplify common tasks, and Database Objects enable you to access all the objects in the database, such as users, tables, sequences, and triggers.

A report contains nodes representing the report's Data Model, Web Source, Paper Layout, and Paper Parameter Form. You can double-click the icon beside any of these nodes to open the associated window, or open the node to view its components. You learn about individual components within each of these windows later in the chapter.

Modifying the Report Appearance in the Paper Design Window

The Paper Design window shows how the report will look when you print the report on paper. Currently, the report needs some modifications to enhance its appearance. The title appears in a Courier font, which is different than the font of the other report items. Some of the report columns are too wide, other columns are too narrow, and the data wraps to multiple lines. The report should display the Northwoods University logo, rather than the default "Your Company Inc." logo. The following sections describe how to use the Paper Design window to modify the report's appearance so it looks like the formatted report in Figure 7-10.

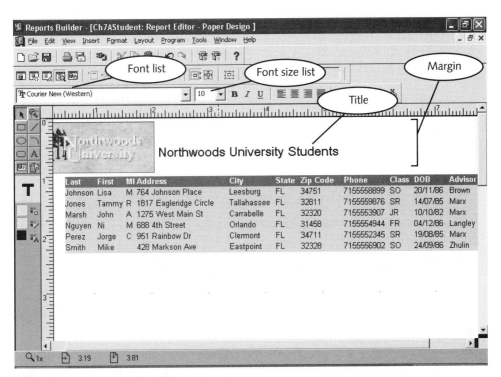

Figure 7-10   Formatted Student report

Modifying the Report Title

The report title appears in the report **margin** (see Figure 7-10), which is the area on the page beyond where the report data appears. The report margin can contain boiler-plate text and graphics, such as a company logo, the date the report was created, the current report page number, or all of these elements. The top margin on the Student report displays the logo from the predefined template and the report title (Northwoods University Students). First you change the title position and font size and style.

To modify the report title:

1. In the Object Navigator, with the CH7ASTUDENT report selected, click **Program** on the menu bar, and then click **Run Paper Layout** to open the Paper Design window.

2. In the Paper Design window, select the **report title**. Selection handles appear around the report title. Drag the title down and to the left so that its left edge is about **0.25** inches from the right edge of the logo, and so that the title is centered vertically between the top and bottom edges of the logo.

TIP

You can make fine adjustments to an object's position by selecting the object and then moving it using the arrow keys on the keyboard.

3. With the title still selected, open the **Font** list on the toolbar (see Figure 7-10), and select **Arial (Western)**. Open the **Font Size** list (see Figure 7-10), and select **16**. If necessary, resize the box that displays the title text so all of the text appears.

4. Save the report.

Adjusting the Column Widths

Next, you adjust the data column widths. Figure 7-7 shows that the Address column wraps to multiple lines. The report would be easier to read if these columns were wider. Also, some of the report fields, such as the Last and First fields, are wider than they need to be. To adjust a column's width in the Paper Design window, you select the column and then drag to make it wider or narrower.

To modify the column widths and positions in the Paper Design window:

1. In the Paper Design window, select any student last name data value to select the values as a group. Selection handles appear around all of the last name values.

2. Select the center selection handle on the right side of any of the last name values, and then drag the mouse pointer toward the left edge of the screen display to make the column narrower. Adjust the width so the column is as narrow as possible, but so that none of the names wrap to the next line.

3. Repeat Step 2 for the student first name data values.

4. Select any of the middle initial values, then drag the column to the left edge of the screen so the middle initial field values appear beside the first name field values.

5. Adjust the remaining column widths and positions so the report columns look like Figure 7-10. You may need to move the advisor field values to the right edge of the screen to make the report field area wider.

6. Save the report.

Applying Format Masks to Report Columns

In Reports Builder, you can specify format masks for fields that display NUMBER and DATE data. You cannot specify format masks for fields that display CHAR and VARCHAR2 data. To change the format mask of a column, you select the column in the Paper Design window, open the column's Property Inspector, and then enter the desired format mask. The **Property Inspector** is a window that displays properties and associated values for report objects, and is similar to the Forms Builder Property Palette.

The format mask property allows you to choose from a list of predefined format masks, or enter a customized format mask. Now you modify the format mask of the column that displays each student's date of birth.

To change the date field's format mask:

1. Click any of the DOB data values. Selection handles appear around all of the data values to indicate that the column is selected.

2. Right-click, and then click **Property Inspector**. The Property Inspector opens.

3. Locate the Format Mask property under the Field node. Select **MM/DD/RR** from the drop-down list as the property value and then close the Property Inspector. The dates appear in the new format mask. Adjust the column widths if necessary so each date appears on a single line.

Importing a Graphic Image

Next, you replace the default template logo with the Northwoods University logo. You delete the default logo, and then import the graphic image of the Northwoods University logo. You import graphic images into reports just as you import graphic images into forms.

To replace the template logo with the Northwoods University logo:

1. Select the template logo, and then press **Delete**. The template logo no longer appears.

2. To import the Northwoods University logo, click **Insert** on the menu bar, and then click **Image**. The Import Image dialog box opens. Click **Browse**, navigate to the Chapter07 folder on your Data Disk, select **Nwlogo.jpg**, click **Open**, and then click **OK**. The logo image appears in the Paper Design window.

3. Verify that the **logo image** is selected and resize it so it appears as in Figure 7-10. Adjust the position of the report title as necessary, and then save the report.

Closing and Reopening Reports

Sometime during this chapter you may want to take a break, so you need to learn how to close a report, and then reopen the report when you are ready to resume working. Now you close your report file, close Reports Builder, and then start Reports Builder again. You reopen your report file, and display the report in the Paper Design window again.

To close and then reopen Reports Builder:

1. Click **File** on the menu bar, and then click **Close**. The Reports Builder Object Navigator window appears, which shows the report components in a hierarchical tree.

2. Close Reports Builder, and then start it again. The Welcome to Reports Builder dialog box opens. Select the **Open an existing report** option button, and then click **OK**.

3. In the Open dialog box, navigate to the Chapter07\Tutorials folder on your Solution Disk, select **Ch7AStudent.rdf**, and then click **Open**. The report appears in the Object Navigator. Maximize the Object Navigator window, and maximize the Reports Builder window.

4. To open the form in the Paper Design window, click the **Run Paper Layout** button 📇 on the toolbar, and then connect to the database as usual. The report appears in the Paper Design window.

VIEWING THE REPORT AS A WEB PAGE

Reports Builder allows developers to make reports available in formats that appear in a Web browser. (You learn about these formats in the section titled "Creating Dynamic Web Pages in Reports Builder" later in this chapter.) If users ultimately are going to view a report in a Web browser, you should preview the report as a Web page to confirm that its formatting is correct. To preview a report as a Web page, you click the Run Web Layout button 📇 on the Reports Builder toolbar. This action creates the Web page source code for the report. The **Web page source code** is a file with an .htm extension that contains the Hypertext Markup Language (HTML) commands and text to represent the report content and formatting.

Hypertext Markup Language (HTML) is a page-layout language consisting of commands that define the appearance of a Web page. HTML files are text files that have an .htm or .html extension. When your Web browser displays a Web page, it translates the report data and layout into .htm file commands, and then displays the output in a browser window.

When you click the Run Web Layout button 📇 to view a report in a Web browser, Reports Builder displays the report using the default report formatting, which is how the report appears in the Paper Design window before you perform any custom formatting, such as changing the column widths or substituting an alternate graphic image for the logo. To view formatting changes in the Web page output, you must preview the report using either a Paginated HTML or a Paginated HTMLCSS format. A **Paginated HTML format** displays the output one page at a time. A **Paginated HTMLCSS format** displays the output one page at a time, and uses a Cascading Style Sheet (CSS) to define formatting styles. A Cascading Style Sheet is a special file that defines formatting specifications that you can apply to multiple Web pages. Now you view the report using both the default Web page format and the Paginated HTMLCSS format.

To view the report as a Web page:

1. In the Object Navigator, click the **Run Web Layout** button 📇 to generate and display the report as a Web page. The report appears in a browser window, as shown in Figure 7-11. (Maximize the window if necessary.) Note that the report appears in its default format, with the default template logo. Note also that the Web page .htm filename appears in the browser's Address field.

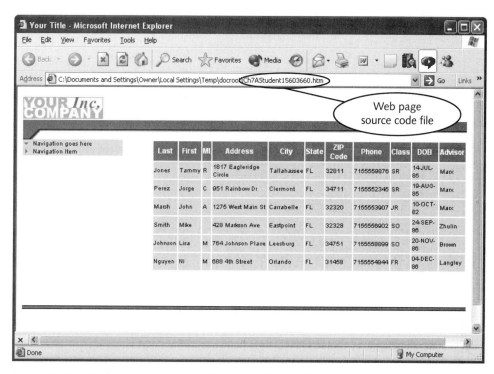

Figure 7-11   Viewing a report as a Web page (default format)

2. Close the browser window.

3. To view the Web page using the Paginated HTMLCSS format, click the **Run Paper Layout** button to open the Paper Design window. Click **File** on the menu bar, point to **Preview Format**, and then click **Paginated HTMLCSS**. The report appears in a browser window, and should look similar to Figure 7-12. (Your formatting may be slightly different.) Note that the report retains its custom formatting.

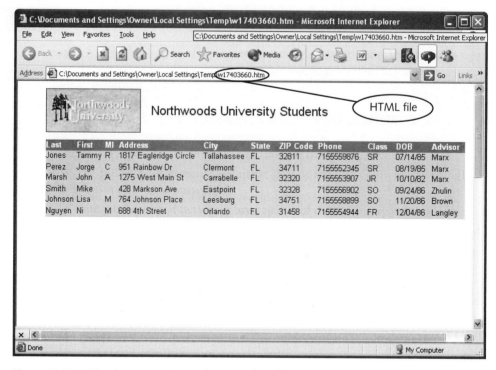

Figure 7-12 Viewing a report as a formatted Web page

4. Close the browser window.

5. Close the report in Reports Builder.

CREATING A MASTER-DETAIL REPORT

Recall that you can use Reports Builder to create reports that show master-detail data relationships in which one record has many associated detail records through a foreign key relationship. In this section, you create the Northwoods University Course report in Figure 7-13.

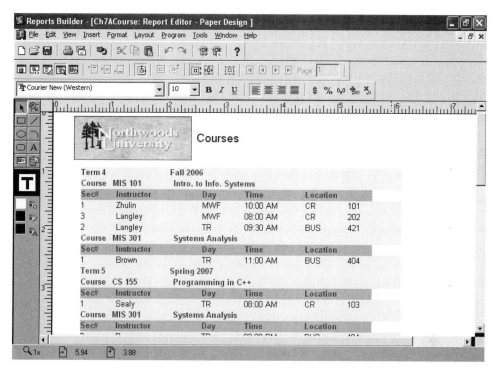

Figure 7-13   Northwoods University Course report

This report displays term IDs and term descriptions, and the call IDs and course names of all courses taught during each term. The report also lists information about each course, including the section number, instructor, day, time, and location. This report involves two master-detail relationships. A term can have multiple courses, so in this relationship, the term data is the master record, and the course data is the detail record. Each course can have multiple course sections, so in this second relationship, the course data is the master record, and the course section data is the detail record.

To create a master-detail report, you use the Report Wizard to specify the report style and data values, just as with the tabular report you created earlier. With a master-detail report, the report's SQL query must retrieve all the master and detail values that appear in the report. When you specify a report SQL query that contains master-detail relationships, the Report Wizard displays the Groups page, which allows you to define the master and detail groups. The following sections describe how to create the report and define the report master and detail groups.

Specifying the Style and Data

To create a new report in the Object Navigator, you select the Reports node, and then click the Create button ✚. A dialog box opens that provides the option of creating the report manually or creating the report using the Report Wizard.

To create the master-detail report:

1. In the Object Navigator, make sure that the Reports node is selected, and then click the **Create** button ✚ to create a new report. The New Report dialog box opens. Make sure that the **Use the Report Wizard** option button is selected, and then click **OK**.

2. When the Welcome page appears, click **Next**. The Report Type page appears. Make sure that the **Create both Web and Paper Layout** option button is selected, and then click **Next**. The Style page appears.

Recall that there are two report styles for creating master-detail reports: group left, in which the master records appear on the left side of the report and the detail records appear in columns to the right of the master records (see Figure 7-2); and group above, in which the detail records appear below the master records (see Figure 7-3). For the Course report, you use the group above style. Now you specify the report title and style, and select to use a SQL query to retrieve the report data.

To specify the title, style, and data source:

1. Type **Courses** in the Title field.

2. Select the **Group Above** option button, and then click **Next**. The Type page appears.

3. Make sure that **SQL Query** is selected, and then click **Next**. The Data page appears.

The report's SQL query must retrieve all master and detail records that appear in the report. Therefore, the SQL query retrieves the term, course, and course section data. Next, you specify the report's SQL query.

To specify the report's SQL query:

1. On the Data page, type the following query in the Data Source definition field:

```
SELECT term.term_id, term_desc, course.course_no, ↵
course_name,
sec_num, f_last, c_sec_day, c_sec_time, bldg_code, room
FROM term, course, course_section, faculty, location
WHERE term.term_id = course_section.term_id
AND course.course_no = course_section.course_no
AND course_section.f_id = faculty.f_id
AND course_section.loc_id = location.loc_id
```

2. Click **Next**. Because this query retrieves data with master-detail relationships (one term might have multiple courses, and one course might have multiple course sections), the Groups page appears.

Using the Groups Page to Specify Master-Detail Relationships

The Report Wizard Groups page specifies how the master-detail data values appear in the report. Data in a master-detail report has multiple levels. This report shows term information in the top level, course information in the next level, and course section information in the most detailed level. In a master-detail report, each data level represents a **group**. The top-level (master) group is Level 1, the next level is Level 2, the next level is Level 3, and so forth.

To specify report groups, you move the fields for each group from the Available Fields list to the Group Fields list. When you first select a field in the Available Fields list and move it into the Group Fields list, Reports Builder automatically creates a Level 1 heading, and places the selection in Level 1. To add additional fields to the Level 1 group, you select one of the fields currently in the Level 1 group, select the new field from the Available Fields list, and then add the new field to the Group Fields list. Figure 7-14 shows the Groups page after you select the term_id field to create the Level 1 group, and then add the term_desc field to the Level 1 group.

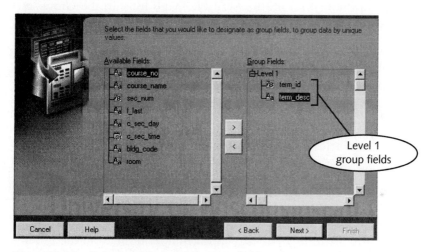

Figure 7-14   Selecting the Level 1 group fields

To create a Level 2 group, you select the Level 1 group in the Group Fields list, and then add a field from the Available Fields list that is to be in Level 2. When you move the selection to the Group Fields list, the Level 2 heading appears, with the selection under it. Figure 7-15 shows the Groups page after you create the Level 2 group, and add to it the course fields.

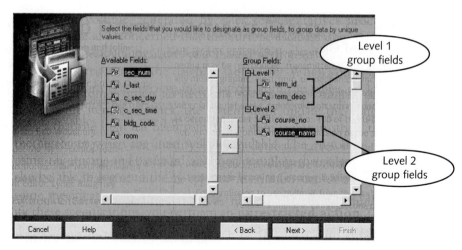

Figure 7-15   Creating the Level 2 group

If you move all of the fields from the Available Fields list into the Group Fields list, an error occurs. So, you must leave the most detailed data fields in the Available Fields list.

Now you create the Level 1 report group. This group contains the term ID and term description data values.

To specify the Level 1 report group:

1. Make sure that **term_id** is selected in the Available Fields list, and then click the **Move one item to target** button > . The Level 1 heading appears in the Group Fields list, with the term_id field below it.

2. Now you add the term_desc field to the Level 1 group. To add another field to Level 1, you first select an existing Level 1 field in the Group Fields list, so if necessary, select **term_id** in the Group Fields list.

3. Make sure that **term_desc** is selected in the Available Fields list, and then click > . The term_desc field appears under term_id in the Level 1 group definition, as shown in Figure 7-14.

HELP

If term_desc appears under a Level 2 heading, select it, click the Move one item to source button < to remove it from the Group Fields list, and then repeat Steps 2 and 3.

Next, you create the Level 2 group. Recall that to create a new group, you must select its parent group in the Group Fields list, select the first field of the new data group from the Available Fields list, and then click the Move one item to target button > . For this report's Level 2 group, you select Level 1 in the Group Fields list and then select the course_no field in the Available Fields list.

To create the Level 2 group:

1. Select **Level 1** in the Group Fields list, make sure that **course_no** is selected in the Available Fields list, and then click the **Move one item to target** button ⟩ . The Level 2 heading appears in the Group Fields list, with course_no under it.

2. To add the course_name field to the Level 2 group, make sure that **course_no** is selected in the Group Fields list, select **course_name** in the Available Fields list, and then click ⟩ . The completed Groups page looks like Figure 7-15.

Recall that the most detailed data fields must remain in the Available Fields list. Because all of the remaining fields in the Available Fields list reference a specific course section, they are in the Level 3 group, so the group specifications are complete.

The order in which the fields appear in the Group Fields list and Available Fields list reflects the order that the fields appear on the report. If you want to change the order of the fields, you select a field that is higher in the list and drag it down to a lower position.

To finish the report, you must specify the display fields, totals fields, field labels, and template. Then, you modify the report's appearance in the Paper Design window by repositioning the margin labels.

To finish the report:

1. On the Groups page, click **Next**. The Fields page appears.

2. On the Fields page, click the **Move all items to target** button ⟫ so that all of the report fields appear in the Displayed Fields list. Click **Next**. The Totals page appears.

3. Because none of the data fields are to appear as totals, do not select any fields. Click **Next**. The Labels page appears.

4. Modify the field labels and widths as follows. (Some of the fields are not intended to have labels, so "(deleted)" indicates that you should delete the current label value.)

Fields and Totals	Labels	Width
term_id	**Term**	8
term_desc	(deleted)	10
course_no	**Course**	5
course_name	(deleted)	10
sec_num	**Sec#**	4
f_last	**Instructor**	10
c_sec_day	**Day**	5
c_sec_time	**Time**	7
bldg_code	**Location**	5
room	(deleted)	6

5. Click **Next**. The Template page appears. Make sure that the **Predefined template** option button is selected, make sure that the **Beige** template is selected in the Predefined template list, click **Next**, and then click **Finish**. The report appears in the Paper Design window. Save the report as Ch7ACourse.rdf in the Chapter07\Tutorials folder on your Solution Disk.

6. Change the report title font to **14-point Arial (Western)**, and reposition the title so that it looks like the title in Figure 7-13. Note the relative positions of the group fields. The Level 2 fields appear under the Level 1 fields, and the Level 3 fields appear under each Level 2 field.

7. To display the Northwoods logo on the report, delete the current logo, click **Insert** on the menu bar, click **Image**, navigate to the Chapter07 folder on your Data Disk, select **Nwlogo.jpg**, click **Open**, and then click **OK**. Resize and reposition the image as necessary so your report looks like Figure 7-13.

8. The Time data field currently displays date values because it does not use a format mask that displays the time components of the dates. To change the format mask of the Time data field, select any time value, right-click, click **Property Inspector**, select the **Format Mask** property, open the list, select **HH:MI AM**, and then close the Property Inspector. The Time values appear as shown in Figure 7-13.

9. Save the report as **Ch7ACourse.rdf** in the Chapter07\Tutorials folder on your Solution Disk, and then close the report in Reports Builder.

REPORT TEMPLATES

When you create many reports that have a similar appearance in terms of fonts, graphics image, and background color, it is useful to create a custom template to specify the report appearance. Along with giving all of your reports a similar appearance, a custom template keeps you from having to perform the same formatting tasks over and over again. In the following sections, you learn how to create a custom template and how to apply custom templates to reports.

Creating a Custom Template

Many companies have design standards that specify the appearance of reports in terms of background colors, font sizes and styles, and graphic image enhancements. To enforce standards and save time in report development, you can apply custom templates to reports. A **custom template** defines the font sizes and styles for a report's title, column headings, and data values. The custom template also defines text and background colors and boilerplate objects, such as graphic images. When you create a custom template, Reports Builder stores the template definition in a template definition file that has a .tdf extension.

Next, you create a custom template for Northwoods University reports by modifying the existing Beige predefined template. The new custom template displays the Northwoods logo rather than the default logo, and displays the date the report was generated. First, you open the predefined template file and save the file using a different name.

To open and save the template file:

1. In the Object Navigator, click the **Open** button on the toolbar, navigate to the Chapter07 folder on your Data Disk, open the **Files of type** list and select **All Files (*.*)**, select **rwbeige.tdf**, and then click **Open**. RWBEIGE appears under the Templates node in the Object Navigator.

2. Click **File** on the menu bar, click **Save As**, and save the template file as **Northwoods.tdf** in the Chapter07\Tutorials folder on your Solution Disk. The template name changes to NORTHWOODS.

If the error message "REP-0069 Internal error" appears, click OK, and continue working.

The Object Navigator shows that a template has five components: Data Model, Paper Layout, Report Triggers, Program Units, and Attached Libraries. In this book, you work with the Paper Layout view, which defines the appearance of template objects.

This book does not explicitly address creating report triggers, program units, or attached libraries in templates because these are advanced Reports Builder topics.

The Paper Layout Template Editor Window

The Paper Layout Template Editor is an environment within the Paper Layout window that you use for editing templates. The Paper Layout Template Editor is similar to the Forms Builder Layout Editor, and represents objects symbolically so you can specify object types and relationships. You cannot view or edit a template in the Paper Design window, because the template defines how a report looks within a specific template. Now you open the template in the Paper Layout Template Editor.

To open the template in the Paper Layout Template Editor:

1. Double-click the **Paper Layout** button 📝 under the NORTHWOODS template in the Object Navigator.

2. The report template appears in the Paper Layout Template Editor, as shown in Figure 7-16.

The Template Editor is similar to the Forms Builder Layout Editor: it has a tool palette, toolbars, painting region, and status line. The painting region has rulers to help you position report components. There are two areas within the painting region in a report template: the margin, where the report title and boilerplate objects, such as the logo, appear; and the **body**, which contains the report data. When you edit a template in the Template Editor, you can edit the objects in only one area at a time. The Template Editor toolbar has a Margin button 🔳 that enables you to toggle between the margins and the body for editing. Currently, 🔳 is not pressed (see Figure 7-16), so the report margins are not visible.

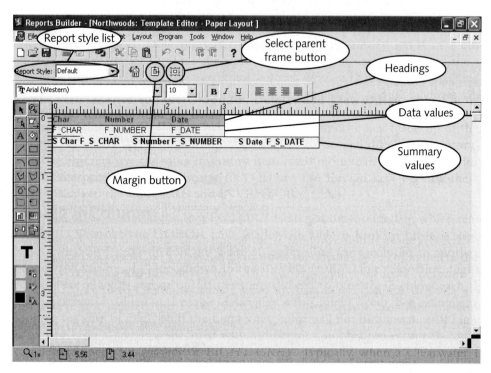

Figure 7-16   Paper Layout Template Editor

The Report Style list (see Figure 7-16) allows you to select different report styles, and modify the template for each style. Currently, the Report Style list value is Default, which specifies the template properties for all report styles.

On the Paper Layout Template Editor toolbar, the Insert Date and Time button 📅 enables you to insert the current date and time in the report, and the Insert Page Number button 🔢 enables you to insert the current page number. These buttons open dialog boxes that specify the format of the date or page number and its placement position. The 🔢 button is available only when the Margin button 🔳 is pressed and the margins are open for editing.

Editing Template Margins

Next, you open the margins for editing. You delete the current logo, and replace it with the Northwoods University logo. You also specify to display the current page number on the bottom report margin.

To edit the template margins:

1. Click the **Margin** button [icon] on the toolbar to open the margins for editing. Select the default logo, and then press **Delete** to delete the logo.

2. Click **Insert** on the menu bar, and then click **Image**. Click **Browse**, navigate to the Chapter07 folder on your Data Disk, select **Nwlogo.jpg**, click **Open**, and then click **OK**. The Northwoods University logo appears on the template. Resize the logo so it fits in the top margin area.

3. To display the page number on the bottom margin, click the **Insert Page Number** button [icon]. The Insert Page Number dialog box opens.

4. Make sure that **Bottom–Center** is selected for the page number position, make sure that the **Page Number Only** option button is selected, and then click **OK**. The placeholder for the page number appears on the bottom report margin, with the text "Page &<Page Number>." Select the page number placeholder if necessary, and change its font to **Arial (Western)** with a font size of **8-point** and in **italics**.

5. Click the **Save** button to save the template file.

Editing the Template Body

To edit attributes of the report body, which displays the report data values, you cancel the Margin button [icon] selection to open the body for editing. The report body has two types of attributes: default and override. **Default attributes** define the default visual attributes for all report styles, which can include object placement, font types and sizes, background colors, and so forth. **Override attributes** define attributes for individual report styles, such as tabular reports or form letters. When you apply a template to a report, the template applies all default attributes to the report. The template applies override attributes only to report styles for which you define the override attributes.

To define override attributes, you select the target report style in the Report Style list, which appears in the Paper Layout Template Editor window on the top-left edge of the toolbar. Then you change the attributes for the selected report styles. Whenever you create or modify a template, you must modify that template for every report style to which you apply the template.

The default report template body shows placeholders for the report field headings, field data values, and summary values. Figure 7-16 highlights the heading, data, and summary placeholders. For example, the field headings appear as individual placeholders with the titles Char, Number, and Date. You can modify properties, such as the background color, or font size, style, or color, of individual template placeholder fields. For example, you could select the Char placeholder in Figure 7-16 and change its background color to specify that the first report field heading appears with a different background color.

Each set of placeholders appears within a **frame**, which is an object that encloses similar objects within a report. In Figure 7-16, the heading placeholders are within a frame, the data value placeholders are within a second frame, and the summary value placeholders are within a third frame. To modify the properties of all of a specific placeholder type, you select the frame that encloses the individual placeholders. Then, you change the frame property values as necessary. To select the frame, you select one of the individual placeholders, and then click the Select Parent Frame button ⬚. A **parent frame** is the frame that directly encloses an object.

Next, you modify the Tabular and Group Above report styles. You add override attributes to specify that the field headings appear on a background color of dark green and that the field heading text appears in yellow.

To modify the template body:

1. Click the **Margin** button ⬚ to cancel its selection and open the template body for editing.

2. Open the **Report Style** list to display the different report styles, and select **Tabular**.

3. To change the background color of the field headings, select the **Char** placeholder, and then click the **Select Parent Frame** button ⬚. Selection handles appear around all of the field headings to indicate that their enclosing frame is selected. Note on the tool palette that the **Fill Color** tool ⬚ shows the current color as the same color that appears as the background of the column headings.

4. Select ⬚, and then select a **dark green square** on the color palette. The background color of the heading placeholders appears dark green.

5. To change the color of the field heading text, you must change the text color in the individual field placeholders. Select the **Char** placeholder, press and hold the **Shift** key, and then select the **Number** and **Date** placeholders. The three field heading placeholders appear selected.

6. Select the **Text Color** tool ⬚ on the tool palette, and then select a **yellow square** on the color palette. The heading placeholder text appears in yellow.

7. To change the body template properties for the Group Above report style, open the **Report Style** list and select **Group Above**. Note that for this report style, the field headings still appear with a light green background and the field heading text is still blue.

8. Repeat Steps 3 through 6 to modify the field heading background color and text color for the Group Above report style, as shown in Figure 7-17.

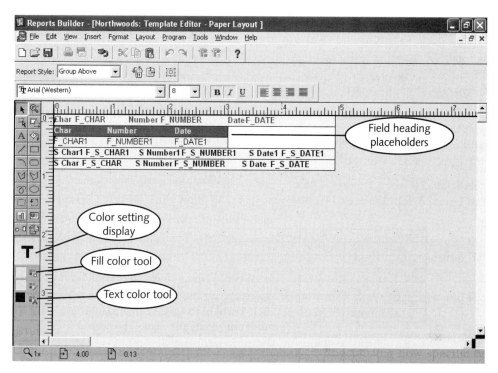

Figure 7-17   Color modification for Group Above report style field headings

9. Save the template file, and then close the template file.

APPLYING CUSTOM TEMPLATES TO REPORTS

Recall that you use the Report Wizard Template page to select a report template. You can select a predefined template, you can select a custom template, or you can specify not to use a template. You can modify an existing report's template by opening the Report Wizard in reentrant mode, and changing the template specification. If you apply a template to a report and then later apply another template, Reports Builder removes the original template formatting, and replaces it with the formatting in the most recent template. The following sections describe how to apply custom templates to reports by specifying the path to the template file, and how to register a custom template in Reports Builder so the custom template appears in the predefined template list.

Applying Templates by Specifying the Template Filename

To apply a custom template file to a report by specifying the template filename, you select the Template file specification option button on the Report Wizard Template page, and enter the full folder path and filename of the template file, including the drive letter. This approach provides a quick and easy way to apply custom templates to forms.

The disadvantage of this approach is that the template file must be available at the specified file location whenever you open the report file in Reports Builder.

Now you open the Ch7ACourse.rdf report that you created earlier in the lesson, start the Report Wizard in reentrant mode, and apply the Northwoods.tdf template file to the report using the template file specification.

To apply the template to the report using the template file specification:

1. Open **Ch7ACourse.rdf** from the Chapter07\Tutorials folder on your Solution Disk. The report appears in the Object Navigator. Save the file as **Ch7ACourse_TEMPLATE.rdf**.

2. Click **Tools** on the menu bar, and then click **Report Wizard** to open the Report Wizard in reentrant mode so you can modify the report template.

3. Select the **Template** tab to open the Template page. (If necessary, click **Next** until the Template tab appears.)

4. Select the **Template file** option button, click **Browse**, navigate to the Chapter07\Tutorials folder on your Solution Disk, select **Northwoods.tdf**, and then click **Open**. Click **Finish** to save your changes and close the Report Wizard. The report appears in the Paper Design window using the formatting specified in the new template.

5. Note that the report now displays the column headings using the template background and text colors. Because you added the Northwoods logo to the report earlier, the report also now displays two Northwoods logos, one on top of the other. Delete one of the logos.

6. Note also that the report now displays page numbers at the bottom of the page. (You might need to scroll to the bottom edge of the window to see the page numbering.) Save the report, and then close the report.

Registering Custom Templates in Reports Builder

Another way to apply a custom template to a report is to register the custom template so it appears in the Predefined Templates list on the Template page in the Report Wizard. The advantage of registering a template is that you don't have to specify the path to the template file, and the file does not always have to be available at the specified location when the report opens.

Registering a custom template is a two-step process. First, you modify the Developer user preferences file, so the custom template appears in the Predefined Templates list. Then, you must copy the template file to the Reports Builder templates folder.

Modifying the User Preferences File

To register a custom template, you need to modify the Developer user preferences file. The **user preferences file** is a text file that specifies configuration information for a specific user for most of the Developer utilities, including Forms Builder and Reports Builder. This information includes user preferences such as whether the Wizard Welcome pages appear, or which format masks appear in the Property Palette or Property Inspector format mask lists.

You register a custom template file by modifying the cauprefs.ora user preferences file, which is in the Developer_Home folder on your workstation. (The Developer_Home folder is the folder in which you specified to install Developer10*g* during the installation process.) Every time Reports Builder opens, the application reads the user preferences file and configures the environment based on the file contents. Every time you close Reports Builder, the application writes modified preference information back to the file. Therefore, you must close Reports Builder before you manually modify this file—otherwise, Reports Builder overwrites your changes the next time it closes. You also cannot open Reports Builder and then open the Report Wizard in reentrant mode immediately after you modify the cauprefs.ora file, because the Wizard overwrites your changes with the settings it saved from the previous report.

All users who run Reports Builder on the same workstation share the same user preferences file. Before modifying the file, you first make a backup copy of the existing cauprefs.ora file and place it in the Chapter07\Tutorials folder on your Solution Disk. After you finish modifying and using the new cauprefs.ora file, you replace the modified file with the original file.

To make a backup copy of the user preferences file:

1. Close Reports Builder.

2. Start Windows Explorer, and navigate to the **Developer_Home** folder on your workstation.

If you followed the installation instructions for Developer10*g* provided with this book, your Developer_Home folder will be C:\Oracle\OraDev. If you cannot find this folder, ask your instructor or technical support person for the path to your Developer_Home folder.

3. Copy **cauprefs.ora** to the Chapter07\Tutorials folder on your Solution Disk.

The user preferences file contains sections defining preferences for every report style. You have to specify a separate template description and filename for each report style (tabular, group above, group left, and so forth). For now, you specify a predefined custom template for the tabular report style only. To modify the user preferences file to display a custom template name in the Predefined Templates list in the Report Wizard, you need to specify the **template description**, which is the description that appears in the Predefined Templates list, such as "Beige." You also need to specify the **template filename**, which is the name of the associated template .tdf file. Now you examine the user preferences file to see how you specify custom template descriptions and corresponding filenames.

To examine the user preferences file:

1. Start Notepad, click **File** on the menu bar, click **Open**, navigate to the **Developer_Home** folder, open the **Files of type** list, and then select **All Files**. Select **cauprefs.ora**, and then click **Open**. The file opens.

2. Scroll down until you see the `Reports.Tabular_Template_Desc =` command, as shown in Figure 7-18. (Notice that the properties after `Reports.` appear in alphabetical order.)

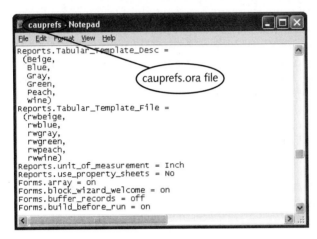

Figure 7-18 Predefined template specifications in the user preferences file

The `Reports.Tabular_Template_Desc` = command specifies the descriptions that appear in the Predefined Templates list for tabular-style reports. The `Reports.Tabular_Template_File` = command specifies the names of the corresponding template file-names. The first description in the description list corresponds with the first filename, the second description corresponds with the second filename, and so forth. Now you will add the description "Northwoods" to the description list, and then add the filename "Northwoods" to specify that the name of the template file is Northwoods.tdf. Note that you do not specify the .tdf extension in the filename.

To add the commands to the user preferences file to register the Northwoods custom template:

1. In Notepad, modify the beginning of the command that defines the template descriptions so it includes the Northwoods template description. The command should appear as follows. (The following modified command does not show all of the template descriptions in the list.)

```
Reports.Tabular_Template_Desc =
(Northwoods,
Beige,
Blue,
```

2. Modify the beginning of the command that defines the template filenames so it appears as follows. (The following modified command does not show all of the template filenames in the list.)

```
Reports.Tabular_Template_File =
(Northwoods,
rwbeige,
rwblue,
```

3. Save the file, and then exit Notepad.

Copying the Template File to the Reports Builder Templates Folder

The last step in registering the custom template is to place a copy of the template .tdf file in the default Reports Builder templates folder. In Developer10*g*, the default templates folder is Developer_Home\Reports\Templates. Now you copy the Northwoods.tdf file from the Chapter07\Tutorials folder on your Solution Disk to the default templates folder. Then you start Reports Builder, create a new report that displays information about Northwoods University terms, and apply the new predefined template to the report. (Recall that you cannot start the Report Wizard in reentrant mode immediately after modifying the user preferences file, because the wizard overwrites the changes.)

To copy the template file to the default templates folder, create a report, and then apply the template:

1. In Windows Explorer, copy **Northwoods.tdf** from the Chapter07\Tutorials folder on your Solution Disk to the Developer_Home\reports\templates folder.

2. Start Reports Builder. When the Welcome to Reports Builder window appears, make sure that the **Use the Report Wizard** option button is selected, click **OK**, and then click **Next**.

If a message appears stating there is an error in the user preferences file, click OK, and close Reports Builder. Start Notepad, open the cauprefs.ora file from the default templates folder, make sure that you made the modifications exactly as described in the previous set of steps, save the file, close Notepad, and then repeat Steps 1 and 2. If the error occurs again, click OK, close Reports Builder, start Notepad again, open the saved cauprefs.ora file from the default templates folder, save the file on your Solution Disk, repeat the steps to modify the file, save the file, close Notepad, and repeat Steps 1 and 2.

3. Accept the default selection on the Report Type page, and click **Next**. On the Style page, type **Terms** for the report title, make sure that the **Tabular** option button is selected, and then click **Next**.

4. Accept the default data source on the Data Source page, and click **Next**. Type **SELECT * FROM term** in the **Data Source definition** field, and then click **Next**. Connect to the database in the usual way.

5. On the Fields page, click the **Move all items to target** button ⟨ » ⟩ to include all of the fields in the report, and click **Next**. Do not select any values on the Totals page, and click **Next**. Accept the default values on the Labels page, and click **Next**.

6. On the Template page, the Northwoods template should appear in the Predefined template list. Make sure that **Northwoods** is selected, click **Next**, and then click **Finish**. The report appears in the Paper Design window, and is formatted using the Northwoods template. Maximize the Paper Design window and the Reports Builder window.

7. Save the report as **Ch7ATerm.rdf** in the Chapter07\Tutorials folder on your Solution Disk.

The Template page displays thumbnail images of the predefined templates so that when the user selects a template, the user can view the template before applying it to a report. Currently, when the selection in the Predefined template list is Northwoods, an image with the text "Image Not Available" appears. To create a thumbnail image of a custom template that appears when you select a template from the Predefined template list, you make a screenshot of a representative sample area of a report that uses the template, save the image using the .bmp file type, and name the image file *template_filename*.bmp. (In this filename, *template_filename* is the name of the template file, so the thumbnail image filename for a file that shows a representation of the Northwoods template would be Northwoods.bmp.) Then, you copy the image file to the default template file directory. The image then appears on the Templates tab when you select the template from the Predefined template list. You must exit Reports Builder and then restart it to be able to view the thumbnail image.

Next, you copy an image file named Northwoods.bmp that shows a representative sample of the Northwoods.tdf template into the Reports Builder templates folder. Then you start Reports Builder, open the Ch7ATerm.rdf report file, start the Report Wizard in reentrant mode, and view the thumbnail image.

To create and view the thumbnail image:

1. Exit Reports Builder.

2. In Windows Explorer, copy **Northwoods.bmp** from the Chapter07 folder on your Data Disk to the Developer_Home\Reports\Templates folder on your workstation.

3. Start Reports Builder. When the Welcome to Reports Builder window appears, select the **Open an existing report** option button and click **OK**, navigate to the Chapter07\Tutorials folder on your Solution Disk, select **Ch7ATerm.rdf**, and click **Open**.

4. To open the Report Wizard in reentrant mode, click **Tools** on the menu bar, and then click **Report Wizard**. Select the **Template page**. The Northwoods template should be selected in the Predefined templates list, and a thumbnail image of the template should appear.

5. Click **Cancel** to close the Report Wizard, save the report and then close Reports Builder.

Recall that you need to restore the backup copy of the user preferences file to the Developer_Home folder on your workstation. You should also delete your Northwoods.tdf and Northwoods.bmp files from the Reports Builder templates folder so they won't be used by other students.

To restore the backup user preferences file and delete the other files:

1. In Windows Explorer, copy **cauprefs.ora** from the Chapter07\Tutorials folder on your Solution Disk to the Developer_Home folder on your workstation hard drive to restore the original user preferences file.

2. Navigate to the Developer_Home\Reports\Templates folder, delete **Northwoods.tdf** and **Northwoods.bmp**, and then close Windows Explorer.

7

SUMMARY

- ❑ A report is a summary view of database data that users can view on a screen or print on paper. Reports retrieve database data using SQL queries, perform mathematical or summary calculations on the retrieved data, and format the output to look like invoices, form letters, or other business documents.

- ❑ Reports Builder can display Oracle 10*g* data in a variety of report styles. A commonly used report style is tabular, which presents data in a table format with columns and rows. Other commonly used report styles are group left and group above, which display master-detail relationships in which one master record can have several associated detail records through a foreign key relationship. In a group left report, each master item is listed on the left side of the report, and the multiple detail items appear to the right of the master item. In a group above report, each master item appears above the multiple detail lines.

- ❑ To create a report, you specify the data that is to appear in the report, select the report style, and then configure the report properties and layout. You can use the Report Wizard to specify the report data and report style. You can open the Report Wizard in reentrant mode to modify properties of existing reports.

- ❑ The Paper Design window shows how the report is going to appear when you print it on paper. You can use the Paper Design window to change the report's appearance by repositioning report objects, changing font sizes and styles, and resizing the widths of report columns.

❐ The Reports Builder Object Navigator shows report components as a hierarchical tree. You can use the Object Navigator to view and modify components, or to open different windows within the Reports Builder environment.

❐ When you view a report as a Web page, Reports Builder creates a static HTML representation of the report contents. To display formatting changes in the HTML representation, you must view the report in either Paginated HTML or Paginated HTMLCSS format.

❐ To create a master-detail report using the Report Wizard, the report SQL query must retrieve all of the data that appears in the report. You use the Wizard Groups page to specify the master and detail data groups.

❐ A report template defines the report appearance in terms of fonts, graphics, and fill colors in selected report areas. A report template has a margin, which contains boilerplate objects such as a logo image, the date the report was generated, and page numbers. The template also has a body region, which displays the report data.

❐ Reports Builder stores template-formatting definitions in template definition files, which have a .tdf extension. You can create a custom template to specify formatting characteristics.

❐ You use the Paper Layout Template Editor window to edit custom template files. This window represents objects symbolically to highlight their types and relationships.

❐ When you create a custom template, default attributes define properties that appear in all report styles that use the template. Override attributes define properties that are unique to a specific report style.

❐ To use the Report Wizard Template page to apply custom templates to reports, you can either specify the path to the template file, or register the custom template so it appears in the Predefined templates list.

❐ To register a template file, you modify the Developer user preferences (cauprefs.ora) file so it contains both the template description and filename. You also must place the template file in the default templates folder.

REVIEW QUESTIONS

1. Which report format displays the detail data as a series of columns and rows, without any grouping?

2. In a grouped report, the data is grouped based on the contents of the _____ record.

3. To retrieve data from a database table to populate a report, the SQL query retrieving the data must already be stored in the database. True or False?

4. You can use the Paper Design window to edit the report format generated by the Report Wizard. True or False?

5. Explain how to widen the column of a report.

6. If you click the Run Web Layout button, the resulting file displayed in the Web browser actually contains _____ specifying the format of the report.

7. When you display a report using the Paginated HTML format preview option, commands stored in a Cascading Style Sheet (CSS) are used to format the report. True or False?

8. If you create a report that displays each faculty member and each of his or her advisees, which database table contains the master records?

9. Suppose you use a template that specifies gray as the background color to create a report. After the report is generated, you change the background color to white. Which color is actually displayed, the color specified by the template or the background attribute you specified later?

10. By default, a(n) _____ appears in the upper-left corner of a report created by the Report Generator.

7

MULTIPLE CHOICE

1. When you view a report as a Web page, Reports Builder creates a _____ .
   a. static HTML file
   b. Java server page
   c. Java applet
   d. dynamic XML file

2. To create a master-detail report, you must select the _____ report style.
   a. master-detail
   b. group above
   c. multi-matrix
   d. tabular

3. When you create a template, which section of the report is *not* stored in the template file?
   a. font color of headings
   b. filename of logo and its placement in the report margin
   c. location of page numbers that appear on the report
   d. data displayed in body of report

4. User preferences are stored in the _____ file.
   a. upref.ora
   b. cauprefs.ora
   c. template.ora
   d. u_settings.ora

5. To include a thumbnail of a template to display as a preview, you must:

   a. Store the thumbnail image in the Images folder.

   b. Name the file the same filename as the template file.

   c. Include the filename of the thumbnail in the Thumbnail property of the template.

   d. all of the above

6. The _____ is a window that displays the properties and their values of report objects.

   a. Property Palette

   b. Property Editor

   c. Property Inspector

   d. Report Navigator

7. To display an image as a logo at the top of a report, you must:

   a. Paste the image into the report.

   b. Export the image from the Image Generator.

   c. Include the image in the Object Navigator.

   d. Import the image into the report.

8. To display a student's telephone number as (999) 999-9999, you must specify the appropriate _____ property.

   a. visual display

   b. number format

   c. format mask

   d. Web page tag

9. The user preferences file is read when _____.

   a. a report is previewed

   b. a new report is created

   c. Report Builder is opened

   d. the Web browser is opened

10. Which of the following does *not* appear in the report margin?

   a. database data

   b. title

   c. page number

   d. logo

PROBLEM-SOLVING CASES

The following cases reference the sample database of Clearwater Traders (Figure 1-14). Run the Ch7Clearwater.sql script in the Chapter07 folder on your Data Disk to refresh the sample databases. All required files are in the Chapter07 folder on your Data Disk. Save all solution files in the Chapter07\Cases folder on your Solution Disk. Save all SQL commands in a file named Ch7AQueries.sql.

1. In this case, you first create a predefined template for Clearwater Traders. Then you create a report listing Clearwater Traders customers, as shown in Figure 7-19, and apply the template to the report.

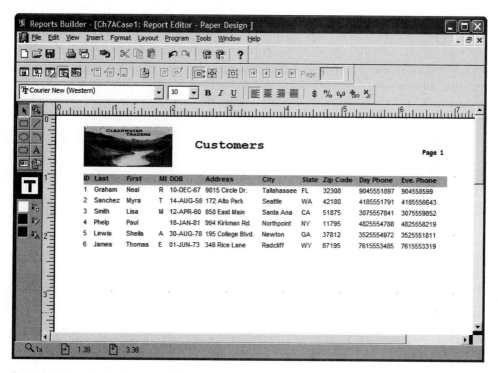

Figure 7-19   Customers report

a. Open the **rwpeach.tdf template file**, and save it as **Clearwater.tdf**.

b. Replace the current logo with the **Clearwater Traders logo** (Clearlogo.jpg).

c. Insert the current page number in the top-right corner of each report page, just under the date, using the format "Page *current_page_number*." Do not display the total number of pages. Right-justify the page number so its right edge aligns with the right edge of the report body. Format the page number using an **8-point italic Comic Sans MS (Western)** font.

d. For the tabular, group left, and group above report styles, change the background of the column headings to a **light gray shade**, and the text color to **black**. For the same report styles, change the data field background color to **white**.

e. Create a new report named **Ch7ACase1.rdf** that displays the fields from the Clearwater Traders CUSTOMER table shown in Figure 7-19. Change the report title to **Customers**, use descriptive column headings, and apply the **Clearwater.tdf** template file by specifying the path to the template file from the Chapter07\Cases folder on your Solution Disk. Modify the report formatting as necessary so the finished report looks like Figure 7-19.

2. Create an incoming shipment report for Clearwater Traders that lists item description, inventory ID, item size, and color values, and then lists the shipment ID and date expected for all incoming shipments that have not yet been received for that item. Format the report as shown in Figure 7-20. Apply the **Clearwater.tdf** template you created in Case 1. (If you did not create the template, apply the **Peach** predefined template, delete the current logo, and replace it with the **Clearwater Traders logo**, which is stored in the Clearlogo.jpg file.) Save the report as **Ch7ACase2.rdf**.

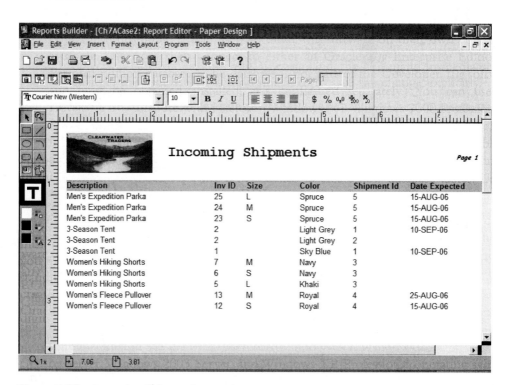

Figure 7-20 Incoming Shipments report

3. Create a report that makes mailing labels for the customers in the Clearwater Traders database, as shown in Figure 7-21. (*Hint*: You can create mailing labels only by using a paper layout.) Format the labels using a **10-point Arial (Western)** font. Do not apply a template to the report. Save the report as **Ch7ACase3.rdf**.

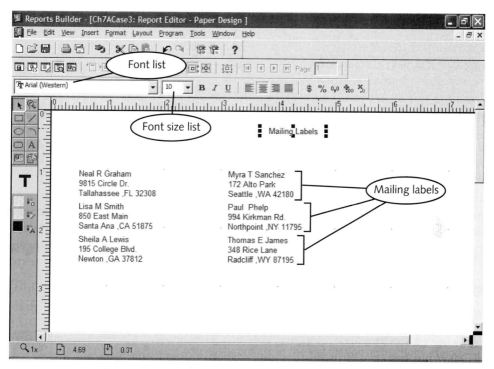

Figure 7-21   Mailing Labels report

4. Create a report using the Group Above style that displays each customer's order, along with the items contained on the order. You should include the customer's first and last name and ID, along with each order number, order date, and the description, quantity, and price of each item. The customer and order information should be listed on the report as separate groups. Include page numbers at the bottom of each report page. Use **Customer Orders** as the report title, and format the report to give it a professional appearance. Save the completed report as **Ch7ACase4.rdf**.

5. Modify the Customer Order report from the previous case so it displays data using the group left style. Make the appropriate modifications to the column widths to accommodate the displayed data. Save the modified report as **Ch7ACase5.rdf**.

◀ LESSON B ▶

After completing this lesson, you should be able to:
- Describe the components of a report
- Modify report components
- Modify the format of master-detail reports
- Create parameters to allow the user to customize report data

REPORT COMPONENTS

In Lesson A, you used the Report Wizard to create reports, and you learned how to use the Paper Design window to make minor formatting changes such as resizing field widths. Sometimes, however, you need to modify the output that the Report Wizard generates more extensively. For example, you might want to format a report so that blank space appears between individual master records in a master-detail report, or you might want to specify where data appears on a report to create documents such as invoices or transcripts.

To customize report output, you need to become familiar with the underlying components of a report and learn how to modify these components directly. In this lesson, you learn more about the Data Model, which specifies the data that the report displays; the Paper Layout view, which displays the report components as symbolic objects; and the report frames, which group related report objects. To explore these report components, you first work with the Northwoods University Location report in Figure 7-22.

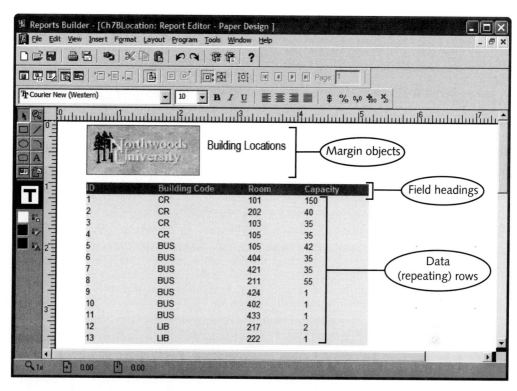

Figure 7-22   Northwoods University Location report

The Location report is a tabular report that shows information from the LOCATION table in the Northwoods University database. Now you start Reports Builder and open the report file.

To start Reports Builder and open the file:

1. Start Reports Builder. Select the **Open an existing report** option button, and then click **OK**.

2. Navigate to the Chapter07 folder on your Data Disk, select **Location.rdf**, and then click **Open**. The Object Navigator window opens. Maximize the Object Navigator window, maximize the Reports Builder window, and then save the file as **Ch7BLocation.rdf** in the Chapter07\Tutorials folder on your Solution Disk.

3. In the Object Navigator, click the **Run Paper Layout** button 🇶, and connect to the database in the usual way. The report appears in the Paper Design window, as shown in Figure 7-22.

Note that the report title and logo are in the top margin. The body contains a row with the field headings and multiple rows showing data from each record in the LOCATION table. The information in the report body comes from the report's Data Model, which the next section describes.

The Data Model Window

The Data Model window shows the report's SQL query and the associated record groups. A **report record group** is a set of records that represents the data fields that a query retrieves. A simple tabular report has a single record group, and a master-detail report has multiple record groups. Now you open the Data Model window for the Northwoods University Location report. Because this is a simple tabular report, it has a single record group.

To open the Data Model window:

1. Click the **Data Model** button on the Reports Builder toolbar.

NOTE
You can also open the Data Model window by clicking View on the menu bar, pointing to Change View, and then clicking Data Model, or by double-clicking the report's Data Model icon 🖃 in the Object Navigator.

2. The Data Model window opens, as shown in Figure 7-23.

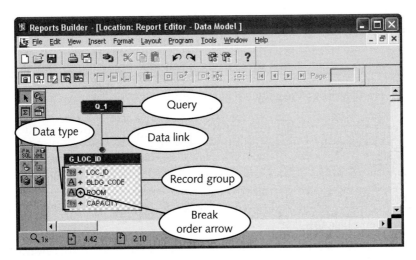

Figure 7-23 Data Model window

The Data Model window has a tool palette for creating queries and report data fields. It also has a toolbar for navigating to other windows within Reports Builder, and for opening and saving files and running and printing reports. In Figure 7-23, there are three objects in the Data Model window: Q_1, which represents the SQL query that retrieves the report data; G_LOC_ID, which represents the record group that this query creates; and the line between Q_1 and G_LOC_ID, which is the **data link** that shows that the Q_1 query determines the data that appears in the G_LOC_ID record group.

The Q_1 query retrieves the LOC_ID, BLDG_CODE, ROOM, and CAPACITY fields, and the G_LOC_ID record group contains these fields. In the Data Model, each field is called a column. The general format for the name of a record group is G_*first_query_field*. Because LOC_ID is the first column the query retrieves, the record group name is G_LOC_ID.

The icon in front of each column name indicates the field data type, and the break order arrow beside the icon indicates the column break order, which controls the order in which the column data values appear.

Record Group Column Properties

The components in the report Data Model have properties that you can view and modify using the Property Inspector. Now you open the Property Inspector for the LOC_ID column in the G_LOC_ID record group and view its properties.

To open the Property Inspector for the LOC_ID record group column:

1. In the Data Model window, select the **LOC_ID** column in the G_LOC_ID record group. The fieldname appears with a black background to show that it is selected.

2. Right-click, and then click **Property Inspector**. The LOC_ID column's Property Inspector opens.

The Name property (LOC_ID) is the same as the corresponding database field. The Column Type property describes the type of data that the column displays. Table 7-1 summarizes the different record group column types.

Column Type	Description
Database - Scalar	Discrete data value retrieved from a database table
Summary	Data value calculated by applying a summary function (such as SUM, AVG, or COUNT) to scalar report columns
Formula	Data value calculated by applying a user-defined formula to values in scalar report columns

Table 7-1   Report record group column types

The LOC_ID Column Type value is Database - Scalar, which indicates that this column displays a discrete value from a database table. The Datatype and Width properties describe the data type and maximum width of the data column, respectively. The Value if Null property allows the developer to substitute a different value for the data field if the retrieved data value is NULL. The XML Settings property node specifies properties of the column if it was imported from an XML file. (XML, Extensible Markup Language, defines a standard way to structure and store data in text files, and is often used to store and retrieve data in Web-based applications.)

Modifying the Report's SQL Query

You can modify the report's SQL query in the Data Model window. Now you open the query and modify it by adding an ORDER BY clause to order the records by BLDG_CODE.

To modify the query in the Data Model window:

1. Close the Property Inspector, and then double-click **Q_1** in the Data Model window. The SQL Query Statement dialog box opens, showing the report's SQL query.

2. To modify the query, add the following ORDER BY clause as the last line of the query, and then click **OK**:

    ```
    ORDER BY bldg_code
    ```

The Data Model window appears again.

3. Click the **Run Paper Layout** button 🗋 on the toolbar to view the result of your change. The data records now appear sorted by building codes.

TIP

You can also change the order in which the records appear by opening the Report Wizard in reentrant mode and modifying the query on the Data page.

4. Click the **Data Model** button 🗋 to reopen the Data Model window, and then save the report.

Creating a Group Filter to Control Report Data

Sometimes you need to limit the data that a report query retrieves. If a report retrieves a very large number of records, it takes a long time for the report to appear. To limit the number of records that the report query retrieves, you can create a **group filter**, which is a structure that uses some criteria to limit the number of records that a report query retrieves. To create a group filter, you assign a value to the Filter Type property in the report record group's Property Inspector. You can set the record group Filter Type property value to First, which specifies that the report displays a specific number of records from the beginning of the retrieved data set; Last, which specifies that the report displays a specific number of records from the end of the retrieved data set; or PL/SQL, which allows you to write a PL/SQL function that evaluates each individual record to determine whether or not it appears in the report.

TIP

Alternately, you can modify the report's SQL query by creating a search condition to limit the number of records that the report retrieves, but you may not know ahead of time that a query will retrieve many records. Also, you can structure a SQL query search condition to retrieve a specific fixed number of records, such as 10 or 20 or 200 records.

TIP

Group filters that you define using PL/SQL functions retrieve data values very slowly. When you need to filter data based on retrieved values, you should perform the filtering using a SQL search condition in the report query.

Now you create a group filter to display the first 10 Northwoods University locations that the data set retrieves. You set the G_LOC_ID record group Filter Type property to First and specify to display the first 10 records. Then you view the result of the group filter by running the report in the Paper Design window.

To create and test a group filter:

1. In the Data Model window, click the border of the **G_LOC_ID** record group to select the record group. Selection handles appear on the borders of the record group.

2. Right-click, and then click **Property Inspector** to open the record group Property Inspector.

3. Select the **Filter Type** property, open the property list, and select **First**.

4. Change the Number of Records property value to **10**, select another property to apply the change, and then close the Property Inspector.

5. Click the **Run Paper Layout** button 🔳 on the toolbar to open the Paper Design window. The report now displays only the first 10 records from the LOCATION table.

6. Save the report.

Understanding Report Objects

After you create a report using the Report Wizard, you may need to reposition report items manually to improve the report appearance. For example, you may need to move related fields (such as first name, middle initial, and last name) so they appear beside one another on the report. Master-detail reports often are more attractive and easier to understand if blank space appears between the data for each new master record value. To reposition report items, you need to understand the objects that make up the report layout, and how the objects in the report layout relate to the report record group.

When you examine the Location report in the Paper Layout window (see Figure 7-24), note that the report displays rows that show the data from each record in the LOCATION table. These data rows are called **repeating rows**, because each row shows the same data fields (LOC_ID, BLDG_CODE, ROOM, CAPACITY), but with different data values. Now you view the report layout components in the Paper Layout window. Recall that the Paper Layout window shows the report items as objects, rather than as actual data values. (You used the Paper Layout Template Editor window in Lesson A to modify report template objects.) You can use the Paper Layout window to modify the report structure.

To view the report layout objects in the Paper Layout window:

1. Click the **Paper Layout** button 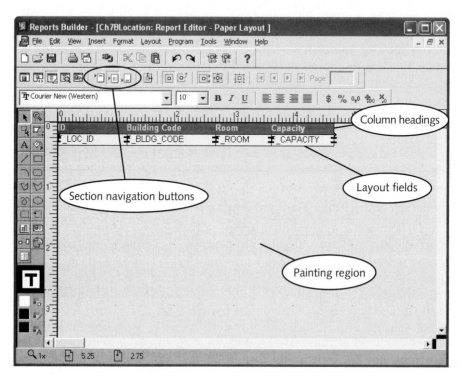 on the toolbar.
2. The report objects appear in the Paper Layout window, as shown in Figure 7-24.

Figure 7-24 Paper Layout window

The Paper Layout window is similar to the Paper Layout Template Editor window and to the Forms Builder Layout Editor. When you open a template .tdf file, the file appears in the Template Editor. However, when you open a report, the file appears in the Paper Layout window. The Paper Layout window has a tool palette, toolbars, painting region, and status line. In addition, the top toolbar has three section navigation buttons that allow you to move to different sections of the report.

A report has three sections: header, main, and trailer. The **header** section is an optional page (or pages) that appears at the beginning of the report and can contain text, graphics, data, and computations to introduce the report. The **main** report section usually has multiple pages, and contains the report data and computations. The report **trailer** section, like the header section, is an optional page (or pages) that appears at the end of the report. The trailer can include summary data, or for a report with hundreds of pages, it could contain text to mark the end of the report.

The first button in the section navigation button group is the Header Section button ▸▢, which enables you to activate the report header and make it available for editing. The second button is the Main Section button ▸▤, which enables you to activate the report body. The third button is the Trailer Section button ,▢, which allows you to activate the report trailer.

Recall that a report has margins that contain boilerplate objects such as the company logo, the report title, and the report page numbers. You can click the Edit Margin button ▣, which is also called the Margin button in the Template Editor, to open the margins portion of the active report section. (You used the Margin button ▣ earlier to toggle between the body and the margins of the report template.)

Within the painting region, the Paper Layout window displays **layout fields** that correspond to the record group columns in the report Data Model. The report displays the Data Model's retrieved data values in the corresponding layout fields. The default name for a report layout field is F_*column_name*. Figure 7-24 displays layout fields named F_LOC_ID, F_BLDG_CODE, F_ROOM, and F_CAPACITY. The Paper Layout window also displays a column heading for each field.

Recall that a report uses frames to group similar objects. You work with report frames in the Paper Layout window. The following section describes report frames in detail.

Report Frames

As you learned in Lesson A, **frames** are containers for grouping related report objects so that you can set specific properties for a group of objects, rather than having to set the property for each item. For example, all of the report column headings are in a frame, so you can easily apply the same background color to all of the headings. Individual frames are not always visible in the Paper Layout, because they are on top of each other. Figure 7-25 shows a schematic representation of the frames in the Location report.

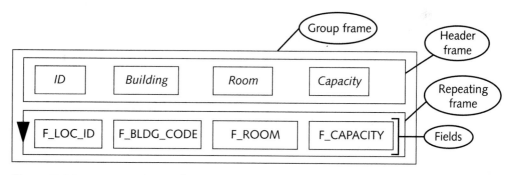

Figure 7-25   Location report frames

Each record group in a report has a corresponding **group frame**, which encloses a repeating frame and an optional header frame. A **repeating frame** encloses repeating data rows, and is designated with a down-pointing arrow on its left side. Each repeating data row contains multiple data fields. When you create a report using the Report Wizard, the Wizard creates a field for each column in the report record group, and assigns to each field the default name F_*column_name*. Figure 7-25 shows that a repeating frame encloses the form layout fields. Repeating frames have **variable sizing**, which means that they shrink or stretch vertically depending on how many data records appear in the report. For example, if the Location report query retrieves 13 records, then the repeating frame stretches to display all 13 records.

A **header frame** encloses all of the column headings for a record group. In Figure 7-25, a header frame encloses the column headings ID, Building, Room, and Capacity. The Report Wizard creates a header frame only when report values appear in a tabular format.

If you move an object outside its enclosing frame, an error message appears in Reports Builder when you view the report in the Paper Design window. For example, if you move the F_LOC_ID field outside its repeating frame, an error message appears. Furthermore, objects in a frame must be totally enclosed by their surrounding frames, or an error message appears when you view the report in the Paper Design window. In Figure 7-25, the header frame and repeating frame must be completely enclosed by the group frame. If one of these frames is outside the group frame or if its borders overlap the borders of the group frame, an error message appears.

The Report Wizard derives frame names from the names of their associated record groups. Table 7-2 shows the general format that the Report Wizard uses for naming report frames, as well as the names of the frames in the Location report. (Recall that in the Location report, the record group name is G_LOC_ID.)

Frame Type	Default Name Format	Frame Name in Location Report
Group	M_*record_group*_GRPFR	M_G_LOC_ID_GRPFR
Header	M_*record_group*_HDR	M_G_LOC_ID_HDR
Repeating	R_*record_group*	R_G_LOC_ID

Table 7-2 Default report frame names

To become familiar with working with frames, you now select the different frames in the Location report. It is difficult to select individual frames in a report, because they are on top of each other. The best way to select a specific report frame in the Paper Layout window is to select an item that is in the frame and then select the item's parent frame by clicking the Select Parent Frame button 🔲 on the Paper Layout toolbar. Recall that a parent frame is the frame that directly encloses an object. In Figure 7-25, the header frame directly encloses the individual column headings, so the header frame is the parent frame of each column heading. Similarly, the group frame directly encloses the header frame, so the group frame is the header frame's parent frame. First, you select the report header frame.

To select the header frame:

1. In the Paper Layout window, make sure that the Edit Margin button ⧉ is not pressed and that the report body is available for editing. (Your screen should look like Figure 7-24, and the report margins should not be visible.)

2. Click the **ID** column heading so that selection handles appear around its perimeter, and then click the **Select Parent Frame** button ⊡ on the toolbar to select the ID column heading's parent frame, which is the header frame. Selection handles appear around all the column headings.

3. Click **Tools** on the menu bar, and then click **Property Inspector**. The Property Inspector opens, confirming that the header frame is selected and that the frame name property is M_G_LOC_ID_HDR.

TIP

You cannot open the frame Property Inspector by right-clicking and then clicking Property Inspector because right-clicking deselects the frame.

7

4. Close the Property Inspector.

Recall that the group frame is the parent frame of the header frame. Because the header frame is currently selected, clicking the Select Parent Frame button ⊡ again selects the group frame. Now you select the group frame and open its Property Inspector.

To select the group frame and open its Property Inspector:

1. In the Paper Layout window, with the M_G_LOC_ID_HDR header frame currently selected, click the **Select Parent Frame** button ⊡. This selects the header frame's parent frame, which is the group frame. Selection handles appear around all the report objects.

2. Click **Tools** on the menu bar, and then click **Property Inspector**. The Property Inspector opens for the group frame, and shows the frame's name as M_G_LOC_ID_GRPFR.

3. Close the Property Inspector.

Finally, you select the report's repeating frame. To do this, you select one of the report layout fields, and then click the Select Parent Frame button ⊡ to select the layout field's parent frame, which is the repeating frame.

To select the report's repeating frame:

1. In the Paper Layout window, select **F_LOC_ID**. Selection handles appear around the layout field.

2. Click the **Select Parent Frame** button ⊡ to select the field's parent frame, which is the repeating frame. Selection handles appear around all the data fields.

3. Click **Tools** on the menu bar, and then click **Property Inspector**. The Property Inspector for the repeating frame opens. The repeating frame's name is R_G_LOC_ID. The Source property is G_LOC_ID, which is the record group that is the source of the repeating frame's data. Do not close the Property Inspector.

You can change frame properties on the frame Property Inspector. For example, you might want to change the frame background color, specify that the frame items print on a new page, or specify special printing instructions, such as leaving a set amount of blank space between rows in a repeating frame. Next, you modify some of the properties of the R_G_LOC_ID repeating frame. You modify the repeating frame so that only five records appear per page. Then you increase the vertical spacing between each record so that 0.25 inches of blank space appear between each data row.

To modify the repeating frame properties:

1. On the R_G_LOC_ID repeating frame Property Inspector, change the Maximum Records per Page property value to **5**.

2. Change the Vert. Space Between Frames value to **.25**, click another property to save the change, and then close the Property Inspector.

3. Click the **Run Paper Layout** button 🏗 to view the report in the Paper Design window. The first five records appear on the first report page, with 0.25 inches of blank space between each row, as shown in Figure 7-26.

If red dashed lines appear in the Paper Design window, it means you currently have a frame selected in the Paper Layout window. Click anywhere on the report to deselect the frame.

Because you modified the report to display five records per page, your report now has multiple pages. When a report has more than one page, the Paper Design paging buttons are enabled (see Figure 7-26). The Next Page button ▶ allows you to move through the report pages, one page at a time. When the Paper Design window displays any report page except the first page, the Previous Page button ◀ is enabled, which allows you to move through the report pages in reverse order. The Last Page button ▶| displays the last page of the report, and the First Page button |◀ displays the first page of the report. The Page field displays the page number of the current report page. To jump to a specific report page number, you can type the page number in the Page field and then press Enter. In the following steps, you move through the report pages using the report paging buttons and Page field. Then you close the report and save your changes.

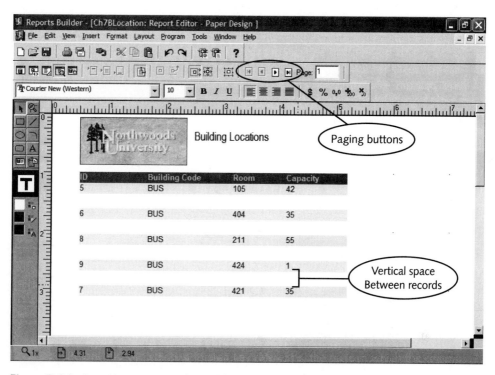

Figure 7-26   Location report with modified repeating frame properties

To move through the report pages, and then close the report:

1. Click the **Next Page** button ▶ on the toolbar. The next five records appear, and the Page field displays page 2.

2. Click the **First Page** button ◀ to navigate to the first page of the report.

3. Highlight the **1** currently displayed in the Page field, type **2**, and then press **Enter**. The second report page appears.

4. Close the report, and click **Yes** to save your changes.

COMPONENTS OF A MASTER-DETAIL REPORT

Now that you understand how to view and modify the objects of a single-table report, you are ready to learn how to modify the objects of a master-detail report to make it more readable. Reports with master-detail relationships have multiple record groups and multiple group frames. To learn about master-detail report objects, you work with the Northwoods University Class List report in Figure 7-27.

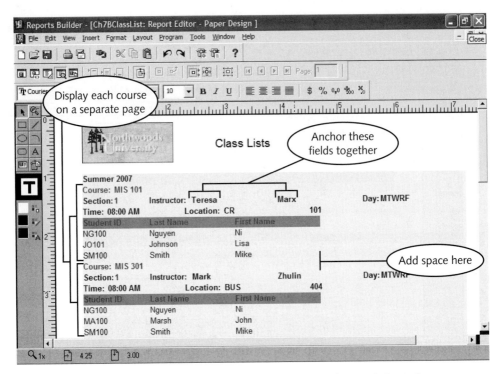

Figure 7-27 Northwoods University Class List report (partial output shown)

The Class List report is a master-detail report that provides class lists for all of the classes offered during the Summer 2007 term at Northwoods University. The report output includes the term description and detail records showing the course number for all courses offered during the term. For each course number, detail records show information about the section number, the instructor's first and last name, and the day, time, building code, and room for each course offered during the term. For each section number, detail records show the student ID, last name, and first name of each student enrolled in the course. Therefore, the report has a total of three master-detail relationships: a term has multiple courses; a course has multiple course sections; and a course section has multiple students. Now you open the report file, save the report using a different filename, and then view the report in the Paper Design window.

To open, save, and view the master-detail report:

1. In the Object Navigator, open **ClassList.rdf** from the Chapter07 folder on your Data Disk, and save the report as **Ch7BClassList.rdf** in the Chapter07\Tutorials folder on your Solution Disk.

2. Click the **Run Paper Layout** button 🖳 to display the report in the Paper Layout window. The report opens in the Paper Design window, as shown in Figure 7-27.

Figure 7-27 highlights some of the problems with the report's current format. Because the class list needs to be distributed to individual instructors, the information for each course should appear on a separate page. The report would be easier to understand if the course section data fields (section number, instructor first and last name, day, time, and location building code and room) were stacked on top of each other vertically, rather than spread horizontally on the page. Fields that should appear adjacent to each other, such as the instructor first and last name, should be anchored together instead of being spaced to accommodate the largest possible field width. Also, the report would look better if there were blank lines between the course section and student information, and if the student fields were indented on the page. To make these changes, you need to reposition the data fields, which requires modifying the frames that the Report Wizard generates. You also need to modify some properties of the report frames. First, you examine the report's master-detail Data Model.

Master-Detail Data Model

The Data Model in a master-detail report is more complex than the Data Model for a single-table report because it contains multiple record groups. Recall that when you use the Report Wizard to create a master-detail report, you use the Groups page to create groups that define the data levels within the report. Each group that you create on the Report Wizard Groups page represents a record group in the report Data Model. For each master-detail relationship, the master records are in one record group, and the detail records are in a separate record group. Next, you examine the report Data Model.

To examine the Data Model:

1. Click the **Data Model** button ⬚ on the toolbar to open the Data Model window. The report record groups appear. The report contains four separate record groups. (Some of the record groups might currently appear off-screen.) Now you move the record groups so that they are all visible on the screen. To move a record group, you select it and then drag it to the desired position.

2. Select **Q_1**, which represents the report SQL query, and move it to the top-left corner of the Data Model painting region, as shown in Figure 7-28.

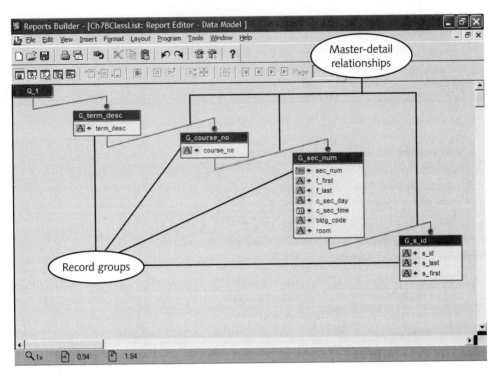

Figure 7-28   Master-detail data model

3. Select **G_term_desc**, which represents the Level 1 record group, and move it so that it is to the right and slightly below Q_1, as shown in Figure 7-28.

4. Move the other record groups so they appear as shown in Figure 7-28.

If a scroll bar appears on the bottom or right side of one of the record groups, it means that the record group is not large enough to display all of its columns. Select the record group by clicking it, and then drag the right-center handle horizontally to the right or drag the bottom-center handle vertically to the bottom so that all fields appear and the scroll bar no longer appears.

Note that the Data Model displays four record groups: G_term_desc, G_course_no, G_sec_num, and G_s_id. The Data Model groups the report record groups according to master-detail relationships, and each data link between two record groups represents a master-detail relationship. The highest-level (Level 1) group contains the term description, which is Summer 2007. Because the term has multiple course number values, the second (Level 2) record group contains the course numbers. One course number might have several sections, so all of the data fields that are unique for a given course section (sec_num, f_first, f_last, c_sec_day, c_sec_time, bldg_code, room) appear in the Level 3 G_sec_num record group. Each section number might have several students, so the student fields (s_id, s_last, s_first) appear in the Level 4 (most detailed) G_s_id record group.

Master-Detail Report Frames

To make the report formatting modifications suggested in Figure 7-27, you have to reposition the data fields within the frames. You also need to modify properties of individual frames. Now you open the Paper Layout window and examine the report frames.

To examine the report frames:

1. Click the **Paper Layout** button ⬛ on the toolbar.

2. The master-detail report layout appears in the Paper Layout window, as shown in Figure 7-29. (If the report margins appear, click the Edit Margin button ⬛ to open the report body for editing.)

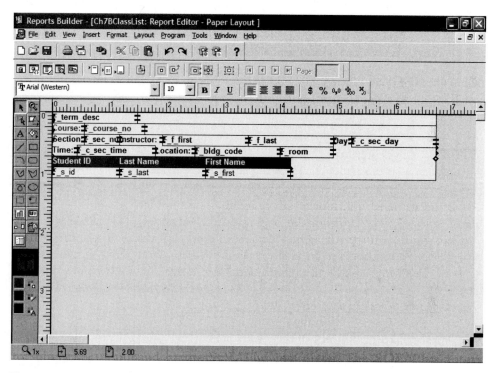

Figure 7-29   Master-detail report Paper Layout

Currently, it is difficult to interpret the report frame relationships because many of the frames are stacked on top of each other. You need to modify the sizes of some of the frames so you can move items around in the frames and create space between different record group items. You begin by examining a series of schematic diagrams that explain the frame structure of the report. Then you examine the individual report frames in the Paper Layout window.

Master-Detail Group Frame Relationships

Recall that in a report, each record group has an associated group frame. The Class List report contains four separate record groups, so it has four separate group frames. The group frames for the more detailed record groups are nested inside the group frames for the less detailed record groups. Recall that in this report, the Level 1 record group (G_term_desc) shows information for a term description. The Level 2 record group (G_course_no) shows information for a particular course number. The Level 3 record group (G_sec_num) shows information for a specific course section associated with that term and course number. The Level 4 record group (G_s_id) lists the students who are in a specific course section.

The group frame associated with the G_s_id record group is enclosed by the group frame associated with the G_sec_num record group. The group frame associated with the G_sec_num record group is enclosed by the group frame associated with the G_course_no record group, and the group frame associated with the G_term_desc record group encloses all the other group frames. Figure 7-30 shows the relationships among the report group frames.

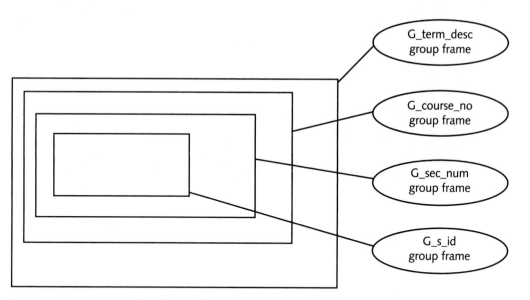

Figure 7-30   Master-detail group frame relationships

Frames Within the G_s_id Group Frame

Next, you examine the relationships of the frames within the G_s_id group frame. Recall that when record group fields appear in a tabular format, a header frame encloses the column headings, a repeating frame encloses the layout fields, and a group frame encloses the header frame and repeating frame. Figure 7-31 shows a schematic diagram of the frames that the Report Wizard creates for the G_s_id record group. These frames appear in the Paper Layout window, but are hard to interpret because the header frame and repeating frame are stacked directly on top of each other, and the group frame is directly on top of these two frames.

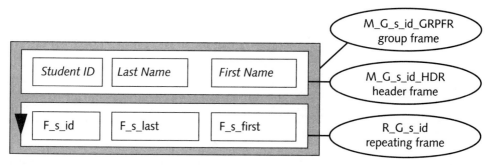

Figure 7-31   G_s_id record group frames

The layout fields that display the student data (F_s_id, F_s_last, F_s_first) are enclosed by a repeating frame named R_G_s_id. The column headings (Student ID, Last Name, and First Name) are enclosed by a header frame named M_G_s_id_HDR. Both the repeating frame and the header frame are enclosed by a group frame named M_G_s_id_GRPFR, which is shaded in the figure.

Frames Within the G_sec_num Group Frame

Recall from Figure 7-30 that the G_s_id group frame lies within the G_sec_num group frame. Figure 7-32 shows that the G_sec_num group frame contains a repeating frame that encloses the record group's headings and layout fields, and also contains the G_s_id group frame. (The G_sec_num record group does not contain a header frame because the frame does not display data in a tabular format.) In the figure, all group frames appear shaded.

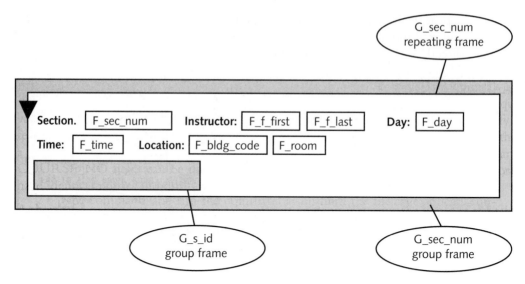

Figure 7-32   G_sec_num record group frames

Note that a down-pointing arrow appears on the left side of the G_sec_num repeating frame.

Frames Within the G_course_no and G_term_desc Group Frames

Recall from Figure 7-30 that the G_course_no group frame encloses the G_sec_num group frame, and the G_term_desc group frame encloses the G_ course_no frame. Figure 7-33 shows that the G_course_no and G_term_desc group frames also contain repeating frames that enclose their associated layout fields and column headings. Because these frames do not display tabular data, they do not contain header frames. In the figure, all group frames appear shaded.

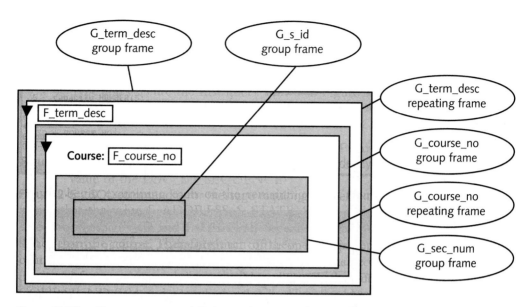

Figure 7-33   G_course_no and G_term_desc record group frames

Selecting Frames in the Paper Layout Window

To become familiar with the frame relationships and gain more experience selecting specific frames, you examine the frames in the Paper Layout window. Recall that these frame relationships are hard to see because many of the frames appear directly on top of each other. Recall also that the best way to select a frame is to select an item inside the frame, and then click the Select Parent Frame button 🔲. Because a frame and its parent frame sometimes are directly on top of each other, there may be no visual change on your screen when you click 🔲. To keep track of the current frame selection, you can open the frame Property Inspector immediately after you select the frame and check the frame name.

To select a specific repeating frame, you select a data item in the frame, and then click 🔲 to select its parent frame. To select a group frame, you select a data item in the frame, click 🔲 to select its repeating frame, and then click 🔲 again to select the repeating frame's parent frame, which is the group frame. Now you use this process to select the G_term_desc repeating frame and group frame.

To select the G_term_desc repeating and group frames:

1. In the Paper Layout window, select the **F_term_desc** layout field, and then click the **Select Parent Frame** button 🔲 to select the field's parent frame, which is the G_term_desc repeating frame (R_G_term_desc). Because this repeating frame encloses all of the report items, selection handles appear around all the report items.

2. Click 🔲 again to select the repeating frame's parent frame, which is the G_term_desc group frame (M_G_term_desc_GRPFR). There is no visible change in the selection handles that appear in the Paper Layout window because the group frame is directly on top of the repeating frame.

3. To view the selected frame name, click **Tools** on the menu bar, and then click **Property Inspector**. The Name property value appears as M_G_term_desc_GRPFR, which indicates you selected the G_term_desc group frame.

4. Close the Property Inspector.

Modifying Master-Detail Report Properties

At this point you should understand the structure of the master-detail report. You are now ready to begin making the formatting changes suggested in Figure 7-27. Next, you learn how to print record group data values on separate pages and reposition the report items.

Printing Report Records on Separate Pages

The first change you make to the report format is to make the data for each course print on a separate page. To create a page break between sets of repeating records, you must open the Property Inspector for the repeating frame that contains the records that you want on each separate page, and change its Maximum Records per Page Property to 1. For this report, you want to print each course number on a separate page. To do this, you select the repeating frame that encloses this field, which is R_G_course_no, open its Property Inspector, and change its Maximum Records per Page property to 1. Then you view the report in the Paper Design window and examine the result.

TIP

Sometimes you have to experiment to determine which repeating frame's Maximum Records per Page property must be changed to make report page breaks appear as desired.

To make the data for each course section number appear on a separate page:

1. In the Paper Layout window, select the **F_course_no** field, and then click the **Select Parent Frame** button 🔳 to select the R_G_course_no repeating frame.

2. Select **Tools** on the menu bar, and then click **Property Inspector**. Confirm that you have selected R_G_course_no, and then change the Maximum Records per Page property to **1**. Click another property to save the change, and then close the Property Inspector.

3. Click the **Run Paper Layout** button 🔲 to display the report in the Paper Design window. The report output appears with the course data and class list for MIS 101 Section 1 on the first report page.

4. Click the **Next Page** button ▶ to view the next class list. The class list for MIS 301 Section 1 appears.

5. Click ▶ again. The class list for MIS 441 Section 1 appears.

6. Save the report.

Repositioning Report Objects

The next formatting task is to reposition the report objects to improve the report appearance. Now you move the course section-specific fields and labels (section number, instructor, day, time, and location) so they stack on top of each other. You must leave some blank space between the course section fields and the student fields, and indent the student fields. When you reposition report objects, be careful not to move objects outside their enclosing frames. If you move an object outside its enclosing frame, the report layout is no longer consistent with the report Data Model, and an error message appears when you try to view the report in the Paper Design window.

To help avoid this error, Reports Builder has a feature called **confine mode**, which determines whether you can move objects outside their enclosing frames. When you enable confine mode, you cannot move an object outside its enclosing frame. When you disable confine mode, you can freely drag objects anywhere on the Paper Layout painting region. You enable confine mode by clicking the Confine On button 🔲 on the Paper Layout toolbar. You disable confine mode by clicking the Confine Off button. It is always safest to leave confine mode enabled, because it is difficult to locate an object that you have accidentally moved outside its enclosing frame and place it back into the correct frame.

TIP   Another way to enable confine mode is to right-click anywhere on the Paper Layout painting region, and then click Confine Mode On. To disable confine mode, right-click anywhere in the Paper Layout painting region, and then click Confine Mode Off.

As you reposition report objects within frames, you also need to understand flex mode. When you enable **flex mode**, an enclosing frame automatically becomes larger when you move an enclosed object beyond the enclosing frame's boundary. To enable flex mode, you click the Flex On button 🔲 on the Paper Layout toolbar. To disable flex mode, you click the Flex Off button. Flex mode overrides confine mode: when flex mode is enabled, Reports Builder automatically resizes frames, regardless of whether confine mode is enabled or disabled.

TIP   Another way to enable flex mode is to right-click anywhere on the Paper Layout painting region, and then click Flex On on the pop-up menu so that Flex On is checked. To disable flex mode, right-click anywhere in the Paper Layout painting region, and then click Flex Off.

A problem with enabling flex mode is that when you move a report layout field, flex mode automatically resizes all the surrounding frames. Sometimes this causes the outermost frames to extend beyond the boundaries of the report body, which generates an error when you view the report in the Paper Design window. Flex mode works well when you need to make a frame longer, because there is usually sufficient extra space on the length of the report page to resize all the surrounding frames. However, when you need to make a frame wider, it is best to leave flex mode off and resize the frame manually, so you can ensure that report frames do not extend beyond the right edge of the report body.

Now you move the STUDENT layout fields (F_s_id, F_s_last, F_s_first) so they appear lower on the report. You must make sure that flex mode is enabled, so when you move the fields, their enclosing frames become longer.

To enable flex mode and then move the fields:

1. Click the **Paper Layout** button ▣ to open the Paper Layout window, and make sure that the **Flex On** button ▣ is pressed. To confirm that flex mode is enabled, right-click in the Paper Layout painting region below the report objects, and confirm that Flex On appears selected.

2. Select the **F_s_id** layout field, press and hold **Shift**, and then select **F_s_last** and **F_s_first** to select all three fields as a group. Drag the columns down until their bottom edges are 1.5 inches from the top of the report, as shown in Figure 7-34. Notice that as you drag the columns down, their parent frame (R_G_s_id) automatically resizes to accommodate the new column locations.

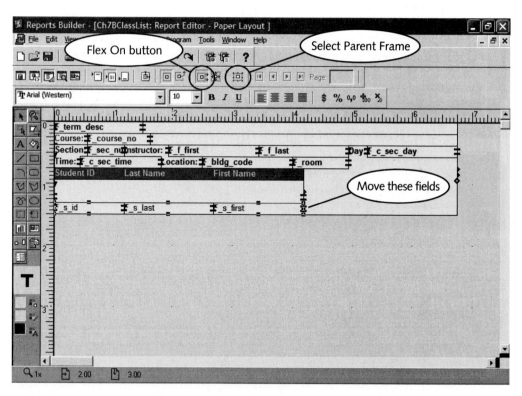

Figure 7-34 Making a report frame longer

3. Click the **Run Paper Layout** button ▣ to view the report in the Paper Design window, and note the new positions of the student data fields. There is now about 0.5 inch of white space between each student record as a result of vertically enlarging the repeating frame and moving the student columns lower in the frame.

4. Click **File** on the menu bar, and then click **Revert** to revert the report back to its state the last time you saved it. Click **Yes** to undo your changes. The Object Navigator window opens.

5. Double-click the **Paper Layout** icon ▣ to reopen the Paper Layout window.

Next, you examine the effects of using flex mode when you make report frames wider. You move a report column to the left edge of the report, which causes the enclosing frames to extend horizontally. Recall that this sometimes causes the report frames to extend beyond the report body boundary, as you see next.

To examine the effects of enabling flex mode when making report frames wider:

1. Confirm that flex mode is enabled, select **F_s_first**, and then drag it to the right edge of the report so the left edge of F_s_first is 5½ inches from the left edge of the report, as shown in Figure 7-35. Note that the frames enclosing F_s_first resize as you drag the data column to the right edge of the report.

7

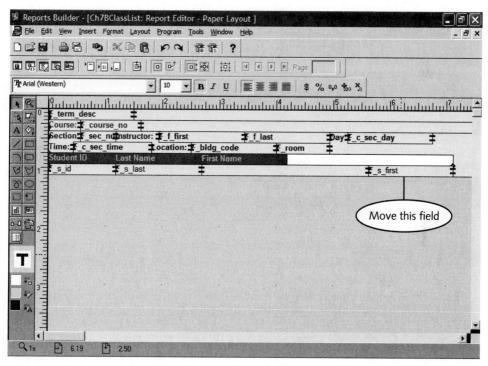

Figure 7-35   Making a report frame wider

2. Click the **Run Paper Layout** button ▣ to view the report in the Paper Design window. The error message "REP-1212: Object 'M_G_term_desc_GRPFR' is not fully enclosed by its enclosing object 'Body'" appears. This error message indicates that one of the report objects extends beyond the edges of the report body.

3. Click **OK**, and then click the **Paper Layout** button 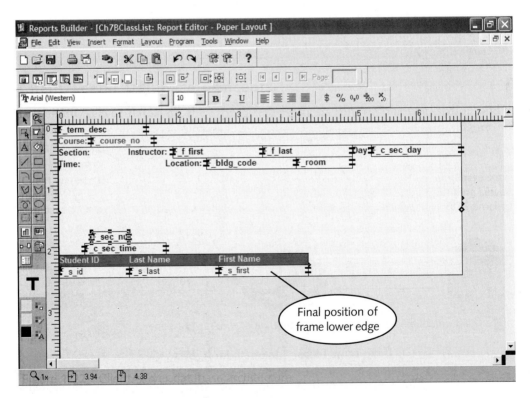 if necessary to open the Paper Layout window. Scroll to the right edge of the report. You see that some of the outer report frames now extend beyond the report body boundaries, which generates the error.

4. Click **File** on the menu bar, click **Revert**, and then click **Yes** to undo your changes. The report appears in the Object Navigator.

5. Double-click the **Paper Layout** icon to reopen the Paper Layout window.

Now that you understand the intricacies and peculiarities of confine mode and flex mode, you are ready to reposition the report objects. First, you make sure that flex mode is enabled, and then make the R_G_sec_num repeating frame and its surrounding frames longer, so that there is room to stack the course section fields vertically.

To resize the R_G_sec_num repeating frame and its surrounding frames:

1. In the Paper Layout window, make sure that **flex mode** is enabled, then select **F_sec_num** and drag it to the lower edge of the report so that the bottom edge of the outermost frame is about 2½ inches from the top of the report, as shown in Figure 7-36.

Figure 7-36   Resizing the report frames

2. Click the **Run Paper Layout** button 🕎 to view the report in the Paper Design window and confirm that all objects are properly within their enclosing frames.

3. Click the **Paper Layout** button 📝 to reopen the Paper Layout window, and then save the report.

Next, you reposition the report fields and labels. You indent the student fields and vertically stack the course section fields and labels. First, you select the student group frame and move it to the right so that the student fields are indented. There is enough room in the frames that enclose the student group frame, so you do this with flex mode disabled. (If you moved the group frame to the right with flex mode enabled, the surrounding frames would become wider and possibly extend beyond the boundaries of the report body.) Then you reposition the course fields and labels.

To indent the student group frame and reposition the course fields and labels:

1. In the Paper Layout window, click the **Flex Off** button 🔅 to disable flex mode. Select **F_s_id**, and then click the **Select Parent Frame** button 🗔 to select the student repeating frame. Click 🗔 again to select the student group frame.

2. Press the **right arrow key** to indent the group frame so that its left edge is about 0.25 inches from the left edge of the report body, as shown in Figure 7–37.

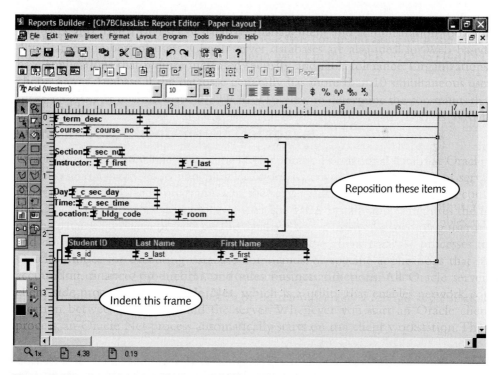

Figure 7-37   Repositioning the report fields and labels

TIP

If you try to use the mouse to drag the frame to a different position, the frame becomes deselected. Always use the arrow keys to move frames in the painting region.

3. Reposition the rest of the report fields and labels as shown in Figure 7-37.

4. View the report in the Paper Design window to confirm that the items appear correctly. Save the report, and then click the **Paper Layout** button 📝 to view the report in the Paper Layout window again.

Adjusting the Spacing Between Report Columns

Another formatting task is to adjust the spacing between report fields so the data values that should appear next to each other (such as the instructor's first and last name) are not separated by blank space. You have probably noticed that some of the report fields have horizontal tick marks on their right and left edges. These marks reflect the field's **elasticity**, which determines whether a field's size is fixed on the report, or whether the field can expand or contract automatically, depending on the height and width of the retrieved data value. Figure 7-38 shows the different elasticity options and associated indicators on report field horizontal and vertical borders.

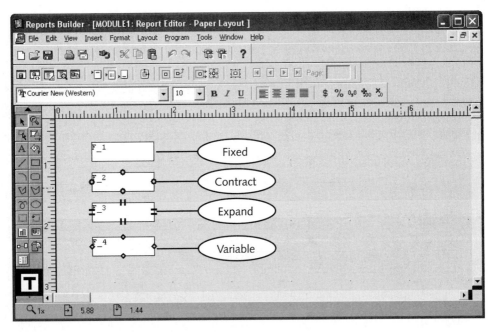

Figure 7-38   Report field elasticity indicators

The markings on the top and bottom edges indicate the **horizontal elasticity**, which specifies how wide the field can be. The markings on the right and left edges indicate the **vertical elasticity**, which specifies how tall the field can be. **Fixed** elasticity means that the field has the exact size shown in the Paper Layout window, and Reports Builder truncates extra text. **Contract** elasticity means that the column contracts if the data value is smaller than the layout size, but Reports Builder truncates wider data values. **Expand** elasticity means that the field expands automatically to accommodate wider data values, but narrower data values still occupy the entire space shown in the layout. **Variable** elasticity means that the report field contracts or expands as needed to fit the data value. A field can have **mixed elasticity**, which means that the vertical and horizontal elasticity can be different. By default, data fields that the Report Wizard creates have vertically expanding and horizontally fixed elasticity. By default, if a data value is wider than the width of the field, the field expands vertically, and the additional characters wrap to the next line. By default, if you manually draw a data field on the Paper Layout display, both the vertical elasticity and horizontal elasticity are fixed.

To make adjacent report fields, such as the instructor first and last name, appear directly next to each other on the report regardless of the width of individual data values, you open the field Property Inspector, and change the Horizontal Elasticity property value to Variable. Now you change the Horizontal Elasticity property for the F_f_first, F_f_last, F_bldg_code, and F_room fields to Variable. You select the fields as an object group, and then change the elasticity property using an intersection Property Inspector. (An intersection Property Inspector is similar to an intersection Property Palette, and allows you to change property values of several objects in one operation.)

To adjust the field elasticity:

1. In the Paper Layout window, select **F_f_first**, press and hold **Shift**, and then select **F_f_last**, **F_bldg_code**, and **F_room** to select the fields as an object group.

2. Right-click, and then click **Property Inspector** to open the intersection Property Inspector. Scroll down to the General Layout property node, and change the Horizontal Elasticity property value to **Variable**.

3. Close the Property Inspector, and then click anywhere on the blank region of the Paper Layout painting region to deselect the fields. Note that a diamond symbol now appears on the top and bottom edges of all the selected data fields to indicate that the horizontal elasticity is variable.

4. Click the **Run Paper Layout** button 📇 to view the report in the Paper Design window. The Instructor's name and location fields now appear right next to their respective labels.

5. To add a blank space to separate the instructor first and last name, select **Marx** in the Paper Design window, and then press the **right arrow key** to move the field toward the right edge of the screen. The first and last names now appear separated by a blank space.

6. To add a blank space to separate the building code and room number, select **101** in the Paper Design window, and then press the **right arrow key** to move the field toward the right edge of the screen. The building code and room number now appear separated by a blank space, and the formatted report appears as shown in Figure 7-39.

7. Save the report.

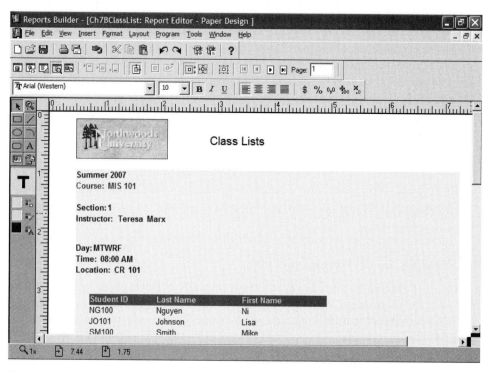

Figure 7-39   Formatted Class Lists report

REPORT PARAMETERS

A **report parameter** specifies how a report appears, or specifies the report's behavior when it runs. Reports have two types of parameters: system parameters and user parameters. **System parameters** specify properties that control how the report appears in the user display and how the report application environment behaves. Examples of system parameters include the currency symbol that appears, and whether the print dialog box opens when the user prints the report. **User parameters** allow the user to select values that specify the data that the report displays. For example, a user parameter in the Class List report would allow the user to select a specific term ID value from a list, and the report would then display class lists for that term only.

System Parameters

Table 7-3 summarizes the report system parameters and describes what the parameters specify.

Parameter Name	Specifies
BACKGROUND	Whether the report should run in the foreground or the background of the current application
COPIES	The number of report copies that are made when the report is printed
CURRENCY	The symbol for the currency indicator, such as "$"or "£"
DECIMAL	The symbol for the decimal indicator, such as "."
DESFORMAT	The output device format, which indicates whether to create PDF or HTML output when sending a report to a file
DESNAME	The name of the output device, such as the filename, printer name, or e-mail userid
DESTYPE	The type of device to which to send the report output, such as screen, file, e-mail, or printer
MODE	Whether the report should run in character or bitmap mode
ORIENTATION	The report print direction (landscape or portrait)
PRINTJOB	Whether the Print Job dialog box should appear before the report runs
THOUSANDS	The symbol for the thousands indicator, such as ","

Table 7-3   Report system parameters

You can modify report system parameter values by opening the Property Inspector of the target system parameter in the Object Navigator. To open a system parameter Property Inspector, you open the report Object Navigator, open the Data Model node, open the System Parameters node, and then open the Property Inspector for the system parameter you wish to modify. In most cases, you accept the default system parameter values.

User Parameters

The user selects a user parameter value from a parameter list. A **parameter list** shows possible values that can appear in a report query's search condition. After the user selects a user parameter value, Reports Builder inserts the value into the query search condition. The report then retrieves data based on the selected search condition value.

Currently, the Northwoods University Class List report shows class lists for the Summer 2007 term only. The report would be more flexible if users could select a term as a user parameter from a parameter list, and the report would display the class lists for the selected term value. You use the following steps to modify a report so it includes a user parameter:

1. Create the user parameter.
2. Create the parameter list.
3. Modify the report query so it uses the parameter as a search condition.

Creating a User Parameter

To create a user parameter, you open the Object Navigator, select the User Parameters node in the report Data Model, and then click the Create button ✚. Then, you open the user parameter Property Inspector and configure the user parameter by specifying the parameter object name and the parameter data type.

The **parameter object name** specifies how Reports Builder internally references the parameter. The parameter object also specifies the label that appears beside the parameter list on the parameter form on which the user selects the parameter value. Therefore, the parameter object name should describe the contents of the parameter list. The parameter object name appears in the Object Navigator in all capital letters, but Reports Builder displays the parameter name in the parameter list heading using mixed-case letters, and replaces underscores with blank spaces. For the Northwoods University Class List report, you name the user parameter TERM_DESCRIPTION in the Object Navigator, and it appears in the parameter list heading as "Term Description."

The **parameter data type** must be the same as the data type of the database field for which the parameter serves as a search condition. You should use number fields rather than character fields for user parameters, because sometimes Oracle 10*g* applications add blank spaces to pad out character fields, making an exact match between the parameter value in the search condition and the data value in the database field difficult to achieve. Therefore, you use TERM_ID for the user parameter, and display the associated term description (TERM_DESC) in the parameter list.

Now you create a user parameter that you associate with the TERM_ID field. After you create the user parameter object, you change the object name. You also change the user parameter data type to NUMBER, because the parameter data type must match the data type of the TERM_ID field in the database.

To create the user parameter:

1. Save the Ch7BClassList.rdf file as **Ch7BClassList_PARAM.rdf** in the Chapter07\Tutorials folder on your Solution Disk. (If you did not create the Ch7BClassList.rdf file earlier in the lesson, a copy of this file is saved as ClassList_DONE.rdf in the Chapter07 folder on your Data Disk.)

2. Open the report node in the Object Navigator, expand the Data Model node, and then select the User Parameters node.

3. Click the **Create** button ✚ to create a new user parameter object.

4. Select the new parameter if necessary, and then click it again so that the background of the parameter name turns blue. Change the name of the new user parameter to **TERM_DESCRIPTION**.

5. To open the user parameter's Property Inspector, select **TERM_DESCRIPTION**, right-click, and then click **Property Inspector**.

6. Make sure that the Datatype property value is **Number**. Change the Width property to **6** to match the maximum width of the TERM_ID field in the TERM table. Do not close the Property Inspector.

Creating the Parameter List

The next step is to specify the values that appear in the parameter list. To do this, you configure the List of Values property in the user parameter Property Inspector. The List of Values property defines the list of values from which the user can select the parameter value. You can enter a static list of predetermined input values, or you can enter a SQL query that dynamically retrieves a list of values from the database. In the following exercise, you enter a SQL query that retrieves term IDs and term descriptions. The first field that the query returns must be the user parameter field (TERM_ID). Reports Builder uses this value as a search condition in the report's data model query. When you configure the parameter list, you select the option of not displaying TERM_ID in the user parameter list of values, so the selection list that the user sees displays only the second query field, which is the term description (TERM_DESC). Now you configure the user parameter list of values.

To configure the parameter List of Values property:

1. In the TERM_DESCRIPTION Property Inspector, select the **List of Values** property, and then click the **More** bar that appears. The Parameter List of Values dialog box opens.

2. Select the **SELECT Statement** option button.

3. Type the query in Figure 7-40 in the SQL Query Statement field.

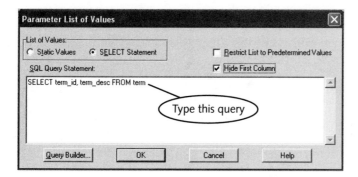

Figure 7-40   User parameter List of Values specification

4. To configure the list so it changes dynamically as users add new values to the TERM table, clear the **Restrict List to Predetermined Values** check box.

5. To hide the TERM_ID values in the parameter list, check the **Hide First Column** check box. Your completed Parameter List of Values dialog box should look like Figure 7-40.

6. Click **OK**, close the Property Inspector, and then save the report.

Modifying the Report Query

Now you need to modify the report query so it retrieves only the records for the parameter selection. Currently, the query search condition that specifies which term records are retrieved is `AND term_desc = 'Summer 2007'`. First you change the search field to TERM_ID. (In the search condition, you must qualify the TERM_ID field with the TERM tablename using the syntax `term.term_id`, because TERM_ID exists in multiple tables in the query's FROM clause.) Then you set the search value so it references the user parameter. To reference a user parameter in a query, you preface the parameter name with a colon, using the syntax *:parameter_name*. You reference the TERM _DESCRIPTION user parameter as `:TERM_DESCRIPTION`.

To modify the report query:

1. In the Object Navigator, double-click the **Data Model** icon 🔳 to open the Data Model, and then double-click **Q_1** to view the Data Model query.

2. Modify the query by replacing the line `AND term_desc = 'Summer 2007'` with the following search condition:

   `AND term.term_id = :TERM_DESCRIPTION`

3. Click **OK** to save the changes and close the dialog box, and then save the report.

Using the Parameter Form

When a report has a user parameter, the parameter form appears before the report appears. The **parameter form** allows the user to specify values for system and user parameters. For user parameters, the user can directly enter a value for the user parameter, or open the parameter list and select a value. Now you run the report and view the report parameter form.

To run the report and view the parameter form:

1. Click the **Run Paper Layout** button 🖳 to run the report.

2. The report parameter form opens and displays a Term Description list for entering a Term Description parameter value or selecting a value from the parameter list, as shown in Figure 7-41.

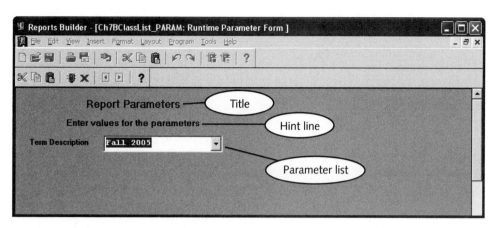

Figure 7-41   Report parameter form

The parameter form displays a title and hint line, and a list of one or more parameters with associated prompts. Because you specified a list of values for the term description parameter, the parameter appears as a list. If you do not specify a list of values for a parameter, the parameter appears as a text field, and the user can enter a value. To run the report and display the values in the Paper Design window, you select a parameter value, and then click the Run Report button 󰞸 on the toolbar. To close the parameter form without running the report, you click the Cancel Run button 󰅖. Now you select the Spring 2007 term from the parameter list and run the report.

To select the parameter value and run the report:

1. Open the parameter list, select the **Spring 2007** term, and then click the **Run Report** button 󰞸 to display the report.

2. The class lists for the Spring 2007 term appear.

No report data values appear if you select Fall 2005, Spring 2006, or Summer 2006 in the parameter list because there are no course section records in the database for these terms.

Customizing the Report Parameter Form

You can customize the appearance of the report parameter form. You can change the text that appears in the title and hint line, reposition the form items, and add additional text and graphics to enhance the appearance of the parameter form. You can also specify to display one or more system parameters on the parameter form, and allow the user to modify the system parameter values at runtime. For example, you might want to allow the user to select the DESTYPE, which specifies whether the report is routed to the screen or to a printer.

The Parameter Form Builder

You use the Parameter Form Builder utility to customize the parameter form. Now you open the Parameter Form Builder.

To open the Parameter Form Builder:

1. Click **Tools** on the menu bar, and then click **Parameter Form Builder**.

2. The Parameter Form Builder dialog box opens, as shown in Figure 7-42.

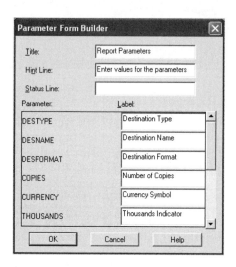

Figure 7-42   Parameter Form Builder dialog box

The Parameter Form Builder allows you to specify general properties for the report parameter form. The parameter form title is the top text line that appears on the parameter form. The hint line is the second line of text that appears on the parameter form, and the status line is the third line of text that appears on the parameter form. (In the parameter form in Figure 7-42, no text appears on the status line.)

The Parameter Form Builder also lists all of the system parameters that the user can modify at runtime. To specify to display a system parameter at runtime, you select the system parameter in the Parameter Form Builder dialog box so the parameter name appears on a black background. To change the name of the prompt that appears beside the text box for the system parameter, you type the new prompt in the text box in the Parameter Form Builder dialog box. You can also use the Parameter Form Builder to change the prompt that appears beside user parameters, and to hide user parameters on the parameter form that you wish to disable.

Now you customize the Class List report parameter form by changing the parameter form title, hint line, and status line. You also select the DESTYPE system parameter value and allow the user to specify the report destination (screen, printer, and so on) at runtime. Then you run the form and view the parameter form changes.

To customize the Class List report parameter form:

1. In the Parameter Form Builder dialog box, change the Title field value to **Northwoods University**.

2. Change the Hint Line field to **Class List Report Parameters**.

3. Change the Status Line field to **Enter parameter values**.

4. To allow the user to select the report destination, select **DESTYPE** in the Parameter list so its background appears black.

5. Scroll down to the bottom of the Parameter list, and note that the Term_Description user parameter you previously created is selected for display. This indicates that the user parameter list will appear on the parameter form when you run the report.

6. Click **OK** to save your changes, and close the Parameter Form Builder dialog box. The parameter form appears in the Reports Builder Paper Parameter Form window.

7. Save the report.

The Paper Parameter Form Window

You use the Reports Builder Paper Parameter Form window to configure the layout of custom parameter forms that you create using the Parameter Form Builder. You can use this window to edit the appearance of the objects that the Parameter Form Builder generates, and to add additional boilerplate text, images, or shapes to customize the parameter form appearance. You open the Paper Parameter Form window by clicking the Paper Parameter Form button 🖼 on the toolbar.

Next, you customize the parameter form in the Paper Parameter Form window. You move the existing parameter form objects so they appear lower on the parameter form, and display the Northwoods logo on the parameter form.

To customize the parameter form:

1. In the Paper Parameter Form window, click **Edit** on the menu bar, and then click **Select All**. All of the parameter form objects appear selected. Move the objects downward on the form to the positions shown in Figure 7-43.

2. To import the Northwoods logo, click **Insert** on the menu bar, and then click **Image**. Click **Browse**, navigate to the Chapter07 folder on your Data Disk, select **Nwlogo.jpg**, click **Open**, and then click **OK**. The logo appears on the parameter form. Modify the image size and appearance so your parameter form looks like Figure 7-43, and then save the report.

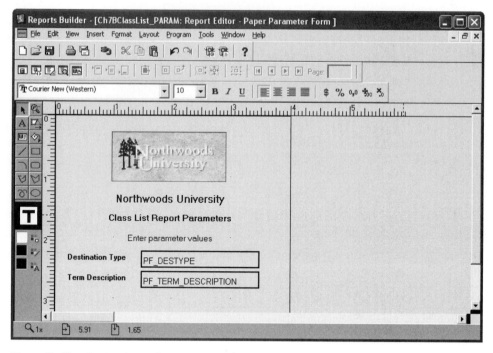

Figure 7-43   Customizing the parameter form appearance

Now you run the report and view the customized parameter form. You specify to send the report output to the screen, and select Summer 2007 for the term description parameter value.

To run the report, view the customized parameter form, and select parameter values:

1. Click the **Run Paper Layout** button 📰. The customized parameter form appears.

2. Make sure that **Screen** is selected in the Destination Type field. Open the **Term Description** list, select **Summer 2007**, and then click the **Run Report** button 📰 to run the report. The report appears in the Paper Design window, and shows the records for the Summer 2007 term.

3. Close the report in Reports Builder. If necessary, click **Yes** to save the changes to the report.

4. Close Reports Builder and all other open applications.

SUMMARY

- ❑ When you create a report, Reports Builder organizes the data fields that the SQL query retrieves into one or more report record groups. The Data Model window shows the report query, the record groups it generates, and the data links between the query and the record groups.

- ❑ In the Data Model window, you can modify the report query, and you can modify the properties of individual record groups and individual record group columns.

- ❑ In a master-detail report, Reports Builder places the master record columns and the detail record columns in separate record groups.

- ❑ You can create a group filter to limit the number of records that a report query retrieves. The group filter can limit the number of retrieved records based on a specific limit value, or based on a PL/SQL function that evaluates the individual data values in each record.

- ❑ The Paper Layout window displays a layout field for each Data Model column, and also displays column headings. You use the Paper Layout window to view and manipulate report items as objects rather than as actual data values. You also use the Paper Layout window to select and manipulate report frames.

- ❑ Frames are containers for grouping report record group data items and column headings. Reports can have header frames, repeating frames, and group frames. Header frames contain all the column headings for a record group and are created only when report values appear in a tabular format. Repeating frames enclose the layout fields that display the actual data values. Group frames group associated header and repeating frames together. An object's parent frame is the frame that directly encloses the object.

- ❑ In a report with master-detail relationships, frames for lower-level record groups are nested inside frames for higher-level record groups.

- ❑ When confine mode is enabled, you cannot move an object out of its enclosing frame. When flex mode is enabled, an enclosing frame automatically becomes larger when you move an enclosed object beyond the enclosing frame's boundary.

- ❑ Field elasticity determines whether a field's size is fixed on the report, or whether it can expand or contract automatically, depending on the size of the retrieved data values.

- ❑ To make data values appear directly next to each other on a report regardless of the width of the individual data fields, you must change the fields' Horizontal Elasticity property to variable.

- ❑ Reports not only have system parameters, which specify properties concerning how the report display looks and how the environment works, but also user parameters, which specify a search condition that controls the data that appears in the report.

- ❑ To enable a user parameter, you create the parameter in the Object Navigator, create the parameter list, and then modify the report SQL query so it uses the parameter in a search condition.

7

❏ The report parameter form allows the user to enter values for system and user parameters. You can use the Parameter Form Builder utility to customize the appearance of a report's parameter form, and the Paper Parameter Form window to modify the parameter form appearance.

Review Questions

1. You can change the properties of report components through the _____.
2. You can use a group filter to execute a WHERE clause and retrieve only those columns that meet the stated search condition. True or False?
3. Identify the three sections in a report.
4. _____ are containers for grouping related report objects.
5. If you want each set of repeating data to appear on a different page, set the _____ property of the repeating frame to 1.
6. Explain how variable elasticity works in a report.
7. If flex mode is enabled, you can make a frame larger by simply dragging one of the objects it contains beyond its current boundary. True or False?
8. What is the advantage of having a user parameter in a report?
9. You can only use a static list of values for the user's choices when creating a user parameter. True or False?
10. What is the appropriate syntax when referencing a user parameter in a SQL query?

Multiple Choice

1. Which of the following properties determines whether a field size is fixed on a report or can be adjusted to accommodate the displayed data?
   a. confine
   b. flex
   c. elasticity
   d. size to fit
2. A _____ limits the number of records that a report query retrieves.
   a. group filter
   b. join condition
   c. system parameter
   d. format mask
3. Which of the following is *not* a frame found in a report?
   a. header
   b. footer
   c. group
   d. repeating

4. Which of the following windows allows you to view the report's SQL query and associated record groups?

   a. Paper Design

   b. Web Layout

   c. Property Inspector

   d. Data Model

5. To view the actual data values displayed in the report, you should view the report through the _____ window.

   a. Paper Layout

   b. Web Layout

   c. Paper Design

   d. Data Model

6. When _____ is enabled, you cannot move an object outside its enclosing frame.

   a. confine mode

   b. flex mode

   c. protected mode

   d. none of the above

7. A(n) _____ allows the user to customize the data that appears in the report.

   a. Data Model

   b. user parameter

   c. system parameter

   d. LOV

8. Which of the following symbols indicates that the field's elasticity is assigned the value of contract?

   a. °

   b. =

   c. «

   d. ◊

9. By default, a parameter form does *not* include which of the following elements?

   a. logo

   b. title

   c. hint line

   d. parameter list

7

10. Which of the following determines whether a report is automatically displayed on a monitor or printed to a printer?

    a. PRINTJOB

    b. DESOUTPUT

    c. MODE

    d. DESTYPE

PROBLEM-SOLVING CASES

Cases reference the Clearwater Traders (Figure 1-24) sample database. All required files are stored in the Chapter07 folder on your Data Disk. Save all solution files in the Chapter07\Cases folder on your Solution Disk.

1. Create the Clearwater Traders Inventory report shown in Figure 7-44. This report lists each item category in the Clearwater Traders database and the corresponding item descriptions. The report also lists in detail the corresponding inventory ID, color, size, price, and quantity on hand.

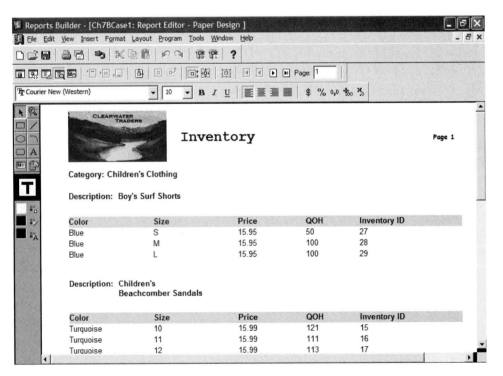

Figure 7-44 Inventory report

a. Use the Report Wizard to create a report that displays the data values in Figure 7-44. If you created the Clearwater.tdf custom template in Case 1 in Lesson A, apply the Clearwater.tdf template to the report. If you did not create the template, use a different template, and replace the default logo with the Clearwater logo (Clearlogo.jpg). Save the report as **Ch7BCase1.rdf**.

b. Format the report so that **0.25** inches of blank space appears between the Category and the Description fields, and so that **0.25** inches of blank space appears between the Description field and the inventory column headings. Configure the report so that each category appears on a separate report page.

2. Create a schematic diagram that shows all the report objects and frames for the Inventory Report created in Case 1. This diagram should be similar to Figure 7-33, except that it should show the detailed contents for all report frames. Label the layout fields, repeating frames, header frames, and group frames using the Report Wizard default names. Save it as **Ch7BCase2.doc**.

3. Create an Order report that displays each order placed by a customer. The report should include each customer's name and address; and for each order, the order number, order date and description of items purchased along with the appropriate quantity and price. In addition, add the formula **OL_QUANTITY * INV_PRICE** to the SELECT statement of the underlying report query and assign **EXT_PRICE** as the column alias. Include the calculated value (extended price) on the report for each item ordered.

When formatting the report, make certain you include a company logo, an appropriate report title, and group the data based on the customer's ID and the order number. Either apply a template or manually format the sections of the report to provide a professional appearance. Save the report as **Ch7BCase3.rdf**.

4. Modify the Order report from Case 3 so that each customer's orders are displayed on a separate page. Add a total through the Report Wizard to determine the total amount due for each order, based on the extended price of each item ordered. Save the report as **Ch7BCase4.rdf**.

5. Create a tabular Sales report that displays the customer ID, order number, and total amount due for each order in a particular month. Also provide the total amount of sales for the month by totaling the amount due of each order. Include a user parameter form to allow users to determine the appropriate month for the report (*Hint*: Include a LIKE statement with the user parameter in the underlying SQL statement to retrieve the indicated month). Save the report as **Ch7BCase5.rdf**.

◀ LESSON C ▶

After completing this lesson, you should be able to:

♦ Display image data in a report
♦ Manually create queries and data links
♦ Create summary columns
♦ Create formula columns
♦ Create reports that display formatted data in a Web browser window

DISPLAYING IMAGE DATA IN REPORTS

Recall that you can store image data in an Oracle 10g database using the LOB (Large Object) data types. You can retrieve and display image data directly in a report by using the Report Wizard to create a report based on a table that has a field with the BLOB data type and that contains image data. Recall also that the F_IMAGE field in the FACULTY table in the Northwoods University database contains data that represents a photograph of each faculty member. In this section, you learn how to create the Faculty report in Figure 7-45, which displays the image of each faculty member.

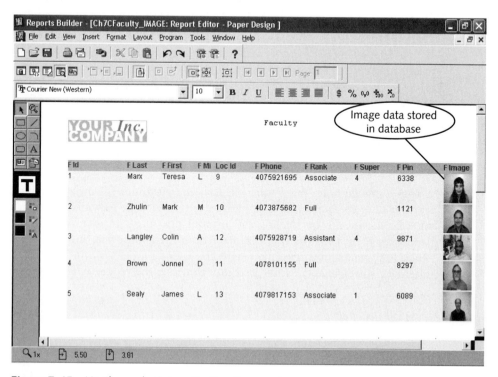

Figure 7-45 Northwoods University Faculty report

598

Loading the Image Data

Before you can create the Faculty report, you must run the script to refresh the case study database tables. You must also load the image data for the FACULTY table into your database. To load the image data, you use the Faculty Image form in Figure 7-46. (You learn how to create a form to load and display image data in Chapter 10. For now, you use this form to load image data.)

Figure 7-46   Northwoods University Faculty Image form

The Faculty Image form is a data block form that allows you to retrieve an existing FACULTY record. When you first run the form, your FACULTY table does not contain any image data, so the image field for each record is blank. You enter into the File field the path to the item image file on your workstation's file system, and then click Load Disk Image to load the image onto the form. You then click the Save button on the Forms Services toolbar to save the image data in the database.

To refresh the database tables and load the image data:

1. Start SQL*Plus, run the **Ch7Northwoods.sql** script from the Chapter07 folder on your Data Disk, and then exit SQL*Plus.

2. Start the Forms Builder OC4J Instance.

3. Start Forms Builder, and open **Faculty_IMAGE.fmb** from the Chapter07 folder on your Data Disk. Save the file as **Ch7CFaculty_IMAGE.fmb** in the Chapter07\Tutorials folder on your Solution Disk.

4. Run the form, and log onto the database when you are prompted to do so.

5. Click in the **Faculty ID** text box, click the **Enter Query** button 🖼, and then click the **Execute Query** button 🖼 to retrieve all of the table records. The record for item Faculty ID 1 (Teresa Marx) appears. The image does not appear, because you have not yet loaded it into the database.

6. In the Image Filename field, type the drive letter, full folder path, and filename for the Marx.jpg image file that is stored in the Chapter07 folder on your Data Disk. For example, if your Data Disk is stored in the C:\OraData folder on your workstation, the entry would be **C:\OraData\Chapter07\Marx.jpg**. (Your drive letter or folder path may be different.)

7. Click **Load Disk Image**. The image appears on the form.

8. Click the **Save** button on the Forms Services toolbar to save the image in the database. The confirmation message "Transaction complete. 1 records applied and saved." may appear, indicating that the image was saved in the database.

9. Click the **Next Record** button ▶ to view the next record. The data values for Faculty ID 2 (Mark Zhulin) appear. Type **C:\OraData\Chapter7\Zhulin.jpg** in the File field. (If your Data Disk files are saved in a different location, type that location instead.) Click **Load Disk Image** to load the image onto the form, and then click the **Save** button on the toolbar to save the record.

10. Repeat Step 9 for the other item records, using the following image files for each record:

Item ID	Image Filename
3	Langley.jpg
4	Brown.jpg
5	Sealy.jpg

11. Close the browser window and close Forms Builder.

Creating and Configuring the Report Image Layout Fields

When you use the Report Wizard to create a report, all data fields automatically appear as text. When a column contains multimedia data, such as image data, the data appears as the text characters "MM" to represent multimedia data. To display the actual images, you must modify the field by opening its Property Inspector and changing its File Format property to Image. Now you use the Report Wizard to create a tabular report based on the Northwoods University FACULTY table. Then you configure the layout field that displays the item images so it displays the image data.

To create and configure the Faculty report:

1. Start Reports Builder. Make sure the **Use the Report Wizard** option button is selected, click **OK**, and then click **Next**.

2. Make sure the **Create both Web and Paper Layout** option button is selected, and then click **Next**.

3. On the Style page, type **Faculty** in the Title field, make sure that the **Tabular** option button is selected, and then click **Next**. On the Data Source page, make sure that **SQL Query** is selected, and then click **Next**.

4. On the Data page, type **SELECT * FROM faculty** for the report query, and then click **Next**. When the Connect dialog box opens, connect to the database as usual.

5. On the Fields page, click the **Move all items to target** button to select all of the table fields for the report, and then click **Next**.

6. The report does not display any totals for any of the data fields, so do not select any fields on the Totals page, and then click **Next**.

7. On the Labels page, accept the default column labels, and then click **Next**.

8. On the Template page, make sure that the **Beige** predefined template is selected, click **Next**, and then click **Finish**. The report appears in the Paper Design window. Note that the layout fields in the F Image column appear as the text "MM."

NOTE

If necessary, adjust the width of the report columns to display each row on a single line.

9. Save the report as **Ch7CFaculty_IMAGE.rdf** in the Chapter07\Tutorials folder on your Solution Disk.

10. Select any of the MM layout fields, right-click, and then click **Property Inspector**. The F_F_IMAGE Property Inspector opens.

11. Under the Column property node, change the **File Format** property value to **Image**. The Report Progress dialog box opens while Reports Builder reformats the report layout. When the formatting completes, close the Property Inspector. The report appears in the Paper Design window and displays the image data, as shown in Figure 7-45.

12. Save the report, and then close the report.

CREATING REPORT QUERIES AND DATA LINKS MANUALLY

Earlier in the chapter, you created reports in which the report data model contained master-detail relationships. To do this, you retrieved all of the report data in a single query, and then used the Report Wizard Groups page to specify the master-detail relationships and create the associated master-detail record groups and data links. Another way to create reports that display master-detail relationships is to create a query manually that retrieves the master records, create a query manually that retrieves the detail records, create a data link between the two queries, and then create layout fields to display the query data. To gain experience creating queries and data links manually, you create the Northwoods University Faculty Class Schedule report in Figure 7-47.

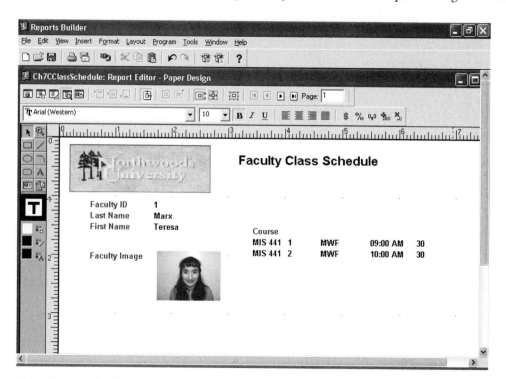

Figure 7-47   Northwoods University Faculty Class Schedule report

This report displays the faculty ID, last name, first name, and faculty image from the FACULTY table. It also shows the associated courses from the COURSE_SECTION table. The report Data Model contains two separate queries: the first query retrieves the data values from the FACULTY table and the second query retrieves the faculty member's class schedule from the COURSE_SECTION table.

To create this report, you first open a report file that currently displays the records from the FACULTY table. Now you open the report, save the report file using a different file-name, and view the report.

To open, save, and view the report:

1. In Reports Builder, open **ClassSchedule.rdf** from the Chapter7 folder on your Data Disk, and save the file as **Ch7CClassSchedule.rdf** in the Chapter07\Tutorials folder on your Solution Disk.
2. Click the **Run Paper Layout** button to view the report in the Paper Design window, as shown in Figure 7-48. (Your report may display a different data value, depending on the internal order of the records in your database.) Currently, the report displays the data for each FACULTY record on a separate report page. Note that the report does not yet display the class schedule.

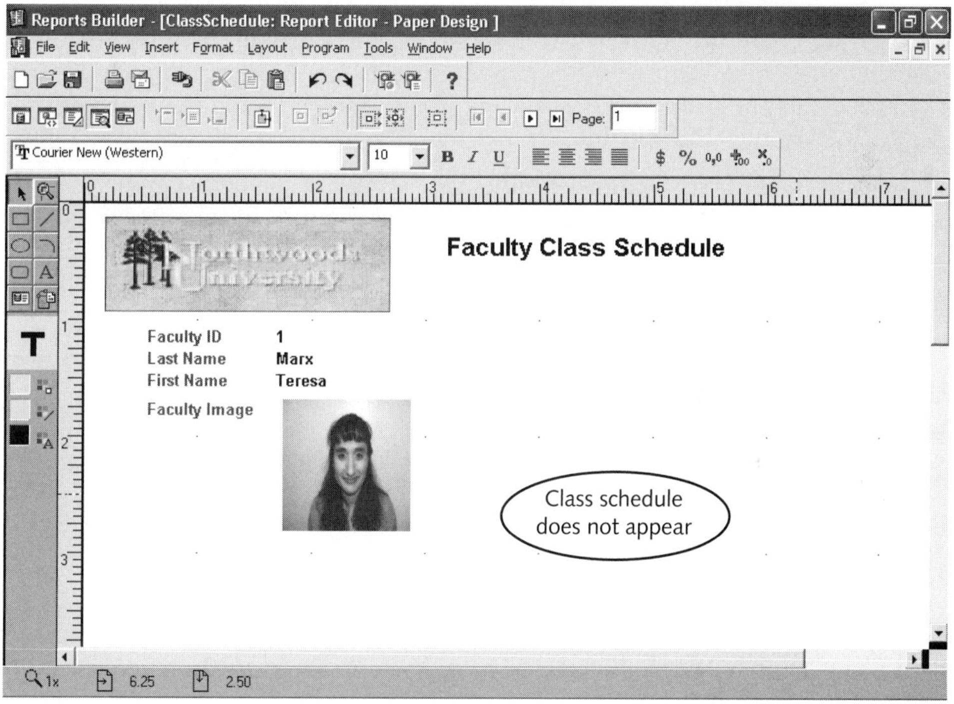

Figure 7-48 Current Faculty Class

If the message "Unable to display image data" appears when you try to view the report, the item image data is not available in the FACULTY table in your database. To load the image data, run the Faculty_IMAGE.fmb form in Form Builder, and load the image data from the associated image files using the steps in the section titled "Loading the Image Data." All required files are in the Chapter07 folder on your Data Disk.

Creating Queries and Data Links Manually

To create a query manually, you open the Data Model window, select the SQL Query tool ![SQL] on the Data Model tool palette, click in the painting region, and then type the SQL query. Then, you manually create data links to represent master-detail relationships between queries. In a data link, the master side of the relationship is called the **parent**, and the detail side of the relationship is called the **child**. You can create the following types of data links:

- **Query to Query**—Defines the data link at the query level. The primary key of the parent record group is a foreign key in the child record group. You might create this type of link if you wanted to add the name of the courses to the class schedule.

- **Group to Group**—Defines the data link at the record group level. One of the columns in the parent record group, which is not necessarily the record group's primary key, has a foreign key relationship with one of the columns in the child record group. You create a Group to Group data link to join the COURSE_SECTION data to the FACULTY data.

- **Column to Column**—Defines at the column level by linking two identical columns that do not necessarily have a foreign key relationship

Sometimes deciding what type of data link to use is a challenging process. Try different data link types until you get the result you want.

To create a data link, you select the Data Link tool ![Data Link] on the Data Model tool palette, and then draw a link from the parent object to the child object. To create a Query to Query link, you draw the link from the parent query to the child query. To create a Group to Group link, you draw the link from the parent record group title bar to the child record group title bar. To create a Column to Column link, you draw the link from the parent column to the child column.

Creating the Query and Data Link

The next step for completing the Faculty Class Schedule report is to display the courses taught by the faculty members. You could modify the query in the Report Wizard so it retrieves the contents of the COURSE_SECTION table, but doing that would reformat the report layout. Instead, you manually create a query in the report Data Model that retrieves all the records from the COURSE_SECTION table, and then create a data link between the F_ID column in the new record group to the F_ID column in the existing record group. Because F_ID is the primary key in the FACULTY table and a foreign key in the COURSE_SECTION table, you create a Query to Query link. The query that retrieves the FACULTY records is the parent query, and the query that retrieves the COURSE_SECTION records is the child query.

To create the query and data link manually:

1. Click the **Data Model** button to open the Data Model window, select the **SQL Query** tool on the tool palette, and then click the painting region on the right side of the existing Q_1 query. The SQL Query Statement dialog box opens.

NOTE

If the Data Model tool palette is not open, click View on the menu bar, and then click Tool Palette.

2. Type **SELECT course_no, f_id, term_id, sec_num, c_sec_day, c_sec_time, max_enrl FROM course_section WHERE term_id = 5** in the SQL Query Statement field, and then click **OK**. A new query named Q_2 and its associated record group named G_course_no appear in the painting region.

NOTE

When you create a query that has a new column with the same name as an existing column, Reports Builder appends a number to the column name to ensure that every Data Model column has a unique name.

3. Reposition the queries so they appear as shown in Figure 7-49.

7

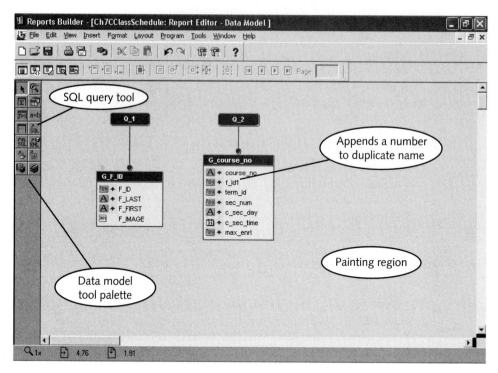

Figure 7-49   Query added manually to report

4. To create the data link, select the **Data Link** tool 🖳 on the Data Model tool palette, select **Q_1**, which is the parent query, and draw a link from Q_1 to Q_2. The data link appears as shown in Figure 7-50, and links F_ID in query Q_1 to f_id1 in query Q_2.

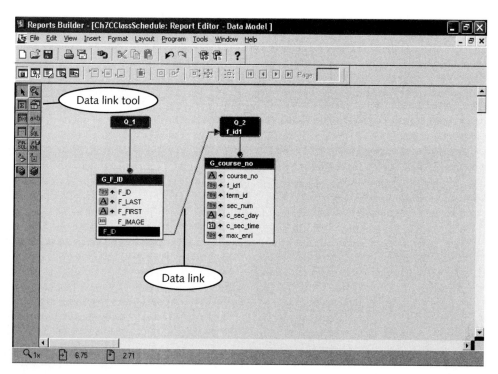

Figure 7-50   Manually linked queries

NOTE

When you use the Data Link tool 📊 to create a data link, you always draw the link from the parent object to the child object.

> 5. Save the report.

Creating Layout Fields Manually

To display values from a query that you create manually, you draw new layout fields in either the Paper Design or Paper Layout window, and then set the field Source property to the name of the new Data Model column for each column. Because a faculty member may teach more than one class, the fields must be enclosed in a repeating frame. Repeating frames automatically expand or shrink depending on the number of courses the G_COURSE_NO retrieves. Now you open the Paper Layout window and set the repeating frame Source property equal to the name of the record group that provides its values, which is G_COURSE_NO. Finally, you draw layout fields in the repeating frame to represent each of the G_COURSE_NO record group columns. You also modify each field's Source property to be the name of the associated record group column.

To draw the repeating frame and fields to display the faculty's class schedule:

1. Click the **Paper Layout** button 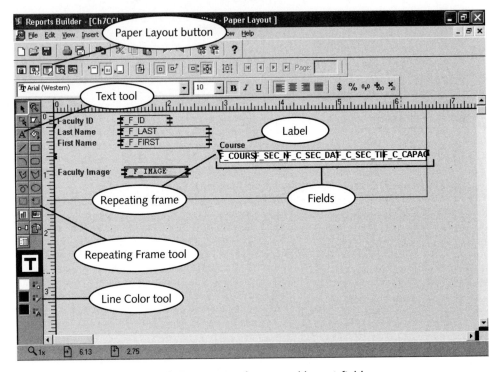 to open the Paper Layout window.

2. To create a label for the repeating frame, select the **Text** tool [A] on the tool palette, click the painting region in the area on the right side of F_FIRST, and then type **Course**, as shown in Figure 7-51. If necessary, change the label font to **10-point bold Arial (Western)**.

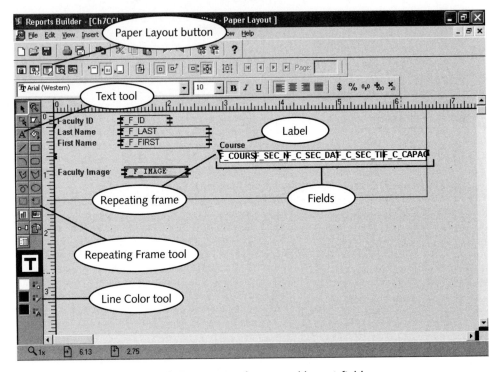

Figure 7-51   Creating the label, repeating frame, and layout fields

3. Select the **Repeating Frame** tool on the tool palette, and draw the repeating frame shown in Figure 7-51.

4. Double-click the **repeating frame** to open its Property Inspector, change the Name property value to **R_G_COURSE_NO** and the Source property value to **G_course_no**. Close the Property Inspector.

5. Select the **Field** tool on the tool palette, and draw five fields inside the repeating frame, as shown in Figure 7-51. These fields display the course number, section number, date and time, and maximum capacity for each course.

6. Open the Property Inspector for each of the fields, change the properties as follows, and then close the Property Inspector.

	Field 1	Field 2	Field 3	Field 4	Field 5
Name	F_COURSE_NO	F_SEC_NUM	F_C_SEC_DAY	F_C_SEC_TIME	F_C_CAPACITY
Source	course_no	sec_num	c_sec_day	c_sec_time	max_enrl
Format Mask				HH:MI AM	

7. Save the report, and then click the **Run Paper Layout** button 🏳 to view the report in the Paper Design window. The report should appear as shown in Figure 7-47. (Your report may display the records in a different order.)

8. If the repeating frame or fields appear outlined, select the frame and the fields as an object group, select the **Line Color** tool ▧ on the tool palette, and then click **No Line**.

9. If the new fields appear in a different font than the rest of the report fields, select the fields as an object group, open the **Font** list and select **Arial(Western)**, and then open the **Font Size** list and select **10**.

10. Save the report, and then close the report in Reports Builder.

CREATING REPORTS THAT DISPLAY CALCULATED VALUES

Sometimes you need to display calculated values on reports. For example, every Northwoods University student receives a student grade report each term that summarizes not only his or her total credits but also term and cumulative grade point averages. To display calculated values in a report, you can create **formula columns**, which display values that PL/SQL functions calculate using report data field values as input parameters. You can also create **summary columns**, which perform summary functions (such as SUM, AVG, or MAX) on report data fields.

To learn about the different report column types, you work with the Northwoods University Student Grade report in Figure 7-52.

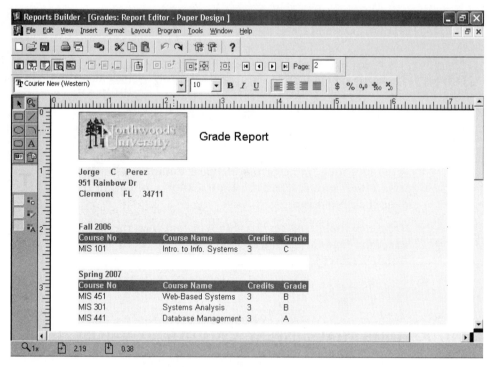

Figure 7-52   Second page of Student Grade report

The student's name and address appear at the top of the grade report, followed by the term description and a list of each course, the course credits, and the grade received for the course. The report sums course credits and calculates the student's grade point average (GPA) for each term. Total credits for all terms and the student's cumulative GPA appear at the end of the report. Now you open the student grade report and save it using a different filename.

To open and save the grade report:

1. In Reports Builder, open **Grades.rdf** from the Chapter07 folder on your Data Disk, and save the file as **Ch7CGrades.rdf** in the Chapter07\Tutorials folder on your Solution Disk.

2. In the Object Navigator, click the **Run Paper Layout** button 🕮 to view the report in the Paper Design window.

Note that the final report formatting has not yet been done: the student's first and last names are not adjacent to each other, the address is not formatted correctly, and there are no blank lines between the student address and the term and course information. You do not want to perform the final formatting tasks yet, because you use the Report Wizard in reentrant mode to create summary columns. Recall that whenever you make a modification using the Report Wizard, all custom formatting—such as changing the size and spacing of fields in frames or adjusting the elasticity property of data fields—is lost. Whenever you create a report that contains calculated report columns, always make sure that all of the report values appear correctly before you perform the final formatting.

Currently, the grade report displays the student name and address, the term description, and the course and course grade data. In the following sections, you create calculated columns that calculate and display the total student credits; total student grade points, which are used to calculate student GPA but do not appear on the finished report; and student GPA. Recall that reports use the terms column and field interchangeably. Summary and formula columns might display only a single value, such as the credits and GPA values, and not appear in a columnar format, but are still called columns.

Before you create the calculated report columns, you open the report Data Model and become familiar with the report record groups. You need to understand the structure of the report data, because you work with the report record groups and frames to make the required modifications.

To view the Data Model:

1. In the Paper Design window, click the **Data Model** button ⊞ on the toolbar. The Data Model view opens, as shown in Figure 7-53.

2. Save the report.

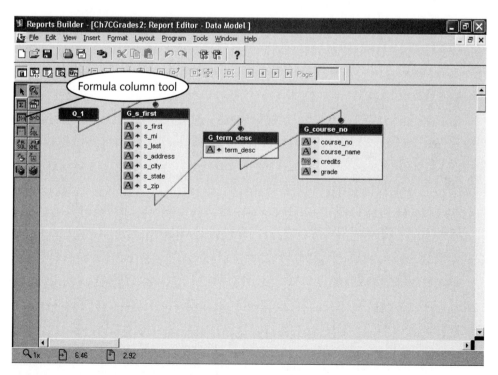

Figure 7-53   Data Model view of Student Grade report

The Data Model shows that the report has three record groups. The Level 1 group, G_s_first, contains the student information. The Level 2 group, G_term_desc, shows the term description, and the Level 3 group, G_course_no, displays the course and grade information. The following sections describe how to create columns to display the calculated report data values.

CREATING FORMULA COLUMNS

Recall that a function is a self-contained program block that returns a single value. PL/SQL programmers can create **user-defined functions**, which are functions that use PL/SQL commands to calculate a return value. A formula column displays a value that a user-defined function returns as a result of performing computations on report data values. For example, you might want to create a formula column to calculate and display a Northwoods University student's grade point average based on his or her course grades and course credits.

To create and display a formula column on a report, you perform the following tasks:

1. Create the formula column in the report Data Model.

2. Write the user-defined function that returns the calculated value for the formula column.

3. Create a layout field in the report to display the formula column value.

Creating a Formula Column in the Report Data Model

To create a formula column in the Data Model, you click the Formula Column tool 🔣, which is on the Data Model tool palette, on the record group that is to contain the formula column. You must be careful to place a formula column in the same record group as the columns that the formula function uses in its calculations, because the formula column calculates a value for each record in the record group.

Northwoods University (like most educational institutions) awards grade points for courses. Grade points are numeric values that represent the grade a student earns in a course times the number of credits the course is worth. Northwoods University awards course grade numeric values as follows: A = 4 grade points, B = 3, C = 2, D = 1, and F = 0.

Northwoods University calculates student grade point averages using the following formula:

$$\frac{\text{SUM}(course\_credits \, * \, course\_grade\_value)}{\text{SUM}(course\_credits)}$$

To calculate grade point averages, you must first calculate the course grade points for each course. The **course grade points** are the product of *course_credits* times *course_grade_value* for each course. For example, if student Tammy Jones receives a grade of "A" in a 3-credit course, she receives 4 points for the "A" times 3 credits, or 12 grade points for the course. You create a formula column to calculate the grade points for each course. Then, you create a summary column to sum the grade points for all courses. Because the report calculates course grade points using the CREDITS and GRADE data fields, you place the formula column that calculates the course grade points for each course in the G_course_no record group.

Now you create the new formula column in the G_course_no record group, and change the name of the new formula column. Formula column names usually appear with the identifier CF, so after you create the formula column, you change its Name property value to CF_course_grade_points.

To create and rename the formula column to calculate course grade points:

1. In the Data Model window, select the **Formula Column** tool [img] on the Data Model tool palette.

TIP

If the tool palette does not appear in your Data Model window, click View on the menu bar, and then click Tool Palette.

2. Click the mouse pointer on the G_course_no record group. A new formula column named *CF_1* appears in the record group. Select and then resize the record group if necessary so that all of the data columns are visible.

3. Select *CF_1*, right-click, and then click **Property Inspector**. Change the new column's Name property to **CF_course_grade_points**. Do not close the Property Inspector.

Creating the Formula Column Function

Every formula column has an associated PL/SQL function that returns the value that the formula column represents. Figure 7-54 shows the general syntax for a formula column function.

```
function column_nameFormula return data_type is
  return_value data_type;
  other variable declarations;
begin
  commands to calculate function return value
  return(return_value);
end;
```

Figure 7-54   General syntax for a formula column function

NOTE

In Figure 7-54, the reserved words function, begin, return, and end appear in lowercase letters because that is how the words appear in the PL/SQL Editor when you create the function.

In this syntax, the `function` keyword signals that the code block is a function. *Column_name*`Formula` specifies the name of the function, and links the function with the formula column. *Column_name* is the name of the formula column, so the function name for the CF_course_grade_points formula column function would be `CF_course_grade_pointsFormula`. The next command declares *return_value*, which is a variable that represents the value that the function returns.

`Begin` signals the beginning of the body of the program function. The function body calculates the function return value, based on columns in the report's record groups, including data columns, summary columns, and other formula columns. To reference a column in the PL/SQL formula column function, you preface the column name, as it appears in the Data Model, with a colon. For example, you reference the credits column as `:credits`. In a formula column function, you can only reference columns that are in the same record group as the formula column, or that are in higher record groups in the Data Model. Finally, the function uses the `return` command to return the value that *return_value* represents.

Now you create the PL/SQL function to calculate the course grade points for a single record in the G_course_no record group. You work with the PL/SQL Editor, which you used in Forms Builder in Chapters 5 and 6 to create form triggers. The PL/SQL Editor displays the template for the function, which contains the `FUNCTION` command that defines the function, and the `BEGIN` and `END` commands that define the function body and end the function. You enter commands in the function body to create an IF/ELSIF decision structure to evaluate the course grade, and then calculate the associated course grade points by multiplying the grade points by the credits.

To write the PL/SQL function to calculate the course grade points:

1. In the CF_course_grade_points Property Inspector, select the **PL/SQL Formula** property so that a button appears, and then click the button. The PL/SQL Editor opens, and displays the template for the formula column function.

2. Type the commands in Figure 7-55 in the PL/SQL Editor to calculate the grade points for a single course section record.

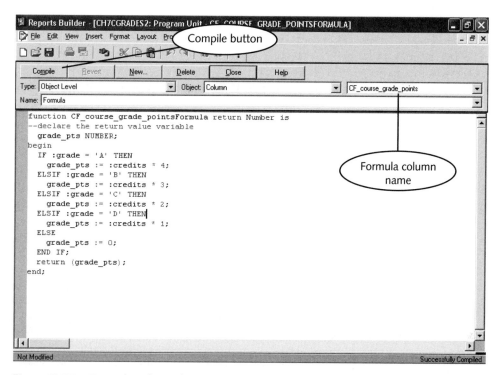

Figure 7-55   Formula column function to calculate course grade points

3. Click **Compile** at the top of the window to compile the code, correct any syntax errors, close the PL/SQL Editor, and then close the Property Inspector.

4. Save the report.

Creating a Layout Field to Display the Formula Column

Currently, the formula column exists in the Data Model but does not appear on the report. To display a formula column on a report, you open the Paper Layout window, and use the Field tool 🔳 to create a new report layout field. You must place the layout field in the same repeating frame as the source values for the formula. Then, you set the new field's Source property to the name of the formula column.

Recall that the columns that provide the data values in the grade points formula are GRADE and CREDITS, which are in the G_course_no record group. Therefore, the field that displays the formula column must be in the R_G_course_no repeating frame. Next, you open the report Paper Layout window, make the R_G_course_no repeating frame wider, draw a new layout field in the frame, and then assign the CF_course_grade_points formula column as the layout field's data source.

Recall from Lesson B that to make a report frame wider, you enable flex mode, and then move an item inside the frame to enlarge the surrounding frames automatically. Now you open the Paper Layout window, enable flex mode, and make the R_G_course_no frame wider.

To make the R_G_course_no frame wider:

1. Click the **Paper Layout** button ![icon] to open the Paper Layout window.

2. Make sure that the **Edit Margin** button ![icon] is not pressed, so the report body is open for editing. (When the report body is open for editing, the report margins do not appear in the Paper Layout window.)

3. Confirm that the **Confine On** button ![icon] is pressed and that the **Flex On** button ![icon] is pressed.

4. Select **F_grade**, and then click the **Select Parent Frame** button ![icon] to select the R_G_course_no repeating frame.

5. Click the *center* selection handle on the right edge (do *not* use the selection handle in the lower-right corner) of the R_G_course_no frame, and drag it toward the right edge of the screen display so the repeating frame is about four-inches wide, as shown in Figure 7-56.

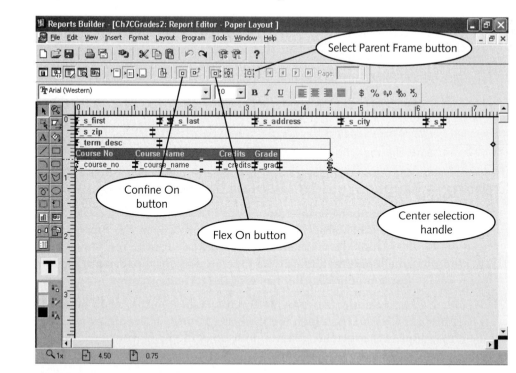

Figure 7-56   Widening the R_G_course_no frame

Make sure that the outside group frame remains on the report body. Otherwise, an error appears later when you run the report in the Paper Design window.

CAUTION

6. To confirm that the report runs with no errors, click the **Run Paper Layout** button 🗐 to view the report in the Paper Design window.

If an error message appears, click File on the menu bar, and then click Revert. Click Yes to undo your changes, and then repeat Steps 1 through 6.

HELP

7. Save the report.

Now you draw in the R_G_course_no repeating frame the layout field that displays the value of the grade points for each course section. Then you modify the field properties. You change the fieldname and assign its data source to be the CF_course_grade_points formula column.

To draw the layout field and modify its properties:

1. Click the **Paper Layout** button 🖼 to open the Paper Layout window, select the **Field** tool 🔠 on the tool palette, and then draw a new field inside the R_G_course_no repeating frame, as shown in Figure 7-57. Make sure that the field is totally enclosed by the frame.

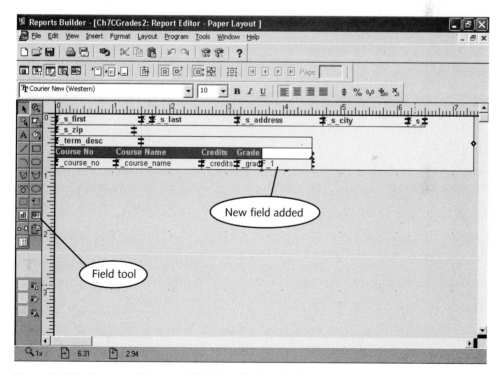

Figure 7-57   Drawing the layout field to display the formula column

2. Make sure that the new field is selected, right-click, and then click **Property Inspector**. Change the Name property value to **F_course_grade_points**.

3. Select the **Source** property, open the list, and select **CF_course_grade_points** as the field data source. Then close the Property Inspector.

4. Click the **Run Paper Layout** button 🐾 to view the report in the Paper Design window. Jorge Perez's transcript on page 2 of the report should look like Figure 7-58 and display the course grade point formula column values as shown.

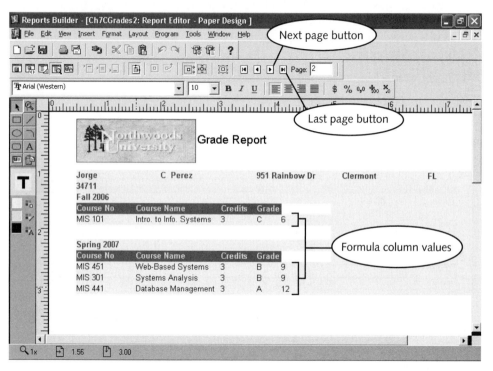

Figure 7-58   Viewing the formula column values

If a border appears around the formula column values, select the formula column field in the Paper Design window, select the Line Color tool 🖉 on the tool palette, and then click No Line. If the field has a different background color than the report, select the formula column field, select the Fill Color tool ■□, and then click No Fill. If the field has a different font name or size than the other report fields, select the formula column field, open the Font list at the top of the Paper Design window, and select Arial (Western). Open the Font Size list, and select 10.

HELP

If an error message appears stating that a display field references a data column at a frequency below its group, this means that you placed the field in the wrong frame. For example, you cannot place a column that displays each course's grade points (which the function calculates using data values from individual course section records) in the repeating frame for the term, because there are multiple course section records in a term.

5. Save the report.

CREATING SUMMARY COLUMNS

A summary column returns a summary value, such as the sum or average, of a series of data fields in a repeating frame. You can create summary columns for any report data field using the Totals page in the Report Wizard. You can also create summary columns manually. The following sections describe how to create summary columns using both approaches.

Creating a Summary Column Using the Report Wizard

To create a summary column using the Report Wizard, you select the field to be summarized and the summary function, such as Sum or Average, on the Report Wizard Totals page. When you create a summary column using the Report Wizard, the wizard automatically modifies the report Data Model, and creates fields to display the summary column on the report layout. Figure 7-59 shows the Report Wizard Totals page.

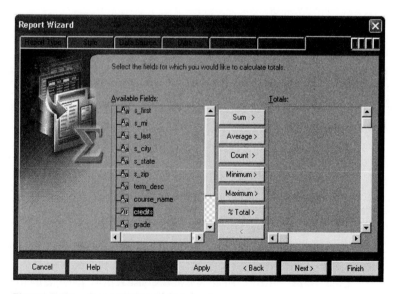

Figure 7-59   Report Wizard Totals page

The Totals page displays all of the report fields in the Available Fields list. To create a report summary column, you select the field to be summarized in the Available Fields list, and then click the button that describes the summary function you would like to apply to the selected field.

TIP

If you select a field that has a character or date data type, the Sum, Average, and % Total buttons are disabled, because these functions apply only to numerical data.

When you use the Report Wizard to create a summary column, the wizard not only creates a summary column in each record group above the record group that contains the source field, but also a summary column that summarizes the data for the overall report. In this report, the summary column source field is CREDITS, which is in the G_course_no (Level 3) record group. The Report Wizard creates a summary column that sums the credits at the term level (Level 2), which corresponds to the G_term_desc record group, and at the student level, which corresponds to the G_s_first (Level 1) record group. It also creates a summary column that summarizes the credits for all students in the report. Now you open the Report Wizard in reentrant mode, and create a summary column to sum total credits for each student for each term, and for each student for all terms.

To open the Report Wizard in reentrant mode and create a summary column:

1. Save the report as **Ch7CGrades_SUM.rdf** in the Chapter07\Tutorials folder on your Solution Disk.

2. Click **Tools** on the menu bar, and then click **Report Wizard** to open the Report Wizard in reentrant mode. The Report Wizard opens, with tabs for each page across the top of the window.

3. Select the **Totals** tab. The Totals page appears. Scroll down, if necessary, in the Available Fields list, and select **credits**.

4. Click the **Sum** button. Sum(credits) appears in the Totals list, indicating that the Report Wizard created a summary column to sum the CREDITS field.

5. Click **Finish** to close the Report Wizard and save the change. In the Paper Design window, note that the total credits appear for each term and for each student.

HELP

Note that the formula column values no longer appear. Recall that when you open the Report Wizard in reentrant mode and apply a change, all custom formatting objects are lost.

6. Click the **Last Page** button ▶| to view the last report page, and note the field that displays the total credits for the entire report, which has the value 36. Because the summary column that sums the credits for all students is not needed in this report, you hide it.

7. Right-click the field that displays **36**, and then click **Property Inspector**. Set the Visible property to **No**, and then close the Property Inspector.

8. Select the last **Total** label, and then press **Delete** to delete the label.

9. Save the report, and then close the report in Reports Builder.

Creating a Summary Column Manually

When you use the Report Wizard to create a summary column, the Report Wizard eliminates all of your custom formatting and deletes custom objects, such as the field you created to display the course grade points. An alternative to creating a summary column using the Report Wizard is to create a summary column manually. Creating a summary column manually allows you to control the summary column placement and frequency, and enables you to retain other custom objects. Creating a summary column manually is similar to creating a formula column, and requires that you perform the following tasks:

1. Create the summary column in the report Data Model, and modify the summary column properties.

2. Create a layout field to display the summary column on the report.

Creating and Configuring a Summary Column in the Data Model

To create a summary column in the report Data Model, you click the Summary Column tool ☒ in the record group that is to contain the summary column. You must create the summary column in the record group that is one level higher than the columns it sums, because the summary column displays a single value that summarizes multiple values in the lower-level record group.

After you create the summary column, you modify its properties to specify the column name, the report field that it summarizes, and the summary function (SUM, AVG, and so forth). Usually, you preface the names of summary columns that you create manually with CS, followed by the name of the column being summed. For example, the name of a summary column that sums term credits would be CS_credits.

Now you open the Ch7CGrades.rdf file you were working on earlier, and manually create a summary column that sums the values that the CF_course_grade_points formula column calculates for student grade points for each term. Because the formula column appears in the G_course_no record group, you place the summary column in the record group that is one level higher, which is the G_term_desc record group. Summary columns have a Reset At property, which specifies the record group level at which the summary column is reset to zero and resumes summing. You want to calculate the sum of grade points for each term, so you reset the summary column at the G_term_desc record group. Now you open the file, create the summary column in the Data Model, and modify its properties.

To open the file and create the summary column in the Data Model:

1. Open the **Ch7CGrades.rdf** file from the Chapter07\Tutorials folder on your Solution Disk. (If you did not create this file earlier in the lesson, a copy of the file is stored as Grades_DONE.rdf in the Chapter07 folder on your Data Disk.) Save the file as **Ch7CGrades_MANSUM.rdf** in the Chapter07\Tutorials folder on your Solution Disk.

2. In the Object Navigator, double-click the **Data Model** icon 📧 to open the Data Model window. Select the **Summary Column** tool ⅀ on the tool palette. Click the mouse pointer at the bottom of the **G_term_desc** record group. A new summary column named *CS_1* appears in the G_term_desc record group.

3. Select *CS_1*, right-click, and then click Property Inspector. Change the Name property value to **CS_CF_course_grade_points**.

4. Select the **Source** property, and then open the list. The list displays only the numerical fields that appear in the G_course_no record group. Select **CF_course_grade_points**.

5. Because you want the summary column to calculate the total grade points for each term, open the **Reset At** list, and then select **G_term_desc**.

6. Close the Property Inspector and then save the report.

Drawing the Summary Column Layout Field

To display the summary column value on the report, you draw a layout field on the report, and set its data source as the name of the summary column. You must place the layout field in the repeating frame that corresponds to the record group that contains the summary column. Because the summary column is in the G_term_desc record group, you draw the field so that it appears in the R_G_term_desc repeating frame. Then you modify the field name and set its Source property value as the summary column.

To create the summary column layout field:

1. Click the **Paper Layout** button 📝 to open the Paper Layout window. Confirm that **confine mode** and **flex mode** are enabled.

2. Select the **F_term_desc** field, and then click the **Select Parent Frame** button 🔲 to select the R_G_term_desc repeating frame. Open its Property Inspector to confirm that you have selected the correct frame, and then close the Property Inspector.

3. Click the mouse pointer on the center selection handle on the bottom edge of the frame, and make the frame longer by dragging its bottom edge about 0.25 inches toward the bottom of the screen display (see Figure 7-60).

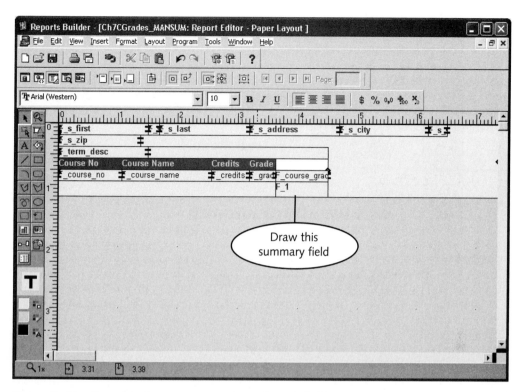

Figure 7-60   Drawing the summary column layout field

4. Select the **Field** tool on the tool palette. Draw a new field directly below the F_course_grade_points field, as shown in Figure 7-60.

5. Open the new field's Property Inspector, and change its name to **F_CS_CF_course_grade_points**.

6. Select the **Source** property, open the list, select **CS_CF_course_grade_points**, and then close the Property Inspector.

7. Click the **Run Paper Layout** button to view the report in the Paper Design window and confirm that the new summary column sums the credit point values correctly, as shown in Figure 7-61. The summary column should appear once per term.

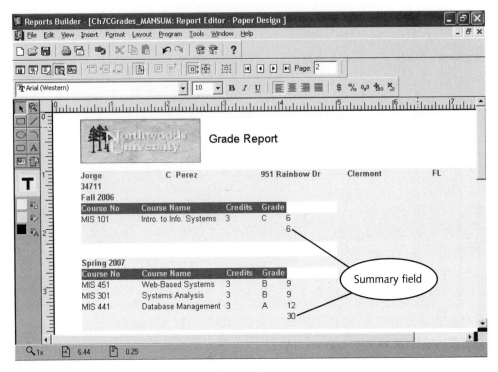

Figure 7-61   Viewing the summary column values

8. Save the report, and then close the report in Reports Builder.

In the student grade report, the student grade points per course and the sum of the student grade points per term are intermediate totals that do not appear on the final report. When you are creating a report with calculated columns that represent intermediate values that do not appear on the final report, it is a good practice to display the intermediate values during development to confirm that the values are correct. For example, Figure 7-61 allows you to view the summed grade points per term and confirm that the summed field values are correct. To suppress this intermediate value in the final report display, you can delete the field on the Layout Model, because the calculated column still exists in the report Data Model. Or, you can change the display field's Visible property value to No in the field Property Palette.

DISPLAYING FORMATTED REPORTS AS WEB PAGES

Recall that when you click the Run Web Layout button 🔾 on the Reports Builder toolbar, Reports Builder displays the current report in a browser window. The report appears in the default report format and does not display any post-Report Wizard formatting changes, such as resized column widths or alternate logo images. If you view the report using either the Paginated HTML or Paginated HTMLCSS format, Reports

Builder shows how the report looks after you make formatting changes. When you view the report in the Paginated HTML or Paginated HTMLCSS format, Reports Builder translates the report into a static Web page, which means that the content is fixed at the time the developer, which in this case is Reports Builder, generates the page. If the database data changes, the static Web page does not change to reflect the new data values.

To distribute formatted reports to users, you can distribute the static .htm Web pages to display a snapshot of the database at a current point in time. However, in Web-based applications, it is desirable to allow users to generate and view reports that reflect the current contents of the database. The following sections describe how the Oracle 10*g* Application Server dynamically generates Web pages that contain current database data, and how you can create and modify these reports.

Using the Oracle 10*g* Application Server to Generate Reports Dynamically

7

Recall that a Web server is a computer that is connected to the Internet and runs special Web server software. Web servers store the files that people can access via the Internet. When a browser connects to a Web server and requests a Web page, the Web server processes the request and sends the Web page to the browser. To generate forms and reports that appear as Web pages, you use the Oracle 10*g* Application Server. The **Oracle 10*g* Application Server (OAS)** is the Web server Oracle Corporation uses to deliver Web-based Oracle 10*g* database applications to users. The OAS runs a process called a Report Server to generate reports dynamically. The **Report Server** process queries the database when the user requests the report, and generates a dynamic Web page that displays the current database data. (Recall that a dynamic Web page is a Web page whose contents change based on user inputs or retrievals from data sources such as databases.) Figure 7-62 illustrates how the user's browser, the OAS, and the database server interact when the user requests a dynamic Web page that contains report data.

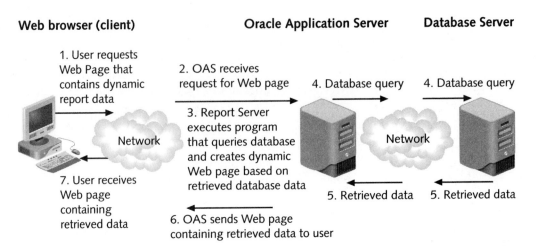

Figure 7-62   Creating a dynamic Web page that displays database data

To create a dynamic Web page that retrieves database data, the user first requests the Web page from the OAS. When the OAS receives this request, it forwards the request to its Report Server process, which starts a program that queries the database, retrieves the data, and then creates an HTML file that contains the data values. The OAS then returns the HTML file to the user, and the formatted data values appear in the user's browser.

Creating Reports that Appear as Dynamic Web Pages

To create a report that the OAS can deliver to users as a dynamic Web page, you create the report using the techniques you have learned in this chapter, and then save the report in either a report definition file (RDF) or a Java Server Page (JSP) format. So far, you have been saving all your reports in the RDF format. You can also save a report as a **Java Server Page (JSP)**, which is a text file with a .jsp extension that contains HTML commands for creating a Web page, along with Java program commands for adding dynamic content to the Web page.

Reports in either the RDF or JSP format contain commands to retrieve data values from an Oracle 10*g* database and then display the data values as a formatted report. The difference between the two formats is in how the OAS processes them to create a dynamic Web page to represent the report. The format in which you save the report depends on how the Web administrator configures your OAS.

When you create a report and save it in either the RDF or JSP format, the report appears in a browser window with the default formatting that you see when you click the Run Web Layout button 🔧 on the Reports Builder toolbar. There are two techniques for modifying the appearance of Web reports. The first technique is to use the Web Source window to modify the default formatting. The second technique is to open an existing HTML file in Reports Builder that contains the formatting for the report, and use Reports Builder to add commands to retrieve and display the report data. Before you can learn how to use these techniques to modify the report output, you need to understand the structure of HTML documents.

Overview of HTML Documents

An HTML document is a text file with an .htm or .html extension that contains formatting symbols, called **tags**, which define how a Web page appears in a Web browser, and **elements**, which represent the content that appears on the Web page, such as text and graphic images. You enclose tags in angle brackets (< >) using the following general syntax: `<tag_name>element</tag_name>`. The first tag is the **opening tag**, and the second tag, in which a front slash (/) precedes *tag_name*, is the **closing tag**. *Tag_name* specifies a particular formatting symbol, and *element* specifies the item, such as text or a graphic image, that the tag formats.

The code in Figure 7-63 shows the basic structure of an HTML document.

```
<HTML>
  <HEAD>
    <TITLE>Web_page_title</TITLE>
  </HEAD>
  <BODY>
     Web_page_body_elements
  </BODY>
</HTML>
```

Figure 7-63   Basic HTML document structure

The `<HTML>` opening and closing tags specify that the enclosed text is an HTML document. The `<HEAD>` opening and closing tags enclose the document's header section. The **header section** contains information about the Web page. The header section defines *Web_page_title*, which is the text that appears in the title bar of the user's browser window when the page appears. You enclose the *Web_page_title* text in the opening and closing `<TITLE>` tags. The header section can also contain information that the browser uses for processing the Web page.

Web_page_body_elements is the set of text, graphics, and other elements that comprise the content of the Web page, as well as the HTML tags that format the Web page content. You enclose these elements in the opening and closing `<BODY>` tags.

HTML documents often contain **comments**, which are text descriptions that do not appear on the Web page, but explain the document contents. An **opening comment tag** uses the syntax `<!--`, which is an opening angle bracket, then an exclamation point, followed by two hyphens. A **closing comment tag** uses the syntax `-->`, which is two hyphens followed by a closing angle bracket.

Viewing the Report Web Source File

When you create a new report in either RDF or JSP format, Reports Builder automatically generates a **Web source file** that contains the HTML tags that define the report appearance, and the program commands that create the dynamic report elements. A Web report embeds program commands using tags that are similar to HTML tags, and are prefaced by `rw:`. For example, the first program tag embedded in the Web source of a report is `<rw:report id="report">`. The Web source file is the same for reports in either format: if you save a report in the RDF format, its Web source file is the same as when you save the same report in the JSP format.

To become familiar with the Web source file that Reports Builder generates, you work with the Northwoods University Terms report in Figure 7-64.

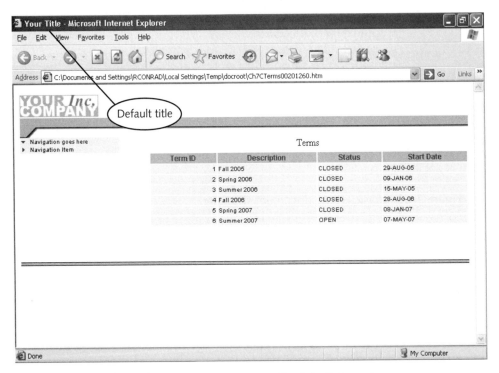

Figure 7-64 Northwoods University Terms report in default format

The Terms report displays data from the Northwoods University TERM table. Figure 7-64 shows the report output in a browser window. This report was created using the Report Wizard, and no changes have been made to the report format, so the report appears in the default format shown. Now you open the Terms report, view the report in the browser window, and view the report's Web source file.

To open and view the report in a Web browser, and then view its Web source code:

1. Open **Terms.rdf** from the Chapter07 folder on your Data Disk, and save the file as **Ch7CTerms.rdf** in the Chapter07\Tutorials folder on your Solution Disk.

2. Double-click the **Web Source** icon 🔲. The Web Source window opens, as shown in Figure 7-65.

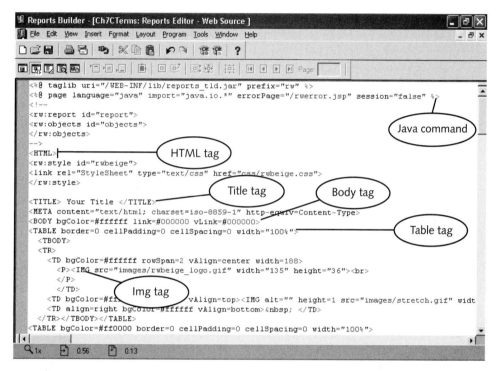

Figure 7-65   Web Source window

In the Web Source window, the HTML tags appear enclosed in angle brackets, and the Java commands appear enclosed in the <% and %> command **delimiters**. In Figure 7-65, note the <HTML> tag that signals the beginning of the HTML document body, the <TITLE> tag, and the <BODY> tag.

The HTML tags and elements define the Web page in Figure 7-64 that displays the report. Note that the text within the <TITLE> tag is "Your Title," which is the title that appears in the browser window in Figure 7-64. The <TABLE> tag defines the table in which the data values appear, and the first <IMG> tag defines the default logo image that appears on the Web page.

The Web source file has a section called the **data area**, which contains the commands that dynamically retrieve and display the report data. These commands follow an HTML comment with the text <!--Data Area Generated by Reports Developer -->. Now you view the data area in the Terms report Web source file.

To view the data area in the Web source file:

1. In the Web Source window, scroll down in the file until you find the HTML comment that appears as follows:

   `<!--Data Area Generated by Reports Developer -->`

2. The data area appears below the comment, as shown in Figure 7-66.

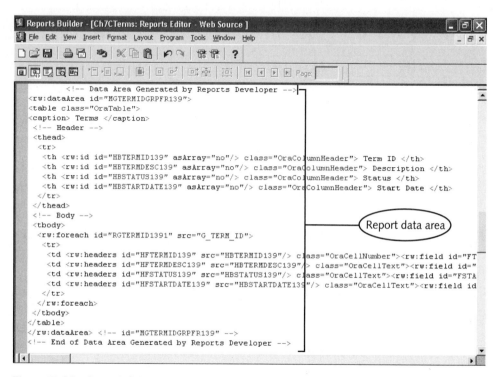

Figure 7-66   Report data area

Modifying the Default Appearance of a Web Report

Recall that by default, a report appears in a browser window with the formatting shown in Figure 7-64, and that there are two techniques for modifying the appearance of Web reports. The first technique is to use the Web Source window to modify the default formatting, and the second technique is to open an existing HTML file in Reports Builder that contains the formatting for the report, and use Reports Builder to add commands to retrieve and display the report data. The following sections illustrate these techniques.

Modifying the Report Web Source Code

Recall that you can modify the default appearance of a report by changing the HTML tags in the Web Source window. To gain experience customizing a report in the Web

Source window, you now modify the HTML document title. Then you run the form in the Web Layout window, and view the change.

CAUTION Do not modify the commands in the data area, or the report will not run correctly.

To modify the HTML commands and run the report:

1. Scroll to the top of the Web Source window, place the insertion point after the opening **<TITLE>** tag, delete the **Your Title** text, and replace it with **Northwoods University**. The **<TITLE>** tag should appear as follows:

   `<TITLE>Northwoods University</TITLE>`

2. Save the report, and then click the **Run Web Layout** button 🔲 to view the report in a Web browser window. Maximize the browser window if necessary. The report appears as shown in Figure 7-67, and shows the modified browser window title.

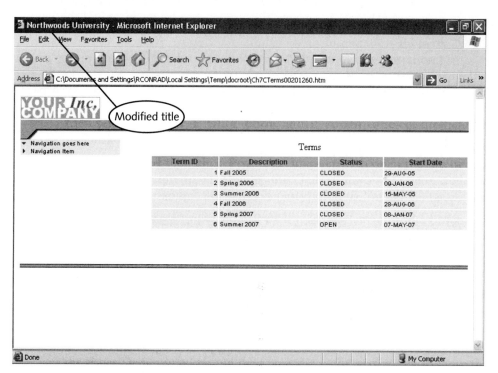

Figure 7-67   Report with modified browser window title

3. Close the browser window, and then close the report in Reports Builder.

You could further modify the HTML tags in the Web source window to change the report format. For example, you could change the logo image, colors, and column widths, add a title, or add links to other Web pages. However, be careful not to change the values that appear in the data area, or the report data will not appear correctly.

Adding Dynamic Report Data to an Existing Web Page

Recall that another way to develop reports that display formatted data is to add the commands that dynamically retrieve and display the report data to an existing HTML document. This approach allows you to add dynamic data to existing Web pages, or create new reports that retain the look and style of existing Web pages on a Web site.

To add report data to an existing HTML document, you perform the following tasks:

1. Open the HTML document in Reports Builder.
2. Open the Data Model window, and manually create a record group that contains the data that the report is to display.
3. Use the Report Block Wizard to create the data area in the Web page by inserting a report block in the document. A **report block** is a series of Java commands that retrieve and format data values in one of the report styles that can appear in a browser window.

To gain experience adding report data to an existing Web page, you open an existing HTML document that defines a Web page displaying the Northwoods University logo and information about the university. You add commands to this Web page to display information currently in the FACULTY database table.

Opening an HTML Document in Reports Builder

When you open an HTML document file in Reports Builder, Reports Builder automatically adds commands to the HTML source code to display the document as a dynamic Web page report. Now you open an HTML document named Northwoods.htm, which contains the HTML tags and elements to represent the Web page in Figure 7-68. You save the file in the RDF file format, and view the commands that define the Web page in the Web Source window.

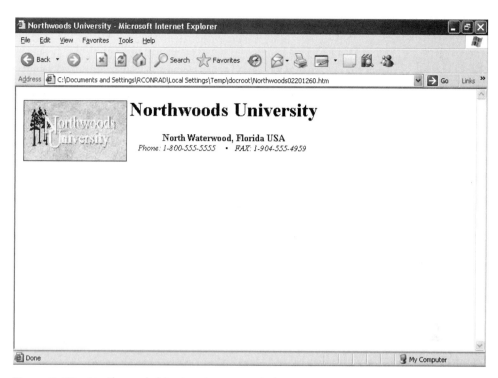

Figure 7-68   Northwoods.htm in a browser window

To open the HTML document and view its source code:

1. In Reports Builder, open **Northwoods.htm** from the Chapter07 folder on your Data Disk.

2. Click the **Run Web Layout** button 🗔 on the Reports Builder toolbar. The Web page appears in the browser window, as shown in Figure 7-68. Maximize the browser window if necessary. The document does not currently display any report data. Later in this section, you add commands that retrieve and display report data on the Web page.

If the Northwoods University logo image does not appear on the Web page, make sure that the NWlogo.jpg file is in the C:\OraData\Chapter07 folder on your workstation. If your Data Disk is on an alternate drive letter or has an alternate folder path, close the browser window, double-click the Web Source icon 🗔 in the Object Navigator, and find the path that specifies the location of the image file, which is currently set as C:\OraData\Chapter07\NWlogo.jpg. Change the path to represent the location of the file on your Data Disk, and then repeat Step 2.

3. Close the browser window.

4. In the Object Navigator, double-click the **Web Source** icon ▣ to open the Web Source window. The HTML document source code appears as shown in Figure 7-65. Note that Reports Builder has automatically added commands to convert the HTML document to a dynamic Web report.

5. Save the file as **Ch7CNorthwoods_HTML.rdf** in the Chapter07\Tutorials folder on your Solution Disk.

Creating the Report Query

The next step for adding report data to an existing Web page is to create the query that retrieves the data that the report displays. Now you open the Data Model window and manually create a query that retrieves all of the fields from the Northwoods University FACULTY table.

To create the report query manually:

1. Click the **Data Model** button ▣ on the toolbar. Currently, the Data Model is empty, because the report does not contain any queries or record groups.

2. Select the **SQL Query** tool ▣ on the tool palette, and then click anywhere in the painting region. The SQL Query Statement dialog box opens.

3. Type **SELECT * FROM faculty** in the SQL Query Statement field, and then click **OK**. The new query and associated record group appear in the Data Model window.

4. Save the report.

Inserting the Report Block into the Web Source Code

Recall that the report block represents a series of Java commands that retrieve and format data values in one of the report styles that can appear in a browser window. To insert a report block, you open the Web Source window, and place the insertion point at the position in the source code where you want the data area to appear. Then you start the Report Block Wizard, which generates the report block commands. The Report Block Wizard is similar to the Report Wizard, and has the following pages:

- **Style page**—Allows you to specify the title that appears on the report and select the report style. Recall that only the tabular, group left, group above, matrix, or matrix with group report styles can appear in a browser window, so the other report styles are disabled.

- **Groups page**—Allows you to select the record group that appears in the report and specify if the records repeat across the report horizontally, or down the report vertically

- **Fields page**—Allows you to select the query fields that appear in the report

- **Labels page**—Allows you to specify the report field labels and widths
- **Template page**—Allows you to select a report template

Next, you open the Web Source window, and position the insertion point where the report data area should appear. Then you start the Report Block Wizard and configure the report block.

To add the report block to the Web source code:

1. In the Data Model window, click the **Web Source** button 🔲 to open the Web Source window.

2. To specify the position in the Web page where Reports Builder is to add the commands to display the report data, place the insertion point on the blank line after the HTML comment `<!--Insert Web page content here -->`.

3. Click **Insert** on the menu bar, and then click **Report Block**. The Report Block Wizard opens, and displays the Style page. Type **Faculty** in the Title field, make sure that the **Tabular** option button is selected, and then click **Next**.

4. The Groups page appears. The Available Groups list displays all record groups within the report's Data Model. The current Data Model contains only the G_F_ID record group, which is selected. Click the **Down** button to select the record group and specify to display the records vertically down the report page, and then click **Next**.

5. The Fields page appears. Click the **Move all items to target** button ⟩⟩ , select the **F_IMAGE** field, and then click the **Move one item to source** button ⟨ to remove the F_IMAGE field from the final report. Repeat the removal process for the **F_PIN**, **F_SUPER**, and **F_RANK** fields, and then click **Next**.

6. The Labels page appears. Change the report labels and widths as follows, and then click **Next**.

Field	Label	Width
F_ID	**Faculty ID**	13
F_LAST	**Last Name**	12
F_FIRST	**First Name**	12
F_MI	**MI**	4
LOC_ID	**Location**	10
F_PHONE	**Phone**	12

7. The Template page appears. Because the report already contains the formatting from the original HTML document, you do not apply a template. Select

the **No template** option button, and then click **Finish**. Note that the commands to specify the report block now appear in the Web source code.

8. Save the report, and then click the **Run Web Layout** button 🖫. The formatted Web page appears in the browser window and displays the data values, as shown in Figure 7-69.

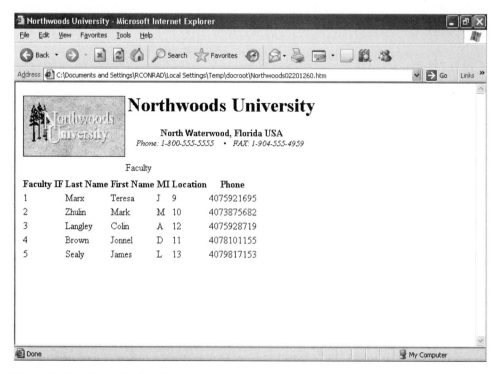

Figure 7-69   Formatted Web page showing data values

9. Close the browser window, and then close the report in Reports Builder.

10. Close Reports Builder.

You have learned how the OAS generates dynamic Web pages to display report data. You have also learned how to modify the appearance of these Web pages in the Reports Builder Web Source window and how to add report data to an existing Web page. In Chapter 8, you learn how to create integrated applications that allow users to work with forms that display reports in browser windows.

SUMMARY

- You can use the Report Wizard to create a report that displays data from a table that contains image data in an LOB data field. To display the image data, you modify the layout field's File Format property to Image.

- You can manually create queries and data links in the Data Model view and then create layout fields to display the values that the queries retrieve. Data links that you create manually can be joined using Query to Query, Group to Group, or Column to Column links.

- You can create formula and summary columns to display calculated values based on report Data Model column values. Formula columns display a value that a PL/SQL function returns after performing mathematical computations on report data. Summary columns summarize data fields using group functions, such as SUM, AVG, or COUNT.

- To create a formula column, you create the column in the Data Model, write the formula function in PL/SQL, and then create a field on the report layout to display the calculated value. You must place a formula column in the record group that contains the columns that the formula references.

- To display a formula column on a report, you create a new report data field that is in the same repeating frame as the source values for the formula and that uses the formula column as its data source.

- When you use the Report Wizard to create a summary column, it creates a summary field that displays the summary value in each group level above the summarized field's record group in the report. It also creates a summary field that displays the summary value for the entire report.

- To create a summary column manually, you create the summary column in the Data Model, draw a field on the report layout to display the summary column value, and then set the new field's source property as the summary column name. The summary column must appear in the report Data Model in a record group that is above the record group of the column being summarized.

- The Oracle 10*g* Application Server (OAS) is the Web server Oracle Corporation uses to deliver Web-based Oracle 10*g* database applications to users. The OAS Report Server process queries the database when the user requests the report, and generates a dynamic Web page that displays the current database data.

- The OAS can dynamically generate reports that are saved in either the binary report definition file (RDF) format, or in the Java Server Pages (JSP) format. All reports have an associated Web source file, which contains the HTML tags and elements that define the report appearance in a Web browser.

7

❑ When you create a report in Reports Builder, the report appears in its default report format in the browser window. To modify the report appearance, you can change the HTML tags and elements in the Web source file.

❑ To create reports that appear on existing Web pages, you can add a report block to an existing HTML document.

REVIEW QUESTIONS

1. Program commands inserted into a Web source file by the Report Builder are prefaced by `rw:`. True or False?

2. What is the difference between Query to Query and Column to Column data links?

3. When manually adding data to a section of a report, if there might be several records displayed in that section, you must insert the field through a(n) _____ frame.

4. If you want to include a user-defined function in a report, after inserting the field, which property must you change to specify the function to use for the calculation?

5. If you make changes to a report using the Report Wizard in reentrant mode, all custom formatting currently in the report will be lost. True or False?

6. You can use the _____ tool to create a query in the Data Model to retrieve data for a report.

7. Suppose you want to create a summary column using the Report Wizard, but the field being summarized contains text data. Which summary functions are available for this type of data?

8. In a dynamic Web page, the database is queried each time the page is displayed, so any record changes committed after the Web page was created are displayed. True or False?

9. What are the valid report formats for a Web page in Oracle 10*g*?

10. When you are creating a data link using the Data Link tool, you always draw the link from the _____ object to the _____ object.

MULTIPLE CHOICE

1. In a data link, the master side of the relationship (which is normally the one side of a 1:M relationship) is called the _____.

   a. parent

   b. detail record

   c. primary

   d. child

2. To display an image in a report, you change the File Format property of the associated _____ to Image.

   a. data model column

   b. record group

   c. layout field

   d. repeating frame

3. When you create a data link manually, you always draw the link from the _____ column to the _____ column.

   a. parent, child

   b. child, parent

   c. left, right

   d. right, left

4. A _____ column displays a value that a PL/SQL function calculates.

   a. formula

   b. summary

   c. calculated

   d. data

5. You use a _____ data link to associate two columns that do not have a foreign key relationship.

   a. Query to Query

   b. Group to Group

   c. Column to Column

   d. Manual

6. _____ is the Web server that Oracle Corporation uses to generate and distribute forms and reports as dynamic Web pages.

   a. Reports Server

   b. Apache Web server

   c. OC4J Instance

   d. Oracle 10*g* Application Server

7. When you are inserting a Report Block into a Web page's source code to display records, you should add the code immediately beneath the _____ tag.

   a. <TITLE>

   b. <BODY>

   c. <!-- Insert Web page content here -->

   d. <!--Data Area Generated by Reports Developer-->

7

8. To view your report in a Web browser, click the _____ button.

   a. Run in Browser

   b. Run Web Layout

   c. Execute Web Page

   d. Display

9. When using a formula column function, the result displayed in the report is the value specified by the _____ command within the PL/SQL function.

   a. begin

   b. return

   c. IF

   d. function

10. If you have two queries in a report's data model and they both include the column MYVAL, the name of the column in the second query is automatically changed to _____ .

    a. MYVAL_A

    b. MYVALA

    c. MYVAL_1

    d. MYVAL1

PROBLEM-SOLVING CASES

The following cases use the Clearwater Traders (Figure 1-25) sample database. All required files are in the Chapter07 folder on your Data Disk. Store all solutions in the Chapter07\Cases folder on your Solution Disk.

1. Create the report shown in Figure 7-70 that displays data about inventory items at Clearwater Traders. You need to load the image data in the database before you can create the report.

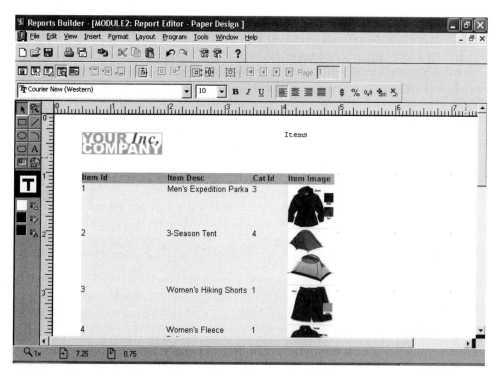

Figure 7-70   Clearwater Traders Items report

a. Start the OC4J Instance and Forms Builder, and open the **Item_IMAGE.fmb** form file. Run the form, and load the image data for each inventory item. Be sure to specify the complete path to the image file, including the drive letter and folder path, in the Image Filename field. Use the following filename for each inventory item:

Item IDImage	Filename
1	**Parka.jpg**
2	**Tents.jpg**
3	**Shorts.jpg**
4	**Fleece.jpg**
5	**Sandals.jpg**
6	**Surfshorts.jpg**
7	**Girlstee.jpg**

b. Use the Report Wizard to create the report. Use a tabular report style, and format the report as shown in Figure 7-70. Save the report in a file named **Ch7CCase1.rdf**.

2. Modify the Clearwater Traders Item report from the previous case so each item is displayed on a separate page. Save the modified report in a file named **Ch7CCase2.rdf**.

3. In this case you create a report displaying information about the current inventory items, including available colors, sizes, price, and quantity, as shown in Figure 7-71.

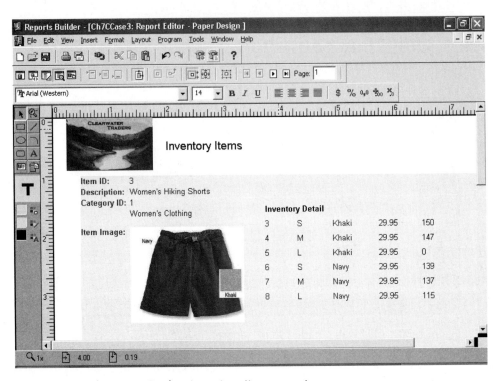

Figure 7-71   Clearwater Traders Inventory Items report

a. Open the **Items.rdf** file from the Chapter07 folder on your Data Disk. Save the file as **Ch7CCase3.rdf** in the Chapter07 folder on your Solution Disk. Run the report and notice that the category description and the inventory detail are not included on the report.

b. Create the necessary query and Query to Query data link to add the category description to the report. Then include the field through the Paper Layout window.

c. Create the necessary query and Column to Column data link to display the inventory detail from the INVENTORY table. Then include the records by creating a repeating frame and adding the appropriate columns through the Paper Layout window. Include the section label and field names shown in Figure 7-72.

d. Save and run the completed report.

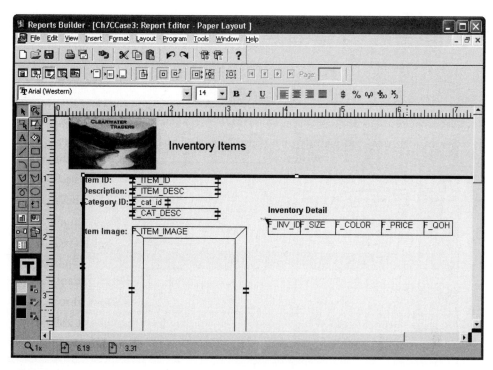

Figure 7-72   Clearwater Traders Inventory Paper Layout

4. Create a report that displays customer order information for Clearwater Traders, as shown in Figure 7-73. The top section of the report should display the customer name and address, as well as the order date, order ID, and payment method. The detail section should show the item ID, item description, size, color, order price, and quantity, and calculate the extended total. The summary section at the bottom of the report should show a subtotal that sums all of the extended totals, calculates sales tax and shipping and handling, and sums the subtotal, tax, and shipping and handling to create a final order total. Format the report as shown in Figure 7-73. Apply the **Beige.tdf** template file, and replace the default image with the **Clearwater.jpg** file from the Chapter07 folder on your Data Disk. Save the report as **Ch7CCase4.rdf**.

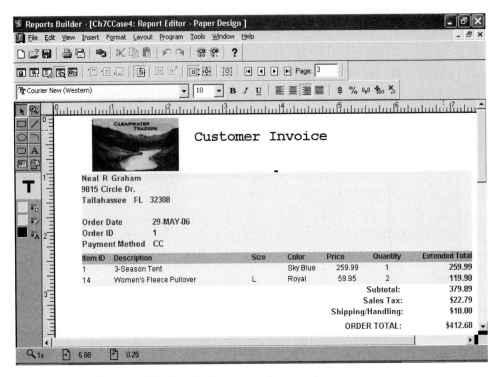

Figure 7-73 Clearwater Traders Customer Invoice report

a. Create the report using the Report Wizard to display the required data fields. Each customer order should appear on a separate report page.

b. Create a formula column to calculate the extended total, which is the order price times quantity. Create a summary column to calculate the subtotal, which is the sum of all the extended totals. Create a formula column that calculates the tax as 6% of the subtotal for Florida residents and as 0% for residents of all other states.

c. Create a formula column to calculate shipping and handling as follows: If the order subtotal is less than $25, the shipping cost is $5.00; if the subtotal is $25 or more, but less than $75, the shipping cost is $7.50; if the subtotal is $75 or more, the shipping cost is $10.00.

d. Create a column to calculate the order total, which is the sum of the subtotal, tax, and shipping and handling.

5. Create a dynamic Web page that displays the current inventory available from Clearwater Traders. The initial Web page displaying the company logo and address is stored in the Clearwater.htm file available in the Chapter07 folder of your Data Disk. Modify the file by retrieving the inventory information and inserting the appropriate report block to display the data. Specify the labels and column widths in the Report Block wizard using the example data shown in Figure 7-74 as a guide. Save the completed file as **Ch7CCase5.htm**.

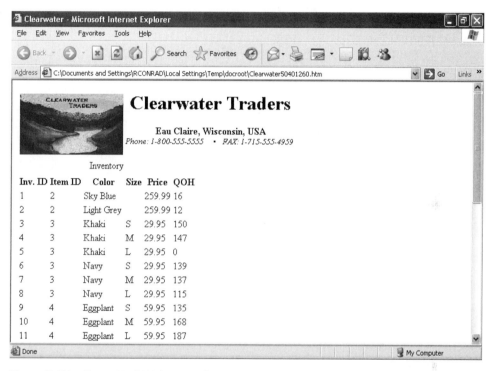

Figure 7-74   Formatted Web page showing current inventory (partial output shown)

8

CREATING AN INTEGRATED DATABASE APPLICATION

◀ LESSON A ▶

After completing this lesson, you should be able to:

- ♦ Understand the steps for developing a database application
- ♦ Design a database application interface
- ♦ Use timers in a Forms Builder application to create a splash screen
- ♦ Create form templates to ensure consistency across application modules
- ♦ Learn how to reference application components in an integrated database application
- ♦ Understand how to open and close form modules in a multiple form application
- ♦ Learn how to display a report in a database application

In Chapter 6, you learned that when you develop a new database system, you begin by identifying not only the business processes that the database system will support, but also the data items required by these processes. Then you create the database tables and load the data. Finally, you develop the forms and reports to manage the data and support the business processes. This chapter describes how to combine the individual form and report components into a single integrated system. To learn how to build an integrated system, you create a single Forms Builder form module that serves as the entry point for the individual system components.

NOTE

This chapter focuses on integrating form and report components into a single application. Web application deployment and security issues are advanced Developer10g topics that are beyond the scope of this book.

DEVELOPING AN INTEGRATED DATABASE APPLICATION

An integrated database application is made up of a variety of individual forms and reports that you combine into a single application. Developing database applications usually involves the following five phases:

1. **Design**—The first phase involves creating the specifications for the application components. These specifications are based on stated user needs, input from project managers and development team members, and user requests for enhancements or bug fixes.

2. **Module development**—During the next phase you create the individual form and report modules.

3. **Module integration**—Once all the individual modules are created, you then integrate them into a single application.

4. **Testing**—This phase has two stages: unit testing and system testing. **Unit testing** involves testing the individual form and report modules to confirm that they work correctly as single applications. **System testing** evaluates whether the modules work correctly when you integrate them into the rest of the system.

5. **Deployment**—Finally, the integrated modules must be packaged into an installable format that you can deliver to customers.

For large database applications, developers usually repeat these phases iteratively and do not necessarily follow this sequence exactly. For example, during the module integration phase, developers might determine that the system requires an additional module, which requires going back to the module development phase. In this chapter, you design the integrated application and learn how to integrate the application modules. (You learned how to create the individual modules in previous chapters.) This chapter does not address testing or deployment.

The Northwoods University Integrated Database Application

In this chapter, the tutorial exercises illustrate how to create an integrated database application for Northwoods University. For this application, the main processes that the system supports are student and faculty services. The system must also provide data block forms for managing the data in the individual database tables, and for creating reports to help managers direct operations. (Previous chapters have described the design and development process for the individual application modules.)

When you create a complex database application, it is a good practice to create separate form modules, rather than combine all of the project forms into a single .fmb file with many different canvases. This approach enables project team members to develop, test, and debug each module independently before integrating the modules into the final application. Smaller modules are easier to work with and make it easier for project teams with multiple team members to split up the development effort, because different team members can simultaneously work on different form files. Smaller modules also create smaller form files, which load faster in Web-based applications.

TIP

When you are designing individual form modules, keep in mind that data blocks with master-detail relationships must be in the same form and cannot be split between multiple form modules.

When you are ready to combine all of the individual application components into the integrated database application, you should place all application files in a single folder, which is called the **project folder**. The project folder should also contain any graphic image files that the applications display. Table 8-1 summarizes the form and report files that compose the Northwoods University database application.

8

Module File	Description
Student_logon.fmb	Custom form that allows students access to the system after verification
Faculty_logon.fmb	Custom form that allows faculty access to the system after verification
Course.fmb	Data block form that displays and edits COURSE data records
Location.fmb	Data block form that displays and edits LOCATION data records
Location.rdf	Report that displays LOCATION information
Terms.rdf	Report that displays TERM information

Table 8-1   Module files in the Northwoods University database application

Next, you start Windows Explorer and create the project folder on your Solution Disk. Then you copy the application files from the Data Disk to the project folder. You also start SQL*Plus, and run the scripts to refresh the case study database tables.

To create the project folder, copy the application files to the folder, and refresh the case study database tables:

1. Start Windows Explorer. Create a folder named **NorthwoodsProject_DONE** in the Chapter08\Tutorials folder on your Solution Disk.

2. Copy the contents of the NorthwoodsProject folder in the Chapter08 folder on your Data Disk to this new folder, and then exit Windows Explorer.

3. Start SQL*Plus, log onto the database, and run the **Ch8Northwoods.sql** script file from the Chapter08 folder on your Data Disk. Then close SQL*Plus.

Integrating the Database Application

An integrated database application should have a single entry point from which users access all of the component forms and reports. Having a single application entry point makes the application easier to use and maintain. In addition, a single entry point ensures that users do not start multiple copies of the same form, which consumes system resources. Similarly, the database application should have a single exit point. Users become frustrated if they click an Exit button to exit a form, and instead exit the entire application.

To integrate individual Oracle form and report modules into a single application, you create a form module, called the **main form**, from which users access all of the individual application components. The main form can have a **splash screen**, which is a front-end entry screen that introduces the application. Then, the application displays the main form screen. The main form screen display often has a **switchboard**, which consists of command buttons that enable users to access the most commonly used forms and reports quickly and easily.

All application components should also be accessible through pull-down menus for users who prefer to use the keyboard rather than the mouse pointer. Pull-down menus also provide access to features used less frequently than the ones on the switchboard and to features that have multiple levels of choices. For example, you could have a pull-down menu titled Reports that contains another menu that lists different reports the user could access. Also, you can use different pull-down menus to restrict users from accessing specific forms, because you might not want all users to be able to insert, update, and delete data using certain forms.

Figure 8-1 shows the design for the main form screen display for the Northwoods University database system, which provides the system's single entry and exit point. Users navigate to all application forms from the main form, and they exit the application from this form. The top section of the design shows the pull-down menu structure. The bottom section of the design shows the system switchboard with a graphic image to enhance the application's appearance.

Users can click the Student Logon and Faculty Logon buttons on the switchboard to access the form modules that allow user verification and subsequent access to the system. Users can click the View Campus Locations button to view a report that displays building and room information.

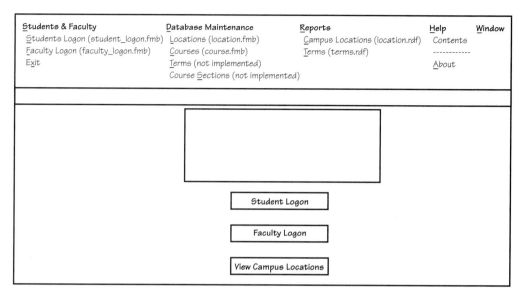

Students & Faculty	Database Maintenance	Reports	Help	Window
Students Logon (student_logon.fmb)	Locations (location.fmb)	Campus Locations (location.rdf)	Contents	
Faculty Logon (faculty_logon.fmb)	Courses (course.fmb)	Terms (terms.rdf)	------------	
Exit	Terms (not implemented)		About	
	Course Sections (not implemented)			

Student Logon

Faculty Logon

View Campus Locations

Figure 8-1   *Main form screen design*

The first pull-down menu item in Figure 8-1 is Students & Faculty. This menu contains selections for logging into the system and exiting the form. (Note that in the design sketch in Figure 8-1, the name of the form file that the menu selection calls appears beside the selection name. The filename does not appear in the completed menu.) The Database Maintenance menu item provides access to the basic forms that let users insert, update, and delete records in several tables in the database. Note that forms for two of the selections, Terms and Course Section, have not been implemented yet. This chapter describes how to develop application placeholder elements called **stubs**, which are programs or messages that handle undeveloped system features.

The Reports menu item provides access to the Campus Location and Terms reports. The final two menu items on the Northwoods University database application menu are Help and Window. The Help menu has two second-level selections: Contents, which provides access to a topics list for the Help database, and About, which gives details about the application. The Window menu selection allows users to move between windows in a multiple-window application.

TIP

Most Windows applications have Help as the last menu choice, but Forms Builder automatically places Window as the last pull-down menu selection.

Next, you start Forms Builder, and create the main form module, which is a custom form that integrates the rest of the application modules. You then save the form in a file that you store in the project folder.

To start Forms Builder and create the main form module:

1. Start the OC4J Instance.

2. Start Forms Builder. Click the **Connect** button 🔟, and connect to the data-base. A new form module appears in the Object Navigator.

3. Change the form module name to **MAIN**.

4. To create the switchboard canvas, select the Canvases node, and then click the **Create** button ➕ to create a new canvas. Change the name of the new canvas to **MAIN_CANVAS**.

5. To configure the main application window, open the Windows node, and change the form window name to **MAIN_WINDOW**. Select the MAIN_WINDOW node, right-click, and then click **Property Palette**. Change the Title property value to **Northwoods University**, and then close the Property Palette.

6. Save the form as **Ch8AMain.fmb** in the Chapter08\Tutorials\ NorthwoodsProject_DONE folder on your Solution Disk.

Next you create a control block that contains the switchboard command buttons. Then, you create the buttons and import the graphic image.

To create the command buttons and import the graphic image:

1. In the Object Navigator, select the Data Blocks node, and click the **Create** button ➕ to create a new block. Select the **Build a new data block manually** option button, and then click **OK**. The new block appears under the Data Blocks node. Change the block name to **MAIN_BLOCK**.

2. Double-click the **Canvas** icon ▣ beside MAIN_CANVAS to open MAIN_CANVAS in the Layout Editor.

3. To create the Student Logon button, confirm that **MAIN_BLOCK** is selected in the Block list in the Layout Editor, select the **Button** tool ▣ on the tool palette, and draw a button on the canvas in the approximate position of the Student Logon button in Figure 8-1.

4. Right-click the new button, click **Property Palette**, change the following properties, and then close the Property Palette.

Property	Value
Name	STUDENT_LOGON_BUTTON
Label	Student Logon
Width	100
Height	16

5. Select the **Student Logon** button if necessary, click the **Copy** button 🗐 on the toolbar, and then click the **Paste** button 📋 two times to create the Faculty Logon and View Campus Locations buttons. Move each pasted button to the approximate positions shown in Figure 8-1.

TIP

When you paste a button, Forms Builder places the pasted button directly on top of the copied button.

6. Open the new button Property Palettes, modify the property values as follows, and then close the Property Palettes.

Name	Label
FACULTY_LOGON_BUTTON	**Faculty Logon**
VIEW_CAMPUS_LOCATIONS_BUTTON	**View Campus Locations**

7. To import the graphic image that appears in the background of the canvas, click **Edit** on the menu bar, point to **Import**, and then click **Image**. Click **Browse**, navigate to the Chapter08\Tutorials\NorthwoodsProject_DONE folder, select **Trees.gif**, click **Open**, and then click **OK**.

8. Make sure that the image is selected and resize and reposition the image as needed so that the image completely fills the canvas without overlapping.

8

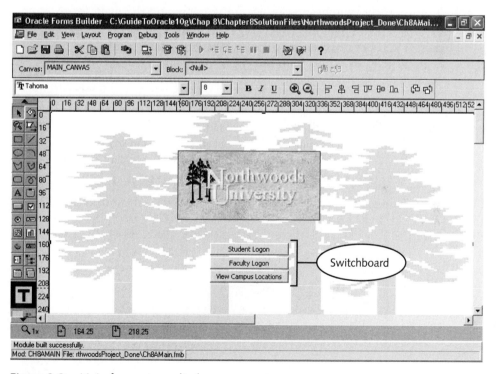

Figure 8-2 Main form screen display

CAUTION

If any part of the image is off the canvas surface, an error will occur when you run the form.

9. To import the logo, click **Edit** on the menu bar, point to **Import**, and then click **Image**. Click **Browse**, if necessary navigate to the Chapter08\Tutorials\ NorthwoodsProject_DONE folder, select **NWlogo.jpg**, click **Open**, and then click **OK**.

10. Reposition and resize the logo as needed to match the logo shown in Figure 8-2. Your form should look like Figure 8-2.

11. Save the form, and then run the form to view it in the browser window.

12. Close the browser window.

CREATING A SPLASH SCREEN

Recall that a splash screen is the first image that appears when you run an application. To implement the splash screen for the Northwoods University database application, you create two separate windows in the main application form. When the user starts the program, the splash screen window shown in Figure 8-3 appears for five seconds. Then the application focus switches to the second window, and the main form screen appears.

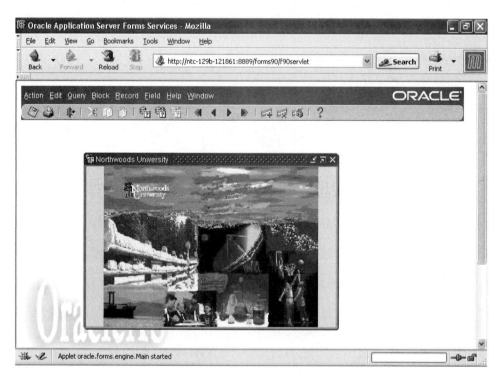

Figure 8-3   Splash screen window

NOTE

To implement the splash screen, you create the splash screen window, canvas, and data block, and configure the canvas to display the splash screen image. Then, you create a PRE-FORM trigger that displays the splash screen image and sets a timer. When the timer expires, a form-level trigger associated with the WHEN-TIMER-EXPIRES event executes, which sets the application's focus to an item on the main form canvas. This hides the splash screen window and displays the main form canvas.

Creating the Splash Screen Window, Canvas, and Data Block

To create the splash screen, you first create and configure the splash screen window, canvas, block, and canvas items. Splash screen windows typically do not fill the entire screen, so you also change the window size and adjust the X and Y Position property values so that the window appears centered on the screen.

To create the splash screen window, canvas, and block:

1. In the Object Navigator, select the Windows node, and then click the **Create** button ➕ to create a new window in the form. Change the name of the new window to **SPLASH_WINDOW**.

2. Select the SPLASH_WINDOW node, right-click, and then click **Property Palette**. Change the Title property value to **Northwoods University**. Scroll down to the Physical property node, change the following properties, and then close the Property Palette.

Property	Value
X Position	**80**
Y Position	**40**
Width	**320**
Height	**200**

3. To create the new canvas, select the Canvases node, and then click ➕. Change the name of the new canvas to **SPLASH_CANVAS**.

4. To create the new block, select the Data Blocks node, and then click ➕. Select the **Build a new data block manually** option button, click **OK**, and then change the name of the new block to **SPLASH_BLOCK**.

5. Save the form.

Because you have resized SPLASH_WINDOW to 320 pixels wide by 200 pixels high, you must also resize SPLASH_CANVAS to the same size. In addition, you must confirm that the canvas Window property is SPLASH_WINDOW, which specifies that the canvas appears in SPLASH_WINDOW.

To configure SPLASH_CANVAS:

1. Right-click the **Canvas** icon ▥ beside SPLASH_CANVAS, click **Property Palette**, and change the Width to **320** and the Height to **200**.

8

2. Confirm that the Window property value is **SPLASH_WINDOW**. If it is not, select the **Window** property, open the list, and select **SPLASH_WINDOW**.

3. Close the Property Palette, and save the form.

Creating the Splash Screen Canvas Image

The canvas that appears in a splash screen window usually displays boilerplate objects, such as text, shapes and lines, and graphic images. The splash screen in the Northwoods University database application is going to display a graphic image. You can use two approaches to display a graphic image on a form. The first approach is to create the image as a boilerplate image, as you did when you imported logos into forms in previous chapters. This approach stores the image inside the .fmb file when you design the form. However, this approach does not work for a splash screen. The splash screen canvas is going to display only boilerplate objects, and Forms Services does not display a canvas that contains only boilerplate objects and that does not contain any block items. To display a graphic image on a splash screen, you must use a second approach that creates a dynamic image item on the splash screen canvas.

A **dynamic image item** is a block item that displays an image loaded by a trigger while the form is running. Because the dynamic image item is a block item rather than a boilerplate item, Forms Services displays the canvas. Next, you create an image item on SPLASH_CANVAS. (You create the trigger to load the file image data into the form image item a little later.) To create an image item, you use the Image Item tool 🖼 on the tool palette to draw the image on the canvas. Then, you configure the image item's properties. You change the image Sizing Style property value to Adjust, so that the item automatically adjusts the image size to fit the splash screen correctly. You modify the image item size so that it is the same size as the canvas, and change its X and Y positions so that it is in the top-left corner of the canvas.

To create an image item for SPLASH_CANVAS and modify its properties:

1. Double-click the **Canvas** icon 🖼 beside SPLASH_CANVAS to open the canvas in the Layout Editor. If necessary, select **SPLASH_BLOCK** from the Block list.

2. Select the **Image Item** tool 🖼 on the tool palette, and draw an image item that is about the same size as the canvas and just inside the canvas borders.

3. Right-click the new image item, and then click **Property Palette**. Change the Name property value to **SPLASH_IMAGE**.

4. To specify that the graphic size automatically adjusts to the same size as the image item on the canvas, select the **Sizing Style** property, open the list, and select **Adjust**.

5. To position SPLASH_IMAGE in the upper-left corner of the canvas, change the X Position and Y Position properties to **0**, if necessary.

6. To make the image item the same size as the canvas, change the Width property to **320**, and the Height property to **200**.

7. Close the Property Palette, and save the form.

Loading the Splash Image

To load an image file into an image item, you use the READ_IMAGE_FILE built-in procedure, which has the following syntax:

```
READ_IMAGE_FILE('filename', 'file_type', 'item_name');
```

In this procedure, *filename* is a text string that specifies the complete path and filename of the graphic art image file. If you specify the filename as a literal value, such as 'c:\myfilename.tif,' you must enclose the value in single quotation marks. You can also specify *filename* as a variable value that references a text string specifying the path and filename. *File_type* is a character string that represents the image file type. Legal values include BMP, PCX, PICT, GIF, CALS, PCD, and TIF, and the value must be enclosed in single quotation marks.

TIP

The different image file types reflect both the graphics art applications that create the files and the image compression techniques that allow the images to occupy less file space. Bitmap (.BMP) files are usually uncompressed, whereas PCX, GIF, and TIF files use different compression methods. PICT and PCD files are made by specific graphics applications, and CALS files are used to compress black-and-white images for fax transmissions. Almost all popular graphics applications support one or more of these types of files.

Item_name is a text string that represents the name of the form image item in which the image data is to appear. You specify this value in single quotation marks, using the format '*block_name.item_name*'. For the image item you just created, the parameter value would be 'splash_block.splash_image'.

Recall that you can create a PRE-FORM trigger to execute when the form first loads. Now you create a PRE-FORM trigger to load the image into the image item on the splash canvas.

To create a PRE-FORM trigger to load the splash image:

1. Click **Window** on the menu bar, and then click **Object Navigator** to switch to the Object Navigator.

2. Select the Triggers node under the MAIN form module, right-click, point to **SmartTriggers**, and then click **PRE-FORM**. The PL/SQL Editor opens.

3. Type the following command to load the image file. (The path to your image may be different if you are storing your Solution Disk files on a different drive letter or with a different folder path.)

8

```
READ_IMAGE_FILE('C:\OraSolutions\Chapter08\Tutorials\
NorthwoodsProject_DONE\NWsplash.tif', 'TIFF',
'splash_block.splash_image');
```

The text string that specifies the filename parameter in the READ _IMAGE_FILE procedure must appear all on one line in the PL/SQL Editor, or an error will occur.

CAUTION

4. Compile the trigger, correct any syntax errors, and then close the PL/SQL Editor.

Displaying the Splash Window

Recall that when a form contains multiple canvases, the canvas that appears when the form first opens is the canvas whose associated block items appear first in the Data Blocks list in the Object Navigator. For the splash window and splash canvas to appear first when the main form opens, SPLASH_BLOCK must appear first in the Data Blocks list. Now you view the form Data Blocks list, and confirm that SPLASH_BLOCK appears first.

To confirm that SPLASH_BLOCK appears first in the Data Blocks list:

1. In the Object Navigator, select the Data Blocks node, and then click the **Collapse All** button 🗗 to close all data block nodes.

2. Open the Data Blocks node, and confirm that the form data blocks appear in the following order: SPLASH_BLOCK, MAIN_BLOCK. If the blocks do not appear in this order, select **SPLASH_BLOCK**, and move it to the top of the list under the Data Blocks node.

Hiding the Splash Window

You use a form timer to control how long the splash image appears and to signal when to hide the splash window. A **form timer** is a form object that you create using program commands. You use timers to control time-based events, such as displaying objects for a set time interval or creating animated objects. When you create a timer, you specify the time interval that elapses before the timer expires. When the timer expires, a form-level trigger named WHEN-TIMER-EXPIRED fires. Every form timer fires the same WHEN-TIMER-EXPIRED trigger. As a result, you cannot create and use multiple timers that expire at different times, because all timers fire the same trigger. The following sections describe how to create, modify, and delete timers, and how to create a WHEN-TIMER-EXPIRED trigger.

Creating a Timer

To create a timer, you use the following syntax to declare a form timer object in the declaration section of a form trigger:

```
DECLARE
      timer_id TIMER;
```

In this syntax, *timer_id* can be any legal PL/SQL variable name. It uses the TIMER data type, which creates a new form timer object. After you declare the timer object, you call the CREATE_TIMER function in the program body to specify the timer properties. The CREATE_TIMER function has the following syntax:

```
timer_id := CREATE_TIMER('timer_name', milliseconds,⤶
iteration_specification);
```

This function requires the following parameters:

- **Timer_id**—A previously declared variable of data type TIMER
- **Timer_name**—Specifies the name of the timer object; its value can be any legal Oracle variable name, and is enclosed in single quotation marks.
- **Milliseconds**—A numeric value that specifies the time duration, in milliseconds, until the timer expires. When the timer expires, it calls a form-level trigger named WHEN-TIMER-EXPIRED. To create a timer that expires in five seconds, you specify the milliseconds value as 5000.
- **Iteration_specification**—Specifies if the timer should be reset immediately after it expires. Valid values are REPEAT, meaning it should be reset immediately and start counting down again, and NO_REPEAT, meaning it should stay expired. You can use the REPEAT option to create animated graphics that appear repeatedly. Because the splash screen appears only once, you use the NO_REPEAT value.

Next you modify the PRE-FORM trigger to declare the timer object, and create and set the timer.

To modify the PRE-FORM trigger to create and set the timer:

1. In the Object Navigator, select the PRE-FORM node, right-click, and then click **PL/SQL Editor** to open the trigger in the PL/SQL Editor. Modify the PRE-FORM trigger by adding the shaded commands in Figure 8-4 to declare, create, and set the timer.

```
DECLARE
      splash_timer_id  TIMER;
BEGIN
      READ_IMAGE_FILE('C:\OraSolutions\Chapter08\Tutorials\ ⤸
      NorthwoodsProject_DONE\NWsplash.tif', 'TIFF', ⤸
      'splash_block.splash_image');
      splash_timer_id := CREATE_TIMER('splash_timer', 5000, NO_REPEAT);
END;
```

Figure 8-4   Commands to create and set the splash screen timer

> 2. Compile the trigger, correct any syntax errors, and then close the
>   PL/SQL Editor.

> 3. Save the form.

Hiding the Splash Window When the Timer Expires

Recall that when the timer expires, a form-level trigger named WHEN-TIMER-EXPIRED fires. If the form does not contain a WHEN-TIMER-EXPIRED trigger, nothing happens. If this trigger exists, then the code in the trigger executes. When the timer in the PRE-FORM trigger expires, you want the application to display the main form window. To show a different window in a multiple-window application, you use the SHOW_WINDOW built-in. To hide the current window, you use the HIDE_WINDOW built-in. These built-ins have the following general syntax:

```
SHOW_WINDOW('window_name');
HIDE_WINDOW('window_name');
```

In this syntax, *window_name* specifies the window's Name property, as it appears in the Object Navigator. After the SHOW_WINDOW built-in executes, a command must execute the GO_ITEM built-in to move the form insertion point to an item on that window in order for the window to have the application focus and be visible.

Next, you create the WHEN-TIMER-EXPIRED trigger, add commands to show MAIN_WINDOW, add commands to hide SPLASH_WINDOW, and use the GO_ITEM procedure to switch the application focus to STUDENT_LOGON_BUTTON. Then you run the form and test the splash screen.

To create the WHEN-TIMER-EXPIRED trigger and run the form:

> 1. In the Object Navigator, select the Triggers node, and then click the **Create** button ➕ to create a new form-level trigger. The MAIN:Triggers dialog box opens.

> 2. Scroll down the list, select **WHEN-TIMER-EXPIRED**, and then click **OK**. The PL/SQL Editor opens.

> 3. Type the following commands to create the trigger:

```
SHOW_WINDOW('MAIN_WINDOW');
HIDE_WINDOW('SPLASH_WINDOW');
GO_ITEM('main_block.student_logon_button');
```

4. Compile the trigger, correct any syntax errors, close the PL/SQL Editor, and then save the form.

5. Run the form. The splash screen should appear for five seconds, as previously shown in Figure 8-3, and then the main form window should open. The splash window should appear centered in the Forms Services window. If the splash window does not appear centered, close the browser window, and adjust the X and Y position of SPLASH_WINDOW until the splash window appears centered in the browser window.

If the error "FRM-30041: Position of item places it off of canvas" appears, then your splash image is not entirely on the surface of SPLASH_CANVAS. Open the SPLASH_CANVAS in the Layout Editor and adjust the size and position of the image so that it is within the bounds of the canvas.

If SPLASH_WINDOW opens, but no image appears, double-check to make sure that the NWsplash.tif image file is in the Chapter08\Tutorials\ NorthwoodsProject_DONE folder on your Solution Disk, and that you correctly specified the path to the file in the PRE-FORM trigger. An error message will appear in the status line showing the location from which the application is trying to load the image file.

6. Close the browser window.

ENSURING A CONSISTENT APPEARANCE ACROSS FORM MODULES

A large database application can include hundreds of different form modules that many different form developers create. It is important that all of the forms have a consistent look and feel, both to make the forms appear as a polished and integrated application and to reduce user training time and frustration. For example, an application will confuse users if in some forms, they must press Ctrl+L to open LOV displays, whereas other forms contain command buttons to open LOV displays. Similarly, the application will confuse users if some forms display an Exit button for returning to the application switchboard, whereas other forms display a Return button. Two ways to standardize the appearance of multiple forms in an application are to use template forms and to use visual attribute groups.

Template Forms

A **template form** is a generic form that includes standard form objects, such as graphics, command buttons, and program units that appear on every form in an application. You store the template form in a location that is accessible to all developers. When you create a new form in an application, you base the new form on the template form. Having a template form speeds up the form development process, because you and your fellow developers do not have to make the same modifications, such as resizing the window or importing the corporate logo, each time you create a new form. Having a template form also ensures a consistent form appearance, because all developers build from the same template.

You already learned how to create a report template in Chapter 7, which creates a standard appearance for application reports.

TIP

To create a new form based on a template form, you click File on the menu bar, point to New, and then click Form Using Template. When you create a form based on a template form, Forms Builder prompts you to save the new form using a different filename, so you do not accidentally overwrite the original template form.

Next, you create a template form for the Northwoods University database. Figure 8-5 shows the template form interface design.

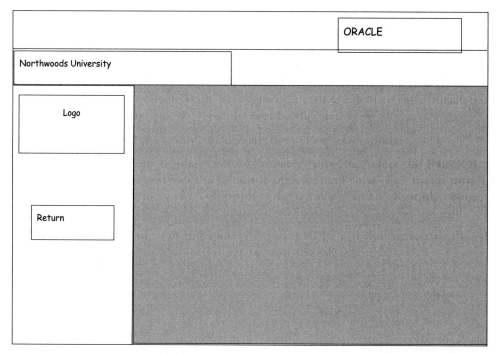

ORACLE

Northwoods University

Logo

Return

Figure 8-5   Northwoods University template form design

The canvas has a medium gray background color. The background of the left one-third of the canvas is a lighter color than the rest of the canvas. The Northwoods University logo appears in the top-left corner, and command buttons for the specific application appear under the logo. Under the form-specific buttons, the template has a Return button that exits the current form and displays the main form. The window title is "Northwoods University." The area on the right side of the form displays form objects, such as text items and radio buttons, which are specific for each individual form.

To create the template form:

1. In the Object Navigator, select the MAIN node, and then click the **Collapse All** button ⊟ to collapse the MAIN form objects.

2. Select the top-level Forms node, and then click the **Create** button ✚ to create a new form. Change the form module name to **NORTHWOODS_TEMPLATE**.

3. To create the new canvas, select the Canvases node, and then click ✚. Change the canvas name to **TEMPLATE_CANVAS**.

4. To configure the template window, open the Windows node, and change the form window name to **TEMPLATE_WINDOW**. Select the TEMPLATE_WINDOW node, right-click, and then click **Property Palette**. Change the Title property value to **Northwoods University**, and then close the Property Palette.

5. Save the form as **Northwoods_Template.fmb** in the NorthwoodsProject_ DONE folder on your Solution Disk.

Next, you configure the canvas and add the template canvas objects in Figure 8-5. You change the canvas background color to a medium gray. You create a trigger for the Return button that exits the form. To exit a form, you use the EXIT_FORM built-in, which has the syntax EXIT_FORM;.

To configure the canvas and add the template canvas objects:

1. In the Object Navigator, right-click the TEMPLATE_CANVAS node, and then click **Property Palette**. Select the **Background Color** property, click the **More** button ⬚ , and select a **medium gray square**. Then close the Property Palette.

2. Double-click the **Canvas** icon ▣ beside TEMPLATE_CANVAS to open the form canvas in the Layout Editor.

3. To create the boilerplate rectangle, select the **Rectangle** tool ▢ on the tool palette, and draw a rectangle as shown in Figure 8-6 that defines the area that contains the logo and return button. The rectangle should start at the top-left corner of the canvas, and should extend the entire length of the canvas. The rectangle's bottom edge should be even with the bottom edge of the canvas.

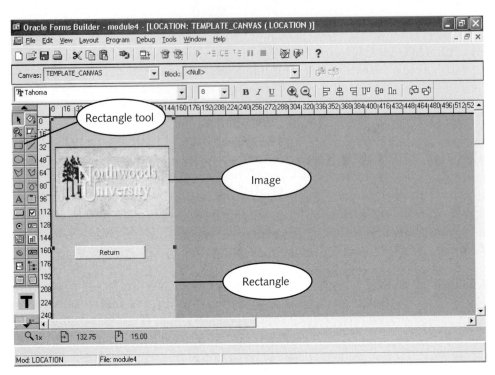

Figure 8-6 Northwoods University template form

Do not place the rectangle so it extends beyond the canvas surface or an error will occur when you run the form.

CAUTION

4. Select the rectangle if necessary, select the **Fill Color** tool ![icon] on the tool palette, and change the rectangle fill color to a lighter shade of gray than the canvas background color. The filled rectangle should appear slightly lighter than the rest of the canvas.

Do not use bright or saturated colors on screen displays, because they can cause eyestrain.

TIP

5. Make sure that the rectangle is still selected, select the **Line Color** tool ![icon] on the tool palette, and then click **No Line**.

6. To import the Northwoods University logo, click **Edit** on the menu bar, point to **Import**, and then click **Image**. Click **Browse**, navigate to the Chapter08\ Tutorials\NorthwoodsProject_DONE folder on your Solution

Disk, select **NWlogo.jpg**, click **Open**, and then click **OK**. The logo appears on the canvas. Resize and reposition the logo as shown in Figure 8-6.

7. To create the Return button, select the **Button** tool ▣ on the tool palette, and draw the button as shown in Figure 8-6. Double-click the button to open its Property Palette, change the button properties as follows, and then close the Property Palette:

Name	**RETURN_BUTTON**
Label	**Return**
Width	**90**
Height	**16**

8. To create the button trigger, select the button, right-click, point to **SmartTriggers**, and then click **WHEN-BUTTON-PRESSED**. Type the following command, compile the trigger, debug it if necessary, and then close the PL/SQL Editor.

```
EXIT_FORM;
```

9. Click **Window** on the menu bar, and then click **Object Navigator** to open the Object Navigator. Note that when you created the Return button, Forms Builder automatically created a new control block that contains the button. Change the new block name to **TEMPLATE_BLOCK**, and then save the form.

Visual Attribute Groups

The form template ensures a uniform overall appearance for application forms, but developers can still specify different properties for individual block items. For example, one developer might use a 10-point Arial font for text items or command button labels, while another might use an 8-point font. To ensure a standard appearance for block items, you create a **visual attribute group**, which is a form object that defines object properties, such as text item colors, font sizes, and font styles. After you create a visual attribute group, you assign the visual attribute group to the Visual Attribute Group property of form windows, canvases, and items.

To create a visual attribute group, you create a new visual attribute group object in the Object Navigator, and then specify its properties. In the group's Property Palette under the General property node, the Type property defines the type of object attributes to which the attribute properties apply. The possible values are Common, which means that the specified properties apply to all attributes in the object; Prompt, which means that the properties apply only to object prompts; and Title, which means that the defined properties apply only to object titles. For example, for a visual attribute group that defines the appearance of both the prompt and data values of a text item, you would create a Common type visual attribute group. To create a visual attribute group that defines only the appearance of the text item's prompt, you would create a Prompt type visual attribute group.

8

The visual attributes group's Color property nodes specify the foreground (text) color, background color, and fill pattern. The Font property node specifies the font size, style, and appearance. Next, you create a visual attribute group to specify the properties of text items in Northwoods University forms.

To create the visual attribute group for form text items:

1. In the Object Navigator window, select the Visual Attributes node under the NORTHWOODS_TEMPLATE form module, and then click the **Create** button ➕. A new visual attribute group object appears.

2. Select the new visual attribute group, right-click, and then click **Property Palette**. Change the Name value to **TEXT_ITEM_VISUAL_ATTRIBUTES**.

3. Make sure that the Visual Attribute Type property value is **Common**.

4. To specify that the text appears in a dark blue color, select the **Foreground Color** property, click the **More** button [...], and then select a **dark blue square**.

5. To specify that the text appears on a white background, select the **Background Color** property, click [...], and type **white** in the property.

6. Select the **Font Name** property, click [...], select **Arial**, and then click **OK**. Select the **Font Size** property, delete the current value, and then type **8**.

7. Close the Property Palette, save the form, and then close the form in Forms Builder. The MAIN form module should still appear in the Object Navigator.

Creating a Form Based on a Template Form

Next, you create a new form based on the Northwoods template. You use the Data Block and Layout Wizards to create a new data block and layout based on the LOCATION table in the Northwoods University database.

To create a new form using the template:

1. In the Object Navigator, click **File** on the menu bar, point to **New**, and then click **Form Using Template**. The Open dialog box opens, prompting you to select the template form. Select **Northwoods_Template.fmb**, and then click **Open**. A new form appears in the Object Navigator. Note that the new form contains objects in the Data Blocks, Canvases, and Visual Attributes nodes. These are the objects that you created in the template form.

You cannot create a new form module based on a template form if the template form file is currently open in Forms Builder.

TIP

2. Change the form module name to **LOCATION**.

3. To create the new data block, select the Data Blocks node, click the **Create** button ✚, make sure the **Use the Data Block Wizard** option button is selected, and then click **OK**. When the Data Block Wizard Welcome page appears, click **Next**.

4. When the Type page appears, make sure that the **Table or View** option button is selected, and then click **Next**. On the Source page, click **Browse**, select the **LOCATION** database table, and click **OK**. Click the **Move all items to target** button ⟩ to select all of the table fields for the data block, leave the Enforce data integrity check box cleared, and then click **Next**.

5. When the Master-Detail page appears, click **Next** because you do not want to create a master-detail relationship. Accept LOCATION for the data block name, and click **Next**. When the Finish page appears, make sure that the **Create the data block, then call the Layout Wizard** option button is selected, and then click **Finish**.

6. When the Layout Wizard Welcome page appears, click **Next**. Accept the default values on the Canvas page, and then click **Next**. (Recall that you are basing the new form on the template form, and the template form's canvas name is TEMPLATE_CANVAS.)

7. On the Data Block page, click ⟫ to select all of the data block fields for the layout, and then click **Next**. On the Items page, accept the default prompt values, and then click **Next**.

8. On the Style page, make sure that the Form option button is selected, and then click **Next**. On the Rows page, type **Location** for the frame title, leave the Records Displayed value as 1, click **Next**, and then click **Finish**.

9. In the Layout Editor, the new data block objects are not visible because they appear below the existing form template objects. Scroll to the bottom of the window, select the **Location** frame, and drag it to the top of the canvas. Resize the frame, and format the text item labels so your form looks like Figure 8-7.

8

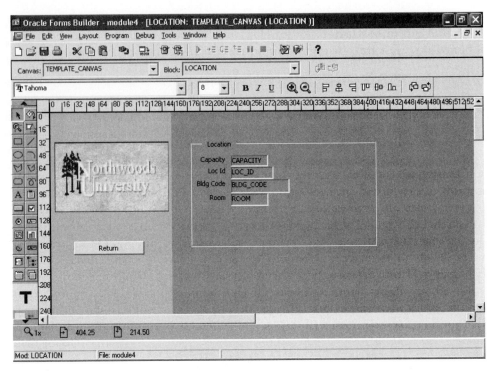

Figure 8-7 Location form based on template form

To modify the text item prompts so they appear on a single line, select the prompt, click the prompt again to open the text for editing, and delete the hard returns that wrap the text to a new line. To make the text items appear in a single column, make the frame narrower and longer.

10. Save the form as **Location.fmb** in the NorthwoodsProject_DONE folder on your Solution Disk.

Next, you apply the visual attribute group to the form text items. You select all of the text items as an object group, and open the intersection Property Palette. Then, you change the Visual Attribute Group property value to the name of the custom visual attribute group that you created earlier.

To apply the visual attribute group to the form text items:

1. In the Layout Editor, select the **LOC_ID** text item, press and hold the **Shift** key, and then select **BLDG_CODE**, **ROOM**, and **CAPACITY** so all of the text items are selected as an object group. You do not need to select the item prompts, because their properties are defined with their associated text items.

2. Click **Tools** on the menu bar, and then click **Property Palette** to open the intersection Property Palette for the text item object group.

3. Scroll down to the Visual Attributes property node, select the **Visual Attribute Group** property, open the list, and select **TEXT_ITEM_VISUAL_ATTRIBUTES**.

4. Close the Property Palette, and save the form. Note that the form text items now appear as dark blue text on a white background, in an 8-point Arial Regular font, as specified in the custom visual attribute group.

5. Run the form, click the **Enter Query** button 🔎, and then click the **Execute Query** button 🔎. Step through the table records to view the records, and then close the browser window.

6. Close the LOCATION form in Forms Builder.

REFERENCING APPLICATION COMPONENTS

When you create an integrated database application, the main form application uses program commands to open other form modules and to run report modules. The application also references files that provide graphic image data for images such as the splash screen image. The following sections describe approaches for referencing application components.

Using the Path Specification Approach

In Developer10*g* database applications, you can use the **path specification approach**, and reference application components such as graphic images, forms, and reports by specifying the complete path, including the drive letter and folder path, to the application file. (You must use the path specification approach to reference dynamic image items such as the image on the splash screen.) The path specification must include the drive letter, folder path, and filename, including the file extension. This approach works well for a development team that has standardized procedures for saving all of the project files to a specific location on a file server. For example, the development team that creates the Northwoods University database application could decide to configure their workstations so their M: drive connects to a file server that contains the NorthwoodsProject_DONE folder.

The path specification approach makes it difficult to move the application to a different storage location because you must change the path specifications in all commands that reference the form or report files. To make it easier to move an application that uses path specifications, it is useful to create a **global path variable**, which is a variable that references a text string specifying the complete path to the drive and folder where you store all the application files. The global path variable enables you to place the path information in a single location. If you move the application files to a new location, you only have to change the specification in the global path variable, rather than changing the path statements in many different locations.

The global path variable value needs to be visible to multiple forms. Therefore, this variable must be a global variable. A **global variable** is a variable that you create in the code of one form, and that you can subsequently reference in any other open form. A global variable remains in memory during the entire duration of the user's database session, and is available to other forms, even when the form that created the global variable closes. A global variable is a character variable, with a maximum length of 255 characters.

To create a global variable and assign a value to it, you use the following syntax:

```
:GLOBAL.variable_name := variable_value;
```

Variable_name must follow the Oracle naming standard. Notice that to create a new global variable, you simply assign a value to it. You do not need to declare a global variable explicitly. If you want to create a global variable that has a NUMBER or DATE data type, you must assign the data value as a character, and then convert the value to the desired data type using the Oracle SQL TO_DATE or TO_NUMBER conversion function. You reference a global variable using the syntax :GLOBAL.*variable_name*.

Suppose that the Northwoods University database application files are stored in the \OraSolutions\Chapter08\Tutorials\NorthwoodsProject_DONE folder on the C: drive on your workstation. You would use the following command to assign that folder location to a global variable named project_path:

```
:GLOBAL.project_path := 'C:\OraSolutions\Chapter08\
Tutorials\NorthwoodsProject_DONE\';
```

Note that the global path variable specification includes a trailing back slash (\ ) in the file path. Also note that you enclose the path specification in single quotation marks, and that the command must appear on a single line in the PL/SQL Editor.

To reference the global path variable and specify the complete path to a file that is stored in the application project folder, you concatenate the path variable to the filename. For example, the following expression references the NWsplash.tif image file that is stored in the C:\OraSolutions\Chapter08\Tutorials\NorthwoodsProject_DONE folder:

```
:GLOBAL.project_path || 'NWsplash.tif'
```

If you write a program command that references a global variable value that has not yet been assigned a value, an error occurs.

Next, you add a command to the PRE-FORM trigger to create a global path variable. The command assigns to the global path variable a text string that specifies the drive letter and folder path where you store your project application files. You must specify the drive letter where your folder is stored, along with the complete path specification, including all subfolders. Then, you modify the READ_IMAGE_FILE command that loads the dynamic image onto the splash screen canvas so it uses the global path variable. Finally, you run the form and confirm that the splash screen image appears correctly.

To create the global path variable in the PRE-FORM trigger:

1. In the Object Navigator, open the MAIN form module node.

2. Open the Triggers node under the MAIN form module. Select the PRE-FORM node, right-click, and then click **PL/SQL Editor**. The PRE-FORM trigger opens in the PL/SQL Editor.

3. To create the global path variable, add the following command as the first command in the trigger:

```
:GLOBAL.project_path := 'C:\OraSolutions\Chapter08\↵
Tutorials\NorthwoodsProject_DONE\';
```

The text string that specifies the global path variable value cannot contain any blank spaces in the folder names, and you cannot break the text string across multiple text editor lines.

Be sure to change the path specification as necessary to reflect the exact location of the folder that contains your tutorial solutions.

4. Modify the READ_IMAGE_FILE command as follows so it uses the global path variable to reference the dynamic image, then compile the trigger, correct any syntax errors, and close the PL/SQL Editor.

```
READ_IMAGE_FILE(:GLOBAL.project_path ||↵
'NWsplash.tif', 'TIFF', 'splash_block.splash_image');
```

5. Save the form, and then run the form. The splash screen image should appear as before.

6. Close the browser window.

Referencing Forms and Reports Using Module Names

When you create a form or report, you assign a module name to the form or report in the Object Navigator. The main application form can reference any form or report using the module name, provided that either:

- The developer stores the form or report file in the default form or report folder.

- The form or report is available on an Oracle Application Server (OAS).

The **default form folder** is a folder on the client workstation in which Developer10*g* always looks for form files when a command references a form in an integrated database application. Similarly, the **default report folder** is a folder on the client workstation in which Developer10*g* always looks for report files when a command references a report in an integrated database application. Developer10*g* specifies these folder paths in the system registry when it installs Forms Builder and Reports Builder. (The **system registry** is a database that Windows operating systems use to store system configuration information. The registry consists of keys, which are folders that contain related values. Related values are variables that have corresponding data entries.)

The FORMS90_PATH registry value specifies the default form folder path. This path has the default value of C:*oracle_developer_home*\forms90. The REPORTS_PATH registry value specifies the default report folder path. This path has the default value of C:*oracle_developer_home*\REPADM61\srw. When you are developing new database applications, it is convenient to store form and report files in the default form and report folders.

NOTE

The Forms Server of the Oracle Application Server 10g still uses the Forms 9.0 underlying engine.

OPENING AND CLOSING FORMS IN AN INTEGRATED DATABASE APPLICATION

Recall that in an integrated database application, the main form serves as the entry and exit point for all other application forms and reports. You write commands for the switchboard buttons and menu selections that call specific forms and reports. You can also create form command buttons to allow users to close forms explicitly. The following sections describe the commands to open and close forms.

Opening Forms in an Integrated Database Application

Forms Builder provides a number of built-in procedures that enable you to open one form from another. These procedures can reference forms using path specifications or by using form module names for forms that are in the default form folder. Table 8-2 summarizes these procedures. The form that calls the second form is the parent form, and the form that is called is the child form.

Procedure Name	Description
CALL_FORM	Opens a child form and immediately switches the application focus to the child form; has options for hiding the parent form, displaying the menu from the parent form, and passing a parameter list from the parent form
OPEN_FORM	Opens a child form, with the option of not immediately changing the application focus to the child form; has an option for creating a new database session for the child form or using the parent form's database session
NEW_FORM	Opens a child form and exits the parent form

Table 8-2   Built-in procedures to open a child form from a parent form

The following sections describe the built-in procedures.

CALL_FORM

The CALL_FORM procedure opens a child form and immediately switches the application focus to the child form. You use the CALL_FORM procedure when the user clicks a switchboard button or selects a menu item, and expects to see the child form immediately. The CALL_FORM procedure has the following syntax:

```
CALL_FORM('form_specification', display,
switch_menu, query_mode, parameter_list_id);
```

This procedure uses the following parameters:

- *Form_specification* is a character string that specifies the child form. If the child form is in the default form folder, *form_specification* is the name of the child form module as it appears in the Object Navigator. If the child form is not in the default form folder, *form_specification* specifies the full path and filename, including the drive letter, to the child form's .fmx file. *Form_specification* always appears enclosed in single quotation marks. For example, the *form_specification* value for the STUDENT_LOGON form that is stored in the default form folder is 'STUDENT_LOGON'. The *form_specification* value for the Student_logon.fmx file that is stored in the C:\OraSolutions\Chapter08\ Tutorials\NorthwoodsProject_DONE folder on your Solution Disk is 'C:\OraSolutions\Chapter08\Tutorials\NorthwoodsProject_ Done\Student_Logon.fmx'.

TIP

You could specify the *form_specification* value using the global path variable as follows: :Global.project_path || 'Student_logon.fmx'.

- *Display* specifies whether the parent form is hidden or not hidden by the child form. Valid values are HIDE and NO_HIDE. You specify the display parameter value as NO_HIDE when you want the two forms to appear side by side on the screen. The default value is HIDE.

- *Switch_menu* specifies whether the child form displays the same pull-down menus as its parent form, or displays different menus. Valid values are NO_REPLACE (pull-down menus are inherited from the calling form) and DO_REPLACE (different pull-down menus are used). The default value is NO_REPLACE. You learn how to replace pull-down menus in Lesson B of this chapter. For now, the calling form uses the standard Forms Services menu selections, and you accept the default NO_REPLACE value.

- *Query_mode* specifies whether the child form runs in normal mode, in which the user can insert, update, or delete values, or in query mode, in which the user can only view data. Valid values are NO_QUERY_ONLY, which runs the child form in normal mode, and QUERY_ONLY, which runs the child form in query mode. The default value is NO_QUERY_ONLY.

- *Parameter_list_id* specifies the identifier for an optional parameter list that the parent form can use to pass data values to the child form.

All of the parameters are optional except for *form_specification*. The following command opens the STUDENT_LOGON form that is stored in the default form folder, and accepts the default values for the other procedure parameters:

```
CALL_FORM('STUDENT_LOGON');
```

If you decide to specify a value for any of the optional parameters, you must specify values for all of the preceding optional parameters in the list. For example, if you specify a value for *switch_menu*, you must also specify a value for *display*. However, you do not need to include values for the subsequent optional parameters if you are willing to accept the default values. For example, the following command opens the STUDENT_LOGON form from the default folder, hides the parent form, and replaces the default menus:

```
CALL_FORM('STUDENT_LOGON', HIDE, DO_REPLACE);
```

OPEN_FORM

The OPEN_FORM procedure opens a child form and gives the developer the option of not immediately changing the application focus to the child form. Many developers prefer this procedure because it allows the user to multitask between parent and child forms in the application. The OPEN_FORM procedure has the following syntax:

```
OPEN_FORM('form_specification', activate_mode, ↵
session_mode,parameter_list_id);
```

This procedure uses the following parameters:

- *Form_specification* is the same parameter as in the CALL_FORM procedure.

- *Activate_mode* specifies whether the application focus switches to the child form or is retained by the parent form. Values can be ACTIVATE, which switches the application focus to the child form, and NO_ACTIVATE, which retains the application focus in the parent form. The default value is ACTIVATE.

- *Session_mode* specifies whether the child form uses the same database session as the parent form, or whether the child form starts a new database session. Values can be NO_SESSION, which specifies that the child form shares the parent form's session, and SESSION, which specifies that the child form starts a new session. The default value is NO_SESSION. When a child form is opened in the same session as the parent form, the DBMS commits all uncommitted values in both forms when a COMMIT command executes. A benefit of opening a child form in a separate session is that whenever a COMMIT command is issued, transactions are committed only in the form in which the COMMIT command executes. This allows the developer to control how the forms commit transactions.

■ *Parameter_list_id* specifies the identifier of an optional parameter list that the parent form can use to pass data values to the child form.

As with the CALL_FORM procedure, you can omit all of the parameters except *form_specification*. And, as with CALL_FORM, if you specify one of the optional parameters, you must specify values for all of the optional parameters that precede it. The following command opens the STUDENT_LOGON form that is stored in the default form folder, retains the form focus in the parent form, does not start a new database session, and passes a parameter list named my_list to the child form:

```
OPEN_FORM('STUDENT_LOGON', NO_ACTIVATE, NO_SESSION, ↵
my_list);
```

NEW_FORM

The NEW_FORM procedure opens a child form, and immediately exits the parent form. You use this procedure to allow users to open application forms without going through the main form switchboard or menus. For example, you might allow the user to open a form and create a new Northwoods University COURSE record, and then immediately open a second form and create a new COURSE_SECTION record. The advantage of NEW_FORM is that it prevents the user from opening multiple forms at once, which consumes system memory. However, because users should access all application forms from the main form, we discourage using the NEW_FORM procedure.

The NEW_FORM procedure has the following syntax:

```
NEW_FORM('form_specification', rollback_mode,
query_mode, parameter_list_id);
```

Form_specification, *query_mode*, and *parameter_list_id* are the same parameters as in the CALL_FORM built-in. The *rollback_mode* parameter specifies whether the DBMS automatically commits or rolls back uncommitted records in the parent form. Values can be TO_SAVEPOINT, which specifies that the DBMS rolls back all uncommitted transactions to the last savepoint; NO_ROLLBACK, which specifies that the DBMS does not roll back any uncommitted transactions; and FULL_ROLLBACK, which specifies that the DBMS rolls back all uncommitted transactions and restores the database to its state at the beginning of the database session. The default value is TO_SAVEPOINT.

Closing Forms in an Integrated Database Application

Up to this point, you have closed forms by closing the browser window. Forms Builder also provides a number of built-in procedures that enable you to create a form command button that allows the user to close a form in an application. Table 8-3 summarizes the procedures that allow you to close forms.

Procedure Name	Description
CLOSE_FORM	Closes the specified form, which might not be the current form
EXIT_FORM	Closes the current form, and provides options for committing or rolling back the uncommitted data

Table 8-3   Procedures to close forms

The CLOSE_FORM procedure closes a specific form, which might not be the form the user is currently viewing. This procedure allows developers to control the open forms in an application. For example, a form trigger could determine if the user has opened two instances of the same form, and automatically close one of the forms.

The CLOSE_FORM procedure has the following syntax:

```
CLOSE_FORM('form_identifier');
```

Forms Builder creates a unique form ID for every open form. The *form_identifier* parameter is the form_ID or name of the form module as specified in the Object Navigator. *Form_identifier* appears enclosed in single quotation marks. In this book, you use the form module name for the *form_identifier* value.

The EXIT_FORM procedure closes the current form. You use this procedure to create a trigger for an Exit button that the user clicks to close a form. The EXIT_FORM procedure uses the following syntax:

```
EXIT_FORM (commit_mode, rollback_mode);
```

This procedure has two optional parameters. *Commit_mode* specifies how the DBMS handles uncommitted form data. Valid values are ASK_COMMIT, which causes the form to ask the user to save uncommitted changes; DO_COMMIT, which automatically commits unsaved data; and NO_COMMIT, which automatically discards uncommitted changes. The default value is ASK_COMMIT. *Rollback_mode* specifies if the DBMS automatically commits or rolls back uncommitted records in the parent form. The *rollback_mode* parameter is also used in the NEW_FORM procedure, and was described earlier.

Creating Triggers to Open Forms

Next, you create and test the triggers for the buttons on the application switchboard that call the Student Logon form and Faculty Logon form. Because it is a good practice for every application to have a single entry and exit point, the Northwoods University database application uses the CALL_FORM procedure to call the forms named STUDENT_LOGON and FACULTY_LOGON, while keeping the main form open. When the user clicks the Return button on either child form, the EXIT_FORM procedure closes the child form and returns execution to the main form. The trigger commands use the global path variable to call the forms in the NorthwoodsProject_DONE folder on your Solutions Disk. The commands accept the default values for the optional parameters in the CALL_FORM procedure.

To create and test the button triggers:

1. In the Object Navigator, click **Tools** on the menu bar, and then click **Layout Editor**. Select **MAIN_CANVAS** from the list if necessary to open the canvas in the Layout Editor. Click **OK**.

2. Select the **Student Logon** button, right-click, point to **SmartTriggers**, and then click **WHEN-BUTTON-PRESSED** to create a new button trigger. Type the following command in the PL/SQL Editor:

   ```
   CALL_FORM(:GLOBAL.project_path || 'Student_logon.fmx');
   ```

3. Compile the trigger, correct any syntax errors, and then close the PL/SQL Editor.

4. Select the **Faculty Logon** button, right-click, point to **SmartTriggers**, and then click **WHEN-BUTTON-PRESSED** to create a new button trigger. Type the following command in the PL/SQL Editor:

   ```
   CALL_FORM(:GLOBAL.project_path || 'Faculty_logon.fmx');
   ```

8

5. Compile the trigger, correct any syntax errors, and then close the PL/SQL Editor.

6. Save the form, and then run the form. When the main form appears, click **Student Logon**. The Student Logon form (Student_logon.fmx) opens.

TIP

If the Student Logon form does not open, check the PRE-FORM trigger to confirm that you correctly entered the path for the global path variable to the folder that stores your project files. Also make sure you entered the path to the .fmx file correctly in the Sales button trigger, and confirm that the Student_logon.fmx file is in the project folder.

7. Click **Exit** to exit the Student Logon form and return to the main form application.

8. Click **Faculty Logon** to open the Faculty Logon form (Faculty_logon.fmx). Click **Return** to exit the form and return to the main form application.

9. Close the browser window.

DISPLAYING A REPORT IN AN INTEGRATED DATABASE APPLICATION

Recall that a report displays a summary view of database data at a specific point in time. Reports can run as stand-alone applications or can appear within integrated database applications. To display a report in an integrated database application, you install and start the local report server, and then configure the main application form so it generates the report as an HTML file and displays the report in a browser window.

Installing and Starting the Local Report Server

Recall from Chapter 7 that a report server is a process that runs in conjunction with the Oracle Application Server 10*g*, queries the database when the user requests a report, and dynamically generates a Web page that displays the current database data. To view dynamic report data while you are developing a new integrated database application, you create and then start a **local report server**, which is a server process that runs on your client workstation. The following sections describe how to accomplish these tasks.

Installing a Local Report Server

To install a local report server, you run an Oracle 10*g* utility named Rwserver, and pass to it parameters that instruct it to install a new local report server and assign a name to the local report server. To pass the following parameters to the Rwserver utility, use this command:

```
-install server_name tcpip
```

The install parameter instructs the utility to install the local report server. *Server_name* specifies the name of the new local report server, and tcpip instructs the report server to use the TCP/IP protocol to enable communication between the report server and the database application. (The **TCP/IP protocol** is a set of rules that networked computers use to communicate.)

To start the Rwserver utility and install a local report server, you can start a command prompt session, navigate to the *oracle_developer_home*\bin folder, and type the following command:

```
rwserver -install server_name tcpip
```

An easier way to start and run the Rwserver utility is to create a shortcut to the utility, and configure the shortcut so it passes the correct parameters to the utility. (In the Windows operating system, a **shortcut** is a small file that starts a target program, and optionally passes parameters to the target program.) In the following exercise, you configure the shortcut to pass parameters to and install a new local report server named localrepserver. Then you run the shortcut and install the local report server.

To create the shortcut to the Rwserver utility and install the local report server:

1. Start Windows Explorer, and navigate to the *oracle_developer_home*\bin folder on your workstation. If necessary, click the **Name** column heading to sort the files by filename. Scroll down the list, right-click **rwserver.exe**, and then click **Create Shortcut**. A new shortcut named Shortcut to rwserver.exe appears in the folder.

2. Change the shortcut name to **StartRepServer.exe**, and then move the shortcut to the Chapter08\Tutorials folder on your Solution Disk.

3. To configure the new shortcut, right-click **StartRepServer.exe** in the Chapter08\Tutorials folder on your Solution Disk, and then click **Properties**. The StartRepServer.exe Properties dialog box opens.

4. To create the parameters that are passed to the utility to specify to install the local report server and assign its name, make sure that the **Shortcut** tab is selected, click the insertion point at the end of the command in the Target field, press the **spacebar**, and then type the following parameters: `-install localrepserver tcpip`.

5. Click **OK** to save your changes and close the dialog box.

6. To install the local report server, double-click **StartRepServer.exe**. A Reports Server dialog box opens asking if you want to install localrepserver as an NT service. Click **Yes**. A dialog box appears with the message "Server 'oracle_developer_homeReports [localrepserver]' installed. Go to Control Panel Services to start it up." Click **OK**.

If a dialog box opens with the message "Unable to install server 'oracle_developer_homeReports [localrepserver]'", it is probably because someone has already installed a report server with this name on your workstation. If this happens, proceed to the next section titled "Starting the Local Report Server."

Starting the Local Report Server

The Rwserver utility installs the local report server as a Windows service. A **Windows service** is a process that runs in the main memory of a workstation and is accessible to all users who log onto the workstation. To start the local report server, you must open the Services control panel on your workstation, and start the service.

To start the local report server service:

1. Click **Start** on the Windows taskbar, click **Control Panel**, double-click **Administrative Tools** (in Classic View), and then double-click **Services**. The Services control panel window opens.

If you are using Windows 2000, click Start, point to Settings, click Control Panel, double-click Administrative Tools, and then double-click Services. Use these instructions throughout the chapter to open the Services window.

2. Scroll down in the Services list until you find the *oracle_developer_home*Reports [localrepserver] service. (You may need to resize the Name column to view the service name.) The service Status column should appear blank, indicating the service is not yet started.

If the Status value appears as Started, as shown in Figure 8-8, it means that the service is already running, so skip the next step, which starts the service.

HELP

If the service does not appear in the list, repeat Steps 3–6 in the previous set of steps until you successfully install the new report server.

HELP

3. Select *oracle_developer_home*Reports [localrepserver], right-click, and then click **Start**. A Service Control window opens for a few moments, and then the service status appears as Started, as shown in Figure 8-8.

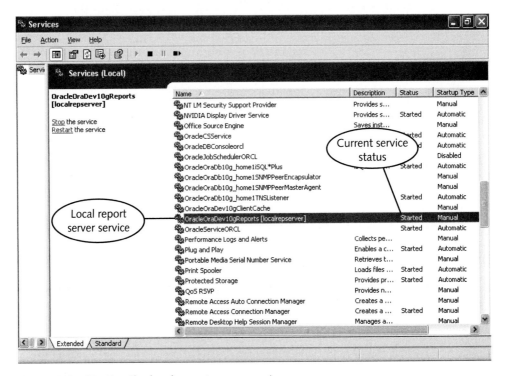

Figure 8-8   Starting the local report server service

4. Close the Services window, and then close the Administrative Tools (or Control Panel, if you are using Windows 2000) window.

Creating a Report Object

To display a report in an integrated database application, you create a report object in the main application form. A **report object** is a top-level Forms Builder object in the Object Navigator window that represents a Reports Builder report file. You create a new report object by selecting the Reports node in the Object Navigator tree, and then clicking the Create button ➕. When you create a new report object, you can specify to create a new report or to base the report object on an existing report. When you opt to create a new report, Forms Builder starts Reports Builder and allows you to configure the new report. You have the option of creating a new report based on an existing data block in the current form. For example, you can create a data block form that contains a data block based on the Northwoods University LOCATION table. Then, you can create a new report object, and base the associated report on the data in the existing LOCATION data block. When Reports Builder starts, the new report automatically contains the LOCATION data block fields.

After you create a report object, you configure its properties using the report object Property Palette. Table 8-4 summarizes important report object properties.

Property	Description	Example Values
Filename	Specifies the report filename. This value can be a complete path specification, including the drive letter and folder path. Or, it can specify a report file that is in the default report folder.	C:\OraSolutions\Chapter08\ Tutorials\ClearwaterProject_DONE\ Inventory.rdf
Execution Mode	Specifies whether the report executes in Runtime or Batch mode. Runtime specifies that the report executes without user interaction, and Batch specifies that the user can interact with the report.	Runtime, Batch
Communication Mode	Specifies how the form communicates with the report. Synchronous specifies that program execution control returns to the form only after the report closes. Asynchronous specifies that control returns to the form while the report is still running, and users can multitask between the form and the report.	Synchronous, Asynchronous

Table 8-4   Report object properties

Property	Description	Example Values
Data Source Data Block	Specifies the data block that provides the report data. (This property applies only to reports that you create based on a data block.)	Block name, as it appears in the Object Navigator
Report Destination Type	Specifies the report output destination.	Screen, File, Printer, Preview, Mail, or Interoffice
Report Destination Name	Specifies the name of an output file, printer, or e-mail user account or distribution list to which the report is sent.	Possible values can be a filename, printer name, e-mail user account, or distribution list
Report Destination Format	Specifies the report output format.	PDF, HTML, HTMLCSS, RTF, DELIMITED
Report Server	Specifies the name of the report server that generates the report.	localrepserver

Table 8-4   Report object properties (continued)

You can also configure the properties of a report object dynamically in a form trigger using the SET_REPORT_OBJECT_PROPERTY built-in, which has the following syntax:

```
SET_REPORT_OBJECT_PROPERTY('object_name',
property_name, value);
```

In this syntax, *object_name* is the name of the report object, as it appears in the Object Navigator. *Property_name* references the property name to be set, and *value* is the new property value. Table 8-5 summarizes the *property_name* and *value* specifications for the SET_REPORT_OBJECT_PROPERTY built-in. Note that the *property_name* value that you specify in the SET_REPORT_OBJECT_PROPERTY built-in is not necessarily the same as the property name that appears in the Property Palette.

Property	SET_REPORT_ OBJECT_PROPERTY *property_name*	SET_REPORT_ OBJECT_PROPERTY *value*	Example Command
Filename	REPORT_FILENAME	Report filename, enclosed in single quotation marks	SET_REPORT_OBJECT_PROPERTY ('LOCATION_REPORT', REPORT_FILENAME, :GLOBAL. project_path || 'Location.rdf');
Execution Mode	REPORT_EXECUTION_ MODE	BATCH or RUNTIME	SET_REPORT_OBJECT_PROPERTY ('Location_report', REPORT_EXECUTION_MODE, RUNTIME);

Table 8-5   SET_REPORT_OBJECT_PROPERTY property names and values

Property	SET_REPORT_ OBJECT_PROPERTY *property_name*	SET_REPORT_ OBJECT_PROPERTY *value*	Example Command
Communication Mode	REPORT_COMM_ MODE	SYNCHRONOUS or ASYNCHRONOUS	SET_REPORT_OBJECT_PROPERTY ('Location_report', REPORT_COMM_MODE, SYNCHRONOUS);
Data Source Data Block	REPORT_SOURCE_ BLOCK	Block name, enclosed in single quotation marks	SET_REPORT_OBJECT_PROPERTY ('Location_report', REPORT_SOURCE_BLOCK, 'LOCATION_BLOCK');
Report Destination Type	REPORT_DESTYPE	PREVIEW, FILE, PRINTER, MAIL, CACHE, or SCREEN	SET_REPORT_OBJECT_PROPERTY ('Location_report', REPORT_DESTYPE, FILE);
Report Destination Name	REPORT_DESNAME	Report destination filename, printer name, e-mail user account, or e-mail distribution list, enclosed in single quotation marks	SET_REPORT_OBJECT_PROPERTY ('LOCATION_REPORT', REPORT_DESNAME, :GLOBAL.project_path || 'Location.htm');
Report Destination Format	REPORT_DESFORMAT	PDF, HTML, HTMLCSS, RTF, or DELIMITED enclosed in single quotation marks	SET_REPORT_OBJECT_PROPERTY ('Location_report', REPORT_DESFORMAT, 'HTML');
Report Server	REPORT_SERVER	Report server name, enclosed in single quotation marks	SET_REPORT_OBJECT_PROPERTY ('Location_report', REPORT_SERVER, 'localrepserver');

Table 8-5   SET_REPORT_OBJECT_PROPERTY property names and values (continued)

8

In the following exercise, you create a report object, and configure the report object properties using the report Property Palette. You specify localrepserver as the report server, specify the report object to run in Runtime mode, and generate the report output to a file using an HTML destination format. You base the report object on the Location.rdf file in the NorthwoodsProject_DONE folder on your Solution Disk.

To create and configure a report object:

1. In the Object Navigator, select the Reports node, and then click the **Create** button ✚ to create a new report object. The MAIN: New Report dialog box opens.

2. Select the **Use Existing Report File** option button, click **Browse**, navigate to the NorthwoodsProject_DONE folder on your Solution Disk, select **Location.rdf**, click **Open**, and then click **OK**. The new report object appears under the Reports node.

684 Chapter 8 Lesson A   Creating an Integrated Database Application

3. Change the new report object name to **LOCATION_REPORT**.

4. Select **LOCATION _REPORT**, right-click, and then click **Property Palette**. Change the following report property values, close the Property Palette, and then save the form.

Property	Value
Execution Mode	**Runtime**
Report Destination Format	**HTML**
Report Server	**localrepserver**

Displaying the Report Object

To display the report object, you create a form trigger for the button that the user clicks to display the report, which is the View Campus Locations button on the main application form. This trigger contains commands to configure the report filename and output filename dynamically. It also contains a command to run the report and generate an HTML output file, along with commands to display the HTML file in a browser window.

Dynamically Configuring the Report Filename and Output Filename

When you created the report object, you specified the report filename. The report filename currently uses the absolute path specification to the project folder, which is currently the NorthwoodsProject_DONE folder on your Solution Disk. If you move the project folder to a new location, you need to open the report object Property Palette, and change the Report Filename property. This procedure is time consuming and prone to error for an application that contains many reports. To make your database application more portable, you use the SET_REPORT_OBJECT_PROPERTY built-in in the trigger to reset the report object filename. You use the global path variable to specify that the report filename is in the project folder.

Currently, the report object does not specify the report destination filename property. You use the SET_REPORT_OBJECT_PROPERTY built-in to specify the destination filename as a file named Location.htm in the project folder. Next you create a trigger for the View Campus Locations button on the main application form, and add the commands to specify the report filename and report output destination filename dynamically.

To create the trigger to specify the report filename and destination filename dynamically:

1. Click **Tools** on the menu bar, and then click **Layout Editor**. Make sure that *MAIN_CANVAS* is selected in the MAIN: Canvases list, and then click **OK**.

2. Select the **View Campus Locations** button, right-click, point to **SmartTriggers**, and then click **WHEN-BUTTON-PRESSED**. The PL/SQL Editor window opens.

3. Type the following commands to set the report filename and destination file-name dynamically:

```
SET_REPORT_OBJECT_PROPERTY('LOCATION_REPORT',
REPORT_FILENAME, :GLOBAL.project_path ||↵
'Location.rdf');
SET_REPORT_OBJECT_PROPERTY('LOCATION_REPORT',↵
REPORT_DESNAME, :GLOBAL.project_path || 'Location.htm');
```

4. Compile the trigger, and correct any syntax errors. Do not close the PL/SQL Editor.

Commands to Generate the Report

To instruct the local report server to generate a report and create the report output file, you use the RUN_REPORT_OBJECT built-in, which is a function that has the following syntax:

```
report_var := RUN_REPORT_OBJECT('report_object_name');
```

In this syntax, *report_var* is a previously declared variable with the VARCHAR2 data type that uniquely identifies the generated report. *Report_object_name* is the name of the report object as it appears in the Object Navigator. You use the following commands to declare a variable named rep_result to represent the *report_var* variable, and generate the INVENTORY_REPORT object:

```
DECLARE
 rep_result VARCHAR2(30);
BEGIN
 rep_result := RUN_REPORT_OBJECT('LOCATION_REPORT');
END;
```

Now you add the commands to the View Inventory button trigger to declare a variable named rep_result as the *report_var* variable, and generate the report.

To add the commands to generate the report:

1. In the PL/SQL Editor, add the shaded commands in Figure 8-9 to declare the variable, and generate the report.

```
DECLARE
      rep_result   VARCHAR2(30);
BEGIN
      SET_REPORT_OBJECT_PROPERTY('LOCATION_REPORT',↵
      REPORT_FILENAME, :GLOBAL.project_path || 'Location.rdf');
      SET_REPORT_OBJECT_PROPERTY('LOCATION_REPORT',↵
      REPORT_DESNAME, :GLOBAL.project_path || 'Location.htm');

      rep_result := RUN_REPORT_OBJECT('LOCATION_REPORT');
END;
```

Figure 8-9 Commands to generate the report

2. Compile the trigger, and correct any syntax errors. Do not close the PL/SQL Editor.

Commands to Display the Report in a Browser Window

The final step is to create the commands that display the report output file in a browser window. To open a new browser window and display an HTML file, you use the WEB.SHOW_DOCUMENT built-in, which has the following syntax:

```
WEB.SHOW_DOCUMENT ('URL', 'target');
```

In this syntax, *URL* represents the Uniform Resource Locator of the HTML file that displays the report. A **Uniform Resource Locator (URL)** is a string of characters, numbers, and symbols that specify a Web page HTML document. In this command, you specify the URL as a call to the report writer servlet, passing the name of the report file to it. In addition, you include parameters to identify the destination format, the destination type, and the Oracle user authentication information.

Target specifies the properties of the browser window in which the form loads the document. Possible values include '_self', which specifies to load the HTML document in the current browser window, and '_blank', which specifies to load the HTML document in a new blank browser window. *Target* must be in lowercase letters, and enclosed in single quotation marks.

You would use the following command to display in a new blank browser window a report output file named Location.rdf that is stored in the workstation's C:\OraSolutions folder using a file URL:

```
WEB.SHOW_ DOCUMENT( '/reports/rwservlet?&report='|| ↵
:GLOBAL.project_path || ' Location.rdf'||'&desformat= ↵
html&destype=cache&userid=scott/tiger@orcl','_blank');
```

Next, you add the command to the form trigger to display the report output file in a new blank browser window. The command uses the global path variable to display the report output file that is stored in the project folder. Then you run the form and display the report.

To add the command to display the report, then run the form and display the report:

1. In the PL/SQL Editor, add the shaded command in Figure 8-10 to display the finished report in a browser window.

```
DECLARE
        rep_result   VARCHAR2(30);
BEGIN
        SET_REPORT_OBJECT_PROPERTY('LOCATION_REPORT',
        REPORT_FILENAME, :GLOBAL.project_path || 'Location.rdf');
        SET_REPORT_OBJECT_PROPERTY('LOCATION_REPORT',
        REPORT_DESNAME, :GLOBAL.project_path || 'Location.htm');
        rep_result := RUN_REPORT_OBJECT('LOCATION_REPORT');

        WEB.SHOW_DOCUMENT('/reports/rwservlet?&report='||
        :GLOBAL.project_path || 'Location.rdf'||'&deformat=
        html&destype=cache&userid=scott/tiger@orcl','_blank);

END;
```

Figure 8-10   Command to display the report output file in a new browser window

2. Compile the trigger, correct any syntax errors, and then close the PL/SQL Editor.

3. Save the form, and then run the form. The splash screen appears, and the main form application appears.

4. Click **View Campus Locations**. The report appears in the browser window, as shown in Figure 8-11. Maximize the browser window if necessary.

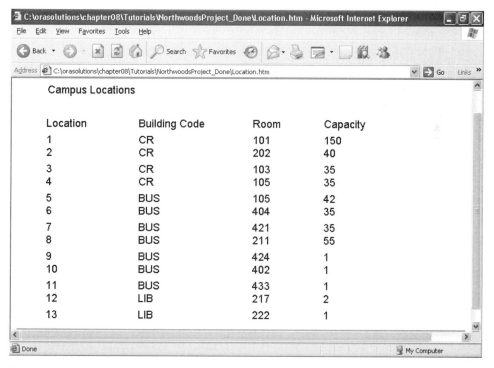

Figure 8-11   Viewing the report in the browser window

If your report does not appear, make sure that the local report server is installed and started, using the instructions in the section titled "Starting the Local Report Server."

If your report still does not appear, open Windows Explorer, and confirm that the report server is generating the Location.htm output file and storing the file in the NorthwoodsProject_DONE folder on your Solution Disk. If the Location.htm file is not present, double-check the commands in the View Campus Locations button trigger to make sure the command that specifies the destination filename is correct and the command that displays the destination filename in the browser is correct.

5. Close the browser window that displays the report, and then close the browser window that displays the main application form.

6. Close the Ch8AMain.fmb form in Forms Builder, click **Yes** to save your changes, and then close Forms Builder and all other open applications.

SUMMARY

◻ Developing a database application requires the following phases: design, module development, module integration, testing, and deployment. Unit testing involves testing individual modules, and system testing evaluates the integration of the modules.

◻ When you use Forms Builder to create a complex database application, it is a good practice to separate the system into individual form modules that developers independently create, test, and debug. Then, the developers integrate the individual form modules into a single application.

◻ To integrate individual forms and reports into a single application, you place all module files in a central project folder. You create a main form module from which users access all of the individual application components from a single entry and exit point. This main form module has a switchboard to access the most commonly used components and menus for accessing all system components.

◻ A splash screen is the first image that appears when you run an application. A timer controls how long the splash screen window appears and when the main form application window appears.

◻ To create a splash screen, you load the splash image, and set a timer in the form's PRE-FORM trigger. Then, you create a WHEN-TIMER-EXPIRED trigger that displays the main application window after a preset time interval.

◻ A template form contains standard form objects that appear on every form in an application. After you create a template form, you use it as the starting point when you create all other application forms.

◻ A visual attribute group is a form object that defines object properties, such as text item colors, font sizes, and font styles. You can assign visual attribute groups to form windows, canvases, and items.

❏ To reference forms and reports in integrated database applications, you can use path specifications or you can store the forms in the default form folder and the reports in the default report folder. If you use path specifications, you should create a global path variable to make the application more portable.

❏ A form global variable is a form variable that you create in one form and can then reference in any other open form. A global path variable represents the path to the project folder. If you change the project folder location, you only have to modify the path specification in the global path variable.

❏ To open one form from another in an integrated application, you use the CALL_FORM, OPEN_FORM, or NEW_FORM procedure. The CALL_FORM procedure opens a child form and switches the application focus to the child form. The OPEN_FORM procedure opens a child form and does not immediately change the application focus to the child form. The NEW_FORM procedure opens a child form and immediately exits the parent form.

❏ To exit a form programmatically, you use the CLOSE_FORM or EXIT_FORM procedure. The CLOSE_FORM procedure closes a specific form, and the EXIT_FORM procedure closes the current form.

❏ To view a report while you are developing an integrated database application, you install and start the local report server, create a report object in the main application form, and then write commands to run the report object and display the output in a browser window.

❏ A report object is a top-level form object that represents a Reports Builder report file. You configure the report object to specify the report file it represents and the properties that the local report server uses to generate and display the report. You can also configure report object properties dynamically using the SET_REPORT _OBJECT_PROPERTY built-in.

❏ To display a report, you configure the report object to generate its output to a destination file that uses an HTML format. Then you create a form trigger that contains commands to generate the report and display the output file in a browser window.

REVIEW QUESTIONS

1. Identify the phases of database application development.
2. A switchboard consists of radio buttons and check boxes that enable users to set their application preferences. True or False?
3. When creating menu items, you can use _____, or placeholder elements, for elements that have not yet been developed.
4. Which trigger identifies the action to take when a timer expires?
5. A global variable exists in one form, but can be referenced by other forms. True or False?
6. In the command :GLOBAL.page_title := 'Northwoods University';, which component is the variable name?

7. In the CALL_FORM and OPEN_FORM procedures, the _____ parameter is the only required parameter.

8. The LocRepserver utility is required to install the local report server on a client machine. True or False?

9. By default, when the local report server is installed it starts automatically every time the computer is restarted. True or False?

10. Use the _____ built-in procedure to open a new browser window and display an HTML file.

MULTIPLE CHOICE

1. Which of the following testing stages focuses on whether the modules work correctly after they have been integrated into the system?

   a. unit testing

   b. module testing

   c. system testing

   d. project testing

2. When you are creating a main form, a _____ should contain the switchboard command buttons.

   a. macro

   b. control block

   c. button module

   d. frame

3. Which of the following is a *correct* statement?

   a. When creating a splash screen, you must make certain the image item is not larger than the canvas.

   b. When creating a splash screen, you must make certain each side of the image item extends at least one inch past the canvas.

   c. You can use the AFTER_SPLASH_SCREEN trigger to specify where the focus is placed after the splash screen is closed.

   d. Both a and c

4. If you create a timer to control the display of a splash screen, use the _____ function to specify the timer properties.

   a. TIMER_PROPERTIES

   b. CREATE_TIMER

   c. SHOW_WINDOW

   d. DISPLAY_SPLASH

5. To reference a user-defined set of object properties for formatting text items in a template, select the _____ property from the text item's Property Palette.

    a. Visual Attribute Group

    b. Group Format

    c. Text Group Format

    d. Grouping

6. To facilitate file movement without having to change the path reference in all triggers and other items, you can use a _____, and only that single path reference will have to be updated with the new file location.

    a. global path variable

    b. static path reference

    c. template

    d. all of the above

8

7. Which of the following procedures opens a child form, and at the same time exits or closes the parent form?

    a. CALL_FORM

    b. OPEN_FORM

    c. CLOSE_FORM

    d. NEW_FORM

8. A local report server is a _____ process that runs on your client workstation.

    a. user

    b. server

    c. shared

    d. developer

9. Which of the following properties specifies that the report object should run without user interaction?

    a. Communication Mode

    b. Report Destination Type

    c. Report Server

    d. Execution Mode

10. To load an HTML document into the current browser window, you should assign the value of _____ to the Target property of the WEB.SHOW _DOCUMENT procedure.

   a. _self

   b. _top

   c. _current

   d. _blank

PROBLEM-SOLVING CASES

The cases refer to the Clearwater Traders (Figure 1-25) sample database tables. Place all solution files in the specified folder in the Chapter08\Cases folder on your Solution Disk.

1. In this case, you create a project folder for an integrated database application for Clearwater Traders. Then you create a template form for the application, and use the template form to create a form that displays data from the INVENTORY table.

   a. Create a folder named **ClearwaterProject_DONE** in the Chapter08\Cases folder on your Solution Disk that will be the project folder, and copy all the files from the Chapter08\ClearwaterProject folder on your Data Disk to the new project folder.

   b. Create a form named **CLEARWATER_TEMPLATE**, and save the form as **Clearwater_Template.fmb** in the new project folder.

   c. Create a window named **TEMPLATE_WINDOW**, a canvas named **TEMPLATE_CANVAS**, and a block named **TEMPLATE_BLOCK**. Change the window title to **Clearwater Traders**.

   d. Create a boilerplate rectangle on the left edge of the canvas, as shown in Figure 8-12. Fill the rectangle with a light gray color.

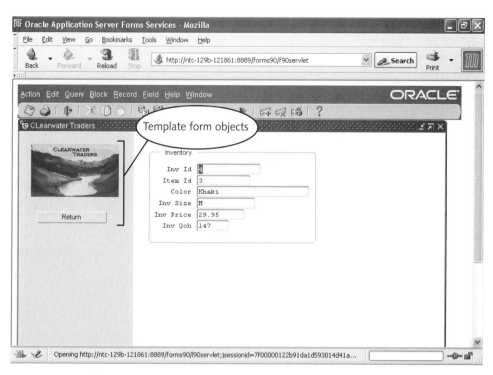

Figure 8-12   Clearwater Traders template form

e. Import the **Clearwater Traders logo** (Clearlogo.tif) from the
    ClearwaterProject_DONE folder, and place it on the template canvas as shown.

f. Create the **Return** button shown on the canvas, and create a **WHEN–BUTTON-
    PRESSED** trigger that exits the current form.

g. Create a visual attribute group named **TEXT_ATTRIBUTES** that formats text
    items using an 8-point Courier New font. Specify that the text items appear on a
    white background.

h. Save the template form.

2. Create a new form named **INVENTORY_FORM** that is based on the
    CLEARWATER_TEMPLATE form. Save the new form as **Inventory.fmb** in
    the ClearwaterProject_DONE folder. Create a data block and layout based on the
    INVENTORY table, as shown in Figure 8-12. Apply the TEXT_ATTRIBUTES
    visual attribute group to all form text items.

3. In this case, you create the main application form for the Clearwater Traders inte-
    grated database application. The main application form allows users to access sales
    and receiving information using the switchboard in Figure 8-13.

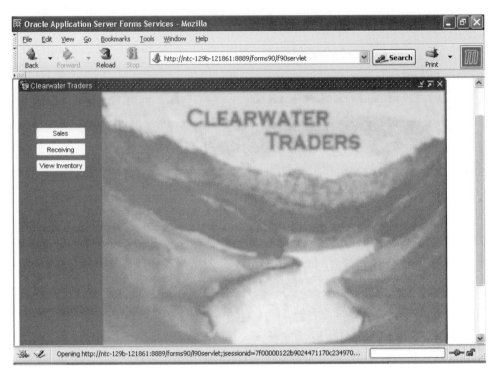

Figure 8-13   Clearwater Traders main application form

    a. Create a new form named **Main.fmb** that will be the main application form, and save the form in the project folder.

    b. Create a PRE-FORM trigger that initializes a global path variable to specify the location of the project files.

    c. Create a form splash screen that displays the Clearsplash.tif image that is stored in the project folder.

    d. Format the main form application canvas as shown in Figure 8-13. Use the Clearimage.gif image stored in the project folder as the background image on the canvas.

4. Create the **Sales** and **Receiving** switchboard buttons for the previously created Main.fmb form as shown in Figure 8-13. Create a trigger for the Sales button so that when the user clicks the button, the form that is stored in the Sales.fmx file in the project folder opens and displays current sales information. Create a trigger for the Receiving button so that when the user clicks the button, the form that is stored in the Receiving.fmx file in the project folder opens and displays received orders. Use the global path variable in all commands. Save the modified form.

5. Create a report object in the previously created Main.fmb form that is associated with the Inventory.rdf report in the project folder. Create the **View Inventory** switchboard button shown in Figure 8-13. Then create a trigger for the View Inventory button so that when the user clicks the button, the report object appears in a browser window. Use the global path variable to specify the location of the report filename and the report output filename, and also to display the report output in the browser window. Save the modified form.

◀ LESSON B ▶

After completing this lesson, you should be able to:

♦ Create custom pull-down menus
♦ Display custom pull-down menus in form modules
♦ Write program commands to control menu items
♦ Create context-sensitive pop-up menus

Database applications usually have a menu bar that contains pull-down menus that allow a user to click a top-level choice and display related selections. Many applications have **pop-up menus**, which appear when the user right-clicks the mouse pointer. Pop-up menus are context sensitive, and different menu selections appear, based on the current screen selection or mouse pointer location. In this lesson, you learn how to add these menu components to database applications.

CREATING CUSTOM PULL-DOWN MENUS

Currently, whenever you display a form in the Forms Services window, the default Forms Services pull-down menu selections appear on the menu bar. Figure 8-14 shows the default menu selections.

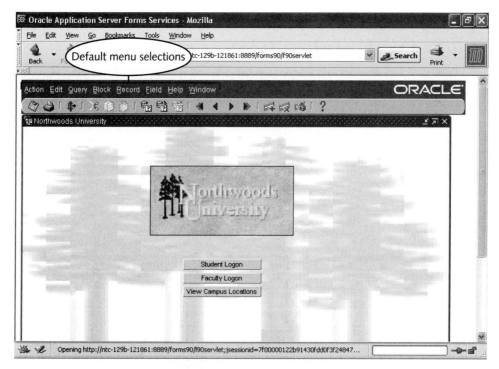

Figure 8-14   Default Forms Services menu selections

Recall from Lesson A that when you create an integrated database application, the user should be able to access all application components using menu selections. The default Forms Services menu selections are appropriate for data block forms, but you need to create a custom menu for the main application form in an integrated database application.

To replace the default Forms Services pull-down menu choices with custom pull-down menu choices, you create a menu module. A **menu module** is a module that you create in Forms Builder that is independent of any specific form. Forms Builder saves a menu module in your workstation's file system as a design file with an .mmb extension and as an executable file with an .mmx extension. You attach the executable (.mmx) menu file to a form module in the form module Property Palette, and the custom menu selections appear when you run the form.

A menu module contains one or more menu items. A **menu item** is a set of menu selections that appear horizontally on the menu bar. For example, in Figure 8-1 the

Student & Faculty, Database Maintenance, Reports, Help, and Window selections constitute a menu item. A menu module can have multiple menu items, so you can create several different menu items in a single menu module, and then programmatically specify the menu item that appears when you attach the menu module to a form. This reduces the number of menu modules that you have to create. The following sections describe how to create a menu module and its associated menu items, and how to create menu selections in a menu item.

Creating a Menu Module and Menu Item

To create a new menu module, you select the Menus node in Forms Builder, and then click the Create button ✚. To create a new menu item within a menu module, you select the Menus node within a menu module, and click ✚. Now you start Forms Builder, and create a new menu module and a new menu item for the Northwoods University database application.

To start Forms Builder and create a menu module and menu item:

1. Start the OC4J Instance.

2. Start Forms Builder. The menu module is independent of any form in an application, and you do not need to open or create a form while you are working with a menu module. Select **MODULE1**, and then click the **Delete** button ✖ to delete the default form module.

3. To create the menu module, select the Menus node, and then click the **Create** button ✚. Rename the new menu module **NW_MENU**.

4. To create a new menu item in the menu module, select the Menus node under NW_MENU, and then click ✚. Rename the new menu item **MAIN_MENU**. Your Object Navigator window should look like Figure 8-15.

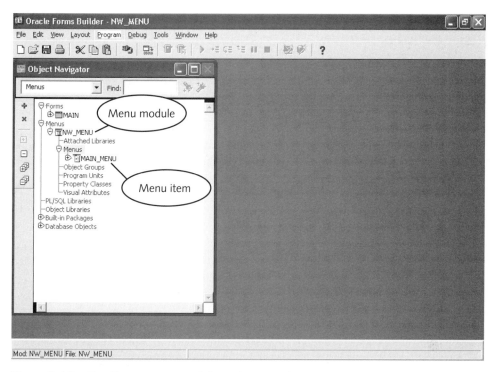

Figure 8-15   Creating a menu module and a menu item

5. Click **File** on the menu bar, click **Save**, and then save the menu module as **NW_MENU.mmb** in the NorthwoodsProject_DONE folder on your Solution Disk.

Creating Menu Selections Within a Menu Item

The next step is to specify the pull-down menu selections that appear on the menu item. Figure 8-16 shows the components of a menu.

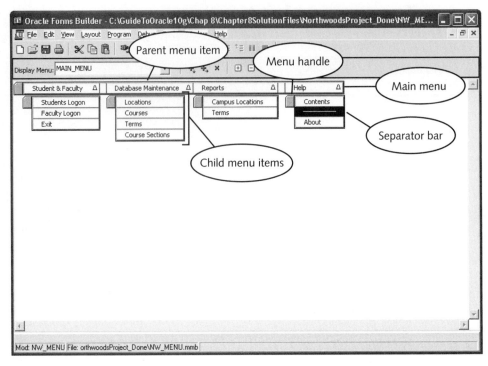

Figure 8-16   Menu components

The **main menu** corresponds to a menu item in the Object Navigator and defines the **parent menu items**, which are the menu selections that appear horizontally on the menu bar. Each parent menu item has an individual menu that contains its **child menu items**, which are the choices that appear when the user clicks the parent menu item. Both the main menu and each individual menu have a **menu handle**, which allows you to detach the menu and move it to a different location in the Menu Editor.

A child menu item can have an associated child submenu. For example, you could click a top-level parent menu item, point to a second-level child menu item to view its menu items, and then select one of the third-level child menu items. You can nest menu items to as many levels as necessary, but it is a good design practice to nest menu items no deeper than three levels.

When the user clicks a child menu item that has no lower levels, a menu code trigger runs. A **menu code trigger** is a PL/SQL program that is associated with a child menu item and contains commands to perform actions such as calling another form, calling a report, or clearing the form.

To create parent and child menus, you use the Forms Builder **Menu Editor**, which allows you to visually define the menu bar structure and create the underlying triggers that execute when a user selects a menu choice. To open the Menu Editor, you double-click the Menu Module icon 🔲 or double-click the Menu Item icon 🔲.

When the Menu Editor first opens, a default parent menu item appears in the top-left corner. You create the new parent menu items from left to right, using the default parent menu item as the starting point. To create the first new parent menu item, you select the default parent menu item and click the Create Right button ⊕ on the Menu Editor toolbar to create the new parent menu item that appears on the right side of the default parent menu item. You can continue to click ⊕ to create parent menu items from left to right across the top of the menu bar.

To create a new child menu item, you select the associated parent menu item, and then click the Create Down button ⊕ on the toolbar to create the first child menu item. To create the next child menu item, you select the child menu item that you just created, and then click ⊕ again. To change the label of a menu item, you select the item, delete the default label text, and type the desired label text. To delete a menu item, you select the item, and then click the Delete button ✖.

Next, you open the Menu Editor and create the parent and child menu items for the Northwoods University database application, as shown in the design sketch in Figure 8-1. (These menu items also appear in the Menu Editor window in Figure 8-16.) Note that the Window selection, which is the far-right parent menu item in the design sketch in Figure 8-1, does not appear in Figure 8-16, and you should not create it. You do not explicitly create the Window menu selection in a menu item because Forms Builder automatically adds this selection at runtime as the far-right parent menu item.

To open the Menu Editor and create the parent and child menu items:

1. Double-click the **Menu Module** icon 🔳 beside NW_MENU. The Menu Editor opens. The default parent menu item named <New_Item> appears.

TIP

> Another way to open the Menu Editor is to select a menu module or menu item, click Tools on the menu bar, and then click Menu Editor.

2. Verify that <New Item> shows a blue background. If it does not, click it until it does, and type **Student & Faculty** to specify the first parent menu item label.

TIP

> If you press Delete when the menu item is selected and has a black background, you delete the entire menu item rather than just the label text.

3. To create a new parent menu item to the right of the Student & Faculty menu item, click the **Create Right** button ⊕ on the Menu Editor toolbar.

4. Type **Database Maintenance** to change the label of the new parent menu item.

5. Select the **Student & Faculty** parent menu item so its background turns black, and then click the **Create Down** button ⚓ on the Menu Editor toolbar to create a child menu item under the Student & Faculty menu item.

6. Verify that the menu item label has a blue background, and type **Students Logon**.

7. Click ⚓ again to create the second child menu item, and change the new child menu item's label to **Faculty Logon**.

8. Click ⚓ again to create another child menu item, and change the new child menu item label to **Exit**.

9. Select the **Database Maintenance** parent menu item, and then create its child menu items, as shown in Figure 8-16.

10. Repeat Steps 3 through 9 to create all of the parent and child menu items in Figure 8-16. Under the Help parent menu item, just create the Contents and About child menu items. You create the separator bar between the Contents and About child menu items later.

11. Click **File** on the menu bar, and then click **Save** to save the menu module.

In the Menu Editor, you can open individual parent menu items to show their associated child menu items, and close individual parent menu items to hide their associated child menu items. When a parent menu item has child menu items, and the child menu items are currently closed, a Closed Tab ▼ appears on the parent menu item. When a menu item has child menu items, and the child menu items are currently opened, an Opened Tab ⏏ appears on the parent menu item. When a menu item does not have child items, no tab appears on the menu item. Now, you open and close individual parent menu items.

To open and close parent menu items:

1. Notice that each parent menu item with underlying child selections currently has an Opened Tab ⏏ beside its label name. Click ⏏ on the Student & Faculty menu item to hide its child menu items. The child menu items no longer appear, and a Closed Tab ▼ appears on the Student & Faculty menu.

2. Click ▼ on the Student & Faculty menu item to reopen the item's child menu items.

Menu Item Properties

Each parent and child menu item has an associated Property Palette in which you specify properties to customize the item appearance and functionality. Table 8-6 summarizes some of the menu item properties that appear under the Functional property node.

Property Name	Description
Enabled	Specifies whether the menu item is enabled or disabled when the menu first opens
Menu Item Type	Determines how the menu item appears in the menu and whether the item can have associated menu code; allowable values are Plain, Check, Radio, Separator, and Magic
Visible in Menu	Specifies whether the menu item is visible or hidden

Table 8-6   Functional property node menu item properties

The menu item property you work with in this section is the Menu Item Type, which defines how the menu item appears on the menu. A menu item can have the following different menu item type values:

- **Plain**—Displays a text label and has an associated menu code trigger that fires when the user selects the menu item

- **Check**—Specifies a property that users can enable or disable. For example, a check menu selection labeled "Toolbar" specifies whether a toolbar appears or is hidden.

- **Radio**—Specifies a selection in a group of menu selections that behave like radio buttons and in which only one selection can be activated at a time. For example, you could create a radio menu selection that defines the names of different report servers that users can use to generate reports.

- **Separator**—Specifies that the menu selection appears as a separator bar, as shown in Figure 8-16.

- **Magic**—Allows you to specify that the menu selection is one of the following predefined magic types: Cut, Copy, Paste, Clear, Undo, About, Help, Quit, or Window. The magic menu items have built-in functionality supplied by Forms Builder.

By default, all menu items are Plain item types unless you specify otherwise. To change a menu item's type property, you create the menu item, open its Property Palette, and modify the Menu Item Type property. Now you create the separator bar in the Help submenu that separates Contents and About in Figure 8-16. You also change the Exit menu item to a magic item type. The Exit item is a Quit magic type, so the form closes when the user selects Exit.

To create the separator bar menu item and modify the Exit menu item type:

1. Select the **Contents** child menu item, and then click the **Create Down** button ⁌ to insert a new child menu item under Contents.

2. Select the new menu item, right-click, and then click **Property Palette** to open the new menu item's Property Palette.

8

You can also open the Property Palette of a menu item by double-clicking the menu item or by selecting the menu item, clicking Tools on the menu bar, and then clicking Property Palette.

3. Change the name property of the menu item to **SEPARATOR_BAR**.

4. Open the **Menu Item Type** property list, select **Separator**, and close the Property Palette.

5. Double-click the **Exit** menu item to open its Property Palette, change its Menu Item Type value to **Magic**, and its Magic Item value to **Quit**. Then close the Property Palette, and save the menu module.

Closing and Reopening Menu Module Files and Individual Menu Items

Recall that Forms Builder stores menu modules separately from form applications. Therefore, to open a menu module, you need to retrieve it from the file system independently of the form in which it is to appear. For example, suppose you decide to take a break and close your menu module file; you will come back to work on it later. In the following steps, you close the menu module and then reopen it.

To close the menu module file and then reopen it:

1. Click the **Close** button on the Menu Editor window, which is the inner window. The Object Navigator appears.

If a dialog box opens asking if you want to save changes to NW_MENU, you closed the Forms Builder window instead of the Menu Editor window. Click Cancel, and then click the Close button on the inner Menu Editor window.

2. To close the menu module, make sure that the NW_MENU node is selected, click **File** on the menu bar, and then click **Close**. If a dialog box opens asking if you want to save your changes, click **Yes**. The NW_MENU menu module object no longer appears under the Menus node.

3. To reopen the menu module, click **File** on the menu bar, click **Open**, open the **Files of type** list, select **Menus (*.mmb)**, navigate if necessary to your NorthwoodsProject_DONE folder, select **NW_Menu.mmb**, and then click **Open**.

4. To reopen the Menu Editor, double-click the **Menu Module** icon 🔳 beside NW_MENU.

Creating Access Keys

Most pull-down menu selections have an underlined letter in the selection label. This letter is called the menu item's **access key**, and allows the user to open or select the menu item by using the keyboard instead of the mouse pointer. In the Forms Services window, users can press the Alt key plus the access key to open parent menu items. Once a user opens a parent menu item, he or she can select a child menu item simply by pressing the child item's access key, without pressing the Alt key.

The first letter of each menu item label is the default access key. If two menu selections have the same first letter, then the access key opens the first selection. For example, in the Database Maintenance parent menu item in Figure 8-16, the Courses and Course Sections child menu items both start with the letter C. If the user opens the Database Maintenance parent menu item, and then presses C, the Courses menu item is selected. Therefore, it is sometimes necessary to change the default access key, because it already has been used or because another key seems more intuitive. For example, the access key for Exit might be X rather than E because X sounds more like Exit. To override the default access key choice, you type an ampersand (&) before the desired access key letter in the menu label. The ampersand does not appear on the menu when the menu appears in Forms Services. To make an ampersand appear within a menu label, you need to type the ampersand twice. For example, to create the menu label Students & Faculty, you would type the label as Students && Faculty.

Next, you add the access key definitions shown in the menu design sketch in Figure 8-1. You also change the label for the Student & Faculty menu item to Students & Faculty Logon by entering the ampersand two times.

To add the ampersand to the Student & Faculty label and specify the menu access keys:

1. In the Menu Editor, open the Student & Faculty menu item for editing by selecting it and then clicking it again, and change the label to Students && Faculty Logon.

2. Open the following menu items for editing, and change their labels as follows. Your modified menu labels should appear as shown in Figure 8-17.

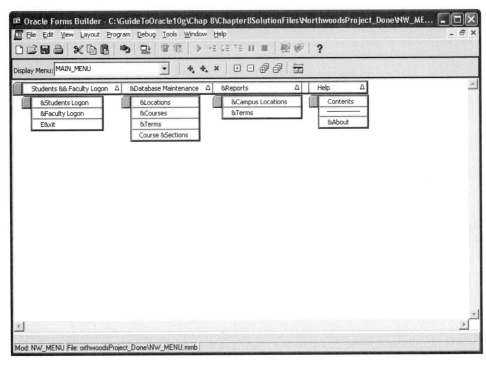

Figure 8-17 Specifying menu access keys

3. Save the menu module.

Creating Menu Code Triggers

Recall that child menu items have associated menu code triggers, which are PL/SQL programs that execute when the user selects the child menu item. Menu code triggers can contain PL/SQL commands that call built-in procedures or functions that do not reference specific form items, such as CLEAR_FORM or LIST_VALUES. Menu code triggers can also contain module navigation commands, such as the CALL_FORM and OPEN_FORM commands to navigate to other forms, and the RUN_REPORT_OBJECT command to run a report object. Menu code commands can reference global variable values, such as the global project path that specifies the location of the application files. However, menu code triggers cannot contain commands that directly reference specific form items, such as GO_ITEM(:Campus Locations_block.loc_id). This limitation exists because the menu is not associated with a specific form, so the form item is not visible to the menu module. If you try to reference a specific form block item in a menu code trigger, a compile error occurs when you compile the menu code trigger. In addition, every child menu item must have an associated menu code trigger, or an error occurs when you compile the menu module.

To create a menu code trigger for a child menu item, you select the menu item in the Menu Editor, right-click, and then click PL/SQL Editor. This opens the PL/SQL Editor, in which you specify the trigger code. Now you implement the menu code triggers for the child menu items that open other forms by executing the CALL_FORM built-in.

To create the menu code triggers that open other forms:

1. In the Menu Editor, right-click the **Students Logon** menu item, and then click **PL/SQL Editor**. The system automatically creates a menu code trigger for the menu item, and you can enter the code that executes when the user selects the menu item.

2. Type the following command to call the Student Logon form. Log onto the database in the usual way, compile the trigger using the Compile PL/SQL code button 🖹, correct any syntax errors, and then close the PL/SQL Editor.

   ```
   CALL_FORM(:GLOBAL.project_path || 'Student_logon.fmx');
   ```

3. Create menu code triggers with the associated code for the following child menu items:

Child Menu Item	Code
Faculty Logon	`CALL_FORM(:GLOBAL.project_path || 'Faculty_logon.fmx');`
Locations	`CALL_FORM(:GLOBAL.project_path || 'Location.fmx');`
Courses	`CALL_FORM(:GLOBAL.project_path || 'Course.fmx');`

4. Compile the triggers using the 🖹 and then save the menu module.

You also need to create a menu code trigger for the Reports Campus Locations child menu item that displays the Campus Locations report. The trigger for the Reports Campus Locations child menu item executes the same commands that you created in the View Campus Locations button in the main application form in Lesson A. (Recall that these commands configure the report object, generate the report output file, and then display the report output file in a browser window.) Next, you open a file named NorthwoodsMain.fmb that contains the completed form that you created in Lesson A. You save the file as Ch8BMain.fmb in the project folder, and copy the commands from the View Campus Locations button trigger to the menu code trigger for the Reports Campus Locations child menu item.

To copy the View Campus Locations button commands to the Reports Campus Locations child menu trigger:

1. Close the Menu Editor. Click **File** on the menu bar, click **Open**, navigate to the Chapter08 folder on your Data Disk, and open **NorthwoodsMain.fmb**. Save the file as **Ch8BMain.fmb** in the NorthwoodsProject_DONE folder on your Solution Disk.

2. Open the Triggers node under the form module, select the PRE-FORM node, right-click, and click **PL/SQL Editor**. If necessary, change the command that assigns the :GLOBAL.project_path variable so it specifies the location of the NorthwoodsProject_DONE folder on your Solution Disk. Then compile the trigger and close the PL/SQL Editor.

3. Click **Tools** on the menu bar, click **Layout Editor**, make sure that **MAIN_CANVAS** is selected, and then click **OK**.

4. Select the **View Campus Locations** button, right-click, and then click **PL/SQL Editor**. Select all of the code in the source code pane, click the **Copy** button 📋 to copy the code, and then close the PL/SQL Editor.

5. Click **Window** on the menu bar, and then click **Object Navigator** to open the Object Navigator. Double-click the **Menu Module** icon 🔳 beside NW_MENU to open the Menu Editor.

6. Select the Reports **Campus Locations** child menu item, right-click, and then click **PL/SQL Editor** to create the menu code trigger. Verify that the insertion point is in the source code pane, and then click the **Paste** button 📋. The copied code appears in the source code pane.

7. Compile the trigger using the Compile PL/SQL code button 🔲 and *not* the Compile Module button 🔲, correct any syntax errors, close the PL/SQL Editor, and save the menu module.

If you used the Compile Module button, at this point you would see a menu compile error message such as the one in Figure 8-18, because every lowest-level child menu item must have a menu code trigger. To complete the menu code triggers, you need to create stubs for the child menu selections that do not yet have menu code. The stubs display a message informing the user that the selection is not yet available. You create stubs for the Terms and Course Sections child menu items under the Database Maintenance parent menu item, the Terms item under the Reports parent menu item, and the Contents and About items under the Help parent menu item.

To create the menu code stubs for the unimplemented menu items:

1. Create a menu code trigger for the **Terms** (under Database Maintenance) menu item by opening the PL/SQL editor and entering the following code:

```
MESSAGE('Selection not yet implemented.');
```

2. Compile the trigger using the Compile PL/SQL code button 🔲, correct any syntax errors, and then close the PL/SQL Editor.

3. Repeat Steps 1 and 2 for the Course Sections, Terms (under Reports), Contents, and About child menu items, and then save the menu module.

DISPLAYING A MENU MODULE IN A FORM

To display a menu module in a form, you first compile the menu module. Then you attach the compiled menu module file to the form by opening the form module Property Palette, and referencing the compiled menu module file in the form module's Menu Module property value. The following sections describe how to accomplish these tasks.

Compiling a Menu Module

When a form module displays a menu module, it uses a compiled version of the menu module file, which has an .mmx file extension. To compile a menu module, you open the menu module in the Menu Editor, and then click the Compile Module button 🖳 on the Forms Builder toolbar. Or, you select the menu module in the Object Navigator, and then click 🖳 . This compiles the menu module and saves the compiled file in the same folder that stores the menu design .mmb file. Every time you modify the menu module, you must recompile the .mmb file into a new .mmx file. Otherwise, the form displays the former .mmx file, and does not show the most recent changes. Now you compile the menu module and generate the executable menu file.

To compile the menu module:

1. In the Menu Editor, click the **Compile Module** button 🖳 . The message "Module built successfully" appears on the status line, and indicates that Forms Builder successfully compiled the menu module.

2. After you successfully compile your menu module, close the Menu Editor.

If a menu compile error message such as the one in Figure 8-18 appears, it indicates that you forgot to create a menu code trigger for one of the child menu items.

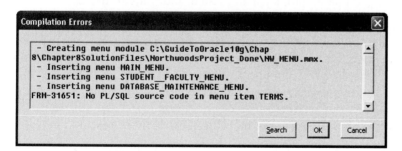

Figure 8-18   Menu module compile error

The "No PL/SQL source code in menu item TERMS" message indicates that the Terms menu item is missing its menu code trigger. Recall that every child menu item must have a menu code trigger. If you encounter a similar error, click OK, add the necessary trigger code under the menu selection indicated in the error message, and then recompile the menu file.

Attaching the Menu Module to the Form

To attach a custom pull-down menu module to a form, you open the form module Property Palette and reference the compiled menu module file in the form module's Menu Module property value. The default value for the form Menu Module property is DEFAULT&SMARTBAR. The DEFAULT specification indicates that the form displays the default Forms Services pull-down menus. The SMARTBAR specification indicates that the Forms Services window displays the **smartbar**, which is the Forms Services toolbar that you use for working with data block forms. Because the main module form is not a data block form, you do not display the smartbar.

There are two approaches for referencing a menu module in a form. The first approach is to save the compiled menu module in the project folder, and change the Menu Module property value to the complete path to the compiled menu module .mmx file, including the drive letter and folder path. This approach is satisfactory when you are developing the menu module, but is not practical when you deploy the finished project to users. The folder path must always reflect the exact location of the menu module in the user's file system. If the user moves the menu files to a new location, the developer must modify the Menu Module property in every form that displays a menu to reflect the new menu file location. (You cannot use the global path variable in the Menu Module property value.)

The second approach is to copy the compiled menu module file to the default form folder, and change the Menu Module property value to the menu module filename. If you place the menu module file in the default form folder, you do not need to include the drive letter or folder path of the menu module file, which makes it easier to move the menu files to a different location in the user's file system. However, you must always ensure that the compiled menu module file is in the default form folder. You should use this approach when the menu module is completed and tested, and ready to deploy to users.

Now you attach the NW_MENU.mmx file to the main application form. Because the project is in the development stage, you reference the compiled menu module that is stored in the project folder. You also change the form module Console Window property to specify that the form module appears in MAIN_WINDOW. You must change the Console Window property when you attach a custom menu to a form, or the status line does not appear on the Forms Services window. Then you run the form and confirm that the menu module appears in the Forms Services window.

To attach the menu module to the main application form and then run the form:

1. In the Object Navigator, select the MAIN form module node, right-click, and then click **Property Palette**.

2. Change the value of the Menu Module property to the full path and filename of your newly generated menu module. If your NorthwoodsProject_DONE

folder is stored on the C drive in the OraSolutions\Chapter08\Tutorials folder, the full path is **C:\OraSolutions\Chapter08\Tutorials \NorthwoodsProject_DONE\NW_MENU.mmx**.

3. To display the status line in the Forms Services window, select the **Console Window** property, open the list, and select **MAIN_WINDOW**. Then close the Property Palette.

You cannot specify the menu module location using the :GLOBAL. *project_path* variable, so if you change the location of the project files, you need to change the menu module path in the form Property Palette.

4. Save the form, and then run the form. The splash screen appears, and then the main application form appears and displays the custom menu selections, as shown in Figure 8-19.

8

Figure 8-19   Main application form with custom menu selections

HELP

If only the "Window" selection appears, close the browser window, then confirm that the NW_MENU.mmx file is in the project folder, and that you specified the path to the NW_MENU.mmx file correctly in the Menu Module property in the MAIN form's Property Palette.

5. Test all of your pull-down menu choices to confirm that they call the correct form or report, and that the "Selection not yet implemented." message appears for unimplemented selections. Note that all of the called forms display the custom menu selections also.

The data block forms (Locations, Items, and Campus Locations) will not work correctly because they require the default Forms Services menu module and the smartbar. You learn how to restore the default menus to these forms later in the lesson.

6. Test all of the pull-down menu choices using the access keys to confirm that the access keys work correctly. Recall that to open a top-level menu selection, you press Alt plus the access key, and that to execute a child-level menu selection, you simply press the access key.

If the Campus Locations report does not appear, make sure that the local report server is installed and started on your workstation. To install and start the local report server, refer to the section titled "Installing and Starting the Local Report Server" in Lesson A.

7. Close the browser window.

Specifying Alternate Pull-down Menus in Called Forms

If you opened the form module Property Palettes of the child forms in the Northwoods University database application (Student_logon.fmb, Faculty_logon.fmb, Location.fmb, Course.fmb, and Location.fmb), you would discover that the Menu Module value is the default value, which is DEFAULT&SMARTBAR. However, every time you call a child form from the main form module, the child form displays the custom menu module that appears in the parent form. To enable a child form to display its own menu module, rather than its parent's menu module, you need to modify the CALL_FORM command that calls the child form. Recall that the syntax for the CALL_FORM command is as follows:

```
CALL_FORM('form_specification', display, switch_menu,
query_mode, parameter_list_id);
```

Recall that *display* specifies whether or not the parent form is hidden when the child form appears. Valid values are HIDE (the parent form is not visible) and NO_HIDE (the parent form remains visible). *Switch_menu* specifies whether the child form inherits its pull-down menus from the parent form, or displays its own menu module, as specified in the child form's Menu Module property. Valid *switch_menu* values are NO_REPLACE (the child form inherits pull-down menus from the parent form) and DO_REPLACE (the child form displays different pull-down menus). Next, you modify the CALL_FORM commands in the menu code triggers that call the application's data block forms, which are the Locations and Courses. You add the HIDE parameter to specify that the parent form is not visible, and you add the DO_REPLACE parameter to specify that the child form displays its own menus. Then you run the form and confirm that the child forms display their own menus.

To modify the CALL_FORM commands in the menu module, and then run the form:

1. In the Object Navigator, double-click the **Menu Module** icon ⊞ beside **NW_MENU** to open the Menu Editor.

2. Select the **Locations** menu selection, right-click, and then click **PL/SQL Editor** to open the menu code trigger. Add the following boldface code to the command, and then compile the trigger, correct any syntax errors, and close the PL/SQL Editor.

   ```
   CALL_FORM(:GLOBAL.project_path || 'Location.fmx', ↵
   HIDE, DO_REPLACE);
   ```

3. Repeat Step 2 for the Courses menu selections.

4. Save the menu module, compile the menu module, and then close the Menu Editor.

HELP

If an error message stating "Error writing file NW_MENU.mmx" appears, press Ctrl+Alt+Delete, click Task Manager to open the Task Manager, select the Processes tab, select the ifWeb90.exe process, and then click End Process.

5. Select the **MAIN** form module in the Object Navigator, and then run the form. Click **Database Maintenance** on the menu bar, and then click **Locations**. The Locations form appears, and shows the default Forms Services window menu selections and toolbar buttons. Click **Return** to return to the main application form.

6. Close the browser window.

8

USING PROGRAM COMMANDS TO CONTROL MENU ITEMS

Forms Builder provides built-in procedures to access and modify menu properties while a form is running. Table 8-7 summarizes the Forms Builder menu built-ins.

Built-in Name	Description	Example
FIND_MENU_ITEM	Returns the object ID for the specified menu item	`FIND_MENU_ITEM ('NEW_ORDERS');`
GET_MENU_ITEM_ PROPERTY	Retrieves the current value of a menu item, such as the label or whether the item is enabled or visible	`GET_MENU_ITEM_PROPERTY ('NEW_ORDERS', LABEL);`
REPLACE_MENU	Replaces the current menu module in all application windows with a different menu module	`REPLACE_MENU ('ALT_CW_MENU');`
SET_MENU_ITEM_ PROPERTY	Dynamically modifies the property of a menu item	`SET_MENU_ITEM_ PROPERTY('NEW_ORDERS', LABEL, 'New Sales Orders');`

Table 8-7   Forms Builder menu built-in programs

To use these built-ins, you must reference menu items using their system-assigned names. When you create new menu items in the Menu Editor, Forms Builder automatically assigns descriptive names to the items. You can determine the system-assigned names of individual menu items by viewing the menu structure in the Object Navigator. For example, Figure 8-20 shows the structure of the Students & Faculty Logon menu item in the Object Navigator. Forms Builder assigned the name STUDENT_FACULTY_MENU to the parent Students & Faculty Logon menu item, and assigned names to the child menu item names as shown.

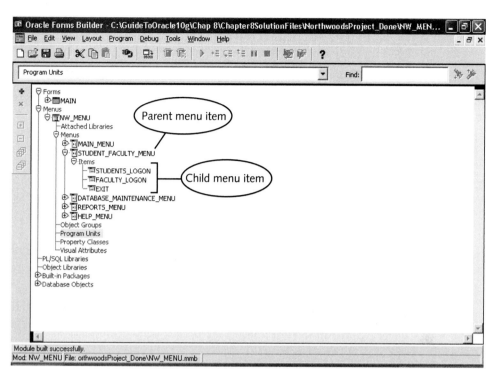

Figure 8-20   System-assigned menu item names

The SET_MENU_ITEM_PROPERTY built-in is useful for dynamically changing menu item properties, such as disabling a menu item. This built-in has the following general syntax:

```
SET_MENU_ITEM_PROPERTY('menu_name', property, value);
```

This syntax uses the following parameters:

- *Menu_name* is the menu item name, specified in the format *parent_menu_name.menu_item_name*. Note that the value is enclosed in single quotation marks. For example, you would use the expression 'STUDENT_ FACULTY_ MENU.FACULTY_LOGON' to reference the Faculty logon menu item.

- *Property* is the name of the property to be modified. This value might be different from the property name in the Property Palette, so you should open the Forms Builder online Help system to determine the exact syntax for referencing the property that you want to modify.

- *Value* is the desired property value.

Now you use the SET_MENU_ITEM_PROPERTY built-in to change a menu property dynamically at runtime. Suppose that a user opens the Student logon form for the Northwoods University database application, and then opens the Students & Faculty

Logon parent menu item, and selects Students Logon. This action opens the Student Logon form again, which does not make sense in the application, and causes the Student Logon form to open multiple times, which consumes workstation memory. To avoid this error, you disable the Students Logon menu item when the user opens the Student Logon form.

To disable a menu item in a specific form, you execute the SET_MENU_ ITEM_PROPERTY built-in in the form's PRE-FORM trigger. In the Northwoods University database application, you configure the built-in to set the Students Logon menu item's Enabled property to False. You also modify the form's Return button to reenable the menu item when the user exits the form.

To modify the Student Logon form to disable and then reenable the Students Logon menu item:

1. In the Object Navigator, select the MAIN form module node, and then click the **Collapse All** button 🗗 to close the form components.

2. Select the **NW_MENU** node, and then click 🗗.

3. To confirm the name of the Students Logon menu item and its parent menu item, open the **NW_MENU** node, open the **Menus** node, open the **STUDENT_FACULTY_MENU** node, and then open the **Items** node to view the **STUDENTS_LOGON** node. Your screen should appear similar to Figure 8-20.

Your system-assigned menu item names may be slightly different.

4. Open the **Student_logon.fmb** form from the NorthwoodsProject_DONE folder on your Solution Disk.

5. Select the Triggers node under the STUDENT_LOGON node, right-click, point to **SmartTriggers**, and then click **PRE-FORM**. The PL/SQL Editor opens.

6. To disable the Students_Logon menu item in the form, add the following command to the trigger:

```
SET_MENU_ITEM_PROPERTY('STUDENT_FACULTY_MENU.STUDENTS_↵
LOGON', ENABLED, PROPERTY_FALSE);
```

7. Compile the trigger, correct any syntax errors, and then close the PL/SQL Editor.

8. In the Object Navigator, open the Data Blocks node under the STUDENT_LOGON form module, open the TEMPLATE_BLOCK node, open the Items node, select the EXIT_BUTTON node, right-click, and then click **PL/SQL Editor**. The Exit button trigger opens in the PL/SQL Editor.

9. To reenable the Students Logon menu item when the form closes, add the following command as the first line of the trigger:

```
SET_MENU_ITEM_PROPERTY('STUDENT_FACULTY_MENU.STUDENTS_
LOGON', ENABLED, PROPERTY_TRUE);
```

10. Compile the trigger, correct any syntax errors, and then close the PL/SQL Editor.

Before you can test the commands to confirm that they disable and then reenable the Students Logon menu selection, you must recompile the Student Logon form module. Recall that the CALL_FORM command in the main application form calls the Student_logon.fmx file. If you modify the Student_logon.fmb form, but don't recompile the form, then the main application form will call the existing Student_logon.fmx file, which does not contain your recent changes. Forms Builder automatically generates this file when you run the Student_logon.fmb form. Or, you can recompile the form module by selecting the module in the Object Navigator, and then clicking the Compile Module button 🔲 on the Forms Builder toolbar. Now you recompile the STUDENT_LOGON form module, and then run the MAIN application and confirm that the menu item is disabled and then reenabled.

To recompile the STUDENT_LOGON form, and then run the MAIN application and confirm that the menu item is disabled and then reenabled:

1. Select the STUDENT_LOGON node, and then click the **Compile Module** button 🔲 on the toolbar. Forms Builder recompiles the form module.

2. Select the MAIN node, and then run the form. The splash screen appears, and then the main application form opens.

3. Click **Students & Faculty Logon** on the menu bar, and then click **Students Logon**. The Student Logon form opens.

4. Click **Students & Faculty Logon** on the menu bar, and note that the Students Logon selection is now disabled.

HELP   If an error message with the text "Cannot find menu item: Invalid ID" appears, it is because you did not correctly specify a menu item name in the SET_MENU_ITEM_PROPERTY command. Review the commands you added to the PRE-FORM and RETURN_BUTTON triggers, and make sure *menu_item.menu_name* exactly matches the menu name values that appear in your menu module's Property Palette.

5. Click **Return** on the Student logon form to return to the main form application.

6. Click **Students & Faculty Logon** on the menu bar, and note that the Students Logon selection is enabled again.

7. Close the browser window.

8. In Forms Builder, close the STUDENT_LOGON and MAIN form modules, and the NW_MENU menu module. There should be no open items in the Object Navigator.

POP-UP MENUS

Recall that pop-up menus are context-sensitive menus that appear when the user right-clicks a specific item displayed on the screen. In Forms Builder, pop-up menus are top-level form objects. You associate a pop-up menu with a specific form, and the pop-up menu can appear only in that form. The following sections describe how to create pop-up menus and attach pop-up menus to form items. A pop-up menu is form specific, whereas a pull-down menu module is an independent object that can appear in any form.

Creating a Pop-up Menu

To create a pop-up menu, you select the Popup Menus node in the Object Navigator, and then click the Create button ➕ to create a new Popup menu object. Then you open the object in the Menu Editor, define the menu items, change the menu item labels, and create the menu code triggers.

Now you create a pop-up menu in the Faculty Logon form in the Northwoods University database application. Currently, the Faculty Logon form allows the user to select Faculty ID values using an LOV that appears when the user places the insertion point in the Faculty ID text item and then presses Ctrl+L. As an alternative, you create a pop-up menu that displays the menu selection "List Values" when the user right-clicks the Faculty ID text item. If the user clicks List Values, the LOV display for the Faculty ID LOV opens. First, you open the Faculty Logon form (Faculty_Logon.fmb) in Forms Builder. Then, you create the pop-up menu object, change its name, and specify its menu item label and trigger code.

To create and configure the pop-up menu:

1. Open **Faculty_Logon.fmb** from the NorthwoodsProject_DONE folder on your Solution Disk.

2. Select the Popup Menus node, and then click the **Create** button ➕ to create a new pop-up menu object. Change the object name to **FACULTY_ID_MENU**.

3. Double-click the **Menu Item** icon ▦ beside FACULTY_ID_MENU to open the pop-up menu object in the Menu Editor. The Menu Editor opens, and the default <New Item> menu item appears.

4. Change the <New Item> label to **List Values**.

5. To create the menu code trigger, select the **List Values** menu item, right-click, and then click **PL/SQL Editor**. Type the following command in the source code pane:

```
LIST_VALUES;
```

6. Compile the trigger, correct any syntax errors, and then close the PL/SQL Editor.

7. Close the Menu Editor, and save the form.

Attaching a Pop-up Menu to a Form Object

You can attach a pop-up menu object to a form canvas or to a data block item, such as a text item, command button, or check box. To assign the pop-up menu to a form object, you open the Property Palette of the object to which you want to attach the menu, and change the object's Popup Menu property value to the name of the associated pop-up menu object.

Now you attach the pop-up menu to the F_ID text item. You open the Property Palette of the F_ID text item, and change its Popup Menu property to the name of the pop-up menu, which is FACULTY_ID_MENU. Then you run the form and test the pop-up menu.

To attach the pop-up menu to the form item, and then run the form:

1. Make sure that the FACULTY_LOGON node is selected in the Object Navigator, click **Tools** on the menu bar, and then click **Layout Editor**.

2. Double-click the **F_ID** text item to open its Property Palette.

3. In the Functional property node, open the **Popup Menu** property list, and select **FACULTY_ID_MENU** to attach the pop-up menu to the text item.

4. Close the Property Palette, save the form, and then run the form. The form opens in the browser window.

5. Right-click the **F_ID** text item. The pop-up menu appears, and displays the selection List Values.

6. Click **List Values**. The LOV display for the Faculty ID LOV opens.

7. Click **Cancel** (scroll down if necessary), and then close the browser window.

8. Close the FACULTY_LOGON form module in Forms Builder, then close Forms Builder and all other open applications.

9. Shut down the OC4J Instance.

You have now learned how to design and create individual form and report modules, and how to combine these modules into an integrated database application. Chapters 9 and 10 describe how to enhance the functionality of your applications using stored PL/SQL programs and advanced Developer10*g* features.

SUMMARY

- Integrated database applications usually have pull-down menus, which appear on the menu bar, and pop-up menus, which appear when the user right-clicks a specific form item with the mouse pointer.

- To define custom pull-down menus in a form, you create a menu module, which is independent of any specific form and is stored in the file system as a file with an .mmb extension. A menu module contains one or more menu items, which are sets of menu selections that appear on the menu bar.

- A menu module contains parent menu items, which are the menu selections that appear on the menu bar, and child menu items, which are the choices that appear when the user clicks a parent menu item. A menu item has a Property Palette on which you can specify how the menu item appears and functions.

- Menu items have access keys, which allow the user to open or select the menu item using the keyboard instead of the mouse pointer. The first letter of each menu item label is the default access key. Users can press the Alt key plus the access key to open parent menu items. Users can simply press the access key to select child menu items.

- Every child menu item must have an associated menu code trigger, which is a PL/SQL program that runs when the user selects the item. Menu code triggers can contain built-in commands, which cannot reference specific form items, and module navigation commands. Menu code triggers can also reference global variable values.

- To attach a pull-down menu module to a form, you specify the full path to the menu module's compiled .mmx file as the Menu Module property value in the form Property Palette. Whenever you modify a menu module design (.mmb) file, you must rebuild it so the modifications are reflected in the menu module executable (.mmx) file.

- By default, child forms display the menu module of their parent form. To display a different menu module in a called form, you specify the new menu module filename in the form's Menu Module property, and then call the form using the DO_REPLACE option in the CALL_FORM command.

- Forms Builder provides several built-in procedures that allow you to modify menu properties while a form is running. You can use the SET_MENU_ITEM_PROPERTY built-in to enable and disable menu items dynamically.

- You can create context-sensitive pop-up menus that appear when a user right-clicks a form item or canvas. To create a pop-up menu, you create a Popup Menu object in the form, then create its menu items and associated menu code triggers. You assign a pop-up menu object to an item's Popup Menu property in the target item's Property Palette.

REVIEW QUESTIONS

1. Each menu module in a form can contain only one menu item. True or False?

2. The menu item in the Object Navigator defines the _____ menu items that appear across the menu bar.

3. What are the valid menu item types for a menu item?

4. A menu module is stored in the same file as its associated form application. True or False?

5. What is the default access key for a menu item?

6. A compiled version of a menu module file has _____ as its filename extension.

7. Use the _____ Editor to create parent and child menus.

8. A menu code trigger cannot directly reference a specific form item. True or False?

9. Why do you need to include the Message command for a child menu item that has not been implemented?

10. What is the overall procedure for creating a pop-up menu for a specific form item?

MULTIPLE CHOICE

1. A menu module is saved in a nonexecutable design file having _____ as the file extension.

   a. mmx

   b. mmb

   c. fmb

   d. fmx

2. You can use the _____ to move a parent or child menu to a different location in the Menu Editor.

   a. menu handle

   b. menu trigger

   c. resizing handles

   d. move pointer

3. By default, all menu items are assigned the _____ item type.

   a. Check

   b. Magic

   c. Plain

   d. Radio

4. Which of the following symbols is used to designate a menu access key?

    a. *

    b. /

    c. &

    d. ^

5. To associate a compiled menu module to a form, reference the name of the compiled file as the form's _____ property.

    a. Menu Module

    b. Associate

    c. Lookup

    d. Compiled Reference

6. Which CALL_FORM parameter determines whether the child form inherits its parent's pull-down menus?

    a. display

    b. switch_menu

    c. query_mode

    d. parameter_list_id

7. When the main application form calls a form file for display, what is the extension for the called file?

    a. .fmb

    b. .err

    c. .fmx

    d. .cmp

8. The default value for the Menu Module property is _____.

    a. DEFAULT

    b. SMARTBAR

    c. MAIN

    d. DEFAULT&SMARTBAR

9. Pop-up menus are _____ form objects.

    a. child-level

    b. top-level

    c. menu-level

    d. node-level

10. To open a parent menu item, you must:

   a. Press the access key.

   b. Press the Ctrl key plus the access key.

   c. Press the Esc key.

   d. Press the Alt key plus the access key.

PROBLEM-SOLVING CASES

The cases refer to the Clearwater Traders (Figure 1-25) sample database. Place all solution files in the specified folder in your Chapter08\Cases folder.

NOTE To complete the following cases, you must have completed the cases in Lesson A of this chapter. The following cases must be completed in sequence.

NOTE You also must have created the o_id_sequence in a previous exercise. If you have not, do so now with the following command in SQL*Plus:

```
CREATE OR REPLACE SEQUENCE o_id_sequence
START WITH 500;
```

1. In this case, you create a new menu module that displays the pull-down menu selections for the Clearwater Traders integrated database system. Figure 8-21 shows the menu design.

Sales/Receiving	Database Maintenance	Reports	Help	Window
New Orders (Sales.fmb)	Customer (Customer.fmb)	Inventory (Inventory.rdf)	Help Topics	
Receiving (Receiving.fmb)	Items (Item.fmb)	Order Source Revenue	-----------	
Exit	Inventory (Inventory.fmb)	(not implemented)	About	
	Colors (not implemented)			
	Order Sources (not implemented)			

Figure 8-21

   a. Create a new menu module named **CWMENU**, and save the module as **CWMenu.mmb** in the ClearwaterProject_DONE folder that you created in Case 1, Lesson A, of this chapter. Create a menu item named **MAIN_MENU**, and then create the parent and child menu items shown in Figure 8-21. Specify the menu access keys as shown.

   b. Create menu code triggers that enable the child menu items to call the associated target files. Use the :GLOBAL.project_path variable in the commands. Create stubs that display the message "**Selection not yet implemented.**" for the Colors, Order Sources, Order Source Revenue, Help, and About child menu items.

   c. Configure the Exit child menu item as a magic menu item that automatically exits the form, and then save the menu.

2. Configure the CALL_FORM commands in the menu and switchboard triggers so that the New_Orders.fmb and Receiving.fmb forms inherit the menus on the parent form; and the Customer.fmb, Items.fmb, and Inventory.fmb forms display their own menu modules. Save the modified menu module.

3. Modify the CWMENU module and the main application form (Main.fmb form created in Lesson A) so that when the user opens either the New_Orders.fmb or Receiving.fmb forms from the switchboard or menu, the menu item that calls the form is disabled. Save the modified menu module.

4. Open the Receiving.fmb form and create a pop-up menu named **SHIPMENT_ID_MENU** for the SHIP_ID text item. The menu should contain an item called **List Values** that will access the LOV display for that text item. Save the modified **Receiving.fmb** form.

5. Open the **Main.fmb** form created in Lesson A and attach the completed CWMenu menu module to the main application form. Run the form and test the access keys and make certain the correct form/report is attached to the menu items.

CHAPTER 9

ADVANCED SQL AND PL/SQL TOPICS

◀ LESSON A ▶

After completing this lesson, you should be able to:

♦ Create and use indexes

♦ Work with PL/SQL stored program units

♦ Create server-side stored program units in SQL*Plus

♦ Use Forms Builder to create stored program units

You have learned how to write SQL commands to create database objects and insert, update, delete, and view Oracle 10*g* database data. You have also learned how to write PL/SQL programs to manipulate database data, and how to use Forms Builder and Reports Builder to create database applications. This chapter describes advanced SQL and PL/SQL concepts that improve the performance of your database applications, and make database applications easier to implement and maintain.

DATABASE INDEXES

An index in a book contains topics or keywords, along with page references for the locations of materials associated with each topic or keyword. Although you could search through every page in a book to find a particular topic, consulting the index speeds up the process considerably. Similarly, a **database table index** is a distinct database table that contains data values, along with a corresponding column that specifies the physical locations of the records that contain those data values.

Consider the queries in Figure 9-1 that retrieve records from the Northwoods University COURSE_SECTION table.

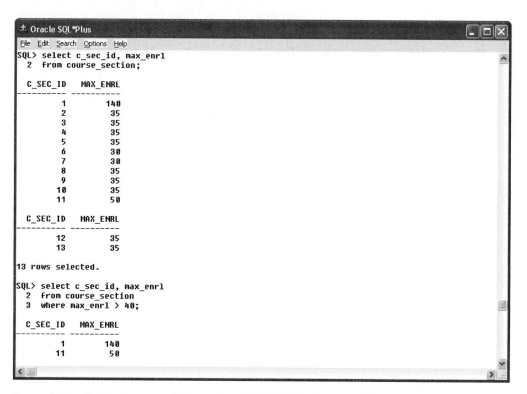

Figure 9-1   Retrieving records from the COURSE_SECTION table

The first query shows the C_SEC_ID and MAX_ENRL fields for all table records, and the second query shows the records in which the MAX_ENRL value is greater than 40. To process the second query, the DBMS must read every record in the table, and then retrieve the ones in which the MAX_ENRL value is greater than 40. In actual practice, the COURSE_SECTION table might contain thousands or tens of thousands of records, so this process might take a long time.

To speed up data retrievals on queries that contain search conditions, you can create a database table index on the search condition fields. Oracle 10*g* bases database table indexes on a table's ROWID field. Every Oracle 10*g* database table contains a field named **ROWID**, which specifies the internal location of the record in the database. Oracle 10*g* encodes ROWID values using an internal data format in which the values specify the physical location of the row within the database. You can retrieve a table's ROWID values by explicitly specifying the ROWID fieldname in the SELECT clause. Figure 9-2 shows the ROWID values for the COURSE_SECTION table.

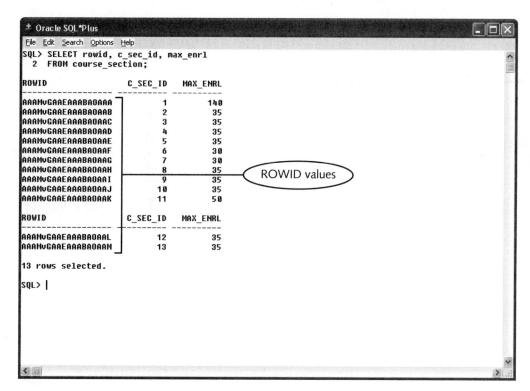

Figure 9-2   COURSE_SECTION table ROWID field values

If you create a SQL query that selects all of the rows in a database table using the syntax SELECT *, the database does not retrieve the values in the ROWID field.

When you create an index on a specific table field, Oracle 10*g* creates a table that stores the ROWID values in one field and the sorted indexed field values in another field. For example, if you create an index on the MAX_ENRL field in the COURSE_SECTION table, the database table index contains the ROWID field and the MAX_ENRL field, sorted in ascending order. Figure 9-3 shows the values that the index would contain.

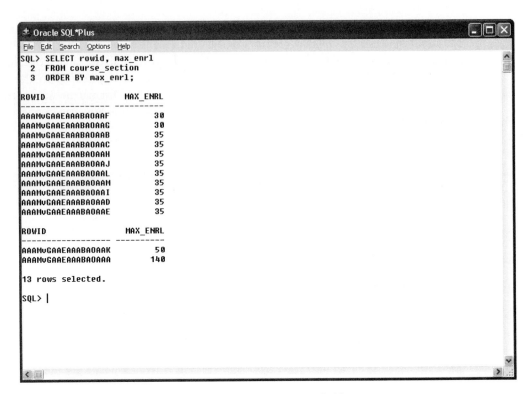

Figure 9-3   Contents of an index on the MAX_ENRL field

When you execute a query that uses an indexed field in a search condition, the DBMS queries the database table index, finds the ROWID values for the required records, and then retrieves the desired records based on the ROWID values. Because the index contains sorted values, the DBMS can search the index more quickly than it can search the original table, and, therefore, it retrieves the records faster.

NOTE

Oracle 10*g* usually stores index values internally in a data structure called a B-tree, which speeds up the search process. A B-tree is an inverted tree that uses a divide-and-conquer search strategy. It has search nodes that contain values that direct the search process and leaf nodes that contain the actual data values. A discussion of the mechanics of a B-tree is beyond the scope of this book, but you can find information on B-trees in most computer science books that describe data structures.

When you create a table, Oracle 10*g* automatically creates an index on the table's primary key. To improve query performance, you should create indexes on columns that users often use in search conditions. A table can have an unlimited number of indexes. However, each time a user inserts, updates, or deletes a record in an indexed table, the DBMS automatically updates the index as well as the original table. This places additional processing overhead on the DBMS.

Creating an Index

You should create an index after you insert all of the data into a table rather than when you first create the table, because it is more efficient for the DBMS to create index entries for all table records at one time, rather than to populate the index for each individual record as the user inserts it. To create an index, you execute the following command in SQL*Plus:

```
CREATE INDEX index_name
ON tablename (index_fieldname);
```

To make it easy to recognize and identify indexes, *index_name* should use the naming convention *tablename_fieldname*. For example, the name of the index on the MAX_ENRL field in the COURSE_SECTION table would be COURSE_SECTION_MAX_ENRL.

TIP

If the index-naming convention makes the object name violate the Oracle naming standard by exceeding 30 characters, you should abbreviate the table and field names.

Next you start SQL*Plus, refresh the case study database tables, and create an index named COURSE_SECTION_MAX_ENRL to index values in the MAX_ENRL field.

To create the index:

1. Start SQL*Plus, and connect to the database.

2. Run the **Ch9Northwoods.sql** script file in the Chapter09 folder on your Data Disk.

3. Start Notepad, and save the new Notepad file as **Ch9AQueries.sql** in the Chapter09\Tutorials folder on your Solution Disk.

4. Type the command in Figure 9-4 to create the index, then copy the command, paste it into SQL*Plus, and execute the command. The confirmation message "Index created" confirms that the DBMS successfully created the index.

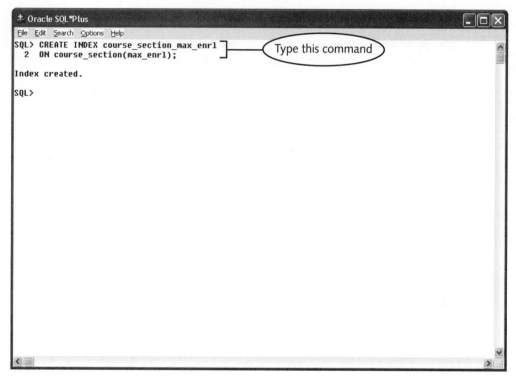

Figure 9-4 Creating an index

NOTE

For the remainder of this lesson, type all SQL and PL/SQL commands in the Ch9AQueries.sql file, copy the commands, and then paste and execute the commands in SQL*Plus.

Creating Composite Indexes

A **composite index** is a database index table that contains multiple sorted columns (up to 16) that the DBMS can use for identifying the row location. You create a composite index to speed up retrievals for queries that contain multiple search conditions joined by the AND or OR operators. Consider the contents of the Northwoods University ENROLLMENT table, which the first query in Figure 9-5 retrieves.

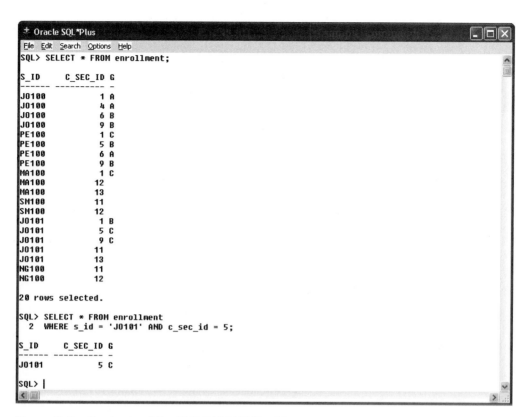

Figure 9-5   Contents of the ENROLLMENT table

In the ENROLLMENT table, the S_ID column stores the student ID values, the C_SEC_ID stores the identifier for each section of a course, and the GRADE field (which is truncated to G in the query output) indicates the grade a student received in a course, if any. Suppose that users at Northwoods University often query the ENROLLMENT table to retrieve the grade a student received in a particular course. The second query in Figure 9-5 retrieves the grade value for student JO101 in the course associated with C_SEC_ID of 5.

To speed up retrievals for this query, you would create a composite index based on the query search conditions. The first query search condition, which is on the S_ID column, is called the **primary search field**. The second search condition, which is on the C_SEC_ID column, is called the **secondary search field**. The index stores the ROWID values, along with the sorted primary and secondary search columns. Figure 9-6 shows the values for a composite index on the ENROLLMENT table, based on the S_ID and C_SEC_ID column.

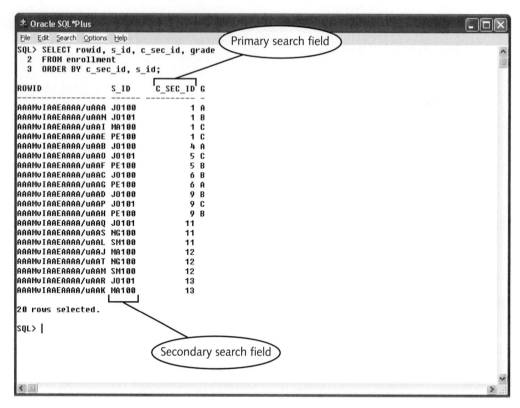

Figure 9-6   Contents of a composite index on the ENROLLMENT table

Note that the index structure makes it very easy for the DBMS to locate a specific S_ID value, and then to locate a specific C_SEC_ID value for that S_ID value.

You use the following syntax to create a composite index:

```
CREATE INDEX index_name
ON tablename(index_fieldname1, index_fieldname2, …);
```

Next, you create the composite index based on the S_ID and C_SEC_ID columns in the ENROLLMENT table.

To create the composite index:

1. Type and execute the command in Figure 9-7 to create the composite index.

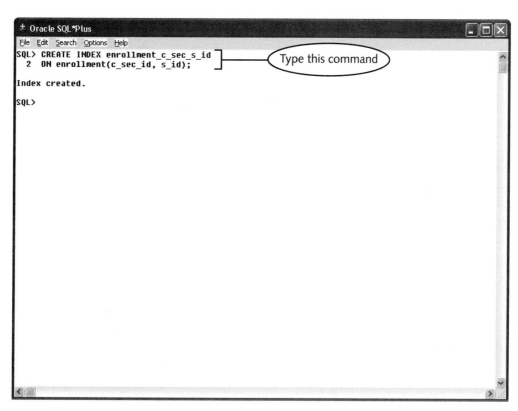

Figure 9-7   Creating a composite index

2. The message "Index created" confirms that the DBMS successfully created the index.

Viewing Index Information Using the Data Dictionary Views

Recall that you can query the data dictionary views to retrieve information about database objects. You can retrieve information about your indexes using the USER_INDEXES data dictionary view. Next, you retrieve the names of the indexes in your user schema.

To retrieve index information:

1. Type and execute the query in Figure 9-8 to retrieve the names of all of the indexes in your user schema.

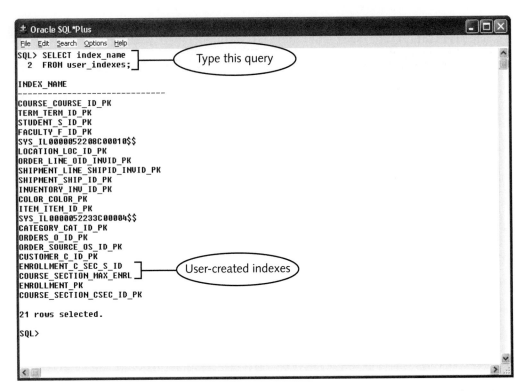

Figure 9-8   Querying the USER_INDEXES data dictionary view (partial output shown)

Your output may be different depending on the tables and indexes that you have created or dropped.

If the column headings for the query repeat, you need to adjust the pagesize for your SQL*Plus environment. Click Options on the menu bar, click Environment, select pagesize from the Set Options list, select the Custom option button, type 80 in the Value field, click OK, and then repeat Step 1.

Dropping an Index

You might need to drop an existing index if applications no longer use queries that are aided by the index or if the index does not improve query performance enough to justify the overhead it creates on insert, update, and delete operations. To drop an index, you execute the following command:

```
DROP INDEX index_name;
```

Now you drop the ENROLLMENT_C_SEC_S_ID index that you created earlier.

To drop an index:

1. Type and execute the following command:

```
DROP INDEX enrollment_c_sec_s_id;
```

2. The message "Index dropped" confirms that the DBMS successfully dropped the index.

Determining When to Create an Index

The need for an index depends on many factors, such as the speed of the database server hardware, the speed of the network, the number of users, and the types of queries that users perform.

In general, Oracle Corporation recommends creating an index on a table field when all of the following conditions occur:

- The table contains a large number of records (a rule of thumb is that a large table contains over 100,000 records).
- The field contains a wide range of values.
- The field contains a large number of null values.
- Application queries frequently use the field in a search condition or join condition.
- Most queries retrieve less than 2% to 4% of the table rows.

In general, Oracle Corporation recommends not creating an index on a table field when one of the following conditions occurs:

- The table does not contain a large number of records.
- Applications do not use the proposed index field in a query search condition.
- Most queries retrieve more than 2% to 4% of the table records.
- Applications frequently insert or modify table data.

The decision to create an index is based on judgment, experience, and running benchmark tests. When you develop database applications, it is a good practice to estimate the maximum number of records that each table might contain, and then create and load enough sample data records so you can execute sample queries, identify performance bottlenecks, and determine where indexes are needed.

OVERVIEW OF PL/SQL STORED PROGRAM UNITS

A program unit is a self-contained group of program statements that can be used within a larger program. (Recall that you created program units in Forms Builder for blocks of code that may be called by multiple form triggers.) It is a good practice to decompose a complex program into smaller program units, because it is easier to conceptualize, design, and debug a small program unit than a large, complex program. When all of the smaller program units work correctly, you can link them into the large program. In addition to making the programming process easier, programming units save valuable programming time because you can reuse them in multiple database applications.

So far in this book, all of the PL/SQL programs that you have created and executed in SQL*Plus are **anonymous PL/SQL programs**, which are programs that you submit to the PL/SQL interpreter and run, but that do not interact with other program units. (You also used PL/SQL commands within form triggers and form program units, but these are not anonymous PL/SQL programs because they exist within a form.) In this chapter, you learn to create stored PL/SQL program units. **Stored PL/SQL program units** are program units that other PL/SQL programs can reference, and that other database users can execute. Stored program units can receive input values from other program units, and pass output values to other program units. You can link stored program units to database application components you create in Forms Builder and Reports Builder to enhance their functionality.

A PL/SQL stored program unit can run on either the database server or on the client workstation. **Server-side program units** are stored in the database as database objects and execute on the database server. **Client-side program units** are stored in the file system of the client workstation and execute on the client workstation. The advantage of using server-side program units is that they are stored in a central location that is accessible to all database users. And, they are always available whenever a user makes a database connection. Users can easily access and run server-side program units without having to locate files on remote workstations. The disadvantage of using server-side program units is that they place all processing on the database server. If the database server is very busy, this can result in applications with very slow response times. If you are creating a program unit that only you or a few co-located users use, and system performance is too slow, then you should create a client-side program unit.

Table 9-1 summarizes the different types of Oracle stored program units, and shows where they are stored and where they execute in the client/server architecture.

Program Unit Type	Description	Where Stored	Where Executed
Procedure	Can accept multiple input parameters, and return multiple output values	Database	Server-side
Function	Can accept multiple input parameters, and can return a single output value	Database	Server-side
Library	Contains code for multiple related procedures or functions	Operating system file	Client-side
Package	Contains code for multiple related procedures, functions, and variables and can be made available to other database users	Database	Server-side
Database trigger	Contains code that executes when a user inserts, updates, or deletes records	Database	Server-side

Table 9-1   Types of Oracle 10g stored program units

The rest of this chapter describes these program unit types in detail, and shows how to create all of these types of stored program units in both SQL*Plus and Forms Builder.

CREATING STORED PROGRAM UNITS

Stored program units can be either procedures or functions. A procedure is a program unit that can receive multiple input parameters and return multiple output values or return no output values. A procedure can also perform an action such as inserting, updating, or deleting database records. A function is a program unit that can receive multiple input parameters, and always returns a single output value. A function works well when you want to compute a single value, because you can assign the return value of the function to a variable within the calling program. A procedure works well when you want to manipulate the values of several variables.

TIP

Stored program units are also called stored procedures, regardless of whether they are procedures or functions.

The following sections describe how to use SQL*Plus to create stored program unit procedures and functions.

Stored Program Unit Procedures

You use the **CREATE PROCEDURE** command to create a stored program unit that is a procedure. Figure 9-9 shows the command's general syntax.

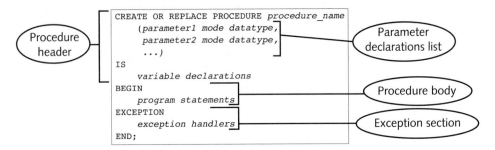

```
CREATE OR REPLACE PROCEDURE procedure_name
    (parameter1 mode datatype,
     parameter2 mode datatype,
     ...)
IS
    variable declarations
BEGIN
    program statements
EXCEPTION
    exception handlers
END;
```

Procedure header

Parameter declarations list

Procedure body

Exception section

Figure 9-9   Syntax to create a stored program unit procedure

Recall that an anonymous PL/SQL program block has three sections: declaration, body, and exception. Instead of a declaration section, a stored program unit has a header section. The **header** defines the program unit name, specifies the parameters that the program unit receives or delivers, and declares the procedure or function variables. The header begins with the **CREATE OR REPLACE PROCEDURE** command. This command instructs the DBMS to create a new procedure or replace an existing procedure. The OR REPLACE clause is optional, but it is a good practice to include it, because an error occurs if you attempt to create a procedure that has the same name as an existing procedure.

Procedure_name defines the name of the program unit. *Procedure_name* must be a unique name within the user's database schema, and must adhere to the Oracle naming standard. The following command defines a procedure named calc_GPA, which calculates and returns the grade point average (GPA) of a student at Northwoods University:

    CREATE OR REPLACE PROCEDURE calc_GPA

Procedures receive input values and deliver output values using parameters, which are variables that pass information from one program to another. The next item in the header defines the program unit **parameter declarations list**, which specifies the program unit parameters and declares their data types. The IS keyword follows the parameter list. You can insert additional variable declarations after the IS keyword to define local variables within the program unit.

The keyword BEGIN marks the beginning of the program unit body, and the keyword EXCEPTION marks the beginning of the exception section. The body specifies the program unit commands, and the exception section specifies the program unit exception handlers. In a stored program unit, the header and body sections are required, and the exception section is optional. The END keyword followed by a semicolon signals the end of the program unit.

Creating the Parameter Declarations List

The program unit parameter declarations list defines the parameters and declares their associated data types. You enclose the parameter declarations list in parentheses, and separate each individual parameter with a comma. *Parameter1, parameter2,* and so on, define the names of each parameter, and must be legal PL/SQL variable names. Note that in the header, you indent the parameter declarations so they align vertically.

Mode defines the **parameter mode**, which describes how the program unit can change the parameter value. Allowable *mode* values are:

- **IN**—Specifies a parameter that is passed to the program unit as a read-only value that the program unit cannot change. If you declare a parameter and omit the *mode* specification, the parameter uses the IN mode by default.

- **OUT**—Specifies a parameter that is a write-only value that can appear only on the left side of an assignment statement in the program unit.

- **IN OUT**—Specifies a parameter that is passed to the program unit, and whose value can also be changed within the program unit.

In the parameter declarations list, *datatype* specifies the data type of each parameter, and can be any legal Oracle 10*g* or PL/SQL data type. When you specify the parameter data type, you do not specify the precision or scale value for a numerical data type, or the maximum width for a character data type. You would use the following command to declare the calc_GPA procedure, and define a parameter declarations list that specifies the student ID and term ID values as IN parameters with the NUMBER data type, and the output GPA value as an OUT parameter with the NUMBER data type:

```
CREATE OR REPLACE PROCEDURE calc_GPA
       (student_id IN NUMBER,
        term_id IN NUMBER,
        student_GPA OUT NUMBER)
```

Creating a Stored Procedure in SQL*Plus

Now you create a stored procedure in SQL*Plus named UPDATE_ENROLLMENT_ GRADE that updates a student's grade for a course in the Northwoods University ENROLLMENT table. (The advantage of creating this stored procedure is that once you create and debug it, you can use it in any application that needs to update the grade column in the ENROLLMENT table, and you do not need to write the commands again.) The procedure accepts three input parameters: the student ID, course section ID, and the relevant grade to be updated.

9

To create a stored procedure in SQL*Plus:

1. In SQL*Plus, type **SET SERVEROUTPUT ON SIZE 4000** and then press **Enter** to initialize the DBMS_OUTPUT buffer. (Recall that you must initialize the DBMS_OUTPUT buffer so you can use the DBMS_OUTPUT.PUT_LINE procedure to display PL/SQL program output.)

2. In Notepad, type the commands in Figure 9-10 to create the stored procedure.

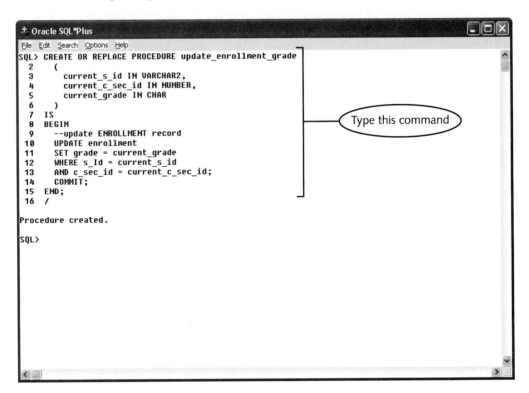

Figure 9-10   Creating a stored procedure in SQL*Plus

3. Copy the commands, paste the copied commands into SQL*Plus, and then run the program by pressing **Enter**. If your program executes successfully, the message "Procedure created" appears, as shown in Figure 9-10.

NOTE

Your database account must have the CREATE PROCEDURE database privilege to create a stored procedure.

HELP

If your program has compile errors, proceed to the next section titled "Debugging Stored Program Units in SQL*Plus."

Debugging Stored Program Units in SQL*Plus

Debugging stored program units in SQL*Plus is similar to debugging any program. The first step is to identify the program line or lines causing the error (or errors). Figure 9-11 shows a stored program unit with compile errors.

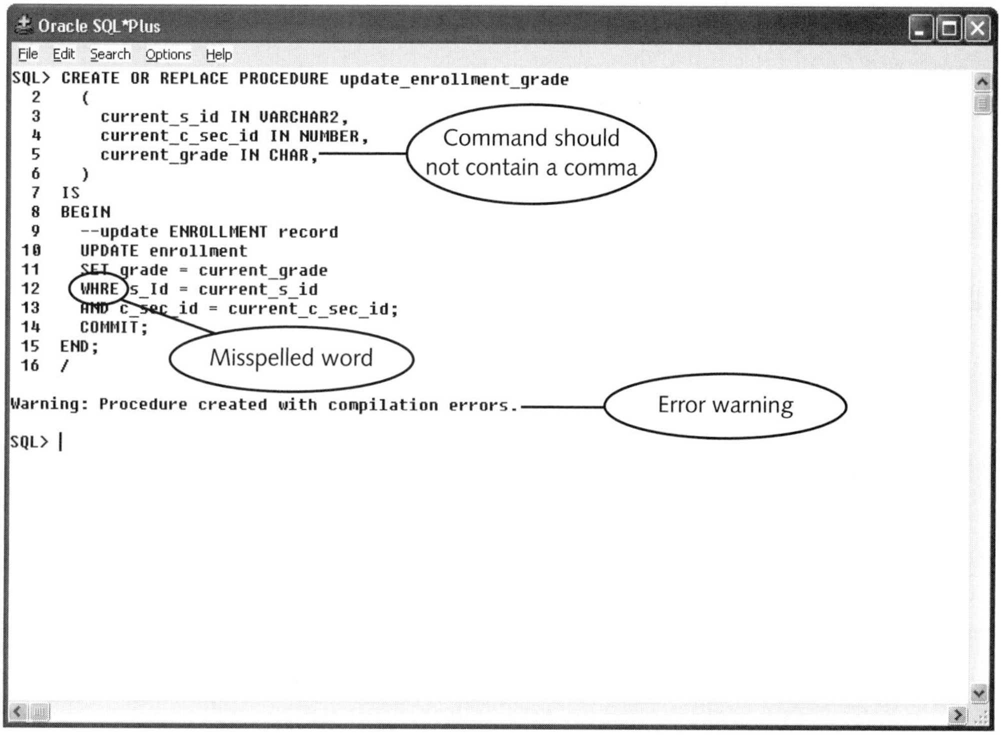

Figure 9-11   Stored program unit with compile errors

When a stored program unit has compile errors, the SQL*Plus interpreter displays only an error warning message. The interpreter does not automatically display compile error messages and line locations, as it does for anonymous program units. Thankfully, the interpreter writes all compile errors to a system table that you can access using the USER_ERRORS data dictionary view. To display a summary listing of compile errors generated by the last program unit that was compiled, you execute the **SHOW ERRORS** command, as shown in Figure 9-12.

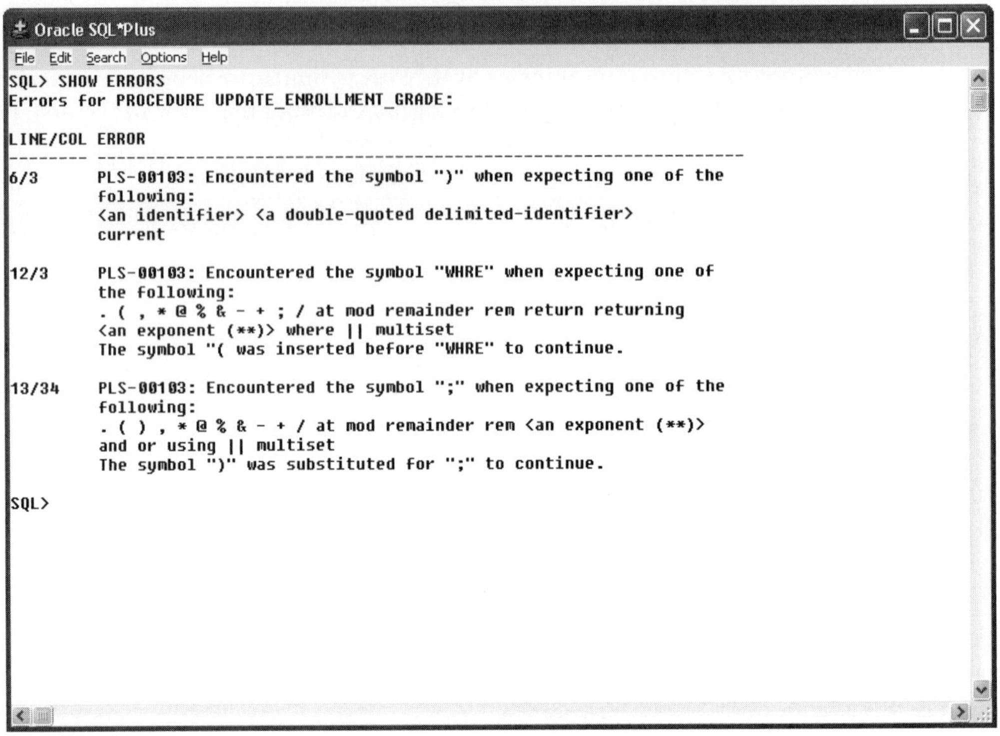

Figure 9-12   Viewing compile errors

In the compile errors listing, the LINE/COL field displays the line number of the error and the character position of the error. The ERROR field shows the error code number and associated error message. The first error message indicates that an error occurred on Line 6, when the interpreter encountered the closing parenthesis. Because the previous command ended with a comma, the interpreter expected another parameter declaration rather than the closing parenthesis. The second error message indicates that the WHERE clause on Line 12 was improperly formed. The last error shown also results from the error on Line 12, but only occurred after Line 13, and the semicolon to terminate the statement, was encountered.

The advice provided in Chapter 4 regarding debugging compile errors in anonymous program blocks also applies to debugging named program blocks. Remember that error locations might not correspond to the line of the actual error, so you might need to isolate the error location systematically by commenting out and modifying suspect lines. Using DBMS_OUTPUT.PUT_LINE commands to display variable values during execution also helps you identify and correct errors.

Calling a Stored Procedure

You have created the UPDATE_ENROLLMENT_GRADE stored procedure, and it exists as a database object in your user schema. You can execute a stored procedure directly from the SQL*Plus command line. Or, you can create a separate PL/SQL program that contains a command to call the stored procedure, and passes the parameter values to the procedure.

Calling a Stored Procedure from the SQL*Plus Command Line

You use the following general syntax to call a stored procedure from the SQL*Plus command line:

```
EXECUTE procedure_name
(parameter1_value, parameter2_value, ...);
```

In this syntax, *procedure_name* is the name of the procedure, as defined in the procedure header. *Procedure_name* is followed by the **parameter value list**, which specifies the values that the calling program passes to the procedure. The values in the parameter value list must be in the same position as the corresponding parameters in the parameter declarations list. For example, the UPDATE_ ENROLLMENT_GRADE procedure's parameter declarations list declares parameters of current_s_id, current_c_sec_id, and current_grade, in that order. The parameter value list in the procedure call must pass an associated value for each parameter, in the correct order. These values can be constants, such as the number 1, or variables that have assigned values, such as current_grade. When you include a variable in the parameter value list, the variable name does not have to be the same as the parameter name in the parameter declarations list.

The following command calls the UPDATE_ ENROLLMENT_GRADE procedure. The current_s_id parameter value is MA100, the current_c_sec_id is 12, and the parameter value B is stored in a variable named current_grade.

```
EXECUTE update_enrollment_grade(MA100, 12, B);
```

Figure 9-13 shows the command that calls the UPDATE_ENROLLMENT_GRADE procedure, and how the parameters in the parameter declarations list map to the parameters in the parameter value list.

```
Procedure Header
PROCEDURE update_enrollment_grade
       (current_s_idINVARCHAR2,current_c_sec_id IN NUMBER, current_grade INVARCHAR2)

Procedure Call: EXECUTE update_enrollment_grade(MA100,12,B);
```

Figure 9-13   Passing parameters to a procedure

It is important to remember that the variables or constants that you pass for each parameter must be in the same order as the parameters appear in the parameter declarations list. The first value in the procedure calling statement is assigned to the first parameter in the procedure declaration; the second value is assigned to the second parameter, and so forth. The parameter value list must also contain a value for every parameter in the parameter declarations list, and the parameters must be of the same data type as the associated declarations in the parameter declarations list. For example, in the parameter declarations list, the current_c_sec_id variable in the UPDATE_ENROLLMENT_ GRADE procedure is declared as having the NUMBER data type. The parameter value list in the command that calls the UPDATE_ENROLLMENT_GRADE procedure must also contain a parameter that has the NUMBER data type as the second value in the list, or an error occurs.

Next you execute the UPDATE_ENROLLMENT_GRADE stored procedure from the SQL*Plus command line, and pass to it the values MA100 for the student ID, 12 for the course section ID, and B for the updated grade. First, you query the database to determine the current grade for this record, so you can later confirm that the stored procedure executed correctly.

To execute the stored procedure from the SQL*Plus command line:

1. Type and execute the first query in Figure 9-14 to retrieve the current grade value for student MA100 in course section 12. The GRADE value should appear as NULL as shown.

2. Type and execute the second command in Figure 9-14, which executes the UPDATE_ENROLLMENT_GRADE stored procedure.

3. Type and execute the third query in Figure 9-14, which retrieves the updated record. The updated grade should appear as a B, which confirms that the stored procedure executed correctly.

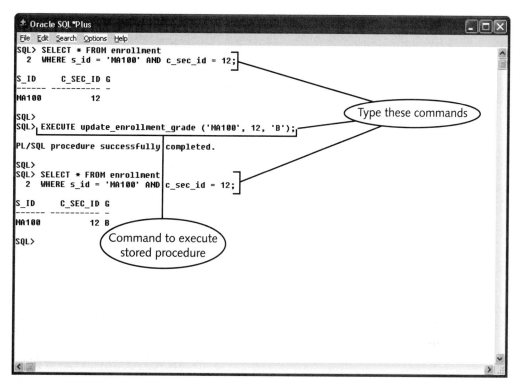

Figure 9-14   Executing a stored procedure from the SQL*Plus command line

Calling a Stored Procedure from a Separate PL/SQL Program

Calling a stored procedure from another stored program unit or from an anonymous PL/SQL program is similar to calling a stored procedure from the SQL*Plus command line, except you omit the EXECUTE command. For example, you use the following command in another PL/SQL program to call the UPDATE_ENROLLMENT_ GRADE stored procedure, and pass to it the parameter values MA100 for the student ID, 12 for the course section ID, and the letter B as the student's grade in the class:

```
update_enrollment_grade(MA100, 12, B);
```

Now you write and execute an anonymous PL/SQL program that contains a command to call the UPDATE_ENROLLMENT_GRADE stored procedure. The program declares a variable named current_update_grade, and sets its value as B. It passes the s_id parameter as the number MA100 and the c_sec_id as 12. Then you query the database again to confirm that the stored procedure updated the GRADE value correctly.

To create and execute an anonymous PL/SQL program to test the stored procedure:

1. To reset the grade assigned to the student back to its initial NULL value, type:

   ```
   update enrollment set grade = NULL
   WHERE s_id = 'MA100' AND c_sec_id = 12;
   ```

2. To verify that the grade value was reset to NULL, type:

   ```
   SELECT * FROM enrollment WHERE s_id = 'MA100' AND ↵
   c_sec_id = 12;
   ```

3. Type the commands in Figure 9-15 to create an anonymous PL/SQL program to call the stored procedure, then execute the commands to call the stored procedure and update the database.

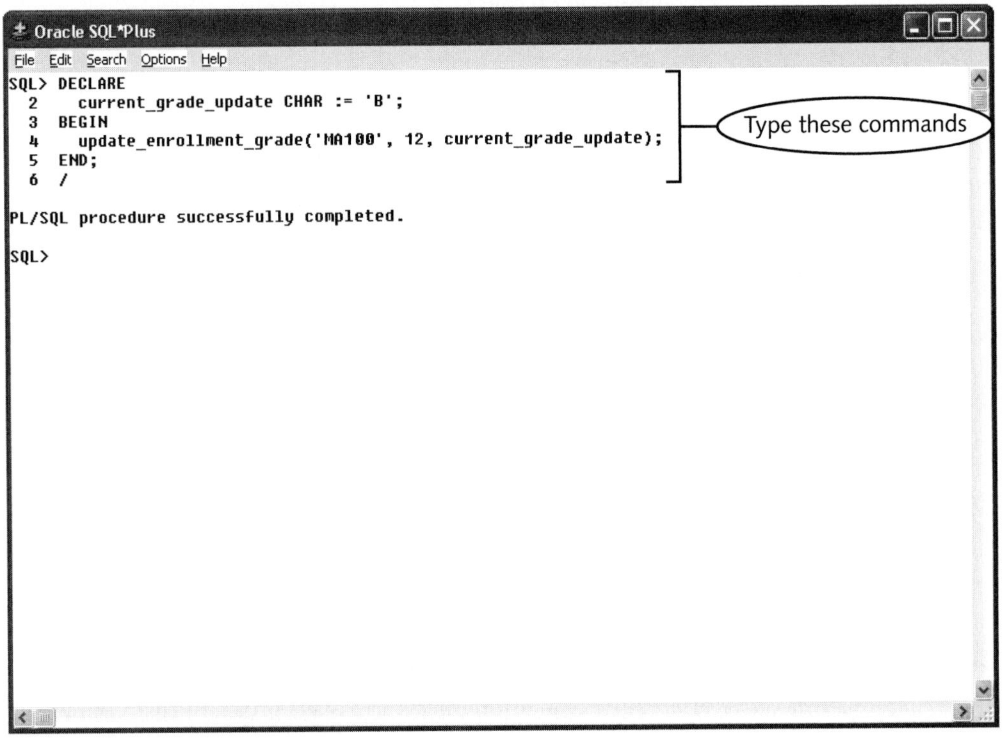

```
± Oracle SQL*Plus
File  Edit  Search  Options  Help
SQL> DECLARE
  2     current_grade_update CHAR := 'B';
  3  BEGIN
  4     update_enrollment_grade('MA100', 12, current_grade_update);     Type these commands
  5  END;
  6  /
PL/SQL procedure successfully completed.

SQL>
```

Figure 9-15   Executing a stored procedure from a PL/SQL anonymous block

4. Type and execute the following query to retrieve the updated GRADE value:

   ```
   SELECT * FROM enrollment
   WHERE S_id = 'MA100' AND c_sec_id = 12;
   ```

The updated value should appear as B, which confirms that the anonymous PL/SQL program successfully called the procedure and passed to it the specified parameter values.

Creating a Stored Program Unit Function

Recall that a function is similar to a procedure, except that it returns a single value to the calling program. Figure 9-16 shows the general syntax for creating a stored program unit that is a function.

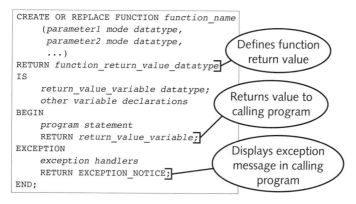

```
CREATE OR REPLACE FUNCTION function_name
    (parameter1 mode datatype,
     parameter2 mode datatype,
     ...)
RETURN function_return_value_datatype     Defines function
IS                                        return value
    return_value_variable datatype;
    other variable declarations           Returns value to
BEGIN                                     calling program
    program statement
    RETURN return_value_variable;
EXCEPTION                                 Displays exception
    exception handlers                    message in calling
    RETURN EXCEPTION_NOTICE;              program
END;
```

Figure 9-16   Commands to create a stored program unit function

To create a stored program unit that is a function, you use the **CREATE OR REPLACE FUNCTION** command, followed by the *function_name* specification. *Function_name* must be unique in the user's database schema, and must follow the Oracle naming standard. The syntax of the function's parameter declarations list is identical to the parameter declarations list of a procedure. After the parameter declarations list, the function header contains the RETURN value specification, which defines the data type of the value that the function returns and declares the variable that represents the value that the function returns. (Because a procedure does not return a value, it does not contain the RETURN value specification.)

Function_return_value_datatype defines the data type that the function returns, and *return_value_variable* declares the variable that represents the function return value. For example, the following function header creates a function named AGE that receives an input value that is a person's date of birth, and returns a number value that is the person's age, based on the current system date:

```
CREATE OR REPLACE FUNCTION age
    (date_of_birth IN DATE)
RETURN NUMBER IS
    current_age NUMBER;
```

The last command in the function body is the RETURN command, which instructs the function to return *return_value_variable* to the calling program. The **RETURN EXCEPTION_NOTICE** command is the last command in the exception section, and instructs the function to display the exception notice in the program that calls the function. (Recall that the exception section is optional in all stored program units.)

Now you create a function named AGE that receives a parameter that is a person's date of birth. The function calculates the person's age by subtracting the date of birth from the current system date and returns the age value as a NUMBER data type. To format the age in years, the function divides the result of the difference between the birth date and the system date, which is in days, by 365.25 and then truncates the decimal portion.

To create a function that returns a person's age:

1. Type and execute the commands in Figure 9-17 to create the AGE function.

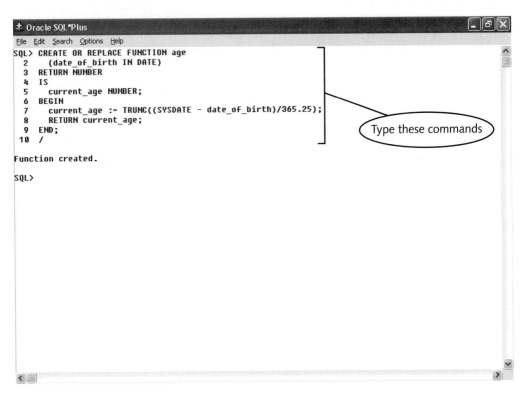

Figure 9-17   Creating a stored program unit function

2. Debug the program if necessary until the message "Function created" appears, indicating you have successfully created the function.

Calling a Function

Recall that a function returns a single data value to the calling program. Calling a function requires assigning the command to call the function to a previously declared variable in the calling program, using the following general syntax:

```
variable_name := function_name(parameter1, ↵
parameter2, ...);
```

As with procedures, the variables or constants that you pass for the parameter values must be in the same order in which the parameters appear in the function declaration.

Now you write an anonymous PL/SQL program that calculates the age of a person born on 07/01/1971 by calling the AGE function and passing to it the date of birth value as a parameter. The program declares a variable named current_age, which is the variable to which it assigns the function return value. The program also declares a variable named current_dob, which is assigned an initial value of 07/01/1971. (The assignment command that initializes the current_dob value uses the TO_DATE function, which converts the character string representation of the date to an internal date value.)

To write the program to call the function:

 1. Type the commands shown in Figure 9-18 to call the AGE function and display the current age of a person born on 07/01/1971.

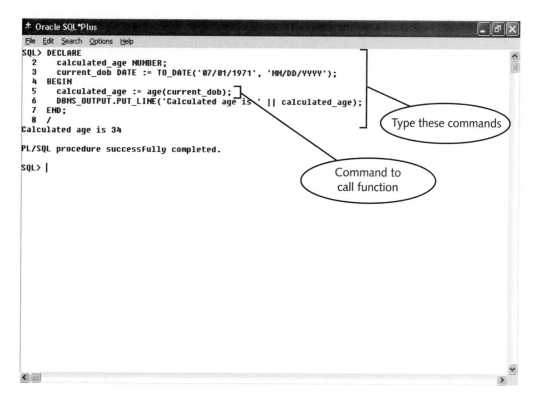

Figure 9-18   Calling a function from an anonymous PL/SQL program

 2. Execute the commands to display the calculated age value. Note that your output will be different, based on your current system date.

Using Forms Builder to Create Stored Procedures and Functions

In Chapter 6, you learned how to create a program unit in Forms Builder. Recall that a program unit is a top-level form object, and you can call a program unit from any trigger within the form. However, the program unit is not visible outside the form in which you create it, and cannot be called by programs outside the form, such as triggers in other forms, or PL/SQL programs that you run in SQL*Plus.

You can also use Forms Builder to develop stored program unit procedures and functions that are visible to other programs and to other database users. You create and test the program unit within a form. When the program unit is working correctly, you then save it as a stored program unit in your database schema. This makes the program unit visible to all of your forms, as well as to other PL/SQL programs that you execute in SQL*Plus, and to other database users. The advantage of using Forms Builder to create stored program units is that it provides an enhanced development and debugging environment. The PL/SQL Editor provides a color-coded editor for entering and debugging program unit commands. It displays compile error messages immediately, which makes it easier to compile and test program units. And, you can use the Forms Debugger to step through program unit commands and view how variable values change.

Creating a procedure or function that is a stored program unit in Forms Builder involves the following steps:

1. Create the procedure or function as a form program unit.

2. Test and debug the form program unit by calling it from commands within a form trigger.

3. Save the form program unit as a stored program unit in the database.

In this section, you create and test stored program unit procedures and functions in Forms Builder using the program unit test form in Figure 9-19.

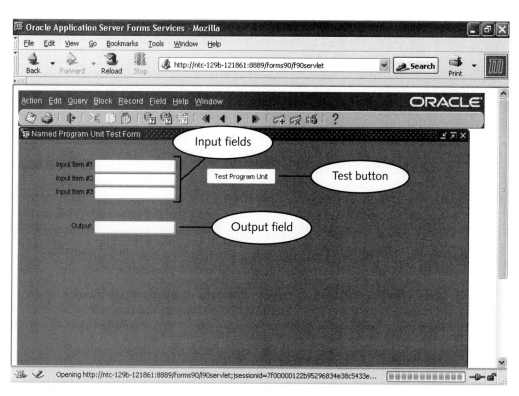

Figure 9-19   Program unit test form

You create a new program unit within the form. To test the program unit, you enter input parameter variables in the input fields. The form contains three input fields, so you can use it to test program units that have a maximum of three input parameters. (If you create a program unit that contains fewer than three input parameters, you do not use the extra input fields. If you create a program unit that contains more than three input parameters, you need to create more input fields on the test form.)

The form also contains a test button. You create a form trigger for the test button that calls your new program unit and passes the form input field values to the program unit. If the program unit returns output values, the form output field displays the results. Now you start the OC4J Instance, start Forms Builder, and open the program unit test form.

To start Forms Builder and open the program unit test form:

1. Start the OC4J Instance.

2. Start Forms Builder, click the **Connect** button 🖰, and connect to the database.

3. Open **Test.fmb** from the Chapter09 folder on your Data Disk. Save the file as **Ch9ATest_PROCEDURE.fmb** in the Chapter09\Tutorials folder on your Solution Disk.

Creating, Testing, and Saving a Stored Program Unit Procedure in Forms Builder

To create a stored procedure in the test form, you create the new procedure as a form program unit, then write a form trigger for the test button that calls the program unit. If the program unit updates the database, you use SQL*Plus to verify that the update was successful. Then, you save the form program unit as a stored program unit in the database.

Creating a Stored Procedure in the Test Form

Recall that to create a new form program unit, you select the Program Units node in the Object Navigator tree, and then click the Create button ✚. You specify the program unit type (procedure, function, or package), and then type the program unit code in the PL/SQL Editor. Now you create a new form program unit procedure named TOGGLE_STATUS. This procedure modifies the STATUS field in the Northwoods University TERM table. If the current STATUS value is "OPEN," the procedure changes the status to "CLOSED." If the current STATUS value is "CLOSED," the procedure changes the status to "OPEN." The procedure receives the TERM_ID value as an input parameter. The procedure uses an implicit cursor to retrieve the current STATUS value, an IF/THEN decision structure to evaluate the current value, and an UPDATE command to modify the status value in the TERM table.

To create the form program unit procedure:

1. In the Object Navigator, click the Program Units node, and then click the **Create** button ✚ to create a new program unit. The New Program Unit dialog box opens.

2. Type **TOGGLE_STATUS** in the Name field, make sure that the Procedure option button is selected, and then click **OK**. The PL/SQL Editor opens, and shows the procedure template.

TIP

The procedure template contains elements that are similar to the procedure header in Figure 9-9, except that it omits the **CREATE OR REPLACE** command. When you create a new procedure in Forms Builder, these commands are implicit within the procedure code.

3. Add the shaded commands in Figure 9-20 to create the procedure.

```
PROCEDURE TOGGLE_STATUS
    (current_term_id NUMBER)
IS
    current_status VARCHAR2(10);
    new_status VARCHAR2(10);
BEGIN
    --retrieve the current status
    SELECT status INTO current_status
    FROM term
    WHERE term_id = current_term_id;
    --evaluate the current status value
    --and set the new status value
    IF current_status = 'OPEN' THEN
        new_status := 'CLOSED';
    ELSE
        new_status := 'OPEN';
    END IF;
    --update the record using the new status value
    UPDATE term
    SET status = new_status
    WHERE term_id = current_term_id;
    COMMIT;
END;
```

Figure 9-20   TOGGLE_STATUS program unit procedure commands

4. Click the **Compile Module** button [icon] to compile the procedure. Correct any syntax errors, close the PL/SQL Editor, and save the form.

Creating a Form Trigger to Test the Program Unit Procedure

The next step is to test the program unit procedure. You create a WHEN-BUTTON-PRESSED trigger for the form test button, and add the command to call the program unit and pass to it the parameter value. Recall that to call a program unit, you specify the name of the program unit, followed by its parameter value list. For the TOGGLE_STATUS program unit, you pass the value of the first input text item as the procedure parameter value. This item is in the TEST block, and is named INPUT_ITEM1, so you reference it as `:test.input_item1`, and use the following command to call the program unit:

```
TOGGLE_STATUS(:test.input_item1);
```

After you create the trigger, you run the form and execute the program unit, then switch to SQL*Plus and execute a query to confirm that the program unit successfully modified the database.

To create and execute the form trigger, then confirm that it successfully updated the database:

1. Click **Tools** on the menu bar, and then click **Layout Editor** to view the form canvas.

2. Select the **Test Program Unit** button, right-click, point to **SmartTriggers**, and then click **WHEN-BUTTON-PRESSED**. The PL/SQL Editor opens for the form trigger.

3. Type the following command to call the program unit and pass the parameter for the TERM_ID value. Then compile the trigger, correct any syntax errors, and close the PL/SQL Editor.

   ```
   TOGGLE_STATUS(:test.input_item1);
   ```

4. Click the **Run Form** button to run the form. The form appears in the Forms Services window.

5. Type **6** in the Input Item #1 field. (This is the ITEM_ID value for the Summer 2007 term. Currently, the STATUS value for this term is "OPEN," so the program unit should change the STATUS value to "CLOSED.")

6. To execute the program unit, click **Test Program Unit**. The message "No changes to save" appears, indicating the program unit successfully updated the database.

7. Switch to SQL*Plus, and type and execute the following query to check the updated STATUS value for TERM_ID 6. The STATUS value should now be "CLOSED."

   ```
   SELECT * FROM term;
   ```

8. To change the STATUS value for TERM_ID 6 back to "OPEN," switch back to the browser window, make sure that 6 still appears in the Input Item #1 field, and then click **Test Program Unit** again.

9. Repeat Step 7. The STATUS value for TERM_ID 6 should now be "OPEN."

10. Switch back to the browser window, and then close the browser window.

Saving the Program Unit as a Stored Procedure in the Database

You have now successfully created and tested a form program unit. The final step is to save the program unit as a stored program unit in the database. This makes the program unit available to all of your forms, and to PL/SQL programs that you execute in other Oracle utilities, such as SQL*Plus. To save the program unit as a stored program unit in the database, you use the Database Objects node in the Object Navigator, as shown in Figure 9-21.

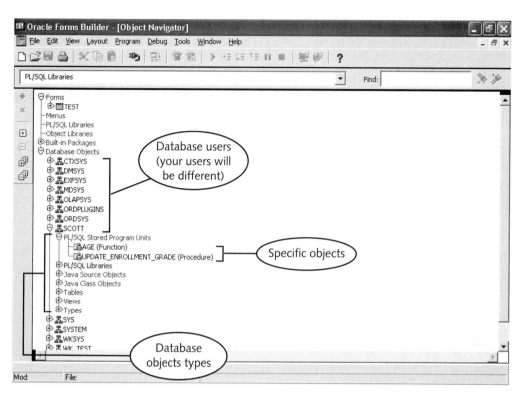

Figure 9-21   Database objects node

The **Database Objects node** is a top-level node within the Object Navigator, and is not associated with a specific form. This node contains child nodes that represent every database user. Each user's node contains nodes that represent database object types, such as stored program units, libraries, tables, and views. When you open an individual object type node, a list displays the objects of that object type that are in the selected user's database schema. Figure 9-21 shows that the SCOTT schema contains PL/SQL Stored Program Unit objects named AGE and UPDATE_ENROLLMENT_GRADE.

To save a form program unit as a stored program unit in the database, you drag the form program unit's Program Unit icon [icon], and drop it onto the PL/SQL Stored Program Units node under your user schema. Before you can save a form program unit as a stored program unit in the database, you must successfully compile the form program unit in Forms Builder. If a form program unit has not been compiled, or if you compile the program unit, then change it and neglect to recompile it, the form trigger appears in the Object Navigator with an asterisk (*) following its name. For example, Figure 9-22 shows that the TOGGLE_STATUS program unit has been changed, but has not yet been recompiled.

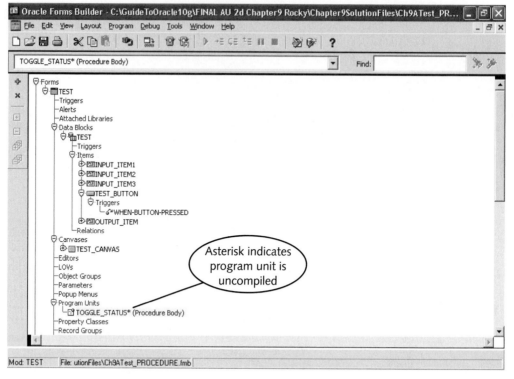

Figure 9-22 Uncompiled program unit in the Object Navigator

If a form program unit appears in the Object Navigator with an asterisk (*) beside its name, you must open the program unit in the PL/SQL Editor, then recompile it before you can save it as a stored program unit in the database.

Now you open the Database Objects node, open your database objects, and save the TOGGLE_STATUS form program unit as a stored program unit in the database. Then you execute the TOGGLE_STATUS program unit in SQL*Plus to confirm that it is available outside the form.

To save the form program unit as a stored program unit in the database, and then test the stored program unit:

1. Click **Window** on the menu bar, and then click **Object Navigator**.

2. Open the Database Objects node, and then open the node for your user schema. (This node has the same name as the username you use when you log onto the database.)

 If the Database Objects node is empty, you need to connect to the database.

3. Open the PL/SQL Stored Program Units node. Your stored program units should resemble the listing shown in Figure 9-21, with the exception of the TOGGLE_STATUS form program unit. (Your stored program units will be different if you did not create the AGE function and UPDATE_ENROLLMENT_GRADE procedure earlier in the lesson, or if you have created other stored program units.)

4. To store the TOGGLE_STATUS form program unit as a stored program unit in the database, scroll up in the Object Navigator if necessary, and view the TOGGLE_STATUS form program unit. Confirm that an asterisk does not appear beside its name, which indicates that the form trigger has been compiled.

5. Select the **Program Unit** icon ⬚ beside TOGGLE_STATUS, drag the icon toward the bottom edge of the screen, and drop it onto the PL/SQL Stored Program Units node under your username. After a moment, TOGGLE_STATUS appears as a stored procedure in your user schema.

6. To test the new stored program unit, switch to SQL*Plus. Type and execute the following command to test the stored program unit and pass to it the value for TERM_ID 6 as the input parameter:

```
EXECUTE toggle_status(6);
```

The "PL/SQL procedure successfully completed" message appears, indicating the procedure ran successfully.

7. To confirm that the procedure updated the database, type and execute the following command to retrieve the STATUS values from the TERM table:

```
SELECT * FROM term;
```

8. The STATUS value for TERM_ID 6 should appear as "CLOSED." Because its last value was "OPEN," this confirms that the procedure successfully modified the value.

9. Switch to Forms Builder, and close the Ch9ATest_PROCEDURE.fmb form. If a dialog box opens asking if you want to save your changes, click **Yes**.

Creating, Testing, and Saving a Stored Program Unit Function in Forms Builder

Recall that a function is a program unit that returns a specific value to a calling program. As with procedures, you can create and debug function program units in Forms Builder, and then save them as stored program units.

Creating a Program Unit Function in Forms Builder

To create a program unit function in a form, you click the Program Units node in the Object Navigator, and then select the Function option button on the New Program Unit dialog box. Now you open the stored program unit test form, save the form using a different filename, and create a new program unit function. The function is named DAYS_BETWEEN, and calculates the number of days between two input dates.

To create a program unit function:

1. In Forms Builder, open **Test.fmb** from the Chapter09 folder on your Data Disk, and save the file as **Ch9ATest_FUNCTION.fmb** in the Chapter09\ Tutorials folder on your Solution Disk.

2. Select the Program Units node, and then click the **Create** button ➕. The New Program Unit dialog box opens.

3. Type **DAYS_BETWEEN** in the Name field, click the **Function** option button, and then click **OK**. The function template appears in the PL/SQL Editor.

The syntax for a function program unit is similar to the syntax for the stored program unit function in Figure 9-16. The header defines the function name, declares its parameter list, and defines the data type and variable that the function returns. The body contains the program statements, and must contain a RETURN command to return the value to the calling program. The optional exception section must contain the RETURN EXCEPTION_NOTICE command to instruct the function to display error handler messages in the calling program.

Now you add the commands to create the function. The function accepts two input variables to represent the input dates, and declares a return value variable named curr_days_between, which has a NUMBER data type and represents the number of days between the two dates. The function calculates the difference between the two dates and returns the difference to the calling program.

To add the commands to create the function:

1. In the PL/SQL Editor, type the shaded commands in Figure 9-23.

```
FUNCTION DAYS_BETWEEN
      (first_date IN DATE, second_date IN DATE)
RETURN NUMBER IS
      curr_days_between NUMBER;
BEGIN
      curr_days_between := second_date - first_date;
      RETURN curr_days_between;
END;
```

Figure 9-23   Commands to create the form program unit function

2. Compile the program unit, correct any syntax errors, close the PL/SQL Editor, and then save the form.

Testing the Program Unit Function

To test the program unit function, you create a trigger for the form test button that accepts the date values that the user enters in the Input Item #1 and Input Item #2 input fields, calculates the difference between the two dates, and displays the function return value in the form output field. Because the input field values have the DATE data type, you must change the Data Type property of the first two text items to DATE, or the text items cannot correctly pass the input values to the program unit function. Now you change the input field data types, and create and test the test button trigger.

To change the input field data types and create and execute the test button trigger:

1. Click **Tools** on the menu bar, and then click **Layout Editor**.

2. To modify the Data Type properties of the first two input fields, click **INPUT_ITEM1**, press and hold the **Shift** key, and then click **INPUT_ITEM2** to select the input fields as an object group.

3. Right-click, and then click **Property Palette** to open the intersection Property Palette. Scroll down to the Data node, change the Data Type property value to **Date**, and then close the Property Palette.

4. To create the test button trigger, click the **Test Program Unit** button, right-click, point to **SmartTriggers**, and then click **WHEN-BUTTON-PRESSED**. Add the following command to call the function, pass to it the values in the first two input fields, and display the result in the output field:

```
:test.output_item := DAYS_BETWEEN
(:test.input_item1, :test.input_item2);
```

5. Compile the trigger, correct any syntax errors, and then close the PL/SQL Editor.

6. Save the form, and then run the form.

7. Type **22–OCT–2006** in the Input Item #1 field and **31–OCT–2006** in the Input Item #2 field, and then click **Test Program Unit**. The function displays the value 9 in the Output field, which is the correct number of days between the two input dates.

8. Close the browser window.

Saving the Program Unit Form as a Stored Program Unit in the Database

The final step is to save the completed function in the database as a stored program unit. As before, you first confirm that the function has been compiled by noting that an asterisk (*) does not appear beside its name. Then you drag the Program Unit icon and drop it onto the PL/SQL Stored Program Units node under your user schema in the Database Objects node. To confirm that the new stored program unit function is available to all of your programs, you create an anonymous PL/SQL program that calls the function and displays the result in SQL*Plus.

To save the function as a stored program unit in the database, and then test the function:

1. In the Object Navigator, click the DAYS_BETWEEN program unit node, drag it toward the bottom edge of the screen, and drop it onto the PL/SQL Stored Program Units node under the node that represents your user schema. The function appears in the list of stored program units.

2. In SQL*Plus, type and execute the commands in Figure 9-24 to create an anonymous PL/SQL program to test the new stored program unit function. The program output should display the value 9, which is the number of days between the two input dates. Close SQL*Plus.

```
DECLARE
      days_in_between NUMBER(3);
      first_date DATE :=
      TO_DATE('22-OCT-2006', 'DD-MON-YYYY');
      second_date DATE :=
      TO_DATE('31-OCT-2006', 'DD-MON-YYYY');       Type these commands
BEGIN
      days_in_between :=
      DAYS_BETWEEN(first_date, second_date);
      DBMS_OUTPUT.PUT_LINE(days_in_between);
END;
/
```

Figure 9-24   Anonymous PL/SQL program to test the stored program unit function

3. Switch back to Forms Builder, close the form, and click **Yes** to save your changes if you are prompted to do so.

4. Close Forms Builder, and shut down the OC4J Instance.

5. Close Notepad, and save the changes to the **Ch9AQueries.sql file**.

SUMMARY

❏ A database table index is a distinct database table that contains specific data field values and the corresponding physical location of their associated database records. You create an index on a field that is frequently used in a query search condition for the purpose of speeding up data retrievals.

❏ An index database table contains the ROWID value for each record, which specifies the physical location of the record in that database, along with sorted data values of the indexed field.

❏ When you create a new database table, the Oracle 10*g* DBMS automatically creates an index on the table's primary key.

❏ A composite index is a database index table that contains multiple sorted columns that the DBMS can use for identifying row locations for queries that contain multiple search conditions joined by the AND or OR operators.

❏ You can view the names of your database table indexes by querying the USER_INDEXES data dictionary view.

❏ Anonymous PL/SQL programs are not available to other users, cannot be called by other program units, and cannot receive input parameters. Stored PL/SQL program units are database objects that other programs can call, and that can receive input parameters.

❏ Stored program units can be server-side program units, which you store in the database as database objects and execute on the database server, or client-side program units, which you store in the file system of the client workstation and execute on the client workstation.

❏ Oracle 10g stored program units include procedures, functions, libraries, packages, and triggers.

❏ Procedures can manipulate multiple data values, and functions return a single data value. Procedures receive and deliver output values using parameters. You can call a procedure or function from the SQL*Plus command line or from another PL/SQL program.

❏ You call a function using a command in another PL/SQL program that assigns the function return value to a program variable.

❏ You can use the Forms Builder PL/SQL Editor to develop stored program units that are visible to other programs and to other database users. You create and test the program unit within a form, and then save it as a stored program unit in your database schema.

REVIEW QUESTIONS

1. What is the purpose of an index?

2. You use the EXECUTE command to call a stored procedure. True or False?

3. A(n) _____ is a database index table based on multiple search criterion frequently referenced by users.

4. What are the valid modes in the parameter declaration list of a procedure?

5. How do you call a function?

6. The _____ identifies the values that are passed to the procedure from the calling program.

7. When you create a check constraint, an index is automatically built on the column(s) referenced by the search condition. True or False?

8. A function is listed under which node of the Object Navigator?

9. The _____ value specifies the physical location of the record in the database.

10. You must specify the USING clause to reference a specific index during the execution of a query. True or False?

MULTIPLE CHOICE

1. Which of the following commands creates a stored procedure?

   a. CREATE OR REPLACE STORED PROCEDURE

   b. CREATE OR REPLACE FUNCTION

   c. CREATE OR REPLACE PROCEDURE

   d. none of the above

2. Which of the following commands is used to remove an index from the database?

   a. DELETE INDEX

   b. DROP INDEX

   c. ALTER DATABASE REMOVE INDEX

   d. REMOVE INDEX

3. A(n) _____ can insert, update, or delete database records.

   a. procedure

   b. function

   c. index

   d. all of the above

4. An index is automatically created for the _____ of a table.

   a. most frequently used search condition

   b. unique constraint

   c. primary key

   d. NULL constraint

5. A(n) _____ can return a maximum of one output value when it is executed.

   a. procedure

   b. trigger

   c. package

   d. function

6. Which of the following statements is *correct*?

   a. You should only create an index if most queries return at least 20% of the table rows in the results.

   b. Do not create an index for a table containing more than 100 rows because it slows down DML operations.

   c. You should create indexes on columns that are frequently referenced by search conditions in an application query.

   d. You can create indexes only on columns that contain a large number of NULL values.

7. Which of the following statements retrieves information useful for debugging stored program units?

   a. `DISPLAY ERRORS;`

   b. `SELECT errors FROM USER_UNITS;`

   c. `SHOW PROCEDURE ERROR;`

   d. none of the above

8. Which of the following symbols indicates that a program unit needs to be compiled or recompiled?

   a. #

   b. *

   c. ^

   d. !

9. Which section of a function defines the value that is returned to the calling program?

   a. header

   b. body

   c. exception section

   d. execution section

10. Which of the following is the default parameter mode in a program unit?

   a. IN OUT

   b. IN

   c. OUT

   d. RETURN

9

PROBLEM-SOLVING CASES

These cases refer to the Clearwater Trading company (Figure 1-25) database. Save all solution files in the Chapter09\Cases folder on your Solution Disk. Run the Ch9Clearwater.sql script in the Chapter09 folder on your Data Disk to refresh the sample databases before beginning work on the Cases.

1. Use SQL*Plus to create a stored program unit function named ADD_DAYS that returns a date value representing a specific number of days after a given date. The function should receive input variables of a date, and an integer representing the number of days to add to the date. Save the function code in a file named Ch9ACase1.sql.

2. Using the ADD_DAYS function created in the previous case, write an anonymous PL/SQL program that retrieves the date that Clearwater Traders shipment ID 3 is expected, and that uses the ADD_DAYS function to calculate and display dates that are 30, 45, and 60 days after the expected date. Add the anonymous PL/SQL program beneath the original code for the function and name the file Ch9ACase2.sql. Format the output as shown in Figure 9-25.

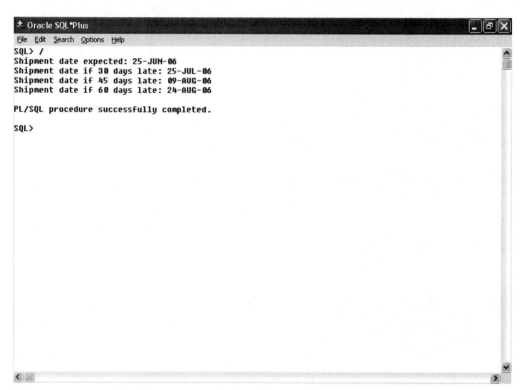

Figure 9-25   Results of anonymous PL/SQL program

3. Create a procedure named UPDATE_INVENTORY that updates the INV_QOH column in the INVENTORY table based on an inventory ID and quantity amount specified by a calling program. The new inventory amount should reflect the previous inventory amount plus the amount of inventory received in the shipment or sold to customers. The previous inventory amount is the amount already stored in the INVENTORY table, whereas the amount of inventory received is a parameter passed from the calling program. Save the procedure in a file named Ch9ACase3.sql.

4. Create an anonymous PL/SQL program that calls the UPDATE_INVENTORY procedure created in Case 3. Execute the anonymous PL/SQL program and update the INVENTORY table to reflect that three units of the item with inventory ID 1 were sold. (*Hint*: Pass a negative value to the procedure.) Execute a SELECT statement before and after execution to verify that INV_QOH is updated correctly. Save the program unit in a file named Ch9ACase4.sql.

5. In this case, you use Forms Builder to create a stored procedure that inserts a new record into the Clearwater Traders SHIPMENT table, and then inserts a corresponding record into the SHIPMENT_LINE table.

   a. In SQL*Plus, create a new sequence named SHIP_ID_SEQUENCE that starts at 300. Save the command in a file named Ch9ACaseQueries.sql. (If you created this sequence in a previous chapter, drop the sequence and create it again.)

   b. In Forms Builder, open Test.fmb from the Chapter09 folder on your Data Disk, and save the file as Ch9ACase5.fmb. Then create a new form program unit procedure named UPDATE_SHIPMENT, and specify that the procedure receives input parameters corresponding to the inventory ID and shipment quantity expected.

   c. Add a command to the UPDATE_SHIPMENT procedure that inserts a new record in the SHIPMENT table. The command retrieves the next SHIP_ID_SEQUENCE value and uses the current system date for the SHIP_DATE_EXPECTED value.

   d. Add a command to the procedure to insert a new record into the SHIPMENT_LINE table that uses the current sequence value for SHIP_ID and the inventory ID and shipment quantity values that were input as parameters. Specify that the SL_DATE_RECEIVED value is NULL.

   e. Write a trigger for the form test button to confirm that the procedure works correctly. Use the test form input fields to enter the INV_ID and SL_QUANTITY values dynamically. Then use SQL*Plus to confirm that the procedure successfully inserts the data values.

   f. Save the procedure as a stored program unit in your database schema.

9

CALLING STORED PROGRAM UNITS FROM OTHER STORED PROGRAM UNITS

PL/SQL database applications can become very large and involve many lines of code. To make these applications more manageable, you often decompose applications into logical units of work, and then write individual program units for each logical unit. By placing all of the code in stored program units that you store in the database, the code is in a single location. All developers can access the program units, and do not need to rewrite program units that already exist. If a program unit needs to be updated, a developer updates the stored program unit, rather than updating all of the applications that use the stored program unit.

To integrate all of the individual program units, you must be able to call program units from other program units and pass parameter values to the called program units. For example, when a new course is offered at Northwoods University, you must insert a record into the COURSE table that contains general information about the course, and then insert the associated records for each section of the course in the COURSE_SECTION table. It is logical to break this into two procedures: one to insert the course information, and another to insert the course section information.

Calling a stored program unit from another stored program unit is similar to calling a stored program unit from an anonymous PL/SQL program: you execute a PL/SQL command that specifies the name of the called program unit, followed by its parameter value list. The main difference is that you must create the called program unit before you create the calling program unit. If you try to compile a program unit that calls a program unit that does not yet exist, a compile error occurs.

Now you use Forms Builder to create a stored program unit procedure named CREATE_NEW_COURSE. This procedure receives as input parameters the course number, course name, number of credit hours, term ID, section number, faculty ID, day and start time, and then inserts the relevant course information into the COURSE table. Then, the procedure calls a second procedure named CREATE_NEW_COURSE_SECTION and passes to it the values for the term ID, section number, faculty ID, date and time. The second procedure inserts a new record into the COURSE_SECTION table. To create and test these program units, you use the Course Information form in Figure 9-26, which contains text items for the course input parameters.

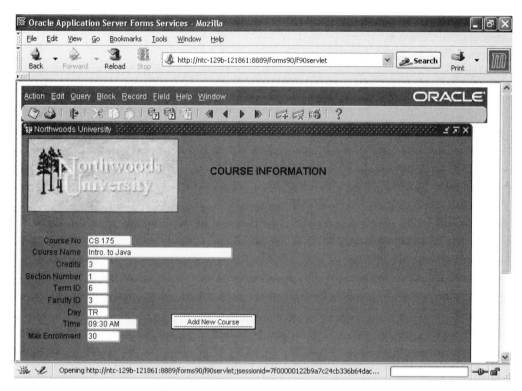

Figure 9-26   Course Information form

To begin, you run the scripts to refresh your database tables. Then, you create the sequence to generate the course section ID surrogate key. Then you open the Course Information form in Forms Builder, and create the CREATE_NEW_COURSE_SECTION procedure as a form program unit. This procedure receives the term ID, section number, faculty ID, and date and time as input parameters, and inserts the values into the database.

To refresh your database tables, create the sequence, open the Course Information form, and create the called procedure:

1. Start SQL*Plus, log onto the database, and type **SET SERVEROUTPUT ON SIZE 4000** to initialize the SQL*Plus output buffer.

2. Refresh your database tables by running the **Ch9Northwoods.sql** scripts in the Chapter09 folder on your Data Disk.

3. Start Notepad, and save the new file as **Ch9BQueries.sql** in the Chapter09\ Tutorials folder on your Solution Disk. Then type the following command to create the sequence, copy the command, and paste and execute the command in SQL*Plus. (If you already created this sequence in a previous chapter, drop the existing sequence and create it again.)

```
CREATE SEQUENCE c_sec_id_sequence
START WITH 14;
```

NOTE

From this point forward in this lesson, type all SQL commands in the Ch9BQueries.sql file in Notepad, copy the command, paste the command into SQL*Plus, and then execute the command.

4. Start the OC4J Instance.

5. Start Forms Builder, and connect to the database.

6. Open **COURSE_INFO.fmb** from the Chapter09 folder on your Data Disk, and save the form as **Ch9BCourseInfo.fmb** in the Chapter09\Tutorials folder on your Solution Disk.

7. To create the new program unit procedure, click the Program Units node in the Object Navigator, and then click the **Create** button ✚.

8. Type **CREATE_NEW_COURSE_SECTION** in the Name field, make sure that the Procedure option button is selected, and then click **OK**. The procedure template appears in the PL/SQL Editor.

9. Add to the template the shaded commands in Figure 9-27. Compile the code, debug it if necessary, close the PL/SQL Editor, and save the form.

```
PROCEDURE CREATE_NEW_COURSE_SECTION
   (current_course_sec_id IN NUMBER,
    current_course_no IN VARCHAR2,
    current_term_id IN NUMBER,
    current_sec_num IN NUMBER,
    current_f_id IN NUMBER,
    current_day IN VARCHAR2,
    current_sec_time IN DATE,
    current_max_enrl IN NUMBER)
IS
BEGIN
   --insert new course section
   INSERT INTO course_section(c_sec_id, course_no, term_id,
      sec_num, f_id, c_sec_day, c_sec_time, max_enrl)
   VALUES (current_course_sec_id, current_course_no,
      current_term_id, current_sec_num, current_f_id,
      current_day,
   TO_DATE(current_sec_time, 'DD-MON-YYYY HH:MI AM'),
      current_max_enrl);
   COMMIT;
END;
```

Figure 9-27   Procedure to insert a new course section

Next, you create the CREATE_NEW_COURSE procedure. This procedure receives the course number, course name, and number of credit hours as input parameters. It retrieves the next C_SEC_ID_SEQUENCE value and saves the value as a local variable named current_c_sec_id, and then inserts a new COURSE record based on the input parameters and sequence value. It then calls the CREATE_NEW_COURSE_SECTION procedure, and passes to it the course section parameter values. After creating the procedure, you create a form trigger for the Add New Course button that calls the CREATE_NEW_COURSE procedure, and passes to it the form text item values. Finally, you run the form and test the procedures.

To write the CREATE_NEW_COURSE procedure and form trigger, and run the form:

1. Select the Program Units node in the Object Navigator, and then click the **Create** button ➕ to create a new program unit. Create a new procedure named **CREATE_NEW_COURSE**.

2. Add to the procedure template the shaded commands in Figure 9-28. Compile the procedure, debug it if necessary, and then close the PL/SQL Editor.

```
PROCEDURE CREATE_NEW_COURSE
   (current_course_no IN VARCHAR2, current_course_name IN VARCHAR2,
    current_credits IN NUMBER, current_sec_num IN NUMBER,
    current_term_id IN NUMBER, current_f_id IN NUMBER
    current_day IN VARCHAR2, current_sec_time IN DATE,
    current_max_enrl IN NUMBER)
IS
    current_course_sec_id IN NUMBER;
BEGIN
    --retrieve next C_SEC_ID_SEQUENCE value
    SELECT c_sec_id_sequence.NEXTVAL
    INTO current_course_sec_id
    FROM dual;
  --insert the COURSE record
  INSERT INTO course VALUES (current_course_no,current_course_name,
            current_credits);
  --call the procedure to insert the COURSE_SECTION record
  CREATE_NEW_COURSE_SECTION(current_course_sec_id,current_course_no,
      current_term_id, current_sec_num, current_f_id, current_day,
current_sec_time, current_max_enrl);
END;
```

Figure 9-28 Procedure to insert a new order and call the procedure to insert a new order line

3. Click **Tools** on the menu bar, and then click **Layout Editor**. To create the form trigger to call the procedure, click the **Add New Course** button, right-click, point to **SmartTriggers**, and then click **WHEN-BUTTON-PRESSED**. Type the following command to call the procedure and pass to it the form text item values:

```
create_new_course(:course.course_no, :course.course_name,
:course.credits, :course.section_number,
:course.term_id, :course.f_id, :course.sec_day,
:course.sec_time, :course.max_enrl);
```

4. Compile the trigger, debug it if necessary, close the PL/SQL Editor, and save the form.

5. Run the form, type the input values in Figure 9-26 to specify a new customer order, and then click **Add New Course** to call the form procedures. The "No changes to save" message appears on the message line, indicating that the procedures successfully inserted the records.

6. Close the browser window.

7. To confirm that the procedures inserted the records into the database, switch to SQL*Plus, and type and execute the queries in Figure 9-29. The query output should show the new COURSE record, and the corresponding COURSE_SECTION record.

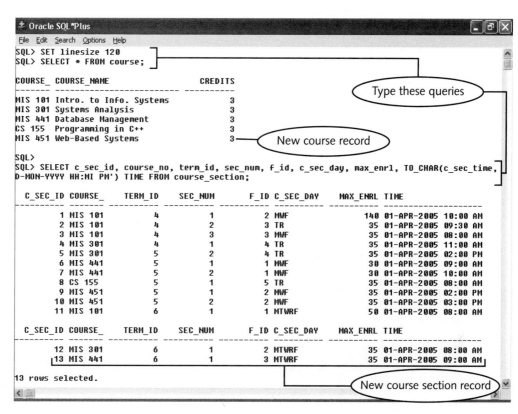

Figure 9-29 Viewing the new records in SQL*Plus

If your C_SEC_ID value in the COURSE_SECTION table is different from 14, use the C_SEC_ID value that appears in the results of the first query as the search condition in the second query.

PL/SQL LIBRARIES

A **PL/SQL library** is an operating system file that contains code for multiple related procedures and functions. You can attach a PL/SQL library to a form or report, and triggers within the form or report can then reference the library's procedures and functions. You can also attach a library to a second library, which makes the attached library's program units available to the second library. You usually store a PL/SQL library in the file system of the client workstation. The library design file has a .pll extension, which stands for "PL/SQL Library." After you create and debug a library .pll file, you compile the library into a library executable file, which has a .plx extension, which stands for "PL/SQL Library Executable."

You can create a library object in the database that references a library .pll file on a shared network drive. You can also store PL/SQL libraries in the database, provided that your DBA runs specific scripts to create database tables in the SYSTEM database schema to store and manage server-side libraries. In this book, you store all PL/SQL libraries in the client workstation file system.

The advantage of using a library over multiple stored program units is that the library places the commands for multiple related program units in a single location that developers can access and use. The library code always executes on the client workstation, which offloads processing from the database server and improves application performance.

The following sections describe how to use Forms Builder to create PL/SQL libraries, add program units to libraries, attach libraries to forms, and modify library program units.

Creating a PL/SQL Library

You can use Forms Builder to create libraries and then add form program units and stored program units to the library. Figure 9-30 shows an example of a PL/SQL library in Forms Builder that contains multiple program units.

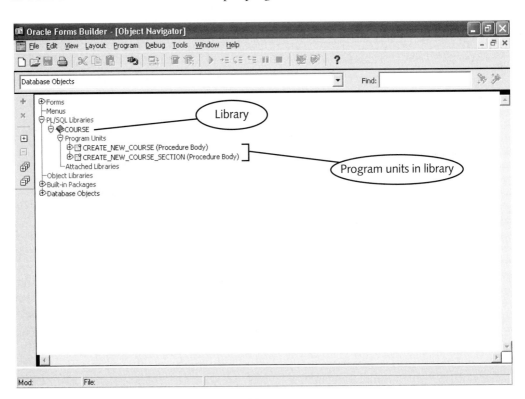

Figure 9-30   PL/SQL library in Forms Builder

To create a new PL/SQL library, you select the PL/SQL Libraries node in Forms Builder, click the Create button ✚, and then save the library in your workstation's file system. You then add completed compiled program units to the library. These program units can be form program units or stored program units in the database. If you add to a library a program unit that calls another program unit, you must be sure to include the called program unit in the library, and recompile the calling program unit after you add it to the library. This enables the system to verify that the called program unit is available within the library.

After you create a library, you save the library .pll file in the file system of your workstation or on a shared network drive. Then you select the library node, and click the Compile Module button 🖳 to create the library executable (.plx) file. If you modify one or more program units within a library, you must always remember to recompile the library .plx file or the form or report to which you attach the library will use the previous library version.

Next, you create a new library named COURSE, which contains program units for working with Northwoods University courses. After you create the library, you add to it the program units you just created in the Course Information form.

To create the library and add the program units:

1. Switch to Forms Builder, click **Window** on the menu bar, and then click **Object Navigator**.

2. Select the PL/SQL Libraries node, and then click the **Create** button ✚. A new library node appears, and has a default name, such as LIB_002 (your default name may be different). Note that the library has a Program Units child node that represents the program units the library contains, and an Attached Libraries node that represents other libraries attached to this library.

3. To change the library name, make sure the new library node is selected, and then click the **Save** button on the toolbar. Navigate to the Chapter09\ Tutorials folder on your Solution Disk if necessary, change the filename to **COURSE.pll**, and click **Save**. The new library node now appears as COURSE in the Object Navigator.

4. To add form program units to the library, make sure that the CREATE_ NEW_COURSE and CREATE_NEW_COURSE_SECTION program units are compiled and asterisks (*) do not appear beside their names. Then select **CREATE_NEW_COURSE**, press and hold the **Shift** key, and select **CREATE_NEW_COURSE_SECTION**.

5. Drag the program units toward the bottom edge of the screen, and then drop the program units onto the Program Units node under the COURSE library node. Note that an asterisk appears beside the CREATE_NEW_COURSE program unit. This is because it references the CREATE_NEW_COURSE_ SECTION program unit. Recall that when you move a program unit that

9

calls other program units to a new object, you must recompile the calling program unit.

6. Double-click the **Program Unit** icon [icon] beside CREATE_NEW_ COURSE to open the program unit in the PL/SQL Editor. Compile the program unit, and then close the PL/SQL Editor. Note that the asterisk no longer appears beside the program unit's name.

7. Select the COURSE library node, and then click the **Save** button to save the library. Your completed library should look like Figure 9-30.

8. Make sure that the COURSE library node is selected, and then click the **Compile Module** button [icon] to compile the library and create the library .plx file.

9. To close the library in Forms Builder, make sure that the COURSE library node is selected, click **File** on the menu bar, and then click **Close**. The library no longer appears in the Object Navigator.

If an error message with the text "Unable to fetch record from table TOOL_MODULE" appears, click OK to close the message, open Windows Explorer, delete the existing COURSE.plx file in the Chapter09\Tutorials folder on your Solution Disk, and then repeat Step 7. If the error persists, save the form, close the form, and close Forms Builder. Delete the COURSE.plx file again, restart Forms Builder, connect to the database, and open the library. Recompile the library program units if necessary, and then recompile the library.

10. Close the COURSE_INFO form in Forms Builder. If a dialog box opens asking if you want to save your changes, click Yes.

Attaching a Library to a Form

After you create a library, you can attach the library to a form or report, and then reference the library program units in form triggers and report formula column functions just as if the program units were directly in the form or report. To attach a library to a form, you select the form's Attached Libraries node, click the Create button [icon], and specify the PL/SQL library .pll filename. The form actually uses the compiled library .plx file, and looks for it in the same folder location as the .pll file you specify. If the .plx file does not exist, the form automatically compiles the .pll file and creates the associated .plx file.

When you attach the library, a message appears asking if you want to store folder path information for the library. If you include path information for a library, the library .pll file must always be available at the specified folder path and filename. If you do not include path information for a library, you must store the library .pll file in the default form folder. (Recall from Chapter 8 that the default form folder is the folder on the client workstation in which Developer10*g* always looks for form files when a command references a form in an integrated database application. By default, this folder is the C:\oracle_developer_home\forms90 folder on your workstation.)

Recall that the original Course Information form did not contain the program units for adding new courses and course sections. Next, you attach the COURSE library to the form, and specify not to remove the folder path information.

To attach a PL/SQL library to a form:

1. In Forms Builder, open **COURSE_INFO.fmb** from the Chapter09 folder on your Data Disk, and save the form as **Ch9BCourse_Info_LIB.fmb** in the Chapter09\Tutorials folder on your Solution Disk. Note that the form does not contain any program units.

2. To attach the library to the form, click the form's Attached Libraries node, and then click the **Create** button ➕. The Attach Library dialog box opens.

3. Click **Browse**, navigate if necessary to the Chapter09\Tutorials folder on your Solution Disk, select **COURSE.pll**, click **Open**, and then click **Attach**. A message appears advising you that the library file contains a nonportable path specification, and asking if you want to remove the path. To save the library path information, click **No**. The COURSE library appears under the Attached Libraries node.

When you attach a library to a form, the form stores a reference to the program unit code, and does not store the actual code. Because the library code is stored in a central location, it can be modified and thus kept consistent across all applications. Next you examine the attached library in the form, create a form trigger to reference the library program units, and insert a new course. Then you test the form to make sure the trigger successfully references the library program units.

To examine the attached library, then create and test a form trigger to reference the library program units:

1. Open the COURSE node under the Attached Libraries node in the Object Navigator, then open each library program unit node, so your screen display looks like Figure 9-31. The Specification node under a library program unit specifies the program unit parameter names and data types, so the developer knows what parameter values he or she needs to pass to the library program unit.

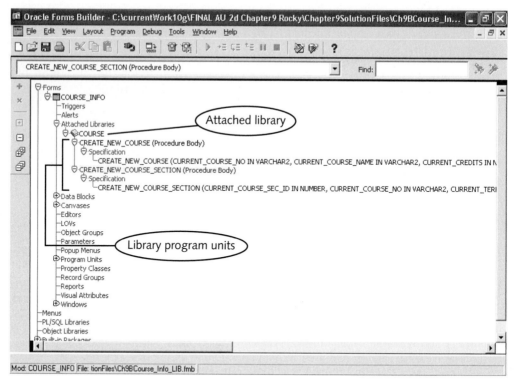

Figure 9-31   Attaching a library to a form

2. Click **Tools** on the menu bar, and then click **Layout Editor**. To create the form trigger to call the procedure, click the **Add New Course** button, right-click, point to **SmartTriggers**, and then click **WHEN-BUTTON-PRESSED**. Type the following command to call the procedure and pass to it the form text item values:

```
create_new_course(:course.course_no, :course.course_name,
:course.credits, :course.section_number,
:course.term_id, :course.f_id, :course.sec_day,
:course.sec_time, :course.max_enrl);
```

3. Compile the trigger, debug it if necessary, and then close the PL/SQL Editor.

4. Save the form, and then run it.

5. Type **MIS 150** in the Course No field, **Computer Concepts** in the Course Name field, **3** in the Credits field, **1** in the Section Number field, **6** in the Term ID field, **3** in the Faculty ID field, **MWF** in the Day field, **10:00 AM** in the Time field, **30** in the Max Enrollment field, and then click **Add New Course**. The "No changes to save" message appears on the message line, indicating that the procedures successfully inserted the records. If a dialog box with the message "FRM-40401: No changes to save" appears, click **Yes**.

6. Close the browser window.

7. To confirm that the form inserted the COURSE record into the database, switch to SQL*Plus, and type **SELECT * FROM course;**. The new record for MIS 150 appears.

8. To confirm that the form inserted the COURSE_SECTION record into the database, type **SELECT * FROM course_section WHERE course_no = 'MIS 150';**. The new COURSE_SECTION record for the course appears.

Modifying Library Program Units

The program units you add to libraries are completed and debugged, and should not require modifications. However, if a program unit within a library requires maintenance, you open the library in Forms Builder, modify the program unit as necessary, compile the program unit within the library, and then recompile the library to create a new .plx file. All forms that contain the library as an attached library object then use the modified code.

Next, you modify the CREATE_NEW_COURSE_SECTION program unit in the COURSE library so it displays a custom message when the program unit inserts the new course section record. Then you modify the Course Information form so it suppresses the default system messages, and run the form to confirm that it uses the modified library.

To modify a library program unit, modify the form, and then run the form:

1. Switch back to Forms Builder. To open the COURSE library, click **File** on the menu bar, click **Open**, navigate if necessary to the Chapter09\Tutorials folder on your Solution Disk, select **COURSE.pll**, and click **Open**. The COURSE library appears under the PL/SQL Libraries node.

2. Open the Program Units node under the library. The library program units appear.

3. To modify the program unit, double-click the **Program Unit** icon 🔳 beside **CREATE_NEW_COURSE_SECTION**. The program unit appears in the PL/SQL Editor.

4. Type the following command after the last program unit command:

```
Message('New course and course section created.');
```

5. Compile the program unit, debug it if necessary, and then close the PL/SQL Editor.

6. Save the library, and then recompile the library.

If an error message with the text "Unable to fetch record from table TOOL_MODULE" appears, click OK to close the message, open Windows Explorer, delete the existing COURSE.plx file in the Chapter09\Tutorials folder on your Solution Disk, and then repeat Step 6. If the error persists, save the form, close the form, and close Forms Builder. Delete the COURSE.plx file again, restart Forms Builder, connect to the database, and open the library. Recompile the library program units if necessary, and then recompile the library. Then open CH9BCourse_Info_LIB.fmb again.

7. To suppress the default system messages in the Course Information form so the custom message appears, click the Triggers node under the form module, right-click, point to **SmartTriggers**, and click **PRE-FORM**. Type the following command, compile the trigger, debug it if necessary, close the PL/SQL Editor, and save the form.

   `:SYSTEM.MESSAGE_LEVEL := 25;`

8. Run the form. Type **MIS 200** in the Course No field, **Micro Computer Apps** in the Course Name field, **3** in the Credits field, **1** in the Section Number field, **6** in the Term ID field, **2** in the Faculty ID field, **MWF** in the Day field, **3:00 PM** in the Time field, **30** in the Max Enrollment field, and then click **Add New Course**. The custom message appears on the message line, indicating that the procedures successfully inserted the records.

9. Close the browser window.

10. Close the library in Forms Builder. If a dialog box opens asking if you want to save your changes, click **Yes**.

If an error appears stating that another process may be using the COURSE.pll library file, press CTRL+ALT+Del to start the Task Manager, select the Processes tab, select the ifweb90.exe or ifbld90.exe process if either is listed, click End Process, click Yes, close the Task Manager, and then click OK.

11. Close the Course Information form in Forms Builder. If a dialog box opens asking if you want to save your changes, click **Yes**.

PACKAGES

Recall that you create stored program units to make application code more modular, easier to maintain, and reusable across multiple database applications. In the previous section, you learned how to create PL/SQL libraries to make program units available to applications. Another way to make PL/SQL program units available to multiple applications is by creating a package. A **package** is a code library that contains related program units and variables. A package is stored in the database and executes on the database server. (Recall that a library is stored in the client or server file system, and executes on the client workstation.) After you create a package and grant other users the privilege to use the package, any PL/SQL program can reference the package procedures and functions.

Packages have more functionality than PL/SQL libraries. You can create variables in packages, and multiple database applications can access and reference these variables. Packages can also contain definitions for explicit cursors, which multiple applications can use to retrieve and process database data. Packages are also more convenient to use than PL/SQL libraries. You store packages in the database, so you do not need to worry about specifying a file or folder path to the package location. The package procedures and functions are available without explicitly attaching them to a form or report, as you must do with a library.

Oracle 10*g* uses built-in packages to provide enhanced functionality in its version of SQL. A package named STANDARD defines all built-in functions and procedures, such as ADD_MONTHS, and defines the properties of the database data types.

A package has two components: the package specification and the package body. The following sections describe these components.

9

The Package Specification

The **package specification**, also called the **package header**, declares package objects, including variables, cursors, procedures, and functions that are in the package and are to be made public. When an object is **made public**, program units outside the package can reference the package's objects.

All of the variables that you have used so far are private variables, which are visible only in the program in which they are declared. As soon as the program terminates, the memory that stores the private variables is made available for other programs to use, and programs can no longer access the variable values. You can use a package to declare **public variables**, which are visible to many different PL/SQL programs. Public variable values remain in memory even after the programs that declare and reference them terminate. You declare a public variable in the DECLARE section of a package, using the same syntax you use to declare a private variable. You reference public variables in programs in the same way you reference private variables.

 A public variable is similar to a form global variable.

TIP

Figure 9-32 shows the general syntax for a package specification.

```
PACKAGE package_name
IS
    public variable declarations
    public cursor declarations
    public procedure and function declarations
END;
```

Figure 9-32   General syntax for a package specification

The keyword PACKAGE specifies that the code block is a package. *Package_name* identifies the package, and must adhere to the Oracle naming standard. You can declare the package objects (variables, cursors, procedures, and functions) in any order. A package does not have to contain all four object types; for example, a package can consist of just variable declarations, or it can consist of just procedure or function declarations.

You declare variables and cursors in packages using the same syntax you use to declare these items in other PL/SQL programs. To declare a procedure in a package, you specify the procedure name, followed by the parameters and variable types, using the following syntax:

```
PROCEDURE procedure_name
(parameter1 parameter1_data_type,
parameter2 parameter2_data_type, ...);
```

To declare a function in a package, you specify the function name, parameters, and return variable type, as follows:

```
FUNCTION function_name
(parameter1 parameter1_data_type,
parameter2 parameter2_data_type, ...)
RETURN return_datatype;
```

The commands for the package procedures and functions appear in the package body, which you learn about in the next section.

The Package Body

The **package body** contains the commands to create the program units that the package specification declares. You must always create the package specification before the package body, or an error occurs. The package body is optional, because sometimes a package contains only variable or cursor declarations, and no procedure or function declarations.

Figure 9-33 shows the general syntax for the package body.

```
PACKAGE BODY package_name
IS
    private variable declarations
    private cursor specifications
    program unit commands
END;
```

Figure 9-33   General syntax for a package body

Package_name in the package body must be the same as *package_name* in the package specification. Variables that you declare at the beginning of the package body are private to the package, which means that they are visible to all modules in the package body but are not visible to modules outside of the package body.

Each package program unit has its own declaration section and BEGIN and END statements. Each program unit that you declare in the package body must have a matching program unit forward declaration in the package specification, with an identical parameter list. Variables that you declare within individual procedures and functions are visible only to the individual procedures or functions in which they are declared.

You can create a package using either SQL*Plus or Forms Builder. The following sections illustrate each approach.

Creating a Package Using SQL*Plus

To create a package using SQL*Plus, you execute commands at the SQL*Plus command line first to create the package specification and then to create the package body. The following sections describe how to accomplish these tasks.

Creating the Package Specification in SQL*Plus

When you create a package specification in SQL*Plus, you use the **CREATE OR REPLACE PACKAGE** command, followed by the package specification. The DBMS automatically stores the package specification as a database object. Now you use SQL*Plus to create a package specification for a package named COURSE_PACKAGE that contains the procedures to insert new course and course sections in the Northwoods University database. Earlier in this lesson, you created a procedure named CREATE_NEW_COURSE that inserts a new course row into the COURSE table and then calls the CREATE_NEW_COURSE_SECTION procedure, which inserts a new record into the COURSE_SECTION table. In this section you modify these program units and add them into COURSE_PACKAGE. In the package, you declare the values for the term ID, section number, faculty ID, date, and time as public variables, and the program that calls the package initializes these variables rather than passing the variable values as parameters. (Recall that public variables can be referenced by any procedure.) Because you have been inserting many records in the COURSE and COURSE_SECTION tables, you first run the Ch9Northwoods.sql script to refresh your database tables.

To use SQL*Plus to create the package specification:

1. In SQL*Plus, run **Ch9Northwoods.sql** from the Chapter09 folder on your Data Disk to refresh the database tables.

2. Switch to Notepad, and type the commands in Figure 9-34 to create the package specification.

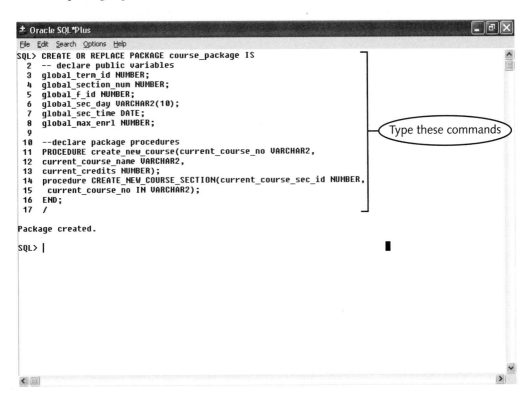

Figure 9-34   Creating a package specification in SQL*Plus

3. Copy the commands, switch to SQL*Plus, paste the copied commands, and then execute the commands. Debug the commands if necessary. The message "Package created" confirms that the DBMS successfully created the package specification.

Creating the Package Body in SQL*Plus

To create a package body in SQL*Plus, you place the **CREATE OR REPLACE PACKAGE BODY** command before the package body specification. The package body must contain commands to define each procedure or function that appears in the package specification, and the header for the procedure or function must exactly match the procedure header or function in the package specification. Now you create the COURSE_PACKAGE package body in SQL*Plus. To avoid errors, you copy the procedure declarations from the package specification, and paste them into the package body. Then you add the commands that specify the body of each procedure.

To create the package body in SQL*Plus:

1. Switch to Notepad, and type the following as the first command of the package body:

   CREATE OR REPLACE PACKAGE BODY course_package IS

2. To ensure that the procedure declarations exactly match the procedure declarations in the package specification, find the commands in the package specification that declare the CREATE_NEW_COURSE and CREATE_NEW_COURSE_SECTION procedures, copy the commands, and then paste the copied commands after the command you typed in Step 1.

3. Modify the pasted commands and add the shaded commands so your package body looks like Figure 9-35. Copy the code, execute it in SQL*Plus, and debug the code if necessary. The message "Package body created" confirms that the DBMS successfully created the package body.

9

```
CREATE OR REPLACE PACKAGE BODY course_package IS
 PROCEDURE CREATE_NEW_COURSE
        (current_course_no IN VARCHAR2,
         current_course_name IN VARCHAR2,
         current_credits IN NUMBER)
 IS
         current_course_sec_id NUMBER;
 BEGIN
         --retrieve next C_SEC_ID_SEQUENCE value
         SELECT c_sec_id_sequence.NEXTVAL
         INTO current_course_sec_id
         FROM dual;
         --insert the COURSE record
         INSERT INTO course
         VALUES(current_course_no, current_course_name, current_credits);
         --call the procedure to insert the COURSE_SECTION record
         CREATE_NEW_COURSE_SECTION(current_course_sec_id, current_course_no);
 END CREATE_NEW_COURSE;

 PROCEDURE CREATE_NEW_COURSE_SECTION (current_course_sec_id IN NUMBER,
 current_course_no IN VARCHAR2)
 IS
 BEGIN
         --insert new course section
         INSERT INTO course_section(c_sec_id, course_no, term_id, sec_num,
         f_id, c_sec_day, c_sec_time, max_enrl)
         VALUES (current_course_sec_id, current_course_no, global_term_id,
         global_section_num, global_f_id, global_sec_day,
         TO_DATE(global_sec_time, 'DD-MON-YYYY HH:MI AM'), global_max_enrl);
         COMMIT;
 END CREATE_NEW_COURSE_SECTION;
 END course_package;
 /
```

Figure 9-35   Commands to create the package body

Referencing Package Objects

The first time a program references a package object, the DBMS loads the package into the database server's main memory and allocates space to store the package's variable values. From then on, the package remains in memory and is available to all users who have the privilege to execute the package.

To reference a package item, you must preface the item with the package name, using the syntax *package_name.item_name*. For example, to assign the value 6 to the global_term_id variable in COURSE_PACKAGE, you use the following command:

```
COURSE_PACKAGE.global_term_id := 6;
```

To call a procedure or function that is within a package, you preface the procedure or function name with the package name. For example, you use the following command to call the CREATE_NEW_COURSE procedure within the COURSE_PACKAGE and pass to it variable values representing the course number, course name, and credit hours:

```
COURSE_PACKAGE.CREATE_NEW_COURSE
(current_course_no, current_course_name, current_credits);
```

You have used the DBMS_OUTPUT built-in package to display output in SQL*Plus: the package name is DBMS_OUTPUT, and the procedure name is PUT_LINE, so you call the procedure using the command `DBMS_OUTPUT.PUT_LINE('message text');`.

When you create a new package, the package exists in your user schema, and you are its owner. To grant other users the privilege to execute a package, you use the general syntax `GRANT EXECUTE ON package_name TO username;`.

Next, you write an anonymous PL/SQL program to test COURSE_PACKAGE. The program initializes the values of global_term_id to 6, global_section_num to 1, global_f_id to 4, global_s_date to MW and global_sec_time to 4:00 PM. Then, the program calls the CREATE_NEW_COURSE procedure from within the package. After you run the program, you query the database to confirm that the package correctly inserted the records.

To create and test the program to reference the package items:

1. Type the commands in Figure 9-36 to initialize the package variables and call the package procedure.

9

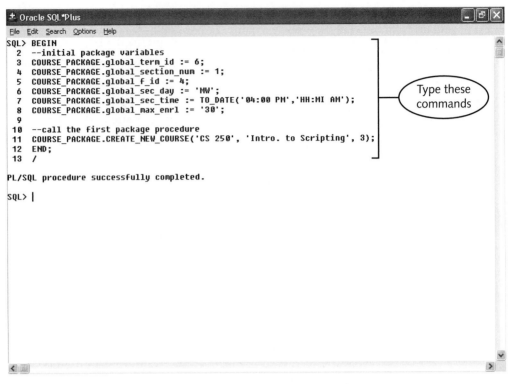

Figure 9-36 Anonymous PL/SQL program to reference package items

2. Execute the commands, and debug them if necessary. The message "PL/SQL procedure successfully completed" indicates that the program successfully referenced the package items.

3. Type and execute the queries in Figure 9-37 to confirm that the new order and order line records were inserted correctly.

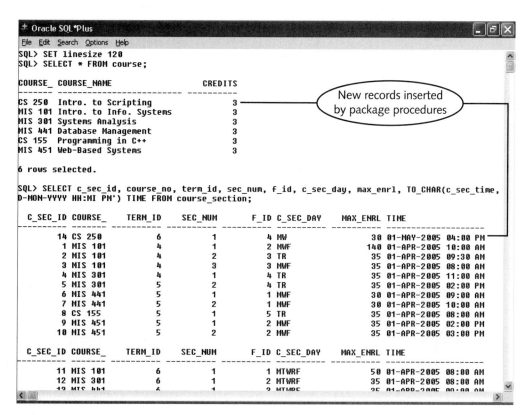

Figure 9-37 Records inserted by package procedures

NOTE

Your c_sec_id sequence value may not match the one in Figure 9-37.

Creating a Package in Forms Builder

To create a package in Forms Builder, you create the package specification and package body as program units within a form. After you test and debug the package, you store the package in the database.

Creating the Package Specification in Forms Builder

To create a package specification in Forms Builder, you create a new program unit of type Package Spec, and then create the package specification in the PL/SQL Editor, using the general syntax in Figure 9-32.

Now you open the program unit test form from the Chapter09 folder on your Data Disk, and create a new package specification in Forms Builder for a package named FUNCTION_PACKAGE. This package contains the AGE and DAYS_BETWEEN functions that you created in Lesson A to calculate a person's age based on his or her date of birth and to calculate the number of days between two known dates.

To create the package specification:

1. Switch to Forms Builder, and open **Test.fmb** from the Chapter09 folder on your Data Disk. Save the file as **Ch9BTest_PACKAGE.fmb** in the Chapter09\Tutorials folder on your Solution Disk.

2. In the Object Navigator, click the Program Units node, and then click the **Create** button ⊕. The New Program Unit dialog box opens.

3. Type **FUNCTION_PACKAGE** in the Name field, click the **Package Spec** option button, and then click **OK**. The PL/SQL Editor opens.

4. Type the commands in Figure 9-38 to create the package specification. Compile the commands, debug them if necessary, and then close the PL/SQL Editor and save the form.

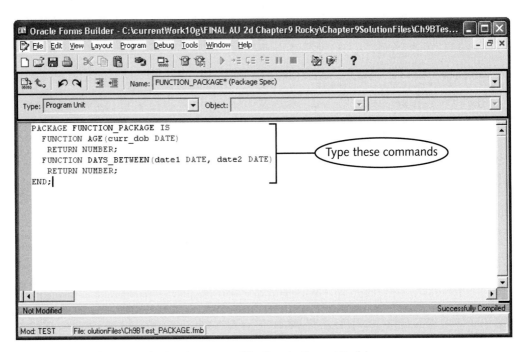

Figure 9-38   Creating the package specification in Forms Builder

Creating the Package Body in Forms Builder

To create a package body in Forms Builder, you create a new program unit of type Package Body, and then enter the package body commands in the PL/SQL Editor. Recall that the package body must have the same name as the package specification, and that you must have already created the package specification before you can create the package body.

Now you create and compile the FUNCTION_PACKAGE body, and then create a form trigger for the test button that calls one of the package functions. The trigger calls the DAYS_BETWEEN function, and displays the number of days between the current system date and January 1, 2003. To test the package, you write a form trigger that references the package items. Recall that to reference a package item, you use the general syntax *package_name.item_name*.

To create the package body, create a form trigger, and test the package:

1. To create the package body, click the Program Units node, and then click the **Create** button ✛. Type **FUNCTION_PACKAGE** in the Name field, click the **Package Body** option button, and then click **OK**.

2. Type the commands in Figure 9-39 to create the package body.

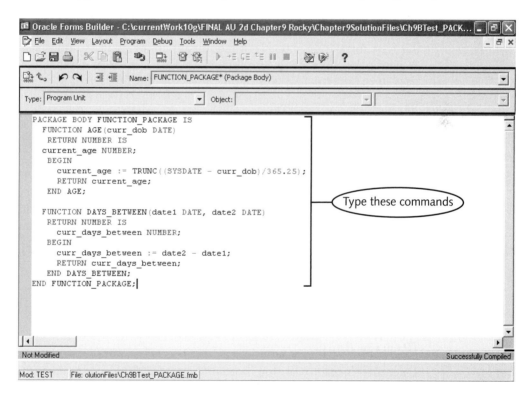

Figure 9-39   Creating the package body in Forms Builder

3. Compile the package body, correct any syntax errors, and then close the PL/SQL Editor.

4. To create the form trigger to test the package, click **Tools** on the menu bar, and then click **Layout Editor**. Click the **Test Program Unit** button, right-click, point to **SmartTriggers**, and then click **WHEN-BUTTON-PRESSED**.

5. Type the following command to call the DAYS_BETWEEN function in the package, and display the result in the form output field:

```
:test.output_item := FUNCTION_PACKAGE.DAYS_BETWEEN
(TO_DATE'01/01/2003', 'MM/DD/YYYY'), SYSDATE);
```

6. Compile the trigger, correct any syntax errors, and then close the PL/SQL Editor.

7. To change the data type of the form output field so it displays a formatted numerical value that does not display any decimal fractions, select **OUTPUT_ITEM**, right-click, and then click **Property Palette**. Scroll down to the Data property node, change the Data Type property value to **Number** and the Format Mask property value to **999,999,999**, and then close the Property Palette.

8. Save the form, and then run it.

9. The test form trigger does not require any input values, so do not type any values in the input fields, and then click **Test Program Unit**. The number of days since January 1, 2003, appears in the output field.

10. Close the browser window.

Saving the Package in the Database

Recall that you ultimately save packages in the database to make the packages' objects available to other database users and to all of your PL/SQL programs. To save a package that you create in Forms Builder in the database, you drag the package specification and package body to the Object Navigator Database Objects node, and drop the program units onto the PL/SQL Stored Program Units node under your database schema. Now you save the FUNCTION_PACKAGE specification and body in the database.

To save the package in the database:

1. Click **Window** on the menu bar, and then click **Object Navigator**.

2. Click **FUNCTION_PACKAGE (Package Spec)** under the Program Units node, press and hold **Shift**, and then click **FUNCTION_PACKAGE (Package Body)**.

3. Drag the objects toward the bottom edge of the screen, and drop them onto the PL/SQL Stored Program Units node under your database schema. The package specification and body appear among your stored procedures.

4. Close the TEST form in Forms Builder, and save any changes, if necessary.

DATABASE TRIGGERS

Database triggers are program units that execute in response to the database events of inserting, updating, or deleting a record. Database triggers are different from form triggers, which are programs that execute in response to form events such as clicking a button.

Database triggers are useful for maintaining integrity constraints. For example, when you delete a course from the Northwoods University COURSE table, you might create a trigger to automatically delete all associated records in the COURSE_SECTION table. Triggers are also useful for creating auditing information. For example, you might want to record the username of every user who modifies the GRADE field in the Northwoods University ENROLLMENT table.

Database triggers are similar to all PL/SQL program units in the sense that they have declaration, body, and exception sections. However, one difference between triggers and other program units is that triggers cannot accept input parameters. Triggers and program units also differ in the way they execute. You must explicitly execute program units by typing the program unit name at the command prompt or calling them from another procedure. In contrast, a trigger executes only when its triggering event occurs.

9

Database Trigger Properties

Developers define database triggers based on the type of SQL statement that causes the trigger to fire, the timing of when the trigger fires, and the level at which the trigger fires.

Trigger timing defines whether a trigger fires before or after the SQL statement executes, and can have the values BEFORE or AFTER. A BEFORE trigger allows you to capture the data values in the database before the trigger statement executes. For example, you might create a BEFORE trigger that fires when a user updates the GRADE field in the Northwoods University ENROLLMENT table to record the grade value before it is updated. You might create an AFTER trigger to update the student's grade point average after the grade is changed.

The **trigger statement** defines the type of SQL statement that causes a trigger to fire, which can be INSERT, UPDATE, or DELETE. For example, you create a DELETE trigger to automatically delete all associated COURSE_SECTION records when the user deletes a COURSE record, or you use an UPDATE trigger to create an audit trail when a user updates the ENROLLMENT table.

The **trigger level** defines whether a trigger fires once for each triggering statement or once for each row affected by the triggering statement, and can have the values ROW or STATEMENT. **Statement-level triggers** fire once, either before or after the SQL triggering statement executes. For example, you would use a statement-level trigger to create an audit trail that records the username and system date and time when a user updates the ENROLLMENT table. This trigger creates a single audit record, regardless of how many ENROLLMENT rows the user's command updates. **Row-level triggers** fire once for each row affected by the triggering statement. In the audit trail example, you would use a row-level trigger to record exactly which rows in the ENROLLMENT table the user changes and record the GRADE values before and after the change.

You can reference the value of a field in the current record both before and after the triggering statement executes. To reference a value in the trigger body before the triggering SQL statement executes, you use the syntax :OLD.*fieldname*. For example, to reference the value of the GRADE field before it is updated, you would use the statement :OLD.grade. To reference a field value after the triggering statement executes, you use the syntax :NEW.*fieldname*. Therefore you use the statement :NEW.grade to reference the grade value after the SQL statement executes and updates the grade.

Creating Database Triggers

Figure 9-40 shows the general syntax to create a database trigger.

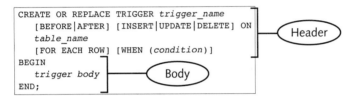

```
CREATE OR REPLACE TRIGGER trigger_name
   [BEFORE|AFTER] [INSERT|UPDATE|DELETE] ON
   table_name
   [FOR EACH ROW] [WHEN (condition)]
BEGIN
   trigger body
END;
```

Header

Body

Figure 9-40   General syntax to create a trigger in SQL*Plus

The **database trigger header** specifies the trigger's properties, including the trigger timing (BEFORE or AFTER), SQL statement type (INSERT, UPDATE, or DELETE), the database table with which you associate the trigger, and the trigger level (statement-level or row-level). The **trigger body** contains the PL/SQL code that executes when the trigger fires.

The Database Trigger Header

In the database trigger header, the command **CREATE OR REPLACE TRIGGER** specifies to create a new trigger. *Trigger_name* must follow the Oracle naming standard. A trigger can have the same name as existing PL/SQL program units in a database schema, because trigger names are stored in a namespace that is different from the namespace that other Oracle database objects use. (A **namespace** is a place in the database that stores object identifiers for a user's database objects.) The BEFORE or AFTER command specifies the trigger timing, and the INSERT, UPDATE, or DELETE command specifies the statement type that causes the trigger to fire. To create a trigger that fires for multiple statement types, you join the statement types using the OR operator. The FOR EACH ROW clause specifies that the trigger is a row-level trigger. To create a statement-level trigger, you omit the FOR EACH ROW clause.

For example, the following trigger header defines a trigger named TEST_TRIGGER that fires after the execution of either an UPDATE or a DELETE statement in the ENROLLMENT table, and fires once for each row that is updated or deleted:

```
CREATE OR REPLACE TRIGGER test_trigger
AFTER UPDATE OR DELETE ON
ENROLLMENT
FOR EACH ROW;
```

The **WHEN** clause in the trigger header is optional, and you use it only for row-level triggers. It specifies that the trigger fires only for rows that satisfy a specific search condition. For example, in the trigger to record when a grade value is updated or deleted, you would want the trigger to fire only when the current value of the GRADE field is changed, but not when the grade is initially inserted. To do this, you would specify the following WHEN clause:

```
WHEN OLD.grade IS NOT NULL;
```

Note that in the WHEN clause, you omit the colon before the OLD or NEW qualifier.

The Database Trigger Body

The database trigger body contains the commands that execute when the trigger fires. The trigger body is a PL/SQL code block that contains the usual declaration, body, and exception sections. It cannot contain transaction control statements, such as COMMIT, ROLLBACK, and SAVEPOINT. You can reference the NEW and OLD field values only in a row-level trigger.

Creating Database Triggers to Leave an Audit Trail for the Northwoods University ENROLLMENT Table

Now you create a statement-level and a row-level trigger to leave an audit trail for the Northwoods University ENROLLMENT table and track when users insert, update, and delete table records. When you create a trigger that leaves an audit trail, you create one or more tables to store the audit trail values. Possible audit trail information can include when an action occurs, who performs the action, and the nature of the action.

Figure 9-41 shows the audit trail tables for the ENROLLMENT table.

ENRL_AUDIT

ENRL_AUDIT_ID	DATE_UPDATED	UPDATING_USER
NUMBER(5)	DATE	VARCHAR2(30)

ENRL_ROW_AUDIT

ENRL_ROW_AUDIT_ID	S_ID	C_SEC_ID	OLD_GRADE	DATE_UPDATED	UPDATING_USER
NUMBER(5)	VARCHAR2(6)	NUMBER(6)	CHAR(1)	DATE	VARCHAR2(30)

Figure 9-41 ENROLLMENT audit trail tables

For the ENROLLMENT table audit trail, whenever one or more ENROLLMENT records change, a statement-level trigger inserts a single audit trail record into the ENRL_AUDIT table. The table has a primary key field named ENRL_AUDIT_ID. The DATE_UPDATED field stores the date of the change, and the UPDATING_USER field stores the username of the user who makes the change. (You use the Oracle 10*g* USER system variable to retrieve the username of the current database user.)

The row-level trigger inserts an audit trail record into the ENRL_ROW_AUDIT table that records individual record changes. The ENRL_ROW_AUDIT table has a primary key field named ENRL_ROW_AUDIT_ID that stores the table's primary key. The S_ID field stores the student ID, and the C_SEC_ID field stores the course section ID. The OLD_GRADE field stores the old grade value of the record, the DATE_UPDATED field stores the date the change was made, and the UPDATING_USER field stores the username of the user who made the change.

Next, you create the ENRL_AUDIT and ENRL_ROW_AUDIT tables in your database schema by running a script file named TriggerTables.sql that contains the commands to create the audit trail tables. This script also contains commands to create sequences named ENRL_AUDIT_ID_SEQUENCE and ENRL_AUDIT_ROW_ID_SEQUENCE, which automatically generate values for the primary keys for the new trigger tables.

To create the ENRL_AUDIT and ENRL_ROW_AUDIT tables and sequences:

1. In SQL*Plus, run the **TriggerTables.sql** script in the Chapter09 folder on your Data Disk.

2. The "Table created" and "Sequence created" confirmation messages should appear twice, indicating that the script successfully created the tables and sequences.

You can create database triggers in SQL*Plus and in Forms Builder. The following sections illustrate creating triggers in both environments.

Creating a Database Trigger in SQL*Plus

Now you use SQL*Plus to create the statement-level database trigger that fires whenever a user inserts, updates, or deletes a record in the ENROLLMENT table. This trigger follows the general syntax in Figure 9-40. The trigger timing is AFTER. The statement type is INSERT, UPDATE, or DELETE, and the table is ENROLLMENT. When the trigger fires, it inserts a record into the ENRL_AUDIT table that retrieves the next value in the table's sequence and inserts the date the change was made, along with the username of the user who made the change. After you create the trigger, you change the grade in the ENROLLMENT record for S_ID 1 and C_SEC_ID 1 from A to B, and query the ENRL_AUDIT table to confirm that the trigger fired correctly and inserted the audit trail record.

To create and test the statement-level trigger:

1. Type the commands in Figure 9-42.

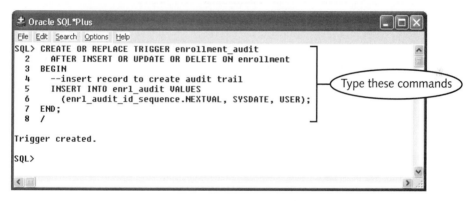

Figure 9-42   Creating a statement-level trigger in SQL*Plus

2. Execute the commands, and debug them if necessary. The message "Trigger created" confirms that the DBMS successfully created the trigger.

> **NOTE**
> To create a trigger, you must have the CREATE TRIGGER database privilege, and you must be the owner of the table associated with the trigger.

3. Type and execute the following command to update the GRADE value in the ENROLLMENT table for S_ID NG100 and C_SEC_ID 11:

```
UPDATE enrollment
SET grade = 'B'
WHERE s_id = 'NG100' AND c_sec_id = 11;
```

The message "1 row updated" should appear.

4. Type and execute **SELECT * FROM enrl_audit;** to verify that the trigger inserted the audit trail record into the ENRL_AUDIT table. The output should show a record of the update, including the ENRL_AUDIT_ID value generated by the sequence, the current date, and your username.

Note that this trigger fired only once, when one row was updated. It would also fire only once if multiple rows were updated, because it is a statement-level trigger.

Creating a Database Trigger in Forms Builder

To create a database trigger in Forms Builder, you open the Database Objects node in the Object Navigator, open the node for your user schema, open the Tables node to display your database tables, open the node that represents the table to which the trigger is attached, select the Triggers node, and then click the Create button. Then you use the Database Trigger dialog box in Figure 9-43 to specify the trigger properties.

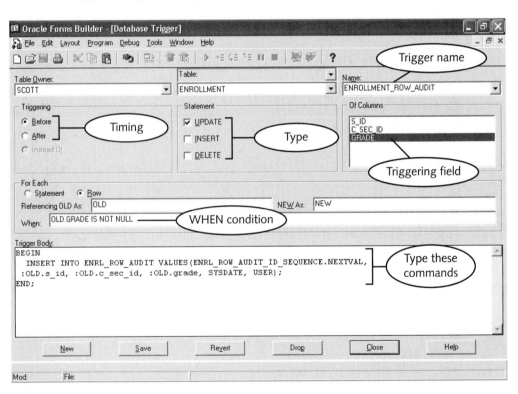

Figure 9-43   Database Trigger dialog box

The Database Trigger dialog box allows you to specify the trigger properties. The Table Owner field displays the table owner, and the Table field shows the table to which the trigger is attached. The other fields on the dialog box allow you to specify the trigger name, triggering statement, timing, and level. The Database Trigger dialog box also allows you to enter the PL/SQL commands that make up the trigger body. The Referencing OLD As and NEW As fields allow you to enter alternate aliases for OLD (field values before the trigger fires) and NEW (field values after the trigger fires), in case you have database tables named OLD or NEW.

Now you use Forms Builder to create a row-level trigger named ENROLLMENT_ ROW_AUDIT that records information in the ENRL_ROW_AUDIT table (see Figure 9-41). The trigger timing is BEFORE, because it records the current (old) values of S_ID, C_SEC_ID, and GRADE. The statement type is UPDATE, and the table is ENROLLMENT. The WHEN clause specifies that the trigger fires only when the user changes the grade value. When the trigger fires, it inserts a record into the ENRL_ROW_AUDIT table, which records the current (old) values of S_ID, C_SEC_ID, and GRADE, as well as the date the change was made and who made the change.

To create the trigger in Forms Builder:

1. Switch to Forms Builder, and open the following nodes in the Object Navigator: Database Objects, *your_username*, Tables, ENROLLMENT, and Triggers. (Some of these nodes may be open already.) The ENROLLMENT_ AUDIT trigger that you just created in SQL*Plus should appear.

HELP

If an error occurs, close Forms Builder, start it again, and connect to the database.

2. Make sure that the Triggers node is selected, and then click the **Create** ✛ button. The Database Trigger dialog box in Figure 9-43 opens. The Database Trigger dialog box currently shows the properties for the ENROLLMENT_ AUDIT trigger you created earlier.

3. To create a new database trigger, click **New** on the lower-left area of the dialog box. The dialog box field values show the default values for a new database trigger.

4. Delete the default trigger name in the Name field, and type **ENROLLMENT_ROW_AUDIT** for the trigger name.

5. Be sure the Before option button is selected to specify the trigger timing, check the **UPDATE** check box to specify the trigger statement, and select **GRADE** in the Of Columns list to specify that the trigger fires when the user updates the GRADE field.

6. Click the **Row** option button to specify that this is a row-level trigger. The trigger's type and action specifications should appear as shown in Figure 9-43.

To finish creating the trigger, you specify the WHEN condition, specify the OLD and NEW aliases, and enter the trigger body code that executes when the trigger fires. Because you don't have tables named OLD or NEW, you use the default alias names of OLD and NEW.

To enter the trigger aliases and trigger body code:

1. Type **OLD** in the Referencing OLD As field, and type **NEW** in the NEW As field, as shown in Figure 9-43.

2. To specify the WHEN condition, type **OLD.GRADE IS NOT NULL** in the When field. Recall that in the WHEN condition, you do not preface the OLD reference with a colon.

3. Click the insertion point in the blank line after the BEGIN command in the Trigger Body field, and type the following commands. Note that in the trigger body code, you preface the OLD references with a colon.

```
INSERT INTO ENRL_ROW_AUDIT VALUES
(ENRL_ROW_AUDIT_ID_SEQUENCE.NEXTVAL,
:OLD.s_id, :OLD.c_sec_id, :OLD.grade, SYSDATE, USER);
```

4. Your completed database trigger specification should look like Figure 9-43. Click **Save** to create the trigger.

5. Click **Close** to close the Database Trigger dialog box. The new trigger appears in the Object Navigator window under the ENROLLMENT table.

Now you switch to SQL*Plus and test the new trigger. You change the grade for S_ID NG100 and C_SEC_ID 11 from B to A to fire the trigger, and then query the ENRL_ROW_AUDIT table to confirm that the trigger inserts the audit trail record providing row-level information.

To test the row-level trigger:

1. Type and execute the following query to update the ENROLLMENT table and fire the row-level trigger:

```
UPDATE enrollment
SET grade = 'A'
WHERE s_id = 'NG100' AND c_sec_id = 11;
```

2. Type and execute the following query to verify that the trigger inserted the audit trail record in the ENRL_ROW_AUDIT table. The output should show a record of the update, including the ENRL_ROW_AUDIT_ID value generated by the sequence, the student ID and course section ID, the former grade value, the current date, and your username.

```
SELECT * FROM enrl_row_audit;
```

The advantage of creating triggers in Forms Builder is that the Database Trigger dialog box allows you to specify trigger properties using option buttons and check boxes. The disadvantage of creating triggers in Forms Builder is that it is difficult to locate and debug errors in the trigger body because the Database Trigger dialog box provides very cryptic error messages.

Disabling and Dropping Triggers

After you create a new trigger, it automatically fires every time its triggering event occurs. If you no longer need a trigger, you drop it from the database using the following general command:

```
DROP TRIGGER trigger_name;
```

You can also **disable** a trigger, which allows the trigger to still exist in the user's database schema, but causes it no longer to fire when the triggering event occurs. When you first create a trigger, it is automatically enabled. The SQL command to disable or enable a trigger is:

```
ALTER TRIGGER trigger_name [ENABLE | DISABLE];
```

Next, you disable the ENROLLMENT_ROW_AUDIT trigger and then enable it again. Then, you drop the trigger.

To disable, reenable, and drop the trigger:

1. Type and execute the ALTER TRIGGER commands in Figure 9-44 to disable and then reenable the ENROLLMENT_ROW_AUDIT trigger.

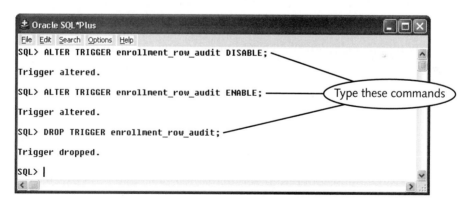

Figure 9-44 Disabling, enabling, and dropping a trigger

2. Type the DROP TRIGGER command in Figure 9-44 to drop the trigger.

Viewing Information About Triggers

You can use the USER_TRIGGERS data dictionary view to retrieve information about triggers. This data dictionary view contains columns describing the trigger name, type, triggering events, owner of the database table or other objects referenced in the trigger, trigger status, and so forth. Now you execute a query that lists the names, types, and events of your database triggers.

To query the USER_TRIGGERS data dictionary view:

1. Type and execute the command in Figure 9-45 to display information about your database triggers. (Your output will be different, based on the triggers in your particular database installation.)

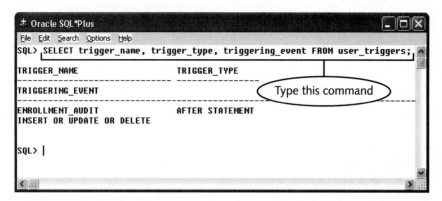

Figure 9-45   Displaying information about database triggers

2. Close SQL*Plus, Forms Builder, and Notepad, and save the changes to the **Ch9BQueries.sql** file. Then shut down the OC4J Instance.

SUMMARY

◘ To make applications more manageable, you often decompose them into logical units of work, write individual program units for each work unit, and call program units from other program units.

◘ To call a stored procedure from another stored procedure, you execute a PL/SQL command that specifies the name of the called procedure, followed by its parameter value list. You must create the called procedure before you create the calling procedure.

◘ A PL/SQL library is a collection of procedures and functions that you can attach to a form or report. You can save a library in a .pll file on a client workstation or a database server. You can store a reference to the library .pll file in the database as a database object.

- To create a PL/SQL library, you create a new library object in Forms Builder, and then add program units to the library. To use a library, you attach the library to a form or report, and then reference the library program units just as if they were program units within the form or report. The form or report only references the library, and does not contain the actual library code.

- A package is a code library that is stored in the database. A package provides a way to manage large numbers of related programs and variables and to create public variables and program units. Package objects include variables, cursors, procedures, and functions. You can reference a package object from any PL/SQL code block, provided you have privileges to execute the package.

- A package specification declares the public package objects. A package body contains code for program units declared in the package specification. You can create a package specification and body in either SQL*Plus or Forms Builder.

- Database triggers execute in response to the database events of inserting, updating, or deleting a record. To create a trigger, you specify the type of SQL statement that causes the trigger to fire; the trigger timing, which specifies if the trigger fires before or after the SQL statement executes; and the trigger level, which specifies whether the trigger fires only once per statement or once for every record that the SQL statement changes.

- The trigger header specifies the trigger properties. The trigger body specifies the PL/SQL commands that execute when the trigger fires.

- You can create a database trigger either in SQL*Plus or using the Forms Builder Database Trigger dialog box. The Database Trigger dialog box allows you to specify trigger properties easily, but makes debugging the code in the trigger body difficult.

REVIEW QUESTIONS

1. What are the valid timing values for a trigger?

2. When creating packages, the package specifications are optional, but the package body is required. True or False?

3. When you create a trigger, it is automatically _____ and fires if the triggering event occurs.

4. After a trigger fires, you can reference NEW and OLD field values if the trigger was a _____ trigger.

5. Which DML operations can be used to specify when a trigger should fire?

6. Packages should always be saved in the appropriate form. True or False?

7. Variables that are declared within individual procedures are visible only to the procedure within which they are declared. True or False?

8. Where is a package executed, on the client or database server?

9. A(n) _____ is similar to a package, but is explicitly attached to a form or report and executed on the client workstation.

10. The inclusion of a WHEN clause in a trigger is only appropriate for a(n) _____-level trigger.

MULTIPLE CHOICE

1. Which of the following commands is not permitted in a database trigger?

   a. INSERT

   b. SELECT

   c. COMMIT

   d. DELETE

2. A _____ trigger fires only once, regardless of how many rows are changed.

   a. row-level

   b. statement-level

   c. database-level

   d. form-embedded

3. When creating package procedures, the actual commands for the package procedure appear in the:

   a. package body

   b. package specifications

   c. calling form

   d. calling report

4. The file containing the PL/SQL Library Executable file has the extension _____.

   a. .plx

   b. .pll

   c. .ocr

   d. .lib

5. When you attach an existing library to a form, you need to specify the file version with the _____ extension.

    a. .plx

    b. .pll

    c. .ocr

    d. .lib

6. You can retrieve information about triggers from the _____ view.

    a. V$TRIGGERS

    b. USERTRIGGER

    c. USER$TRIGGER

    d. USER_TRIGGERS

7. When you create a library design file, it is automatically assigned the extension _____ .

    a. .plx

    b. .pll

    c. .ocr

    d. .lib

8. Which of the following *cannot* accept input values?

    a. procedures

    b. functions

    c. triggers

    d. none of the above—they all accept input values

9. In Forms Builder, if a(n) _____ appears next to the procedure name, it has been modified and should be recompiled.

    a. !

    b. @

    c. /

    d. *

9

10. When you create a package in Forms Builder and decide that you want to make it available to other database users and programs, you should place the package under the _____ node in the Object Navigator.

   a. Forms -> Library

   b. Forms -> Program Units

   c. Database Users -> PL/SQL Stored Program Units

   d. Database Objects -> PL/SQL Stored Program Units

PROBLEM-SOLVING CASES

The cases refer to the Clearwater Traders (Figure 1-26) database. Save all solution files in the Chapter09\Cases folder on your Solution Disk. Run the Ch9Clearwater.sql script in the Chapter09 folder on your Data Disk to refresh the sample databases before beginning work on the Cases.

1. In this case, you use SQL*Plus to create stored procedures that process incoming shipment items that Clearwater Traders receives. The first stored procedure updates the SHIPMENT_LINE table and then calls a second stored procedure that updates the quantity on hand (QOH) of the item in the INVENTORY table. Save all commands in a text file named Ch9BCase1.sql.

   a. In SQL*Plus, create a stored procedure named UPDATE_INV_QOH that receives input parameters of an inventory ID and an update quantity, and then uses these values to update the INV_QOH field in the INVENTORY table.

   b. In SQL*Plus, create a second procedure named UPDATE_SHIPMENT_LINE that receives input parameters of an existing shipment ID and inventory ID. The procedure should update the associated SHIPMENT_LINE record based on the input values, using the system date as the date received. It should then call the UPDATE_INV_QOH procedure to update the associated inventory quantity on hand using the shipment line quantity.

   c. Write an anonymous PL/SQL program to call the UPDATE_SHIPMENT_LINE procedure, and pass to it values to show that the units of inventory ID 5 were received for shipment ID 3. Execute the anonymous PL/SQL program, and then use SELECT statements to display the updated rows from the two tables to verify the procedures worked correctly.

2. In this case, you create a PL/SQL library that contains the procedures you created in Case 1. Then you attach the library to the test form, and create a trigger in the test form to call the library procedures.

   a. Open Test.fmb from the Chapter09 folder on your Data Disk, and save the form as Ch9BCase2.fmb. Create a PL/SQL library named Ch9BCase2, and save the library file as Ch9BCase2.pll. Add to the library the UPDATE_INV_QOH and UPDATE_SHIPMENT_LINE procedures that you created in Case 1.

    b. Attach the Ch9BCase2 library to the test form.

    c. Create a form trigger for the test button that allows the user to input the values for the shipment ID and inventory ID in the form input fields, and then calls the UPDATE_SHIPMENT_LINE procedure.

3. In this case, you create a package named SHIPMENT_PKG that contains procedures associated with the Clearwater Traders database for creating new shipments and shipment lines when orders are placed with suppliers and for processing incoming shipments when orders are received.

    a. In Forms Builder, open Test.fmb, and save the form as Ch9BCase3.fmb.

    b. Create a package specification named ORDER_PACKAGE that declares a variable named CURR_SHIPMENT_ID as a public variable, and declares the procedures and associated input variables shown in Table 9-2.

Procedure Name	Input Variables	Procedure Description
ADD_SHIPMENT	CURR_DATE_EXPECTED	Inserts a new SHIPMENT record using CURR_SHIPMENT_ID and the procedure input variables
ADD_SHIP_LINE	CURR_INV_ID, CURR_QUANTITY	Inserts a new SHIPMENT_LINE record using CURR_SHIPMENT_ID and the procedure input variables
UPDATE_SHIP_LINE	CURR_INV_ID, CURR_QUANTITY	Updates a SHIPMENT_LINE record using CURR_SHIPMENT_ID and the procedure input variables; uses the system date for the SL_DATE_RECEIVED field value
UPDATE_INV_QOH	CURR_INV_ID, CURR_QUANTITY	Updates an INVENTORY record using the procedure input variables

Table 9-2   Case procedures and variables

    c. Create a package body containing commands for the specified procedures.

    d. Store the package specification and body as stored program units in the database.

    e. Modify the label of the test button's label to Add Shipment. Then create a form trigger for the button that calls the package's ADD_SHIPMENT and ADD_SHIP_LINE procedures to add shipment ID 30, for 100 units of inventory ID 3. Use the ADD_MONTHS function to set the date expected as one month from the current system date. (The user will not enter any values in the form input fields.)

f. Create a new button with the label Update Shipment. Then create a form trigger for the button that uses the package's UPDATE_SHIP_LINE and UPDATE_INV_QOH procedures to record that 25 units of inventory ID 2 were received for shipment ID 2.

4. In this case, you use SQL*Plus to create a series of triggers that automatically update the quantity on hand field in the Clearwater Traders INVENTORY table by the correct amount whenever an ORDER_LINE record is inserted, updated, or deleted. Save all source code in a text file named Ch9BCase4.sql.

a. Create a trigger named QOH_INSERT that is associated with the ORDER_LINE table and subtracts the order quantity of the specified inventory item from the INV_QOH field in the INVENTORY table when a new item is sold to a customer.

b. Create a trigger named QOH_UPDATE that is associated with the ORDER_LINE table and correctly adjusts the INV_QOH field in the INVENTORY table whenever the OL_QUANTITY field is updated for the associated inventory item.

c. Create a trigger named QOH_DELETE that is associated with the ORDER_LINE table and correctly adjusts the INV_QOH field in the INVENTORY table when a record is deleted from the ORDER_LINE table.

ADVANCED FORMS BUILDER TOPICS

After completing this chapter, you should be able to:

♦ Create form items that display boilerplate items, calculated data, and images

♦ Create forms that allow users to load and display image data

♦ Create form lists that display static data values

♦ Configure forms that retrieve and manipulate large data sets

♦ Control data block relationships

In the previous chapters, you learned how to create and use data block and custom forms, and how to integrate forms to create a database application. This chapter explores advanced Forms Builder topics—creating forms that display and manipulate different types of data in different ways, and creating reusable form objects that make it easier to develop forms.

CREATING NON-INPUT FORM ITEMS

So far, most of the form items you have created (text items, radio groups, check boxes, and so forth) allow users to input and change data values. Forms Builder also supports a number of items that display data but do not allow the user to change the displayed value. Table 10-1 summarizes the different types of non-input form items and identifies the tool on the Layout Editor tool palette that you use to create the item.

Item Type	Description	Usage	Layout Editor Tool
Display Item	A read-only text item that does not allow user input	Provide calculated fields and read-only data fields	Display Item tool
Boilerplate Objects	Frames and shapes (rectangles, circles, lines, and so forth)	Enhance form appearance	Frame tool Rectangle tool Ellipse tool Line tool
Boilerplate Text	Form text that appears directly on the form	Provide form titles and other nondata text	Text tool

Table 10-1   Non-input form items

To practice creating non-input form items, you work with the Northwoods University Student Grade form in Figure 10-1.

This data block form displays data from the Northwoods University STUDENT_GRADES_VIEW database view. It contains additional **display items** that display text data that the user cannot modify. The form has a display item that shows each student's earned grade points, calculated as the course credit hours times the grade points. A summary display item sums the grade points of all completed courses. A boilerplate rectangle and text label describe and emphasize the summary display item. Now you refresh your case study database tables, then start Forms Builder and open the form, which does not yet contain the display or boilerplate items.

To refresh your database tables, and then open the form:

1. Start SQL*Plus, log onto the database, and run the **Ch10Northwoods.sql** and **Student_Grades_View.sql** scripts in the Chapter10 folder on your Data Disk.

2. Start the OC4J Instance.

3. Start Forms Builder, and connect to the database. Open **Student_Grades.fmb** from the Chapter10 folder on your Data Disk, and save the file as **Ch10Student_Grades.fmb** in the Chapter10\Tutorials folder on your Solution Disk.

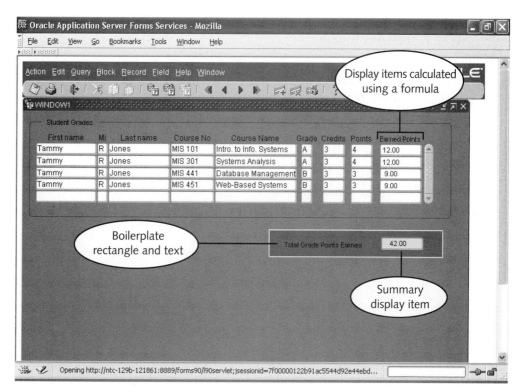

Figure 10-1   Form with calculated display items

Creating Display Items

A display item is a text box that displays data values that the user cannot change directly. Display items can be used to display calculated values based on form data values.

To create a display item, you select the Display Item tool on the Layout Editor tool palette, and then draw the display item on the form canvas. Next you open the display item's Property Palette and configure its properties so it displays the desired data. Now you create the display item that shows the calculated value for each completed course. You create this display item in the STUDENT_GRADES_VIEW data block. Currently, the STUDENT_GRADES_VIEW data block uses a tabular-style layout, with five records appearing on the canvas at one time. When you draw the new display item in the data block, five display items appear, one for each record. You modify the properties of the new display item by changing its name, data type, maximum length, format mask, and prompt.

To create the display item to display the calculated item values:

1. Click **Tools** on the menu bar, and then click **Layout Editor**. The form canvas appears.

2. Select the **Student_Grades** frame, and make it wider by dragging the center selection handle on the right edge of the frame toward the right edge of your screen display.

3. Select the **scrollbar**, and move it toward the right edge of your screen display so that there is an empty space wide enough for the Earned Points column.

4. Open the **Block** list at the top of the Layout Editor window, and select **Student_Grades_View** to place the new display item in the **Student_Grades_View** block.

5. Click ⊞ on the tool palette, and draw a rectangle that is the same height as the other text items, and as wide as the Earned Points field in Figure 10-1. Because the display item is in the **Student_Grades_View** block, and the block displays five records at a time, five individual display items appear, even though you drew only one display item.

TIP   Note that the Display Item tool ⊞ shows a text box with a gray background, while the Text Item tool ⊞ shows a text box with a white background.

6. Resize and reposition the display items so they look like the Earned Points column display items in Figure 10-1.

7. To change the background color of the display items, select the display items, select the **Fill Color** tool ⊞, and select any **white square** but the second **white square** in the top row.

8. Double-click any of the display items on the Layout Editor canvas to open the display item Property Palette, and then modify the following properties (do *not* close the Property Palette):

Property	Value
Name	**EARNED_POINTS_DISPLAY_ITEM**
Data Type	**Number**
Maximum Length	**6**
Format Mask	**99.99**
Prompt	**Earned Points**
Prompt Attachment Edge	**Top**
Prompt Alignment	**Center**

The next step is to specify how the display item calculates the value it displays. To do this, you change the Calculation Mode property, which can have one of three values: None, which indicates the item is not a calculated value; Formula, which indicates the item's value is calculated by a PL/SQL formula; or Summary, which indicates the item's value is calculated by a summary operation (such as SUM or AVG) on a single form item. The Earned Points display item contains values calculated using a PL/SQL formula. Display items that display calculated values always base the calculations on the data values that appear in form items such as text items. In the code for the PL/SQL formula, you reference form items within the formula using the block name and item name, separated by a period and prefaced with a colon. In the Student Grade form, the formula calculates the display item value as the product of the CREDITS and POINTS text items, which are in the STUDENT_GRADES_VIEW block. The formula appears as follows:

:STUDENT_GRADES_VIEW.CREDITS * :STUDENT_GRADES_VIEW.POINTS

A display item that displays a calculated value does not have to be in the same data block as the source items used in the calculation formula.

Now you change the calculation properties of the display item, then run the form, retrieve all of the records from the STUDENT_GRADES_VIEW view for Tammy Jones, and view the calculated earned point values.

To configure the display item calculation properties and run the form:

1. In the EARNED_POINTS_DISPLAY_ITEM Property Palette, scroll up to the Calculation property node, select the **Calculation Mode** property, open the list, and select **Formula**.

2. Select the **Formula** property, click the **More** button ⬚ , type the following formula, and then click **OK**.

   ```
   :STUDENT_GRADES_VIEW.CREDITS * :STUDENT_GRADES_VIEW.↵
   POINTS
   ```

3. Close the Property Palette, and save the form.

4. Run the form, click the **Enter Query** button 🔍, type **Tammy** in the S_First field and **Jones** in the S_Last field, and then click the **Execute Query** button 🔍 to retrieve the records from the STUDENT_GRADES_VIEW view. The records appear, along with the calculated values.

5. Close the browser window.

Next, you create the summary display item that shows the total grade points earned. This item uses the Summary calculation mode. You must specify the name of the item being summarized, the data block that contains the summary item, and the summary function (such as SUM or AVG) that the display item uses. You place the summary item in the same data block as the source items. You must change the summary item's Number of Items Displayed property value to 1, or a separate summary item will appear for each record displayed. You must also change the data block's Query All Records property to Yes, or a compile error will occur when you run the form. Now you create and configure the summary display item.

To create the summary display item:

1. In the Layout Editor, select the **Display Item** tool ▣ on the tool palette, and draw a new display item in the approximate position of the summary display item in Figure 10-1.

2. Open the new display item's Property Palette, modify the item properties as follows, and then close the Property Palette:

Property	Value
Name	**SUMMARY_ITEM**
Data Type	**Number**
Maximum Length	**9**
Format Mask	**9,999.99**
Calculation Mode	**Summary**
Summary Function	**Sum**
Summarized Block	**STUDENT_GRADES_VIEW**
Summarized Item	**EARNED_POINTS_DISPLAY_ITEM**
Number of Items Displayed	**1**

3. Click **Window** on the menu bar, and then click **Object Navigator** to open the Object Navigator. If necessary, drag the SUMMARY_ITEM node so it appears as the last item in the Items list.

4. If necessary, modify the data block's Query All Records property by double-clicking the **Data Block** icon ▤ beside the STUDENT_GRADES_VIEW data block, scrolling down to the Records property node, and changing the Query All Records property value to **Yes**. Then close the Property Palette.

5. Save the form, and then run the form. Click the **Enter Query** button ▦, type **Tammy** in the First name field and **Jones** in the Last name field for the search condition, and then click the **Execute Query** button ▦. The records and calculated values appear, along with the summary item value, which should display 42.00 as the total grade points earned by the student Tammy Jones.

HELP

If any display item in the Earned Points column is not wide enough to show its entire value, close the browser window, select any one of the display items, and make the display items wider. If the summary display item is not wide enough to show its entire value, close the browser window, select the summary display item, and make it wider.

6. Close the browser window.

Creating the Boilerplate Rectangles and Text

Recall that boilerplate items are form items that do not display database data, but are used to enhance form functionality and appearance. You created boilerplate frames around radio groups in a previous chapter. To finish the Student Grade form so it looks like Figure 10-1, you need to create the boilerplate objects. When you place boilerplate objects, such as text and shapes, on top of other boilerplate objects, the objects appear in the order in which you create them. For example, if you create the boilerplate text for the summary display item (Total Grade Points Earned) and then draw the boilerplate rectangle on top of the text, the text will be hidden. To change the order of boilerplate objects, you select the object, and then click the Bring to Front button ⊞ or the Send to Back button ⊞. Next, you draw and format the boilerplate rectangle around the summary item and add the boilerplate text.

To create and format the boilerplate objects:

1. Click **Tools** on the menu bar, and then click **Layout Editor**. Select the **Rectangle** tool ▢ on the tool palette, and draw the boilerplate rectangle shown in Figure 10-1.

2. Select the **rectangle**, click **Layout** on the menu bar, point to **Bevel**, and click **Outset**.

3. With the rectangle still selected, select the **Fill Color** tool ▣ on the tool palette, and select a **gray square** that is slightly darker than the canvas background color.

TIP
Always use subtle colors when creating form enhancements that involve colors. Intense, saturated colors look less professional than subtle shades and can cause eyestrain. Also, always ensure that there is sufficient contrast between foreground text and background colors.

4. Select the **Text** tool ▣A▣ on the tool palette, click in the rectangle at approximately the point where the boilerplate text starts in Figure 10-1, and type **Total Grade Points Earned**. Reposition and resize the boilerplate items as necessary so your form looks like Figure 10-1.

5. Save the form, run the form, click the **Enter Query** button ▣, type **Tammy** in the S_First field and **Jones** in the S_Last field as the search condition, and then click the **Execute Query** button ▣ to retrieve all of the STUDENT_GRADES_VIEW records for the student Tammy Jones. The completed form should look like Figure 10-1.

6. Close the browser window, and then close the form in Forms Builder.

CREATING A FORM TO LOAD AND DISPLAY GRAPHIC IMAGES

You can use image items in a form to display graphic images that enhance the appearance of the form or to retrieve images that are stored in the database or file system. There are two ways to display a graphic image on a form: as a static imported image, which incorporates the image data into the form design (.fmb) file and compiles it into the .fmx file; or as a dynamic image, which loads the image data from the workstation file system into the form at runtime. Usually, you use static imported images to add graphic enhancements that stay the same regardless of the data that currently appears on the form. You have been using static images in your forms to display company logos.

You use dynamic images to display images that are retrieved from the database or file system while the form is running, or to retrieve and display large static images that do not appear every time you run the form. You created a dynamic image to display the splash screen image in the integrated database application in Chapter 8. In this chapter, you learn how to use a dynamic image within a data block to store image data in the database. Specifically, you learn how to store and display images for the items in the Northwoods University FACULTY table using the Faculty form in Figure 10-2.

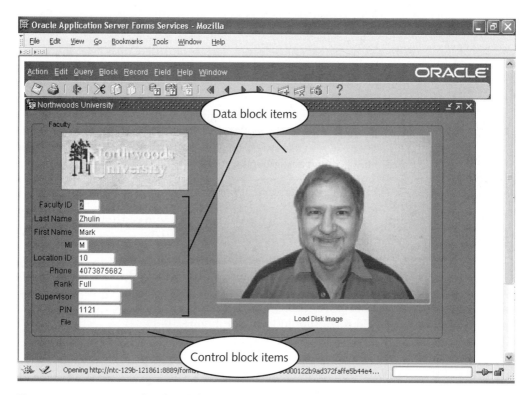

Figure 10-2   Form to load and display image data

The Faculty form contains a data block that displays the fields in the Northwoods University FACULTY table. If a faculty member's image is already loaded in the database, the form automatically displays the image when the user retrieves the associated FACULTY record. If the image is not yet in the database, the user can load the image by typing the image filename in the File field, and then clicking the Load Disk Image button. The File field and Load Disk Image button are in a form control block.

To create this form, you first open a file named Faculty.fmb stored in the Chapter10 folder on your Data Disk. Currently, this form contains no data blocks. You create a data block and layout for the FACULTY table in the Northwoods University database, and then you create the control block items to load the file image.

Creating the Data Block and Layout

Next, you use the Data Block and Layout Wizards to create the data block and layout for the FACULTY table fields. Recall from previous chapters that the FACULTY table has a field named F_IMAGE that is a BLOB (binary large object) data type, and that this field currently contains a locator for the BLOB image. (Recall that a locator is a structure that contains information about the BLOB type and points to the alternate memory location where the BLOB image data will eventually be stored.) Because the F_IMAGE field has the BLOB data type, the Layout Wizard uses an image item to display the F_IMAGE field.

10

To open the form and create the data block and layout:

1. In Forms Builder, open **Faculty.fmb** from the Chapter10 folder on your Data Disk, and save the form as **Ch10Faculty.fmb** in the Chapter10\ Tutorials folder on your Solution Disk.

2. In the Object Navigator, select the Data Blocks node, click the **Create** button ➕ to create a new data block, confirm that the Use the Data Block Wizard option button is selected, and then click **OK**.

3. On the Data Block Wizard Welcome page, click **Next**. Confirm that the Table or View option button is selected, and click **Next**. Click **Browse**, select the **FACULTY** table as the data block source, and then click **OK**. When the FACULTY fields appear in the Available Columns list, note that the data type icon for F_IMAGE is the Binary icon ▦, which indicates that the column has one of the LOB data types.

4. Select all of the fields in the FACULTY table to be included in the data block, click **Next**, accept FACULTY as the data block name, click **Next**, and then click **Finish**.

5. Click **Next** on the Layout Wizard Welcome page. On the Canvas page, make sure FACULTY_CANVAS is the selected display canvas, and click **Next**.

6. On the Data Block page, select all of the available items for display in the layout, and click **Next**.

7. On the Items page, change the prompts and widths as follows, and then click **Next**. (You delete the prompt for the F_IMAGE field.)

Name	Prompt	Width	Height
F_ID	**Faculty ID**	28	14
F_LAST	**Last Name**	120	14
F_FIRST	**First Name**	120	14
F_MI	**MI**	14	14
LOC_ID	**Location ID**	47	14
F_PHONE	**Phone**	74	14
F_RANK	**Rank**	68	14
F_SUPER	**Supervisor**	54	14
F_PIN	**PIN**	54	14
F_IMAGE	(deleted)	258	203

8. On the Style page, be sure that the Form option button is selected, and click **Next**. On the Rows page, type **Faculty** in the Frame Title field, accept the other default values, click **Next**, and then click **Finish**. The layout appears in the Layout Editor.

9. Select the frame, open its Property Palette, change the Update Layout property to **Manually**, and then close the Property Palette.

10. Resize the frame, reposition the form items so your canvas looks like Figure 10-3, and then save the form.

Creating the Control Block and Control Block Items

Now you create the form control block, the File text item, and the Load Disk Image button. Recall that the Load Disk Image button's form trigger loads into the current record's F_IMAGE field the graphic image from the file whose path specification appears in the File field. When the user clicks the Save button on the Forms Services toolbar, the form saves the data for the current image in the database.

To load the specified file image into the form image item, you use the READ_IMAGE_FILE built-in procedure, which has the following general syntax:

```
READ_IMAGE_FILE (filename, 'file_type',
'item_name');
```

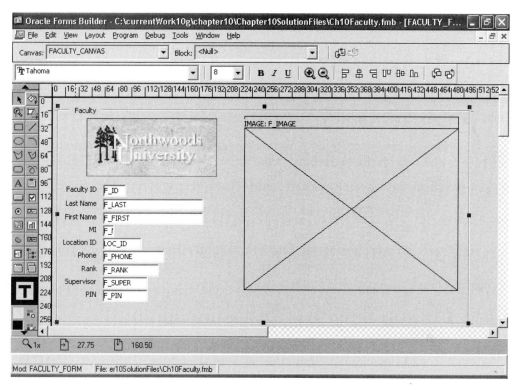

Figure 10-3   Repositioning the form items

This command uses the following parameters:

- *Filename*—The complete specification of the graphic image file, including the drive letter and folder path. You can pass this value to the procedure as a variable, form text item value, or literal character string. When you pass the value as a literal character string, you must place the value in single quotation marks. For example, the complete specification for the Marx.jpg file stored in the Chapter10 folder on your Data Disk might be 'C:\OraData\ Chapter10\Marx.jpg'. The filename specification is not case sensitive.

- *File_type*—Identifies the image file type. You pass this value as a character string in single quotation marks. Legal values are the following file types: ANY, BMP, CALS, GIF, JFIF, JPEG, OIF, PCD, PCX, PICT, RAS, TIFF, or TGA. If ANY is specified, Forms Builder attempts to determine the file type by examining the image. It is a good practice to specify the file type rather than using the ANY value, because making Forms Builder determine the file type slows down performance.

10

TIP

The different image file types depend on the graphics art application used to create the file and how the image is compressed to make it occupy less space. Most popular graphics applications support one of these types of files, which are identified by the file extension.

- *Item_name*—The name of the image item in which the file image appears. You specify this value as a character string using the syntax `'block_name.item_name'`. Note that you do not preface *block_name* with a colon (:).

Now you create the control block, the File text item, and the Load Disk Image button. You also write the button trigger to allow the user to load the file image into the form.

To create the control block and control block items, and run the form and load the image:

1. Click **Window** on the menu bar, and then click **Object Navigator**.

2. To create the control block, select the Data Blocks node, and then click the **Create** button ✚. Select the **Build a new data block manually** option button, and then click **OK**.

3. Right-click the new data block, and then click **Property Palette**. Change the Name property value to **IMAGE_CONTROL_BLOCK** and the Database Data Block property value to **No**, and then close the Property Palette.

4. To specify the block navigation order, select the IMAGE_CONTROL_BLOCK data block node, drag it toward the bottom edge of the screen, and drop it on the FACULTY data block node so that the control block appears as the second block in the data block list.

5. Click **Window** on the menu bar, and then click **FACULTY_FORM: FACULTY_CANVAS (FACULTY)** to reopen the Layout Editor.

6. To create the File text item in the control block, open the **Block** list and select **IMAGE_CONTROL_BLOCK**. Then select the **Text Item** tool 🔲, and draw the File text item shown in Figure 10-2.

7. Open the text item Property Palette, change the property values as follows, and then close the Property Palette:

Property	Value
Name	**FILE_ITEM**
Maximum Length	**300**
Prompt	**File**

8. In the Layout Editor, adjust the prompt position if necessary.

9. To create the button, make sure that IMAGE_CONTROL_BLOCK is selected in the Block list. Then select the **Button** tool 🔲, and draw the Load Disk Image button shown in Figure 10-2.

10. Open the button Property Palette, change the button Name value to **LOAD_IMAGE_BUTTON** and the button Label value to **Load Disk Image**, and then close the Property Palette.

11. Right-click the button, point to **SmartTriggers**, and then click **WHEN-BUTTON-PRESSED**. In the PL/SQL Editor, type the following commands to load the file data into the form item image, and then move the form focus to the F_ID text item so the user can select the next record.

```
READ_IMAGE_FILE(:image_control_block.file_item,
'ANY', 'FACULTY.F_IMAGE');
GO_ITEM('FACULTY.F_ID');
```

12. Compile the trigger, correct any syntax errors, close the PL/SQL Editor, and then save the form.

Now you run the form, retrieve the data records, and load the image for Faculty ID 1 (Teresa Marx).

To run the form and load the image:

1. Run the form, click the **Enter Query** button 🔍, and then click the **Execute Query** button 🔍. The data for Faculty ID 1 (Teresa Marx) appears. The image does not yet appear, because you have not yet loaded the image into the database.

2. Place the insertion point in the File field, and then type the drive letter and folder path to the Marx.jpg image file in the Chapter10 folder on your Data Disk. For example, if your Data Disk is stored in the OraData folder on your C: drive, you type C:\OraData\Chapter10\Marx.jpg.

3. Click **Load Disk Image**. The image appears in the form image item.

HELP

If the image does not appear, make sure that you typed the path to the Marx.jpg image file correctly, and that the file is actually saved at that location in your file system.

4. To save the image in the database, click the **Save** button 💾 on the toolbar. The confirmation message "Transaction complete: 1 records applied and saved" appears, indicating that the DBMS updated the record.

5. Close the browser window.

When you display image items on a form, the displayed image should be about the same size as the form image item. However, different images can have different sizes. To enable all images to appear correctly in the form item image, you need to adjust the image item's Sizing Style property. This property can have one of two values: Crop, which means that images that are too large for the image item will be cropped or cut off; or Adjust, which means that the image is scaled to fit correctly within the image item. Now you change the F_IMAGE item's Sizing Style property value to Adjust, so that Forms Builder automatically adjusts the file image to the size of the form image item. Then you run the form again and insert the images for the rest of the items in the FACULTY table.

To adjust the image item's Sizing Style, and insert the additional images:

1. In the Layout Editor, select **IMAGE: F_IMAGE**, right-click, and then click **Property Palette** to open the image's Property Palette. Under the Functional property node, select the **Sizing Style** property, open the list, and select **Adjust**. Then close the Property Palette.

2. Save the form, and then run the form. Click the **Enter Query** button , and then click the **Execute Query** button to retrieve all of the table records. The data values for Faculty ID 1 appear, along with the image of the item that you inserted before. Note that the File field does not display a value. You do not store the image file path specification in the database.

3. Click the **Next Record** button to display the next table record. The values for Item ID 2 (Mark Zhulin) appear. The image data does not appear, because you have not yet loaded it into the database.

4. Place the insertion point in the File field, and type the full path specification to the **Zhulin.jpg** file in the Chapter10 folder on your Data Disk. If your Data Disk is in the OraData folder on your C: drive, this value would be C:\OraData\Chapter10\Zhulin.jpg.

5. Click **Load Disk Image**. The image appears on the form, as shown in Figure 10-2.

6. Click the **Save** button on the toolbar to save the image in the database. The confirmation message appears, confirming that the record was updated.

7. Click to display the next table record, which is Item ID 3 (Colin Langley). Change the File field value so it specifies the location of the **Langley.jpg** file in the Chapter10 folder on your Solution Disk, click **Load Disk Image** to load the image, and then click on the toolbar to save the image.

8. Repeat Step 7 to load the image data for the remaining FACULTY records using the filenames in Figure 1-26.

9. Close the browser window, save any changes if necessary, and then close the form in Forms Builder.

CREATING STATIC LISTS IN FORMS

A **list item** is a data block item that displays a list from which users can select entries to provide form inputs. You should use a form list item to allow the user to select from a limited number of choices that do not change very often. Use lists instead of radio buttons when there are more than five choices or when there is a limited amount of space on the form to display radio buttons. Form list items display **static lists**, which means that the developer specifies the list values when he or she creates the form. Form list items cannot retrieve database values.

> Lists are useful for enabling the user to select values from a predefined list when the values are not linked to a database table. If you want to create a list of selections for database items that frequently change, such as foreign key references, create a list of values (LOV).

Forms Builder supports three types of static list items, as illustrated in Figure 10-4.

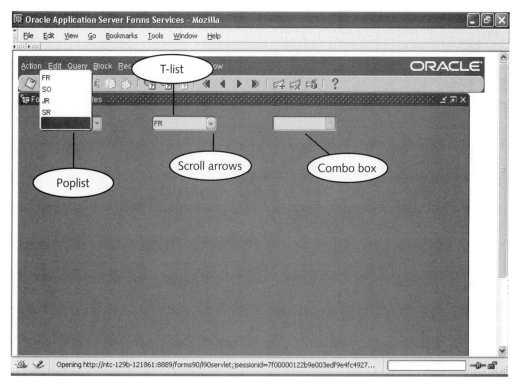

Figure 10-4   Forms Builder list item types

A **Poplist** is a drop-down list that the user opens when needed. When the user opens the list, the drop-down box displays up to 10 items at once. If the list contains more than 10 items, the Poplist displays a vertical scrollbar to allow the user to scroll and view all of the list items. When the user selects a value, the value appears in the list selection box, and the list closes.

A **T-List**, which is also called a **Text List**, always displays the current selection. It has up and down scroll arrows on its right edge, and the user can use these arrows to scroll through the list items sequentially and select a different value. You should not use T-Lists for lists with more than five to ten selections, because it becomes tedious to scroll through every value.

A **Combo box** is similar to a Poplist, except the user has the option of entering a value if the desired value does not appear in the list. The following sections describe how to create and configure a form list item.

Creating a List Item

Recall that a static list item cannot dynamically retrieve database data values. However, you can use a static list item to provide values for form text items if the values do not change, or change infrequently. For example, you might use a static list to provide values for the Northwoods University FACULTY table F_RANK (rank) field, in which the values (Full, Associate, Assistant, and Instructor) do not change very often.

To create a list item to provide input values for a database table field on a data block form, you use the Data Block and Layout Wizards to create the form, and use the Layout Wizard Data Block page to specify that the field's Item Type is a list item. Or, you can convert an existing text item to a list item by opening the text item's Property Palette, and changing the Item Type property value to List Item. To create a list item in a control block on a custom form, you draw the list item on the canvas using the List Item tool ▣ on the Layout Editor tool palette.

To gain practice creating and configuring a list item, you create a data block form that displays values from the Northwoods University FACULTY table. You specify that the F_RANK field appears as a Poplist list item that displays choices for the faculty rank value, as shown in Figure 10-5.

When the user works with the form, he or she can open the Poplist to display allowable values for the F_RANK field. Now, you open a form named Faculty.fmb in the Chapter10 folder on your Data Disk. Currently, this form does not contain any data blocks. You add a data block to the form, and specify to display the F_RANK field as a Poplist list item.

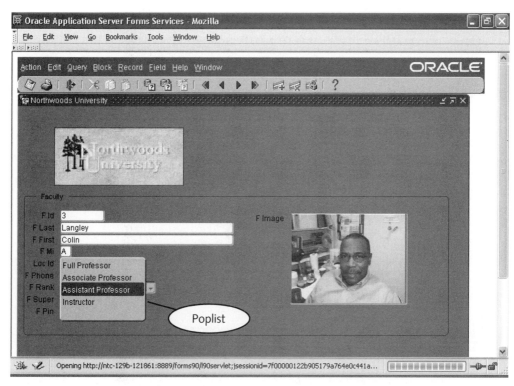

Figure 10-5   Form that uses a Poplist to display faculty rank selection

To create a data block form that uses a list item:

1. In Forms Builder, open **Faculty.fmb** from the Chapter10 folder on your Data Disk, and save the form as **Ch10Faculty_POPLIST.fmb** in the Chapter10\Tutorials folder on your Solution Disk.

2. To create the new data block and layout, select the Data Blocks node, click the **Create** button ✚, confirm that the Use the Data Block Wizard option button is selected, and then click **OK**.

3. On the Data Block Wizard Welcome page, click **Next**. Confirm that the Table or View option button is selected, click **Next**, click **Browse**, select the **FACULTY** table as the data block source, and then click **OK**.

4. Select all of the fields in the FACULTY table to be included in the data block, click **Next**, accept FACULTY as the data block name, click **Next**, and then click **Finish**.

5. Click **Next** on the Layout Wizard Welcome page. On the Canvas page, make sure FACULTY_CANVAS is the selected display canvas, and click **Next**.

6. On the Data Block page, click the **Move all items to target** button ⟩⟩ to select all of the available items for display in the layout.

7. To specify that the F_RANK item appears as a Poplist list item, select **F_RANK** in the Displayed Items list, open the **Item Type** list, select **Pop List**, and then click **Next**.

8. On the Items page, accept the default prompts and widths, and then click **Next**.

9. On the Style page, be sure that the Form option button is selected, and click **Next**. On the Rows page, type **Faculty** in the Frame Title field, accept the other default values, click **Next**, and then click **Finish**. The layout appears in the Layout Editor.

10. Change the frames' Update Layout property to Manually and reposition the form items so your canvas looks like Figure 10-6.

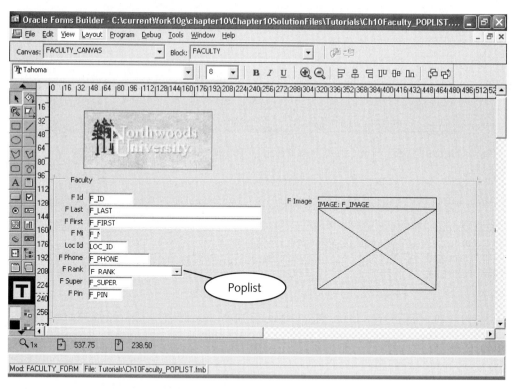

Figure 10-6   Creating and resizing the Poplist

11. Select the **F_RANK** Poplist item, resize it so it is visible, as shown in Figure 10-6, and then save the form.

Configuring a List Item

After you create a list item, you configure it using its Property Palette. You can modify the List Style property to specify that the list is a Poplist, T-List, or Combo box. You use the Elements in List property to open the List Elements dialog box in Figure 10-7 and specify the list items.

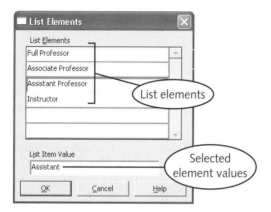

Figure 10-7   List Elements dialog box

The List Elements dialog box contains a List Elements list in which you specify the text of the list selections. When the user selects a list element, the list item has the value represented by the selected element. Sometimes, you want the list to display values that are different from the underlying data that the list item represents. For example, you might have the F_RANK list display actual faculty ranks (such as Full Professor), and have the underlying value that the list item represents be the rank's corresponding abbreviated form (such as Full). To specify an alternate value for a list element, you select the list element in the List Elements list, and type the alternate value in the List Item Value field.

Now you open the F_RANK list item Property Palette, and specify the list elements. You enter the items in the order they appear in Figure 10-7, and then you run the form and test the list.

TIP   Forms Builder always saves list item values using mixed case letters, regardless of how you enter them in the List Item Value field. For example, if you enter a list item value as "FULL," Forms Builder always translates it to "Full."

To add the list elements to the F_RANK list item, and then run the form:

    1. Double-click the **F_RANK** list item to open its Property Palette. Note that the Item Type property value is List Item, and the List Style property value is Poplist.

2. Select the **Elements in List** property, and then click the **More** button More... . The List Elements dialog box opens.

3. Delete the text in the first item in the List Elements list, and then type **Full Professor**.

4. Place the insertion point in the List Item Value field, delete the current text, and type **Full**.

5. Place the insertion point in the second item in the List Elements list, and type **Associate Professor**.

6. Place the insertion point in the List Item Value field, delete the current text, and type **Associate**.

7. Place the insertion point in the third item in the List Elements list, and type **Assistant Professor**.

8. Place the insertion point in the List Item Value field, delete the current text, and type **Assistant**. Complete the appropriate entry for Instructor. When you select a different list element (such as Assistant Professor), its associated List Item Value (Assistant) should appear in the List Item Value field, as illustrated in Figure 10-7.

9. Click **OK** to close the List Elements dialog box and save your changes, and then close the Property Palette and save the form.

10. Run the form, click the **Enter Query** button, enter **3** in the F_id field, and then click the **Execute Query** button. The third FACULTY record (Colin Langley) appears. Note that the F_Rank field value appears as "Assistant Professor," although the stored data value is "Assistant."

NOTE

For the F_Image to display properly, you need to change the F_Image's Sizing Style property to Adjust, as discussed earlier in this chapter.

11. Open the **F_Rank** list. The four faculty rank choices (Full Professor, Associate Professor, Assistant Professor, and Instructor) appear.

12. Select **Associate Professor**. Associate Professor appears in the F_Rank field.

13. Click the **Save** button to save the record. The confirmation message appears, confirming that the DBMS successfully updated the database.

14. Close the browser window.

You can modify the style of an existing list by changing the List Style property in the list Property Palette. Now you change the F_RANK list item to a Combo box, then run the form, and enter Adjunct for the F_RANK value, which is a value that is not in the list.

To change the list to a Combo box and test the list:

1. Open the **Property Palette** for the F_RANK list, select the **List Style** property, open the list, and select **Combo Box**. Then close the Property Palette.

2. Save the form, then run the form.

3. Click the **Enter Query** button 🔲, and then click the **Execute Query** button 🔲 to retrieve all of the FACULTY records. The first FACULTY record appears.

4. Open the **F_RANK** list. The current list items appear. Do not select a value. Close the list. The current value in the rank text box, which is Associate, is highlighted.

5. Type **Adjunct** in the F_Rank field to overwrite the current value, then click the **Save** button. The confirmation message appears, indicating the change was successful.

6. Close the browser window, and then close the form in Forms Builder.

USING FORMS WITH LARGE DATA SETS

10

So far, all of the forms you have created are associated with databases that contain only a few sample records. Production databases can contain thousands, or hundreds of thousands, of data records. Forms that retrieve large data sets process very slowly. For example, suppose that the Northwoods University database contains 100,000 records in the ENROLLMENT table. Depending on factors such as current database activity and network speed, a query that retrieves all of the records into a simple data block form based on the ENROLLMENT table might take several minutes. This performance is significantly worse in a master-detail form that retrieves additional information, such as student and course information for each enrollment record. Form developers need to estimate the maximum number of records that form queries might retrieve, and then design their applications to ensure that form performance remains satisfactory.

Some approaches for improving data retrieval performance in forms are:

- Create indexes on fields used in search conditions.

- Encourage users to count **query hits**, which are the number of records a query will retrieve, before actually retrieving the records. The Oracle 10g DBMS can locate and mark a record for retrieval much faster than it can actually retrieve and display the data values.

- Limit the number of records that queries retrieve by requiring users to enter one or more search conditions to filter the retrieval set.

- Configure form LOVs so that they allow the user to filter data to handle large retrieval sets more efficiently.

Chapter 9 describes how to create and use indexes. The following sections explore counting query hits, requiring search conditions, and configuring LOVs for efficiency.

Counting Query Hits

To count the number of records that a data block form query retrieves before the form actually retrieves the records, you place the form in Enter Query mode. Before you execute the query, you click Query on the menu bar, and then click Count Hits. The DBMS then determines how many records the query will retrieve, but does not yet actually retrieve the records. To illustrate this feature, you use a data block form that displays data from the ENROLLMENT table in the Northwoods University database.

To open the form and determine the number of query hits:

1. In Forms Builder, open **Enrollment.fmb** from the Chapter10 folder on your Data Disk, and save the form as **Ch10Enrollment.fmb** in the Chapter10\Tutorials folder on your Solution Disk.

2. Run the form, and then click the **Enter Query** button to place the form in Enter Query mode. Do not execute the query.

3. To determine the number of query hits, click **Query** on the menu bar, and then click **Count Hits**. The message "FRM-40355: Query will retrieve 20 records" appears in the status bar. Click the **Cancel Query** button to cancel the query.

4. To count the number of query hits that a different query retrieves, click to place the form in Enter Query mode, place the insertion point in the Grade field, and type **C** to determine how many students have earned a grade of C in a course. Do not execute the query yet.

5. To determine the number of query hits, click **Query** on the menu bar, and then click **Count Hits**. The message "FRM-40355: Query will retrieve 4 records" appears in the status bar. Click the **Execute Query** button to run the query and retrieve the records.

6. Close the browser window.

When users unknowingly execute a query that retrieves a large number of records, they are then stuck for several minutes (or even hours) waiting for the data to appear. To avoid this scenario, you can train users to always count query hits prior to actually executing the query.

Limiting Retrievals by Requiring Users to Enter Search Conditions

Another way developers can limit query retrievals is by configuring forms so users must always enter a search condition before executing a query. You can create a form trigger that requires users to enter a search condition for any query that can potentially retrieve a large number of records. You implement the trigger as a block-level trigger that is associated with the PRE-QUERY event, which occurs just before the data block form retrieves the records. The trigger tests if the user has entered a search condition in any of the form text items. If not, the trigger displays a message advising the user to enter a search condition, and then abandons the query by raising the FORM_TRIGGER_FAILURE exception. This is a built-in exception that halts processing of the current transaction. To raise a built-in exception, you use the syntax RAISE *exception_name;*.

Next you create a PRE-QUERY trigger that requires users to specify a query search condition in the Enrollment form. Then you run the form, and confirm that the trigger fires and prompts the user to enter a search condition.

To create and test a trigger to require the user to enter a search condition:

1. In the Object Navigator, open the Data Blocks node, open the ENROLLMENT data block node, select the Triggers node under ENROLLMENT, and then click the **Create** button ✚ to create a new trigger for the ENROLLMENT data block. The ENROLLMENT:Triggers dialog box opens.

2. Select **PRE-QUERY** from the list by typing **PRQ**, and then click **OK**. The PL/SQL Editor opens.

3. Type the trigger commands in Figure 10-8, compile the trigger, debug it if necessary, and then close the PL/SQL Editor.

```
IF (:enrollment.s_id IS NULL) AND
(:enrollment.c_sec_id IS NULL) AND
(:enrollment.grade IS NULL) THEN
  MESSAGE ('Please enter a query
search condition.');
  RAISE FORM_TRIGGER_FAILURE;
END IF;
```

Figure 10-8   PRE-QUERY trigger to require the user to enter a search condition

4. Save the form, and then run the form. Click the **Enter Query** button, and then click the **Execute Query** button. The message "Please enter a search condition" appears on the message line.

TIP
You could also display an alert advising the user to enter a search condition, which would probably more readily obtain the user's attention than a message on the message line.

5. Type **JO101** in the Student ID text item, click [icon] to execute the query, and then click OK. The record for Student ID JO101 appears.

6. Close the browser window, save any changes if necessary, and then close the form in Forms Builder.

The Enrollment form query executes regardless of the search condition that the user specifies, as long as the user enters a search condition in one of the form fields. You could customize the form to require the user to enter a search condition in a specific field by modifying the PRE-QUERY trigger in Figure 10-8 so that it confirms that a specific text item value is not NULL.

Configuring LOVs to Handle Large Retrieval Sets

Users sometimes have to endure performance delays or system malfunctions when they open an LOV display that retrieves a large data set. An LOV must retrieve all records before the LOV display appears.

When you open an LOV display that displays values from one of the sample databases, the LOV display appears almost instantly, because the LOV retrieves only a few records. However, if the LOV retrieves thousands or tens of thousands of records, a long delay occurs before the LOV display appears and is ready for use. To avoid this delay, you can configure LOVs that may retrieve many records to always require the user to specify a search condition before retrieving any records. To learn how to configure LOVs, you use the custom Enrollment form in Figure 10-9 that allows Northwoods University records management staff to enter student grades.

When this form opens, the staff member opens an LOV associated with the Student ID text item and then opens an LOV associated with the Course Section ID text item. The staff member then enters the grade the student earned for that course, and then clicks Save to save the updated record. Now you open and run the form to review its operation.

To open and run the form:

1. Open **CustomEnrollment.fmb** from the Chapter10 folder on your Data Disk, and save the form as **Ch10CustomEnrollment.fmb** in the Chapter10\Tutorials folder on your Solution Disk.

2. Run the form. The form appears in the browser window.

3. Place the insertion point in the Student ID field, and press **CTRL+L** to open the LOV associated with the field. The LOV display appears, and shows the student IDs and first and last names. Select the student whose ID is MA100.

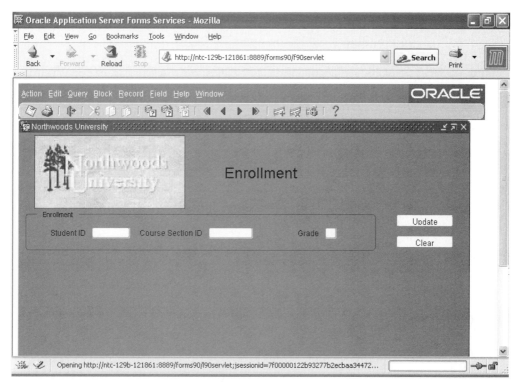

Figure 10-9 Custom form to post student grades

4. Place the insertion point in the Course Section ID field, and press **CTRL+L** to open the LOV associated with the field. The LOV display appears, and shows the course section IDs, course names, course numbers, and grades.

5. Select the Database Management course. The course section ID and grade values appear on the form.

6. Type **C** in the Grade field, and then click **Update**. An alert box appears confirming that the database will be updated. Click **Yes**. The message "Student grade posted" appears in the message line.

7. Close the browser window.

Currently, the LOV associated with the Student ID field has the default configuration settings. Because Northwoods University could potentially have tens of thousands of students, you need to configure the LOV so it does not attempt to display every student when the records clerk opens the LOV display. Now you use the Forms Builder LOV Wizard to open the LOV in reentrant mode and examine the LOV properties that you can configure to help manage large data sets.

To open the LOV Wizard in reentrant mode:

1. In the Object Navigator, open the LOVs node, select the STU_LOV node, click **Tools** on the menu bar, and then click **LOV Wizard**. The LOV Wizard opens in reentrant mode, and tabs appear at the top of the Wizard pages.

2. Click the **Next** button until the Advanced page appears, as shown in Figure 10-10. This page allows you to configure parameters to enable the LOV to handle large data sets. Do *not* close the LOV Wizard.

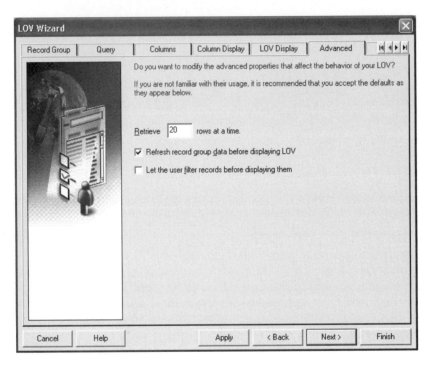

Figure 10-10   LOV Wizard Advanced page

On the LOV Wizard Advanced page, you can specify several LOV properties. The following sections describe these properties.

Record Group Fetch Size

The Retrieve field on the LOV Wizard Advanced page specifies the value for the LOV **Record Group Fetch Size property**, which determines the number of records that the Oracle 10*g* DBMS fetches in each query processing cycle. Oracle 10*g* query processing involves three phases:

- **Parse phase**—The DBMS checks the query syntax to confirm that the syntax is correct and that the user has adequate system and object privileges for the requested database objects.

- **Execute phase**—The DBMS prepares to retrieve the requested records.

- **Fetch phase**—The DBMS retrieves the records and returns them to the user.

You can specify the maximum number of records that the LOV query retrieves in each cycle. For queries that retrieve large data sets, you improve performance by making the record group fetch size larger than the default value. The RECORD GROUP FETCH SIZE can be set to 0, and then Forms calculates the optimized FETCH SIZE, depending on the size and type of the column(s) under consideration.

Automatic Refresh

The Refresh record group data before displaying LOV check box on the LOV Wizard Advanced page specifies the LOV **Automatic Refresh property**, which determines whether the LOV display refreshes each time the user opens the LOV. By default, this check box is checked, so the LOV retrieves its display values from the database each time the user opens the LOV. If the check box associated with this property is cleared, the LOV retrieves the records only once, when the user opens the LOV display the first time. The form stores the values in the client workstation's main memory. When the user opens the LOV subsequent times, the display shows the saved values from the first retrieval. This box should be cleared for an LOV in which the values do not change very often and that could retrieve a large data set.

Filter Before Display

The Let the user filter records before displaying them check box specifies the LOV **Filter Before Display property**, which determines whether or not records appear before the user filters the display using a search condition. If the check box associated with this property is checked, no records appear in the LOV display without first allowing the user to enter a search condition in the Find text box. If this check box is cleared, which is the default value, all records that the LOV query retrieves appear in the LOV display when it first opens. This box should be checked for an LOV that could retrieve a large number of records.

You can configure an LOV's Record Group Fetch Size, Automatic Refresh, and Filter Before Display properties on the LOV Wizard Advanced page. Or, you can specify the Record Group Fetch size on the Property Palette of the record group associated with the LOV, and the Automatic Refresh and Filter Before Display properties on the LOV Property Palette.

Next, you modify the STU_LOV properties to accommodate a potentially large data set. Because the records clerk knows the name of the student whose grade is being posted, you modify the STU_LOV Filter Before Display Property value to allow the records clerk to filter the data before the LOV display appears, and to only display the current student's data. (You do not need to modify the other properties; the LOV retrieves only a few records because the user filters the data.)

10

An LOV allows the user to search for data values in the first column of the LOV display. To display the field on which users will most likely search so it appears first in the LOV display, you place the search field first in the LOV SQL query, and Columns display on the LOV Columns page in the LOV Wizard. Now you modify STU_LOV so the student last name (S_LAST) is the first column that the LOV display shows, so users can search on this column.

To modify STU_LOV:

1. In the LOV Wizard, check the **Let the user filter records before displaying them** check box to enable the Filter Before Display property.

2. To place the S_LAST column first in the LOV SQL query, select the **Query** tab, and change the query text so it appears as follows:

    SELECT DISTINCT s_last, s_first, s.s_id FROM student s, enrollment e WHERE s.s_id=e.s_id

3. To specify the S_LAST column as the first column that appears in the LOV column display, select the **Columns** tab, and drag **S_ID** to the bottom of the list, so the column order is as follows: S_LAST, S_FIRST, S_ID.

4. Click **Finish** to save your changes and close the LOV Wizard.

5. Save the form.

Now you run the form and open the LOV. When the LOV first opens, it displays only the Find field, which allows the user to enter a search condition before retrieving any LOV records. You can type the wildcard character (%) to retrieve all records. You can type an exact search condition, such as the text string "Marsh" to retrieve only records in which the S_LAST field value is "Marsh." Or, you can type an inexact search condition, such as "M%", which retrieves all records in which the S_LAST field begins with the letter M.

To open the LOV display and enter a search condition:

1. Run the form.

2. Place the insertion point in the Student ID field, and then press **Ctrl+L** to open the LOV. The LOV display appears as shown in Figure 10-11, and requires the user to enter a search condition. Note that the LOV display does not yet display any records.

3. Make sure that the insertion point is in the Find field, and then type **M%** to retrieve all S_LAST values in which the first letter is M.

4. Click **Find**. The record for student John Marsh, who is the only student whose last name begins with the letter M, appears.

5. Click **Cancel** to close the LOV Display, and then close the browser window.

6. Close the form in Forms Builder.

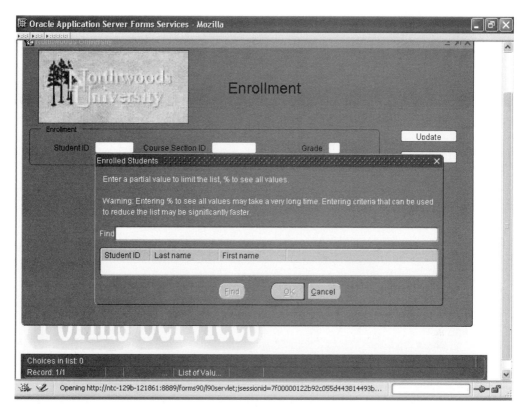

Figure 10-11   LOV display that requires the user to enter a search condition

CONTROLLING DATA BLOCK MASTER-DETAIL RELATIONSHIPS

In Chapter 5, you used the Data Block Wizard to create data block forms with master-detail relationships. Recall that in a master-detail relationship, a master database record can have multiple related detail records that are defined through foreign key relationships. Figure 10-12 shows the Faculty Students form, which is a data block form with a master-detail relationship.

Recall that a record in the Northwoods University FACULTY table can have multiple related student records in the STUDENT table. The F_ID field is the primary key in the FACULTY table and is a foreign key in the STUDENT table. When the user retrieves a specific record in the Faculty frame in the Faculty Students form, which is associated with the master block, the form automatically displays the selected faculty member's students, records in the Students frame, which is associated with the detail block.

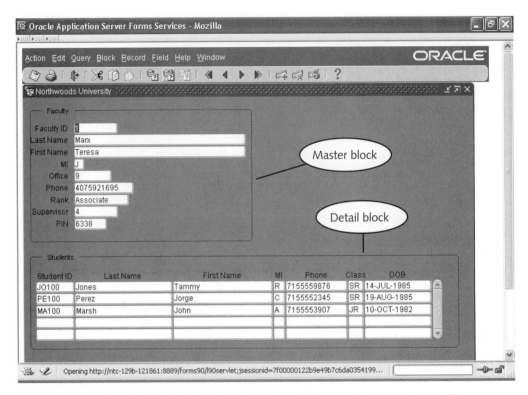

Figure 10-12   Faculty Students data block form with a master-detail relationship

In a master-detail form, the master block and the detail block are **coordinated**, which means that the master record and the detail record are associated by a relationship. When a user performs a **coordination–causing event**, which is any operation that causes the current record in the master block to change, the detail block values also change. In this section, you become familiar with the form objects and triggers that define master-detail relationships, and you learn how to modify these relationships. Now you open the Faculty Students form, save it using a different filename, and then run the form to review how a master-detail form works.

To open and run the master-detail form:

1. In Forms Builder, open **Faculty_Students.fmb** from the Chapter10 folder on your Data Disk, and save the file as **Ch10Faculty_Students.fmb** in the Chapter10\Tutorials folder on your Solution Disk.

2. Run the form, click the **Enter Query** button 🖫 to place the form in Enter Query mode, and then click the **Execute Query** button 🖫 to execute the query. The data values for the first FACULTY record (Faculty ID 1, Teresa Marx) appear as shown in Figure 10-12. Note that the detail frame shows Teresa's associated STUDENT record.

3. Click the **Next Record** button ▶. The data values for the next FACULTY record (Faculty ID 2, Mark Zhulin) appear. Note that the detail block values now show Mark's students' records.

4. Close the browser window.

Recall from Chapter 5 that when you use the Data Block Wizard to create a form that displays a master-detail relationship between two blocks, Forms Builder automatically creates a relation object in the master block. Forms Builder also automatically creates several triggers and PL/SQL program units that enforce coordination between the form master and detail blocks. The following sections describe the relation object and explore relation-handling triggers and program units. You learn how to use commands to modify properties of a master-detail relationship while a form is running.

The Data Block Relation Object

When you create a new master-detail relationship using the Data Block Wizard, Forms Builder creates a relation object in the master block that is named using the following syntax: *masterblock_detailblock*. In the master-detail form you just ran, the master block is FACULTY, and the detail block is STUDENT, so the relation object name is FACULTY_STUDENT. The relation object specifies properties about the relationship, such as the name of the detail block, the SQL join condition that specifies the relationship between the master block and the detail block, and how the form handles deletions of master block records. Next, you open the FACULTY_STUDENT relation object Property Palette, and examine the relation object properties.

To examine the relation object properties:

1. In the Object Navigator, open the Data Blocks node, open the FACULTY data block node, and then open the Relations node. The FACULTY_STUDENT relation appears.

2. Select the FACULTY_STUDENT node, right-click, and then click **Property Palette**. The relation Property Palette opens, as shown in Figure 10-13.

Important relation properties include:

- **Relation Type**—This can be either Join, which specifies that the relationship is based on joining two key fields in the blocks, or Ref, which specifies that the relationship was created using a REF pointer. (A REF pointer is an item in an Oracle 10*g* object table that creates a relationship between two tables by specifying the location of the related field.) In this relationship, the Relation type is Join.

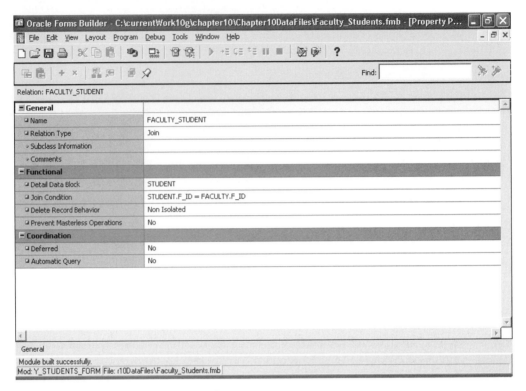

Figure 10-13   Relation Property Palette

- **Detail Data Block**—This specifies the name of the detail block in the relationship. In this relationship, the Detail Data Block is STUDENT.

- **Join Condition**—This specifies the names of the master block item and detail block item on which the blocks are joined. The Join Condition property uses the syntax *detail_block.join_item = master_block.join_item*. In this relationship, the Join Condition is STUDENT.F_ID = FACULTY.F_ID.

- **Delete Record Behavior**—This specifies how deleting a record in the master block affects records in the detail block. Possible values are: Non Isolated, which prevents the user from deleting a master record when associated detail records exist in the database; Isolated, which allows the user to delete a master record when associated detail records exist; and Cascading, which performs a **cascading delete**, in which the form deletes the master record and then deletes all of the associated detail records in the detail block's database table. In this relationship, the current value is Non Isolated, which is the default value.

- **Prevent Masterless Operations**—This specifies whether users can query or insert detail block records when there is no master record in the master block. When this value is set to Yes, the form displays an error message stating that the form cannot insert or query detail records without a parent (master) record present. When this value is set to No, users can manipulate detail blocks independently of master blocks. The default value is No, which is also the current value for this relation.

- **Deferred** and **Automatic Query**—These work together to determine whether the detail block records automatically change when the user selects a new master record, or if the user has to navigate explicitly to the detail block and refresh the detail records. If Deferred is set to No, which is the default value, then when the user selects a new master record, the form fetches the detail records immediately, regardless of the value of Automatic Query. When Deferred is set to Yes and Automatic Query is set to Yes, the form defers fetching the detail records until the user navigates to the detail block. At that point, the form automatically fetches the new records. When Deferred is set to Yes and Automatic Query is set to No, the form does not automatically fetch the detail records, and the user must navigate to the detail block and explicitly execute a query using the EXECUTE_QUERY built-in.

Next, you run the form again and examine how the default Delete Record Behavior and Prevent Masterless Operations relation properties behave. You attempt to delete a master record that has associated detail records. You also try to query the detail block when no master record exists.

To examine the default relation properties:

1. Close the Property Palette, and run the form. Click the **Enter Query** button 📠 to place the form in Enter Query mode, and then click the **Execute Query** button 📠 to execute the query. The data values for the first FACULTY record (Faculty ID 1, Teresa Marx) appear.

2. To try to delete a master record that has associated detail records, make sure that the insertion point is in the Faculty ID text item, and then click the **Remove Record** button 📠 to attempt to delete the master record. The message "Cannot delete master record when matching detail records exist" appears in the message line.

3. Click the **Save** button on the toolbar to try to commit the change. The message "FRM-40401: No changes to save" appears, indicating that the record was not deleted.

4. Click **Block** on the menu bar, and then click **Clear** to clear the form.

5. To try to query the detail block when no master record exists, place the insertion point in the **Student ID** text item, which is in the detail block. Click 📠 to place the form in Enter Query mode, and then click 📠 to execute the query.

10

All of the records in the STUDENT table appear, indicating that you successfully performed a masterless operation.

6. Close the browser window.

Under most circumstances, you should accept the default relationship properties. In almost all situations, you should leave the Delete Record Behavior property set to Non Isolated. It is dangerous to allow users to delete master records or perform cascade deletes, because these settings can allow users to delete records that they do not intend to delete. One possible reason to set the Prevent Masterless Operations property to Yes is to keep users from trying to insert detail records when no corresponding master record exists. For example, if a user tries to insert a new student record for a faculty ID that does not exist in the database, an error will occur. You set the Deferred property to Yes to defer retrieving of detail records to improve form performance if the detail records constitute a large data set.

Relation Triggers and Program Units

When you create a master-detail relation, Forms Builder automatically generates several triggers and program units to manage the relation. Table 10-2 summarizes the relation-handling triggers.

Trigger Name	Scope	Purpose
ON-CLEAR-DETAILS	Form	Clears detail block records
ON-POPULATE-DETAILS	Master Block	Coordinates values in master and detail blocks
ON-CHECK-DELETE-MASTER	Master Block	Prohibits deleting master record when detail records exist
PRE-DELETE	Master Block	Executes cascading deletes

Table 10-2   Relation-handling triggers

The ON-CLEAR-DETAILS and ON-POPULATE-DETAILS triggers are always present in a form with a master-detail relationship. The specific trigger that Forms Builder creates for controlling the deletion of master records depends on the value of the relation's Delete Record Behavior property. When the Delete Record Behavior property value is set to Non Isolated, Forms Builder creates an ON-CHECK-DELETE-MASTER trigger, which fires when the user attempts to delete the master record. The trigger checks for the existence of detail records. If detail records exist, the form does not perform the deletion. When the Delete Record Behavior property is set to Cascading, Forms Builder creates the PRE-DELETE trigger, which executes a cascading delete. When the Delete Record Behavior property is set to Isolated, neither of these triggers are present, and the form deletes master records without trigger intervention.

Table 10-3 summarizes the relation-handling program units. These program units are called by the relation-handling triggers or by one another, and handle the details of the master-detail processing.

Program Unit Name	Purpose	Called By
CLEAR_ALL_MASTER_DETAILS	Clears detail records	ON-CLEAR-DETAILS trigger
QUERY_MASTER_DETAILS	Fetches detail records	ON-POPULATE-DETAILS trigger
CHECK_PACKAGE_FAILURE	Determines if trigger successfully executes, and displays corresponding error message if it does not	ON-POPULATE-DETAILS trigger, CLEAR_ALL_MASTER_DETAILS and QUERY_MASTER_DETAILS program units

Table 10-3   Relation-handling program units

Next, you examine the relation-handling triggers in the Faculty Students form.

To examine the form's relation-handling triggers:

1. In the Object Navigator, open the Triggers node under the FACULTY_STUDENTS_FORM form module. The ON-CLEAR-DETAILS form-level trigger appears, as shown in Figure 10-14.

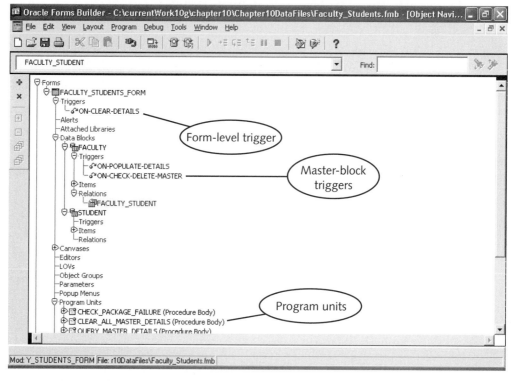

Figure 10-14   Default relation-handling triggers and program units

2. Open the Data Blocks node, and then open the FACULTY data block node, which is the master block. Open the Triggers node under FACULTY to examine the master block's triggers. The ON-POPULATE-DETAILS and ON-CHECK-DELETE-MASTER triggers appear, as shown in Figure 10-14. Recall that by default, the relation's Delete Record Behavior property is set to Non Isolated, which means that the ON-CHECK-DELETE-MASTER trigger fires when the user attempts to delete the master record.

3. Open the Program Units node to display the relation-handling program units. The three relation-handling program units appear, as shown in Figure 10-14. (Recall that if an asterisk appears beside the program unit name, it means that the program unit is not yet compiled. Forms Builder automatically compiles these triggers the next time you run the form.)

Next, you change the relation's Delete Record Behavior value and observe how the form's relation-handling triggers change. First you change the Delete Record Behavior value to Isolated. This allows the user to delete master records freely. When you do this, Forms Builder removes the ON-CHECK-DELETE-MASTER trigger from the form. Next you change the relation's Delete Record Behavior property to Cascading. When you do this, Forms Builder creates a PRE-DELETE trigger for the form, which executes cascading deletes.

To change the Delete Record Behavior property and observe how the triggers change:

1. In the Object Navigator, open the Relations node under the FACULTY master block, right-click the **FACULTY_STUDENT** relation, and then click **Property Palette**. Change the Delete Record Behavior property value to **Isolated**, and then close the relation Property Palette.

2. If necessary, open the Triggers node under the FACULTY master block, and observe that the ON-CHECK-DELETE-MASTER trigger shown in Figure 10-14 no longer appears.

3. Open the FACULTY_STUDENT relation Property Palette again, change the Delete Record Behavior property to **Cascading**, and then close the Property Palette.

4. Open the Triggers node under the FACULTY master block, and observe that Forms Builder has created a PRE-DELETE trigger to handle cascading deletes.

5. Open the FACULTY_STUDENT relation Property Palette again, change the Delete Record Behavior property back to **Non Isolated**, and then close the Property Palette.

6. Open the Triggers node under the FACULTY master block, and observe that Forms Builder has deleted the PRE-DELETE trigger, and has re-created the ON-CHECK-DELETE-MASTER trigger.

Recall that data block forms can display multiple master-detail relationships. For example, one block can be a detail block in one master-detail relationship, and a master block in a second master-detail relationship. When you add a new master-detail relationship to a form that already has an existing master-detail relationship, the new relationship uses the form's existing triggers and program units and does not create new ones, as long as the correct trigger is present to handle the new relationship's Delete Record Behavior property.

Changing Relation Properties Dynamically

You can allow the user to specify the retrieval properties of a relation while a form is running. For example, to improve form performance, a user might prefer not to retrieve detail records automatically every time the master record changes. Recall that you can configure a relation's Deferred and Automatic Query properties to determine whether the detail block records automatically change when the user selects a new master record. The default value of the Deferred property is No, so by default, Forms Builder fetches detail records immediately. You can use program commands to change the values of the Deferred and Automatic Query properties so the user can control the detail block's fetch behavior. To do this, you use the SET_RELATION_PROPERTY built-in, which has the following syntax:

10

```
SET_RELATION_PROPERTY('relation_name',
property, value);
```

This command uses the following parameters:

- *Relation_name*—References the relation in which the property value is to be changed, as it appears in the Object Navigator. *Relation_name* is enclosed in single quotation marks.

- *Property*—References the name of the property to be changed. This value is DEFERRED_COORDINATION for the Deferred property, and AUTOQUERY for the Automatic Query property. (To find the names of other relation properties used in this command, search in the Forms Builder online Help system for the SET_RELATION_PROPERTY built-in.)

- *Value*—Represents the new property value. For the Deferred and Automatic Query properties, this value can be either PROPERTY_ TRUE, which corresponds to Yes in the Property Palette, or PROPERTY_ FALSE, which corresponds to No in the Property Palette. (To find the values of other relation properties used in this command, search in the Forms Builder online Help system for the SET_RELATION_PROPERTY built-in.)

Next, you modify the Faculty Students form so that it has two radio buttons, as shown in Figure 10-15. These radio buttons allow the user to control the master-detail relationship retrieval behavior.

When the user selects the Fetch Detail Records Immediately radio button, the detail records update whenever the master record changes. When the user selects the Fetch Detail Records Later radio button, the detail records do not update until the user navigates to the detail block. First you create a control block that contains the radio buttons, and then you create the radio buttons and associated radio group and configure their properties.

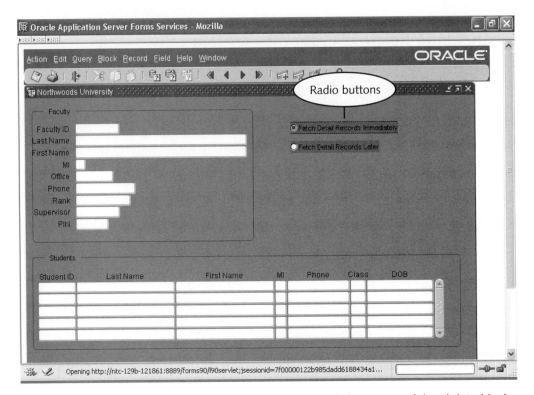

Figure 10-15 Radio buttons to allow the user to control the timing of detail data block retrievals

To create the control block and radio buttons:

1. In the Object Navigator, select the Data Blocks node, and then click the **Create** button ✚ to create a new block. Select the **Build a new data block manually** option button, and then click **OK**.

2. Select the new **data block**, right-click, and then click **Property Palette**. Change the Name property value to **QUERY_CONTROL_BLOCK** and the Database Data Block property value to **No**, and then close the Property Palette.

3. Click **Tools** on the menu bar, then click **Layout Editor**, and confirm that QUERY_CONTROL_BLOCK is selected in the Block list. Select the

Radio Button tool ⊙ on the tool palette, and draw the Fetch Detail Records Immediately radio button on the canvas, as shown in Figure 10-15.

4. Double-click the new **radio button** to open its Property Palette, change its properties as follows, and then close the Property Palette:

Property	Value
Name	**IMMEDIATE_RADIO_BUTTON**
Label	**Fetch Detail Records Immediately**
Radio Button Value	**IMMEDIATE**

5. If necessary, make the radio button larger so its label is entirely visible, and change its fill color to be as close to the same color as the canvas as possible.

6. Select the new **radio button**, click the **Copy** button 📋 on the toolbar, and then click the **Paste** button 📋. The Radio Groups dialog box opens. Accept the default radio group name, and then click **OK**. The pasted radio button appears on top of the existing button.

7. Drag the new **radio button** so it is below the existing button, as shown in Figure 10-15.

8. Double-click the new **radio button** to open its Property Palette, change its properties as follows, and then close the Property Palette:

Property	Value
Name	**LATER_RADIO_BUTTON**
Label	**Fetch Detail Records Later**
Radio Button Value	**LATER**

9. Click **Window** on the menu bar, and then click **Object Navigator**. To configure the new radio group, select the new radio group under QUERY_CONTROL_BLOCK, right-click, and then click **Property Palette**. Change the Name property value to **QUERY_RADIO_GROUP**. Scroll down to the Data node, and set the Initial Value property value to **IMMEDIATE**, then close the Property Palette and save the form.

Next, you need to create the trigger that changes the FACULTY_STUDENT relation's Deferred and Automatic Query properties based on the selected radio button. Recall that when the relation's Deferred property value is No, Forms Builder fetches the detail records immediately. When the relation's Deferred property value is Yes and the Automatic Query property value is Yes, the form waits to fetch the detail records until the user navigates to the detail block. Now you create a trigger associated with the radio group's WHEN-RADIO-CHANGED event that fires whenever the user selects a different radio button. This trigger evaluates the new value of the radio group. If the radio group value is IMMEDIATE, then the first radio button has been selected, and the trigger uses the SET_RELATION_PROPERTY built-in to set the relation's Deferred property to No. If the radio group value is LATER, then the second radio button has been selected, and the trigger sets the relation's Deferred property and Automatic Query property to Yes. After the relation properties are set, the trigger executes the GO_BLOCK built-in to place the insertion point back in the FACULTY block so the user can execute the next query.

To create the trigger to change the relation properties dynamically:

1. In the Object Navigator, select the Triggers node under the QUERY_RADIO_GROUP node, and then click the **Create** button ➕ to create a new trigger. The FACULTY_STUDENTS_FORM: Triggers dialog box opens.

2. Select **WHEN-RADIO-CHANGED** in the list, and then click **OK**. The PL/SQL Editor opens.

3. Type the commands in Figure 10-16, compile the trigger, and correct any syntax errors. Then close the PL/SQL Editor, and save the form.

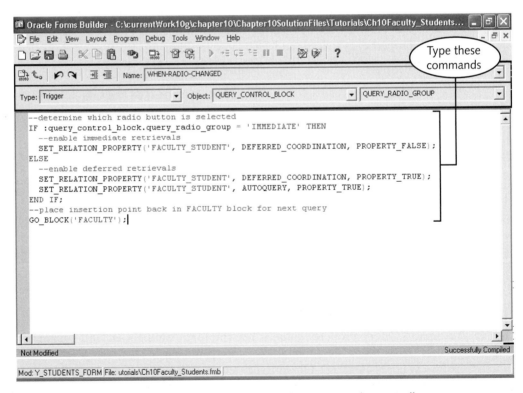

Figure 10-16   Trigger to control relation retrieval properties dynamically

Next, you run the form and confirm that you can use the radio buttons to change the relation retrieval properties. You retrieve all of the form records using the default retrieval action, which populates the detail block as soon as the master block value changes. Then you select the radio button that delays detail block fetches until you navigate to the detail block, and confirm that when you select a new master record, the form delays fetching the detail records until you navigate to the detail block.

To test the trigger to change the relation properties:

1. Run the form, place the insertion point in the Faculty ID text item in the Faculty frame, click the **Enter Query** button 🔳 to place the form in Enter Query mode, and then click the **Execute Query** button 🔳 to execute the query. The data values for the first FACULTY record (Faculty ID 1, Teresa Marx) appear, along with her associated students' information.

2. Click the **Next Record** button ▶ to retrieve the next master record. The values for Faculty ID 2 (Mark Zhulin) appear, along with his associated students' information.

3. Select the **Fetch Detail Records Later** radio button to delay retrieving the detail records, and then click ▶ to retrieve the next master record. The data values for faculty ID 3 (Colin Langley) appear, but note that his students' records do not appear.

4. Place the insertion point in the Student ID text item in the Students frame. After a moment, the detail records appear for the current master record.

5. Place the insertion point in the Faculty ID text item, and then click ▶ to retrieve the next master record. Again, the student records do not appear. Place the insertion point in the Student ID text item to retrieve the student records.

6. Select the **Fetch Detail Records Immediately** radio button to switch back to fetching the detail records immediately. The insertion point automatically moves to the FACULTY block.

7. Click the **Previous Record** button ◀ to retrieve the last master record again. The records for Faculty ID 3 (Colin Langley) appear, and his associated students' information appears immediately.

8. Close the browser window, and click **No** if you are asked if you want to save your changes.

9. Close the form in Forms Builder and save the changes if necessary.

10. Close Forms Builder and shut down the OC4J Instance.

10

Chapter Summary

- A form display item displays text data that the user cannot modify. Display items can display the results of summary or function calculations.

- When you place boilerplate objects, such as text and shapes, on top of other boilerplate objects, the objects appear in the order in which you create them.

- You can display a graphic image on a form two ways: as a static imported image, which incorporates the image data into the form design (.fmb) file and compiles it into the .fmx file; or as a dynamic image, which loads the image data from the workstation file system into the form at runtime. You use static imported images to add graphic enhancements that stay the same regardless of the data that currently appears on the form. You use dynamic images to display images that are retrieved from the database or file system while the form is running, or to retrieve and display large static images that do not appear every time you run the form.

- To create a form that loads and displays dynamic images that the database stores, you base the form on a table that contains a data field with the BLOB data type. To load the images into the database, you create a form trigger that uses the READ_IMAGE_FILE built-in to load the image into the form image item.

- A list item is a data block item that displays a static list from which users can select to provide form inputs. Use a form list item when the user is selecting from a limited number of choices that do not change very often. Use lists instead of radio buttons when there are more than five choices or when there is a limited amount of space on the form to display radio buttons.

- There are three different types of list items. A Poplist is a drop-down list that the user opens when needed. A T-List always displays the current selection, and has up and down scroll arrows on its right edge that enable the user to scroll through the list items sequentially. A Combo box is similar to a Poplist, except the user has the option of entering a value if the desired value does not appear in the list.

- Form developers need to estimate the maximum number of records that form queries might retrieve, and then design their form applications to ensure that form performance remains satisfactory.

- Approaches for improving form data retrieval performance include creating indexes on search fields, training users to count query hits, forcing users to enter search conditions in forms that might retrieve large data sets, and configuring LOVs to allow users to filter data.

- In a master-detail form, the master block and the detail block are coordinated, which means that the master record and the detail record are associated by a relationship. When a user performs a coordination-causing event, which is any operation that causes the current record in the master block to change, the detail block values also change.

❑ When you use the Data Block Wizard to create a form that displays a master-detail relationship between two blocks, Forms Builder automatically creates a relation object in the master block, and creates several triggers and PL/SQL program units that enforce coordination between the form master and detail blocks.

❑ A relation's Delete Record Behavior property specifies how deleting a record in the master block affects records in the detail block. Possible values are: Non Isolated, which prevents the user from deleting a master record when associated detail records exist in the database; Isolated, which allows the user to delete a master record when associated detail records exist; and Cascading, which performs a cascading delete, in which the form deletes the master record and then deletes all of the associated detail records in the detail block's base table.

❑ A relation's Prevent Masterless Operations property specifies whether users can query or insert detail block records when there is no master record in the master block.

❑ A relation's Deferred and Automatic Query properties work together to determine whether the detail block records automatically change when the user selects a new master record or whether the user has to navigate explicitly to the detail block and refresh the detail records.

REVIEW QUESTIONS

1. Describe the difference between a text item and a display item.

2. List three ways to improve form data retrieval performance.

3. What Oracle 10*g* data type should you use when you create database fields to store image data?

4. A form contains a data block named my_block, which contains a text item named my_path that displays the path to an image file, and an image item named my_image. Write the command to load the file specified in the my_path text item into the my_image image item. Assume that the file image is a bitmap.

5. You create a static list with the _____ style to allow the user to enter an alternate list value.

6. To require the user to enter a search condition in an LOV display, you set the LOV _____ property value to Yes.

7. In a form with a master-detail relationship, the master block is named STUDENT, and the detail block is named ENROLLMENT. What is the name of the form relation object, and where is the relation object located?

8. What is a cascading delete?

9. How do you configure the Deferred and Automatic Query properties of a relation object that handles detail block updates by requiring the user to navigate to the detail form, and then explicitly execute an EXECUTE_QUERY command?

10. When does Forms Builder create an ON-CHECK-DELETE-MASTER trigger?

10

MULTIPLE CHOICE

1. A frame that surrounds form objects is an example of a _____.

   a. text item

   b. display item

   c. boilerplate object

   d. graphic object

2. A _____ displays text data that the user cannot change.

   a. text item

   b. display item

   c. boilerplate text object

   d. graphic image item

   e. both a and c

3. A display item can display the result of a _____ that returns a value based on form data values.

   a. PL/SQL function

   b. SQL query

   c. form trigger

   d. both a and b

4. Which tool palette tool do you use to create a display item?

   a.

   b.

   c.

   d.

5. You must create a _____ to load and display image data from the database.

   a. data block form

   b. custom form

   c. static item image

   d. READ_IMAGE trigger

6. You use a(n) _____ to display list values that do not change, and a(n) _____ to display list values that are stored in the database and change frequently.

   a. static list, Poplist

   b. static list, LOV

c. LOV, Poplist

d. Poplist, T-List

7. You should use a(n) _____ only when the list contains no more than five to ten selections.

a. Poplist

b. T-List

c. Combo box

d. LOV

8. You create a(n) _____ PRE-QUERY trigger to require users to enter a search condition before executing a query.

a. form-level

b. block-level

c. canvas-level

d. item-level

9. In the _____ query phase, the DBMS checks the query syntax.

a. fetch

b. execute

c. compile

d. parse

10. When a relation object's Delete Record Behavior property value is _____, the user can delete a master record when associated detail records exist.

a. Non Isolated

b. Deferred

c. Cascading

d. Isolated

PROBLEM-SOLVING CASES

The cases reference the Clearwater Traders sample databases. The data files needed for all cases are stored in the Chapter10 folder on your Data Disk. Store all solutions in the Chapter10\Cases folder on your Solution Disk. Save the text for all SQL commands in a file named Ch10Queries.sql in the Chapter10\Cases folder. Run the Ch10Cleanwater.sql script in the Chapter10 folder on your Data Disk to refresh the tables in the sample database before completing the cases.

1. The Ch10Case1.fmb form currently contains a data block that displays inventory information from the Clearwater Traders database. Modify the form to include the

calculated fields in Figure 10-17 that display the calculated value for each inventory item, and the total value for all items in the current inventory. Add boilerplate text and a rectangle as shown to highlight the item totals. (*Hint*: Calculate the value of an inventory item by multiplying the price of an item by the quantity on hand (QOH). Calculate the total value of the inventory as the sum of the value of each inventory item.) Save the completed form as Ch10Case1_DONE.

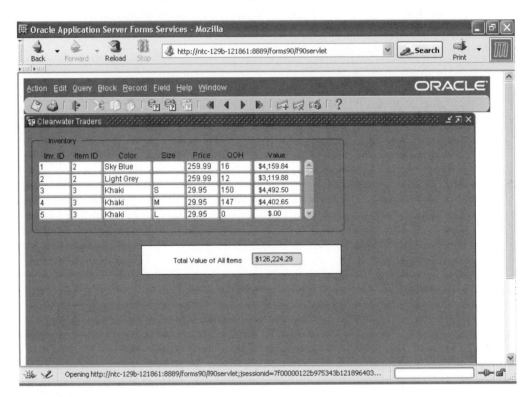

Figure 10-17

2. The form in Figure 10-18 has an LOV associated with the Customer ID field in the Clearwater Traders CUSTOMER table that allows the user to select a customer, and has a second LOV associated with the Order ID field in the Clearwater Traders ORDERS table that allows the user to select an order for that customer. After the user selects a customer and an order, the order line details appear in the view-only Order Line Detail frame. The extended total (Price times Quantity) appears for each order line, along with the order subtotal, sales tax, and final total.

a. Open Ch10Case2.fmb, and save the file as Ch10Case2_DONE.fmb. The form currently displays the items in the Customer Orders frame.

b. Create a view named ORDER_LINE_VIEW that contains the order ID, inventory ID, price, order quantity, item description, color, and size for every record in the ORDER_LINE table. Then, create a data block to display the view as shown in the Order Line Detail frame in Figure 10-18, and create a relationship between the form control block and the data block so that when the user selects an order, the order line values for the selected order appear.

c. Create a display item to show the extended total (Price times Quantity) for each order line.

d. Create display items to calculate the order subtotal (sum of all extended totals), sales tax (6% of the order subtotal), and total order cost (subtotal plus sales tax). Add the boilerplate text items and the filled rectangle object to highlight the summary data as shown.

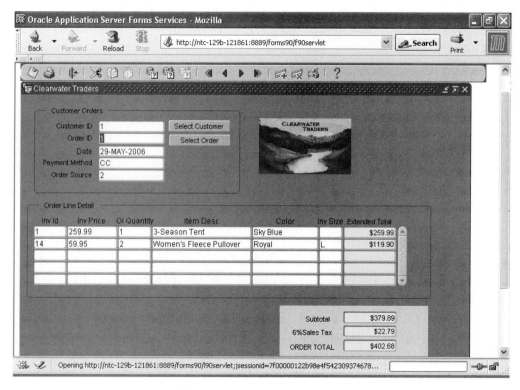

Figure 10-18

3. Create the form in Figure 10-19 that uses a data block form to load and display data and images from the Clearwater Traders ITEM table. Save the form as Ch10Case3_DONE.fmb. To enhance the form's appearance, import the Clearwater Traders logo as a static imported image. (The logo image is stored in the CWlogo.tif file in the Chapter10 folder on your Data Disk.) After you create the form, load each item image into the ITEM table using the image files stored in the Chapter10 folder on your Data Disk. The name of the JPEG image file for each item is similar to the item's description.

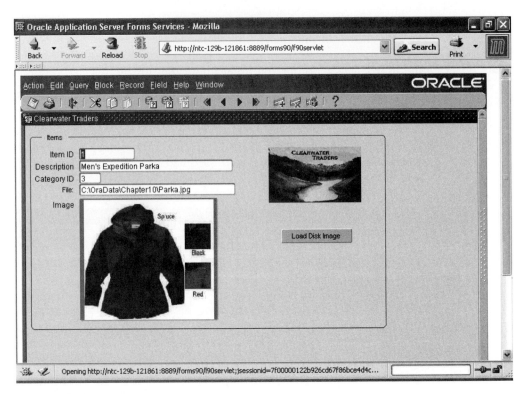

Figure 10-19

4. In this case, you modify the COLOR table in the Clearwater Traders database so it contains a new field that contains a graphic image representing the associated color. Then you create the form in Figure 10-20 that loads and displays the names of the colors and their associated images. Save the completed form as Ch10Case4_DONE.fmb.

   a. In SQL*Plus, modify the COLOR table by adding a new field named COLOR_IMAGE that has the BLOB data type.

   b. Create the form in Figure 10-20 to load and display the color images. After you create the form, load each color image into the COLOR table using the image files in the Chapter10 folder on your Data Disk. To load a specific color image, select the record associated with the color name in the tabular form, and then click the Load Disk Image button. The name of the .tif image file on your Data Disk for each color is the same as the color name.

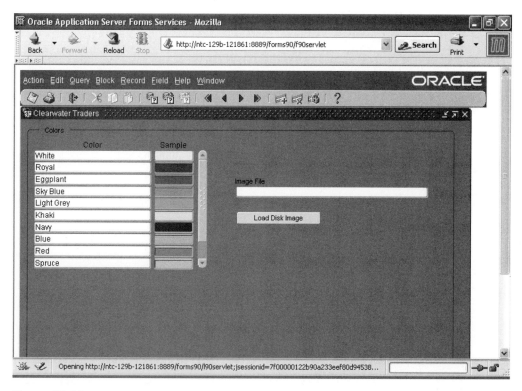

Figure 10-20

5. A data block form based on the Clearwater Traders ORDERS table is saved in a file named Ch10Case5.fmb. The form contains LOVs that allow the user to select values for the Order ID, Customer ID, Payment Method, and Order Source fields. The ORDERS table contains a record detailing order data each time a customer places an order, so this table could contain a large number of records. In this case, you modify the form to manage large data sets. Save the modified form as Ch10Case5_DONE.fmb.

   a. Create a PRE-QUERY trigger to ensure that the user enters a search condition in the Order ID, Customer ID, Date, Payment Method, or Order Source text items. If a search condition exists in any one of these items, execute the query. If no search condition exists in any of these items, display a Stop alert instructing the user to enter a search condition in one of them.

   b. Modify all of the form LOVs so they always allow the user first to filter data before retrieving any values.

   c. Modify the LOVs so the user can search on the following columns in the associated LOVs:

LOV	Search Column
Customer ID	Customer last name
Payment Method	Payment method description
Order Source	Order source description

6. Create the form in Figure 10-21 that displays records from the Clearwater Traders ORDERS table. Create a Combo box to display the values in the O_METHPMT field. Display the following list elements and associated item values in the Combo box list:

List Element	List Item Value
Check	CHECK
Credit Card	CC
Debit Card	DC
Money Order	MO

Save the form as Ch10Case6_DONE.fmb.

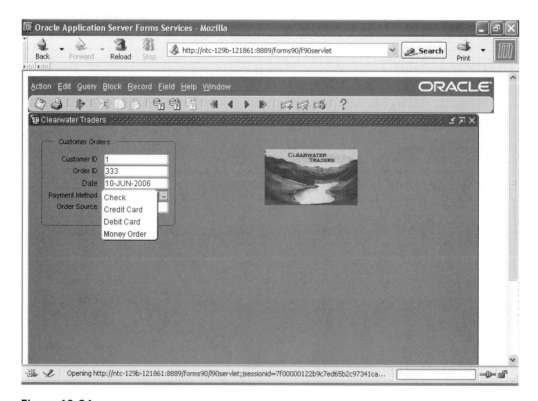

Figure 10-21

7. Figure 10-22 shows a master-detail form that displays Clearwater Traders CUSTOMER records in the master block, and related ORDERS records in the detail block that is saved as Ch10Case7.fmb. In this case, you add the check box that lets the user specify how the form fetches detail records. Save the modified form as 10Case7_DONE.fmb.

a. Create the check box as shown, and configure the check box so it is checked when the form first opens. (*Hint:* You need to place the check box in a control block.)

b. Configure the form so that when it first opens, the form immediately fetches detail records when the user retrieves a master block record.

c. Create a trigger that executes when the user checks or clears the check box. Add code to set the form relation properties according to the user preference regarding how the form fetches detail records. If the check box is checked, fetch detail records immediately. If the check box is cleared, do not fetch detail records until the user places the form insertion point in the detail block.

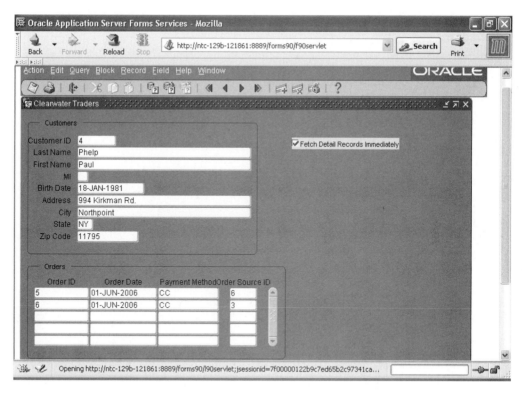

Figure 10-22

11

INTRODUCTION TO ORACLE 10*g* DATABASE ADMINISTRATION

◀ LESSON A ▶

After completing this lesson, you should be able to:

♦ Understand database administration tasks

♦ Install an Oracle 10*g* database, configure client applications, and remove Oracle 10*g* applications

♦ Perform database administration tasks using Oracle 10*g* Enterprise Manager

♦ Understand Oracle 10*g* data storage structures

♦ Work with Oracle 10*g* database instance files

A database system consists of data, the DBMS that manipulates the data, and the database applications that allow users to interact with the data. **Database administration** involves installing, configuring, maintaining, and troubleshooting a database system. A database administrator (DBA) is the person responsible for performing database administration tasks in organizations. In previous chapters, you learned how to perform database developer tasks such as creating and populating database tables and creating database applications. Database developers need to understand basic database administration tasks so they can effectively interact with DBAs and communicate their needs and problems. Sometimes, database developers are also database administrators. In this chapter, you learn about database administration and learn how to use the Oracle Enterprise Manager 10*g* utility to perform basic database administration tasks.

NOTE

This chapter assumes that the Enterprise Manager database administration utility is installed on your workstation. It also assumes you have already installed the Oracle 10g Database (Release 1) on your workstation, or that you are using an existing Oracle 10g Enterprise Edition database over a network. You do not actually install the database in the chapter exercises, but you do learn about the installation process.

OVERVIEW OF DATABASE ADMINISTRATION

Recall from Chapter 1 that a database provides a centralized repository for organizational data. The organizational department that typically manages the database is an information technology (IT) department, sometimes known as the information systems (IS) department. The IT department usually has two roles: a **service role**, to support users as they interact with the database in their daily activities, and a **production role**, to provide users with specific solutions to information management problems, such as database applications or reports. The DBA assists in the service role by performing the following tasks:

- Install and upgrade the DBMS software on the server.
- Optimize database performance by configuring how the database uses storage space in the server's main memory and file system.
- Create and maintain user accounts to control database access.
- Monitor data storage space and allocate additional storage space as needed.
- Start and shut down the database to perform database maintenance tasks.
- Perform database backup and recovery operations

The DBA assists in the production role by performing the following tasks:

- Install and upgrade developer client utilities (such as Forms Builder and SQL*Plus) on developer client workstations.
- Deploy finished database applications to users.
- Assist developers in designing and creating database tables.
- Assist developers in designing and creating form and report components and integrated database applications.
- Assist in testing and debugging new applications.
- Assist in training developers and users.

This chapter focuses on the DBA support role, and describes how to perform DBA support role tasks for an Oracle 10g database.

INSTALLING AN ORACLE 10*g* DATABASE

The Oracle 10*g* DBMS is available in three different editions, depending on your company's needs and budget. The three Oracle 10*g* DBMS editions are:

- **Enterprise Edition**, which is the most powerful DBMS, and is appropriate for installations that require a large number of transactions performed by multiple simultaneous users; Enterprise Edition includes many additional utilities for managing the database and enhancing its functionality.

- **Standard Edition**, which is sufficient for high volume multiple-user installations, but does not include some of the additional utilities in Enterprise Edition.

- **Personal Edition**, which provides a single-user DBMS for developing database applications.

The exact steps for installing an Oracle 10*g* database depend on the specific version you are installing and the operating system of the server workstation. The following sections describe the general concepts that you need to understand as you install an Oracle 10*g* DBMS so you can make appropriate choices and enter correct option values during the installation process.

Oracle 10*g* Folder Structure

A server workstation may contain multiple Oracle products. When you install an Oracle 10*g* database, by default a directory with the structure `C:\oracle\product\10.1.0`, where C is any hard drive, is created; this directory is called *Oracle_Base*, and it is the root of the Oracle 10*g* database directory tree. Each individual product's subfolder is called the product's **Oracle Home**. The \Oracle_Home directory is located beneath C:\Oracle_Base and contains subdirectories for Oracle software executables and network files.

Whenever you install an Oracle product, the installation utility prompts you to enter the name of the product's Oracle Home. Each individual product must be installed in a different Oracle Home. If you do not install each product in a different Oracle Home, the product may not run, and previously installed products may not run correctly. Figure 11-1 shows a workstation file system in which the Oracle Base is named `C:\oracle\product\10.1.0`, and that contains the Oracle Home subfolder named Db_1.

In Figure 11-1, the Oracle Base folder contains the subfolders named admin, Db_1, flash_recovery_area, and oradata. You learn about these folders later in this chapter.

11

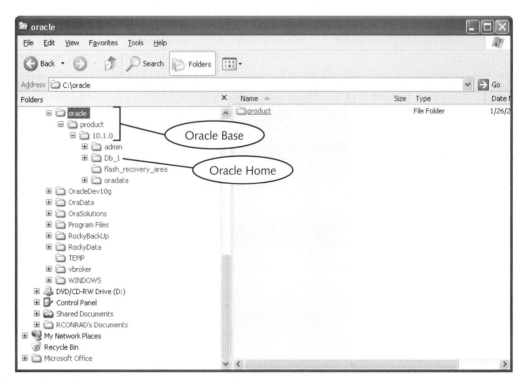

Figure 11-1   Oracle 10*g* folder structure

Database Server System Requirements

Before a DBA begins the database installation process, he or she must ensure that the workstation that runs the DBMS has sufficient main memory and disk space to support the installation. The DBA must also confirm that the server workstation has the correct operating system version. The online documentation provided on the product CD contains the exact system requirements. These requirements change depending on the exact DBMS release number, so you should always check the online documentation before installing the DBMS.

NOTE

To view the system requirements for the Oracle 10*g* DBMS, put the Oracle 10*g* Database Release 1 CD 1 in your CD-ROM drive. When the Oracle Database 10*g* – Autorun window opens, click the Browse Documentation button, click the Documentation link or Documenation tab, click the Oracle Database Installation Guide HTML link, scroll down to the Oracle Database Preinstallation Requirements section, and click the Oracle Database System Requirements link.

Table 11-1 summarizes the Windows-based operating systems and versions that are compatible with all versions of the Oracle 10*g* DBMS.

Operating System	Version
Windows NT with Service Pack 6a or higher	Windows NT Server 4.0 Windows NT Server Enterprise Edition 4.0 Windows NT 4.0 Server, Terminal Server Edition
Windows 2000 with Service Pack 1 or higher	All editions, including Terminal Services and Windows 2000 MultiLanguage Edition (MLE)
Windows Server 2003	Windows Multilingual User Interface Pack is supported
Windows XP Professional	Windows Multilingual User Interface Pack is supported

Table 11-1   Windows operating systems on which you can install the Oracle 10*g* DBMS

To run any Oracle 10*g* DBMS, your server workstation should have at least a 200MHz processor. The server workstation must also have at least 256 MB of main memory; the online documentation recommends at least 512 MB. The virtual memory size should be double the amount of RAM, and 100 MB of temporary disk space and a 256 video adapter are also listed as required in the online documenation.

Oracle recommends installing Oracle components on NTFS (NT File System). The NTFS system requirements listed in Table 11-2 are more accurate than the hard disk values reported by the Oracle Universal Installer Summary screen. The Summary screen does not include accurate values for disk space, the space required to create a database, or the size of compressed files that are expanded on the hard drive.

The hard disk requirements for Oracle Database components include 32 MB required to install Java Runtime Environment (JRE) and Oracle Universal Installer on the partition where the operating system is installed. If sufficient space is not available, installation fails and an error message appears.

Table 11-2 lists the space requirements for NTFS. The starter database requires 720 MB of disk space. The figures in Table 11-2 include the starter database. FAT32 space requirements are slightly larger.

Installation Type	System Drive	Oracle Home Drive
Basic Installation	100 MB	1.5 GB
Advanced Installation: Enterprise Edition	100 MB	1.5 GB
Advanced Installation: Standard Edition	100 MB	1.4 GB
Advanced Installation: Personal Edition	100 MB	1.5 GB

Table 11-2   Hard Disk Space Requirements for NTFS

Oracle 10*g* Universal Installer

To install any Oracle 10*g* product, you use **Oracle Universal Installer (OUI)**. Oracle Universal Installer is a Java-based graphical user interface (GUI) tool that enables you to install and remove Oracle software. OUI is provided on the CD-ROM for all Oracle 10*g* products. When you install a product, the installation process automatically installs Universal Installer on your workstation so you can later modify the product's components or install new products. OUI automatically installs the Oracle version of the Java Runtime Environment (JRE). This version is required to run Oracle Universal Installer and several Oracle assistants. Now you start Oracle Universal Installer and view the Oracle products on your workstation.

To start Oracle Universal Installer and view the Oracle products installed on your workstation:

1. Click **Start** on the taskbar, point to **All Programs**, point to **Oracle – OraDb10g_home1**, point to **Oracle Installation Products**, and click **Universal Installer**. Oracle displays a command window that checks the system requirements, and then the Oracle Universal Installer: Welcome page opens.

2. Click **Installed Products**. The Inventory window opens and shows the installed products in their associated Oracle Homes, as shown in Figure 11-2. (Your screen may look different depending on the products that are installed on your workstation and the names of their associated Oracle Homes.)

Figure 11-2   Universal Installer Inventory window

3. Open the nodes of the Oracle Home products to view the individual utilities and applications within each product.

NOTE To remove a specific application or utility, you check the item's node, and then click Remove.

4. Click **Close** to close the Inventory window.

To use Universal Installer to install a new product, you click Next on the Welcome page. Universal Installer provides a series of pages that lead you through the new product installation process. These pages are specific for each product installation. Universal Installer displays the following pages for the Oracle Database 10*g* installation:

- **Specify File Locations page**—Specify the location of the application source files and the location and name of the Oracle Home to which you wish to install the application.

- **Select Installation Type page**—Choose from among the installation options: Enterprise Edition, Standard Edition, Personal Edition, or Custom.

- **Select Database Configuration page**—Specify the properties of the pre-configured starter database that Universal Installer creates.

- **Specify Database Configuration Options page**—Specify the database name and instance name values that uniquely identify the database, the character set that the database uses to store character data, and the option to load sample demostration tables in the database created.

- **Select Database Management Option page**—Choose Grid Control or Database Control.

- **Specify Database File Storage Option page**—Identify the drive letter and folder path where the database stores the files that contain the actual database data values, and choose Automatic Storage Management or Raw Devices.

- **Specify Backup and Recovery Options page**—Specify whether you enable automated backups after you have installed the software.

- **Specify Database Schema Passwords page**—Enable the entering and confirming of passwords for all of the privileged database accounts.

- **Summary page**—Provides messages and other details about the installation process and the products about to be installed.

11

Specify File Locations Page

Figure 11-3 shows the Universal Installer Specify File Locations page.

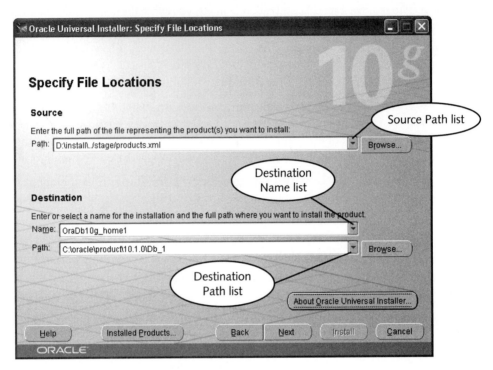

Figure 11-3   Specify File Locations page

The **Source Path list** specifies the location of the application source files. The Source Path list can specify a CD-ROM drive, a network drive location, or a Web server URL that has been configured as a product staging (distribution) site. The stage location is the centralized location where components have been placed for installations. The components are bundled in one stage location and their description file is the products.xml file. OUI will read this file and provide a list of the components available to be installed.

The **Destination Name list** specifies the name of existing Oracle Homes, and allows you to enter a new Oracle Home name for new products. The **Destination Path list** specifies the drive letter and folder path associated with the Oracle Home that the Destination Name list currently displays. If you open the Destination Name list and select a different Oracle Home, the Destination Path then displays the folder path to the newly selected Oracle Home. Figure 11-3 shows that the folder path for the OraDb10g_home1 Oracle Home is c:\oracle\product\10.1.0\Db_1.

Recall that you must install each individual Oracle 10*g* product in a different Oracle Home. If you select an existing Oracle Home and attempt to install a new product in it, the installation may work correctly, but one or both of the products may not run correctly. To specify a new Oracle Home, you type the new Oracle Home name in the Destination Name list, and specify the associated folder path in the Destination Path list. An Oracle Home name must be 1 to 16 characters long, and can include only alphanumeric characters and underscores. The Oracle Home name cannot include special characters or spaces. If the folder path in the Destination Path list does not yet exist, Universal Installer automatically creates the specified file folders.

NOTE

To select an existing drive letter and folder path in the Destination Path list, click Browse, then navigate to the existing drive and folder.

Now you open the Specify File Locations page, examine the Oracle Homes and corresponding path locations on your workstation, and practice specifying a source and destination.

To open the Specify File Locations page:

1. On the Universal Installer Welcome page, click **Next**. The Specify File Locations page appears, as shown in Figure 11-3. (Your page may look different, depending on the Oracle products installed on your workstation.)

2. Open the **Source Path list**, and select one of the existing source path names.

NOTE

You will not be able to open the Source Path list if all previously installed Oracle products were installed from a single source location.

11

3. Open the **Destination Name list** to view your workstation's Oracle Homes, and select a different Oracle Home if one exists. Note that the Destination Path list value changes and displays the selected Oracle Home's drive letter and folder path.

NOTE

You will not be able to open the Destination Name list if your workstation contains only one Oracle Home.

4. Delete the current value in the Destination Name list, and then type **SampleOracleHome** in the Destination Name list. Note that the Destination Path list is cleared.

5. Click **Browse**, navigate to the Chapter11\Tutorials folder on your Solution Disk, and click **OK**. The new Destination Path value appears in the list.

6. Click **Cancel** to exit Universal Installer, and then click **Yes** when you are asked if you really want to exit.

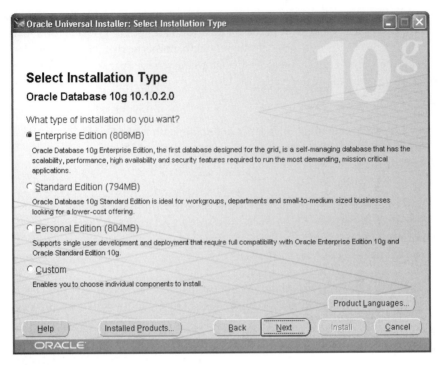

Figure 11-4   Select Installation Type page

Select Installation Type Page

Figure 11-4 shows the Universal Installer Select Installation Type page for the Oracle Database 10g installation.

The option buttons allow you to select the database level (Enterprise Edition, Standard Edition, or Personal Edition). The Custom option button allows you to select individual components for specialized or unique installations. The Product Languages button allows you to choose the language to use for the installed products.

Select Database Configuration Page

Recall from Chapter 1 that the Oracle 10g DBMS provides the programs for managing the database data and delivering database data to database applications. Universal Installer installs the DBMS programs. However, the DBMS is independent of the database that stores the actual data values. Universal Installer can also create a preconfigured database during the installation process. The Database Configuration page (Figure 11-5) allows you to select the preconfigured database that best suits your needs.

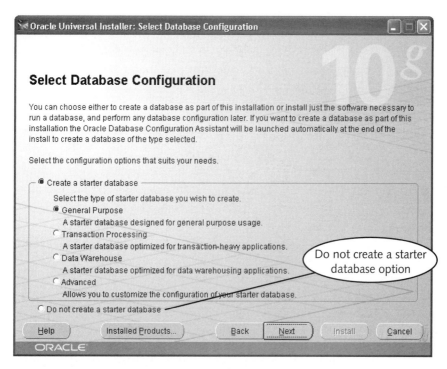

Figure 11-5   Select Database Configuration page

The General Purpose database supports the widest range of applications, from simple transactions to complex queries. The Transaction Processing database is appropriate for an installation that has a large numbers of concurrent users executing simple transactions, such as banking or Internet commerce transactions. The Data Warehouse database is appropriate for data warehouse applications, which contain historical information for researching past transactions such as customer orders, support calls, sales force prospects, or shopping and buying patterns.

The Advanced option button allows the DBA to specify the database configuration parameters. You should select this option only if you are experienced with Oracle installation and can provide specific system and product configuration information. The Advanced configuration takes longer to install and requires more user inputs than the other options. You may also select to not create a preconfigured database. You should select this option if you are an experienced DBA and know how to create and configure a new database. In almost all cases, it is advisable to select one of the preconfigured databases.

Specify Database Configuration Options Page

Recall that with a client/server database such as Oracle 10*g*, the DBMS runs on a server workstation. Client applications connect to the DBMS, request data, and receive data from the DBMS through a network. An Oracle 10*g* database consists of processes that manage the data and files that store the data values. (You learn more about these processes and files later in this chapter.) With Oracle 10*g*, a **database instance** represents a set of processes and memory structures that manipulate the data in a database. Figure 11-6 illustrates the relationships among the database, database instance, and client applications in an Oracle database system

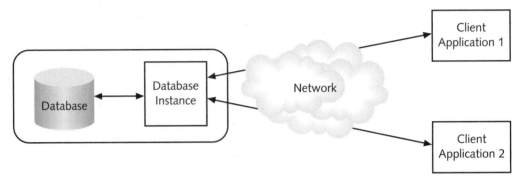

Figure 11-6   Oracle 10*g* database system components

The database instance receives client requests. The instance's processes and memory structures interact with the database to process the client requests and return requested data values or action query confirmation messages to the client applications. A single database server workstation can run multiple Oracle 10*g* database instances, provided it has sufficient main memory and hard drive space.

How does a client application route database requests to the correct Oracle 10*g* database instance? It does so by specifying the global database name and database system identifier. The **global database name** uniquely identifies an Oracle 10*g* database server and distinguishes it from all other Oracle 10*g* database servers in the world. The **database system identifier (SID)** uniquely identifies each database instance on a database server. Figure 11-7 shows the Specify Database Configuration Options page on which you specify the global database name and database SID.

The global database name is based on the database server workstation's Internet Protocol (IP) address or domain name. The Internet Protocol (IP) specifies rules that determine how computers that are connected to the Internet share data. Every computer that is connected to the Internet has a unique **IP address** that specifies the computer's network location. IP addresses are expressed as four numbers (each ranging in value from 0 to 255), separated by periods (or decimal points). An example of an IP address is 137.28.224.52. It is difficult to remember numerical IP addresses, so network servers often have an alternate **domain name**, which consists of meaningful alphanumeric characters that are easy to identify and remember. Examples of domain names are otn.oracle.com and nfl.com.

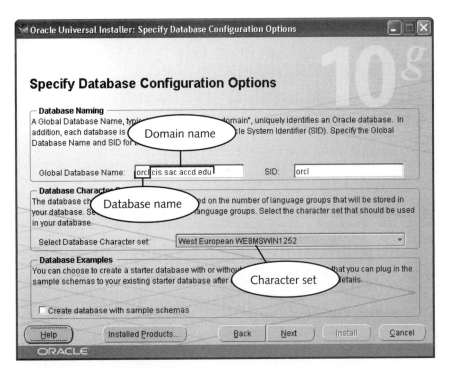

Figure 11-7   Specify Database Configuration Options page

You specify the global database name using the format *database_name.database_domain*. *Database_name* is a text string of one to eight characters that identifies the database. In Figure 11-7, the *database_name* value is orcl. *Database_domain* is a text string of one to 128 characters that specifies the IP address or domain name of the database server. In Figure 11-7, the *database_domain* value is cis.sac.accd.edu.

Recall that the database SID uniquely identifies each database instance on a database server. Usually, the SID is the same as the *database_name* value in the global database name. Note that in Figure 11-7, the SID value is orcl.

NOTE

The Specify Database Configuration Options page also allows you to specify the database character set. Different languages store data using different alphabets and numeric characters. The database character set is determined based on the number of language groups that will be stored in the database. Use the Help button to view the definition of language groups.

The Specify Database Configuration Options page also allows you to specify whether to create the starter database with or without sample schemas. You can plug the sample schemas into your existing database after creation or you can have the Universal Installer do it during installation. These sample schemas are necessary if you want to work on any of the Oracle-provided tutorials.

Select Database Management Option Page

Figure 11-8 shows the Universal Installer Select Database Management Option page for the Oracle Database 10*g* installation. Each Oracle Database 10*g* may be managed centrally using the Oracle Enterprise Manager 10*g* Grid Control, or locally using the Oracle Enterprise Manager 10*g* Database Control. This topic is explained later in this chapter. If you choose Database Control, you may also indicate whether you want to receive e-mail notifications for alerts from the DBMS.

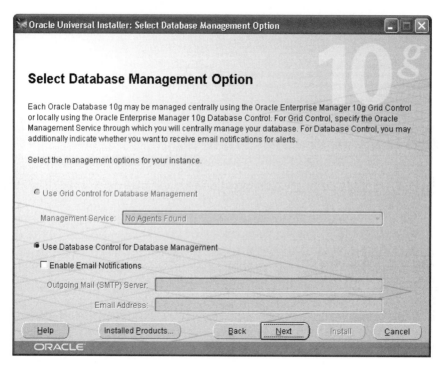

Figure 11-8   Select Database Management Option page

Specify Database File Storage Option Page

In the preceding section, you learned that an Oracle Database 10*g* consists of processes that manage data and files that store data values. To improve database performance, it is good practice to store the files that contain the database process commands on a hard drive different from the drive with the files that contain the actual data values. This setup improves performance, because database operations require a lot of reading from and writing to the server hard drive. Placing the data management processes and the data on separate hard drives distributes the workload between two different hard drives, which enhances performance.

Universal Installer stores the database process files in the Oracle Home folder you specify on the Specify File Locations page (see Figure 11-3). The Universal Installer Specify Database File Storage Option page (Figure 11-9) allows you to specify a different location for the database files that contain the actual data values.

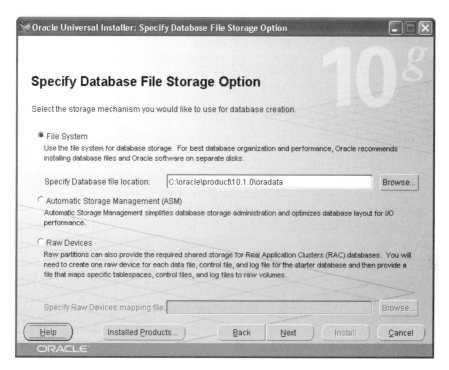

Figure 11-9   Select Database File Storage Option page

You can type an alternate disk drive letter and folder path in the Specify Database file location field on the Specify Database File Storage Option page to specify a different location for the database files. If the folder path does not yet exist, Universal Installer automatically creates it. Or, you can click Browse, and select an existing drive letter and folder path in which to store the database files. By default, Universal Installer stores database files in a folder named oradata that is located in the Oracle Base.

Oracle Database 10*g* makes available the Automatic Storage Management (ASM) system, which simplifies database storage administration and optimizes database layout for I/O performance. To use ASM for database storage, you must create one or more ASM disk groups. Disk groups are managed by a special Oracle instance, called an ASM instance. This instance must be running before you can start a database instance that uses ASM for storage management. If you choose ASM as the storage mechanism for your database, Database Configuration Assistant creates and starts this instance if necessary.

Raw devices are disk partitions or logical volumes that have not been formatted with a file system. When you use raw devices for database file storage, Oracle writes data directly to the partition or volume, bypassing the operating system file system layer. You can sometimes achieve performance gains by using raw devices. However, because raw devices can be difficult to create and administer, and because the performance gains over modern file systems are minimal, Oracle recommends that you choose ASM or file system storage rather than raw devices.

Specify Backup and Recovery Options Page

Figure 11-10 shows the Universal Installer Specify Backup and Recovery Options page for the Oracle Database 10g installation. Before choosing to enable automated backups, make sure that you have sufficient disk space available for storing backup files. If you choose to configure automated backups, Oracle Enterprise Manager schedules a database backup to occur at the same time every day. By default, the backup job is scheduled to occur at 2 a.m. To configure automated backups, you must designate an on-disk storage area for backup files, called the **flash recovery area**. You can use either the file system or an ASM disk group for the flash recovery area.

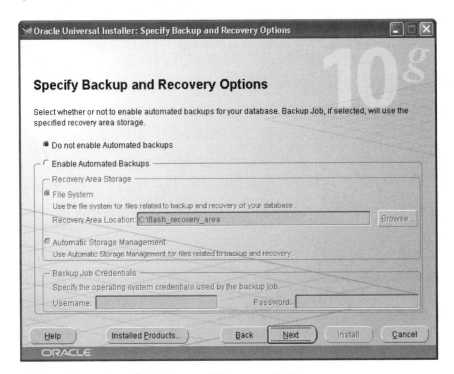

Figure 11-10   Specify Backup and Recovery Options page

Specify Database Schema Passwords Page

Figure 11-11 shows the Universal Installer Specify Database Schema Passwords page for the Oracle Database 10*g* installation. The Starter Database contains preloaded schemas, most of which have passwords that will expire and be locked at the end of install. Specify passwords for the following database administrative accounts (schemas): SYS, SYSMAN, DBSNMP, and SYSTEM. You can use the same password for all accounts, or specify different passwords for each account.

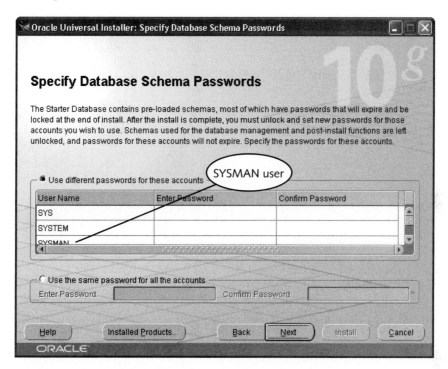

Figure 11-11   Specify Database Schema Passwords page

Summary Page

The Summary page (Figure 11-12) allows you to review the information about the installation settings and software about to be installed. At this point, you click the Install button for the installation to proceed and click the Cancel button to end the installation process. Once the Install button is clicked, the installation process proceeds. If at any time you Cancel the installation process, you need to go though the deinstall process discussed later in the chapter.

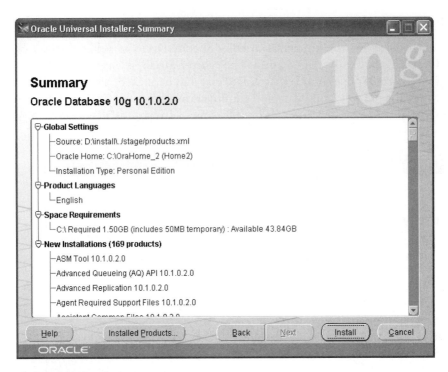

Figure 11-12   Summary page

Configuration Tools Page

Near the end of the Oracle 10*g* database installation process, the Configuration Tools page in Figure 11-13 appears.

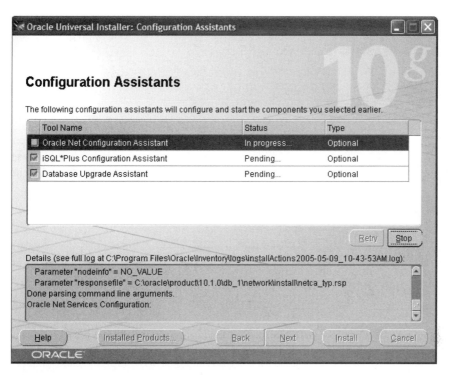

Figure 11-13   Configuration Tools page

The Configuration Tools page shows the installation progress of several optional tools that Universal Installer installs to configure the Oracle 10*g* database. The Tool list has columns that show the Tool Name and Status. The Status column values can be succeeded, in progress, pending, or unsuccessful. Succeeded means that Universal Installer has successfully installed the tool, and in progress means that the installation is in progress. Pending means that the installation is waiting for a user input to continue, and unsuccessful means that the installation failed. The Details field shows comments about the installation.

Two important tools that Universal Installer automatically installs include:

- **Oracle Net Configuration Assistant**—A utility that allows DBAs to configure client workstations so they can connect to database servers using Oracle Net. (Recall from Chapter 1 that Oracle Net is a utility that enables network communications between client applications and the database server.)

- **Oracle Database Configuration Assistant**—Installs the preconfigured database.

If all optional tools install successfully, Universal Installer automatically moves to the next stage of the installation process. If a tool does not install successfully, the DBA can cancel the tool installation process, and the database will still operate correctly. Following the Configuration Tools page is the End of Installation page, which signals the end of the Oracle 10*g* installation process.

CONFIGURING CLIENT APPLICATIONS TO CONNECT TO AN ORACLE 10g DATABASE

To connect to an Oracle 10g database, you type a connect string in the Host String field on the Log On dialog box in SQL*Plus, or in the Database field in the Connect dialog box in Forms Builder.

 NOTE If you are using an Oracle 10g database that is on the same workstation as the SQL*Plus or Forms Builder client application, you can omit the connect string, and the client application automatically connects to the local Oracle 10g database.

The connect string provides Oracle Net the information it needs to create a connection between the client application and a specific database instance. To configure connect strings for client applications that connect to databases that are not on the same workstation as the database server, you use one of the following approaches:

- **Local naming**—Each client application contains a file named tnsnames.ora that contains the configuration information for connecting to the database server.

- **Oracle Internet Directory**—Connection information is stored on a network directory server.

The following sections describe these configuration approaches.

Local Naming

The local naming approach stores connect string and database connection information in a file named tnsnames.ora on the client workstation. Every client application on a workstation has a separate tnsnames.ora file, which is stored in the client application's *Oracle_Home*\network\admin folder. If your computer contains the Developer10g client application and the Enterprise Manager client application, then it contains two separate tnsnames.ora files, one for each client application. Figure 11-14 shows the location of tnsnames.ora on a workstation in which the Developer10g application is stored.

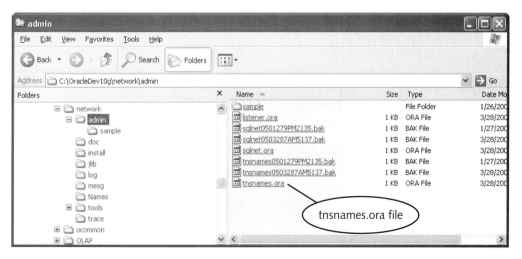

Figure 11-14   Location of the tnsnames.ora file for a client application

To understand the contents of tnsnames.ora, you must understand connect descriptors and service names.

Connect Descriptors

Oracle Net uses a **connect descriptor** to route a client data request to a specific Oracle 10*g* DBMS. A connect descriptor specifies the network communication protocol, the IP address of the database server, and the database instance name. (A **network communication protocol** is a set of rules that enable networked computers to communicate. Most computers that are connected to the Internet use the TCP/IP network communication protocol.)

The structure of an Oracle 10*g* connect descriptor depends on what network communication protocol you use. Figure 11-15 shows the general structure of a connect descriptor for the TCP/IP protocol.

```
(DESCRIPTION=
   (ADDRESS=
      (PROTOCOL=TCP)
      (HOST=server_name)
      (PORT=port_number)
   )
   (CONNECT DATA=
      (SID=database_sid)
   )
)
```

Figure 11-15   Oracle 10*g* TCP/IP connect descriptor structure

In this syntax, *server_name* is the domain name or IP address of the database server. *Port_number* is the port on which the database server process listens for incoming data requests. A **port** corresponds to a memory location on a server. A port is identified by a number, and is associated with a TCP/IP-based server process, such as a database server, e-mail server, or Web server. Every request that a client sends to a server must specify the server's IP address and the port number of the server process to which the message is directed. For example, when you use an e-mail program to retrieve your new e-mail messages, the e-mail program automatically formats the request to include the e-mail server's IP address and the correct port number for the e-mail server process on the server. For an Oracle 10*g* database process, the default port number is 1521. *Database_sid* is the SID that uniquely identifies the database instance. (Recall that you specify the SID on the Specify Database Configuration Options page shown in Figure 11-7 when you install the database.) Figure 11-16 shows an example connect descriptor for a database with the SID "orcl" that is running on a database server at the IP address 137.28.231.52.

```
(DESCRIPTION=
   (ADDRESS=
      (PROTOCOL=TCP)
      (HOST=137.28.231.52)
      (PORT=1521)
   )
   (CONNECT DATA=
      (SID=orcl)
   )
)
```

Figure 11-16   Example connect descriptor

NOTE

The connect descriptor could use the database server's domain name instead of its IP address.

Service Names

To enable client applications to connect to a database server, the DBA sets up a service name for each database to which client applications connect. A **service name** is the connect string that is associated with a connect descriptor, and is the value the remote user types in the Host String field in the Log On dialog box in SQL*Plus or in the Database field in the Connect dialog box in Forms Builder.

The DBA associates service names with connect descriptors in the tnsnames.ora file. The general format of a tnsnames.ora service name entry is as follows:

```
service_name = connect_descriptor
```

Service_name can be any text string up to 255 characters, and can contain any special characters except blank spaces. Usually, a DBA assigns a short, simple, descriptive name to *service_name*, such as "localdb" or "production." Figure 11-17 shows a sample entry in a tnsnames.ora file that assigns the service name "production" to the example connect descriptor from Figure 11-16.

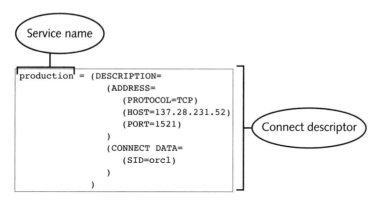

Figure 11-17   Example tnsnames.ora service name definition

Each time a DBA installs a new database, the DBA must update the tnsnames.ora file in the Oracle_Home\network\admin folder of every client application on every client workstation. To do this, the DBA can use the Oracle Net Configuration Assistant utility or manually edit tnsnames.ora using a text editor.

Manually modifying tnsnames.ora every time new client applications or new databases are added can be a daunting task for an organization that has many client workstations that each store many client applications. To address this limitation, Oracle 10*g* supports the Oracle Internet Directory approach for configuring client workstations.

Oracle Internet Directory

Oracle Internet Directory stores connect descriptor and service name information on a directory server called an **Oracle Internet Directory (OID) server**. A directory provides a way to find information. For example, you look up telephone numbers in a telephone directory or building room locations in a building directory. A **directory server** specifies the location of servers on a network. An OID server uses the **Lightweight Directory Access Protocol (LDAP)**, which is a standard protocol for configuring a directory server. LDAP specifies that the directory server contains entries for each server process in a specific format. Figure 11-18 illustrates conceptually how an LDAP directory server stores server information.

11

LDAP directory server

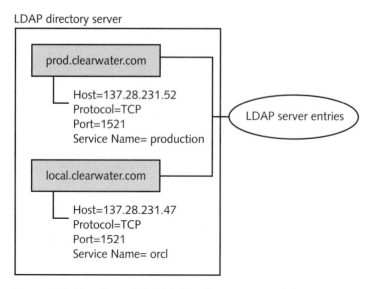

Figure 11-18   Sample LDAP directory server entries

Each LDAP server entry contains a directory listing, which is usually in the form of a URL. The URL then describes database connect descriptor information. It can also store additional information, such as usernames and passwords.

To use an Oracle Internet Directory server, a client database application requests to connect to an Oracle 10*g* database by contacting the OID server and specifying the LDAP server entry for the desired database, as shown in Figure 11-19. The OID server uses the server entry to retrieve information about the database connection, such as the server name, port number, protocol, and SID. The OID server sends this information back to the client, and the client then makes the connect request to the database server.

Creating and configuring an OID server is an advanced database and network administration task, and the steps for doing this are beyond the scope of this book. Oracle Corporation recommends using this approach for managing large database installations with many clients and many database servers.

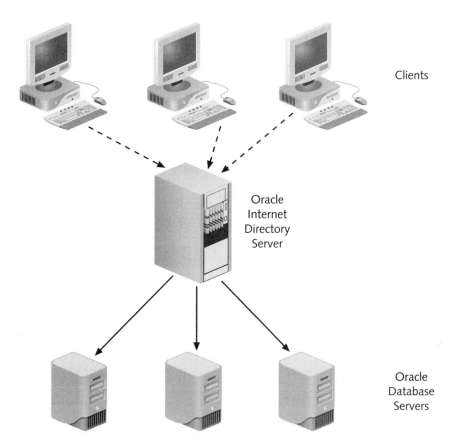

Clients

Oracle
Internet
Directory
Server

Oracle
Database
Servers

11

Figure 11-19   Using OID to create a database connection

REMOVING AN ORACLE 10*g* DATABASE

When you install an Oracle 10*g* database, the installation process automatically writes configuration information into the system registry. (Recall that the system registry is a database that Windows operating systems use to store system configuration information.) This configuration information specifies values such as the names and locations of the database server's Oracle Home and the location of the database files. This information also instructs the system to start the Oracle 10*g* database instance automatically when the computer restarts. You cannot install a new version of an Oracle product in the same Oracle Home until you remove the existing version; you also must remove the existing registry entries.

Recall that you can use Universal Installer to remove Oracle 10*g* applications from a server or client workstation. However, Universal Installer does not remove all of the application files or all of the application configuration information that exists in the system registry. To completely remove an Oracle 10g database and client applications from a workstation, you must perform the following operations:

1. Stop the Oracle services, using the Windows Control Panel.

2. Remove components with Oracle Universal Installer.

3. Manually edit the system registry to remove all Oracle 10g-related entries.

4. Restart the workstation.

5. Check the `Path` environmental variable and remove any Oracle entries.

6. Check the Start menu for any Oracle entries and remove them.

7. Manually delete any existing Oracle 10g-related files and folders. These files are in the Oracle base folder and in the C:\Program Files\Oracle folder.

CAUTION

Editing the Windows registry is a very risky operation. Be sure to consult the Oracle 10g Database Installation Guide for detailed instruction for removing Windows registry entries. If you delete the wrong item, one or more programs may malfunction. In an extreme case, your entire system may malfunction, and you may need to reinstall your operating system. Unless you are very experienced with system registry entries and related Oracle 10g components, you should not attempt to edit your system registry. The alternative to removing these registry entries is to back up all of your data files, then reinstall your operating system.

Using Oracle Enterprise Manager 10g to Perform Database Administration Tasks

NOTE

To perform the exercises in this section, your user account must have the DBA role. If you are using an Oracle 10g Personal database and created the DBUSER user account using the installation instructions provided with the textbook, you granted this role to your account. If you are using an Oracle 10g Enterprise Edition database that you did not install yourself, your instructor or technical support person must grant the DBA role to your user account. Your user account will also have to be set up as an administrator in Oracle Enterprise Manager 10g. This is illustrated in the chapter, and may be preset by your instructor or technical support person.

Recall that important DBA support tasks include configuring how the database uses storage space in the server's main memory and file system, creating and maintaining user accounts, monitoring and allocating data storage space, starting and shutting down the database, and performing backup and recovery operations. (A recovery operation involves restoring the database to a working state after a hardware or software malfunction.) To simplify these database administration support tasks, Oracle provides a Web-based management tool called **Oracle Enterprise Manager (OEM)**. When you choose to create a preconfigured database during the installation, you must select the Oracle Enterprise Manager interface that you want to use to manage the database (see Figure 11-8). Universal Installer installs OEM on the server workstation when you install any Oracle 10*g* database.

OEM has a three-tier architecture, as shown in Figure 11-20.

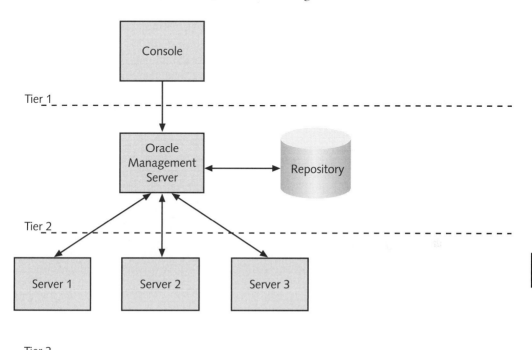

Figure 11-20   Oracle Enterprise Manager architecture

Tier 1 contains the **console**, which is the workstation at which the DBA performs administration tasks. In Oracle Enterprise Manager 10g, this is implemented via a Web browser user interface. Tier 2 contains the Oracle Management Server, which the DBA logs onto when he or she wants to perform administrative tasks. The **Oracle Management Server (OMS)** is an Oracle 10*g* database server that supports database administration tasks in an organization. The OMS interacts with a **repository** that contains information for remotely administering different databases. The OMS then interacts with different database servers, which are in Tier 3. This architecture makes it easier for DBAs to administer multiple databases in an organization's network, because all of the information for each server is stored in the central repository. In addition, this architecture allows DBAs to perform standard administration tasks and then propagate these changes to multiple servers. For example, a DBA could change the backup time schedule on the OMS for all database servers, rather than having to specify the new schedule individually for each server.

There are two ways that you can deploy Oracle Enterprise Manager (see Figure 11-8.) To deploy Oracle Enterprise Manager centrally, you must install at least one Oracle Management Repository and one Oracle Management Service within your environment, and then install an Oracle Management Agent on every computer that you want to manage. This single interface is called Oracle Enterprise Manager Grid Control (or simply Grid Control).

Your other option is to deploy Oracle Enterprise Manager locally on the database system. This local installation provides a Web-based interface called Oracle Enterprise Manager Database Control. Oracle Enterprise Manager Database Control software is installed by default with every Oracle Database installation except Custom.

To start OEM:

1. Open a Web browser and enter the following URL:

   http://localhost:5500/em

2. The Oracle Enterprise Manager Login window opens. Enter the SYSMAN username and the password you specified during installation (see Figure 11-11), as shown in Figure 11-21, and then click **Login**. When you log in to Oracle Enterprise Manager Database Control using the `SYSMAN` user account, you are logging in as the Oracle Enterprise Manager super user. The `SYSMAN, SYS, AND SYSTEM` accounts are automatically granted the roles and privileges required to access all the management functionality provided with Database Control.

CAUTION

You may see a Windows Dialog box asking if you want Windows Password Manager to remember this logon. If necessary, click No.

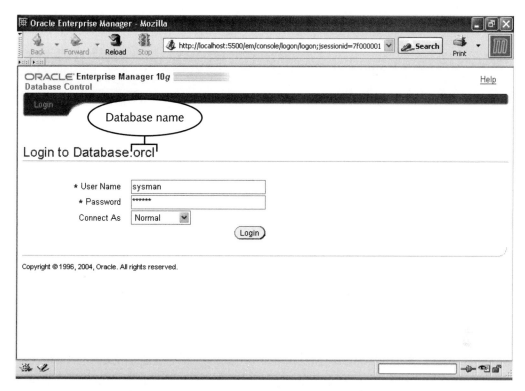

Figure 11-21   Oracle Enterprise Manager 10*g* Login page

NOTE

If you are using an Oracle 10*g* Enterprise Edition database that you did not install yourself, your instructor or technical support person must give you a user account and password.

2. The Oracle Database Licensing Information 10g page may appear. After reading the page, click **I Agree**. The Oracle Enterprise Manager Database Control home page appears, as shown in Firgure 11-22.

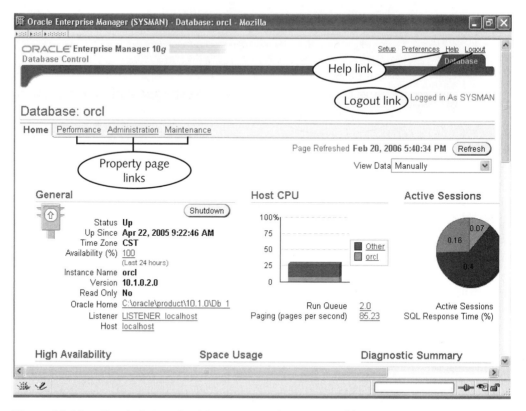

Figure 11-22   Oracle Enterprise Manager Database Control home page

NOTE

Your Oracle Enterprise Manager Databases Control home page may look different from the one in Figure 11-22, but the information will be similar.

Overview of the OEM Console

The OEM Database Control home page in Figure 11-22 displays the **OEM** console, which provides a graphical, Web-based environment that enables DBAs to perform database administration support role tasks for any database server. From the Oracle Enterprise Manager, you can perform administrative tasks such as creating schema objects (tablespaces, tables, and indexes), managing user security, backing up and recovering your database, and importing and exporting data. You can also view performance and status information about your database instance.

You can access context sensitive online help by clicking the Help link, which is displayed on every page. Each page also contains a Help Contents link, which allows you to access a help table of contents. A search facility enables you to search the contents of help.

Managing Navigation in the OEM

A DBA uses OEM to monitor, administer, and maintain an Oracle database. The property page links near the upper-left side of the Database page—Home, Performance, Administration, and Maintenance—access the associated page.

To monitor the performance of the Oracle database system, the DBA clicks the Performance link. To administer tasks such as granting user privileges and storage usage, the DBA clicks the Administration link. To maintain the capability of restoring the database, the DBA clicks the Maintenance link.

NOTE

When you perform the following actions, your screen might not look exactly like the pages illustrated in the figures.

Now you will navigate to the Performance property page. This might take a minute because Oracle must gather current statistical data on your database and produce graphs to display. On the Performance property page, you can view information about sessions CPU usage and other information about the database instance.

To navigate to the Performance property page, and obtain information about sessions CPU usage:

1. Click the **Performance** link on the Database Home Page Navigational Property Pages tabs bar. The Performance property page shown in Figure 11-23 appears.

2. Click the **CPU Used** link to the right of the Sessions: Waiting and Working graph.

3. To return to the Performance property page, click the **Database: *database_name*** link in the upper-left section of the page.

11

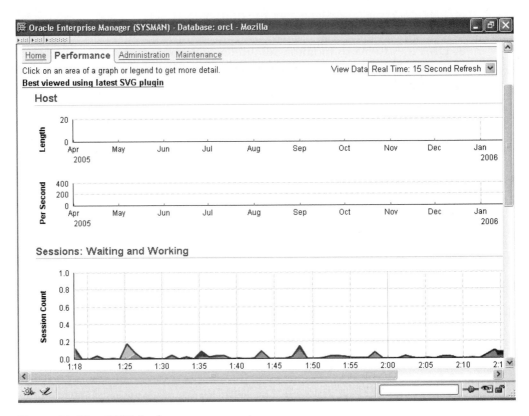

Figure 11-23   OEM Performance property page

Now you will navigate to the Administration property page, and and then you will navigate to information about the users in the database.

1. Click the **Administration** link on the Database Home Page Navigational Property Pages tabs bar. The Administration property page shown in Figure 11-24 appears.

2. Click the **Users** link under the Security heading to open the Users page. Note the buttons on the right side of the page. These allow the DBA to create, edit, and delete database users.

3. To return to the Administration property page, click the **Database:** *database_name* link in the upper-left section of the page.

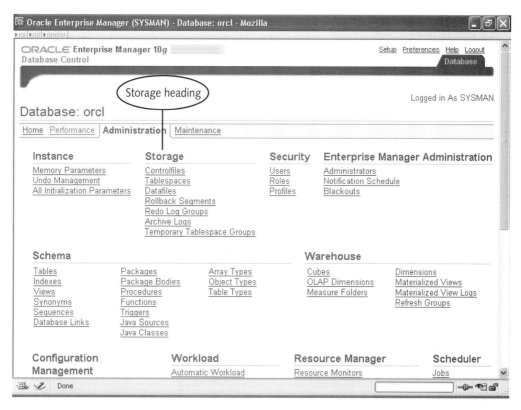

Figure 11-24   OEM Administration property page

Now you will navigate to the Maintenance property page, and you will navigate to the export utility. The export utility is an Oracle utility used to back up your database to an external specially formatted data file.

1. Click the **Maintenance** link on the Database Home Page Navigational Property Pages tabs bar. The Maintenance property page, as shown in Figure 11-25, appears.

2. Click the **Export to Files** link under the Utilities heading to open the Export: Export Type page. Note you can export the entire database, one or more schemas, or one or more tables from a selected schema to a file.

3. To return to the Administration page, click the **Database:** *database_name* link in the upper-left section of the page.

4. Click the **Home** link on the Administration page to return to the OEM home page, as shown in Figure 11-22.

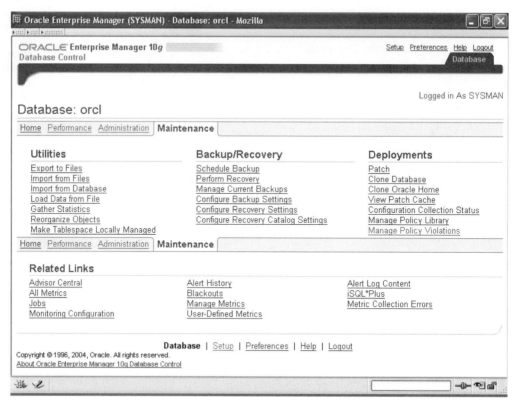

Figure 11-25   OEM Maintenance property page

Connecting to a Database and Performing DBA Tasks

Although you are logged in to the Oracle Enterprise Manager and have DBA privileges, you cannot perform every task that is available in the OEM because you logged in under the default role, which is **Normal**. The DBA must explicitly connect to the OEM by specifying a username, password, and connect role. The username and password identify the DBA. The **connect role** specifies the role for which the DBA is logging in. The connect role values are Normal, which specifies that the DBA will be performing routine maintenance tasks; or SYSOPER or SYSDBA, which allows the DBA to perform administrative tasks, such as shutting down and starting up the database.

To connect to the database as SYSOPER or SYSDBA:

1. Click the **Logout** link in the upper-right corner of the page. A page indicating that you that you have been logged out of Enterprise Manager appears.

2. Click the **Login** button on the page. A Login page similar to the one in Figure 11-21 appears.

3. Type your username and password, select **SYSDBA** from the drop-down list, and then click **Login**. You will likely get an error message, as shown in Figure 11-26, because you have not been assigned SYSDBA privileges.

CAUTION

You may see a Windows dialog box asking if you want Windows Password Manager to remember this logon. Click No.

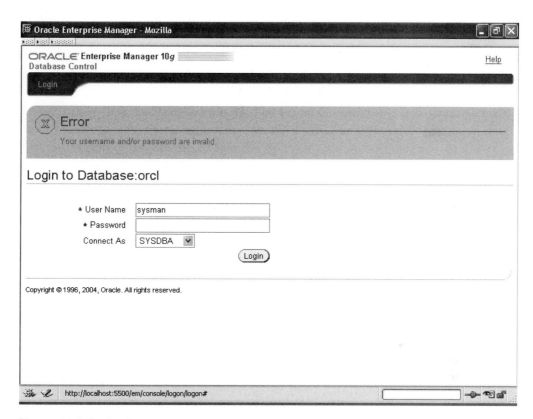

Figure 11-26   Login error message

MANAGING ORACLE 10*g* DATA STORAGE

An Oracle 10*g* database stores and manages data using a variety of data structures. This section describes these data structures and provides instructions for creating and managing data storage structures.

Oracle 10*g* Data Structures

A **data structure** provides a framework to organize data that a computer stores. Tables, lists, and arrays are examples of data structures. Oracle 10*g* databases store data using the data structures shown in Figure 11-27.

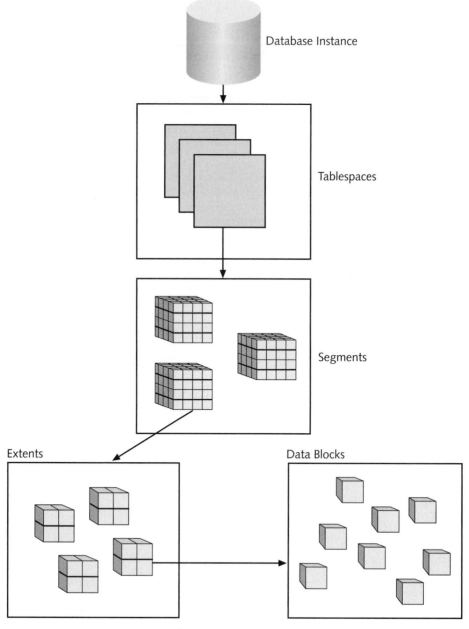

Figure 11-27   Oracle10*g* data structures

The Oracle 10*g* data structures have a hierarchical relationship. Recall that an Oracle 10*g* database instance is a set of processes and memory structures. A database instance stores data in one or more tablespaces. Each tablespace is made up of multiple segments, each segment is made up of multiple extents, and each extent is made up of one or more data blocks. The following sections describe these Oracle 10*g* data structures.

Tablespaces

A **tablespace** stores related database objects. For example, a DBA might create one tablespace to store all of the database objects for the Clearwater Traders database, and a second tablespace to store all of the database objects for the Northwoods University database. By default, an Oracle 10*g* database stores all of the system tables that support the database's functions in a tablespace named SYSTEM, all of the table indexes in a second tablespace named INDX, and all of the user data in a third tablespace named USERS. The database server stores the data for each tablespace in a **datafile**, which is a file with a .dbf extension. The data for each tablespace is stored in a separate datafile in the database server file system. To keep individual datafiles from becoming overly large and hard to manage, a DBA can configure a tablespace so its contents are stored in multiple datafiles. Storing a single large tablespace in multiple datafiles improves database performance if the datafiles are located on different hard drives, because then the same hard drive does not have to service all data requests. Figure 11-28 shows the relationships among tablespaces and datafiles.

In Figure 11-28, the SYSTEM tablespace has an associated datafile named SYSTEM01.DBF, and the INDX tablespace has an associated datafile named INDX01.DBF. Note that the USERS tablespace has two associated datafiles (USERS01.DBF and USERS02.DBF).

You place data that has similar characteristics in the same tablespace. For example, the data in the system tables is constantly being updated and needs to be backed up frequently, so it is logical to place it in a separate tablespace.

A tablespace's **status** determines whether it is available to users. A tablespace's status can be **online**, which means it is available to users, or **offline**, which means that it is unavailable to users. Sometimes a DBA takes a tablespace offline to perform maintenance activities, such as backing up the tablespace's data file(s) or changing the file properties. By placing different types of data in different tablespaces, the DBA can perform maintenance tasks on one tablespace without affecting the activities of the other tablespaces. A DBA can also configure a tablespace so its contents are read-only, which prevents users from adding, modifying, or deleting data.

11

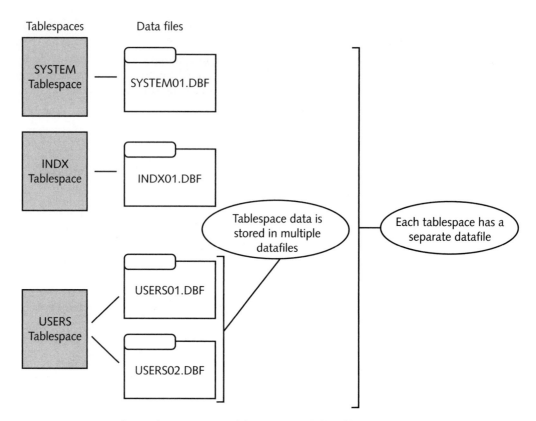

Figure 11-28   Relationships among tablespaces and datafiles

To view a database's tablespaces, you navigate to the Administration property page, and then you navigate to the Tablespaces page. When you click a specific tablespace name, you can view and modify tablespace properties in the tablespace edit page, for example, the names of a tablespace's datafiles and the tablespace's status (online or offline). Now you use OEM to view your database's tablespaces and tablespace properties.

To view your database's tablespaces and tablespace properties:

1. If necessary, login to OEM, and then click the **Administration** link.

2. Click the **Tablespaces** link. A page similar to Figure 11-29 appears. Click the **USERS** tablespace. (If your database does not contain a USERS tablespace, select another tablespace.) The tablespace details appear in the Edit Tablespace page, as shown in Figure 11-30.

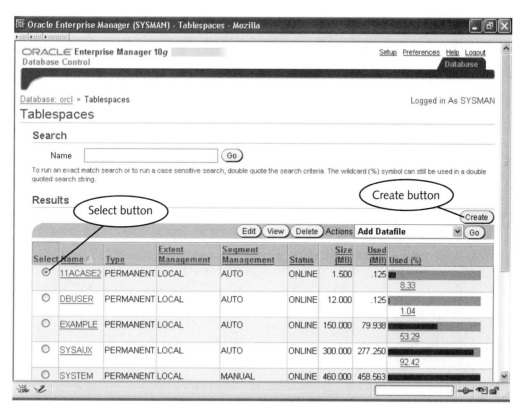

Figure 11-29   Tablespaces page

The Tablespaces page displays all of the database tablespaces, and includes information about tablespace size, space usage, tablespace status, and other information useful to the DBA. You can create, edit and delete tablespaces from this page.

The Edit Tablespace page (see Figure 11-30) displays details about the selected tablespace's datafiles, including the filenames, directory locations on the database server, file sizes, and how much file space is currently occupied by data. You can add a new datafile to the tablespace by clicking Add, typing the datafile name, file directory, and desired size, and then clicking Continue. When you select the option button beside a datafile in the list and click the Edit button, the Edit Datafile property page opens, which allows you to edit properties of the datafile that is currently selected. You can click the Remove button to delete the current datafile selection immediately after you create the datafile.

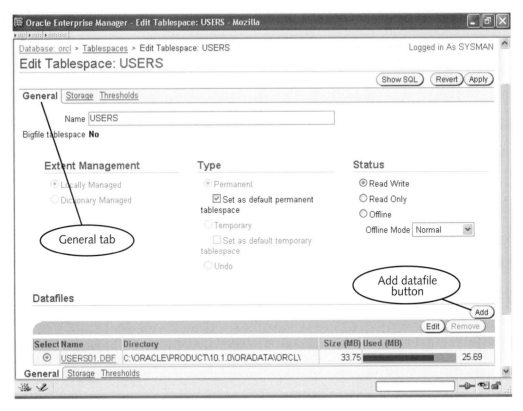

Figure 11-30   Edit Tablespace page

NOTE

The only time you can use Remove to delete a datafile is immediately after you create it, and before you save it by clicking Apply.

The Status column reports the tablespace's status (Online or Offline), and whether the tablespace is read-only. The Type column specifies whether the tablespace is Permanent, Temporary, or Undo. A permanent tablespace stores permanent database objects that must persist across multiple user sessions, such as tables or sequences. A temporary tablespace stores temporary data values that the DBMS uses to process queries, such as intermediate results from a sorting operation. You create a temporary tablespace for users who do not have the required privileges to create permanent database objects, but need a tablespace area that the DBMS can use to store data during query operations. An undo tablespace stores data that the DBMS uses to process rollback operations. (Recall that the DBMS does not make action query transactions permanent until the user issues the COMMIT command. A rollback operation restores the database to its state before the transaction.)

Segments

An Oracle 10*g* database organizes tablespace data within a datafile in multiple segments. Each **segment** stores an individual database object, such as a table or an index. A database table can be **partitioned**, which means that the database stores the table data in multiple different segments. You create partitioned tables when a table contains a lot of records, and you want to store the data across multiple datafiles to improve database performance. You create different partitions using a data field search condition that logically divides the data. Figure 11–31 shows a non–partitioned table and a partitioned table that both store data in the Clearwater Traders ORDERS table. The non-partitioned table stores all of the table data in the same segment. The partitioned table uses different segments to store the data for orders in different years.

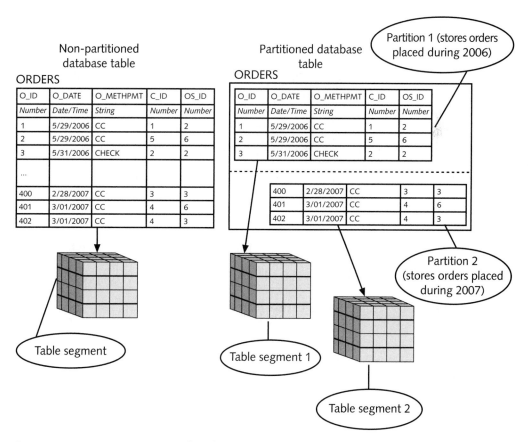

Figure 11-31   Non-partitioned and partitioned database tables and associated segments

Extents

An extent is a contiguous unit of storage space within a segment. The Oracle 10*g* database stores in a single extent a data item, such as a data field, that needs to be retrieved in a single disk input/output (I/O) operation. (Database performance would be much slower if the database server had to use multiple disk I/O operations to retrieve data for a single field.) Figure 11-32 illustrates that each individual field value within a record in the Clearwater Traders ORDERS table is stored in a separate extent within a segment.

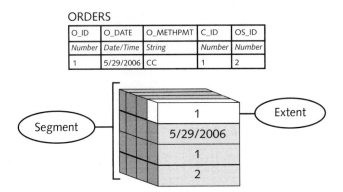

Figure 11-32   Storing individual field values in individual extents

Data Blocks

Recall that an extent is made up of one or more data blocks, as shown in Figure 11-33.

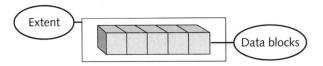

Figure 11-33   Relationship between an extent and its data blocks

In Oracle 10*g* database administration terminology, a **database storage data block** is the smallest storage unit that the database can address. (A database storage data block is different from the form data block that you learned about in Chapter 5 that represents a group of related form items.) A database storage data block corresponds to one or more **operating system blocks**, which are the smallest storage unit that the operating system can address.

By default, an Oracle 10*g* data block stores 8192 bytes. In operating systems that use the NT file system (NTFS) method for storing and managing data, the operating system block is called a cluster, and can store between 512 bytes and 4 KB, depending on the hard drive size and configuration. If the database server's cluster size is 4 KB (4096 bytes), then each data block consists of two operating system blocks.

Figure 11-34 shows that a data block consists of three separate areas: the header, free space, and row data.

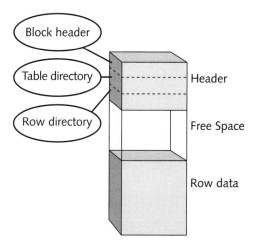

Figure 11-34   Data block components

The header contains information about the data block contents, and is made up of three separate subsections: the block header, the table directory, and the row directory. The **block header** contains general block information, such as the block address within the operating system. The **table directory** specifies which database table's data the data block stores. The **row directory** specifies which rows the block stores. The **row data** stores the actual data values. The **free space** is empty space retained by the block, in case users update the data within the data block, and the updated data occupies more storage space than the original data.

Managing Oracle 10*g* Data Structures

To manage storage structures in an Oracle 10*g* database, the DBA must be able to create and manage tablespaces and their associated datafiles. The following sections describe how to create a tablespace and manage data file extents.

Creating a Tablespace

To create a tablespace in OEM, you navigate to the Administration property page, and then to the Tablespaces page, and then to the Create Tablespace page.

The Create Tablespace page has fields that allow you to specify the tablespace name, which must adhere to the standard Oracle naming rules and must be unique within the database instance. You also specify the tablespace extent management property (Locally Managed or Dictionary Managed), status property (Read Write, Read Only or Offline), and type property (Permanent, Temporary, or Undo).

When you create a new tablespace, you must also specify the name, location, and maximum size of at least one datafile that stores the tablespace data. By default, Oracle 10*g* stores datafiles in the folder path *Oracle_Base*\ORADATA*database_SID*\\. Recall that *Oracle_Base* is the folder in which you place the folders associated with different Oracle Homes. All datafiles are placed in a subfolder named ORADATA, and then placed in a second subfolder that has the name of the current database's SID.

When you create a new tablespace, you can create a new datafile to store the tablespace data. Or, you can reuse an existing datafile that has already been created but is no longer being used. The advantage of reusing an existing datafile is that the datafile space is already allocated and its storage space is marked as being in use on the server hard drive. If you choose to delete an existing datafile rather than reuse it, you need to defragment the hard drive to recover the disk space that it occupied. Now you create a new tablespace and create a new datafile.

To create a new tablespace:

1. If necessary, navigate to the OEM Home page. Click the Administration property page link and then click the **Tablespaces** link that is under the Storage heading. On the Tablespace page click **Create**. The Create Tablespace page appears.

2. Type **DBUSER** in the Name field. Accept the default settings for the Extend Management property, the Type property, and the Status property.

3. To add a datafile for this tablespace, click **Add** in the Datafiles property. The Create Tablespace: Add Datafile page appears, as shown in Figure 11-35.

4. Type **DBUSER01.DBF** into the File Name field. Note that the File Directory field shows the default datafile folder location.

5. Change the File Size field value to **256 KB**.

6. Check the **Automatically extend datafile when full (AUTOEXTEND)** check box. Type **256 KB** into the Increment field, click the **Value option** button to select it, and then change the Maximum File Size field value to **1 MB**.

7. Click **Continue** and the Create Tablespace page appears with the new datafile information displayed.

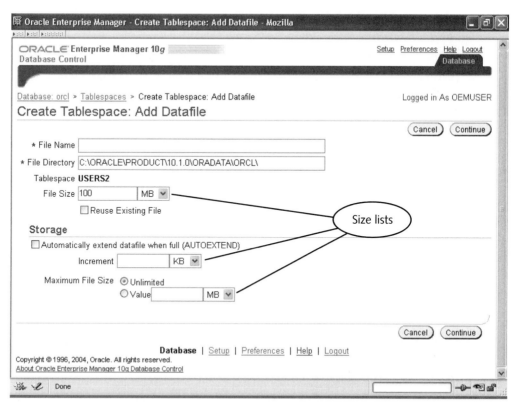

Figure 11-35   Create Tablespace: Add Datafile page

Managing Datafile Extents

Each new datafile has a header that contains information about the datafile and its associated tablespace and one free extent that can store data. As users add new data to the tablespace, the database automatically adds additional extents to the datafile until the datafile reaches its maximum allowable size. When users delete data, the data blocks that store the data are marked as unused, and become available to store new data. The database needs to track the tablespace data blocks that are unused so it can store new data values in the unused data blocks.

Tablespaces in the Oracle 10*g* Release 1 database use **local extent management**, in which the tablespace maintains a bitmap in each datafile that represents whether each data block is free or used. A **bitmap** is a data structure that uses bits, or binary digits 0 and 1, to represent data values. Each bit represents a single data block, and the bit value (0 or 1) specifies whether the data block is free or used. Previous Oracle versions (up to and including Oracle9*i* Release 1) used **dictionary-managed extent management**, whereby the data dictionary in the SYSTEM tablespace tracked all of the free and used data blocks in each tablespace. The advantage of using a locally managed tablespace is that extent management occurs directly in the tablespace, which reduces file I/O operations. If extent management is performed in the data dictionary, the database must constantly consult the data dictionary before adding or updating data.

In Oracle Corporation databases prior to Version 7, the DBA created a datafile for a tablespace and specified the datafile's maximum size. When the datafile reached its maximum size, the DBA needed to create a new datafile. If the datafile was full, users received error messages when they tried to insert or update data, and applications ceased working. To overcome this problem, Oracle Corporation created autoextensible tablespaces. In an **autoextensible tablespace**, when a tablespace datafile becomes full, it automatically grows, or **extends**, in response to new data insertions. The DBA specifies the amount by which the datafile extends when he or she configures the datafile.

Figure 11-36 illustrates how a tablespace extends in response to new data insertions.

CLEARWATER Tablespace

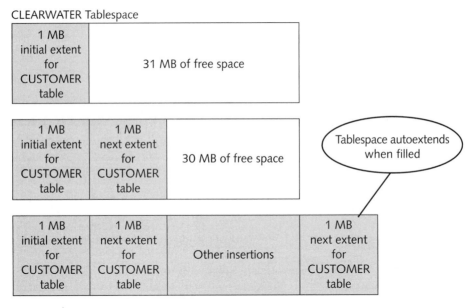

Figure 11-36   Extending a Tablespace in response to new data insertions

In Figure 11-36, a tablespace named CLEARWATER stores data for the Clearwater Traders database tables. Initially, the DBA specifies that the tablespace has a maximum size of 32 MB. When a user creates the CUSTOMER table, the database allocates a 1 MB extent in the tablespace, which is the space that the CUSTOMER table requires. As users insert data into the CUSTOMER table, the initial 1 MB extent becomes filled, and the database creates a 1MB extent to store the new data. The tablespace continues to extend in response to data insertions, until inserted data occupies all 32 MB within the tablespace datafile. As users insert more CUSTOMER records, the tablespace autoextends beyond its 32 MB maximum size, and creates a new 1 MB extent.

Configuring the Tablespace and Datafile Properties

You use the Create Tablespace page in Figure 11-35 to specify how the database manages datafile extents.

In the Extent Management section, note that the Locally Managed option button is selected, which specifies that the tablespace uses local extent management.

The Automatically extend datafile when full (AUTOEXTEND) check box configures a datafile so it extends automatically when it becomes full. The Increment text box allows the DBA to specify the size for all new extents.

You can use SQL commands to create tablespaces and configure their properties. When you use the dialog boxes to configure a new tablespace, OEM generates the associated SQL command. When you click Create on the Create Tablespace page, OEM sends the SQL command to the database instance for processing. You can click Show SQL to view the generated SQL command before you click Create to execute the command.

CAUTION

You must click Show SQL before you click Create to create the tablespace. After you click Create, the SQL command is no longer available for viewing.

Now you view the SQL command that creates the tablespace, and then create the tablespace.

To view the SQL command, then create the tablespace:

1. On the Create Tablespace page, click **Show SQL**. The SQL command text appears, as shown in Figure 11-37.

2. Click **Return** to return to the Create Tablespace page.

3. Click **OK**. The Tablespace page appears with an update message informing you that the tablespace object has been created successfully. Note the new tablespace is included in the list of tablespaces on the page.

4. Click the **Database: *database_name*** link in the upper-left section of the page to return to the Administration page.

Figure 11-37   Viewing the SQL command to create the tablespace

THE ORACLE 10*g* DATABASE FILE ARCHITECTURE

An Oracle 10*g* database is made up of several different files that reside in the file system of the database server workstation. These files include:

- **Parameter file**—Initializes the database specifications and points to the locations of the database control files.

- **Control files**—Contain information about the database tablespaces, datafiles, redo log files, and the current state of the database.

- **Datafiles**—Contain the actual database data values.

- **Redo log files**—Contain rollback information for uncommitted transactions.

Figure 11-38 illustrates the architecture of the Oracle 10*g* database files.

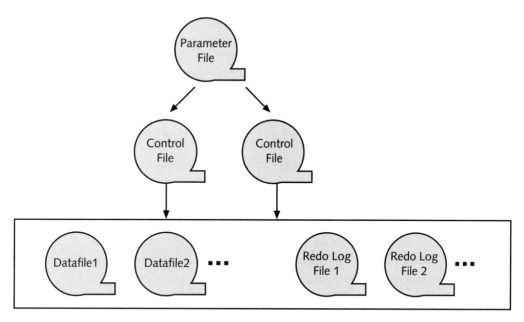

Figure 11-38   Oracle 10*g* database file architecture

The parameter file is the starting point, and contains entries that specify the names and locations of the control files. The control files specify the names and locations of all of the database instance's datafiles and redo log files.

Parameter File

The parameter file is a text file that specifies configuration information about an Oracle 10*g* database instance. When you start an Oracle 10*g* database instance, the process that starts the instance reads the parameter file and uses its information to configure the instance. Important parameter file information includes the names and locations of the control files, and the amount of server main memory that the Oracle 10*g* instance uses to process data retrievals.

The parameter filename is init.ora. By default, Oracle 10*g* stores the parameter file for each database instance in the server's *Oracle_Base*\admin*SID*\pfile folder. For example, suppose that your database instance's *Oracle_Base* folder path is C:\oracle\product\10.1.0\ and its *SID* is orcl. The folder path and filename of the parameter file is c:\oracle\product\10.1.0\admin\orcl\pfile\init.ora. Now you start Windows Explorer, and examine your database instance's parameter file.

To examine the parameter file:

    1. Start Windows Explorer, and navigate to the *Oracle_Base**SID*\pfile folder on your workstation. The init.ora parameter file appears.

2. Start Notepad, and open *Oracle_root\SID\pfile\init.ora*. The parameter file appears in Notepad.

NOTE

Sometimes the database shutdown process appends a number to the init.ora filename and then the filename appears like init.ora.1021001164023, for example. This value allows the database instance to track the parameter file internally. If a number is appended to your init.ora filename, open the numbered file.

3. Scroll down to the File Configuration section, as shown in Figure 11-39, and note the names and location of the control files.

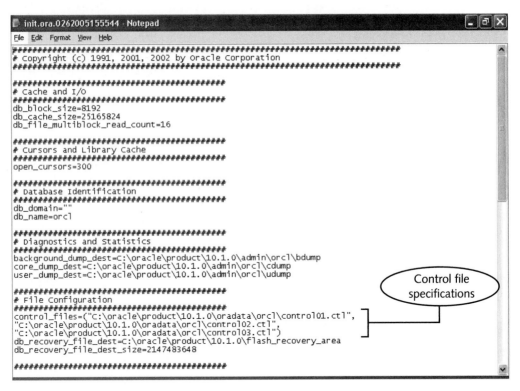

Figure 11-39   Parameter file contents

4. Close Notepad.

Recall that the parameter file is a text file. DBAs can edit the parameter file to modify the database configuration. Because the database instance reads the parameter file when the instance starts, the DBA must shut down the database instance and then restart it to apply the parameter file changes. If an instance's parameter file is not present or becomes corrupted, the process that starts the database instance creates a new parameter file that contains the default database settings.

Control Files

Control files store information about the database structure and state. Specific items in the control files include the global database name, SID, information about the tablespaces and their associated datafiles, and names and locations of redo log files. When you start an Oracle 10*g* instance, the process that starts the database reads the parameter file, finds the locations of the control files, and then reads the control files. By default, Oracle 10*g* stores the control files in the *Oracle_Base*\oradata*SID* folder on the database server. A database server on which *Oracle_Base* is C:\oracle\product\10.1.0 and *SID* is orcl stores the control files in the C:\oracle\product\10.1.0\oradata\orcl folder.

By default, an Oracle 10*g* database instance has three separate control files named CONTROL01.CTL, CONTROL02.CTL, and CONTROL03.CTL. At least one of the control files must be present, or the database instance will not start. Each control file is a mirror image of the others, and contains exactly the same data. The database maintains three copies in case one or more of the files is lost or corrupted. Control files are binary files that can be written to and read by only the database processes. You cannot read or edit a control file directly. Now you view a listing of your database instance's control files in Windows Explorer.

To view a listing of the control files:

1. In Windows Explorer, navigate to the *Oracle_Base*\oradata*SID* folder, and open the *SID* folder.

2. The file listing shows the control files, as shown in Figure 11–40. (You will learn about the other files in this folder shortly.)

11

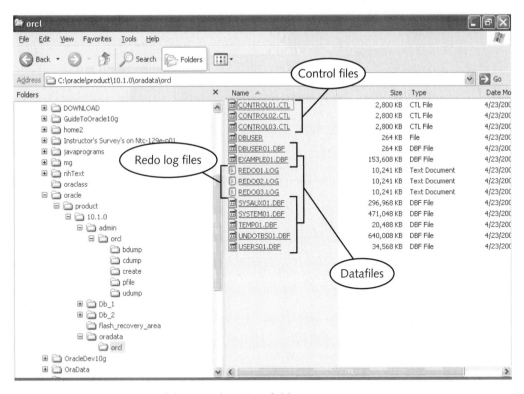

Figure 11-40   Contents of the \oradata\SID folder

You can view properties about control files in OEM by selecting the Controlfiles link under the Storage heading on the Administration page. Now you view the control file information for your database instance.

To view the control file information:

1. Click the **Controlfiles** link under the Storage heading on the Administration page. The control file filenames, file directory, and status appear on the General tab.

2. Click the **Advanced** tab and then database instance information appears on the page.

3. Click the **Database:** *database_name* link in the upper-left section of the page to return to the Administration page.

Datafiles

Recall that datafiles are files with .dbf extensions that store tablespace contents. You created a new datafile when you created a tablespace in a previous exercise. By default, Oracle 10g stores datafiles in the *Oracle_Base*\oradata*SID* folder, which is the same folder that stores the control files. Figure 11-40 shows the datafiles for the orcl database instance.

You can view a listing of a database instance's datafiles in OEM by clicking the Datafiles link under the Storage heading. You can use OEM to create a new datafile and associate it with an existing tablespace. Now you add a new datafile to the DBUSER tablespace you created earlier.

To add and modify the datafiles:

1. Click the **Datafiles** link on the Administration page, and then click **Create**. The Create Datafile page appears.

2. To specify the datafile name, type **DBUSER02.DBF** in the File Name field. Accept the default for the File Directory field.

3. Open the **Tablespace** list, select **DBUSER**, and then click **Select**.

4. Type **10** in the File Size field, and make sure that **MB** is selected in the Size list.

5. Check the **Automatically extend database when full (AUTOEXTEND)** check box, specify that the next extent size is **512 KB**, and make sure that the Unlimited option button is selected.

6. Click **Show SQL** to view the SQL command associated with creating the datafile. Click **Return**.

7. Click **OK**. The Datafiles page appears with the update message "The object has been created successfully." Note that the new datafile appears in the Datafiles list on the Datafiles page.

11

You can also use OEM to modify the properties of an existing datafile. Now you change the maximum size of the existing DBUSER01.DBF datafile to 2 megabytes.

To change the properties of an existing datafile:

1. Select the option button for the *Oracle_Base***ORADATA***SID*\ **DBUSER01.DBF** datafile on the Datafiles page, and then click **Edit**. The Edit Datafile page for the selected datafile appears.

NOTE

You can also open a datafile's Edit Datafile page by opening the Tablespaces page, navigating to the page of the tablespace associated with the datafile, and then clicking Modify by the Datafiles property of the tablespace.

2. Change the *File Size* field value to **2**, and then open the **Size** list and select **MB**.

3. Click **Show SQL** to view the SQL command for changing the datafile size. Click **Return**.

4. Click **Apply** to apply the change.

5. Click the **Database:** *database_name* link in the upper-left section of the page to return to the Administration page.

Redo Log Files

Recall from Chapter 3 that after the user enters all of the action queries in a transaction, he or she can either commit (save) all of the changes or roll back (discard) all of the changes. When the Oracle 10*g* database executes an action query, it updates the data in the datafiles, and also records information to undo the action query's changes. Oracle 10*g* records information to undo action query changes in redo log files. A redo log file has a .log extension. An Oracle 10*g* database stores its redo log files in the *Oracle_Base*\ORADATA*SID* folder. Figure 11-40 shows three redo log files (REDO01.LOG, REDO02.LOG, and REDO03.LOG).

For an INSERT action query, the redo log file contains the ROWID value of the new record. (Recall that every record has a ROWID value that specifies the record's internal location in the database.) For an UPDATE action query, the redo log file contains a **pre-image** of the updated fields, which shows the data values before the update. For a DELETE action query, the redo log file contains an image of how the record looked before it was deleted.

To perform database administration tasks, you do not work with redo log files directly. Instead, you work with rollback segments and redo log groups, which the following subsections describe.

Rollback Segments

An Oracle 10*g* database stores rollback information in redo log files in a data structure called a **rollback segment**. A rollback segment is made up of data blocks configured in a circular fashion, as shown in Figure 11-41.

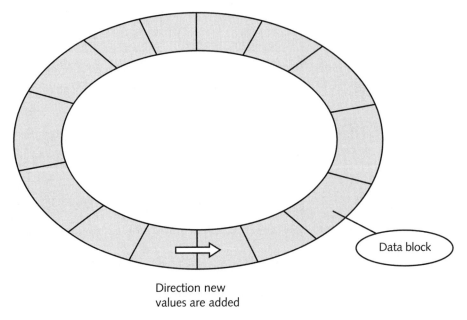

Direction new
values are added

Figure 11-41   Rollback segment

As users perform action queries, the DBMS adds new rollback information to the rollback segment data blocks. By default, an Oracle 10*g* database contains a single rollback segment named SYSTEM. This rollback segment's data blocks are usually stored across multiple redo log files, which keeps any single redo log file from becoming overly large and hard to manage.

Redo Log Groups

An Oracle 10*g* database structures redo log files as redo log groups. A **redo log group** consists of one or more redo log files. Each individual redo log file is called a **member**. Figure 11-42 shows the relationship between redo log groups and members.

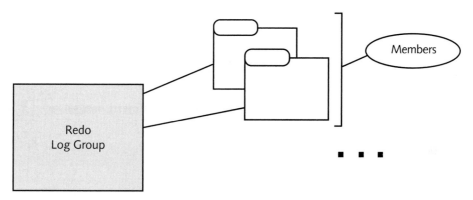

Figure 11-42   Redo log groups and members

By default, an Oracle 10*g* database contains three redo log groups, numbered 1, 2, and 3, which each contain one redo log file, named REDO01.LOG, REDO02.LOG, and REDO03.LOG, respectively. Every Oracle 10*g* database must have at least two redo log groups, so the database can begin writing to the second group when the first group becomes filled with rollback information. A general guideline is that a DBA should create a redo log group for approximately every four users who create action queries.

When all of the data blocks in redo log group 1 are filled, the database performs a **log switch**, which means that it starts writing rollback information to the data blocks in redo log group 2. When all of the data blocks in redo log group 2 are filled, the database performs another log switch, and starts writing to the data blocks in redo log group 3. When all of the data blocks in redo log group 3 are filled, the database begins overwriting the entries in redo log group 1, with the relative certainty that by this time, all users have committed or rolled back their transactions.

Using OEM to Work with Rollback Segments and Redo Log Groups

You can use OEM to create and configure rollback segments and redo log groups. The redo log groups are referenced as log group numbers, starting with number 1. When you create a new redo log group, OEM automatically assigns to it the next larger number.

11

Recall that by default, a new Oracle 10g database instance has three redo log groups, numbered 1, 2, and 3. If you create a new redo log group, OEM automatically assigns to it the number 4. Now you use OEM to examine your database's rollback segment and redo log groups, and create a new redo log group.

To use OEM to work with rollback segments and redo log groups:

1. Click the **Rollback Segments** link under the Storage heading on the Administration page. A single rollback segment named SYSTEM appears.

2. Click the **Database:** *database_name* link in the upper-left section of the page to return to the Administration page.

3. Click the **Redo Log Groups** link. The Redo Log Groups page appears, as shown in Figure 11-43.

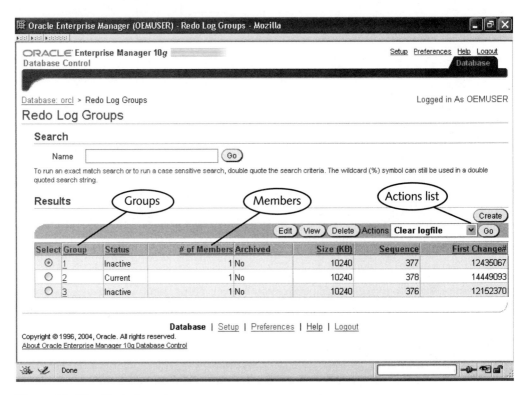

Figure 11-43   Redo Log Groups page

4. Click **Create**. The Create Redo Log Group page appears.

5. Note that the Group # field value is 4. (If your database instance already has a redo log group number 4, accept the default value instead.) Accept the default File Size value of 10240 KB.

6. Change the File Name value to **REDO04.LOG**. (If your Group # is different from 4, use that number in the filename instead.)

7. Click **Show SQL**. Click **Return**.

8. Click **OK**. The update message "The object has been created successfully" appears on the Redo Log Group page.

9. Note that the new redo log group appears in the list of redo log groups, and that its status appears as "Unused," which indicates that the database instance has not yet written any rollback information to it.

In the Redo Log Groups page (see Figure 11-43), a redo log group's status value can be Current, Active, Inactive, or Unused. *Current* means that the database is currently writing rollback information to the redo log group. *Active* means that the redo log group has active (uncommitted) transaction information. *Inactive* means that all of the information in the redo log group is for inactive transactions that have been committed. *Unused* means that the redo log group has just been created.

Recall that when a redo log group's data blocks become filled, the database performs a log switch, and begins writing to the next redo log group. The DBA can **force** a log switch, which causes the database to begin writing to a different redo log group before the current group becomes full. A DBA might force a log switch if he or she wants to create a backup of a redo log group's files. Now you force a log switch so the database instance begins writing to your new redo log group.

11

To force a log switch:

1. Note which redo log group's status has the value "Current."

2. Click the list arrow next to the Actions field. Select **Switch logfile** from the list, and then click **Go**.

3. An update message appears on the page that says "Log group successfully switched." Note that the redo log group whose Status value is "Current" has changed.

4. If necessary, repeat Steps 2 and 3 until your new log group's Status value is "Current."

5. Click the **Logout** link in the upper-right corner of the OEM page.

6. Close any open windows.

SUMMARY

❑ Database administration involves installing, configuring, maintaining, and trouble-shooting a database. A database administrator (DBA) is the person responsible for performing database administration tasks in organizations. A DBA can have a service role, which supports users in daily activities that interact with the database, and a production role, which provides users with specific solutions to information management problems.

❑ There are three editions of the Oracle 10*g* database: Enterprise Edition, for installations that require a large number of transactions performed by multiple simultaneous users; Standard Edition, for high volume multiple-user installations that do not require the additional utilities in Enterprise Edition; and Personal Edition, which provides a single-user DBMS for developing database applications.

❑ Typically, a database server contains a folder called the Oracle Base that contains subfolders for individual Oracle 10*g* products, each of which is called the product's Oracle Home.

❑ When you install Oracle Database 10*g*, a series of Universal Installer pages guides you through the installation process. You must specify the location of the database source files, the database edition, global database name, which uniquely identifies the database server, and SID, which uniquely identifies the database instance on the server.

❑ To configure client applications to connect to an Oracle 10*g* database, you can use local naming, in which each client application contains a file named tnsnames.ora that contains the configuration information for connecting to the database server, or Oracle Internet Directory, which stores connection information on a network directory server.

❑ Oracle Enterprise Manager 10*g*(OEM) is a utility that allows DBAs to perform database administration support tasks, such as creating and configuring database storage structures and user accounts.

❑ A database instance stores data in one or more tablespaces. Each tablespace is made of up multiple segments, each segment is made up of multiple extents, and each extent is made up of one or more data blocks.

❑ The Oracle 10*g* database stores all of the data for a tablespace in one or more datafiles. A datafile can be autoextensible, which means that it automatically grows as users insert more data records.

❑ A tablespace stores related database objects, such as all of the database objects for a specific application. A DBA can take a specific tablespace offline and then create backups or perform other maintenance activities on the tablespace and its datafiles.

❑ A segment stores an individual database object such as a table. To improve system performance for large tables, you can partition a table, and store its objects across multiple segments.

◻ An extent is a contiguous unit of storage space within a segment that needs to be retrieved in a single disk input/output (I/O) operation, such as a data field.

◻ A data block is the smallest storage unit that the database can address, and corresponds to one or more operating system blocks, which are the smallest storage unit that the operating system can address.

◻ The parameter file is a text file that specifies configuration information about an Oracle 10g database instance. When you start an Oracle 10g database instance, the process that starts the instance reads the parameter file and uses its information to configure the instance.

◻ Control files are binary files that store information about the database structure and state, such as the global database name, SID, information about the tablespaces and their associated datafiles, and names and locations of redo log files.

◻ Redo log files record information to undo action query changes through user rollback operations. An Oracle 10g database stores the rollback information in a data structure called a rollback segment. A rollback segment is usually stored across multiple redo log files to keep any single file from becoming too large. Each redo log file is a member of a redo log group.

◻ When all of the data blocks in a redo log group are filled, the database performs a log switch, and starts writing rollback information to the data blocks in another redo log group. The DBA can force a log switch.

11

REVIEW QUESTIONS

1. What is an Oracle Home?
2. What is a database instance?
3. The local naming client configuration approach stores information for connection to a database server in a file named _____.
4. When should a DBA partition a table into multiple segments?
5. Why do data blocks contain free space?
6. What value does a redo log file store when a user inserts a new record?
7. What is a redo log member?
8. Describe the relationship between a redo log group and redo log members.
9. What is a log switch?
10. How can a DBA force a log switch?

11. Creating new database tables is an example of the DBA _____ role, whereas creating new database users is an example of the DBA _____ role.

    a. service, production

    b. SYSOPER, SYSDBA

    c. production, sservice

    d. SYSDBA, SYSOPER

12. The _____ uniquely identifies an Oracle 10g database server and distinguishes it from all other Oracle 10g database servers in the world, and the _____ uniquely identifies each database instance on a database server.

    a. server name, instance name

    b. global database name, connect string

    c. SID, connect string

    d. global database name, SID

13. A _____ specifies the network communication protocol, the IP address of the database server, and the database instance name.

    a. connect descriptor

    b. SID

    c. connect string

    d. service name

14. The tnsnames.ora file contains:

    a. control filenames

    b. service names

    c. global database names

    d. tablespace names

15. The Oracle Internet Directory client configuration approach stores database server configuration information on a:

    a. database server

    b. directory server

    c. file server

    d. database application server

16. An Oracle 10g database must have at least _____ redo log group(s).

    a. one

    b. two

    c. three

    d. none, redo log groups are optional

17. True or False: A database block can consist of exactly one operating system block.

18. True or False: In a default Oracle 10*g* database installation, each redo log file contains a separate rollback segment.

19. True or False: You must install each individual Oracle 10*g* product in a different Oracle Home.

20. Specify the correct Oracle 10*g* data structure (tablespace, segment, extent, data block) to complete the following statements:

    a. A(n) extent is composed of many _____.

    b. A(n) _____ segregates different types of database data.

    c. A(n) _____ stores a contiguous block of data, such as a field value.

    d. A(n) _____ stores a single data object, such as a table.

21. Specify the correct Oracle 10*g* file type (parameter file, control file, datafile, or redo log file) to complete the following statements. (Some statements will have more than one correct answer.)

    a. If the _____ is missing or corrupted, the database instance will not start.

    b. The _____ contains rollback segments.

    c. The _____ points to all of the other files for a database instance.

    d. The _____ contains the actual data values that the database instance stores.

    e. The _____ specifies the database tablespace names.

    f. You can edit the _____ with a text editor.

    g. The _____ is stored in the *Oracle_Base*\ORADATA*SID* folder on the database server.

11

PROBLEM-SOLVING CASES

Save the solution files for all cases in the Chapter11\Cases folder on your Solution Disk.

1. In Notepad, create a file named 11ACase1tnsnames.ora that associates the service name "my_database" with a database server that uses the TCP/IP protocol, and has the following connect descriptor specifications: Server Name = mike.uwec.edu; Port = 1521; SID = test_server.

2. Use OEM to configure a new tablespace named 11ACASE2. Create an associated datafile named 11ACASE201.DBF, and specify that the datafile is initially 1 MB in size. Configure the datafile to automatically allocate new extents that are 512 KB in size. Do not allow the datafile to grow larger than 100 MB in total size. Before you create the tablespace, show the SQL command that creates the tablespace. Copy the SQL command, and paste it into a file named 11ACase2.sql.

NOTE

To complete Case 3, you must have already completed Case 2.

3. Use OEM to add a new datafile named 11ACASE301.DBF to the 11ACASE2 tablespace that you created in Case 2. Configure the datafile so its initial and next size is 512 KB, and so that it automatically extends to an unlimited size. Before you create the datafile, show the SQL command that alters the tablespace, copy it, and paste it into a file named 11ACase3.sql.

NOTE

To complete Case 4, you must have already completed Case 2.

4. Use OEM to modify the 11ACASE201.DBF datafile that you created in Case 2 so all new extents that the database creates are 1 MB in size. Before you apply the change to the datafile, show the SQL command, copy it, and paste it into a file named 11ACase4.sql.

5. Use OEM to create a new redo log group that has the next number within your database instance's redo log groups. Specify that the redo log group has two redo log files of size 4 MB. Name the redo log files 11ACASE501.LOG and 11ACASE502.LOG. Before you create the redo log group, show the SQL command, copy it, and paste it into a file named 11ACase5.sql.

◄ LESSON B ►

After completing this lesson, you should be able to:
- Create and manage user accounts
- Understand an Oracle 10g database instance's memory areas and background processes
- Start and shut down the database
- Understand Oracle 10g database backup and recovery

CREATING AND MANAGING USER ACCOUNTS

Recall that each database user has a user account associated with his or her user schema that is identified by a unique username and password. From a user's viewpoint, an Oracle 10g database consists of the database objects in his or her user schema and the database objects that he or she has privileges to access. An important database administration support task involves creating and managing user accounts to control access to database instances and the objects that a database instance stores and manages.

When you create a new user account, you must specify the following items:

1. General information about the user account, such as the username, password, default tablespace, and temporary tablespace.

2. System privileges the user has in the database.

3. The user's tablespace quota on the database server.

NOTE

You can specify other optional properties for a new user account, but these three items are required. Specifying optional properties is an advanced database administration topic that this textbook does not address.

The following sections describe how to configure new user accounts in an Oracle Database 10g.

Specifying General User Information

To use OEM to create a new user account, you navigate to the Administration property page, click the Users link under the Security heading, and then click Create. This opens the Create User page, and displays the General page in Figure 11-44, which you use to specify general information such as the username, password, and tablespaces.

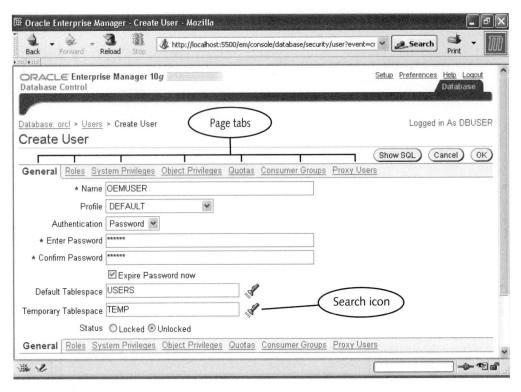

Figure 11-44 Create User page

The General page collects the following account information:

- **Name**—Specifies the username. This value must be unique within the database instance, and must adhere to the standard Oracle naming rules.

- **Profile**—Specifies how the database manages the user's connections and memory usage; the default value is DEFAULT, which is satisfactory for most users.

- **Authentication**—Specifies how the user identifies himself or herself. This textbook uses the Password authentication method, which requires the user to type a password before he or she can log onto the database. Oracle passwords must be 1–30 characters long, can contain any combination of characters or symbols, except blank spaces, and are not case sensitive. The DBA types the initial password in the Enter Password field, and confirms the initial password by retyping it in the Confirm Password field. If the DBA checks the Expire Password Now check box, the user's password expires the first time the user logs onto the database, and he or she is prompted to enter a new password.

- **Default tablespace**—Specifies the user's default tablespace. The **default tablespace** is the tablespace in which the database stores the user's database objects. You should assign new users to the USERS tablespace or to a tablespace associated with applications that they will use. You should never assign the SYSTEM tablespace as a new user's default tablespace because the SYSTEM tablespace is reserved for system objects. You can click the Search icon to select a tablespace from a list of existing tablespaces.

- **Temporary tablespace**—Specifies the tablespace in which the database stores temporary data that it uses for query processing. Temporary data includes intermediate results for queries that perform calculations or sort data. You can click the Search icon to select a tablespace from a list of existing tablespaces.

- **Status**—Specifies whether the account is enabled. If the Locked option button is selected, the account is disabled; if the Unlocked option button is selected, the account is enabled and available for use.

Now you use OEM to begin creating a new user account. You specify the user's name as OEMUSER and accept the default values for the other general specifications.

To begin creating a new user account:

1. Start OEM by opening a Web browser and entering the following URL: **http://localhost:5500/em**

2. The Oracle Enterprise Manager Login window opens. Enter the SYSMAN username and the password you specified during installation (see Figure 11-11) as shown in Figure 11-21, and then click **Login**. When you log in to Oracle Enterprise Manager Database Control using the SYSMAN user account, you are logging in as the Oracle Enterprise Manager super user. The SYSMAN, SYS, AND SYSTEM accounts are automatically granted the roles and privileges required to access all the management functionality provided with Database Control.

NOTE

If you are using an Oracle 10g Enterprise Edition database that you did not install yourself, your instructor or technical support person must give you a user account and password.

3. Click the **Administration** link, click the **Users** link under the Security heading, and then click **Create**. This opens the Create User page, and displays the General page shown in Figure 11-44.

4. Type **OEMUSER** in the Name field.

5. Make sure that **DEFAULT** is selected in the Profile list, and **Password** is selected in the Authentication list.

6. Type **oracle** in the Enter Password field, and type oracle again in the Confirm Password field.

7. Check the **Expire Password now** check box.

8. Click the Search icon next to the Default Tablespace field, and then select **USERS** as the Default tablespace value. (If your database does not have a tablespace named USERS, select another tablespace as the default tablespace, but do not select the SYSTEM tablespace.) Click the **Select** button.

9. Click the next to the Temporary Tablespace field, and then select **TEMP** as the temporary tablespace value. (If your database does not have a tablespace named TEMP, select another tablespace as the temporary tablespace, but do not select the SYSTEM tablespace.) Click the **Select** button.

10. Make sure that the **Unlocked** option button is selected. Do not click OK yet, because you are not finished configuring the user account.

If you accidentally click OK, select OEMUSER on the Users page, and then click Edit.

Specifying System Privileges

There are two types of privileges in an Oracle 10*g* database: system privileges and object privileges. A **system privilege** allows a user to perform a specific task with the Oracle 10*g* database, such as connecting to the database or creating a new table. An **object privilege** allows a user to perform a specific action on a database object, such as selecting data from a table or retrieving a value from a sequence.

You learned how to grant and revoke object privileges in Chapter 3 using the SQL*Plus GRANT and REVOKE commands.

To enable a new user to interact with an Oracle 10*g* database, the DBA must grant system privileges to the new user account. Oracle 10*g* provides system privileges to allow users to create and manipulate database objects such as tables, views, stored procedures, and indexes. For example, the CREATE TABLE privilege enables a user to create new database tables in his or her database schema. Oracle 10*g* enables DBAs to grant privileges with the ANY qualifier, which allows the grantee to modify objects in any user's schema. For example, if a user has the CREATE ANY TABLE privilege, he or she can create a new table in any user's schema.

The System Privileges page in the Create User page in Figure 11-45 lists the user's current system privileges. Clicking Modify provides a list of all available system privileges and allows you to grant system privileges to new users.

Figure 11-45   System Privileges page in the Create User

To grant a privilege, you select the privilege in the Available list and click the Move link. To revoke an existing privilege, you select the privilege in the Granted list and click the Remove link. You can grant a privilege with **Admin Option**, which allows the user with a certain privilege to grant that privilege to other users. By default, newly granted privileges do not have Admin Option. To grant the Admin Option to a privilege, you place a checkmark in the Admin Option check box beside the privilege in the System Privileges list.

Now you open the System Privileges page and grant the CREATE TABLE and CRE-ATE USER system privileges to your new user account. You grant CREATE TABLE with Admin Option and CREATE USER without Admin Option.

To grant system privileges:

1. In the Create User page, click the **System Privileges** link. The System Privileges page in the Create User page, as shown in Figure 11-45, appears.

2. Click Modify. The Modify System Privileges page appears. The available system privileges appear in the Available System Privileges list, as shown in Figure 11-46.

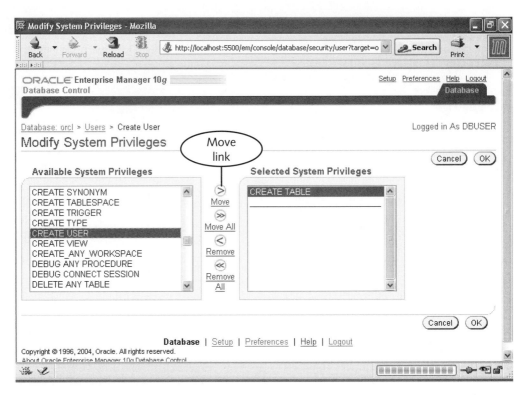

Figure 11-46   Modify System Privileges page

2. Scroll in the Available System Privileges list, select **CREATE TABLE**, and then click the **Move** link. CREATE TABLE appears in the Selected System Privileges list.

3. Scroll in the Available System Privileges list, select **CREATE USER**, and then click the **Move** link. CREATE USER appears in the Selected System Privileges list.

4. Click **OK**. The System Privileges page in the Create User page appears.

5. Check the Admin Option check box beside the CREATE TABLE privilege. Do not click Apply yet, because you are not finished configuring the user account.

Tablespace Quotas

To enable a new user to create new database objects, you must assign to the user account a tablespace quota for the user's default tablespace. A user's **tablespace quota** specifies the amount of disk space that the user's database objects can occupy in his or her default tablespace. If you do not assign a tablespace quota to a new user account, the user will receive an error message when he or she tries to create new database objects. You use the Quota page in the Create User page in Figure 11-47 to assign a tablespace quota to a new user account.

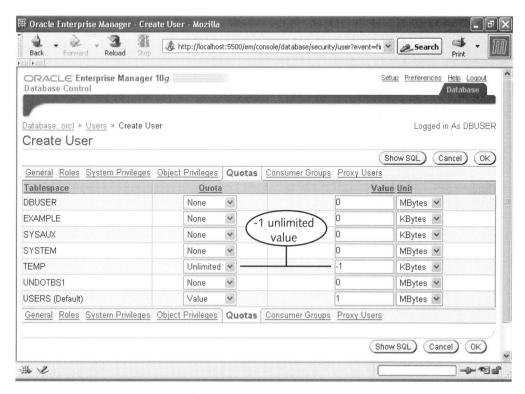

Figure 11-47   Quota page in the Create User page

To assign a tablespace quota to a new user account, you select the user's default tablespace in the Tablespace list. By default, the Quota Size value is **None**. To assign to the user an unlimited quota size, select **Unlimited** from the list box in the Quota field. To assign a specific quota size, select the **Value** from the list box in the Quota field, enter the quota size in the Value field and select a unit size in the Unit field. Now you assign a tablespace quota to the new user account, view the SQL command to create the new user, and then create the new user.

To assign the quota size and create the new user:

1. In the Create User page, click the **Quotas** link. The Quotas page appears.

2. Type **1** in the Value field beside the USERS tablespace. (If you assigned a different tablespace on the General page as the default tablespace, select that tablespace instead.)

3. Verify that **MBytes** is selected in the Unit list.

4. Select **Unlimited** in the Quota field beside the TEMP tablespace (if you assigned a different tablespace on the General page as the default tablespace, select that tablespace instead), and then type **–1** in the Value field.

5. Click **Show SQL**. The SQL command to create the new user appears, as shown in Figure 11-48.

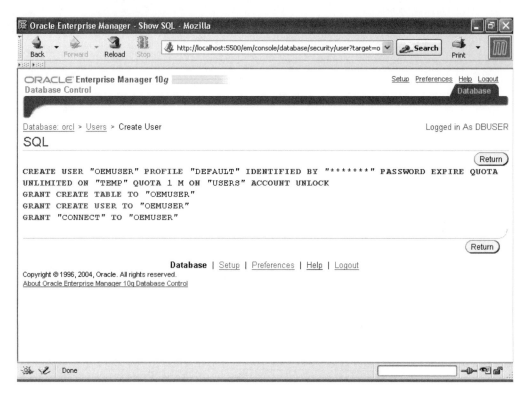

Figure 11-48   SQL page

5. Click **Return**, and then click **OK.** The Users page appears with the update message "The object has been created successfully." Note that the OEMUSER account appears in the Users list on the Users page.

Editing Existing User Accounts

You can use OEM to modify the properties of existing user accounts. You select the user account to be modified on the Users page, and the General page opens, allowing you to change general user information.

You can select other links to modify properties such as the user's system privileges or table-space quota. Now you modify the OEMUSER account and increase its default tablespace quota to 2 MB.

To edit an existing user account:

1. Click the **OEMUSER** link under the UserName column of the Users page.

2. Click the **Quotas** link, change the *Value* field value to **2**, and change the Unit list value to **MBytes** for the **USERS** tablespace (or your user's default tablespace).

3. Click **Show SQL** to view the SQL command for the modification, click **Return**, and then click **Apply** to apply the change.

4. To return to the Administration property page, click the **Database: database_name** link in the upper-left section of the page.

Roles

Suppose that you are working as a database administrator at Northwoods University, and a professor asks you to create database user accounts for 200 students. She asks that you grant the following system privileges to each account: CREATE TABLE, CREATE SEQUENCE, CREATE VIEW, CREATE INDEX, and CREATE PROCEDURE. She also asks you to grant the CREATE TABLE and CREATE VIEW privileges with Admin Option. Being an industrious employee with a strong work ethic, you begin making user accounts and granting system privileges, and finish a few days later. When you inform the professor of your accomplishment, she says, "Great! And by the way, I forgot to tell you that all of the accounts need the CREATE TRIGGER system privilege, also." At this point, you wonder if there is an easier way to assign user account privileges than by opening the System Privileges page for each user account and individually assigning each privilege. There is—through the use of roles.

A **role** is a database object that represents a collection of system privileges that you can assign to multiple users. You create the role and grant to it system privileges. You can then grant the role to database users. If you need to change the privileges of the users to whom the role has been granted, you simply change the role's privileges, and the user accounts automatically reflect the privilege change. The following subsections describe how to create a role and grant a role to a user account.

Creating a Role

To create a new role, you click the Roles link under the Security heading, and then click Create. The Create Role page opens, as shown in Figure 11-49, and displays the General page.

11

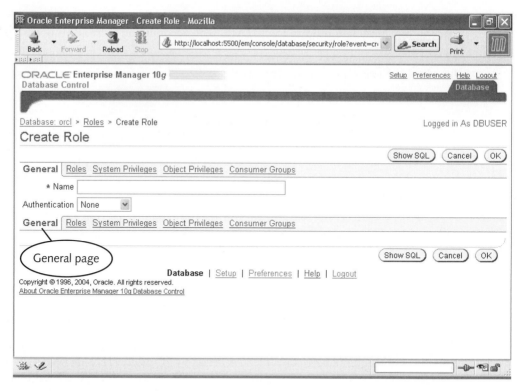

Figure 11-49   Create Role page

The Name field allows you to specify the role name, which must adhere to the standard Oracle naming rules. The General page also allows you to specify the role's authentication method. The default value is *None*, which specifies that the users to whom the roles are granted automatically receive the role's privileges, with no further authentication. You can also specify that users must enter a password or perform other validation measures to receive the role privileges.

A role can inherit privileges from other roles. For example, you might create a role named DEVELOPER that has the privileges needed to perform database development activities such as creating tables, views, and sequences. You might create a second role named DBA_DEVELOPER that has the privileges for these development activities, as well as privileges required for DBA support activities, such as creating tablespaces and users. You would create the DEVELOPER role first, then create the DBA_DEVEL-OPER role, and specify that the DBA_DEVELOPER role inherits the DEVELOPER role privileges. To specify that a new role inherits privileges from an existing role, you use the Modify Roles page, as shown in Figure 11-50.

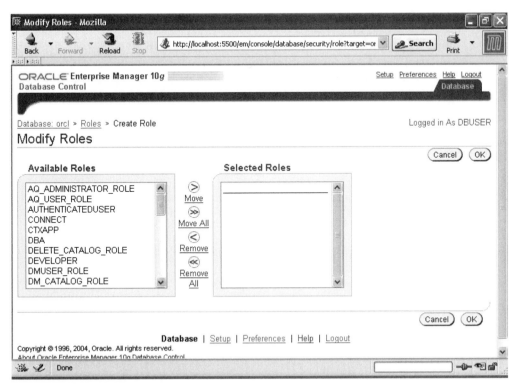

Figure 11-50   Modify Roles page

You use the System Privileges tab in the Create Role page to grant specific system privileges to a role. The Modify System Privileges page in the Create Role page is identical to the Modify System Privileges page in the Create User dialog box in Figure 11-46, except that it grants privileges to a role rather than to a user account. Now you create a new role named DEVELOPER, and grant to it system privileges that allow developers to create different database objects.

To create a role and grant privileges to it:

1. Click the **Roles** link under the Security heading for your database instance, and then click **Create**. The General page of the Create Role page opens.

2. Type **DEVELOPER** in the Name field, and accept the Authentication value of None.

3. Click the **System Privileges** link, click **Modify**, and grant to the role the following system privileges without Admin Option: **CREATE SEQUENCE**, **CREATE TABLE**, and **CREATE VIEW**.

4. Click **OK**. On the System Privileges page of the Create Role page click **Show SQL** to view the SQL command to create the role, click **Return**, and then click **OK**. The update message "The object has been created successfully" message box appears on the Roles page.

Granting a Role to a User Account

To grant a role to a new user account, you open the Role page in the Create User page, and grant the role to the user. To grant a role to an existing user account, you open the user account for editing, and grant the role to the user. Now you grant the new DEVELOPER role to the existing OEMUSER account.

To grant the role to the existing user account:

1. To return to the Administration property page, click the **Database: database_name** link in the upper-left section of the page.

2. Navigate to the **Users** page, and then click the **OEMUSER** link in the UserName column.

3. Click the **Roles** link, click **Modify**, select the **DEVELOPER** role in the Available Roles list, and then click the **Move** link. The DEVELOPER role appears in the Selected Roles list.

4. Click **OK**, click **Show SQL** to view the SQL command to grant the role, click **Return**, and then click **Apply** to apply the role change to the user account.

5. Click the **Database: database_name** link in the upper-left section of the page, to return to the Administration property page.

COMPONENTS OF AN ORACLE 10g DATABASE INSTANCE

Recall from Lesson A that in an Oracle 10g client/server database, a client process that runs on the client workstation connects to a database instance that runs on the database server. The link between the client process and the database instance is called a connection, and it creates a user session. Figure 11-51 illustrates how the client process connects to the server process to create a connection and a user session.

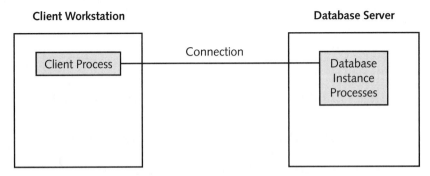

Figure 11-51   Oracle 10g connection architecture

Recall that a database instance consists of a set of processes and associated memory structures that manipulate data in a database's tablespaces. There are several links on the

Administration page of OEM that allow the DBA to view and manage properties of a database instance. For example, you can click the Last Collected Configuration link under the Configuration Management heading for your database instance. This displays the instance General page, as shown in Figure 11-52.

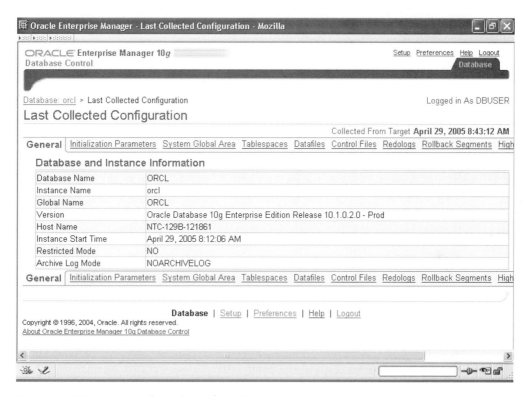

Figure 11-52   Last Collected Configuration page

The General page for a database instance provides information about the database instance, such as the host (server) name and database version. It also provides links to other pertinent information about a database instance, such as the Initialization Parameters link. You use this link to view initialization parameters that the parameter file sets. Now you open the General page for your database instance and view the initialization parameters.

To open the General page and view the initialization parameters:

1. Click the **Last Collected Configuration** link under the Configuration Management heading on the Administration page. The General page appears, and should be similar to Figure 11-52.

2. Click the **Initializaton Parameters** link. The Initializaton Parameters page of the Last Collected Configuration page appears; this page displays the current

values of the system initialization parameters. Unless you are an experienced DBA, you usually accept the default initialization parameter values.

3. Click the **System Global Area** link next to the Initialization Parameters link. This information is discussed in the next section.

To understand the purpose of the parameter values and how to configure them, you must understand the database's server main memory structures and processes.

Oracle 10*g* Server Main Memory Structures

An Oracle 10*g* database instance creates two memory areas in the database server's main memory: the System Global Area and the Program Global Area. The following sections describe these memory areas.

System Global Area

The **System Global Area (SGA)** is a memory area that all database connections use. The purpose of the SGA is to share information among all database processes.

NOTE

The SGA is also sometimes called the Shared Global Area.

In an Oracle 10*g* database instance, the SGA is usually made up of five primary memory areas, as shown in Figure 11-53.

NOTE

The SGA contains other memory areas, but their use and configuration are beyond the scope of this book.

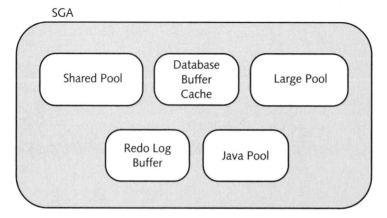

Figure 11-53   Primary memory areas within the System Global Area

The primary SGA memory areas include:

- **Shared pool**—Stores machine-language code and execution plans for frequently used SQL commands. The first time a user executes a SQL command, the database creates a machine-language version and an execution plan for the command. If another user executes the same SQL command, the database instance retrieves the machine-language query and execution plan from the shared pool rather than creating it again, which improves system performance. The shared pool overwrites information for older SQL commands to make room for newer commands. The **shared_pool_size** parameter in the parameter file specifies the shared pool size.

- **Database buffer cache**—Stores data values. When a user inserts or updates data values, the database instance stores the new values in the database buffer cache and writes the values to the datafiles at a later time. When a user retrieves data values, the instance retrieves the data values from the datafiles, stores them in the database buffer cache, and then returns them to the user. If another user retrieves the same data values while the values are still in the database buffer cache, the instance retrieves the values from the database buffer cache rather than retrieving them from the datafiles, which improves system performance. The **db_block_buffers** parameter in the parameter file specifies the number of data blocks allocated to the database buffer cache.

- **Large pool**—Consists of large contiguous memory storage spaces for tasks such as backup and recovery operations. The **large_pool_size** parameter in the parameter file specifies the number of data blocks allocated to the large pool.

- **Redo log buffer**—Stores rollback information for user transactions. As users perform action queries, the database instance writes rollback information in the redo log buffer and then periodically transfers this information to the redo log files. The **log_buffer** parameter in the parameter file specifies the number of main memory data blocks that the database instance allocates for the redo log buffer.

- **Java pool**—Stores machine language representations and execution plans for Java commands used in application programs and database operations. The **java_pool_size** parameter in the parameter file specifies the number of data blocks in the server's main memory that the database instance allocates to the Java pool.

Program Global Area

The **Program Global Area (PGA)** is a memory area that stores information for a specific user connection. The PGA contains two separate memory areas, as shown in Figure 11-54.

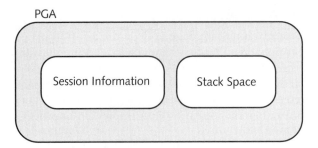

Figure 11-54   Memory areas within the Program Global Area

The **session information** area contains information about the user session, such as the username, time the session started, date of the last activity, and global variable values. The **stack space** contains the values of the variables that the user declares in PL/SQL programs and other programs. The **sort_area_size** initialization parameter in the parameter file specifies the number of data blocks in the server memory that the database instance allocates to the PGA for an individual user session.

Configuring Memory Areas

You can view and edit the sizes of your database instance's SGA and PGA memory areas on the instance Memory Parameters page, as shown in Figure 11-55.

The instance Memory Parameters page shows the current sizes of the shared pool, buffer cache (which combines the database buffer cache and redo log buffer), large pool, and Java pool. The pie chart displays the relative space occupied by each area. The PGA page shows the aggregate PGA target, which is the total memory area that the database instance reserves for PGAs for all instance users. The sum of the SGA and PGA cannot exceed the total server main memory plus the memory required for the operating system and other server processes.

Unless you are an experienced DBA, you usually accept the default sizes for the SGA memory areas. Experienced DBAs configure the SGA memory areas based on their users' needs. If users execute a lot of different SQL commands, the system needs a large shared pool; conversely, if all users execute essentially the same SQL commands over and over, the shared pool can be smaller. If users retrieve large volumes of data, or execute a lot of action queries that generate high amounts of rollback data, then the DBA needs to make the buffer cache area larger. If your system has many users, you might need to increase the PGA size. You can click the Advice button beside the Shared Pool, Buffer Cache, and Aggregate PGA Target fields on the Memory Parameters pages to view a graph and learn more about configuring these memory areas.

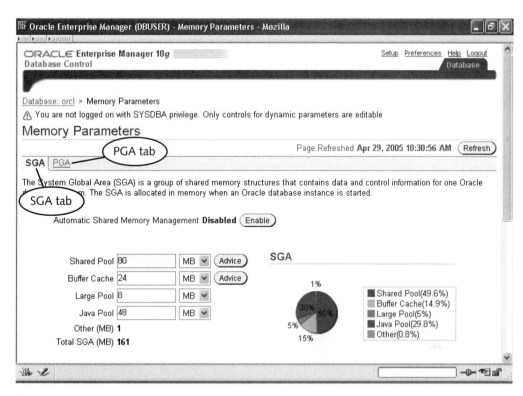

Figure 11-55   Memory Parameters page

Now you will navigate to the SGA page and the PGA page of the Memory Parameters page to view memory properties.

To navigate to the SGA page of the Memory Parameters page:

1. Click the **Database:** *database_name* link in the upper-left section of the page, to return to the Administration property page. The General page appears.

2. Click the **Memory Parameters** link, which is located under the Instance heading. The SGA page of the Memory Parameters page appears and should be similar to Figure 11-55.

3. Click the **Advice** button next to the Shared Pool property. Examine the page and then click **Cancel**.

4. Click the **PGA** link next to the SGA link. The PGA page of the Memory Parameters page appears.

5. Click the **Database:** *database_name* link in the upper-left section of the page, to return to the Administration property page. The General page appears.

You can modify the sizes of memory components on the database instance Memory Parameters page or in the parameter file. After you modify the size of these values, you must then shut down the database and restart it to apply the changes. (You learn how to shut down and restart the database in a later section in this lesson.)

Oracle 10g Background Processes

An Oracle 10g database instance contains a set of background processes to service user requests. For the most part, a DBA cannot directly control these processes. However, it is important to be familiar with them so you understand how the database instance operates. Figure 11-56 illustrates the main Oracle 10g database instance background processes.

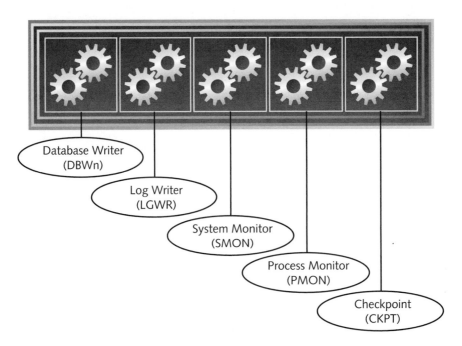

Figure 11-56   Oracle 10g database instance background processes

The Oracle 10g background processes include:

- **Database Writer (DBWn)**—Writes changed data from the database buffer cache to the datafiles. To improve database performance, the database accumulates data changes in the database buffer cache, then writes all of the changes to the datafiles at one time. A database instance can have up to ten separate database writer processes, numbered 0 to 9. The Oracle 10g database references the first process as DBW0, the second as DBW1, and so on. The

db_writer_processes parameter in the parameter file specifies the current number of database processes. By default, the Oracle 10g database has one database writer process. An instance would need multiple processes if it has many users performing multiple simultaneous operations and experiencing performance slowdowns.

- **Log Writer (LGWR)**—Writes redo information from the redo log buffer to the redo log files. As with datafile changes, the database accumulates rollback information in the redo log buffers and writes all of the changes to the redo log files at once to improve system performance.

- **System Monitor (SMON)**—Has three primary functions: to recover lost data after a system hardware or software failure; to deallocate temporary memory areas that the database uses for sort operations; and to manage server disk space by periodically coalescing free space to make larger continuous extents for new data. As users insert, update, and delete data, the server hard drive may become **fragmented**, which means that some extents may become empty and the data becomes scattered across multiple disk locations. **Coalescing free space** defragments the data on the hard drive, and moves all of the stored data so it is in contiguous memory blocks.

- **Process monitor (PMON)**—Monitors and manages individual user sessions. When a user session performs an UPDATE or DELETE action query on a data record, the database locks the record, and other user sessions cannot view the record until the current user session commits or rolls back the transaction. After a commit or rollback, PMON releases the process locks held by the user session. In case of a server hardware or software malfunction, PMON rolls back uncommitted user transactions and releases the session's locks.

- **Checkpoint (CKPT)**—Initiates checkpoints. A **checkpoint** signals the DBWn and LGWR processes to write the buffer contents to the datafiles and redo log files. The **log_checkpoint_interval** parameter in the parameter file specifies the checkpoint interval as a preset amount of time. Checkpoints occur automatically when a redo log file becomes filled, when a log switch occurs, and when the DBA shuts down the database instance. A DBA can force a checkpoint by forcing a log switch.

STARTING AND SHUTTING DOWN THE DATABASE

Database administrators must shut down a database periodically to perform maintenance tasks such as backing up the database, adding new datafiles, modifying the properties of the parameter file, or recovering from a database crash. After the DBA performs the maintenance activity, he or she restarts the database to make it available for new user connections. The following subsections describe how to create an administrative connection for starting and shutting down the database, the startup and shutdown phases, and the steps for starting and shutting down the database.

CAUTION

Shutting down the database makes the database unavailable to other users. You can perform the exercises in this section only if you are using an Oracle 10g Enterprise or Personal database on which you are the *only* user, or which your instructor has configured within your classroom setting so you can perform these steps.

Creating an Administrative Connection

Shutting down a database makes the database unavailable for user connections, so starting and shutting down a database instance are tasks that organizations entrust to only a few database administrators in an organization. To start or shut down a database instance, a DBA must log onto the database using an **administrative connection**. An administrative connection requires the user to have either the SYSDBA or SYSOPER system privilege, and to connect as either a SYSDBA or a SYSOPER. To connect as a SYSDBA or SYSOPER in OEM, you open the *Connect As* list on the Login to Database:*database_name* page (see Figure 11-57), and select SYSOPER or SYSDBA.

NOTE

If you select SYSOPER or SYSDBA in the Connect as list, but enter a username and password for a user account that does not have the SYSOPER or SYSDBA privilege, an error will occur. See Figure 11-26.

NOTE

When you logged onto OEM earlier in the chapter, you selected Normal in the Connect as list, which created a normal (non-administrative) connection.

The SYSDBA or SYSOPER system privileges allow a user to access objects in the SYS tablespace. Table 11-3 summarizes some of the tasks that only users with the SYSDBA and SYSOPER system privileges can perform.

Figure 11-57   Login to Database:*database_name* page as SYSDBA

Task	SYSDBA	SYSOPER
Start the database	X	X
Shut down the database	X	X
Perform database recovery operations	X	X
Recover the database to a specific point in time	X	
Create a new database within an existing database instance	X	
Receive all system privileges with Admin Option	X	

Table 11-3   Tasks that SYSDBA and SYSOPER users can perform

Table 11-3 shows that the SYSOPER privilege is slightly more restrictive than the SYSDBA privilege. SYSOPER users cannot recover the database to a specific point in time, create new databases, or receive all system privileges with Admin Option.

When you install a new database instance, only the SYS user account can access SYS tablespace objects, and only the SYS user account has the SYSDBA and SYSOPER system privileges required to make an administrative connection. To grant either of these system privileges to other users, the DBA who installs the database instance must create an administrative connection by logging onto the database as a SYSDBA using the SYS user account. Because it is not a good idea to use the SYS user account for daily operations, the DBA then grants either the SYSDBA or SYSOPER system privileges to his or her user account and to the user accounts of other DBAs who need to be able to make administrative connections.

Now you logoff as a Normal user and connect to your database instance using the SYS user account, and grant the SYSDBA privilege to your DBUSER account. Then you use OEM to create an administrative connection to your database using your DBUSER account.

To create an administrative connection:

1. Click either **Logout** link on the page, and then click **Login**. The Login to Database page appears.

2. Type **SYS** in the User Name field and type the password you specified during installation (see Figure 11-11) in the Password field, as shown in Figure 11-57.

3. Open the **Connect As** list, select **SYSDBA**, and then click **Login**.

4. Click the **Administration** link, and then open the **Users** link under the Security heading.

5. Click the **DBUSER** link in the UserName column. The General page appears. (If your normal user account has a different username, select the node for that account instead.)

6. To assign the SYSDBA privilege to your user account, select the System Privileges link, click **Modify**, scroll down in the Available System Privileges list, select the SYSDBA privilege, and then click the **Move** link to add SYSDBA to the Selected System Privileges list.

7. Click **OK**, and then click **Apply** to apply the new privilege.

8. To create an administrative connection using your DBUSER account, click either **Logout** link, and then click **Login**. Type **DBUSER** in the User Name field, and your normal password in the Password field. (If you use a user account different from DBUSER, type that account name instead.)

9. Open the **Connect As** list, select **SYSDBA**, and then click **Login**.

Startup and Shutdown States

When you start an Oracle 10*g* database instance, the instance passes through the four states shown in Figure 11-58. When you shut down an Oracle 10*g* database instance, the instance passes through the four states in the reverse order. As the instance starts, it performs the tasks on the lines with the arrows.

Before you start an Oracle 10g database instance, the instance is in the **SHUTDOWN** state. None of its processes are running on the database server, and none of its files are opened. When you first start a database instance, you launch a **startup process**. The startup process first reads the parameter file, and initializes the database parameters. It then starts the background processes, and allocates the memory for the SGA. These steps place the database instance in the **NOMOUNT** state.

From the NOMOUNT state, the startup process opens and reads the database control files, and initializes the database objects. These steps move the database instance to the **MOUNTED** state. From the MOUNTED state, the startup process opens the datafiles and redo log files, performs database recovery operations if necessary, and then starts the Oracle Net server process to service user requests. These steps place the database in the **OPEN** state.

When a DBA starts or shuts down a database instance, he or she can specify to place the database in any of the states shown in Figure 11-58. This allows the DBA to perform specific maintenance activities on the database without completely shutting down the entire database. For example, the DBA could place the database instance in the MOUNTED state to perform operations that require the database processes to be running, but also require the datafiles and redo log files to be closed. Examples of these types of operations include renaming datafiles or adding, dropping, or renaming redo log files. The DBA could place the database instance in the NOMOUNT state to perform backup and recovery activities that require the background processes to be running, but the control file to be closed.

To use OEM to start or shut down a database instance, or to move the database instance to one of the intermediate (NOMOUNT or MOUNTED) states, you create an administrative connection, and then click the Shutdown button (or the Startup button) on the Database Home page (see Figure 11-22.)

To place the database in a different state, you click the option button associated with the target state to which you wish to move the database, and then click Apply.

11

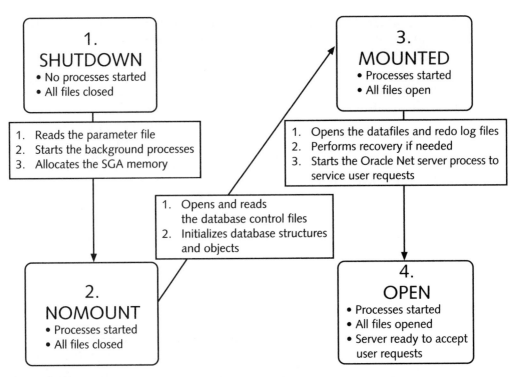

Figure 11-58   Oracle 10g database instance states

Shutdown Options

When shutting down an Oracle 10g database instance, the DBA can specify one of four ways to handle existing user connections:

- **Normal**—The instance does not accept any new connections, but allows current users to finish their transactions and log off normally. The instance shuts down when all users have voluntarily disconnected. When the instance shuts down, it writes all data in the database buffer cache to the datafiles and all data in the redo log buffer to the redo log files. This option leaves the database in a consistent state, so it does not require any recovery operations the next time the database is started. This option is most convenient for users, but is least convenient for the DBA, who must wait until all users log off.

- **Transactional**—The instance does not accept any new connections, and allows users to finish their current transaction. When a user commits or rolls back his or her transaction, the instance disconnects the user. As with the Normal shutdown option, the transactional option writes all buffered data to its associated files and leaves the database in a consistent state. This option is fairly convenient for the DBA. It is less convenient for users, because they cannot continue their work; however, it allows them to finish what they are doing with no data loss.

- **Immediate**—The instance does not accept any new user connections, and immediately terminates current user connections. The database instance automatically rolls back all transactions, so users lose all uncommitted action query changes. This option is most convenient for the DBA, but requires users to re-enter their uncommitted action queries the next time they log onto the database.

- **Abort**—Disconnects all users, stops the instance's processes, and reallocates all server memory. The data in the buffers is not written to the files, and uncommitted transactions are not rolled back. This shutdown option does not leave the database in a consistent state, and the next time the database starts, it will require a recovery operation, which may or may not succeed. This option is equivalent to pulling the database server's plug out of the wall socket, and may result in catastrophic data loss. A DBA should perform an abort shutdown only if the database is malfunctioning and will not shut down using any of the other options.

When a DBA clicks the Shutdown button on the Database Home page, the Startup/Shutdown: Specify Host and Target Database Credentials page in Figure 11-59 appears.

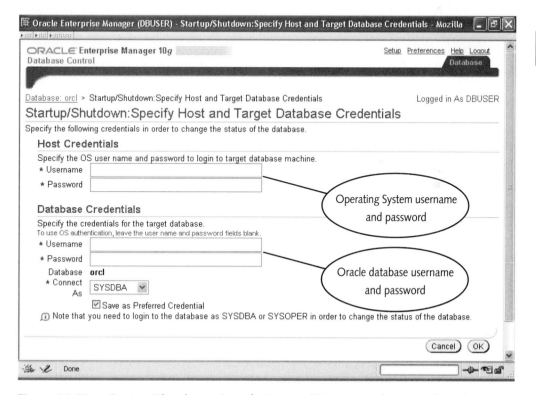

Figure 11-59   Startup/Shutdown: Specify Host and Target Database Credentials page

This page requires you to log in to the machine that is running Oracle, and into the database itself with SYSDBA or SYSOPER privileges. The next screen enables you to shut down the database, or start up the database, specifying options. The DBA selects the desired shutdown option and clicks OK, and OEM shuts down the database.

Startup Modes

When a DBA starts an Oracle 10*g* database instance, he or she can start it in one of two modes: **Unrestricted**, which allows all users to create connections; and **Restricted**, which creates connections only for users who have the RESTRICTED SESSION system privilege.

Most of the time, you start a database in unrestricted mode. You start a database in restricted mode if the database is not functioning properly, and you want to allow only DBAs with the RESTRICTED SESSION privilege to create connections and perform maintenance activities.

When starting the database, the DBA can select from three startup modes on the Startup/Shutdown: Advanced Startup Options page, which is shown in Figure 11-61. The Start the instance option starts the database in the NOMOUNT state; the Mount the database option starts the database in the MOUNT state; and the Open the database starts the database in OPEN state.

If the DBA leaves the Restrict access to database and Force database startup check boxes cleared, the database instance starts in normal mode. If the DBA checks the Restrict access to database check box, the instance starts in restricted mode, and if the DBA checks the Force database startup check box, the instance shuts down the database with abort mode before restarting it. Use this option only if you experience a problem with startup.

On the Startup/Shutdown: Advanced Startup Options page, the Initialization Parameter option allows the DBA to identify which parameter file to use to start up the database. The DBA could specify that the SPFILE, which is the server parameter file, be used in the startup process. This is an alternate parameter file that the database server automatically modifies during operation to improve system performance.

Using OEM to Shut Down and Start a Database Instance

When a DBA shuts down a database instance using the Normal, Transactional, or Immediate shutdown option, the shutdown process performs the following tasks:

1. Writes the contents of the data buffer cache to the datafiles.
2. Writes the contents of the redo log buffer to the redo log files.
3. Closes all files.
4. Stops all background processes.
5. Deallocates the SGA in the server's main memory.

When a DBA restarts a database instance, the startup process performs these tasks in the reverse order. Now you use the OEM Console to shut down and then restart your Oracle 10*g* database instance.

CAUTION

Shutting down the database makes the database unavailable to other users. You can perform these steps only if you are using a Personal Oracle 10*g* database on which you are the only user. If you are using an Oracle 10*g* Enterprise Edition database, you can perform these steps if your instructor or technical support person has configured your system so you do not inconvenience other users by shutting down the database.

To shut down and restart the database instance:

1. Click the **Shutdown** button on the Database Home page, the Startup/ Shutdown: Specify Host and Target Database Credentials page in Figure 11–59 appears.

2. Type the username and password for the computer on which Oracle Database 10*g* is running under the Host Credentials heading.

3. Type **DBUSER** in the username field, and **DBUSER** in the password field under the Database Credentials heading.

4. Verify that SYSDBA is selected in the Connect As field, and that the Save as Preferred Credential checkbox is checked.

11

NOTE

If you get the Error Message: RemoteOperationException: ERROR: Wrong password for user, it is probably because the OS user with whom you are trying to login has not been set up to allow the user to logon as a "Batch Job." To resolve this issue, go to Control Panel, then Administrative Tools, then Local Security Policy. Within Local Policies, go to user Right Assignment. Add the user to Log on as a Batch Job.

5. Click **OK**. The Startup/Shutdown: Confirmation page appears. Click the Advanced Options button and verify that the Shutdown Immediate Mode is selected, as shown in Figure 11–60. Click **OK**.

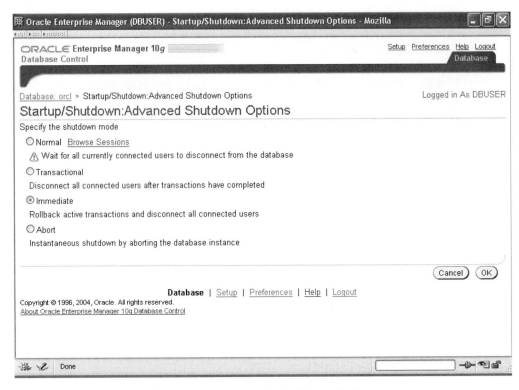

Figure 11-60   Startup/Shutdown: Advanced Shutdown Options page

6. Click **Yes**. The Startup/Shutdown: Activity Information page appears. Because the database is currently being shut down, this operation might take some time.

7. Click the **Refresh** button. The Database: *database name* page appears.

8. To restart the database, click the **Startup** button. The Startup/Shutdown: Specify Host and Target Database Credentials page (see Figure 11-59) appears. Repeat Steps 2, 3, and 4 above.

9. Click **OK**. The Startup/Shutdown: Confirmation page appears. Click the **Advanced Options** button and verify that the Open the database Startup mode is selected, as shown in Figure 11-61. To restart the database instance in

normal mode, make sure the Restrict access to database or Force database startup check boxes are not checked, verify that the Use default initialization parameter option button is selected, and then click **OK**.

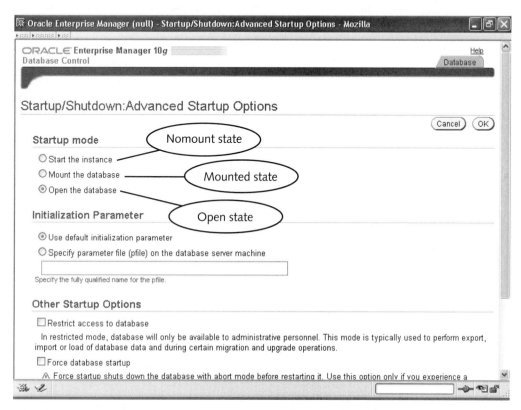

Figure 11-61   Startup/Shutdown: Advanced Startup Options page

10. The Startup/Shutdown: Confirmation page appears. Click **Yes**. The Startup/Shutdown: Activity Information page appears with the message: "Datebase startup is in process. Please wait...." When the database has started, the Login to Database page appears.

11. Click **Login**. Type **DBUSER** in the User Name field, and your normal password in the Password field. (If you use a user account different from DBUSER, type that account name instead.)

12. Open the **Connect as** list, select **SYSDBA**, and then click **Login**.

DATABASE BACKUP AND RECOVERY

CAUTION

You cannot perform the exercises in this section if you are using an Oracle 10*g* Enterprise Edition Database that supports multiple users, because the exercises involve starting and shutting down the database.

Important database administration support tasks include **backup**, which involves creating a copy of the database files, and **recovery**, which involves restoring the database to a working state after a hardware or software malfunction. Database backup and recovery is a complex database administration process, and this textbook provides only a brief introduction to the topic.

Backup operations on an Oracle 10*g* database can be classified as offline backups or online backups. An **offline backup**, which is also called a **cold backup**, requires shutting down the database, then copying all of the database files to an alternate location. An **online backup**, which is also called a **hot backup**, involves backing up critical database files while the instance is running, as well as creating an ongoing archive of database changes so the DBA can restore the database to its state at any time. The following sections describe how to create cold and hot backups, and outline the recovery process.

Creating Offline (Cold) Backups

To create an offline (cold) backup, you shut down the database instance, copy specific database files to an alternate disk location, and then restart the database instance. The alternate disk location should preferably be at a different physical location, so the data will not be lost if the site that houses the database server is destroyed by a natural disaster such as a fire or flood.

When you create a cold backup, you make copies of the following database files:

- Parameter file—Stores the database specifications, and points to the locations of the database control files. Recall that the startup process automatically creates a new parameter file if the existing parameter file is not present. You need to create a backup of the parameter file only if you have modified it with custom parameter settings. Recall that by default, Oracle 10*g* stores the parameter file for each database instance in the server's *Oracle_Base*\admin\ *SID*\pfile folder.

- Control files—Contain information about the database structure. Recall that the control files are mirror images of one another, so you need to copy only one of the control files when you create a cold backup. Recall that by default, Oracle 10*g* stores the control files in the *Oracle_Base*\oradata*SID* folder on the database server.

> - Datafiles—Store the actual data values. You must copy all of the datafiles when you create a cold backup. Recall that by default, Oracle 10*g* stores datafiles in the *Oracle_Base*\oradata*SID* folder, which is the same folder that stores the control files.

You should not make copies of the redo log files for a cold backup. When you shut down the database using the normal, transactional, or immediate shutdown option, all of the user transactions are committed, so the redo log files do not contain any current roll-back information. If you make copies of the redo log files and then use them in a recovery operation, the resulting database might contain inconsistent information.

When you create a cold backup of an Oracle 10*g* database, you must copy the files to a hard drive that has the same operating system block size as the database server. If you do not, then the backup files will have a different block size, and the backup files will not work when you attempt to recover the database.

Now you create a cold backup of your Oracle 10*g* database. You shut down the database instance, copy the database files to your Solution Disk, then restart the database instance.

To create a cold backup of your Oracle 10*g* database:

1. Click the **Shutdown** button on the Database Home page. The Startup/Shutdown: Specify Host and Target Database Credentials page appears.

2. If, in the previous exercise, you left the Save as Preferred Credential check box checked, then all the necessary information is present on the page and you can just click **OK**. If you did not, you will need to type in the usernames and passwords.

3. Click **Yes** on the Startup/Shutdown: Confirmation page. The Startup/Shutdown: Activity Information page appears. The database is currently being shut down; this operation may take some time. Once this operation is complete, click the **Refresh** button to return to the Database Control page.

4. Start Windows Explorer, and create a folder named **OraColdBackup** in the Chapter11\Tutorials folder on your Solution Disk.

5. Navigate to the *Oracle_Base*\admin*SID*\pfile folder on the database server. For example, if your *Oracle_Base* folder path is C:\oracle\product\10.1.0, and your database *SID* is orcl, navigate to C:\oracle\product\10.1.0\admin\orcl\pfile.

6. Copy **init.ora** to the OraColdBackup folder on your Solution Disk.

7. In Windows Explorer, navigate to the *Oracle_Base*\oradata*SID* folder on the database server.

8. Select all of the datafiles and control files, and then copy the files to the OraColdBackup folder on your Solution Disk. (Recall that control files have a .ctl extension, and datafiles have a .dbf extension.) (This operation might take several minutes, because some of these files are quite large.)

11

9. Switch back to OEM and click the **Refresh** button. The Database: *database name* page appears.

10. To restart the database, click the **Startup** button to open the Startup/Shutdown: Specify Host and Target Database Credentials page. If necessary, type the appropriate usernames and passwords, and then click **OK**.

11. Click **Yes** on the Startup/Shutdown: Confirmation page.

12. Click **Login**. Type **DBUSER** in the *User Name* field, and your normal password in the Password field. (If you use a user account different from DBUSER, type that account name instead.)

13. Open the **Connect As** list, select **SYSDBA**, and then click **Login**.

Creating Online (Hot) Backups

Recall that an Oracle 10*g* database instance writes rollback information to redo log files. By default, when all of the redo log files are filled, the instance starts overwriting the oldest entries. To create a hot backup, you must configure the database instance so it saves all redo log file information in archive files. An **archive file** stores the contents of a redo log file after a log switch occurs, which creates an archive of all database transactions. The database instance can use the archive files to re-create all database transactions that have occurred since the last hot backup.

Creating a hot backup is a four-step process:

1. Place the database instance in ARCHIVELOG mode so it automatically creates archive files.

2. Back up the control file.

3. Back up the datafiles.

4. Instruct the database to write the current contents of the redo log files to the archive files, and then create backup copies of the archive log files.

The following sections describe how to accomplish each of these steps.

Placing a Database Instance in ARCHIVELOG Mode

By default, a new Oracle 10*g* database instance runs in NOARCHIVELOG mode. In **NOARCHIVELOG mode**, the instance writes rollback information to redo log files, and when all of the redo log files are filled, the instance starts overwriting the oldest entries. In a database instance that is in **ARCHIVELOG mode**, when a log switch occurs, the instance writes all of the information in the filled redo log file to an archive file. As a result, the instance retains all rollback information in one or more archive files, and the database instance can restore the database to its state at any time.

To use OEM to place a database instance in ARCHIVELOG mode, you create an administrative connection, open the instance Configure Recovery Settings page shown in Figure 11-62, and check the ARCHIVELOG Mode* check box.

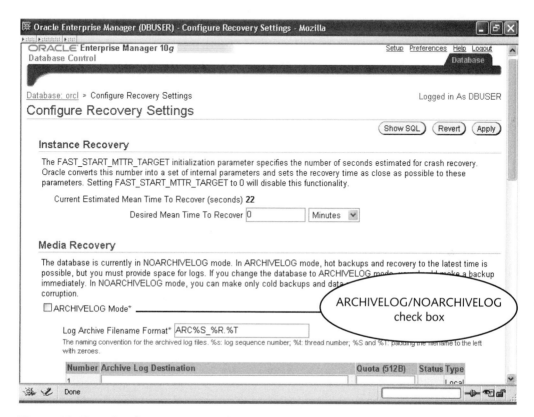

Figure 11-62 Configure Recovery Settings page

The database instance automatically writes the redo log file information to the archive file when a log switch occurs. The db_recovery_file_dest parameter specifies the folder path of the location to which the instance writes the archive log files. By default, this folder path is *Oracle_Base*\flash_recovery_area*SID*\archivelog. For example, if your *Oracle_Base* is C:\oracle\product\10.1.0, and your *SID* is orcl, the folder path would be C:\oracle\product\10.1.0\ flash_recovery_area\orcl\archivelog.

Now you place your database instance in ARCHIVELOG mode. You check the ARCHIVELOG Mode check box on the instance Configure Recovery Settings page, and then click Apply. This **bounces** the database, which means that it shuts down the database instance, modifies the database properties, and then restarts the database instance. In this case, you bounce the database to place it in ARCHIVELOG mode. Then you force a log switch, and view the archive file that the database instance creates.

To place your database instance in ARCHIVELOG mode and view the archive file:

1. On the Database Home page, under the High Availability heading, click the **Disabled** link next to Archiving. The Configure Recovery Settings page, as shown in Figure 11-62, appears.

2. Check the **ARCHIVELOG Mode** check box and click the **Apply** button. The Confirmation page appears with the message "The changes have been made successfully. However, you must restart the database to implement the changes. Do you want to restart the database now?" Click **Yes**.

NOTE

If the ARCHIVELOG Mode check box is disabled, then you have not logged onto OEM using an administrative connection. To create an administrative connection, review the section titled "Creating an Administrative Connection."

If the ARCHIVELOG Mode check box is already checked, skip to Step 5, because your database is already in ARCHIVELOG mode.

3. The Restart Database: Specify Host and Target Database Credentials page opens. If necessary, type the appropriate usernames and passwords, and then click **OK**. The Restart Database: Confirmation page appears, as shown in Figure 11-63.

4. Click **Show SQL**, examine the SQL statements, and then click **Return**.

5. Click **Yes**. The Restart Database: Activity Information page appears. Click **Refresh** and the Database Home page appears.

6. To force a log switch and write the contents of the current redo log file to an archive file, navigate to the Administration page and click the **Redo Log Groups** link under the Storage heading.

7. Select **Switch logfile** in the Actions list, and then click **Go**. The update message "Log group successfully switched" appears.

8. Switch to Windows Explorer, navigate to the folder where the database stores the archive files, and view the new archive file, as shown in Figure 11-64.

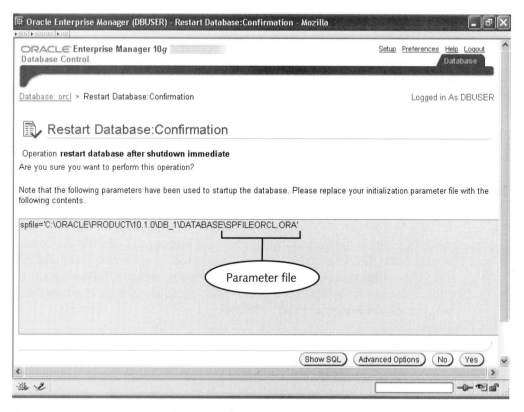

Figure 11-63   Restart Database: Confirmation page

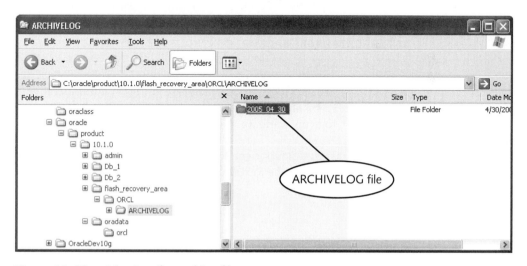

Figure 11-64   Viewing the archive file

Backing Up the Control File

Recall that when you created a cold backup, you shut down the database, and then used Windows Explorer to copy the control files to an alternate disk location. To create a hot backup, you cannot use Windows Explorer to back up the control file directly, because the control file is currently open and in use by the database instance. To back up the control file in a hot backup, you log onto SQL*Plus using an administrative connection, and then execute the following command:

```
ALTER DATABASE BACKUP CONTROLFILE TO TRACE;
```

This command creates a **trace file**, which is a text file that contains information about the operation of a database instance. This particular command creates a trace file that contains the commands to rebuild your control file. The database places the trace file in the *Oracle_Base*\admin*SID*\udump folder on the database server. Now you start SQL*Plus, log on using an administrative connection, and execute the command to create the trace file. Then you copy the trace file to your Solution Disk. To log onto SQL*Plus using an administrative connection, you type the following command in the Connect String field: *service_name* AS SYSDBA. If your database's *service_name* is orcl, you would type orcl AS SYSDBA.

To back up the control file:

1. Switch to Windows Explorer, and create a folder named **OraHotBackup** in the Chapter11\Tutorials folder on your Solution Disk.

2. Start Notepad, and create a new file named **Ch11Queries.sql** in the Chapter11\Tutorials folder on your Solution Disk.

3. In Notepad, type the following command to create the trace file for the control file:

```
ALTER DATABASE BACKUP CONTROLFILE TO TRACE;
```

4. Copy the command.

5. Start SQL*Plus, and log onto the database using an administrative connection. Figure 11-65 shows an example of the SQL*Plus Log On dialog box when a user logs on using an administrative connection.

NOTE

If the error message "Insufficient privileges" appears, it is because your user account does not have the SYSDBA privilege.

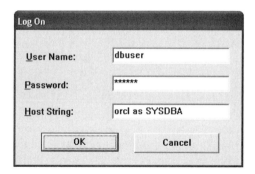

Figure 11-65   Logging onto SQL Plus using an administrative connection

6. Paste the copied command into SQL*Plus, and then press **Enter** to execute the command. The "Database altered" confirmation message appears when the database has finished creating the trace file.

NOTE

For the rest of the lesson, type the SQL*Plus command in Notepad, copy the command, paste the copied command into SQL*Plus, and then execute the command.

7. Switch to Windows Explorer, and navigate to the *Oracle_Base*\admin\ *SID*\udump folder on the database server. (If your *Oracle_Base* is C:\oracle\ product\10.1.0, and your *SID* is orcl, then you navigate to the C:\oracle\ product\10.1.0\admin\orcl\udump folder.)

8. If your Windows Explorer display does not show the dates the files were modified, select **View** on the toolbar, then click **Details** to view the file details. Click the **Date Modified** column one or more times to sort the files by descending dates, so the most recent files appear first in the list. (If you are using Windows 2000, click the Modified column to sort the files by descending dates.)

9. Select the first file in the list, which is the file that was created most recently. This is the trace file that you just created. Copy the trace file to the OraHotBackup folder on your Solution Disk.

Backing Up the Datafiles

To back up datafiles in a hot backup, you complete the following steps:

1. Create an administrative connection in SQL*Plus, then execute a SQL*Plus command that causes the database temporarily to stop modifying the tablespace associated with the datafile(s). (Recall that a tablespace can have one or more associated datafiles.)

11

2. Use Windows Explorer to make a copy of the datafile.

3. Execute a SQL*Plus command to instruct the database to start using the datafile's tablespace again.

You repeat this process for each tablespace you want to back up.

You use the following SQL*Plus command to instruct the database temporarily to stop writing to a tablespace's datafile:

```
ALTER TABLESPACE tablespace_name BEGIN BACKUP;
```

In this syntax, *tablespace_name* is the name of the tablespace whose datafiles you want to back up. After executing this command, you switch to Windows Explorer, and back up the tablespace's datafile(s) to an alternate disk location. Then you execute the following command to instruct the database to resume writing to the tablespace datafiles:

```
ALTER TABLESPACE tablespace_name END BACKUP;
```

Now you back up the datafiles for the DBUSER tablespace in your database instance. (Recall that earlier in the chapter, you created this tablespace and created two datafiles named DBUSER01.DBF and DBUSER02.DBF to store the tablespace data.)

To back up the datafiles for the DBUSER tablespace:

1. Type and execute the following command to instruct the database to stop writing to the datafiles associated with the DBUSER tablespace:

```
ALTER TABLESPACE DBUSER BEGIN BACKUP;
```

The message "Tablespace altered" appears, indicating the command executed successfully.

2. Switch to Windows Explorer, and navigate to the *Oracle_Base*\oradata*SID* folder on the database server. Copy **DBUSER01.DBF** and **DBUSER02.DBF** to the OraHotBackup folder on your Solution Disk. (This might take a few minutes.)

3. When the files are copied, type and execute the following command to make the tablespace datafiles available again:

```
ALTER TABLESPACE DBUSER END BACKUP;
```

The message "Tablespace altered" appears, indicating the command executed successfully.

Backing Up the Archive Files

The final step in creating a hot backup is to instruct the database to archive the redo log files, and then back up the archive files. To instruct the database to archive the redo log files, you create an administrative connection in SQL*Plus, and then execute the following command:

```
ALTER SYSTEM ARCHIVE LOG CURRENT;
```

This command forces a log switch, and then writes the contents of the current redo log file to an archive file. It also ensures that the contents of all redo log files are written to archive files. Next, you back up the archive files for the DBUSER tablespace in your database instance.

To archive the redo log files and back up the archive files:

1. Type and execute the following command:

   ALTER SYSTEM ARCHIVE LOG CURRENT;

   The "System altered" message confirms that the command executed successfully.

2. Switch to Windows Explorer, and navigate to the folder where the database stores the archive files (see Figure 11-64).

3. Select all of the archive files, which are files whose filename starts with the characters ARC. Copy the files to the OraHotBackup folder on your Solution Disk.

4. Close Windows Explorer, SQL*Plus, Notepad, and OEM.

Database Recovery

Recovery is the process of restoring a database from either cold or hot backup files. Recovery is necessary after a database failure caused by a software malfunction, such as a corrupted datafile or control file, or a hardware malfunction, such as a damaged hard drive. Recovery is also necessary if the DBA shuts down the database using the Abort shutdown option. The following subsections describe recovery using cold and hot backups.

Recovery Using a Cold Backup

Recovery using a cold backup allows the DBA to recover the database to its state at the time of the last cold backup. All subsequent changes to the database structure and datafiles are lost. You use the following recovery steps for a system that uses cold backups:

1. Shut down the database, if it is still running.

2. Restore the backed up control file and datafiles to the database server.

3. Restart the database.

Recovery Using a Hot Backup

To recover a database that uses hot backups, you recover the database to the point at which the last hot backup was created. Then, you use the control trace file to generate a new copy of the control file. Next, you restore the backup copies of the datafiles. Finally, you perform **media recovery**, which uses the archive files and current redo log files to update the datafiles to reflect all subsequent changes since the datafiles were backed up. You use the following recovery steps for a system that uses hot backups:

1. Shut down the database, if it is still running.

2. Restore the database from the last hot backup.

3. Restart the database in a restricted session, so only users with the RESTRICTED SESSION privilege can log on.

4. Use media recovery to recover the changes from the archive files, and apply them to the datafiles.

5. Restart the database in an unrestricted session, and make it available to all users.

To perform media recovery, you execute the RECOVER command in SQL*Plus. The **RECOVER** command causes the database to take the contents of the archive files, and apply the information to the datafiles. The RECOVER command has the following syntax:

```
RECOVER DATABASE;
```

Recovery operations tend to be complex and difficult, and usually cannot be performed by novice DBAs. All of the recovery components must be in place and configured correctly, or the database will not run correctly.

Backup Strategies

Recall that cold backups involve shutting down the database and then creating copies of database files. Hot backups involve creating copies of database files while the database is running, and configuring the database to create archive files to record all database changes. An organization's backup strategy might use only cold backups, or might use a combination of both approaches, in which the DBA restores the database to its state at the time of the last cold backup, generates an updated control file, and then performs media recovery to update changed datafiles.

Cold backups are simplest to perform, and are satisfactory for databases whose contents do not change very much and whose users can tolerate downtime while the DBA creates the cold backup files. Hot backups are more complex to perform, but are better suited for dynamic, mission-critical databases whose contents are constantly changing and must be available at all times. Hot backups require the database to run in ARCHIVELOG mode, which consumes system resources and requires extra storage space for archive files. Therefore, systems that rely on hot backups must have adequate database server hardware to support the extra processing and storage needs.

With either backup scheme, the DBA and his or her staff must perform the backup activities correctly and regularly. The frequency with which the DBA performs backups is a critical factor. For cold backups, an infrequent backup schedule results in the loss of large amounts of data. For hot backups, an infrequent backup schedule makes the recovery process take longer. A frequent backup schedule is optimal, but might be impractical.

For large databases, daily backups of all database files are not feasible because it might take all day to copy the database files to an alternate disk location! As a result, the DBA may create **partial backups**, which back up selected tablespaces at certain intervals.

To determine an appropriate organizational backup scheme and frequency, the DBA has to reach an agreement with users and managers regarding the trade-offs of downtime caused by cold backups, versus the extra effort and system resources consumed by hot backups. The organization has to weigh the tradeoffs between data loss or downtime caused by a database failure, against the resources consumed by different backup strategies and schedules. Oracle Corporation and other vendors provide utilities to automate and expedite the backup process, but these utilities can be challenging to configure and deploy.

CHAPTER SUMMARY

- When you create a new user account, you must specify general information about the user account, system privileges the user has in the database, and the user's tablespace quota on the database server.

- When you create a new user account, you specify the account's default tablespace, which is the tablespace in which the database stores the user's database objects, and the account's temporary tablespace, which specifies the tablespace in which the database stores temporary data that it uses for query processing. When you create a new user account, you must grant a tablespace quota in the default tablespace.

- A system privilege allows a user to perform a specific task with the Oracle 10g database, such as connecting to the database or creating and manipulating database objects. If you grant to a user a privilege with Admin Option, the user can then grant the privilege to other users.

- A role represents a collection of system privileges that you can assign to multiple users. If you need to change the privileges of the users to whom the role has been granted, you simply change the role's privileges, and the user accounts automatically reflect the privilege change.

- An Oracle 10g database instance creates two memory areas in the database server's main memory: the System Global Area (SGA) and the Program Global Area (PGA). The SGA is used by all database connections to share information among all database processes. The PGA stores information for a specific user connection.

- The primary areas within the SGA include: the shared pool, which stores machine-language code and execution plans for frequently used SQL commands; the database buffer cache, which stores data values from user insertions, updates, or retrievals; the large pool, which consists of large contiguous memory storage spaces for tasks such as backup and recovery operations; the redo log buffer, which stores rollback information for user transactions; and the Java pool, which stores machine-language representations and execution plans for Java commands used in application programs and database operations.

11

❑ The PGA has a session information area, which contains information about the user session, and a stack space area, which contains the values of the variables that the user declares in PL/SQL programs and other programs.

❑ An Oracle 10*g* database instance contains a set of background processes to service user requests. The background processes include: Database Writer (DBWn), which writes changed data from the database buffer cache to the datafiles; Log Writer (LGWR), which writes redo information from the redo log buffer to the redo log files; System Monitor (SMON), which recovers lost data after a system hardware or software failure, deallocates temporary memory areas that the database uses for sort operations, and manages server disk space; Process Monitor (PMON), which monitors and manages individual user sessions; and Checkpoint (CKPT), which is responsible for initiating checkpoints that signal the DBWn and LGWR processes to write the buffer contents to the datafiles and redo log files.

❑ Checkpoints occur automatically when a redo log file becomes filled, when a log switch occurs, and when the DBA shuts down the database instance. A DBA can force a checkpoint by forcing a log switch.

❑ To start or shut down a database instance, a DBA must log onto the database using an administrative connection, which requires the user to have either the SYSDBA or SYSOPER system privilege and to connect as either a SYSDBA or a SYSOPER.

❑ When you start an Oracle 10*g* database instance, the instance passes through four states: SHUTDOWN, in which none of its background processes are running and all of its files are closed; NOMOUNT, in which the background processes are running and the SGA memory area is allocated; MOUNTED, in which the control file has been opened; and OPEN, in which all processes are running, all files are opened, and the database is ready to accept user connections.

❑ A DBA can shut down a database instance using the following shutdown options: normal, in which the instance does not accept any new connections, but allows current users to continue their transactions and log off normally; transactional, in which the instance does not accept any new connections, and allows current users to continue operations until they complete their current transaction; immediate, in which the instance does not accept any new user connections, and current user connections terminate immediately; and abort, in which all users are immediately disconnected, the instance's processes are immediately stopped, and all server memory is immediately reallocated.

❑ A DBA can start an Oracle 10*g* database instance in one of three modes: unrestricted, which allows all users to create connections; restricted, which limits connections only to users who have the RESTRICTED SESSION system privilege; and read-only, which allows users to read database contents, but does not allow users to perform action queries that modify database contents.

❑ When a DBA shuts down a database instance, the shutdown process writes the contents of the data buffer cache to the datafiles, writes the contents of the redo log buffer to the redo log files, closes all files, stops all background processes, and deallocates the SGA in the server's main memory. When a DBA starts a database instance, the startup process performs these tasks in the reverse order.

❐ Backup involves creating a copy of the database files, and recovery involves restoring the database to a working state after a hardware or software malfunction.

❐ An offline (cold) backup requires shutting down the database, then copying all of the database files to an alternate location. An online (hot) backup involves backing up critical database files while the instance is running, as well as creating an ongoing archive of database changes so the DBA can restore the database to its state at any point in time.

❐ To create a cold backup, you shut down the database instance, copy the parameter file, control files, and datafiles to an alternate disk location, and then restart the database.

❐ To create a hot backup, you place the database instance in ARCHIVELOG mode, which creates archive files that contain the contents of the redo log files. Then you execute SQL*Plus commands to back up the control file and datafiles, and instruct the database to write the current contents of the redo log files to the archive files.

❐ Recovery is necessary after a database failure caused by a software malfunction, such as a corrupted datafile or control file, a hardware malfunction, such as a damaged hard drive, or shutting down the database using the abort shutdown option.

❐ Recovery using cold backups involves shutting down the database, copying the backup files to the database server, and then restarting the database. Recovery using hot backups involves restoring the database using the hot backup, and then performing media recovery to merge the contents of the archive files with the datafiles so the datafiles contain all changes since the last backup.

11

REVIEW QUESTIONS

1. Describe the ramifications when a system privilege is granted without Admin Option.

2. When should you create a role?

3. The _____ is a memory area used by all database connections to share information among all database processes, and the _____ is a memory area that stores information for a specific user connection.

4. When should you start a database instance in restricted mode?

5. Describe the difference between a transactional shutdown and an immediate shutdown.

6. When should you use the abort shutdown option?

7. How do you grant the SYSDBA or SYSOPER privilege to users in a new database installation?

8. When does a checkpoint occur?

9. A(n) _____ backup requires shutting down the database, then copying all of the database files to an alternate location, and a(n) _____ backup involves backing up critical database files while the instance is running.

10. Which database files do you back up when you create a cold backup?

11. What is an archive file?

12. Which background processes perform actions when a checkpoint occurs?

13. To create a new user, the DBA must specify the user's tablespace quota in the _____ tablespace.

    a. SYSTEM

    b. default

    c. temporary

    d. default and temporary

14. A(n) _____ is a database object that represents a collection of system privileges that you can assign to multiple users.

    a. object privilege

    b. administrative connection

    c. privilege with Admin Option

    d. role

15. The _____ is a memory area within the SGA that stores rollback information for user transactions.

    a. PGA

    b. shared pool

    c. redo log buffer

    d. rollback buffer

16. To create an administrative connection, a user must:

    a. log on using the SYS user account

    b. have the DBA privilege with Admin Option

    c. have the SYSDBA or SYSOPER privilege

    d. both a and c

17. You should use the _____ shutdown option when you want to prevent users from losing their uncommitted transactions.

   a. normal

   b. transactional

   c. immediate

   d. abort

   e. either a or b

18. The _____ startup mode allows all users to create connections.

   a. unrestricted

   b. restricted

   c. unmounted

   d. mounted

19. The DBA performs media recovery when recovering a database from a _____ backup.

   a. hot

   b. cold

   c. frequent

   d. NOARCHIVE

20. A database instance must be _____ in order to create hot backups.

   a. in NOARCHIVELOG mode

   b. in ARCHIVELOG mode

   c. shut down

   d. in restricted mode

21. True or False: You can grant a role to a user account only when you first create the account.

22. Write the name of the storage structure (SGA, PGA, shared pool, database buffer cache, or redo log buffer) beside the statement that describes it:

   Stores information about a specific user process

   Stores most recently viewed or changed data values

   Stores compiled SQL queries and PL/SQL program statements

   Stores a record of all changes made to the database

   Stores data that is shared by all user processes

11

23. Write the name of the Oracle process (DBWn, LGWR, SMON, PMON, or CKPT) beside the statement that describes it:

    Periodically signals other processes to start

    Writes contents of the redo log buffer to the redo log files

    Performs rollbacks when requested by a user

    Writes changes in the database buffer cache to the datafiles

    Performs system recovery operations

25. Write the name of the instance state beside the statement that describes it:

    Processes started, control file opened

    No processes started, no files opened

    Processes started, all files opened

    Processes started, no files opened

26. Place the steps that a database instance goes through during the shutdown process in the correct order:

Step Number	Description
_____	Stops all background processes
_____	Closes all files
_____	Deallocates the SGA in the server's main memory
_____	Writes the contents of the redo log buffer to the redo log files
_____	Writes the contents of the data buffer cache to the datafiles

PROBLEM-SOLVING CASES

Save all solutions in the Chapter11\Cases folder on your Solution Disk.

1. Create a new user account with the username 11BCASE1 and initial password "secret." Use the DEFAULT user profile. Specify that the default tablespace is USERS and the temporary tablespace is TEMP. Grant to the user the CREATE TABLE and CREATE VIEW system privileges. Assign the CREATE VIEW system privilege with Admin Option. Assign a disk quota of 10 MB to the default tablespace, and unlimited disk space to the temporary tablespace. Save the SQL command to create the user in a file named 11BCase1.sql.

To complete Case 2, you must have completed Case 1.

2. Create a new role named 11BCASE2ROLE that does not use an authentication method. Grant to the role the DBA role, and the CREATE ANY INDEX and CREATE ANY LIBRARY system privileges. Grant both system privileges with Admin Option. Then grant 11BCASE2ROLE to the 11BCASE1 user account you created in Case 1. Save the SQL commands to complete these tasks in a file named 11BCase2.sql.

3. Find the values of the shared_pool_size, db_block_buffers, large_pool_size, log_buffer, and java_pool_size SGA parameter values for your database instance. Save the parameter names and their associated values in a file named 11BCase3.txt.

4. Create a new user account named 11BCASE4 with the initial password "secret" that can create administrative connections. Use the DEFAULT user profile. Specify that the default tablespace is USERS and the temporary tablespace is system-assigned. Assign to the default tablespace an unlimited tablespace quota. Save the command to create the account in a file named 11BCase4.sql.

11

Appendix

iSQL*Plus

iSQL*Plus User Interface

iSQL*Plus is a component of SQL*Plus that enables you to use a Web browser to perform the same tasks that are available on the SQL*Plus command line. No client machine set-up or installation is required. You simply need a Web browser and an Internet connection to access the Oracle 10*g* database server. This appendix introduces the user-friendly, browser-based iSQL*Plus interface and identifies some of the differences that you may encounter when using iSQL*Plus versus client SQL*Plus.

> The appearance and layout of the iSQL*Plus screens might vary depending on your type of Web browser and browser window size.

Commands Not Supported in iSQL*Plus

There are numerous SQL*Plus commands not supported in the iSQL*Plus user interface. Attempting to use any of the unsupported commands or command options shown in the following paragraphs raises the error message "SP2-0850: Command *command name* is not available in iSQL*Plus." In addition, some commands become obsolete as new versions of Oracle are introduced. You should modify scripts using unsupported or obsolete commands to use alternative commands or options.

The following commands have no context in iSQL*Plus, and therefore raise an SP2-0850: error message:

- ACCEPT
- PASSWORD
- CLEAR SCREEN
- PAUSE

The following SET command variables have no context in iSQL*Plus, and therefore raise an SP2-0850: error message:

- SET EDITFILE
- SET SQLBLANKLINES

- SET TAB
- SET FLUSH
- SET SQLCONTINUE
- SET TERMOUT
- SET NEWPAGE
- SET SQLNUMBER
- SET TIME
- SET PAUSE
- SET SQLPREFIX
- SET TRIMOUT
- SET SHIFTINOUT
- SET SQLPROMPT
- SET TRIMSPOOL
- SET SHOWMODE
- SET SUFFIX

The following commands have security issues on the middle tier, and therefore raise an SP2-0850: error message:

- GET
- SPOOL
- HOST
- STORE

The following SQL buffer editing commands are not relevant in iSQL*Plus, and therefore raise an SP2-0850: error message:

- APPEND
- DEL
- INPUT
- CHANGE
- EDIT
- SAVE

*i*SQL*Plus Versus Client SQL*Plus

Some commands are unsupported in the *i*SQL*Plus interface and some do not function in the same manner. This can be frustrating for those familiar with the client SQL*Plus interface. Table A-1 shows how to implement three common SQL*Plus commands in iSQL*Plus.

SQL*Plus Command	SQL*Plus	iSQL*Plus Alternative
The @ or START command	Calls the script specified by the URL from a Web server and runs the SQL*Plus statements in the script.	Use the load script feature (discussed later in this Appendix).
Spool command	Sends or saves output to a disk storage area. Often used to print or transfer files.	You have one of two options for saving the executed statements and the results from your session: 1. Copy the executed code and results into a text editor or word processor. 2. Save the *i*SQL*Plus Web page by clicking File, and choosing Save As from the File menu. This saves the executed code and the results as an HTML file.

Table A-1   SQL*Plus commands and iSQL*Plus alternatives

iSQL*Plus Navigation

To navigate in iSQL*Plus, you use the global navigation icons. Each are described in the following sections.

Icons

Global navigation icons are displayed on each screen and have three states:

- **Available**—In this state, the icon has a *white* background.
- **Unavailable**—In this state, the icon has a *light brown* background.
- **Active** (when you have navigated to that screen)—In this state, the icon has a *blue* background.

The iSQL*Plus global navigation icons are shown in Table A-2.

Icon	Name	Description
(logout icon)	Logout	Logs you out of the iSQL*Plus session and returns you to the Login screen.
(preferences icon)	Preferences	Opens the iSQL*Plus Preferences screen, on which you can set interface options, system variables, or change your password.
(help icon)	Help	Opens the iSQL*Plus Help in a separate Web browser window.

Table A-2 *i*SQL*Plus Global Navigation Icons

iSQL*Plus Login Screen

You connect to the Login screen from your Web browser using a URL like http://machine_name.domain:port/isqlplus.

To connect to ISQL*Plus:

1. Connect to the Internet, and enter your Oracle 10*g* HTTP Server URL in the browser address area. The format is http://machine_name.domain:port/isqlplus. The URL will be supplied by your instructor.

2. The login screen appears, as shown in Figure A-1. Enter the username and password supplied by your instructor. You can leave the Connection Identifier field blank if you are connecting to the default database. Otherwise, you need to enter the database connection string as included in the server's tnsnames.ora file.

3. Click **Login**. The iSQL*Plus Work screen appears, as shown in Figure A-2.

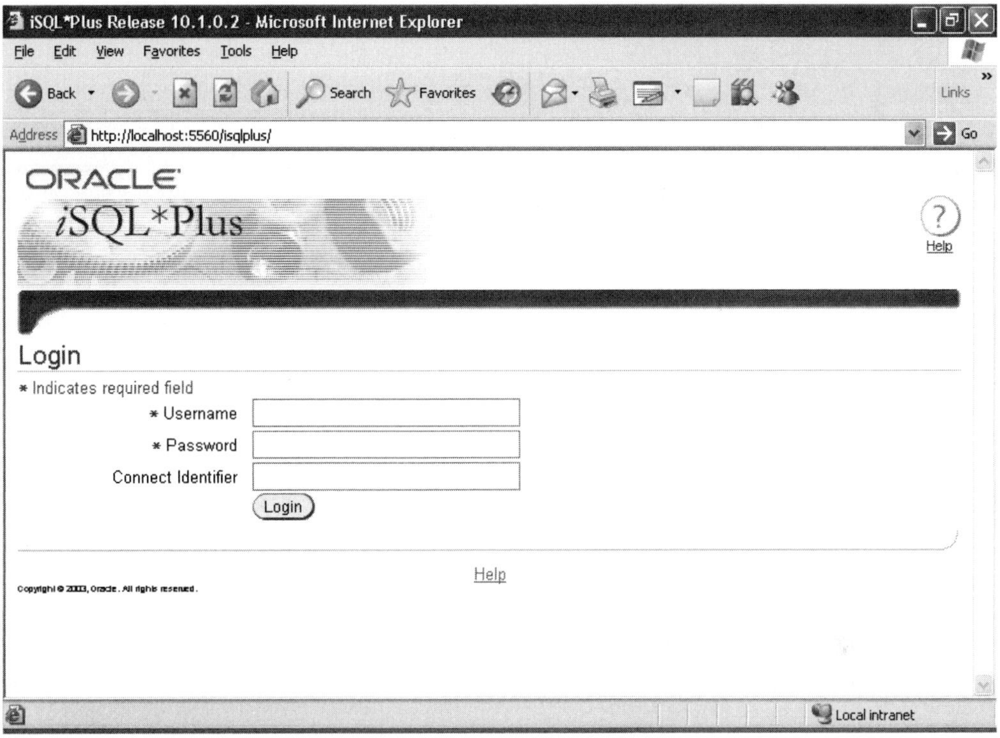

Figure A-1 Login screen

Figure A-2   Workspace

iSQL*Plus WORKSPACE

The Workspace displays the following Input areas and buttons:

- **Enter SQL, PL/SQL and SQL*Plus statements**—Type SQL and PL/SQL statements, or SQL*Plus commands into this Input area. You can resize the Input area in the Interface Configuration screen, which you access from the Preferences screen.

- **Execute**—Executes the contents of the Input area. Depending on your preference settings, the results are displayed in the Workspace or saved to a file.

- **Load Script**—Loads the script specified in the File or URL field into the iSQL*Plus Input area for editing or execution. Clicking this button displays a new page with the following Input areas and button:

  - **URL**—Type the URL of a file to load it into the Input area for editing or execution.

  - **File**—Type the path and filename of a file to load it into the Input area for editing or execution.

- **Browse**—Click this button to search for a script file that you want to load for editing or execution.

- **Save Script**—Saves the contents of the Input area to a file. You are prompted to enter the name of the file. The file extension you choose is for your convenience. It may be useful to identify scripts with an extension of .sql.

- **Clear**—Clears all statements in the Input area, and all displayed output.

- **Cancel**—Interrupts the script that is currently running.

Entering SQL, PL/SQL, and SQL*Plus Statements

You can enter SQL, PL/SQL and SQL*Plus statements by:

- Typing the statements directly into the Enter Statements Input area or work area.

- Copying text from a text editor and pasting it into the Enter Statements area.

- Using the LOAD SCRIPT button to load statements from a file.

To type statements in the work area:

1. Type a SELECT statement or DESCRIBE command directly into the statements Input area, as shown in Figure A-3.

2. Click the **Execute** button to execute the code that is in the statement Input area. Note that the code and results or an error message appear below the Enter Statements Input area. The statement remains available in the Input area so you can easily modify it and execute it again.

3. To maintain a copy of your session, including statements executed and the resulting output, cut and paste to a text editor or word processer. Save the file if you want to keep the copy.

4. To save the statement in a script file to rerun at a later time, click the **Save Script** button. This prompts you for a filename and location. You should use the file extension .sql to identify the file as an SQL script file.

NOTE
A script is a very useful tool. It is a file containing a sequence of commands that you can otherwise enter interactively. The file is saved for convenience, reexecution, and as a backup method. Scripts are often called by operating-system-specific names. In iSQL*Plus, you can execute the script by using the Browse, Load, and Execute buttons.

You can create all your statements in a text file and then test or execute them by copying and pasting them from the text editor to the statement Input area. The text file can be saved as a script. The execution and modification of statements in the script are handled in the same manner discussed in the previous section.

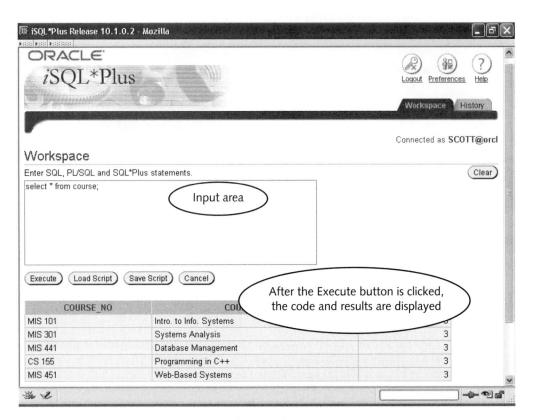

Figure A-3   Entering SQL, PL/SQL, and SQL*Plus Statements

To load statements from a file:

1. Click the **Load Script** button, which is located below the Enter SQL, PL/SQL, and SQL*Plus statements box as shown in Figure A-3. A Choose file dialog box appears, allowing you to select the file you wish to load. This file must be a text file that contains SQL and/or PL/SQL statements.

2. Click the **Browse** button, which is located next to the File box as shown in Figure A-4. A Choose file dialog box appears, allowing you to select the file you want to load. This file must be a text file that contains SQL, SQL*Plus and/or PL/SQL statements.

3. Select the desired file and click **Open**. The file location and name appear in the File text box.

4. Click the **Load** button, and all the text from the file is loaded into the Input area. You can execute or modify the SQL, SQL*Plus and/or PL/SQL statements that are displayed. Note that multiple statements can be entered and executed.

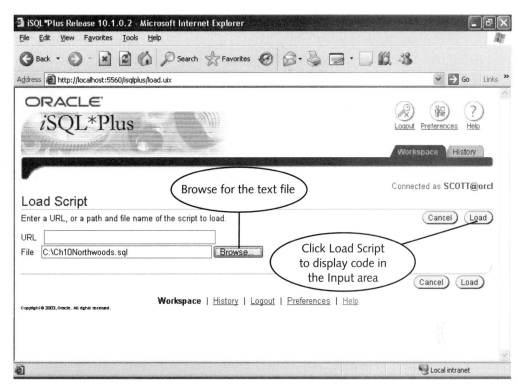

Figure A-4   Load Statements from a File

iSQL*Plus History Screen

The History screen allows you to reload scripts that you previously executed in the same session. A History entry is created each time you execute a new script in the Work screen. The History screen shows the leading 80 characters of the script. Once you load a script from the History, it is moved to the top of the History list, and when the History limit is reached, the earliest scripts are removed. When you exit a session, the History is lost. You can change the default number of entries stored in the History list in the Interface Configuration screen, which you access from the Preferences screen. To access the History screen, which is shown in Figure A-5, click the History tab.

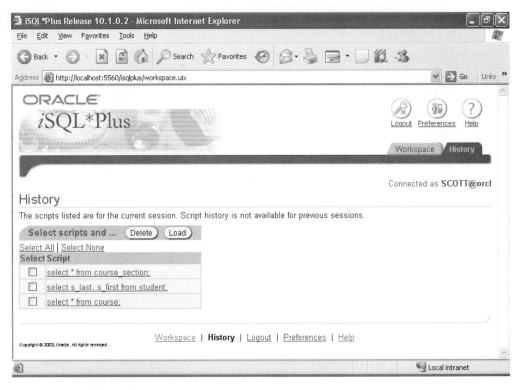

Figure A-5 History Screen

The History screen displays the following buttons:

- **Select scripts and…**—Displays a list of scripts in the History, with the most recently executed at the top. Click the script text to load it into the Input area. Scripts are displayed word for word, so be careful if you have included items that you do not want displayed.

- **Load**—Loads the selected scripts into the Input area of the Work screen.

- **Delete**—Deletes the selected scripts from the History.

- **Select All**—Places a check in all the check boxes next to each script in the history.

- **Select None**—Removes check from all check boxes.

iSQL*Plus Preferences Screen

The Preferences screen, shown in Figure A-6, allows you to set interface and system configurations, and to change your password.

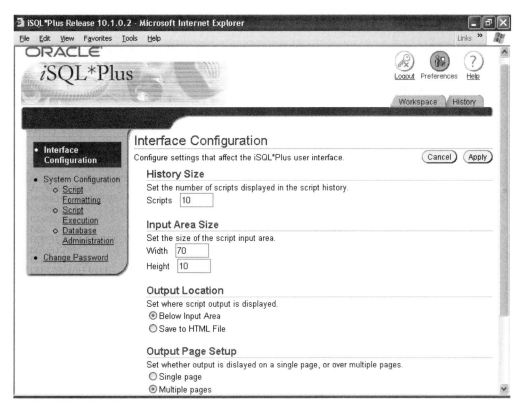

Figure A-6   Preferences screen

Set Interface Configurations

Click Set interface options, or the Go button, to access the Interface Options screen, which is shown in Figure A-7. You use the Interface Configuration screen to:

- Set the History size.

- Set Input area size.

- Determine whether the output is directed to below the Input area or to an HTML file.

- Set up the output page.

The Interface Configuration screen displays the following fields and buttons:

- **History Size**—Enter the number of scripts that are stored in a session. The minimum value allowed is 0, the maximum 100, and the default is 10. If you enter an invalid value, the previous valid value is used.

- **Input Area**—Enter the width in characters, and the height in lines that you want for the Input Area Size. The default width is 70 characters, and the default height is 10 lines.

- **Output**—Select one of two options to set the Output Location for script results generated from your session:

  - **Below Input Area**—When the contents of the Input area are executed, the resulting output is displayed on screen under the Input area. This is the default.

  - **Save to HTML File**—When the contents of the Input area are executed, the resulting output is saved to a file. You are prompted to enter the name of the file. Because the output is in HTML format, it is useful to give the saved output file an .HTM or .HTML extension.

- **Output Page Setup**—Select whether the output should be on a single page or multiple pages. You can specify the number of rows on a page.

Set System Configuration

The System Configuration is categorized by Script Formatting, Script Execution, and Database Administration.

- **Script Formatting**—Allows for the configuration of settings that affect how script output is formatted, and what optional information it contains. Optional information can be line numbers, headings, record counts, and more.

- **Script Execution**—Allows for the configuration of settings that affect how scripts are parsed and executed.

- **Database Administration**—Allows for the configuration of settings related to the database administration tasks of backup and recovery.

Change Your Password

To access the Change Password screen, which is shown in Figure A-7, click Change Password or the Go button on the Preferences screen. If you have logged in with DBA privileges, you can also change the password of other users.

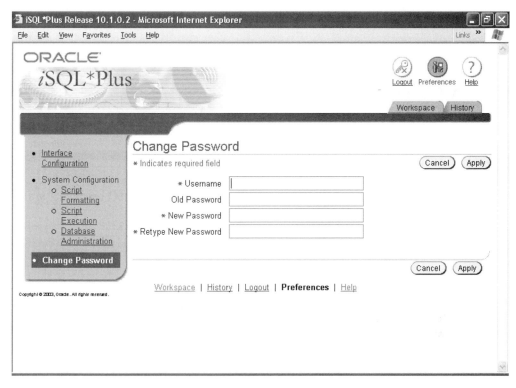

Figure A-7   Change Password screen

The Change Password screen displays the following Input areas and buttons:

- **Username**—Enter the Oracle database account username.
- **New Password**—Enter the new password.
- **Old Password**—Enter the current Oracle database account password.
- **Retype New Password**—Enter the new password again to verify it.

iSQL*Plus Help

The Help screen, which is shown in Figure A-8, can be accessed by clicking the Help icon or the Help link at any time you are working in iSQL*Plus.

This screen allows you to access an abundance of information either by subject content, an index, or by topic searching. Once a topic is located, a printable page option is available. The Help screen appears in a new browser window and does not affect the current session window.

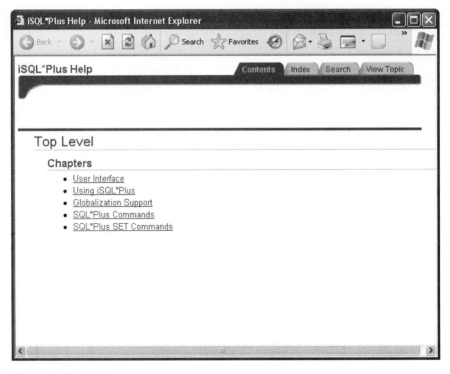

Figure A-8   Help screen

Index

The envelope bound into this text is the Oracle 10*g* Developer Suite, Version 9.0.4.0.1, for Microsoft Windows NT, Windows 2000 Professional or Server, Windows 2003 Server, and Windows XP operating systems.

For complete installation instructions, go to *www.course.com/cdkit*.

YOU MUST REGISTER THE SOFTWARE.

Oracle Technology Network puts you in touch with the online community behind the software that powers the internet. Download the latest development programs and sample code. Engage in discussions with the Web's leading technologists. Keep connected with the latest insights and resources you need to stay ahead. Membership is FREE, and so is the latest development software. Before proceeding to use the software, **you must** register the software at *http://otn.oracle.com/books/* and agree to the Oracle Technology Network Developer License Terms in order to receive the key code to unlock the software. Upon registering the software, you agree that Oracle may contact you for marketing purposes. You also agree that any information you provide Oracle may be used for marketing purposes.